# FUEL CELL SCIENCE

## Wiley Series on Electrocatalysis and Electrochemistry

**Andrzej Wieckowski**, Series Editor

*Fuel Cell Catalysis: A Surface Science Approach*, Marc T. M. Koper

*Electrochemistry of Functional Supramolecular Systems*, Margherita Venturi, Paola Ceroni, and Alberto Credi

*Catalysis in Electrochemistry: From Fundamentals to Strategies for Fuel Cell Development*, Elizabeth Santos and Wolfgang Schmickler

*Fuel Cell Science: Theory, Fundamentals, and Biocatalysis*, Andrzej Wieckowski and Jens Nørskov

# FUEL CELL SCIENCE

## THEORY, FUNDAMENTALS, AND BIOCATALYSIS

Edited by

Andrzej Wieckowski
Jens K. Nørskov

WILEY
A JOHN WILEY & SONS, INC., PUBLICATION

Published by John Wiley & Sons, Inc., Hoboken, New Jersey
Published simultaneously in Canada



***Library of Congress Cataloging in-Publication Data:***

Wieckowski, Andrzej
Fuel cell science : theory, fundamentals, and biocatalysis / edited by Andrzej Wieckowski and Jens Nørskov.
p. cm.
Includes bibliographical references and index.
ISBN 978-0-470-41029-5 (cloth)

Printed in the United States of America

10 9 8 7 6 5 4 3 2 1

## CONTENTS

# FOREWORD

## IS THERE A COMMON "ACTIVITY YARDSTICK" THAT APPLIES TO ALL FUEL CELL ELECTROCATALYSTS?

Thinking what should be the message in the foreword to a book that covers extensively a wide frontier of fuel cell catalysis work, a tempting, albeit somewhat risky, idea kept coming up in my mind: Is it possible to define a common "activity yardstick" that applies to a large number of, if not to all, fuel cell electrocatalysts? Is it possible to make such a generalization when considering the wide variety of catalytic surfaces studied and practiced in low-temperature fuel cells?

When examining the relevant literature, it appears that the more recent searches for active metal electrocatalysts and for active molecular electrocatalysts have had somewhat different priorities. In the case of metal electrocatalysts, the focus has been on tailoring the electronic properties of metal alloy surfaces to achieve an optimized bond strength between the metal surface and the relevant adsorbed intermediates [1]. Such studies have been supported by density functional theory (DFT) calculations, yielding the energies of the bonds between catalytically active surfaces and the likely reaction intermediates [2]. In all such studies, the assumption has been that a complete description of the electrocatalytic process requires consideration of a reactant molecule and a metal surface in contact with water, or aqueous electrolyte. The *electro* element of electrocatalysis has been covered all along by assuming that a change in the interfacial potential difference has an effect on, and only on, the activation energy of any reaction step involving electron transfer. Accordingly, the typical rate expression for an electrocatalytic process takes the form of a product of a preexponential term and a two-component exponential term, with the rate dependence on the electrode potential $E$ fully covered in the exponential term. For a cathodic process within the so-called Tafel regime, the rate expression takes the following general form

$$J(E) = Fk^0 A^*_{\text{cat}}\, C_r^{\gamma} \exp\{-\Delta H^*\text{act}/RT\} \exp\{-(E - E^{\circ}\ \text{cell process})/b\} \qquad (1)$$

where $F$ is the Faraday constant, $k^0$ is a frequency factor, $A^*_{\text{cat}}$ is the overall catalyst surface area per unit electrode cross-sectional area, $C_r$ is the concentration of the reactant molecule at the electrode surface, $\gamma$ is the reaction order $\Delta H^*_{\text{act}}$ is the *chemical component* of the activation energy, and $b$ is the so-called Tafel slope. In the case of molecular electrocatalysts, the more recent achievements in preparation of highly oxygen reduction reaction (ORR)-active, carbon-supported iron complexes [3], resulted from efforts to maximize the overall surface density of effective redox centers, $N^*$. Lefevre et al. [3] showed that an effective center formed when the iron complex was located on a specific, pretailored carbon surface site. The mechanism of electrocatalytic processes taking place at such active surface sites is described in terms of *redox mediation*, where the electrocatalytic activity at a potential $E$ is expected to involve a fraction of $N^*$, $N_{\text{active}}(E)$, defined by

$$N_{\text{active}}(E) = N^* f(E - E^{\circ}_{\text{surface redox}}) \tag{2}$$

For example, in the specific case of ORR catalyzed by a molecular iron complex, a plausible mechanism of ORR at the active complex of iron, X–Fe(II), where X is a surface anchor site and the iron complex is in its reduced form, has been described [3] with a first step involving bonding of dioxygen to the active form of the iron complex, X–Fe(II), assisted by electron transfer from the Fe(II) center:

$$O_2 + \text{X-Fe(II)} = \text{X-Fe(III)-O-O} + e \tag{3a}$$

This step is followed by the completion of the 4e reduction process with regeneration of the active form of the redox system, written in simplified form as follows:

$$\text{X-Fe(III)-O-O} + e + 3e + 4H^+ = \text{X-Fe(II)} + 2H_2O \tag{3b}$$

This mechanism implies that only at a cathode potential sufficiently negative to generate a significant population of the reduced form of the surface redox couple, X–Fe(II), can the rate of the process in Equations (3a) and (3b) rise to a measurable level. In an ideal case where the steady-state population of X–Fe(II) depends on potential according to a simple Nernst equation, the number of active sites at an electrode potential $E$ will be given, for a cathodic process, as

$$N_{\text{active}}(E) = N^*[1/(Z+1)] \tag{4}$$

where $Z = \exp\{(F/RT)\,(E - E^{\circ}_{\text{surface redox}})\}$. Inserting in Equation (1) this dependence of active-site population on electrode potential, the rate expression will take the form

$$J(E) = Fk^0 A^*_{\text{cat}}\, f(E - E^{\circ}_{\text{surface redox}}) C_r^{\gamma} \exp\{-\Delta H^*_{\text{act}}/RT\} \exp\{-(E - E^{\circ}_{\text{cell process}})/b\} \tag{5}$$

where in the simplest case, $f(E - E^{\circ}_{\text{surface redox}}) = 1/(Z+1)$.

The significant difference between Equations (5) and (1) is the appearance in (5) of two sources of rate dependence on electrode potential, associated with two different values of $E^{\circ}$. One is the dependence of the activation energy at an active surface site on the overpotential, $E - E^{\circ}_{\text{cell process}}$, and the second is a dependence of active-site population on $E - E^{\circ}_{\text{surface redox}}$. The former appears in the exponential term of the rate expression, whereas the latter appears in the preexponential term [4].

The tacit assumption behind the use of the simpler expression [Eq. (1)] for processes at metal surfaces is that availability of active sites on metal surfaces does not depend on the electrode potential. This assumption misses, however, a key element of electrocatalysis at metal surfaces [4]. For example, examination of the value of $E^{\circ}_{M/M,ox}$ for metal and metal alloy electrocatalysts of high ORR activity, reveals that "ignition" of the ORR process is tied consistently with the onset of cathodic generation of some minimum surface population (e.g., 1%) of free metal sites on a surface that is fully covered under open-circuit conditions by oxygen species that block ORR. Recognizing that such change in surface composition is required for the onset of the process, one can describe the ORR process at Pt in terms of surface redox mediation, involving the Pt/PtOx surface redox system [4]. ORR ignition requires reduction of a Pt surface oxide, or hydroxide species, for example, according to

$$2\text{Pt}{-}\text{OH}_{\text{surface}} + 2\text{H}^{+} + 2\text{e} = 2\text{Pt}_{\text{surface}} + 2\text{H}_2\text{O} \tag{6a}$$

followed next by reaction of $O_2$ at Pt and with metal sites that become available beyond a threshold potential determined by $E^{\circ}_{M/MeOH}$ according to

$$2\text{Pt}_{\text{surface}} + \text{O}_2 + 2\text{e} + 2\text{H}^{+} = 2\text{Pt}{-}\text{OH}_{\text{surface}} \tag{6b}$$

Continuous repetition of (6a) + (6b) sustains a steady-state rate of a 4e ORR process, taking place at the active (metal) surface sites, with the active site continuously regenerated beyond a threshold potential determined by $E^{\circ}_{M/MeOH}$.

Mediation by a surface redox system is apparently a common feature of a wide variety of electrocatalysts, whether molecular or metallic, and this insight can lead to an attempted definition of a "general key to active electrocatalysts." From Equation (5), an optimum value of $E^{\circ}_{\text{surface redox}}$ will maximize the product of the preexponential and exponential-terms at an electrode potential of technical interest, that is, at a low overpotential-versus-$E^{\circ}_{\text{cell process}}$. Consequently, $E^{\circ}_{\text{surface redox}}$ must not be too far from $E^{\circ}_{\text{cell process}}$, to electroactivate the mediating surface system and thereby ignite the faradaic process at a low overpotential. However, too small a difference between the two standard potentials will mean a small free-energy drive for the reaction of the reactant molecule with the active form of the surface redox system [e.g., reaction (6b)], because the standard free-energy change in that reaction is $F(E^{\circ}_{\text{cell process}} - E^{\circ}_{\text{surface redox}})$. The activation energy of a process like (6b) is expected to be lower, the higher the $(E^{\circ}_{\text{cell process}} - E^{\circ}_{\text{surface redox}})$ difference and, conversely, very close proximity of the two standard potentials will likely result in excessive $\Delta H^{*}_{\text{act}}$. We are looking, therefore, at a need to *optimize the gap between* $E^{\circ}_{\text{surface redox}}$ and $E^{\circ}_{\text{cell process}}$, to satisfy the conflicting requirements of a low overpotential for electrode surface activation and a sufficient free-energy drive for the reaction of the reactant molecule with the active surface site.

On the basis of experimental results reported to date, the optimum value of $(E^{\circ}_{\text{cell process}} - E^{\circ}_{\text{surface redox}})$ for requiring electrocatalytic processes in low-temperature fuel cells is in the range of 300–400 mV. In the case of ORR at unalloyed Pt, for example, $(E^{\circ}_{O_2/H_2O} - E^{\circ}_{Pt/PtOx})$ is near 400 mV and can be lowered further by about

100 mV by alloying [1], resulting in enhanced ORR activity. The rate enhancement derived in this case from lowering of ($E^\circ_{\text{cell process}} - E^\circ_{\text{surface redox}}$) indicates that the beneficial effect of Pt alloying originates from lowering of the *ignition overpotential*, resulting in an increase in the value of the preexponential term in Equation (5) at some given cathode potential $E$. A metal surface where ($E^\circ_{O_2/H_2O} - E^\circ_{M/MOx}$) is either significantly smaller than 300 mV or significantly higher than 400 mV exhibits ORR activity that is lower than that of Pt because it is associated with either high $\Delta H^*_{\text{act}}$ (in the former case), or an excessive ignition overpotential (in the latter case). An aggressive goal for the future would be to minimize further the difference between the two $E^\circ$ values. A surface redox system with $E^\circ_{\text{surface redox}}$ removed less than 300 mV from $E^\circ_{\text{cell process}}$, while, at the same time, securing a low $\Delta H^*_{\text{act}}$ for reaction of the reactant molecule with the active surface site, will enable the onset of significant current at overpotentials lower than those demonstrated to date. The reduction to practice of such a desirable surface function is highly challenging, however, because of the low rates typically associated with processes driven by small changes in free energy.

In summary, at a risk typical for all generalizations, a general rule for active fuel cell electrocatalysts is proposed here, in the hope that it can provide a common ground for the evaluation and development of new electrocatalysts. The rule is based on the recognition that a wide variety of electrocatalytic processes, taking place at either redox-functionalized or metal surfaces, are *surface-redox-mediated*, leading, in turn, to the pursuit of an optimum value for ($E^\circ_{\text{cell process}} - E^\circ_{\text{surface redox}}$) as the guideline for maximizing the electrocatalytic activity. An optimized gap between these two standard potentials best addresses the conflicting demands of a minimum overpotential for surface activation and a high rate of the reaction between the reactant molecule and the active surface site. Since the maximum rate is expected at an *optimal* gap between the $E^\circ$ values, a plot of the rate of the electrocatalytic process versus ($E^\circ_{\text{cell process}} - E^\circ_{\text{surface redox}}$) will obviously take the famous form of a "volcano"; however, this typical shape is now projected and explained in terms of a redox mediation mechanism and the need to optimize the value of ($E^\circ_{\text{cell process}} - E^\circ_{\text{surface redox}}$) to achieve high rates at low overpotential. *Enjoy the book!*

S. GOTTESFELD

## REFERENCES


1. H. A. Gasteiger and N. M. Marković, *Science* **324**(5923), 48–49 (2009).
2. J. Rossmeisel et al., in *Fuel Cell Catalysis: A Surface Science Approach*, M. T. M. Koper, ed., Wiley, Hoboken, NJ, 2009, pp. 57–93.
3. M. Lefevre et al., *Science* **324**(5923), 71 (2009).
4. S. Gottesfeld, in *Fuel Cell Catalysis: A Surface Science Approach*, M. T. M. Koper, ed., Wiley, Hoboken, NJ, 2009, pp. 1–30.

# PREFACE

The book covers some essential topics in the science of fuel cell electrocatalysis [1,2]. It shows an increase in importance of theory and modeling, and the emerging new field of electrocatalysis science: bioelectrocatalysis. It shows a spectacular evolution of the electrocatalysis concepts, from a simple statement of hydrogen evolution/oxidation on platinum to reactions involving advanced nanoengineering and single-crystal surfaces, new methods to study, and complicated chemical moieties (up to enzymes). It is basically a materials/theory volume of chemical physics of fuel cell reactions, including the electron transfer process and structure of the electric double layer, as seen by a new generation of scientists, not necessarily electrochemists. It also shows that operando measurements became possible because of the availability of synchrotron light. It forecasts the work for the future for the current and incoming generation of fuel cell scientists, namely, to use theory and understanding of the process involved (see Chapter 19 and the Foreword), use the operando (advanced in situ) approach, and expect the unexpected from the emerging new field of bioelectrocatalysis. The future is bright and exciting; the combination of the intellectual, high technology, and energy issues makes us strong. We are looking forward.

AW acknowledges the splendid support by the National Science Foundation and the US Army Research Office toward his research in the preparation of this book.

J. Norskov
A. Wieckowski

## REFERENCES

1. S.-G. Sun, P.A. Christensen, and A. Wieckowski, eds., *In-Situ Spectroscopic Studies of Adsorption at the Electrode and Electrocatalysis*, Elsevier, Amsterdam, 2007.
2. A. Wieckowski, E. Savinova, and C. Vayenas, eds., *Catalysis and Electrocatalysis at Nanoparticle Surfaces*, Marcel Dekker, New York, 2003.

***Note***: Color versions of selected figures are available on `ftp://ftp.wiley.com/sci_tech_med/fuel_cell_catalysis`.

# PREFACE TO THE WILEY SERIES ON ELECTROCATALYSIS AND ELECTROCHEMISTRY

The Wiley Series on Electrocatalysis and Electrochemistry covers recent advances in electrocatalysis and electrochemistry and depicts prospects for their contribution to the industrial world. The series illustrates the transition of electrochemical sciences from its beginnings in physical electrochemistry (covering mainly electron transfer reactions, concepts of electrode potentials, and structure of the electrical double layer) to a field in which electrochemical reactivity is shown as a unique aspect of heterogeneous catalysis, is supported by high-level theory, connects to other areas of science, and focuses on electrode surface structure, reaction environments, and interfacial spectroscopy.

The scope of this series ranges from electrocatalysis (practice, theory, relevance to fuel cell science and technology) to electrochemical charge transfer reactions, biocatalysis and photoelectrochemistry. While individual volumes may appear quite diverse, the series promises up-to-date and synergistic reports on insights to further the understanding of properties of electrified solid/liquid systems. Readers of the series will also find strong reference to theoretical approaches for predicting electrocatalytic reactivity by high-level theories such as DFT. Beyond the theoretical perspective, further vehicles for growth are provided by the sound experimental background and demonstration of the significance of such topics as energy storage, syntheses of catalytic materials via rational design, nanometer-scale technologies, prospects in electrosynthesis, new research instrumentation, surface modifications in basic research on charge transfer, and related interfacial reactivity. In this context, one might notice that new methods that are being developed for one specific field are readily adapted for application in another.

Electrochemistry has benefited from numerous monographs and review articles due to its applicability in the practical world. Electrocatalysis has also been the subject of individual reviews and compilations. The Wiley Series on Electrocatalysis and Electrochemistry hopes to address the current activity in both of these complementary fields by containing volumes that individually focus on topics of current and potential interest and application. At the same time, the chapters intend to demonstrate the connections of electrochemistry to areas in addition to chemistry and physics, such as chemical engineering, quantum mechanics, chemical physics, surface science, biochemistry, and biology, and thereby bring together a vast range of literature that covers each topic. While the title of each volume informs of the specific concentration chosen by the volume editors and chapter authors, the integral outcome of the series aims is to offer a broad-based analysis of the total development of the field. The progress of the series will provide a global definition of what electrocatalysis and electrochemistry are concerned with now and how these fields will evolve overtime. The purpose is twofold; to provide a modern reference for graduate instruction and for active researchers in the two disciplines, and to document that electrocatalysis and electrochemistry are dynamic fields that are ever-expanding and ever-changing in their scientific profiles.

Creation of each volume required the editor involvement, vision, enthusiasm and time. The Series Editor thanks all the individual volume editors who graciously accepted the invitations. Special thanks are for Ms. Anita Lekhwani, the Series Acquisitions Editor, who extended the invitation to the Series Editor and is a wonderful help in the assembling process of the Series.

ANDRZEJ WIECKOWSKI
*Series Editor*

# CONTRIBUTORS

**Radoslav R. Adzic**, Brookhaven National Laboratory, Upton, NY 11973

**Matthias Arenz**, Department of Chemistry, University of Copenhagen, Copenhagen, Denmark

**Fraser A. Armstrong**, Department of Chemistry, Oxford University, South Parks Road, Oxford OX1 3QR, United Kingdom

**Renata Bilewicz**, Faculty of Chemistry, University of Warsaw, Pasteura 1, 02-093 Warsaw, Poland

**Nicéphore Bonnet**, Department of Materials Science and Engineering, Massachusetts Institute of Technology, Cambridge, MA 02139

**Fikile R. Brushett**, Department of Chemical and Biomolecular Engineering, University of Illinois at Urbana—Champaign, Urbana, IL 61801

**Juan Pablo Busalmen**, Laboratorio de Bioelectroquímica, INTEMA(CONICET), UNMdP. Juan B. Justo 4302, B7608FDQ, Mar del Plata, Argentina

**Karen Chan**, Department of Chemistry, Simon Fraser University, Burnaby, British Columbia, Canada

**Ismaila Dabo**, Université Paris-Est, CERMICS, Projet Micmac ENPC-INRIA, 6-8 Avenue Blaise Pascal, 77455 Marne-la-Vallée Cedex 2, France

**Michael Eikerling**, Department of Chemistry, Simon Fraser University, Burnaby, British Columbia, Canada

**Abraham Esteve-Nuñez,** Departamento de Química Analítica e Ingeniería Química, Universidad de Alcalá, Madrid, Spain

**Juan Miguel Feliu,** Instituto de Electroquímica, Universidad de Alicante, Apartado de Correos 99, 03080 Alicante, Spain

**Peter Ferrin**, Department of Chemical and Biological Engineering, University of Wisconsin—Madison, Madison, WI 53706

**Dominic Gervasio**, Department of Chemical and Environmental Engineering, University of Arizona, Tucson, AZ 85721

**Hendrik A. Heering**, Leiden Institute of Chemistry, Leiden University, PO Box 9502, 2300 RA Leiden, The Netherlands

**Paul J. A. Kenis**, Department of Chemical and Biomolecular Engineering, University of Illinois at Urbana—Champaign, Urbana, IL 61801

**Michael L. Klein**, Institute for Computational Molecular Science, College of Science and Technology, Temple University, Philadelphia, PA 19122

**Marc T. M. Koper**, Leiden Institute of Chemistry, Leiden University, PO Box 9502, 2300 RA Leiden, The Netherlands

**Carol Korzeniewski**, Department of Chemistry and Biochemistry, Texas Tech University, Lubbock, TX 79409

**Heather J. Kulik**, Department of Materials Science and Engineering, Massachusetts Institute of Technology, Cambridge, MA 02139

**Yanli Li**, Université Paris-Est, CERMICS, Projet Micmac ENPC-INRIA, 6-8 Avenue Blaise Pascal, 77455 Marne-la-Vallée Cedex 2, France

**Nenad M. Markovic**, Materials Science Division Argonne National Laboratory, Argonne, IL 60439

**Nicola Marzari**, Department of Materials Science and Engineering, Massachusetts Institute of Technology, Cambridge, MA 02139

**Manos Mavrikakis**, Department of Chemical and Biological Engineering, University of Wisconsin—Madison, Madison, WI 53706

**Agostino Migliore**, Center for Molecular Modeling, Department of Chemistry, University of Pennsylvania, Philadelphia, PA 19104

**Jens K. Nørskov**, Department of Physics, Center for Atomic-Scale Materials Design, Technical University of Denmark, DK-2800, Lyngby, Denmark

**Marcin Opallo**, Institute of Physical Chemistry, Polish Academy of Sciences, ul. Kasprzaka 44/52, 01-224 Warsaw, Poland

**Odysseas Paschos**, Department of Physics, Technische Universität München, James Franck Strasse 1, D-85748, Garching, Germany

**David E. Ramaker**, Chemistry Department, George Washington University, Washington, DC 20052

**Jan Rossmeisl**, Department of Physics, Center for Atomic-Scale Materials Design, Technical University of Denmark, DK-2800, Lyngby, Denmark

**Christina Roth**, Institute for Materials Science, Technische Universität, Darmstadt, Germany

**Ata Roudgar**, Department of Chemistry, Simon Fraser University, Burnaby, British Columbia, Canada

**Kotaro Sasaki**, Brookhaven National Laboratory, Upton, NY 11973

**Carlo U. Segre**, Physics Division, Illinois Institute of Technology, 3101 S. Dearborn St., Chicago, IL 60616

**Patrick H.-L. Sit**, Center for Molecular Modeling, Department of Chemistry, University of Pennsylvania, Philadelphia, PA 19104 and Institute for Computational Molecular Science, College of Science and Technology,Temple University, Philadelphia, PA 19122

**Eugene S. Smotkin**, Department of Chemistry and Chemical Biology, 417 Hurtig Hall, Northeastern University, Boston, MA 02116

**Ulrich Stimming**, Department of Physics, Technische Universität München, James Franck Strasse 1, D-85748 Garching, Germany and ZAE Bayern Division 1, Walther Meissner Strasse 6, D-85748 Garching, Germany

**Yung-Eun Sung**, World Class University Program of Chemical Convergence for Energy and Environment, School of Chemical and Biological Engineering, Seoul National University, Seoul 151-744, Korea

**Hiroyuki Uchida**, Clean Energy Research Center, University of Yamanashi, 4 Takeda, Kofu 400-8510, Japan

**Miomir B. Vukmirovic**, Brookhaven National Laboratory, Upton, NY 11973

**Mitsuru Wakisaka**, Fuel Cell Nanomaterials Center, University of Yamanashi, 4 Takeda, Kofu 400-8510, Japan

**Jia X. Wang**, Brookhaven National Laboratory, Upton, NY 11973

**Liya Wang**, Department of Chemistry, Simon Fraser University, Burnaby, British Columbia, Canada

**Masahiro Watanabe**, Fuel Cell Nanomaterials Center, University of Yamanashi, 4 Takeda, Kofu 400-8510, Japan

**Andrzej Wieckowski**, Department of Chemistry, University of Illinois at Urbana—Champaign, Urbana, IL 61801

**Holger Wolfschmidt**, Department of Physics, Technische Universität München, James Franck Strasse 1, D-85748 Garching, Germany

**Sung Jong Yoo**, Fuel Cell Center, Korea Institute of Science and Technology, Seoul 136-791, Korea

CHAPTER 1

# Hydrogen Reactions on Nanostructured Surfaces

HOLGER WOLFSCHMIDT and ODYSSEAS PASCHOS

Department of Physics, Technische Universität München, Garching, Germany

ULRICH STIMMING

Department of Physics, Technische Universität München and ZAE Bayern Division 1, Garching, Germany

Hydrogen catalysis is an important scientific field since hydrogen reactions (e.g., hydrogen evolution and hydrogen oxidation) play a key role in electrochemical devices such as fuel cells and electrolyzers. The latter devices have the potential to provide clean and sustainable energy with high efficiencies. This chapter reviews hydrogen catalysis in detail. Details on hydrogen reaction studies from theoretical and experimental perspectives are presented. The former usually complement the results from experimental studies and are used to strengthen them. Various systems that have been explored throughout the years are reviewed. These include model surfaces as well as applied systems. Model catalyst systems comprise Pt and Pd nanoislands deposited on planar surfaces of inert supports, high-quality single-crystal materials, or single nanoparticles created with scanning tunneling microscopy tips. Applied systems consist of metallic nanoparticles deposited on high-surface-area carbon supports. Theory versus experiment, and model versus applied systems are reviewed in detail, and useful insights for hydrogen reactions in these systems are demonstrated

## 1.1 INTRODUCTION

Whereas the nineteenth century was the stage of the steam engine and the twentieth century was the stage of the internal-combustion engine, it is likely that the twenty-first century will be the stage of the fuel cell. Fuel cells have captured the interest of people

*Fuel Cell Science: Theory, Fundamentals, and Biocatalysis*,
Edited by Andrzej Wieckowski and Jens K. Nørskov

around the world as one of the next great energy alternative. They are now on the verge of being introduced commercially, revolutionizing the present method of power production. Fuel cells can use hydrogen as fuel and oxygen or air as oxidant, offering the prospect of supplying the world with clean, sustainable electrical power, heat, and water.

This chapter focuses on hydrogen reactions such as the hydrogen oxidation reaction (HOR) and the hydrogen evolution reaction (HER). These reactions are of utmost importance in developing and improving fuel cell devices. The discussion here is directed principally toward hydrogen electrocatalysis from an experimental as well as theoretical perspective. Starting with an overview on the fundamentals of hydrogen reactions in Section 1.2, studies on single crystals as well-defined and high-quality surfaces are reviewed. An introduction to theoretical work calculating important fundamentals for hydrogen catalysis regarding material properties is then discussed. As predicted by theory, the behavior of nanostructured and bimetallic surfaces differs from that of bulk material. Similar findings supporting the theoretical predictions are shown for large nanostructured surfaces as well as single particles. The section concludes with a short overview of carbon-based catalysts.

The fundamentals of hydrogen reactions are reviewed in Section 1.2. Starting from the general reversible hydrogen reaction, the different reaction pathways suggested by Volmer, Heyrowsky, and Tafel are introduced. Because of the importance of the hydrogen adsorption mechanism and the important findings with new experimental techniques, a short overview of results obtained since the late 1990s is given. An introduction to the correlation between catalytic behavior and the catalyst material significance of this correlation, completes this section using experimental and theoretical calculations, with a conclusion regarding the long-range.

Single crystals and well-defined surfaces play a very important role in surface science. Many scientific contributions are available that study these well-defined surfaces. Section 1.3 introduces the electrochemical behavior toward hydrogen reactions on Pt, Au, and Pd surfaces. The quality of single crystals rapidly increased in the 1990s, resulting in new and different insights. Because of the importance of Pt as a catalyst, the main part of this section focuses on this element. The dependence of the crystallographic orientation toward adsorption as well as electrocatalytic activity is discussed. An introduction to Pd as a catalyst material with the property to absorb hydrogen and Au as an inert support material is the last topic in that section.

Besides experimental work, numerous theoretical calculations for hydrogen catalysis have been performed. Computational methods such as density functional theory (DFT) and Monte Carlo simulations are powerful tools in surface science and catalysis. Theoretical as well as experimental work has been combined in several scientific publications and complement each other well. The first principles of theoretical techniques and theoretical results are shown in Section 1.4. As a main topic, the adsorption behavior of hydrogen is considered and the *d*-band model is introduced. Calculations regarding the hydrogen oxidation reaction and the influence of different reactions pathways are also shown. Theoretical calculations of metals on thin films and supported on various foreign metals are reviewed and are correlated with experimental findings.

The chemical behavior of metal nanoparticles often differs from that of bulk metal. Different effects such as particle size, interparticle distance, and support effects have to be considered in this nanometer-scale regime. Since Pd and Pt are important materials in catalysis, much work was done in the last few decades describing the abovementioned effects. In particular, multilayers, monolayers, and submonolayers of Pd and Pt onto foreign metal supports have shown unexpected behavior. Pd on Au(111) regarding several electrochemical properties introduces this section. Different types of adsorption, absorption, and desorption behavior as well as electrocatalytic activity toward hydrogen reactions are shown and discussed. The deposition of Pd on other supports and the influence of hydrogen reactions hindered by adsorbing foreign adsorbates as well as investigations of Pt overlayers on Au(111) are also discussed. A summary and detailed discussion in Section 1.5 also includes theoretical aspects.

As mentioned above, planar surfaces are thoroughly investigated and serve as widely accepted reference systems with high-quality, reproducible results. For local investigation of small structures, new approaches and setups have to be designed and applied. For this purpose, the electrochemical scanning tunneling microscope (EC-STM) has been modified by several groups in order to create small nanoparticles and nanoparticle arrays and also to investigate corrosion, deposition, dissolution, and reactivity. Due to the tunneling effect, high resolution is achievable and thus leads to a precise technique with atomic resolution. The STM tip can be used in different ways in the electrochemical environment to create and investigate local reactivity of nanostructures. Experimental and theoretical results are compared and are shown to complement each other. Specifically, the activity of a single Pd particle is shown. A discussion of the experimental results of the stability of Pd particles deposited on Au (111) and their reactivity toward HER follows. A summary completes Section 1.6 with a comparison between results obtained at extended Pd nanostructured Au(111) surfaces and single Pd particles.

Section 1.7 presents an overview of studies performed on carbon-based systems. Since carbon has high electrical conductivity, is relatively inexpensive to use, and is highly available, it has been the favored support material for many years. Of the many scientific contributions, only a few can be presented here regarding the mechanism of HER and HOR using metallic nanoparticles with carbon-based supports. The reactivity of these catalysts for hydrogen reactions and CO oxidation is also of major interest. These catalyst systems include glassy carbon, carbon nanofibers, Vulcan, and carbon black for support for metallic nanoparticles, and the more highly oriented and defined pyrolytic graphite (HOPG) are also presented and discussed.

## 1.2 FUNDAMENTALS OF HYDROGEN REACTIONS

### 1.2.1 Hydrogen Catalysis

Over the years a number of studies have been performed in order to investigate the characteristics of hydrogen-related reactions. The general reversible reaction is as follows:

$$H^+ + e^- \leftrightarrow \tfrac{1}{2}H_2 \tag{1.1}$$

Its standard potential is set to 0 V. In the case of proton discharge to form molecular hydrogen the reaction is called a *hydrogen evolution reaction* (HER), while the reverse pathway describes the *hydrogen oxidation reaction* (HOR). However, for the reaction to proceed at sufficient rate, it needs to be catalyzed on an electrode surface. Possible catalyst candidates include various metals such as Pt, Pd, and Ru, as well as enzymes with active centers. Much research focused on finding parameters that influence the activity of materials toward hydrogen electrocatalysis. Even though much progress has been made on this matter, it is still not clearly known how various properties influence the catalytic activity. More details will be given later in this section.

Today it is generally accepted that hydrogen evolution on Pt occurs via two different pathways consisting of every two reaction steps:

Discharge reaction of a proton to form an adsorbed hydrogen atom, known as the *Volmer reaction* [1]:

$$Pt + H^+ + e^- \rightarrow H{-}Pt \tag{1.2}$$

Combination of two adsorbed hydrogen atoms to form molecular hydrogen, known as the *Tafel reaction* [2]:

$$2(H{-}Pt) \rightarrow H_2 + 2Pt \tag{1.3}$$

Combination of an adsorbed hydrogen atom with a proton and an electron to form molecular hydrogen, known as the *Heyrovsky reaction* [3]:

$$H{-}Pt + H^+ + e^- \rightarrow H_2 + Pt \tag{1.4}$$

Two different pathways can occur; the first one is described as a combination of reactions (1.2) and (1.3), known as the *Volmer–Tafel mechanism*. With this mechanism, protons from the solution are discharged on the catalyst surface, forming adsorbed hydrogen atoms. Then, two adjacent adsorbed hydrogen atoms combine to form molecular hydrogen. The second mechanism, known as the *Volmer–Heyrovsky mechanism*, can be described by combining reactions (1.2) and (1.4). A proton from electrolyte solution is discharged on the catalyst surface to form an adsorbed hydrogen atom. This step is followed by combination with another proton and electron to form molecular hydrogen.

Hydrogen oxidation reaction on Pt can be described in a similar way using the reaction pathways in reverse order:

Dissociation of molecular hydrogen into one adsorbed hydrogen atom and immediate discharge of the other atom into proton and electron, similar to the Heyrovsky reaction:

$$H_2 + Pt \rightarrow H{-}Pt + H^+ + e^- \tag{1.5}$$

Adsorption of molecular hydrogen on the catalyst surface in the form of two hydrogen atoms, similar to the Tafel reaction:

$$H_2 + 2Pt \rightarrow 2(H{-}Pt) \tag{1.6}$$

Discharge of an adsorbed hydrogen atom to proton and electron, similar to the Volmer reaction:

$$H{-}Pt \rightarrow Pt + H^+ + e^- \tag{1.7}$$

Similar to hydrogen evolution, the hydrogen oxidation reaction can follow two different pathways. The first mechanism is a combination of reactions (1.5) and (1.7). A hydrogen molecule is positively charged ($H_2 \rightarrow H_2^+$), and immediately one of its atoms is discharged into proton and electron, while the other is adsorbed on the surface of the catalyst. Then the adsorbed hydrogen atom is discharged into proton and electron. The second one is a combination of reactions (1.6) and (1.7). With this mechanism molecular hydrogen is adsorbed on the catalyst surface in the form of two hydrogen atoms, followed by discharge of the atoms into proton and electron.

### 1.2.2 Hydrogen Adsorption Mechanism and Experimental Setups

As was shown above, in both hydrogen reactions (oxidation and evolution) the step of forming a hydrogen adsorbate on the catalyst surface exists in both pathways. Research was performed in order to study the mechanism of hydrogen adsorption on Pt single crystals. Pt is one of the most widely studied catalysts because of its ability to catalyze hydrogen reactions with small overpotentials. Initial studies focused on determining the heat of adsorption of hydrogen on Pt(111) single crystals. An interesting review was published by Markovic and Ross [4], who showed the values for the heat of hydrogen adsorption reported in early years to be inconsistent. Christmann and Ertl [5] reported in 1976 that the value for Pt(111) is approximately equal to 50–60 kJ/mol. However, McGabe and Schmidt [6] in 1977 and Salmeron et al. [7] in 1979 reported higher values, between 70 and 90 kJ/mol. Later, it was found that these corresponded to adsorption of hydrogen on defect sites. Until relatively recently it was accepted that hydrogen tends to adsorb on highly coordinated sites, which for the case of Pt(111) would be the threefold hollow sites. These would lead, though, to a very high coverage of two hydrogen atoms per Pt; therefore, in order to ensure agreement with experimental values, it was accepted that hydrogen occupies the threefold next-nearest-neighbor sites (for details, see Section 1.3). Olsen et al. [8], performing DFT calculations, showed that hydrogen tends to occupy the top sites. Nevertheless, in all cases the values reported were close to each other.

Depending on the overpotential, the adsorbed hydrogen atom on the catalyst surface is referred to as *under-* or *overpotential deposited hydrogen* ($H_{upd}$ or $H_{opd}$). $H_{upd}$ refers to hydrogen atoms adsorbed at potentials positive of the reversible hydrogen electrode (RHE) potential, while $H_{opd}$ occurs at potentials negative of the RHE potential. The state of $H_{upd}$ and $H_{opd}$ depends also on the pH of the electrolyte and

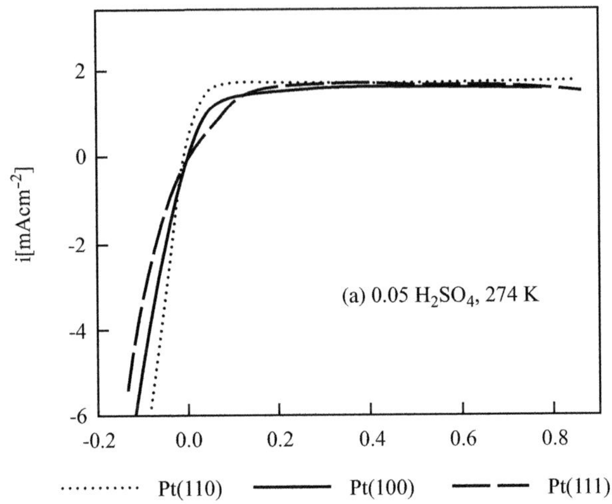


**FIGURE 1.1** Polarization curves for HER and HOR on Pt(*hkl*) in 0.1 M $HClO_4$ at sweep rate 20 mV/s [4].

is generated from either protons or water molecules [4] and can be described by the following reactions:

$$\text{Pt} + \text{H}_3\text{O}^+ + \text{e}^- \rightarrow \text{Pt–H}_{\text{upd}} + \text{H}_2\text{O} \quad (\text{pH} \leq 7) \tag{1.8}$$

$$\text{Pt} + \text{H}_2\text{O} + \text{e}^- \text{Pt–H}_{\text{upd}} + \text{OH}^- \quad (\text{pH} \geq 7) \tag{1.9}$$

There are several possible reasons why reported experimental values sometimes do not agree. The first would be the quality of the single crystal. Crystals having defect sites or impurities adsorbed on their surfaces act differently toward electrochemical reactions. Also, it has been shown that different single crystal faces of Pt have different electrocatalytic rates. Markovic and Ross [4] showed that the activity increases in the order of (111) < (100) < (110) (Fig. 1.1).

Barber et al. [11] showed a slightly different result (Fig. 1.2), where the activity for HER/HOR increases in the order of (100) < (111) < (551) < (110).

However, despite these small differences, it is clearly shown that the activity is strongly affected by the orientation of the Pt single crystal.

The experimental technique also plays a determining role on the results obtained mainly with the appearance of a limiting current density above certain overpotentials for the hydrogen reactions. Especially for the case of HOR on Pt, the exchange current density is high in acidic solutions, but simultaneously, because of the low solubility of hydrogen, the limiting diffusion current is low. Quaino et al. [12] showed that by using Levich–Koutecky analysis the $j(\eta)$ dependence for HOR cannot always be obtained accurately and may be underestimated. Bagotzky and Osetrova [13] were the first to propose an alternative experimental setup that had the potential to solve many issues related to the investigation of HOR. Their setup consisted of Pt microelectrodes

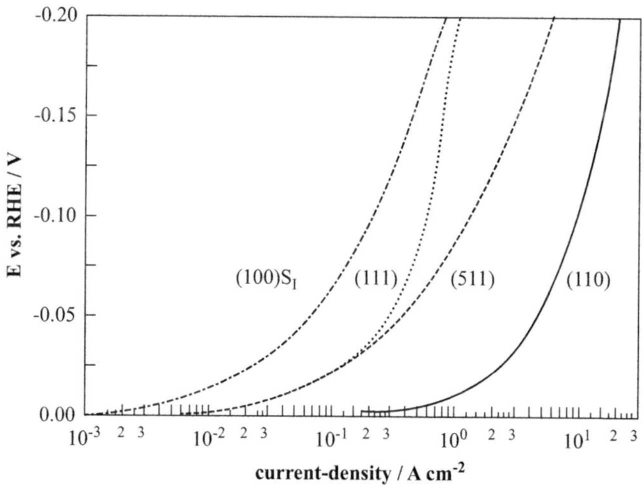


**FIGURE 1.2** Derived Tafel plots, less the diffusion effect, for the Pt (100)$S_I$, (511), (111), and (110) faces as marked on the plot [11].

('A' and 'B') embedded in fused-glass tubes with two different polished surfaces. Because of the small thickness of the electrode, there was an enhanced mass transport of hydrogen. Therefore, values for limiting diffusion current that were one order of magnitude higher than those obtained from RDE setups could be reached. However, the results obtained were affected by the roughness of the microelectrode surface, which can be clearly seen in Figure 1.3.

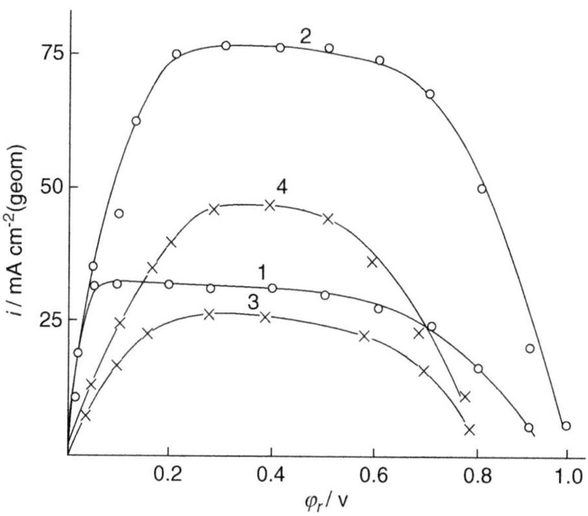


**FIGURE 1.3** Dependence of hydrogen ionization current on potential on microelectrodes with different roughness values *A* and *B* in 0.5 M $H_2SO_4$: (1) *A*, (2) *B* and in 1 M KOH: (3) *A*, (4) *B* [13].

Two electrodes with different roughness values, differentiated by the degree of polishing, were used to study hydrogen oxidation in both acidic and alkaline solutions, and as can be seen, the results are different for the two electrodes. Quaino et al. [14] used a similar setup to study hydrogen oxidation. They were able to demonstrate that at low overpotentials the Tafel–Volmer route dominates the kinetics of HOR. At high overpotentials the Tafel–Volmer effect diminishes while the Volmer–Heyrovsky mechanism becomes dominant.

Also, traces of impurities that can be present in unpurified solutions can compete with the reactions under certain conditions, especially at low current densities [15], resulting in misleading interpretation of the results.

### 1.2.3 Correlation between Activity toward Hydrogen Reactions and Physicochemical Properties of Catalyst Material

In the early years research was focused on finding a relationship between the activity toward hydrogen evolution and oxidation and a property of the catalyst. Conway and Bockris [16] reported a correlation between the exchange current density $j_0$ and the electronic workfunction $f$. *Workfunction* is defined as the energy with which electrons near the Fermi level are bound to the material. According to their study [16], the relationship between $j_0$ and $f$ arises from the dependence of heat of adsorption on $f$. Additionally, they showed that the bond strength between adsorbed hydrogen and metal calculated from Pauling's equation was smaller than the one obtained from experiments. They also, as can been seen in Figure 1.4, using values from the literature, demonstrated that for various metals (e.g., Ta, Mo, W, Cu, Ni, Fe, Rh, Pd, Pt) the logarithm of $j_0$ increases as the heat of adsorption of H decreases, while an opposite trend is observed for Hg, Cd, Pb, and Tl.

For HER, it was also shown (Fig. 1.5) that the logarithm of $j_0$ increases as the $d$ character of the material increases.

The latter was explained by the fact that as the $d$ character increases, more electrons have paired spins and hence require more energy to extract, them causing $\Delta H$ of adsorbed hydrogen atoms to decrease.

Parsons [17] studied the relationship between exchange current density and the ability of the electrode to adsorb atomic hydrogen in terms of the standard free energy $\Delta G_H$. His theoretical studies showed that log $j_0$ reaches a maximum when $\Delta G_H \sim 0$. Even though he mentions a disagreement between experimental and theoretical results (a similar disagreement is also mentioned by Trasatti [18] for the heat of adsorption, the observed trend should still be valid. Metals that adsorb hydrogen weakly ($\Delta G_H$ has positive values), such as Hg, Zn, and Sn, have low exchange current densities. Metals such as Pt that adsorb moderately hydrogen have high values of $j_0$ and metals that adsorb hydrogen strongly, such as Mo, Ta, and W, also have low $j_0$ values.

It was shown that there is dependence between the exchange current density for hydrogen reactions and the workfunction of the catalyst material. However, workfunction values were usually used by electrochemists as obtained from physical experiments. These values were usually measured using nonelectrochemical inputs

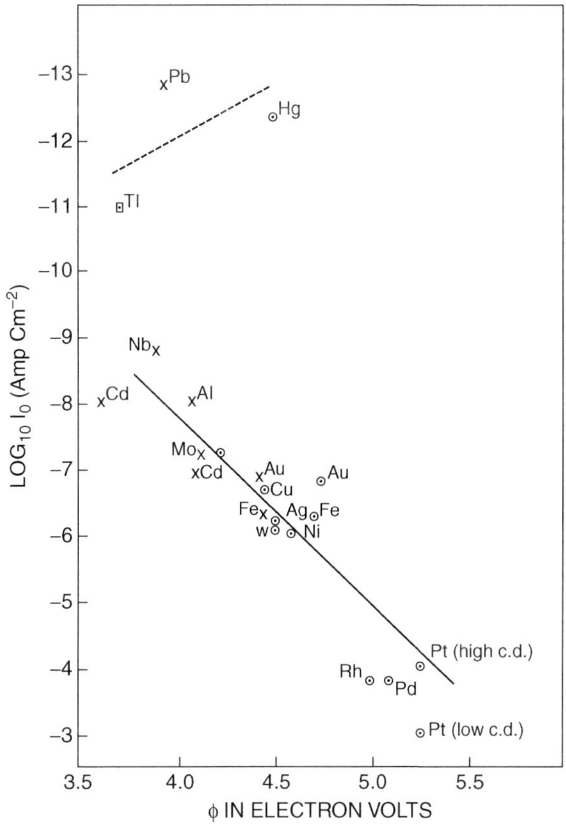


**FIGURE 1.4** Linear dependence of $\log_{10}$ of the exchange current ($i_0$) of HER on the electronic workfunction *f*. Values of $\log_{10} i_0$ are taken from the literature [16].

such as adsorption of gas hydrogen on metals and without taking into consideration the chemical environment surrounding the catalyst. Trasatti [18] published an interesting review on several aspects in order to obtain more accurate data regarding the correlation of hydrogen reactions to physicochemical properties of materials. He argued that the sign of the charge of electrode surface is usually ignored. If the exchange current density $j_0$ is plotted versus the workfunction (Fig. 1.6), then two fairly parallel lines can be obtained.

One line consists of data from transition metals and *sp* metals with positively charged surfaces, while the other includes data from *sp* metals with negatively charged surfaces. It is also noteworthy that the lines are approximately 0.4 eV apart from each other. Trasatti explained the division of materials in these two groups in terms of orientation of water molecules on the catalyst surface. As is shown in Figure 1.7, if the surface is positively charged, then water will be positioned with an oxygen atom toward the metal, whereas in the case of a negatively charges catalyst surface, an opposite orientation is expected.

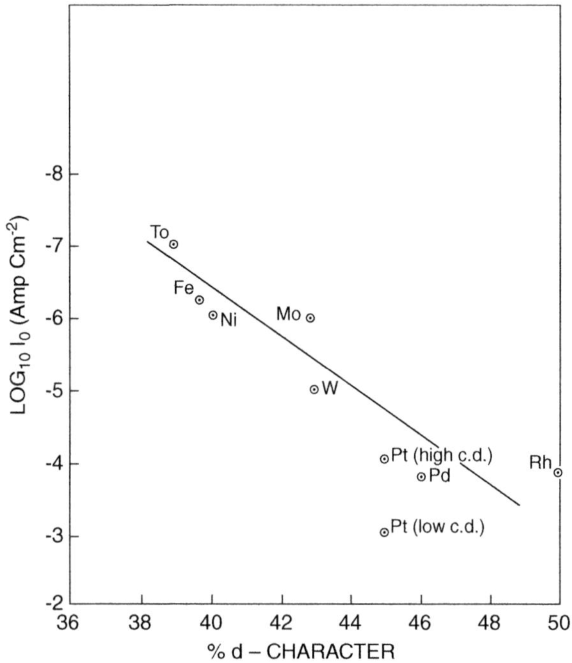


**FIGURE 1.5** $\mathrm{Log}_{10}\, i_0$ for HER as a function of percent *d* character of the metal [16].

Although the plot shows a clear difference between transition and *sp* metals, it includes no information regarding the mechanism of reaction. This information can be factored in only if the exchange current density is plotted versus the heat of adsorption of hydrogen on the metal surface. As mentioned previously, the rate and mechanism of HER depends on the bond strength between the metal and the hydrogen atom (M–H). Parsons reported that it should pass through a maximum, and a similar volcano-shaped curve was reported by Krishtalik and Delahay [19], as shown in Figure 1.8.

As shown in Figure 1.8, Pt is on the top of the volcano curve where the Pt–H bond is neither too strong nor too weak. The general trend observed in the volcano curve is that for several metals, as the bonding energy of hydrogen to the metal increases, the activity also increases, reaching a maximum. Then an opposite trend is observed, where log $j_0$ decreases as the bonding strength of hydrogen to the metal increases.

A similar study was done by Nørskov et al. [20]. Density functional theory (DFT) calculations demonstrated a volcano-type behavior of hydrogen chemisorption energies with respect to exchange current density for hydrogen evolution (Fig. 1.9).

Platinum was again found to be a better catalyst than other metals for HER primarily because hydrogen evolution reaction on Pt is thermoneutral at the equilibrium potential. The findings of this work can be used to predict behavior of other bimetallic systems for HER as well as HOR. The analysis was reported as a new method to obtain H adsorption free energies and understand trends for different systems that are of electrochemical interest.

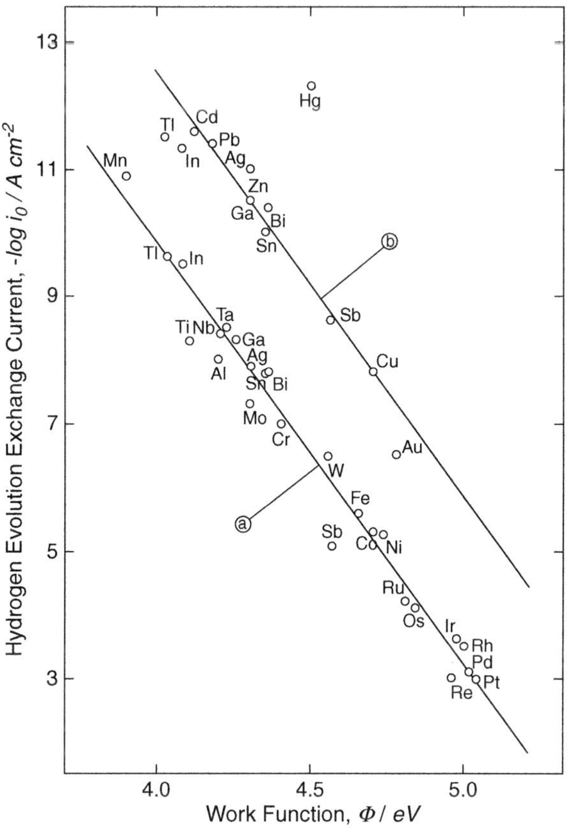


**FIGURE 1.6** Exchange currents for HER versus workfunction of metals [18].

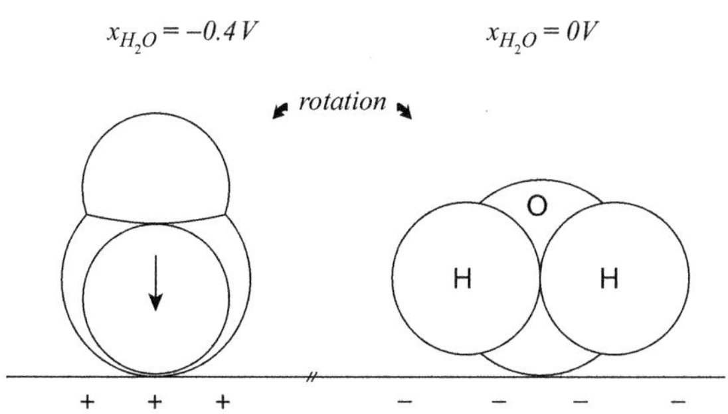


**FIGURE 1.7** Drawing of water molecules in position of maximum and minimum orientation suggested as occurring on positively and negatively charged metal surfaces, respectively [18].

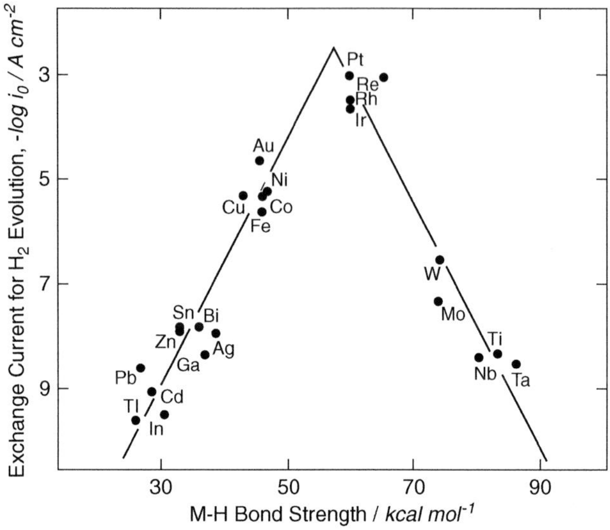


**FIGURE 1.8** Exchange currents for HER versus strength of intermediate metal–hydrogen bond formed during electrochemical reaction [18].

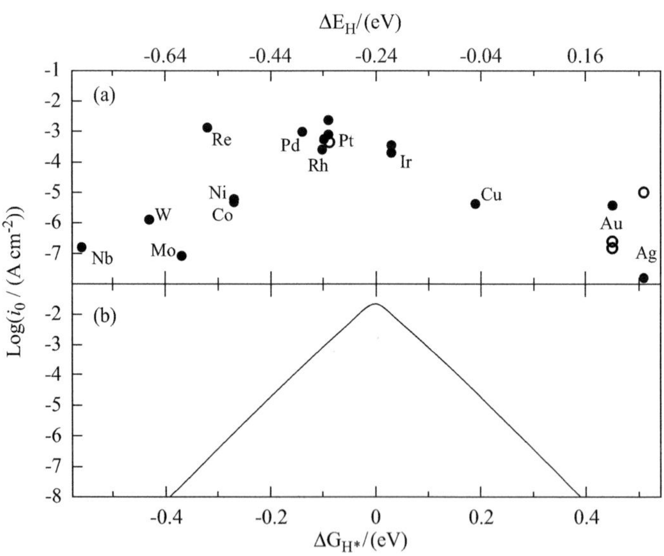


**FIGURE 1.9** (a) Experimentally measured exchange current $\log(i_0)$ for HER over different metal surfaces plotted as a function of calculated hydrogen chemisorption energy per atom, $\Delta E_H$ (top axis). Open symbols indicate data for single crystals. Results obtained for the simple kinetic model are plotted as a function of the free energy for hydrogen adsorption, $\Delta G_{H^*} = \Delta E_H + 0.24\,\text{eV}$ [20].

## 1.3 FUNDAMENTAL STUDIES OF HYDROGEN REACTIONS ON EXTENDED SINGLE-CRYSTAL SURFACES

### 1.3.1 Introduction

Extended well-defined surfaces such as single-crystal surfaces play an important role in surface science by providing accurate and precise systems. Especially in the case of electrochemistry, usage of single crystals has yielded new findings. Hydrogen reactions have been intensively studied have been on polycrystalline as well as single-crystal electrodes. Although the quality of single crystals [21–25] and experimental conditions rapidly increased in the 1990s, fundamental studies started long before [26–29]. Because of the high catalytic activity, Pt and Pd surfaces are of special interest. The main focus of this section is on Pt single crystals and the dependence of crystallographic orientation on electrocatalytic activity. Then Au is also be reviewed, although less work has been performed as compared to Pt. Because of the inertness and low activity of gold and the uncomplicated procedure to prepare Au single crystals, gold is often used as a support for depositing foreign metals, enzymes, and so on. Section 1.3.6 presents a brief introduction to Pd single crystals and Pd overlayers.

### 1.3.2 Pt(*hkl*) Surfaces

Because of the high degree of catalytic activity of Pt toward several reactions such as hydrogen evolution, hydrogen oxidation, oxygen reduction, and methanol oxidation, Pt is used as a catalyst in various applications. Studies of Pt on polycrystalline and single-crystal surfaces began in the 1970s or so. Annealing as a preparation method for Pt single crystals is widely described in the literature [30, 31]. It was also found that the cooling procedure following the annealing process is important for a well-defined structure. By applying the right annealing and cooling procedures, one can observe specific peaks in the hydrogen region in cyclic voltammograms [32–35], and typical surface structures can be observed via STM [36–38].

Typical voltammograms of Pt(111), Pt(100), and Pt(110) in 0.1 M $H_2SO_4$ are shown in Figure 1.10. The following discussion concentrates on important findings on Pt(111). Details for Pt(100) and Pt(110) surfaces can be found elsewhere [37–43]. The characteristic current peak at 0.19 V versus SCE in the voltammogram of Pt(111) in 0.1 M $H_2SO_4$ is caused by a rearrangement in the adsorbed sulfate ion adlayer [44] and can be seen as a parameter for the quality of the single crystal. The ordered $(3 \times 7)$ $R19.1^\circ$ adlayer of these sulfate ions can be observed via STM, which was first demonstrated by Funtikov et al. [45,46]. Comparing the Pt(111) in 0.1 M $HClO_4$ and in 0.1 M $H_2SO_4$ shows no differences in cyclic voltammograms below 0.1 V versus SCE, indicating a weak influence of anions in this region.

Platinum surfaces were studied regarding electrocatalytic activity on polycrystalline electrodes or poorly defined single crystals [26–29]. Well-defined single-crystal surfaces with unique properties and clean experimental conditions, which are absolutely necessary for electrochemical investigations on Pt surfaces, were established in the 1990s. The hydrogen reactions were intensively investigated on polycrystalline as

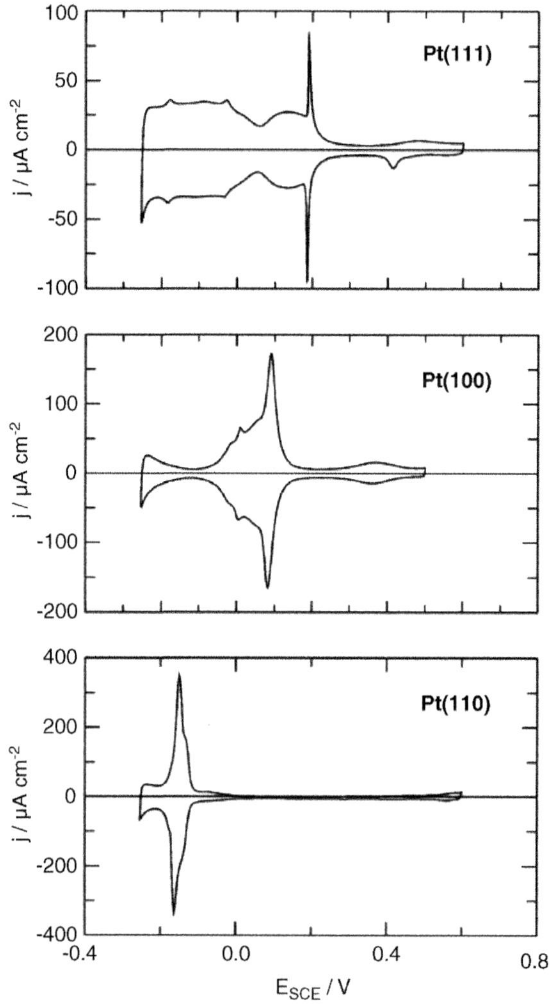


**FIGURE 1.10** Cyclic voltammograms for freshly prepared Pt(111), Pt(100), and Pt(110) in 0.1 M $H_2SO_4$ after cooling in a CO atmosphere and oxidative stripping of the CO adlayer; scan rate 50 mV/s [47].

well as single-crystal surfaces. In early measurments no dependence on crystallographic orientation for electrocatalytic activity was found, which was attributed mainly to poor-quality single crystals and unfavorable experimental conditions. These findings were disproved by Markovic et al. [21,22] and Barber et al. [23,25], who found that the reactivity of different Pt(*hkl*) surfaces toward hydrogen reactions is strongly influenced by surface orientation. It was clearly shown that in alkaline as well as acidic electrolytes, reactivity increases in the order (111) < (100) < (110), as shown in Figure 1.11. In acidic solution the exchange current density of Pt(110) surfaces is 3 times higher than that of Pt(111). The findings are in line with the activation energies

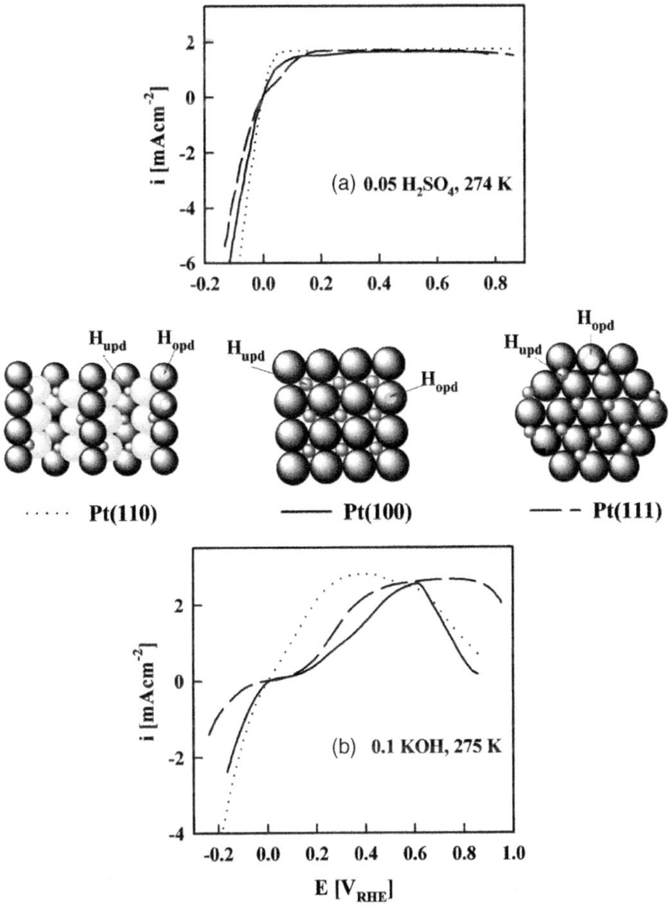


**FIGURE 1.11** (a) Polarization curves for HER and HOR on Pt(*hkl*) in 0.05M $H_2SO_4$ at sweep rate 20 mV/s [2]; (b) polarization curves for HER and the HOR on Pt(*hkl*) in 0.1 M KOH at sweep rate 20 mV/s [1]. Insets show ideal models for the Pt(111)-(1 × 2), Pt(100)-(1 × 1), and Pt(110)-(1 × 1) surfaces; small dots represent the active sites for $H_{upd}$ and $H_{opd}$ [48].

obtained from Arrhenius plots [21] showing the high value of 18 kJ/mol for Pt(111) and the lowest value of 9.5 kJ/mol for the Pt(110) single-crystal surface seen in Figure 1.12.

Arrhenius plots were also used to evaluate the hydrogen evolution in frozen aqueous electrolytes on different polycrystalline metal electrodes [50,51]. Frese et al. [50] investigated HER on polycrystalline Pt electrode in liquid phase and in frozen electrolyte at temperatures ranging between 293 and 176 K. In liquid phase as well as in frozen phase, the Volmer–Tafel mechanism occurs with the Tafel reaction as the rate-determining reaction. An Arrhenius plot (Fig. 1.13) in their paper [50] manifests a linear behavior before and after the melting point of the electrolyte at

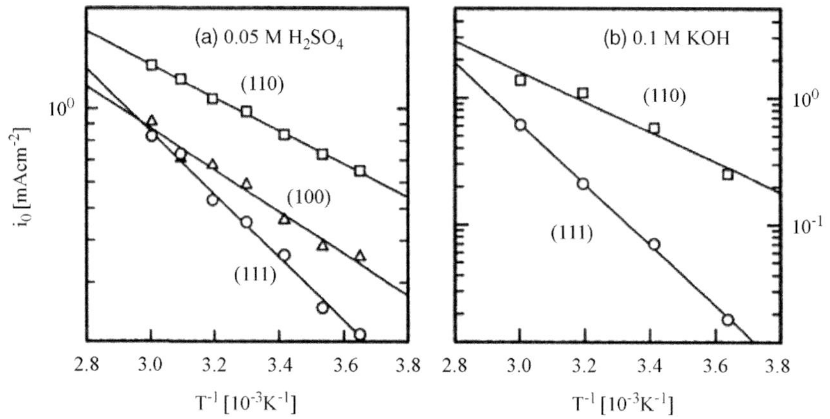


**FIGURE 1.12** Arrhenius plots of the exchange current densities ($i_0$) for HER/HOR on Pt(*hkl*) in (a) acid solution [49] and (b) alkaline solutions [48].

~227 K. The activation energies are in the range of 15 kJ/mol for the liquid electrolyte and 27 kJ/mol in the frozen electrolyte, clearly seen in the change of the slope at the melting point. The value obtained in the liquid electrolyte for the polycrystalline electrode investigated is comparable to the average of the values obtained on single crystals seen in Figure 1.3 [48].

Markovic et al. [22] investigated hydrogen reactions at low temperature to obtain kinetic rates under slow reaction conditions. Here the temperature was varied in the range between 274 and 333 K. In other studies HER and HOR were investigated using a rotating-disk electrode (RDE) setup. Mass-transfer-corrected current densities of HER at 274 K are shown in Figure 1.14. In this approach the exchange current density was determined using micropolarization curves. Using only small overpotentials in

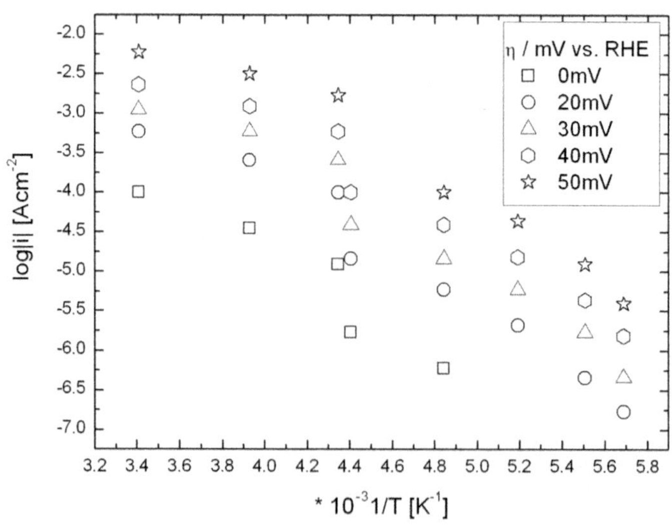


**FIGURE 1.13** Arrhenius plot of polycrystalline Pt (Reproduced from Ref. 50.)

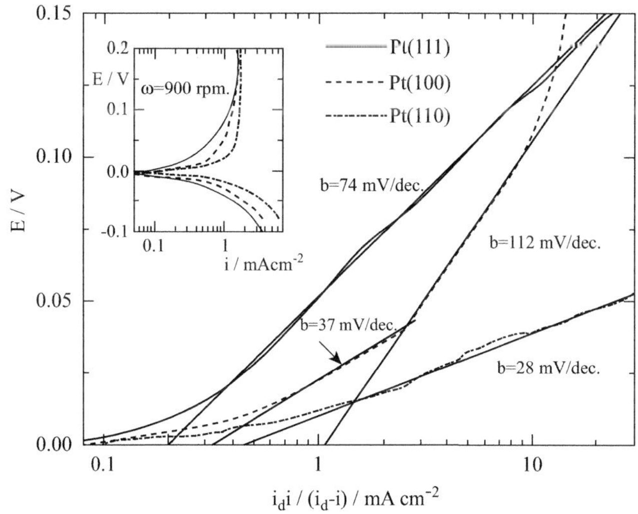


**FIGURE 1.14** Tafel plots of mass-transfer-corrected currents for the HOR on Pt(*hkl*) in 0.05 M $H_2SO_4$ at 274 K. Inset shows polarization curves for HER and HOR at low overpotentials; rotation rate, 900 rpm [22].

the range of ± 10 mV versus the equilibrium potential, one can obtain a reliable value of $j_0$. Therefore the current potential curves were used to determine the slope from which the exchange current densities can be calculated. It was found that different single-crystal surfaces have different Tafel slopes and different exchange current densities ($j_0$). As was found before, the most reactive surface was Pt(110), with the lowest Tafel slope and the highest exchange current density. Interestingly, it was found that for low overpotentials the two branches of the Tafel plot are symmetric, as is shown in the inset of Figure 1.14. Hence, HER and HOR show the same exchange current density. It was added that the evaluation of exchange current density is more or less independent of the rotation speed. The values obtained by the micropolarization curves and the extrapolation in the Tafel plot to overpotentials equal to zero are in good agreement. It was also found that the activity has the same order, Pt(111) < Pt(100) < Pt(110), as mentioned above. At higher temperature the same results were obtained with higher net exchange current densities. The reactivity on Pt(*hkl*) single crystals can be summarized as follows. The reactivity was found to be strongly dependent on the crystallographic orientation of the single crystal. Whereas the activity represented by the exchange current density increases in the order of Pt (111) < Pt(100) < Pt(110), the activation energy decreases in the sequence $\Delta H_{(111)} > \Delta H_{(100)} > \Delta H_{(110)}$. Activity measurements show the same trend for different temperatures with different absolute values.

### 1.3.3 Au(*hkl*) Surfaces

As compared to platinum, gold is also not affected by air and water and hardly assailable by acids or alkalis. Au thin films on mica support are often used in electrochemistry instead of single crystals with comparable results [52,53]. A thin Au film is evaporated on the mica with a very thin chromium interlayer to achieve a good adhesion. Flame annealing is a common procedure for preparing thin Au films and Au single crystals and leads to large (111) oriented terraces.

Cyclic voltammograms for Au(111), Au(100), and Au(110) surfaces in 0.1 M $H_2SO_4$ are shown in Figure 1.15 [27]. The investigations of sulfuric acid with the

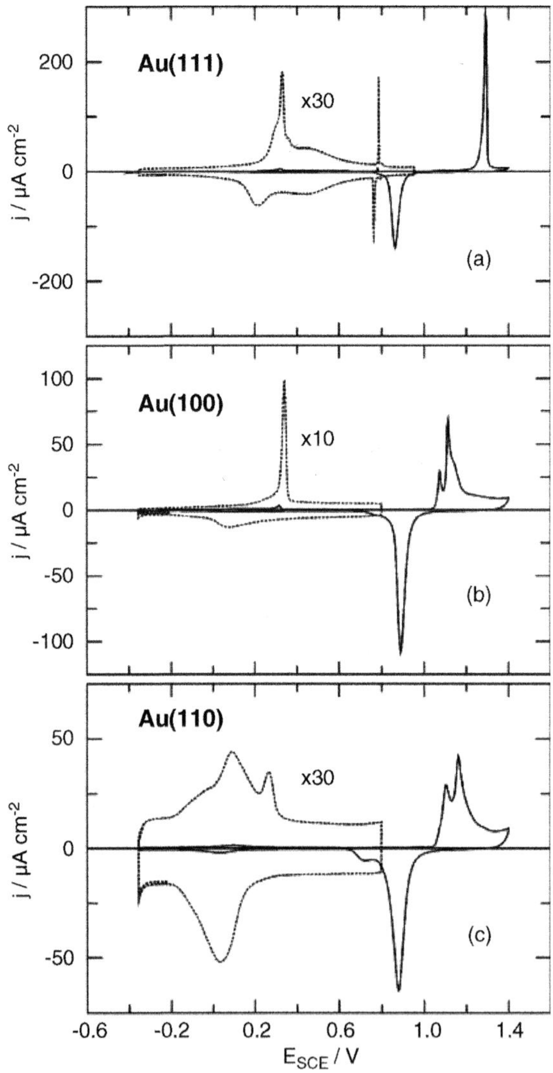


**FIGURE 1.15** Cyclic voltammograms for Au(*hkl*) in 0.1 M $H_2SO_4$, scan rate 10 mV/s. The dotted curves are enlarged as indicated and represent the first cycles in the double-layer region, starting at −0.2 V [47].

typical current peaks in the voltammograms are especially indicative of single-crystal quality. Surface reconstruction of Au(111) and Au(100) is lifted at positive potentials of 0.34 V versus SCE due to adsorption of sulfate ions [54]. This leads to the unreconstructed $(1 \times 1)$ surface. Gold oxide formation begins at potentials higher than 1 V versus SCE. Because of the importance of Au(111) in electrochemistry as a support material, some important properties are summarized. Au(111) has several advantages, such as an easy preparation process and large potential window in the double-layer region. The quality of single-crystal Au(111) can be easily determined by cyclic voltammetry in sulfuric acid and evaluating the peaks attached to the lifting of reconstruction and forming ordered (v3xv7) *R*19.1° sulfate layer [47,55–57]. Defects as well as small terraces will inhibit the well-pronounced peaks at 0.34 V and 0.78 V versus SCE for the abovementioned procedures. Details on investigations on other low-index Au single-crystal surfaces can be obtained elsewhere [54,58–64]

Investigations of Au single crystals with respect to HER are quite rare [65–67] and show a weak dependence on crystallographic orientation in early studies. Comparable to Pt, precise investigations can be done only with high-quality single crystals and high standards for clean experimental conditions. Gonzales et al. [24] investigated the hydrogen evolution in different Au(*hkl*) single-crystal electrodes. To avoid a strong influence of adsorbing ions and to correlate the structure-sensitive reactivity to the different single-crystal electrodes, perchloric acid was used as electrolyte. Using a rotating-disk setup with hanging meniscus, concentration gradients of hydrogen at the interface were avoided. Results similar to those mentioned above for Pt crystal surfaces were obtained. Comparable to Pt, the different Au(*hkl*) surfaces are also structure-sensitive. Figure 1.16 shows that hydrogen evolution is obviously different

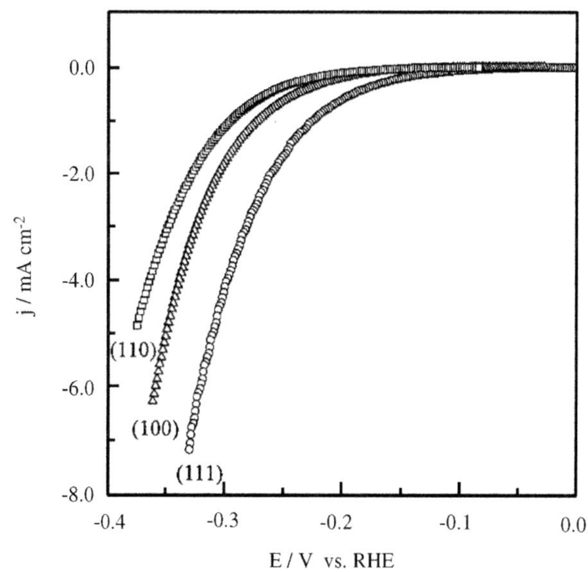


**FIGURE 1.16** Polarization curves for hydrogen evolution on the low-index gold single-crystal electrodes in $N_2$-saturated 0.1 M $HClO_4$; sweep rate 10 mV/s, rotation rate 50 rpm [24].

on the three Au(111), Au(100), and Au(110) surfaces with respect to HER. With this result, it was proved that electrocatalytic activity is definitely dependent on the crystallographic orientation of Au single-crystal surfaces in the following sequence: Au(111) > Au(100) > Au(110). It was suggested that HER activity increases with the atomic density of the surface following the trend of increasing workfunction for the electron [68]. While the activity for hydrogen evolution is also dependent on the crystallographic orientation seen in the case of Pt, the measured net current densities are several orders of magnitude lower for Au.

### 1.3.4 Pd(*hkl*) Surfaces

Palladium, similar to gold and Platinum, is easily deformable and drawable to nearly any kind of profile such as gold and platinum. It is not affected by air and water but dissolves in oxidizing acids and molten alkalis. A unique property of Pd is the high hydrogen absorption ability, which may even change the lattice constant. This behavior complicates the preparation procedure because any kind of hydrogen during annealing would negatively influence the quality. Owing the hydrogen absorption into the bulk Pd, the hydrogen adsorption current region is superposed with the hydrogen absorption current in the cyclic voltammogram. To avoid this problem, thin Pd films that were epitaxially grown on supports such as Au(*hkl*) and Pt(*hkl*) were used to study hydrogen adsorption as well as hydrogen evolution and several other reactions [69–82]. As shown in Figure 1.17 for Pd(111), the potential window of the

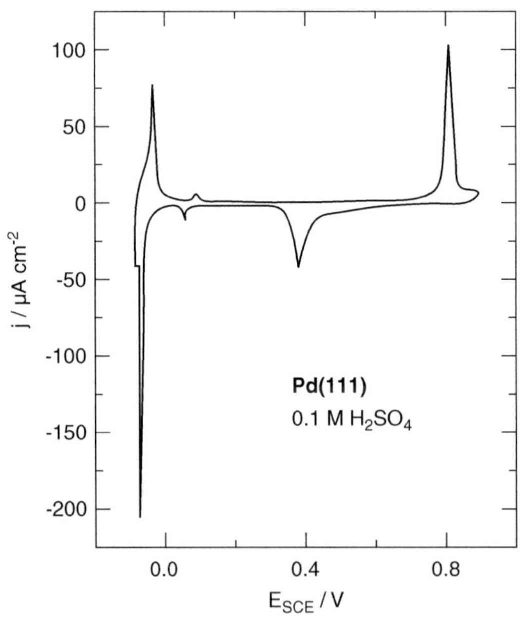


**FIGURE 1.17** Cyclic voltammogram for Pd(111) in 0.1 M $H_2SO_4$; scan rate 10 mV/s [47].

double-layer region is large, and typical current peaks for hydrogen adsorption and oxide formation are pronounced. As mentioned before, Pd is often used as a thin film on different supports; hence the most experimental findings deal with Pd overlayers. Unexpected results for hydrogen reactions of thin Pd layers were obtained relative to electrocatalytic activity, and are reported in later sections

## 1.4 THEORETICAL STUDIES OF HYDROGEN CATALYSIS

### 1.4.1 Introduction

Computational methods such as density functional theory (DFT) or Monte Carlo simulations are powerful tools in surface science and catalysis. More recent improvements provide new insights in adsorption and desorption processes and reaction pathways. These results can be used to develop and design surfaces with specific properties such as new, more effective catalysts. Theoretical predictions often show good agreement with experimental results. Theoretical as well as experimental methods complement each other and therefore allow deeper insights and a comprehensive understanding that could not be explored with only one technique. Up to now theoretical quantum mechanical studies allow qualitative and sometimes even quantitative findings in surface science and catalysis and often play a leading role in new proceedings.

Principles of computational techniques and theoretical results are discussed in Section 1.4.2, The main focus is on the adsorption behavior of hydrogen and a theoretical explanation using the *d*-band center model. Pd on Au(111) is discussed as a special case in Section 1.4.3. This is followed by presentation of results from different theoretical groups from DFT calculations regarding hydrogen oxidation in Section 1.4.4 and the reaction pathway of hydrogen oxidation in Section 1.4.5. Calculations on metal surfaces and thin films of foreign metal on metal support is then reviewed and discussed in comparison to experimental findings.

### 1.4.2 Theoretical Fundamentals of Adsorption Behavior

One of the main objectives for surface scientists is to gain a fundamental understanding of the properties influencing the chemical reactions on surfaces. Hence, chemisorption of different reaction partners to the surface is important to access the abovementioned properties. More recent theoretical investigations show an appropriate, precise, and efficient trend to confirm the experimental results. The main focus is on hydrogen reactions, which are briefly discussed.

A theory of bonding adsorbates on transition metal surfaces was developed by Hammer and Nørskov [83]. In this model the chemisorption energy is determined by the interaction of adsorbate orbitals with surface *sp* and *d* bands. The model is simplified for atomic adsorbates, while the hybridization energy $E_{d\text{-hyp}}$ of the adsorbate with the metal *d* band can be calculated. Constant *sp*-band interaction energies imply that changes in $E_{d\text{-hyp}}$ on various metals are nearly identical to changes in the full chemisorption energies $E_{chem}$. This leads to the conclusion that the Hammer–Nørskov model describes the changes in adsorbate chemisorption energies $E_{chem}$ over different metals that are related to changes of the metal $d$-band centers. Thus the center of the $d$-band has a direct impact on the chemisorption energy $E_{chem}$. Chemisorption energies can be calculated quantitatively with this model, which is, for example, described in References 84–87. Next to the pure metal surfaces, pseudomorphic overlayers deposited on foreign metals also show a change in the chemisorption energies in theoretical calculations. Hydrogen chemisorption on Pd and Re single crystals as well as Pd and Re overlayers were investigated by Pallassana et al. [88,89]. A linear relationship was found between $E_{chem}$ and metal $d$-band center $\varepsilon_d$. Besides adsorption energies, dissociation energies have also been calculated [83]. Summarized trends for chemisorption and thus, activity of well-defined metal surfaces and of overlayers on foreign supports can be calculated. Chemisorption energies and activation energies can be determined for different reactions and give trends for the catalytic behavior.

Next to the more general description of reactivity in terms of chemisorption and activation energy, several theoretical approaches for hydrogen reactions on various catalyst systems are given in the literature [87,90–100]. Hydrogen evolution as well as the hydrogen oxidation play an important role in electrochemical processes and technological applications [101–105]. It has been known for a long time that catalytic activity, which can be represented by the exchange current density plotted versus the hydrogen metal bond strength, is a volcano-shaped curve [106–109]. The shape of the curve is related to the Sabatier principle, which requires a moderate—neither too strong nor too weak—binding energy between surface and reaction partners [110]. Thus, the best catalyst should have intermediate binding energies to the educts as well as to the products. The free hydrogen adsorption energy $\Delta G_H$ is a reliable value for dissipating trends for hydrogen evolution [87,95,109]. Figure 1.18 [94] shows experimental values of exchange current density from the literature [87,95,109] plotted versus theoretical calculated values of $\Delta G_H$. As can be seen, Pt, Pd, Ir, and Ru in the form of bulk crystals or as overlayers on foreign metals are near the top of the volcano curve. Their binding energy to hydrogen is neither too strong nor too weak, therefore yielding good catalyst behavior. Metals or bimetallic surfaces on the left side of the volcano curve bind hydrogen too strongly whereas metals on the right side bind hydrogen too weakly and thus are not adequate catalysts. Although there are some quantitative deviations between different experimental results, a clear trend is visible, which supports the qualitative volcano-shaped behavior. The most reactive catalysts for the hydrogen evolution lie on top of the volcano curve with $\Delta G_H \sim 0$, which is supported by experimental results showing Pt as the best pure metal hydrogen catalyst.

Nørskov et al. [94,96] investigated different alloys using DFT. The solute element was implemented in the host element, and the electrocatalytic properties were determined. It was shown that some alloys have a high degree of activity toward the hydrogen evolution, which is demonstrated by low $\Delta G_H$ values (indicated in dark red in Fig. 1.19). Therefore, $\Delta G_H$ can be used to quantify the hydrogen metal bond strength; values of $\Delta G_H \sim 0$ lead to high HER activity [87]. For this high-throughput method, used to determine catalysts for hydrogen evolution binding energies, activation barriers, preexponential factors, and other parameters were considered. In this

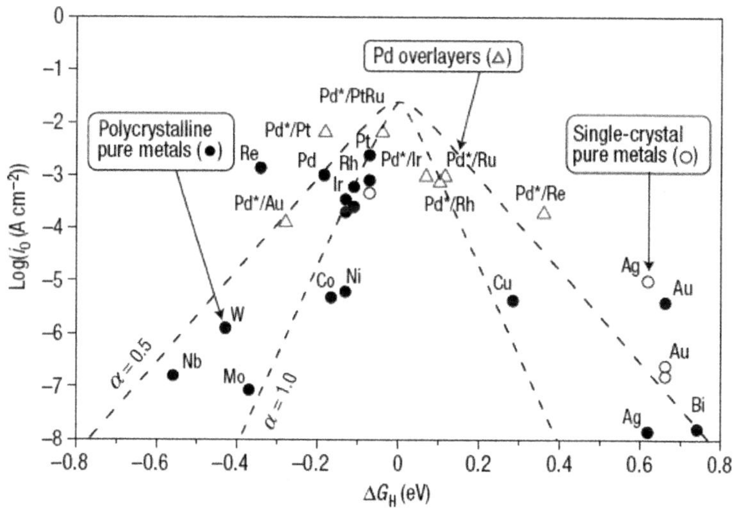


**FIGURE 1.18** Volcano plot for the HER for various pure metals and metal overlayers. $\Delta G_H$ values are calculated at 1 bar of $H_2$ (298 K) and at a surface hydrogen coverage of either $\frac{1}{4}$ or $\frac{1}{3}$ML (i.e., one-fourth or one-third of a monolayer). Experimental data were compiled in the literature [87,95,109]. Where available, computational data are taken from the present work; other computational results are taken from the three references cited above. The experimental data were collected from over 40 years of publications, representing different experimental conditions and surface structures. No corrections for changes in the calculated $\Delta G_H$ values with coverage are included, in contrast to the calculated values for Pd overlayers (denoted by Pd*/ substrate) presented by Greeley et al. [95]. The two curved lines correspond to the activity predictions of simple mean-field, microkinetic models, assuming transfer coefficients ($a$) of 0.5 and 1.0, respectively [12].

case the reduction of protons and the removal of adsorbed hydrogen are balanced. DFT calculations [96] have already shown their power to predict a high activity that is supported by experimental results [94,95]. Hydrogen-binding energies for several bimetallic systems are shown in Figure 1.19. Metal alloys have been evaluated by DFT calculations and were presented in a matrix format. Different promising catalyst systems can be selected and discussed in detail, but instead an overview of trends gleaned from single investigations on several promising candidates should be given. In principle, it can be shown that low solute metal coverages of late-group transition metals have high $\Delta G_H$ values. For early-group metals the opposite trend was found, with free energy decreasing from lower right to upper left of the small matrix inset. These findings are in line with the Hammer–Nørskov model [111,112]. Interestingly, all alloys with Pt, Pd, or Ru are highly active toward hydrogen evolution. Almost all combinations with these three metals as host as well as solute element have $\Delta G_H$ values close to zero. These three metals were also found to be the best catalysts for HER in the Nørskov et al. [87] study, where pure elements were investigated. An astonishing result is that the behavior of a full monolayer of solute metals (shown in Fig. 1.19c) behaves differs from that of a third or two-thirds monolayer.

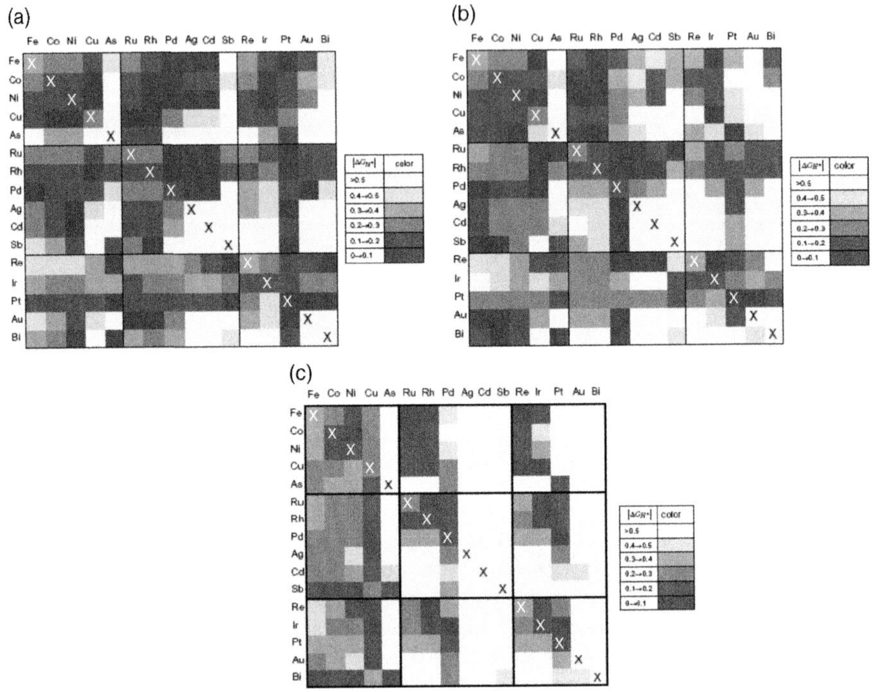


**FIGURE 1.19** Magnitudes of free energies of hydrogen adsorption ($|\Delta G_{\mathrm{H}}|$) at $T = 298$ K, $\theta H_{\frac{1}{3}}$ ML. All free energies are referenced to gas-phase $H_2$. Coverage of solute elements (columns) in the surface layer of host elements (rows) is (a) $\frac{1}{3}$ ML, (b) $\frac{2}{3}$ ML, and (c) 1 ML. The X symbols (tickmarks) on diagonals denote pure elements [96].

This would indicate that submonolayers of solute elements are more reactive than are full monolayers. It was suggested that these findings are due to a strain effect in the lattice constant that strongly influences $\Delta G_{\mathrm{H}}$ and thus, the behavior toward the activity [85]. Although many very promising candidates for catalysts for HER have been presented, only a few are stable under electrochemical conditions. Out of these 768 possibilities, only 30 were predicted to be stable, as they often contain Rh, Ru, Pt, Pd, and Ir, which are known to be active. Other promising candidates containing Rh, As, and Re have to be validated by experiments.

### 1.4.3 Hydrogen Adsorption on Pd Monolayers on Au(111)

After a general overview of chemisorption and adsorption of hydrogen on various metals, a special case is presented in this section. Theoretical calculations of the hydrogen adsorption on Pd on Au(111) [91,92,113] are discussed. Figure 1.20 [113] shows the hydrogen adsorption energy versus the number of Pd overlayers on Au(111) is shown. For comparison, pure Pd is also given on the right side of the figure. According to calculations of lattice constants of Au (4.18 Å) and Pd (3.96 Å), the Pd overlayer is strained by $\sim 5\%$. This expansion leads to higher binding energies of

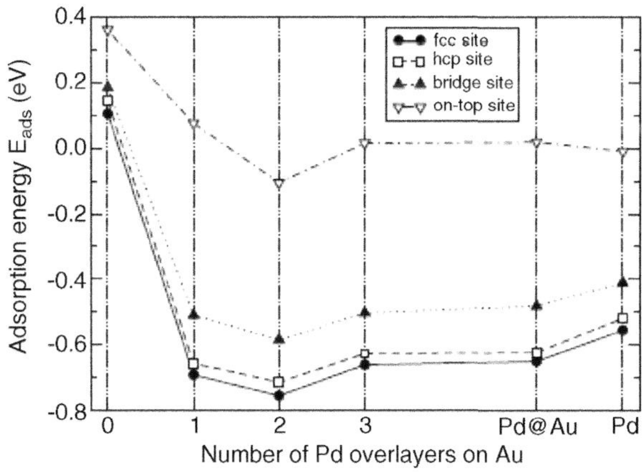


**FIGURE 1.20** Atomic hydrogen adsorption energies as a function of the number of Pd overlayers on Au(111) for different adsorption sites on the (111) at a coverage of $h = 0.25$. The pure Pd substrates with the lateral lattice constant of Au ($a = 4.18$ Å) and Pd ($a = 3.96$ Å) are labeled by Pd@Au and Pd, respectively [113].

hydrogen on Pd according to the *d*-band model. Interestingly, the adsorption energies for one and three monolayers of Pd on Au(111) are similar and do not differ much from those for bulk Pd. Thus, it was suggested that a strain effect due to the support is limited to layers of only a few monolayers' thickness. An unexpected result is that the adsorption energy is maximized for two monolayers of Pd on Au(111), which is due to an extreme position of the *d*-band center. This was explained with the assumption that for only one Pd monolayer, some indirect interactions between the adsorbate and Au(111) influence the activity. This effect is negligible for two monolayers. These results might have to be extrapolated from the solid–vacuum interface to the solid–liquid interface, which requires several modifications. One of the most important issues is the presence of water adlayers on the electrode [114–116]. For less reactive surfaces the water adlayers remain unaffected on the surface and form a stable overlayer. This will change if the surface is reactive and the water may partially dissociate. Although there is an influence of the water adlayer the hydrogen binding energy is reduced by only 100 meV. Because of this weak interaction of water with Pd on Au(111) there will not be a significant influence on the water adlayer. Therefore, it can be concluded that studies at the solid–vacuum interface can give reliable values as well as trends for the adsorption of hydrogen at the solid–liquid interface.

### 1.4.4 Theoretical Fundamentals of HOR

The hydrogen oxidation reaction has been intensively investigated because of its importance in electrochemistry. Interestingly, the rate of hydrogen oxidation varies for different metals over six orders of magnitude. Metals such as mercury have a very slow hydrogen oxidation rate, whereas platinum is highly reactive toward HOR. As mentioned above, the catalytic activity of different materials can be correlated to the workfunction [109], metal hydrogen bonding [87], and the unfilled *d* orbital [117]. More recently developed concepts for gas-phase catalysis were discussed above and mainly in regard to the position of the *d*-band center [90]. The results from gas-phase catalysis calculations can be extrapolated to electrocatalysis, considering important issues such as the interaction with the water adlayer or the solvent. For hydrogen oxidation, a hydrogen molecule is oxidized in two protons and two electrons, whereas the protons are stabilized by a solvent. The electrons are transferred to the metal and the solvated protons desorb into the solution, which will result in a relaxed system [99]. This is induced by lowering the energy of the system when the bonding orbital passes the Fermi level. Santos and Schmickler [98] state that it is important to exercise caution when employing several procedures occurring in the electrochemical environment. As mentioned above, the interaction between water and the proton is very strong; hence, the water molecules have to be considered in the calculations. Furthermore, the electrode potential has not been introduced in the calculations. Additionally, it is not yet possible to consider the solvent fluctuations that change the electronic level past the Fermi level. Different approaches were used to develop a model that accounts for breakage of the electrochemical bond, which was suggested to explain a possible lowering of the activation energy. An important aspect is the interaction of the metal with reaction partners. Therefore, chemisorption, which depends on electronic energy, is of fundamental interest. The chemisorption energy function can thus be calculated and can be used to determine the trend of hydrogen activity for different metals. Up to now it has not been possible to calculate exact values because of insufficient accuracy of available experimental values in the calculation. However, a trend can be established with relative values to estimate whether a catalyst will show good catalytic behavior toward hydrogen oxidation. Santos and Schmickler [98] calculated potential energy surfaces of the bond length between two hydrogen atoms and the generalized solvent coordinate $q$, which corresponds to the solvent configuration. A neutral hydrogen molecule has a solvent coordinate of $q = 0$, whereas two protons have a solvent coordinate of $q = -2$. Figure 1.21 shows two potential energy surfaces for two different catalysts. As can be seen, the reaction pathway starts in a valley where a hydrogen molecule goes through a saddle point and ends in a valley where two protons as reaction products are located. The activation energy determines the saddle point of the curve that has to be conquered. It is obvious that in the case of platinum in Figure 1.21 (c,d) the activation energy is lower than that for *sp* metals (in Fig. 1.21a,b). This leads to the assumption that desorption of hydrogen is the rate-determining step, and not the bond breakage of the hydrogen molecule. Comparing the calculations with different experimental values is a difficult task because of the wide scattering of these values. Different experimental techniques as well as a difficult electrode preparation leads to inconsistent findings. Therefore, it was suggested that the higher measured activity seems to be the most reliable, due to inhibiting effects caused by contamination and slow measurement techniques. Despite these discrepancies in theoretical and experimental values, the theory yields a successful prediction of powerful catalysts.

It can be concluded that two concepts have to be considered: (1) the metal–hydrogen interaction explained by the *d*-band center model and (2) the reaction pathway determined by the strength between the metal and the molecule.

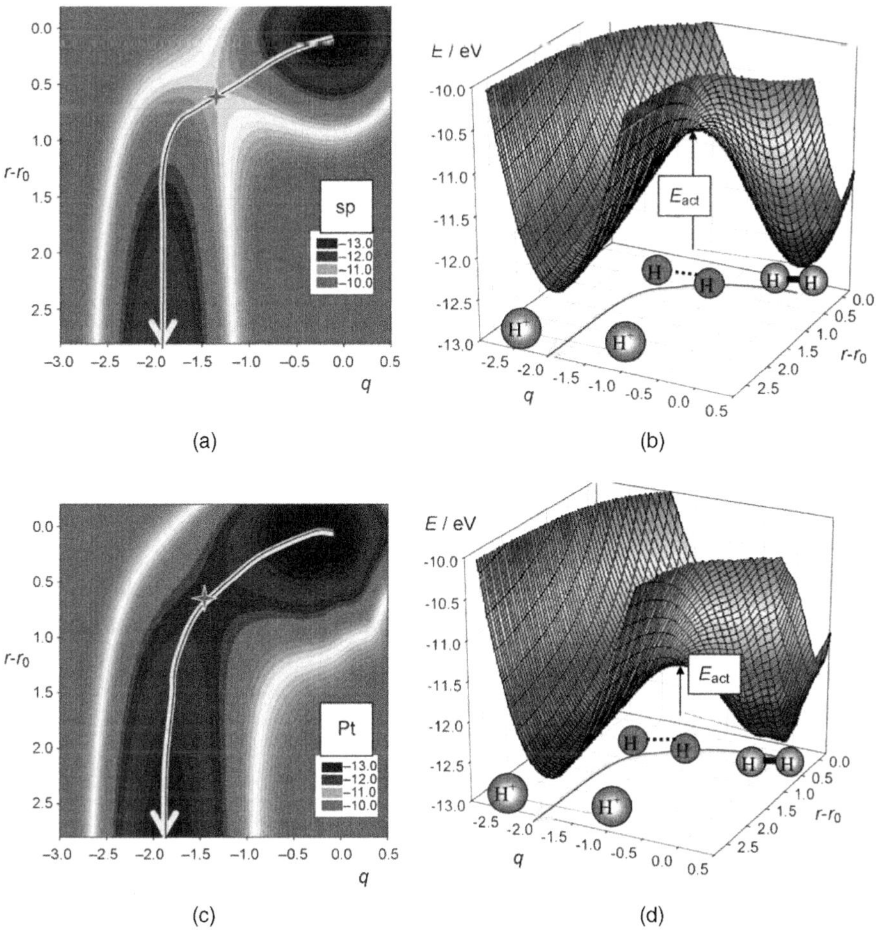


**FIGURE 1.21** Potential-energy surfaces for oxidation of $H_2$ as a function of bond length $r$ and solvent coordinate $q$. The equilibrium bond length is denoted by $r_0$; $r$ is given in atomic units (1 au = 0.529 Å). (a,b) *sp* metal; (c,d) platinum. The red crosses denote the saddle points [98]. (See insert for color representation.)

## 1.4.5 Theoretical Model of Spillover Concept for HER

Eikerling et al. [93] developed a kinetic model of the hydrogen evolution on single Pd particles supported on Au(111). This model is able to explain the experimental finding from STM investigations of Meier et al. [118]. It was shown that hydrogen desorption from Pd surfaces is the rate-determining step. If the desorption is small, adsorbed hydrogen can spill over from the catalyst particle to the Au surface, which acts as hydrogen storage. The principal mechanism of the hydrogen spillover model is shown in Figure 1.22, and the steps are described as follows: (1) the solvated protons diffuse through the electrolyte to the catalyst particle, (2) charge transfer occurs at the catalyst particle and the proton will be reduced to adsorbed hydrogen $H_{ad}$, (3) $H_{ad}$ diffuses on

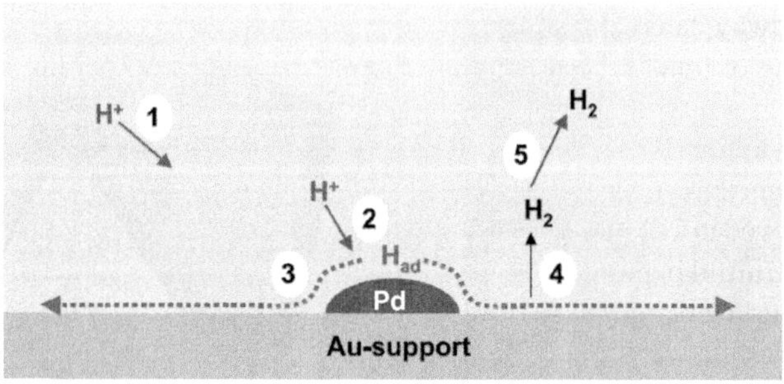


**FIGURE 1.22** Illustration of the processes on the substrate surface involved in the reduction of protons: (1) bulk diffusion of protons, (2) proton discharge and formation of adsorbed hydrogen, (3) spillover and surface diffusion of adsorbed hydrogen, (4) recombination and desorption of molecular hydrogen, and (5) hydrogen diffusion in bulk electrolyte [11].

the Pd particle and spills over to Au(111) support, (4) recombination of two adsorbed hydrogen atoms to an adsorbed hydrogen molecule $H_{2,ad}$ and subsequent desorption of the $H_{2,ad}$ occur, and (5) $H_2$ diffuses in the electrolyte.

It was shown that the diffusion of hydronium ions is much larger for small particles compared to flat surfaces, due to hemispherical diffusion. The diffusion-limiting current can be up to $5 \times 10^5$ A/cm$^2$ for particles of radius 2 nm and a proton concentration of $10^{-4}$ mol/cm$^3$. The experimental setup with an STM tip as local sensor allows elimination of the diffusion limitation due to the small current. Therefore, it can be estimated that the proton discharge and the adsorption at Pd are the rate-determining steps. It was concluded that desorption rates of adsorbed hydrogen on the Pd particle are small and the enhanced current densities can be explained by a hydrogen spillover to the Au surface. For particles with a height of one or two monolayers, turnover rates and small desorption rates were found. The adsorbed atomic hydrogen will expand in diffusion fronts on the Au support and will recombine to hydrogen molecules and subsequently desorb. Larger particles behave in a different way. Their electrocatalytic activity is about two orders of magnitude lower compared to small particles. This means that hydrogen spillover is therefore reduced. This behavior can be explained by an upshift of the *d*-band center and hence a weaker hydrogen bonding to the Pd, which was described by the model of Hammer and Nørskov [83,90]. The spillover model is in good agreement for the hydrogen evolution from experimental work and can be supported by theoretical models. Up to now it has been difficult to apply the same model for the reverse hydrogen oxidation reaction.

### 1.4.6 Summary and Discussion

Modern theoretical approaches provide powerful and fast tools for investigation of fundamentals at the solid–vacuum interface. Comparison with electrochemical experiments indicates that the trends and results are also applicable to the solid–liquid interface and hence to electrochemical issues. Hence chemisorption, adsorption and reaction pathway calculations can be used to gain deeper insights in and improved understanding of fundamental reactions. These findings can be used to determine or modify certain properties and also to systematically enhance the catalytic properties. With modern approaches by theory, findings can be explained more accurately or even predicted. This allows the development of promising catalyst systems to be investigated and proved in a systematic stepwise fashion by experimentalists, which has been shown to be fruitful in more recent studies. Theoretical catalysis on pure metal as well as metal overlayers on foreign substrates has been introduced. Interesting und unexpected results were extracted and will serve as a background for future experimental results.

## 1.5 FUNDAMENTAL STUDIES OF NANOPARTICLES ON SINGLE CRYSTALS

### 1.5.1 Introduction

Small nanostructures and nanoparticles of different materials often show properties different from those of bulk material. Since platinum and palladium are two of the most important catalysts for hydrogen reactions, they have been intensively investigated. The scientific focus is on fundamental processes and reaction mechanisms occurring at these reactive materials, especially their electrocatalytic activity. The reactivity of the catalyst can be influenced by several variables, such as particle size and particle distribution, influence of the support, and the ratio between low and high coordinated atoms of the catalyst. Scientists have already achieved a closer look into the fundamentals of hydrogen reaction, which also allows improvement of catalyst behavior. Therefore, much experimental as well as theoretical work performed to further understand the hydrogen reaction mechanism on well-defined catalyst surfaces was reviewed in Sections 1.3 and 1.4. A special case occurs if Pt and Pd are deposited onto foreign metal supports. Due to lattice strain of the Pt or Pd overlayer, the reactivity toward the hydrogen reactions differs from that of the bulk material. Many experimental studies on monolayers and submonolayers of metals on foreign metal substrates have been performed with interesting findings. However, a complete understanding that describes the overall effect has not been found until now. This section reviews the results on electrocatalytic activity of Pd and Pt multilayers, monolayers, and submonolayers on different single-crystal surfaces and compares them with those for bulk material. The different surfaces were investigated with respect to their electrocatalytic properties toward HER and HOR using various electrochemical methods. Section 1.5.2 describes the behavior of Pd overlayers on different supports. Adsorption and desorption behavior as well as the electrocatalytic activity toward hydrogen reactions of Pd overlayers are described. Investigations of Pt overlayers on Au(111) are shown afterward, followed by a summary and a discussion of all work presented thus far. More recent statements are combined and may provide some new insights into the understanding of different properties influencing activity related to hydrogen reactions.

### 1.5.2 Pd on Single-Crystal Surfaces—Adsorption/Absorption/Desorption of Hydrogen

Palladium deposition was studied regarding deposition mechanisms [119–126] for different metals. Bimetallic electrodes consisting of various metal supports and Pd overlayers have several advantages, such as the relatively easy preparation and well-established electrochemical treatment. By contrast, Pd single-crystal preparation is elaborate because of the necessary absence of hydrogen. Also, the hydrogen absorption into the Pd single crystal must be considered since it can negatively affect the quality of the single crystal. In most cases except for Au(111), Pd grows in one complete monolayer and continues growing in a three-dimensional mode [127]. Pd on Au(111) can grow two-dimensionally in up to two monolayers [119], which is ideal for further investigations. Because of the support, the Pd monolayer behaves differently than bulk Pd. It was shown by Kolb is group [121,127,128] that the adsorption and desorption properties of thin Pd overlayers change with respect to the Pd bulk material. Early studies [128] showed that with increasing Pd overlayer thickness on Au(111), the adsorption behavior of hydrogen approaches that of bulk properties of palladium. This was explained by a change in the lattice of the Pd overlayers due to the Au(111) support material. With increasing thickness of the Pd overlayer on Au(111), the behavior approaches that of Pd(111) massive crystal, indicating the importance of the support. In a study by Kibler et al. [127] on Pd monolayers on different supports, different reaction rates due to lateral strain effects were shown. Several metal electrodes such as Au(111), Re(0001), Ru(0001), Rh(111), Ir(111), Pt/Ru(111), and Pt(111) were used to electrochemically deposit palladium monolayers. The effect on hydrogen adsorption and desorption was investigated. The decorated Pd metal electrodes have different binding energies of Pd to hydrogen caused by the support, as shown in Figure 1.23. Baldauf and Kolb [121] showed that thin Pd overlayer shows a high overpotential for hydrogen absorption. Thus, small hydrogen adsorption currents can be resolved that are typically superposed by absorption currents. Cyclic voltammograms for different Pd metal electrodes are shown in Figure 1.23 [127]. For comparison, a Pd(111) single crystal is also added. Only the Au(111) support has a peak potential that is more positive than that of pure Pd(111). All other supports desorbed hydrogen at much lower potentials compared to palladium. It was suggested by the author that these peaks can be an indicator for the adsorption strength of hydrogen to the metal.

Comparing the results from Figure 1.23 with theoretical calculations from Nørskov et al. [129,130], we see an indicator of direct impact onto the binding energies. The shift of the *d*-band center of Pd overlayers on different supports was calculated assuming the simple model shown in Figure 1.24a. Pd is indicated in the figure, and the Pd atoms are relaxed in the bulk material. Using another support changes the lattice constant of the Pd overlayers and induces a positive or negative strain. It is also shown that a change in the lattice constant affects the *d*-band center in the metal. Figure 1.24b shows the peak potential of the electrode versus SCE for the hydrogen desorption for Pd monolayers on different foreign metal supports. The electrode potentials are plotted versus the *d*-band center shifts from the study by Ruban et al. [130] following a

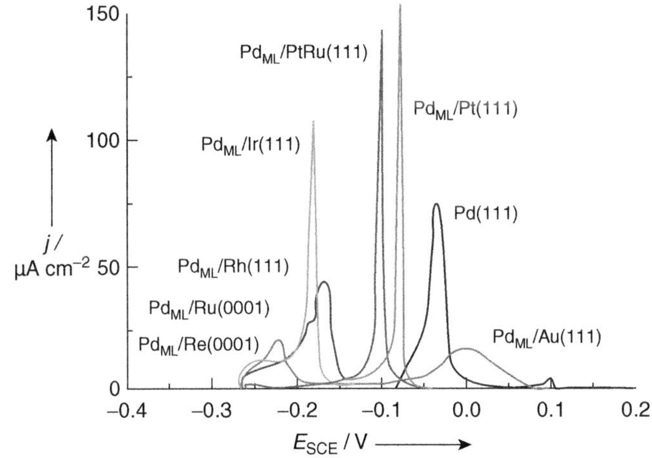


**FIGURE 1.23** Positive voltammetric sweeps for Pd(111) and pseudomorphic palladium monolayers (PdMLs) on seven different single-crystal substrates in 0.1 M $H_2SO_4$, revealing a spectrum for hydrogen desorption. Scan rate $v = 10\,mV/s$ [127].

linear behavior. Kibler et al. [127] showed that the change in electronic behavior of Pd monolayer has two contributions: (1) the lattice constant of the overlayer changes with the different lattice constant of the support and thus induces a shift in the *d*-band center (geometric effect) and (2) there is also a so-called ligand effect that is also caused by a change of the *d*-band center if there is a large electronic interaction between the surface layer and the substrate. However, the systematic change of the *d*-band center leads to the conclusion that the geometric effect is the controlling one. These findings may give rise to the conclusion that changed adsorption behavior due to lattice mismatches will also result in changed reactions or kinetic properties.

### 1.5.3 Pd on Au(111)—HER and HOR

The hydrogen evolution reaction was investigated on Pd deposited on Au(111), and several effects were found for single Pd particles [131–133] as well as large Pd nanostructured Au(111) electrodes [134,135]. Nanoscale effects such as spillover of hydrogen as well as geometric effects can influence the reactivity of Pd supported on Au(111) toward HER. As mentioned in Section 1.5.2, the geometric effects change the binding energy to hydrogen. This effect is due to a strain of the overlayer caused by the support, which has a different lattice constant. This effect was also found from theoretical calculations [127,136–140]. Pandelov and Stimming [134] investigated monolayers and submonolayers of Pd deposited on Au(111) with respect to HER. The Pd overlayer thickness was lowered from four monolayers to about 3.5% of a monolayer, causing an increase of about one order of magnitude for the geometric current density. It was shown that the geometric current density increases with decreasing palladium coverage, as shown in Figure 1.25. The strong enhancement in the case of submonolayers was explained by a hydrogen spillover from the Pd

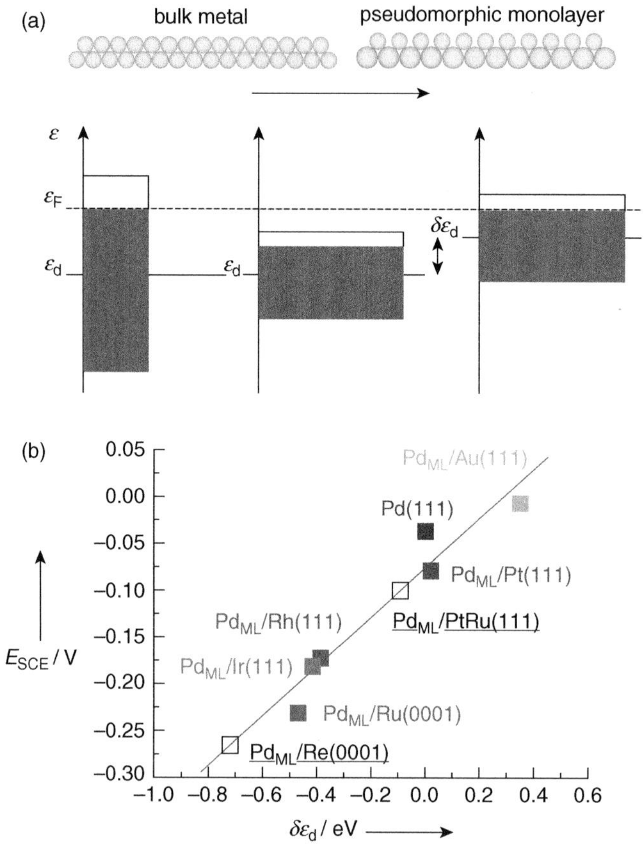


**FIGURE 1.24** (a) The main origin of a shift in the *d*-band center $\varepsilon_d$ is a change in the interatomic distances within an overlayer. If the *d* band of a metal is more than half-filled, an expanded pseudomorphic monolayer will lead to an upshift of $\varepsilon_d$ owing to band narrowing and energy conservation [131]. (b) A plot of the hydrogen desorption potentials versus the shift of the *d*-band center, $d\varepsilon_d$ shows a linear correlation as theoretically predicted [131]. For palladium on the Re(0001) substrate, no hydrogen peak was measured (see text). The hydrogen peak at −0.1 V for palladium on PtRu(111) suggests a shift of the *d*-band center $d\varepsilon_d$ of about −0.1 eV, which has not yet been calculated [127].

catalyst particle on the Au(111) surface. This result indicates that with lower amounts of noble metal, a higher net activity is possible, which is in clear contrast to expectation.

Referring to the current values per active palladium area, a very strong increase in activity with decreasing coverage (Fig. 1.26) is found. Almost three orders of magnitude enhancement in activity comparing four monolayers of palladium with only ∼ 3.5% Pd on Au(111) were determined for two different overpotentials. Different explanations were discussed, whereas no dependence of the current density on the number of edges versus terrace atoms could be found as proposed by Meier

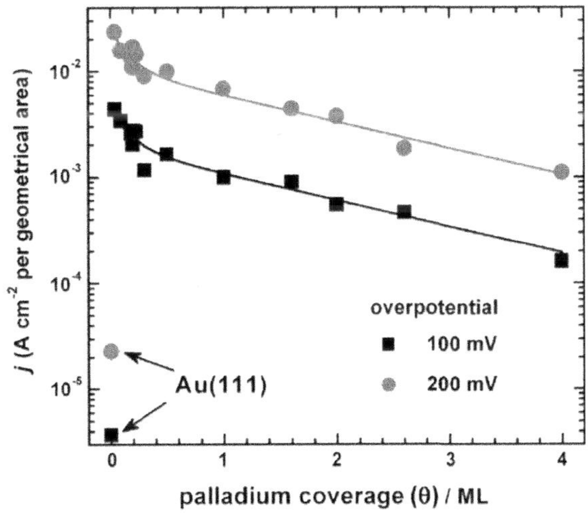


**FIGURE 1.25** Current density of hydrogen reduction as a function of the Pd coverage at 100 and 200 mV overpotential, respectively. The current densities were calculated with respect to the geometric electrode area. The bulk symbols are the measured values, and the solid-line curves are for visual inspection [134].

et al. [132]. While the increasing reactivity for monolayers can be explained by the strain effect in the palladium lattice, the increase for submonolayers of Pd was explained by a spillover mechanism of reaction intermediates. This suggests a direct contribution of the support in the hydrogen evolution.

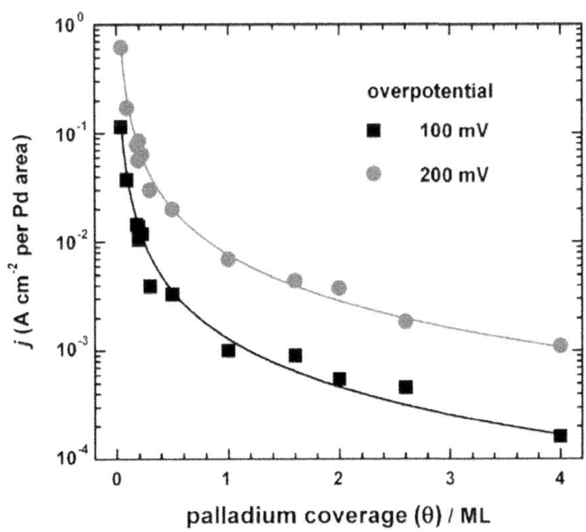


**FIGURE 1.26** Current density of hydrogen reduction as a function of the Pd coverage at 100 and 200 mV overpotential, respectively. The current densities were calculated regarding the Pd area. The bulk symbols are the measured values and the solid curves are a guide for the eye [134].

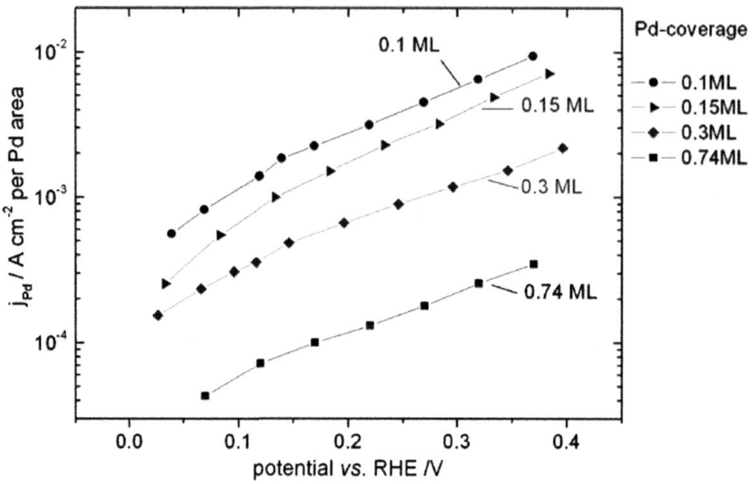


**FIGURE 1.27** Electrocatalytic activity of HOR per Pd area of the Pd/Au(111) system in 0.1 M $HClO_4$. The current density is normalized to the surface area of the deposited Pd [141].

Similar results were found by Wolfschmidt et al. [141] for HOR. A decreasing coverage of Pd on Au(111) shows an increase in the absolute current densities. Referring these values to the active Pd area, a strong enhancement of the specific catalytic activity is found, as shown by Pandelov and Stimming [134] before. In Figure 1.27 the specific current densities for HOR are about two orders of magnitude higher on comparison of 0.74 ML with 0.1 ML of Pd deposited on Au (111). These findings were explained with altered electronic structure of the Pd overlayer. With the shift in the *d*-band center the binding energy to hydrogen is changed and thus can enhance the reaction. The latter explanation is valid for multilayers down to one monolayer and not directly applicable to the case of submonolayers. A spillover model comparable to HER has been considered in the context of a more complex theory. The influence of reactive steps and defects cannot be excluded but does not seem to be an important factor. It is a major challenge to find a general mechanism for HOR on submonolayers.

### 1.5.4 Pd on Pt(111)

Markovic and Ross [142] performed experiments on Pd films on Pt(111) and compared the activity for HER and HOR to the activity of a Pt(111) single crystal. Comparison of pure Pt(111) with a Pd-modified Pt(111) single crystal shows an enhanced reactivity that is clearly seen in Figure 1.28. The catalytic activity for hydrogen reactions was investigated with a RDE at a rotation speed of 1600 rpm. The Arrhenius plot inset in Figure 1.28 shows that the activation energy for hydrogen reactions is only half that for the Pt(111)–Pd electrode compared with Pt(111). For the so-called paradoxical activity of the Pt(111)–Pd surface for HER and HOR, it is important to understand the nature of $H_{upd}$ and the effect on $H_{opd}$ formation. The authors suggested that the

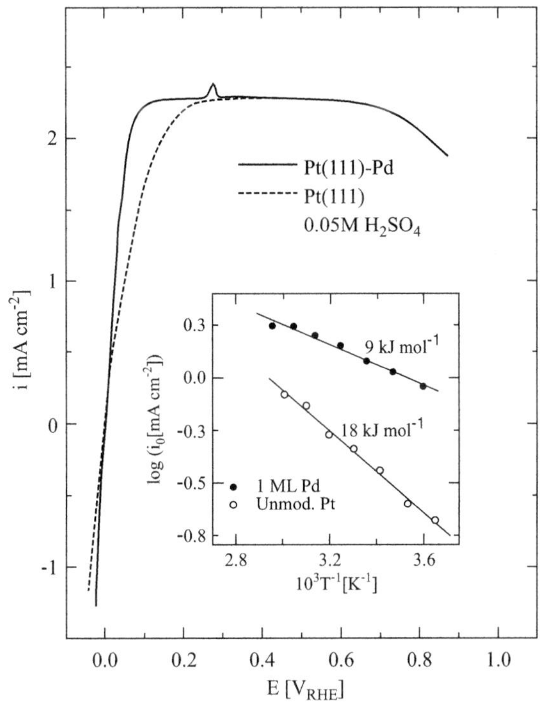


**FIGURE 1.28** Comparison of the HER/HOR on Pt(111)–Pd (solid-line curve) and Pt(111) (dashed-line curve) in 0.05 M $H_2SO_4$ at 278 K. Rotation rate 1600 rpm [inset: Arrhenius plots for HER/HOR on Pt(111)–Pd and Pt(111)] [25].

transition of unreactive $H_{upd}$ into reactive $H_{opd}$, due to high coverages of $H_{upd}$ on Pd/Pt(111), seems to be reasonable. However, it seems uncertain whether the change in adsorption energy has an important role in the enhanced electrocatalytic behavior for the Pd/Pt(111). The importance of the altered adsorption energetics of hydrogen due to electronic effects on Pd on Pt(111) support in contrast to the pure Pt(111) were emphasized later. It was concluded that the electronic effects play an important role in electrocatalysis.

The influence of Cu, Pb, and Bi on the reactivity of different Pt(*hkl*) single-crystal surfaces was also investigated by Markovic and Ross [142]. In these investigations Cu and Pb were deposited as an underpotential monolayer and Bi was deposited as irreversible uncomplete monolayer. Because of overlap of the metal deposition and hydrogen evolution, only hydrogen oxidation was investigated. All metal modified Pt surfaces inhibit the HOR compared to pure Pt(111) surfaces, which is shown in Figure 1.29. These modifications of the Pt surface with overlayers of foreign metals did not result in improvement of the catalytic properties. But some interesting conclusions can be drawn from these inhibiting effects. The activity toward the different overlayers can be expressed as Bi > Pb > Cu deposited on Pt(111). The highly active Bi overlayer becomes less active at higher overpotentials, due to adsorbed OH on adjacent Pt sites.

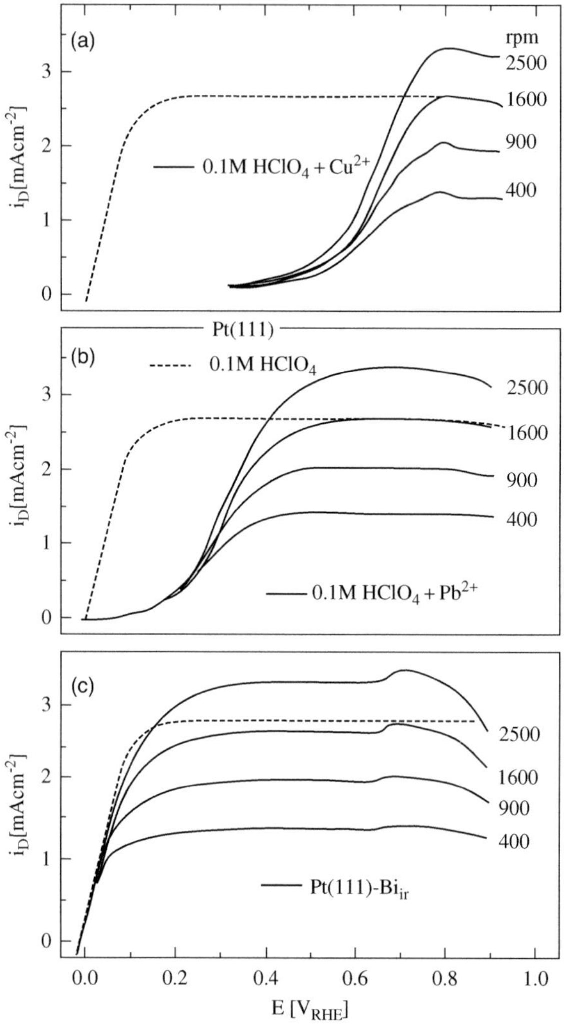


**FIGURE 1.29** (a-c) Polarization curves for HOR on Pt(111) in 0.1 M $HClO_4$ (dotted-line curve), recorded at 1600 rpm. (a) HOR on Pt(111) in 0.1 M $HClO_4$ + $10^{-3}$ M $Cu^{2+}$ (solid-line curve) at sweep rate 20 mV/s; (b) HOR on Pt(111) in 0.1 M $HClO_4$ + $10^{-3}$ M $Pb^{2+}$ (solid-line curve) at sweep rate 20 mV/s; (c) HOR on Pt(111) in 0.1 M $HClO_4$ (dotted-line curve), and on a Pt(111)–$Bi_{ir}$ electrode in 0.1 M $HClO_4$ at sweep rate 20 mV/s [24].

It is suggested that the free Pt sites play a major role for the strong HOR activity in the Bi case. In contrast, Pb and Cu metal adlayers do not influence the energetics of the absorbed hydrogen, but they block the free available Pt sites. The low activity of the Cu-modified Pt(111) surfaces was explained by the strong adsorption of anions, which has a significantly negative influence on HOR.

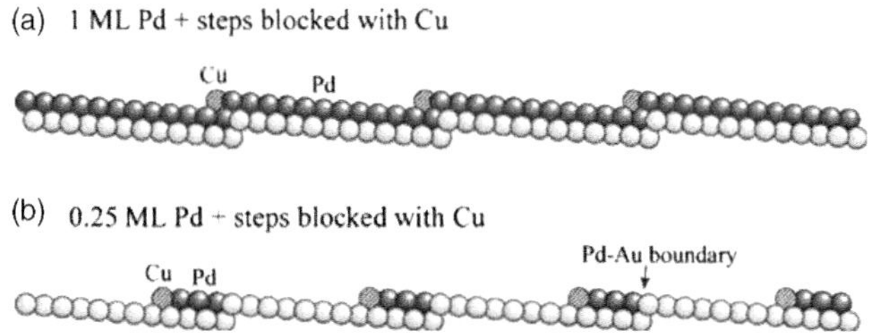


**FIGURE 1.30** Hard-ball models of a Au(665)/Pd electrode where the Pd step sites have been blocked by Cu, modified with (a) 1 ML of Pd and (b) 0.25 ML of Pd [144].

### 1.5.5 Pd on Au(111)—Blocking Active Sites with Cu

Palladium on Au(111) was intensively investigated by the Baltruschat group [144–146]. Different single-crystal surfaces such as Au(111), Au(332), and Au(331) were decorated with Pd to investigate their properties regarding the adsorption of hydrogen. It was found that it is necessary to deposit more than one atomic row of Pd onto the stepped Au surfaces to adsorb hydrogen [145]. Further findings on Au(332)- and Au (665)-decorated single crystals show a strong influence of the steps on the catalytic activity toward hydrogen evolution. This is in line with the findings on the weak hydrogen adsorption according to the Sabatier principle. An interesting effect was found by blocking the active Pd rows on the different stepped Au surfaces with Cu, schematically shown in Figure 1.30. Regarding Pd/Au(665) with Pd steps blocked by Cu, the electrocatalytic activity decreases to values that are in the order of magnitude of a Au(665) covered with more than a full monolayer of Pd. It was concluded that hydrogen spillover seems unreasonable for Pd-covered surfaces, which is in contrast to findings by the Stimming group [134,141,147]. Regarding hydrogen evolution and oxidation, it was found that below one monolayer of Pd on Au(332) the reactivity is independent of coverage. Therefore, it was deduced that the reactivity is caused mainly by the step sites and that contact between the Au and the Pd atoms is necessary. The effect of Pd deposited on terraces is negligible. Furthermore, it was concluded that Cu is able to block the active Pd rows on stepped Au surfaces and hence, to decrease the activity of the Pd atoms regarding hydrogen evolution.

### 1.5.6 Pt on Au(111)

Platinum-modified Au(111) surfaces were investigated by Wolfschmidt et al. [141] in terms of their electrocatalytic properties for HER and HOR. Several scanning tunneling microscopy (STM) images of the investigated nanostructured electrodes are shown in Figure 1.31. Different methods such as integrating the deposition charge, hydrogen underpotential deposition, CO monolayer oxidation, and STM were used to determine the amount of deposited metal. Comparing the different methods allows a detailed and precise access to the exact amount of Pt.

It was shown that monoatomically high Pt islands on Au(111) have a strongly enhanced catalytic activity for both hydrogen reactions. The kinetic current densities

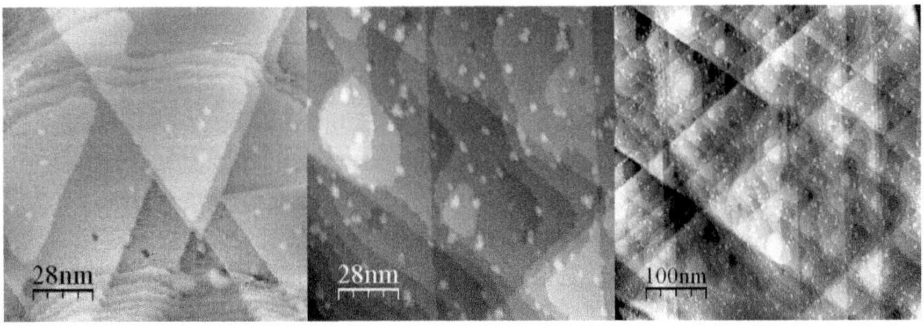


**FIGURE 1.31** STM images of different Au(111) electrode surfaces nanostructured with Pt. The images were recorded in air with $U_{bias} = 100\,mV$ and $I_T = 1\,nA$. The amount of Pt on Au(111) corresponds to (a) 0.025 ML, (b) 0.067 ML, and (c) 0.13 ML of Pt [141].

are shown in a Tafel plot in Figure 1.32. It is obvious that the Pt(111) single-crystal electrode has the highest reactivity for both reactions referred to the simple geometric electrode area. A clear trend toward their electrocatalytic activity for the different amounts of Pt on Au(111) was not found for the geometric current density. Although the activity of the submonolayers is less than the activity of the Pt(111), it can already be seen that the decrease in activity of one order of magnitude is less pronounced than the decrease in coverage, which is more than three orders of magnitude.

The geometric current density referred to the active Pt surface results in the specific current density. Comparing the lower coverages of Pt submonolayers on Au(111) with a Pt(111), single-crystal enhancement factors of up to three orders of magnitude for the specific current densities were found. The catalytic activity thus strongly increases with decreasing coverage for both hydrogen reactions. Although the geometric current density absolutely decreases with decreasing coverage, as discussed above,

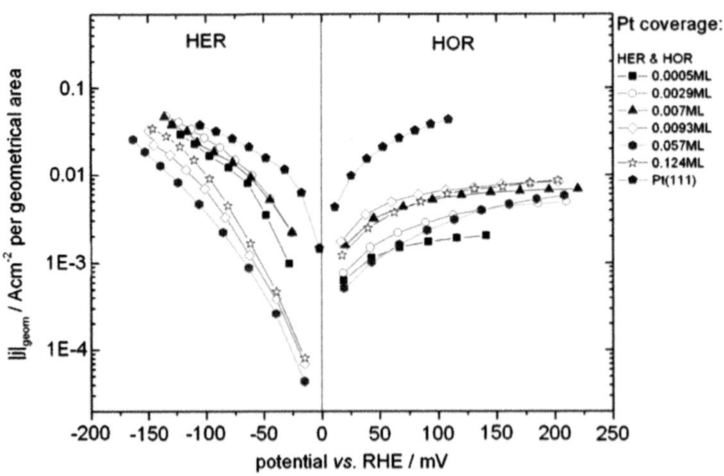


**FIGURE 1.32** Tafel plot of HER and HOR with current density per geometric area for Pt submonolayers on Au(111) and Pt(111) in 1 M $HClO_4$ [141].

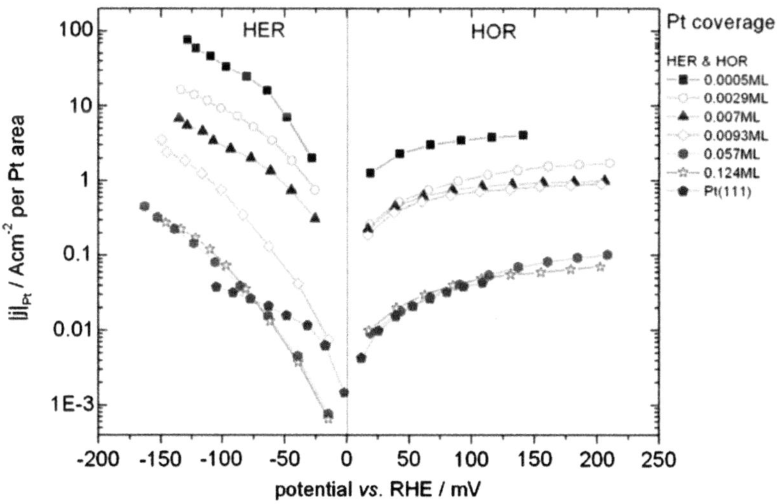


**FIGURE 1.33** Tafel plot of HER and HOR with current density per Pt area for Pt submonolayers on Au(111) and Pt(111) in 1 M $HClO_4$.

the specific current density strongly increases. Because of very high currents, it seems that diffusion effects may play an important role in HOR. Comparing both branches in Figure 1.33, the HOR branch is lower at higher overpotentials compared to lower overpotentials, where HER and HOR are similar. This may lead to the assumption that an impact of diffusion limitation has to be considered although methods were used to avoid the diffusion effects. This is clearly shown in Figure 1.33, where the specific current density is plotted versus the overpotential for several Pt submonolayers on Au (111) and Pt(111). In summary, the specific activity for HER is about three orders of magnitude higher compared to Pt(111) and the specific activity of HOR is more than two orders of magnitude higher compared to Pt(111) (Fig. 1.33). Different explanations for the strong enhancement are given. As already mentioned in the Pd/Au(111) case, there is not one major parameter that causes this behavior. Also, for Pt deposited on Au(111), particle size effect, support effect, and high reactivity defects must be considered. Thus far it is unclear which effect plays the leading role or whether the enhancement is due to a combination of effects.

### 1.5.7 Summary

Adsorption and desorption processes were investigated on single-crystal surfaces by the Kolb's group [121,127,128]. Various indicators were found that underline the importance of Pd overlayer thickness and selection of support. Pd thickness as well as support influence the adsorption and desorption behavior as well as the electrocatalytic activity. The experimental finding were supported by theoretical work from the Nørskov group [129,130,148] and the Gross group [149,150] reported in Section 1.4. The experimental results for HER and HOR on monoatomically high Pd nanoislands deposited on Au(111) indicate a specific electrochemical activity that increases with decreasing coverage [127,134,141]. A similar increase in specific activity was shown for the HER as well as for the HOR for Pt nanostructured Au(111) surfaces [141]. Although the net current densities for HOR are one order of magnitude lower than for HER, the enhancement of specific current density is nearly two orders of magnitude for both reactions. The lower concentration of reactants and thus a diffusion limitation in the case of HOR can affect this behavior. The enhancement was attributed to several effects. For decreasing Pd monolayers down to one monolayer, one has to consider a strain effect of the Pd overlayer according to the Nørskov model. The strong enhancement for the submonolayers was explained by a hydrogen spillover from the Pd particle onto the Au(111) support. Defect sites and steps do not have a big impact on electrocatalytic activity.

Similar results were also obtained by Kibler [127,135] for HER. In contrast to studies by the Stimming group [134,141], the other studies investigated the exchange current densities, and not the specific current densities, and an enhancement of a factor of $\sim$ 4 was found comparing 1 ML with 0.07 ML Pd on Au(111). The importance of the free energy of hydrogen adsorption was underlined, whereas defect sites and steps often play an important role. An influence of the support was also provided with the background of theoretical calculations. Also, the Baltruschat group [144–146] performed several Pd-on-Au(111) experiments. It was suggested that the main part of the electrocatalytic activity is caused by the steps. Pd on the Au (111) terraces is not very active and thus has no influence on enhanced catalytic activity. A model based on a spillover concept from the hydrogen from Pd to the Au (111) support does not seem to be reasonable here, according to Hernandez and Baltruschat [144,145].

Markovic and Ross [142] studied the Pd-on-Pt(111) system. Here also, an improvement of the catalytic activity compared to pure Pt(111) was found. Decorating the Pt(111) surface with Pd enhances the performance toward HER and HOR. This effect can be clearly seen in Figure 1.28. A variation of the thickness was not done, but the positive effect of Pd was also demonstrated in this case.

Blocking the surfaces with different metal adlayers to show an inhibition effect was investigated by Markovic and Ross [142]. All three different metals, Cu, Pd, and Bi, decrease the electrocatalytic activity of Pt toward HOR. Thus, improving the activity with these model surfaces does not seem feasible. In this case, Pt is highly reactive and is inhibited by the metal adlayers.

### 1.5.8 Discussion

Pt/Au(111) [141] as well as the Pd/Au(111) systems [127,134,135,141] show a strong enhanced specific current density toward HER and HOR. Although the net current densities for the Pt on Au(111) are significantly higher than for the Pd on Au(111) system, both show similar behavior. An obvious enhancement of the specific electrocatalytic activity for Pt on Au(111) with decreasing Pt coverage with respect to HER and HOR is clearly seen in Figure 1.33. The specific current densities are approximately three orders of magnitude higher for small coverages of Pt on Au(111) as compared to Pt(111) surface for HER and more than two orders of magnitude higher for HOR. A similar trend was found for the Pd on Au(111) system shown in Figures 1.26 and 1.27. The increase in specific reactivity is also more than two orders of magnitude, whereas the current densities of HOR are lower compared to the current densities of HER. According to the importance of Pt catalyst, the main focus in the discussion is on the Pt results but often refer to the well-investigated Pd/Au(111) system.

For a general understanding of HOR and HER activity of Pd- and Pt-modified Au(111) electrodes, several parameters have to be considered. One parameter is the influence of the substrate on the lattice constant of the Pd or Pt overlayer. According to the Nørskov model, a lateral strain on the metal overlayer causes a shift of the *d*-band center [130,134,137,139,151–154], resulting in a change in the electronic properties of the overlayer. For the Pd/Au(111) system the enhanced activity of Pd monolayers compared to Pd multilayers on Au(111) can be explained by a strain effect of the gold substrate on the Pd layer. Such strain effects gradually decrease with increasing thickness of the Pd layer [133]. This influence is confirmed by the reverse experiment with a support material with a smaller lattice constant. In case of tip-induced Pd particles on Cu(111), no enhancement for HER activity was found [155]. This is in line with expectation and theory, since copper has a smaller lattice constant than does palladium. Furthermore, Pt nanoislands electrodeposited on highly oriented pyrolytic graphite (HOPG) also show no enhanced activity for HER for mono- and submonolayers of Pt [156]. In this case, no influence of the support is expected, due to weak interaction between metal and support. Both results indicate that neither Cu nor HOPG as support enhance the electrocatalytic activity of deposited Pd or Pt particles. Therefore, strain induced by the support material seems to be an important parameter that enhances electrocatylytic activity for HER and HOR. This allows a better understanding of the increase in electrocatalytic activity of deposited Pd and Pt as a function of layer or particle thickness.

Nevertheless, it is important to consider that the most pronounced increase in HER activity was observed for submonolayers of Pt and Pd on Au(111). As is described in Section 1.6, Meier et al. [133] reported about nanoscale effects of single Pd particles that were deposited with an STM tip on Au(111) surfaces. The catalytic activity for HER is drastically depends on the height of the particles and increases by about two orders of magnitude going from particle height of 10 layers to particles of approximately two layers of height. Electronic effects induced by the substrate depending on the height of the deposited particle are also important. The argument of modified electronic properties due to a lattice strain of an isomorphic overlayer is applicable to Pt islands as well as to Pd islands, whether single particles or large nanostructured surfaces. Since the lattice mismatch between gold (0.408 nm) and platinum (0.392 nm) is 4%, and between 4.8% gold and palladium (0.389 nm), it is logical to expect a similar increase in activity based on the Nørskov effect for the Pt/Au(111) as well as for the Pd/Au(111) system. For a further understanding, it has to be considered that tip-induced Pd islands on Au(111) were varied in height. The electrochemically deposited monoatomic Pd islands on Au(111) were varied in coverage; nevertheless, in both cases an enhanced activity is observed. The increasing activity for submonolayers of monoatomically high Pd and Pt with decreasing amounts of deposited metal may be ascribed not only to a lattice mismatch but also to an additional effect. According to the Nørskov model alone, the specific activity of nanoislands of various coverages should be identical.

The spillover model, introduced by Eikerling et al. [147], was described in more detail in Section 1.3. Enhanced catalytic activity of Pd nanoislands is attributed to spillover of atomic hydrogen from the Pd sites to the gold surface. A kinetic model was developed for observed increased activity of single particles on Au(111) surfaces. A spillover of atomic hydrogen from Pd islands with a high coverage of hydrogen to near-hydrogen-free gold surface can energetically be a downhill process [133]. The charge transfer takes place at the Pd islands and the recombination occurring on the Au(111) seems reliable. A potential application of this model to the reverse hydrogen oxidation reaction involving both surface catalyst as well as Au(111) can also be imagined, although it remains difficult to prove this in detail. Thus, the increase of catalytic activity for monoatomic submonolayers of Pd and Pt on Au(111) compared to monolayers could be understood by the involvement of the Au(111) surface as an additional storage of adsorbed atomic hydrogen.

An alternative explanation for the observed increased electrocatalytic reactivity of submonolayers could be based on a possibly higher activity of low coordinated surface atoms such as step sites as proposed by Hernandez and Baltruschat [144,145]. It was found that a Pd monolayer is more reactive than Pd multilayers, which is in good agreement with a strain effect induced by the Au surface. The exchange current density of submonolayers of Pd down to 0.2 ML is independent of Pd coverage. Furthermore, it was demonstrated that codeposition of copper at Pd submonolayers results in an inhibited HER activity. It is suggested that the Pd at the steps controls the kinetics of hydrogen evolution on stepped gold surfaces modified by Pd. In case of Pd nanoislands on Au(111), the reactivity of HER is significantly larger, by a factor of ~5, such as at an overpotential of 50 mV. Analysis of nanostructured Au(111) electrode surfaces reported by Pandelov and Stimming [134] shows no correlation of current density for HER with the number of Pd edge versus Pd terrace atoms on Au(111). The same result was found by Meier et al. [133] from an analysis of single Pd particles on Au (111). Hence, the number of low coordinated atoms such as edge atoms does not seem to be the dominant factor controlling HOR and HER activity on nanostructured Au(111) surfaces. From the experimental results reported, it remains difficult to develop a complete model for HER and HOR on nanostructured Pd/Au(111) and Pt/Au(111) surfaces in spite of the strong effects observed.

Nevertheless, all experimental findings for the extended Pt and Pd nanostructured surfaces presented here tend to show improved catalytic behavior. Although the authors prefer different explanations for their findings, it is not possible at the moment to neglect one of the approaches. Briefly summarized, the kinetics of HER and HOR are obviously influenced by several parameters, and it remains important to understand the origin of these differences. Although Pt is among the best catalysts for hydrogen related reactions, there is still potential for improvement. On the basis of the reported experimental results, it can be stated that the positive effect of Pd and Pt lattice expansion on Au(111) for hydrogen-related reactions is one of the fundamental issues remaining to be understood in more detail.

## 1.6 INVESTIGATIONS OF HYDROGEN-RELATED REACTIONS ON SINGLE Pd PARTICLES

### 1.6.1 Introduction

Planar surfaces have been thoroughly investigated, and most of the reactivity measurements are referred to these extended electrodes [157–163]. In contrast to, nanostructures and nanoparticles often behave differently than bulk material. Therefore novel techniques were developed and applied to investigate the changed properties in the nanometer scale. One possible approach is the scanning electrochemical microscope (SECM), which uses ultramicroelectrodes (UMEs) for investigation. With this technique a resolution of about 100 nm is achievable, which is limited by the geometry of the UME. Meier et al. [164] introduced a technique that uses a tip of an electrochemical scanning tunneling microscope (EC-STM) as probing sensor. Because of the tunneling effect and thus the high precision, atomic resolution is possible, which is several orders of magnitude in geometric precision lower than in other techniques. In this way, the electrocatalytic activity of single Pd particles on Au(111) support toward hydrogen evolution was investigated in situ [164,165]. In this approach the hydrogen evolution at the Pd particle was indirectly measured via the detection of the oxidation of the evolved hydrogen at the tip. Therefore the STM tip was used as a local sensor measuring the activity of a single Pd particle. This approach allows a complete in situ investigation of single particles from characterization to reactivity measurements. In Section 1.6.2 the STM tip is reviewed as a tool for local nanostructuring. The principles of local investigations with the STM tip with their possibilities and limitations are shown in Section 1.6.3. Experimental as well as theoretical principles of particle preparation and stability will follow. Section reviews the results obtained from single Pd particle experiments supported on Au(111). These results are discussed and compared to those for large nanostructured electrodes.

### 1.6.2 STM Tip as a Tool for Local Nanostructuring

Local methods for nanostructuring enable modifications of the substrate on a defined position by a direct impact. This approach to creating nanostructures using macroscopic techniques is called a *top–down approach*, which enables the creation of defects or deposition of clusters on the surface. All these techniques are direct methods, so the structures are directly fabricated in contrast to a mask for indirect deposition. Typical techniques for nanolithography are photolithography, X-ray lithography, electron- and ion-beam lithography, or the scanning probe microscopy techniques such as STM and atomic force microscopy (AFM).

Here different techniques for nanolithography are discussed in detail using STM and EC-STM. In the electrochemical environment there are many different options for using the STM tip to create a cluster or induce a defect. The first attempt uses the tip and forces a contact with the substrate (see Fig. 1.34a). In this way, a defect is created for metal deposition in a second step. A gentle approach is shown in Figure 1.34b that is quite substrate-sensitive. The tip removes only a tarnishing film from the substrate and

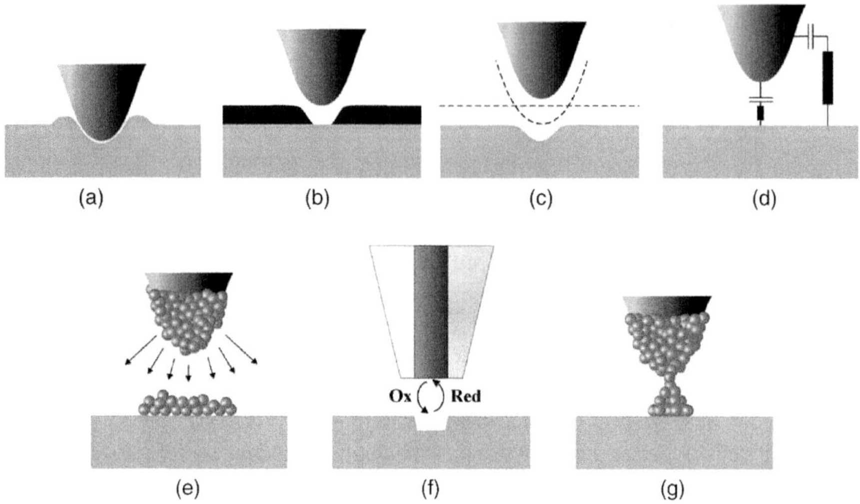


**FIGURE 1.34** Different methods for nanostructuring with an STM tip: (a) mechanical impact; (b) removal of a tarnishing film; (c) double-layer crosstalk; (d) pulsed double-layer charging; (e) burstlike dissolution from the tip; (f) SECM with a nanode; (g) jump-to-contact method [181].

allows metal deposition. Studies at the system Cu on Au with sodium dodecylsulfate films have been reported [166]. The double-layer crosstalk in Figure 1.34c has to be included if the electrochemical double layer of the tip and the substrate interferes. A loss of the potential control of the substrate beneath the tip can result. Different groups investigated the $Cu/Cu^{2+}$ [167] and the $Ag/Ag^{+}$ [168] systems, where dissolution of bulk metal was induced by a very close approach of the tip. The Ertl group [169,170] used the so-called electrochemical machining technique to create structures. An ultrashort pulse was applied between tip and substrate in nanosecond regime for etching processes. Due to the time dependence of the double layer, charging on the upper or lower part this technique is used to set the pulse in such a way that only the tip apex interacts with the substrate potential (Fig. 1.34d). Therefore, it is possible to dissolve metal by oxidizing it in a small area. A quite easy way for tip-induced nanostructures is the burstlike method depicted in Figure 1.34e). The metal on the tip is deposited directly under the tip [171,172]. Schindler et al. [173] explain the effect with a high metal ion concentration after the deposition process that shifts the Nernst potential for the surface under the tip. Figure 1.34f) shows the scanning electrochemical microscopy (SECM). Next to the previously mentioned techniques, SECM can be used for detection of electrochemical processes on a micrometer scale. Especially in the field of electrocatalysis, SECM is a powerful tool for direct investigations of particles. More details will be found in the literature [174,175]. Most previous studies reported in the literature deal with deposition of clusters on substrates induced by the "jump to contact" method (Fig. 1.34g). The tip gets in contact with the substrate without destroying it. This effects a formation of a metal bridge, the so-called connective neck, which will break if the tip is retracted. Different groups use this technique to create large arrays from different metals on different substrates [176–180]. In Section 1.6.3 methods using massive metal tips loaded with another metal from a solution are discussed. The tips consist of platinum with electrodeposited palladium for preparing single Pd nanoparticles, whereas the jump-to-contact method is used.

### 1.6.3 Principles of Single Pd Particle Preparation—Computer Simulation and Experiments

Scanning tunneling microscope tips have been used in many different approaches to generate nanoparticles on well-defined supports [164,165,173,176, 181–188]. Creation of large nanostructured electrodes with particles, single-particle reactivity measurements, and investigation of magnetic properties are several examples of single-nanoparticle investigations using STM. Different methods of using the STM tip for nanostructuring are discussed elsewhere [181] and are reviewed above. In one method reported in the literature [164,165,173,176,183] the tip is loaded with metal by electrochemical deposition from the electrolyte. A tip created by this technique is schematically shown in Figure 1.35a, where the bright atoms are tip atoms and the dark ones are the deposited metal. The tip is moved to the support and brought into contact (Fig. 1.35b). The jump to contact occurs and a connective neck is built (Fig. 1.35c). The tip is subsequently withdrawn and some atoms from the deposited metal on the tip will remain on the surface and yield a nanoparticle (Fig. 1.35d). The particle size is dependent on the distance to the tip, and hence the penetration depth into the surface and, of course, the geometry of the tip. It was shown that with larger movement of the tip in the $z$ direction after building a connective neck, the particle size and hence the number of atoms per particle will increase. Interestingly, with increasing particle size the number of Au atoms also increases and a stable alloy is formed whereas the Au atoms are detached from the surface. Smaller particles generally consist only of Pd atoms.

The abovementioned increase in Au atoms stabilizes the particles as a result of the stable Pd–Au alloys and inhibits the dissolution compared to small particles. This assumption was also theoretically investigated using Monte Carlo simulation. The electrochemical potential $\mu$ of the particles was fixed and the number of atoms was varied. A series of simulations were performed on the basis of computational calculations, which are shown in Figure 1.36. The size and the composition of the particles are plotted versus the distance of the tip after experiencing a jump to contact. Calculations were performed for different electrochemical potentials of Pd. The initial configuration corresponded to the end configuration of the atom dynamic simulation; hence the cluster evolution was followed and the dissolution or growth of the cluster could be determined. It was shown that the particles grow at electrochemical potentials higher than $-3.8$ eV and completely dissolve at smaller values. The value of $-3.8$ eV is smaller than the cohesive energy of Pd of $-3.9$ eV, and thus it is suggested that particles are less stable compared to bulk Pd. Dissolution of particles fits well to the experimental results, which showed that dissolution starts at the base and is fast for small particles with a few atoms.

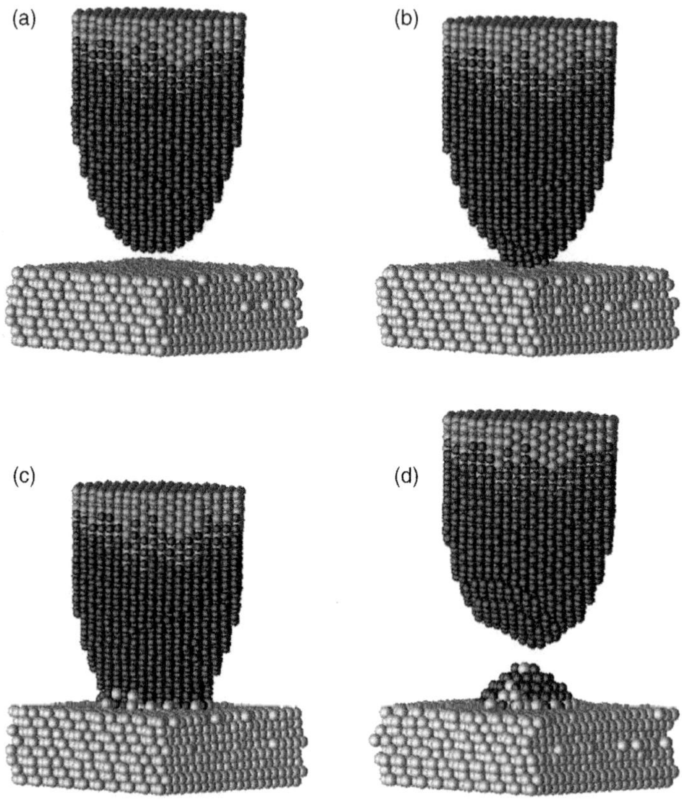


**FIGURE 1.35** Snapshots taken during the generation of a Pd cluster on Au(111): (a) initial state of the system, $\Delta z = +/2.04$ Å; (b) jump to contact, $\Delta z = /0.0$ Å; (c) closest approach, $\Delta z = -6.37$ Å; (d) final state, $\Delta z = 11.04$ Å [177].

Particles with different heights and diameters were also experimentally generated and examined (Fig. 1.37). The differences in geometry are influenced by the folding of the shape of the tip and the shape of the particle. It was found that higher particles also have larger diameters. For a precise evaluation, particle height is plotted versus the approach of the tip toward the surface. At a certain point, cluster height increases with increasing values of $\Delta z$. The jump to contact occurs between $-0.5$ and $-0.6$ nm. Below that the tip is not in contact with the surface, and hence no particle is created. These experimental results agree very well with theoretical ones from Figure 1.36. Although the numbers of atoms shown in Figure 1.36 have been calculated, it is necessary to build a jump to contact to create particles, and it is also obvious that with deeper $z$-direction of the tip, movement particle height increases.

Experimentally generated single Pd particles on Au(111) are shown in Figure 1.38. All particles were produced in situ and characterized using electrochemical scanning tunneling microscopy. As can be seen, the particles differ in diameter and height, which vary with movement of the $z$-direction piezo during particle generation

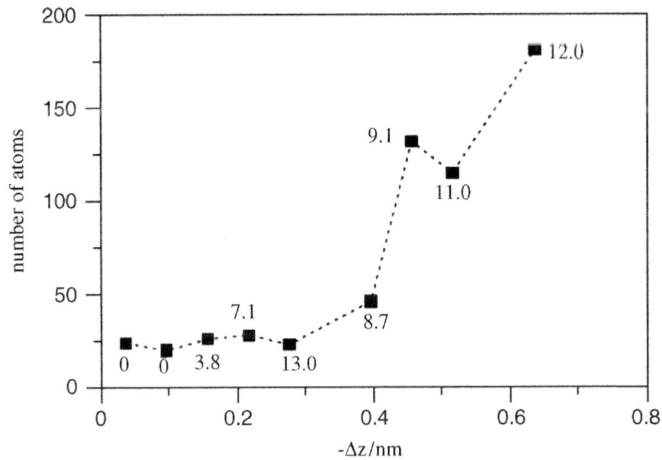


**FIGURE 1.36** Number of atoms in the clusters obtained by computer simulations. The distance of closest approach is referred to as "jump to contact." The numbers close to the points give the percentage of gold atoms in the cluster [177].

and tip geometry. It was shown that the particles are stable for several hours but in some cases they their morphology changes and increasing particle diameter causes an increase in particle height, which is in agreement with the theoretical findings. A detailed description of particle stability and change in morphology follows. The particles are placed on large Au(111) terraces, guaranteeing no influence of defects of the support.

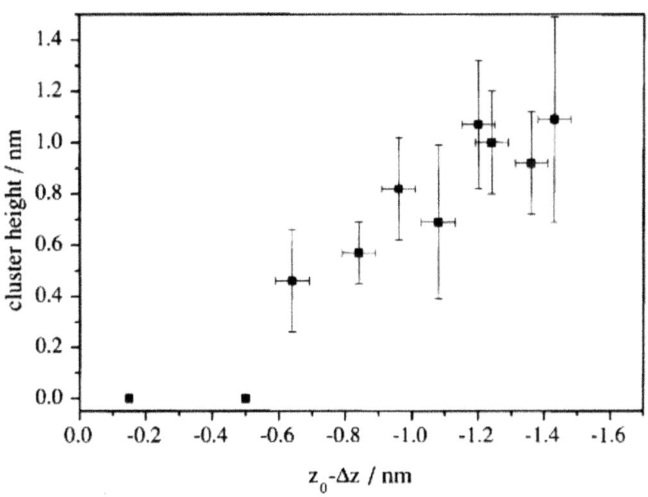


**FIGURE 1.37** Height of Pd clusters on Au(111) as a function of distance of closest approach during generation [177].

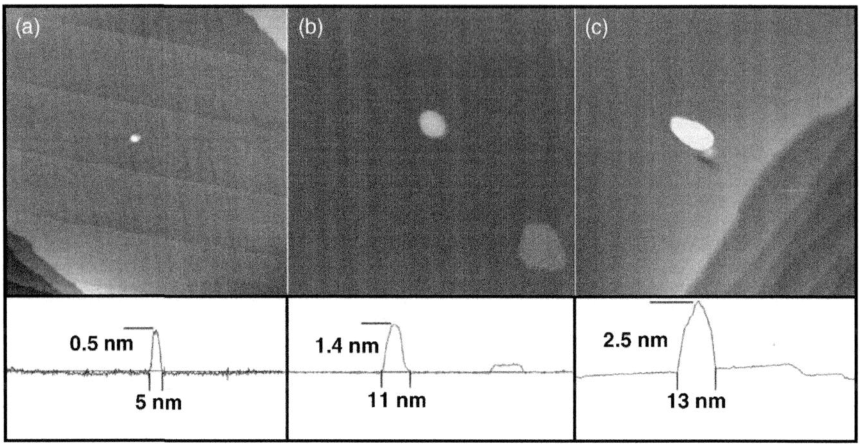


**FIGURE 1.38** In situ STM image of tip-induced palladium particles: (a) one particle of 0.5 nm height; (b) particle of 1.4 nm height; (c) one larger particle of 2.5 nm height. All images are $100 \times 100\,nm^2$. All particles on Au(111) in 0.1 M $H_2SO_4$. $I_{tunnel} = 1$ nA, $U_{WE} = 400$ mV, $U_{tip} = 500$ mV [165].

Experimental and theoretical findings provide a consistent picture of particle generation using electrochemical STM. Increasing movement of the STM tip in the $z$ direction induces increased particle size and particle stability. It was suggested that the stability of large particles is achieved mainly by Au atoms from the support forming stable alloys. In contrast, small particles contain only Pd atoms and thus are easily dissolvable.

### 1.6.4 STM Tip as Local Sensor Investigation of HER

Investigating single nanostructures such as particles of a few nanometers in diameter requires special techniques. Under electrochemical conditions it was demonstrated by Meier et al. [164] that an STM tip can be used as local sensor. With this technique it was possible to investigate single Pd particles supported on Au(111) toward HER. The purely in situ method combines generation of the Pd particles with the tip, characterization of the deposited particles, and subsequent measurement of the particle reactivity. Here, the hydrogen evolution was chosen for investigation because of its importance in electrochemistry. The catalytically active particle on the inert Au(111) support was directly observed with the STM tip. With electrochemical potential control of the support and the tip, it was possible to independently apply appropriate potentials. Therefore, it was feasible to evolve hydrogen on the Pd particle, whereas the tip was at a potential inducing the adjacent oxidation of the evolved hydrogen. The tip was placed ~ 10 nm above the Pd particle and thus collected most of the evolved hydrogen. With the very high sensitivity of the preamplifier in the STM setup, which is normally used to control the tunneling current feedback, it is possible to resolve currents down to several picoamperes. This method is suitable for single particles

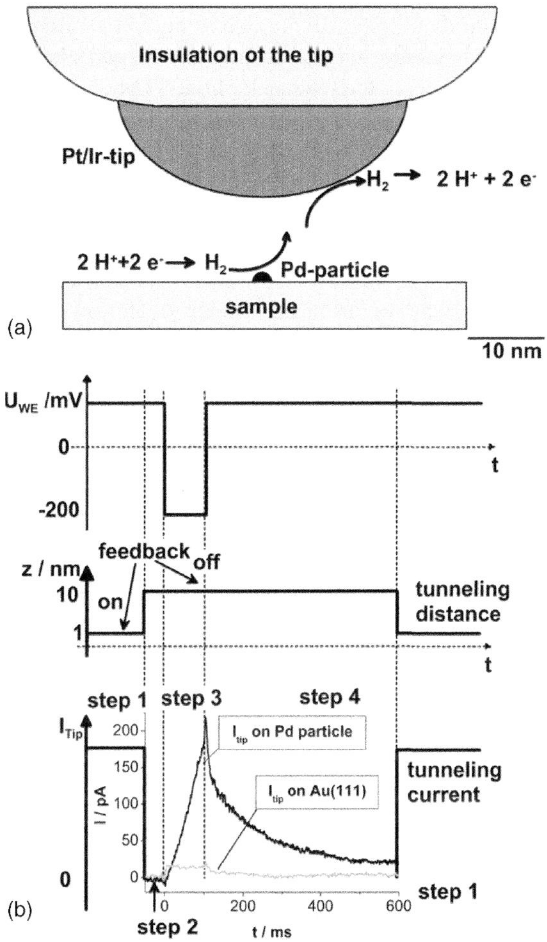


**FIGURE 1.39** (a) Illustration of the proton reduction reactivity measurement; (b) principle of the method: (1) the structure of the palladium particle is characterized; (2) the tip is positioned over the cluster and retracted while the feedback control is switched off; (3) a potential pulse of 100 ms duration is applied to the sample while the tip potential is kept constant and the hydrogen oxidation current on the STM tip is recorded; finally, the feedback control is switched on and the STM tip is back in tunneling mode [164].

and very small coverages. Using higher coverages leads to interferences due to the collection of evolved hydrogen from several particles. The high precision in vertical as well as horizontal resolution due to the nature of the STM using the tunneling effect allows a highly defined positioning measurement system. A detailed description of the local measurement technique with the STM tip is shown in Figure 1.39a, which includes the reaction pathway of HER at the Pd particles and the subsequent oxidation of the evolved hydrogen at the tip. Figure 1.39b is a detailed recording of the applied voltage on the working electrode, the distance of the tip from the support, and the current transient measured at the tip. In the following text all steps are described in detail. The tip was used to create and characterize the Pd particle on the support as a first step. After the scan stopped, the feedback of the STM was turned off and the tip was retracted (step 1). The potentials of the working electrode and the tip were still positive and tunneling current dropped to $\sim 0\,pA$ (step 2). Applying a potential negative of the hydrogen equilibrium potential induced the production of hydrogen from protons in the electrolyte. Protons diffused through the electrolyte to the tip and were oxidized (step 3). The current at the tip is the oxidation current of the evolved hydrogen at the Pd particle. In the last step (step 4) the current was recorded, and decreased because there was no further hydrogen evolution at the particle. This technique allows circumvention of transport limitations, which is not easily achievable for flat surfaces. Assuming a semihemispherical particle and thus hemispherical diffusion, one can estimate the diffusion-limiting current at $\sim 10^5\,A/cm^2$ for a particle with 2 nm radius. Therefore, it can be expected that diffusion limitation is not achieved with a large range of overpotentials.

### 1.6.5 Stability of Pd Nanoclusters on Au(111)—Experimental Results

Palladium nanoparticles were deposited with STM as described before. A contact between the tip and the support builds a connective neck and thus withdrawing the tip yields a particle. Some atoms of the tip stay at the surface and build a particle. For investigation, the particles must be stable on the support. The requirements for performing HER reactivity measurements on deposited Pd particles do not stipulate that no $Pd^{2+}$ ions be in the solution. Especially in this case, potentials are required that are far more negative than the Pd equilibrium potential in a $Pd^{2+}$-free solution. Thus the Pd particles are in a metastable state from a thermodynamic perspective. To gain further insight into the stability of Pd nanoparticles, the dissolution was investigated. Four different particles were regarded and are shown in Figure 1.40a. Particles A, B, and C have the same height but differ in diameter; particle D is larger in height and diameter. Figure 1.40b shows the change in particle height over time at a potential of 330 mV versus RHE. This potential is positive enough to slowly dissolute the Pd particles in the $Pd^{2+}$-free electrolyte. Obviously the change in particle height is not continuous. The decrease in particle height is stepwise with 0.26 nm per step. This value could not be allocated to any lattice constant of Pd or Au. As already seen in theory, the small particles decay faster compared to the big particle, which is still not completely dissolved. It was suggested that the particles consist of a layer structure. Dissolution of particles starts from steps and not from the top, which indicates that the low coordinated atoms are dissolved first. Furthermore, after dissolution of the larger particles, holes in the support were found next to the particles, indicating alloying of the Pd particles with the Au support. It was also concluded that smaller particles mainly containing Pd are less stable compared to larger ones consisting of more Au atoms and forming stable alloys. In summary, the experimental results presented are in line with the theoretical ones described before. Larger particles are more stable than smaller particles as a result of alloying, and small clusters practically consisting of Pd are completely removable from the support.

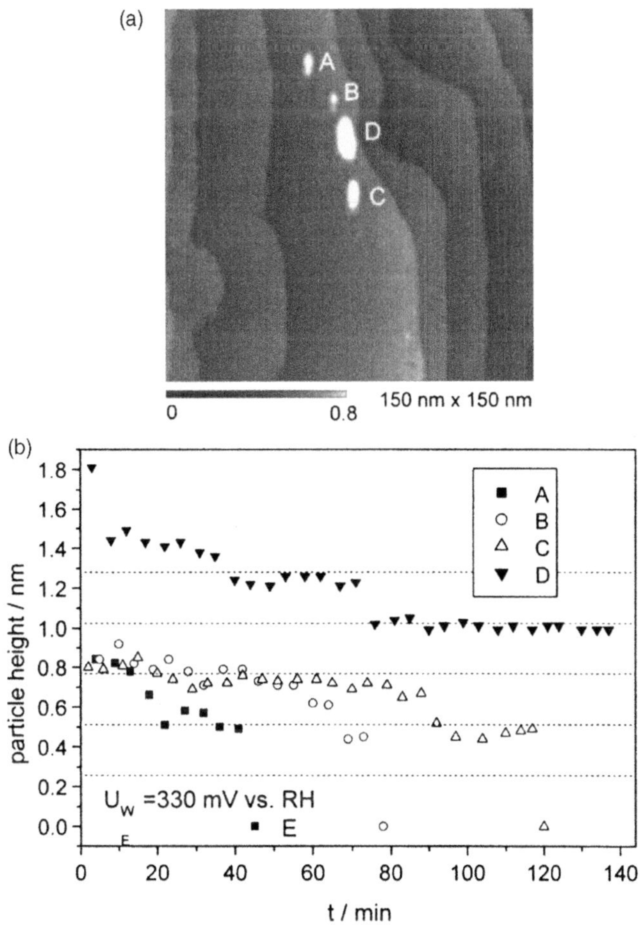


**FIGURE 1.40** (a) STM image of Pd clusters on Au(111) in 0.1 M $H_2SO_4$ immediately after generation, with $I_{tip} = 1.1$ nA, $U_{bias} = 170$ mV; (b) change in particle height with time at 330 mV versus RHE.

### 1.6.6 Reactivity Measurements of Pd Single Particles on Au(111)

Investigating nanoparticles usually means that a large number of particles are considered and the overall behavior is observed. Lowering the number of particles requires a high current resolution setup and sufficient signal-to-background ratio, which is not easily achievable. Also, agglomeration, size distributions, inhomogenities, and diffusion limitations of reaction partners have to be considered. Therefore, a technique investigating only one particle was introduced by Meier et al. [164,165]. As explained in a previous subsection, the tip was used to measure the reaction products. In this case the evolved hydrogen was oxidized at the tip and hence an indirect reactivity was measured. The catalytic activity was shown to be

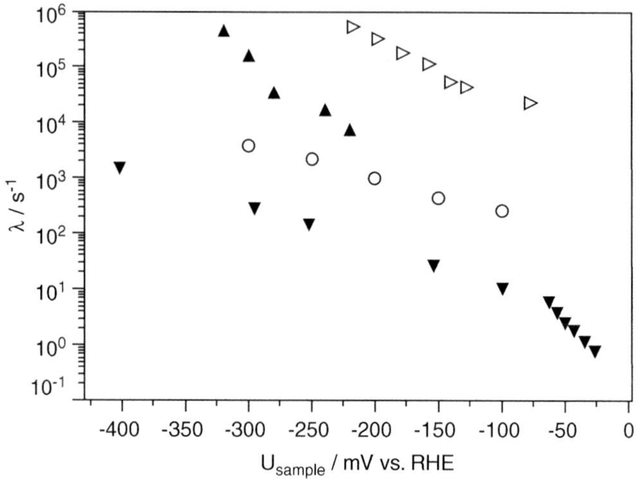


**FIGURE 1.41** Tafel plots of three different Pd particles on Au(111) in 0.1 M $H_2SO_4$: ○, Pd particle with 10 nm height and 200 nm diameter; ▲, Pd particle with 2.5 nm height and 12 nm diameter; ▷, three Pd particles with 0.5 nm height and 6 nm diameter; ▼, polycrystalline Pd [192].

strongly dependent on particle size, especially particle height. Therefore, particles of different height and diameter were investigated. For different overpotentials applied at the single Pd particle, the Au(111) support hydrogen was evolved and the activity was measured. Figure 1.41 shows a Tafel plot for different Pd particles deposited on Au(111). It is obvious that with decreasing particle height, increasing activity is observed. The increase in activity is at least two orders of magnitude, comparing 200 nm with 6 nm in diameter. Detailed study revealed that the current densities, especially for particles with a height of 0.5 nm, are highly reactive, up to two orders of higher magnitude compared to the flat Pd surface. A change in the Tafel slope was also indicated and was explained by surface diffusion effects on the support [189]. The changed Tafel slope also indicates that the reaction mechanism of the HER is different. Whereas a Tafel slope of 118 mV suggests a rate-determining step of the discharge of the protons, a Tafel slope of 29 mV theoretically predicts the recombination of two adsorbed hydrogen atoms as the rate-determining step. Because of the mixed Tafel slope values, it was suggested that recombination becomes more important since the discharge of protons should be rapid. The enhanced activity was discussed regarding several approaches. An influence of low coordinated atoms on reactivity was not found. Charging of the particles due to different workfunctions and a modified double layer were also excluded. Applying the model of the changed *d*-band center [190,191] was suggested to explain the enhancement.

An important relationship was found by correlating the activity with particle height (see Fig. 1.42). A strong increase in activity was observed with decreasing particle height. An influence of the number of low-coordination atoms was excluded.

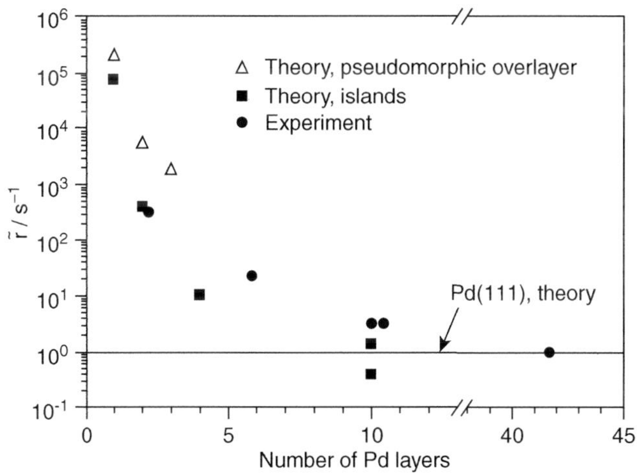


**FIGURE 1.42** Semilogarithmic plot of the relative reaction rate $r$ versus particle height. Full circles represent experimental data ($U_{WE} = 200$ mV); open triangles, theoretical predictions for pseudomorphic overlayers, obtained from DFT calculations; full squares, theoretical predictions for islands, obtained by combining the DFT calculations with molecular dynamics [165].

STM images were evaluated and the ratio of low-coordination atoms to surface atoms was determined. Study of two different effects was suggested: (1) the surface structure of the Pd particle and (2) the hydrogen bonding to the Pd layers. It was shown [193] that Pd grows in (111) orientation on Au(111). Because of the different lattice constants, the Pd overlayers are strained by $\sim 4.8\%$. Up to now it has not really been clear which function should be applied for mismatch versus layer thickness. Molecular dynamics (MD) calculations showed that single-layer islands are pseudomorphic while two Pd monolayers introduce misfit dislocations. It was also shown that for even four monolayers a strain can be observed in the Pd overlayers. Also, the adsorption energy of hydrogen to Pd overlayers in Au(111) was calculated using DFT, whereas the adsorption energy is closely related to the reactivity [165]. In Figure 1.42 shows that the theoretical calculations for islands that are one to three monolayers high are more reactive compared to the Pd(111) bulk. A strong enhancement is observed by increasing the height of islands as well as pseudomorphic overlayers. Experimental findings and the theoretical calculations showed the same trend seen in Figure 1.42. The enhancement in reactivity is caused by a strain of the Pd overlayers induced by the Au support with a larger lattice constant. Therefore, the electronic properties and thus the binding energy to hydrogen are changed.

Similar results were also found on extended Pd nanostructured Au(111) surfaces by Pandelov and Stimming [161], which supports the single-particle approach. Nevertheless, with the advantages of only slight diffusion limitation and the possibility of investigating single particles under electrochemical conditions, this method and the results are unique.

### 1.6.7 Summary

A combination of theoretical and experimental results was reviewed regarding generation, stability, and activity of single Pd nanoparticles on Au(111). It was shown that generation of single particles is dependent on the movement of the STM tip in the *z* direction and the geometry of the tip. With increasing movement toward the surfaces, increasing particles sizes were experimentally obtained and theoretically calculated. Larger particles form Pd–Au alloys with Au atoms from the support, which stabilize the particle and inhibit the dissolution, whereas smaller particles consist only of Pd and are completely dissolvable. The electrochemical STM tip as a probing sensor was used to investigate the HER on single Pd particles in Au. It was found that with decreasing particle height the reactivity strongly increases. This finding is in line with the Nørskov model, which explains the enhanced reactivity due to a shift in the *d*-band center and thus in changed adsorption energy to hydrogen. Similar results were also reported for large Pd nanostructured Au(111) surfaces from Pandelov and Stimming [161]. For this reason two different experimental approaches show the same trend: namely, that with decreasing particle height rather than coverage, the activity toward HER is increased. The method of investigating single particles is a powerful tool for direct investigation of the catalytic activity in the nanometer scale without the limitations due to extended surfaces. Although the experimental approach is sophisticated and time-consuming, new insights into reactivity behavior influenced by the support were found. Together with findings from extended surfaces, an additional argument shows the importance of the support and the influence on reactivity.

## 1.7 STUDIES OF HYDROGEN-RELATED REACTIONS ON CARBON-BASED SYSTEMS

### 1.7.1 Introduction

In this last section of the chapter we present an overview of studies performed on carbon-based systems, for hydrogen-related reactions. Carbon has been the material of choice for many years because of its high electrical conductivity, low cost, and high availability. Carbon is a conventional support used in the present technology of fuel cells. There are many papers, some of which are presented here, devoted to studying HER and HOR mechanisms using metallic nanoparticles on carbon-based supports. Another topic of interest is the activity of these catalysts for CO oxidation in combination with their activity toward HER and HOR.

Carbon monoxide poisoning is a major issue in today's fuel cell technology since it impedes fuel cell commercialization. Fuel cells operating with industrial hydrogen, hydrogen feedstreams derived from carbon-based fuels or alcohols, present deactivation issues of the anode catalyst even with small traces of CO. Therefore, although Pt is shown to be a very good catalyst for HER and HOR, it is deactivated strongly by CO, leading to high overpotentials. Hence, a catalyst is needed that combines the ability to oxidize both CO and $H_2$ effectively.

### 1.7.2 HOR and HER Studies on Carbon-Supported Metallic Nanoparticles

Gasteiger et al. [194] studied $H_2$ and CO electrooxidation on Pt, Ru, and Pt–Ru catalysts using the rotating-disk electrode (RDE) method. Binary platinum catalysts consisting of alloy mixtures of Pt with other metals such as Ru, Rh, and Sn showed enhanced activity toward CO oxidation. These mixtures facilitate the oxidation of CO by providing oxygen-containing species, therefore increasing the active sites for hydrogen oxidation. Pt–Ru alloys are commonly used nowadays as supported catalysts for direct alcohol fuel cells. Gasteiger et al. [194] studied the alloying effect on $H_2$ and CO oxidation using simple but effective sample geometry. They used arc melting to prepare pure Pt, pure Ru, and Pt : Ru (1 : 1 atomic ratio) alloy samples with a small diameter (~6 mm). As seen in Figure 1.43, the sample containing Pt–Ru alloy showed enhanced activity toward CO oxidation compared to pure Pt sample. They also observed that the alloy sample showed identical activity compared to pure Pt sample toward $H_2$ oxidation, as shown in Figure 1.44. Although the pure Ru sample exhibited the optimum behavior toward CO oxidation, its activity toward $H_2$ oxidation was many orders of magnitude lower.

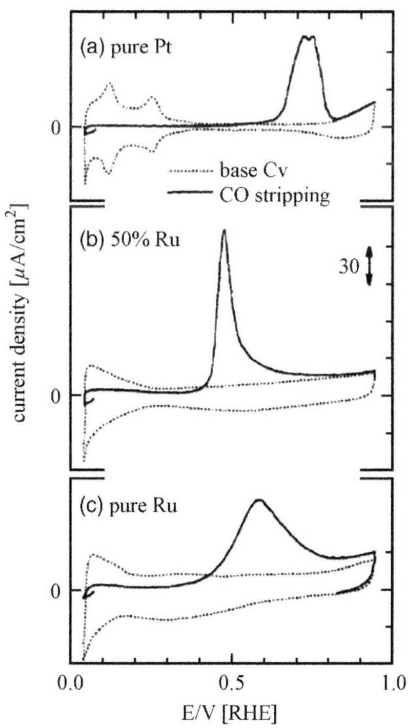


**FIGURE 1.43** CO stripping voltammetry of sputter-cleaned electrodes in 0.5 M $H_2SO_4$ on (a) Pt, (b) Pt–Ru, and (c) Ru. Straight line indicates first positive-going sweep; dashed line, negative-going sweep after stripping of CO, followed by a positive-going sweep. Scan rate was 20 mV/s [194].

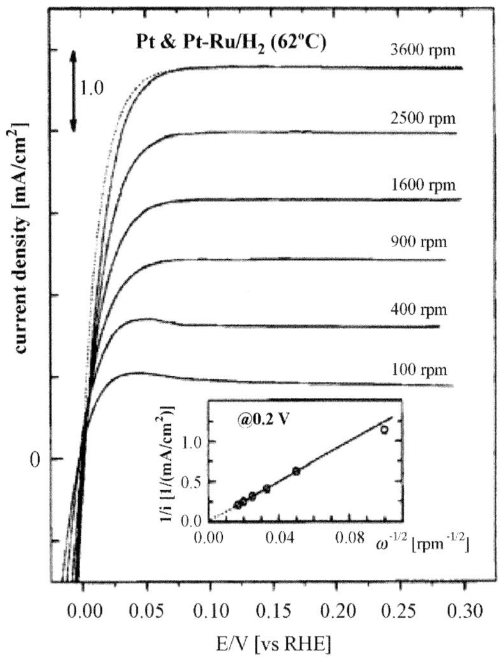


**FIGURE 1.44** Hydrogen reduction and oxidation current densities during positive-going sweeps (20 mV/s) on Pt and Pt–Ru electrodes saturated with $H_2$ at 62 °C [194].

Sasaki et al. [195] investigated bimetallic catalysts consisting of Pt submonolayers deposited on Ru and Au nanoparticles supported on carbon. The catalysts were investigated for hydrogen oxidation as well as oxygen reduction. Deposition of Pt on Ru was carried out by immersing the Ru catalyst on a solution containing Pt, while Pt deposition on Au was achieved by redox displacement of a Cu underpotential deposited monolayer. Two types of samples were prepared and included Pt/Ru and Pt/Au planar surfaces as well as carbon supported catalysts. Representative TEM pictures of the supported PtRu catalysts are shown in Figure 1.45.

The Pt–Ru catalyst showed enhanced activity for hydrogen oxidation in RDE configuration compared to commercial catalysts where CO-contaminated $H_2$ was used as a fuel (Fig. 1.46). The results showed that mass specific activity for $H_2$ oxidation of $PtRu_{20}$ catalyst is 3–4 times larger than for commercial catalysts. Also, CO tolerance tests proved the $PtRu_{20}$ catalysts to be superior to commercial ones.

Chen and Kucernak [196] investigated the kinetics of hydrogen oxidation reaction on single-Pt-particle-supported carbon. The advantage of their method is that they could quantify the mass diffusion properties. They also studied the role of the $H_{upd}$ in the electrochemical oxidation of hydrogen molecules. As discussed previously, HOR involves atomic hydrogen as adsorbed species. The role of adsorbed hydrogen in HOR is not yet fully understood. It is believed that it either could be the intermediate involved in oxidation of dihydrogen or acts as blocking intermediate, hindering the reaction. Their sample preparation included

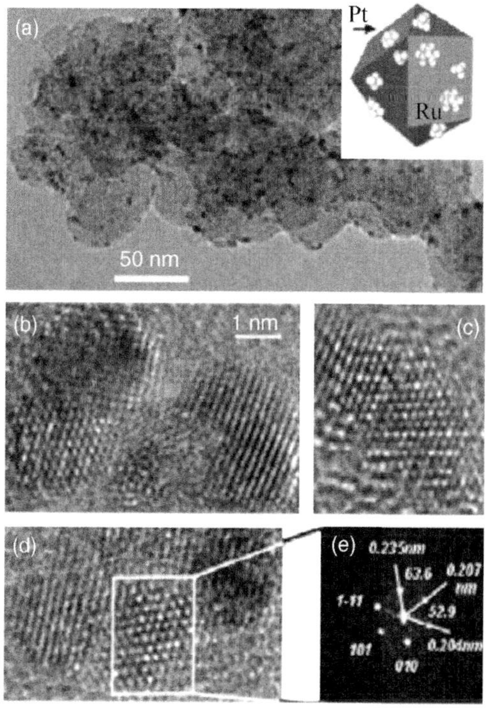


**FIGURE 1.45** TEM micrographs of the $PtRu_{20}$ catalyst prepared by spontaneous deposition of Pt on Ru nanoparticles: (a) low magnification morphology of the metal particles (black dots indicate average size ~2.5 nm) on carbon spheres (average size ~50 nm); (b–d) high-resolution images showing atomic resolved lattice structures; (e) diffractogram obtained from the high-resolution image shown in (d) with measured the angles and lattice spacings that are consistent with hexagonal close-packed (hcp) Ru single-crystal structure [195].

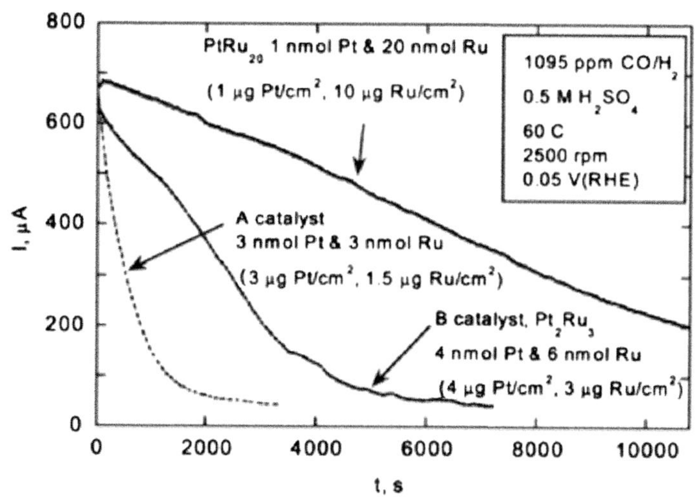


**FIGURE 1.46** Comparison of the CO tolerance of three catalysts based on the current as a function of time for the oxidation of $H_2$ with 1095 ppm of CO at 60°C for the $PtRu_{20}$ and two commercial electrocatalysts at 0.05 V with the loadings indicated in the graph. A considerably larger CO tolerance is seen for the $PtRu_{20}$ catalyst [195].

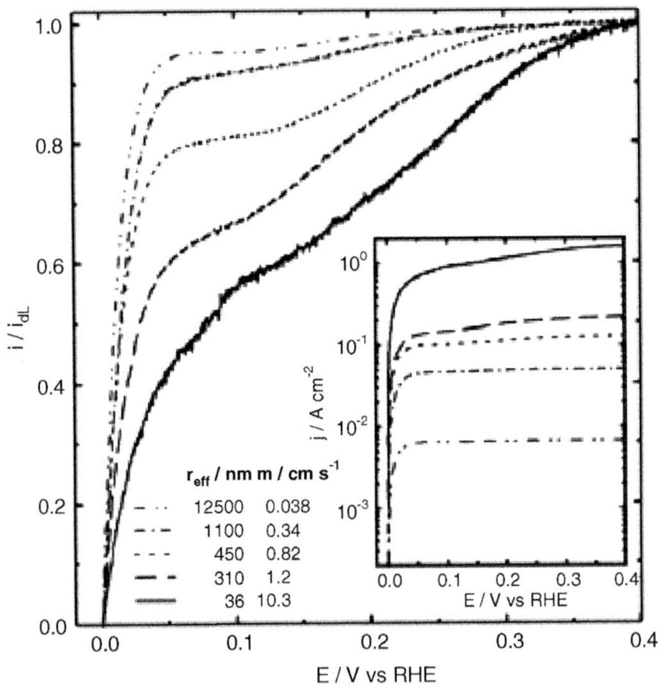


**FIGURE 1.47** Current–potential curves obtained on particle electrodes accompanying the oxidation of hydrogen in 0.1 mol/dm$^3$ $H_2SO_4$ solution saturated with CO free hydrogen gas. The currents are normalized by the limiting values read at about 0.45 V. The numbers in the figure indicate the particle radius and give the mass transport rate for that particle assuming hydrogen diffusion coefficient of $3.7 \times 10^{-5}$ cm/s. Scan rate is 5 mV/s. The inset shows the current density–potential plot for each of these particles [196].

electrochemical etching of carbon fibers, followed by insulation of the latter leaving a very small area exposed (down to a few nanometers). Single Pt particles are then deposited by electrochemical deposition. The results of the study toward HOR are shown in Figure 1.47.

As is shown, the current densities vary almost three orders of magnitude for the range of particles examined. Also, the features of the curves are different from those obtained from RDE experiments. For small particle radii there are two limiting current plateaus indicating two separating processes involved in HOR. As the particle radius increases, the two different limiting currents diminish. The authors explain this behavior to be a result of decreasing the ratio between reaction rate constant and mass transport coefficient. A model is also presented showing that the first plateau is due to the limiting rate of the Tafel reaction. Additionally, they conclude that $H_{upd}$ is a reaction intermediate rather than an inhibitor of HOR.

Arenz et al. [197] examined the catalytic activity of various carbon-supported Pt–Sn catalysts toward oxidation of $H_2$, CO, and $H_2$/CO mixtures. Typical micrographs of the catalysts studied are shown in Figure 1.48.

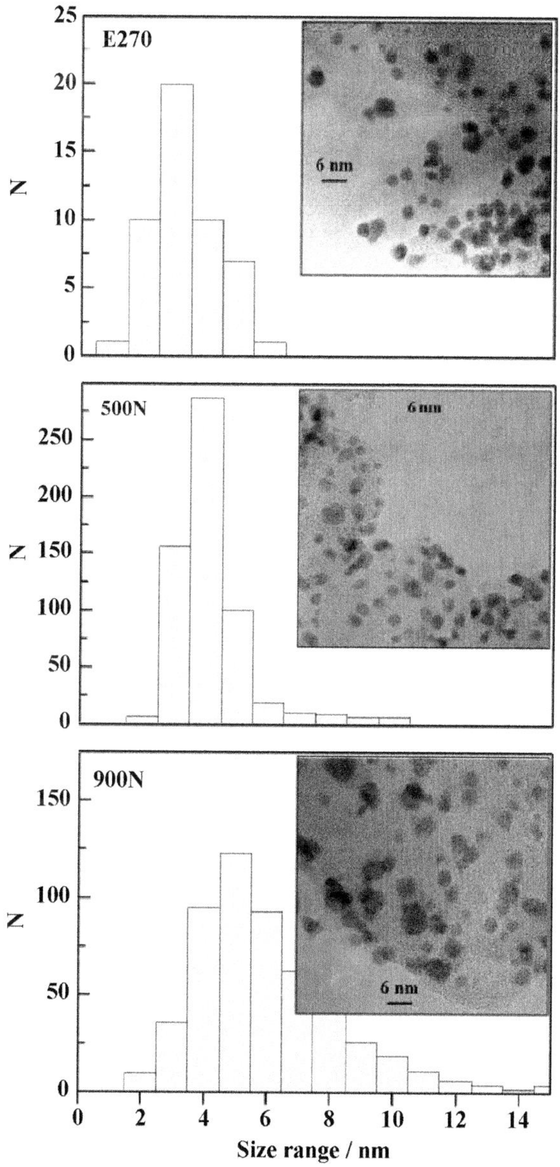


**FIGURE 1.48** TEM micrographs of carbon-supported PtSn high-surface-area catalysts with corresponding histograms of particle size distribution: E270—average particle size 3–4 nm; 500N—average particle size ~4 nm; 900N—average particle size ~6 nm [197].

Polarization curves for $H_2$, CO, and $H_2$/CO oxidation are shown in Figure 1.49, where catalytic activity is expressed as mass activity (moles reactant per unit mass of Pt).

As is shown in Figure 1.49, all three catalysts show similar activities toward $H_2$ oxidation. However, sample E270 having a Pt : Sn atomic ratio of 3 : 1 had the highest mass activity for CO oxidation.

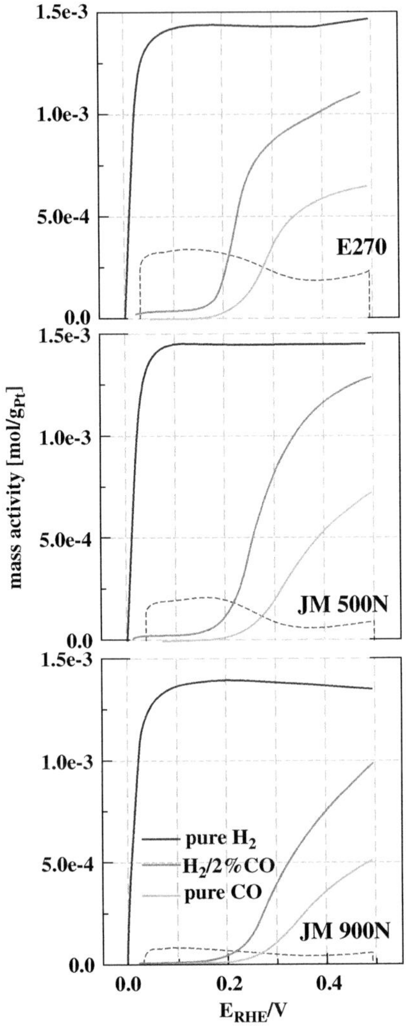


**FIGURE 1.49** Polarization curves for different Pt–Sn/C catalysts recorded in 0.5 M $H_2SO_4$ solution saturated with either pure $H_2$, a $H_2$/2% CO mixture, or pure CO; curves are recorded at 60°C with 1 mV/s and 2500 rpm; dashed curves: CV (au) in argon-purged solution at 60°C, scan rate 50 mV/s [197].

### 1.7.3 HOR and HER Investigations on Highly Oriented Pyrolytic Graphite (HOPG) Systems

Highly oriented pyrolytic graphite is a carbon-based material and has been used in studying electrochemical reactions because of its interesting properties. It is a carbon material with very low surface roughness, and therefore reactions can be studied in a model-like substrate that is closer to the "real" carbon-based catalyst systems. By investigating catalyst particles on HOPG surface, one can exclude the substrate-related effects can be due to weak binding of the metal particles to the substrate. Electrodeposited particles are not epitaxially oriented on a HOPG surface. Furthermore, because of $sp^2$ hybridization of carbon atoms in the HOPG basal plane, there is no possibility for bond formation between hydrogen and the surface. Therefore, hydrogen reactions on a metal-decorated HOPG surface should be affected mainly by morphological parameters, such as particle size and particle area density.

Gimeno et al. [198] utilized STM and electrochemical techniques to investigate the growth mechanism of Pd islands deposited on HOPG from aqueous solutions. They found that the initial stage of growth is under activation control followed by a diffusion

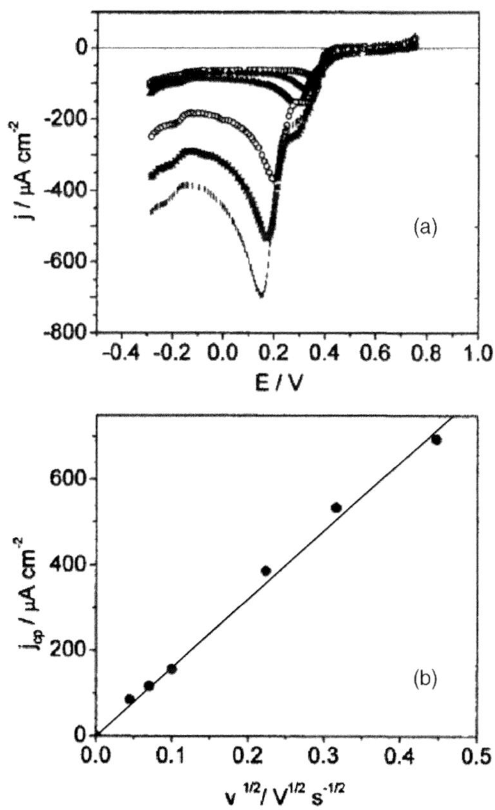


**FIGURE 1.50** (a) Voltammograms obtained for HOPG in aqueous $7.5 \times 10^{-4}$ $PdCl_2$ + 5 $\times 10^{-2}$M $NaClO_4$ + $5 \times 10^{-3}$M $HClO_4$ at different potential scan rates; $u = 2$ mV/s (top curve), 5 mV/s, 10 mV/s, 50 mV/s, 0.1 V/s, and 0.2 V/s (bottom curve); 298 K. (b) plot of cathodic current peak $j_{cp}$ versus $u^{1/2}$ [198].

control. This was clearly demonstrated as shown in Figure 1.50, where the height of the cathodic current peak increases linearly with the scan rate.

Particles formed by the potential step method showed different morphologies, which strongly depended on whether the cathodic potential was above or below the potential of zero charge of the bulk palladium.

Lu and Zangari [199] studied electrodeposited Pt nanoparticles on HOPG. They investigated the different parameters during the deposition process such as the scan rate during deposition and concentration of the Pt precursor in the electrolyte. Another factor investigated was the addition of supporting electrolyte (e.g., HCl and $H_2SO_4$). Supporting electrolyte can influence faradaic processes and in particular, the growth and nucleation. For example, electrolytes would increase overall conductivity, resulting in compression of the double layer and shorten the double-layer charging transient. They found that chlorides in general inhibit the Pt reduction process, due to formation of adsorbed chloride films that occupy available sites for discharge of Pt complexes. On the contrary, addition of sulfate or perchlorate enhances reduction process by compressing the double layer.

Brülle and Stimming [156, 200] investigated the electrodeposition of Pt nanoparticles on HOPG surfaces and their reactivity toward HER. In order to create more active sites, an oxidation step was initially performed prior to deposition. Representative AFM images are shown in Figure 1.51.

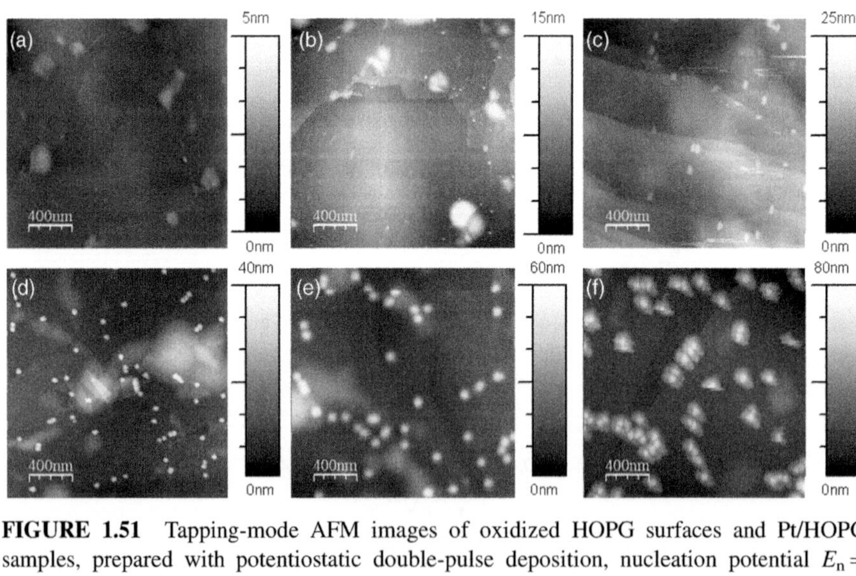


**FIGURE 1.51** Tapping-mode AFM images of oxidized HOPG surfaces and Pt/HOPG samples, prepared with potentiostatic double-pulse deposition, nucleation potential $E_n =$ 30 mV, nucleation time $t_n = 500\ \mu s$, growth potential $E_g = 560$ mV, and different growth times $t_g$: (a) oxidized HOPG surface; (b) $t_g = 100$ s, $Q_{dep} = -2.56$ mC/cm$^2$; (c) $t_g = 150$ s, $Q_{dep} = -5.55$ mC/cm$^2$; (c) $t_g = 300$ s, $Q_{dep} = -13.76$ mC/cm$^2$; (d) $t_g = 450$ s, $Q_{dep} = -19.9$ mC/m$^2$; (e) $t_g = 600$ s, $Q_{dep} = -41.5$ mC/cm$^2$ [200].

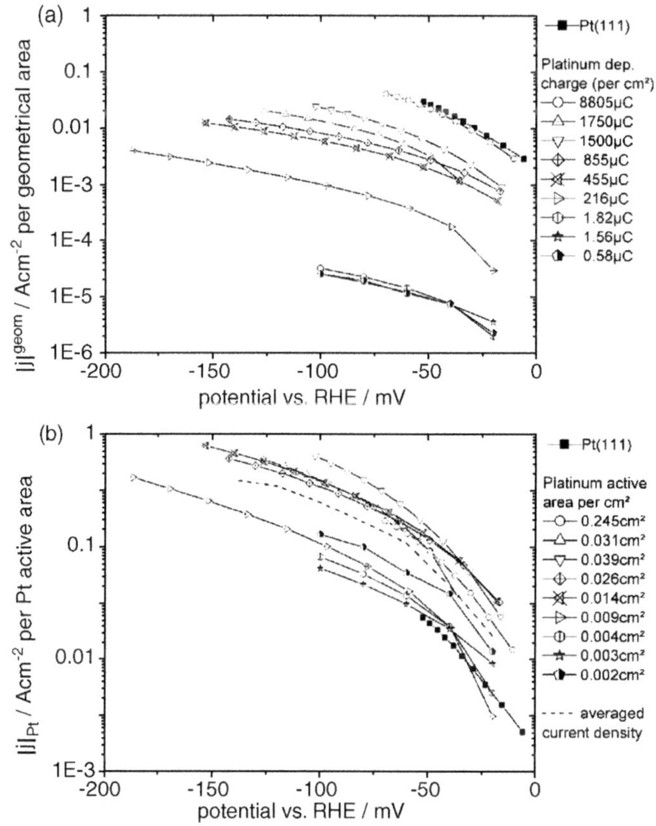


**FIGURE 1.52** Tafel plots for different amounts of platinum on HOPG deposited via single-pulse method, current densities versus geometric area (a); current density normalized to platinum active area obtained from hydrogen ad/desorption charge. An averaged curve, assuming that the variation of the specific current density is due to experimental errors, is represented by the dashed line (b) [156].

HER was studied using potentiostatic pulses, and results are shown in Figure 1.52. As can be seen, the trend observed is that the more Pt is deposited, the higher the activity. For high amounts of deposited Pt, the activity reaches that from a Pt single crystal. If activities are normalized to the Pt active surface area, then they all become comparable within experimental error to Pt single crystal.

Figure 1.53 shows HER current densities normalized to platinum active surface area versus platinum coverage for Pt/Au(111) and Pt/HOPG for very small coverages (Pt surface per geometric area < 0.02). The different behavior of the activities of the two model systems becomes obvious for these small coverages. At a coverage larger than 0.01, the activity regarding HER is similar for both kinds of platinum nanostructured surfaces. For smaller amounts of platinum on Au(111) surfaces, the current densities increase sharply with decreasing coverage. For HER on Pt/HOPG, the

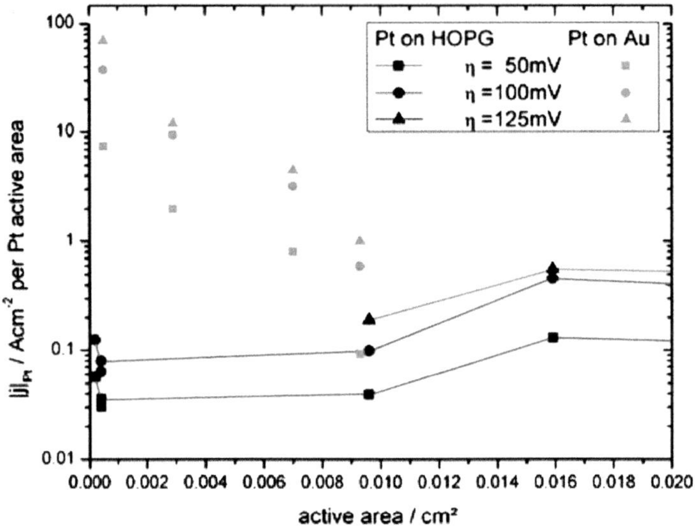


**FIGURE 1.53** Current densities per platinum active area versus platinum active area per substrate ($cm^{-2}$) for Pt/HOPG (single-pulse method) and Pt/Au(111) for different overpotentials ($\eta = 50$ mV, 100 mV, 125 mV) [156].

current densities do not increase but slightly decrease for decreasing coverage. This leads to the conclusion that the enhancement in reactivity for Pt/Au(111) is caused by a mechanism that cannot be applied to the Pt/HOPG model system.

## REFERENCES


1. T. Erdey-Gruz and M. Volmer, *Z. Phys. Chem. Abteil. Chem. Thermodyn. Kinetik Elektrochem. Eigenschaft.* **150**, 203 (1930).
2. J. Tafel, *Z. Phys. Chem. Stochiom. Verwandt.* **50**, 641 (1905).
3. J. Heyrovsky, *Rec. Trav. Chim. Pays-Bas* **46**, 582 (1927).
4. N. M. Markovic and P. N. Ross, *Surf. Sci. Rep.* **45**, 121 (2002).
5. K. Christmann and G. Ertl, *Surf. Sci.* **60**, 365 (1976).
6. R. W. McCabe and L. D. Schmidt, *Surf. Sci.* **65**, 189 (1977).
7. M. Salmeron, R. J. Gale, and G. A. Somorjai, *J. Chem. Phys.* **70**, 2807 (1979).
8. R. A. Olsen, G. J. Kroes, and E. J. Baerends, *J. Chem. Phys.* **111**, 11155 (1999).
9. N. M. Markovic, B. N. Grgur, and P. N. Ross, *J. Phys. Chem. B* **101**, 5405 (1997).
10. J. H. Barber and B. E. Conway, *J. Electroanal. Chem.* **461**, 80 (1999).
11. J. Barber, S. Morin, and B. E. Conway, *J. Electroanal. Chem.* **446**, 125 (1998).
12. P. M. Quaino, M. R. G. de Chialvo, and A. C. Chialvo, *Phys. Chem. Chem. Phys.* **6**, 4450 (2004).
13. V. S. Bagotzky and N. V. Osetrova, *J. Electroanal. Chem.* **43**, 233 (1973).

14. P. M. Quaino, J. L. Fernandez, M. R. G. de Chialvo, and A. C. Chialvo, *J. Mol. Catal. A: Chem.* **252**, 156 (2006).
15. J. O. M. Bockris, *J. Serb. Chem. Soc.* **70**, 475 (2005).
16. B. E. Conway and J. O. Bockris, *J. Chem. Phys.* **26**, 532 (1957).
17. R. Parsons, *Trans. Faraday Soc.* **54**, 1053 (1958).
18. S. Trasatti, *J. Electroanal. Chem.* **39**, 163 (1972).
19. L. I. Krishtalik and P. Delahay eds., *Advances in Electrochemistry and Electrochemical Engineering*, Vol. 7, Interscience, New York, 1970.
20. J. K. Nørskov, T. Bligaard, A. Logadottir, J. R. Kitchin, J. G. Chen, S. Pandelov, and U. Stimming, *J. Electrochem. Soc.* **152**, J23. (2005).
21. N. M. Markovic, S. T. Sarraf, H. A. Gasteiger, and P. N. Ross, *J. Chem. Soc. Faraday Trans.* **92**, 3719 (1996).
22. N. M. Markovic, B. N. Grgur, and P. N. Ross, *J. Phys. Chem. B* **101**, 5405 (1997).
23. J. Barber, S. Morin, and B. E. Conway, *J. Electroanal. Chem.* **446**, 125 (1998).
24. J. Perez, E. R. Gonzalez, and H. M. Villullas, *J. Phys. Chem. B* **102**, 10931 (1998).
25. J. H. Barber and B. E. Conway, *J. Electroanal. Chem.* **461**, 80 (1999).
26. S. Schuldin, M. Rosen, and D. R. Flinn, *J. Electrochem. Soc.* **117**, 1251 (1970).
27. K. Seto, A. Iannelli, B. Love, and J. Lipkowski, *J. Electroanal. Chem.* **226**, 351 (1987).
28. H. Kita, S. Ye, and Y. Gao, *J. Electroanal. Chem.* **334**, 351 (1992).
29. R. Gomez, A. Fernandezvega, J. M. Feliu, and A. Aldaz, *J. Phys. Chem.* **97**, 4769 (1993).
30. J. Clavilier, R. Faure, G. Guinet, and R. Durand, *J. Electroanal. Chem.* **107**, 205 (1980).
31. J. Clavilier, *J. Electroanal. Chem.* **107**, 211 (1980).
32. S. Motoo and N. Furuya, *J. Electroanal. Chem.* **172**, 339 (1984).
33. N. Markovic, M. Hanson, G. McDougall, and E. Yeager, *J. Electroanal. Chem.* **214**, 555 (1986).
34. S. Motoo and N. Furuya, *Ber. Buns. Gesells. Phys. Chem. Chem. Phys.* **91**, 457 (1987).
35. J. Clavilier, A. Rodes, K. Elachi, and M. A. Zamakhchari, *J. Chim. Phys. Physico-Chim. Biol.* **88**, 1291 (1991).
36. J. Clavilier, J. M. Orts, and J. M. Feliu, *J. Phys. Iv* **4**, 303 (1994).
37. J. M. Feliu, A. Rodes, J. M. Orts, and J. Clavilier, *Polish J. Chem.* **68**, 1575 (1994).
38. L. A. Kibler, A. Cuesta, M. Kleinert, and D. M. Kolb, *J. Electroanal. Chem.* **484**, 73 (2000).
39. J. Clavilier, D. Armand, S. G. Sun, and M. Petit, *J. Electroanal. Chem.* **205**, 267 (1986).
40. J. Clavilier, J. M. Feliu, A. Fernandezvega, and A. Aldaz, *J. Electroanal. Chem.* **269**, 175 (1989).
41. E. Vlieg, I. K. Robinson, and K. Kern, *Surf. Sci.* **233**, 248 (1990).
42. N. M. Markovic, B. N. Grgur, C. A. Lucas, and P. N. Ross, *Surf. Sci.* **384**, L805 (1997).
43. S. Horch, H. T. Lorensen, S. Helveg, E. Laegsgaard, I. Stensgaard, K. W. Jacobsen, J. K. Nørskov, and F. Besenbacher, *Nature* **398**, 134 (1999).
44. D. A. Scherson and D. M. Kolb, *J. Electroanal. Chem.* **176**, 353 (1984).
45. A. M. Funtikov, U. Stimming, and R. Vogel, *J. Electroanal. Chem.* **428**, 147 (1997).
46. A. M. Funtikov, U. Linke, U. Stimming, and R. Vogel, *Surf. Sci.* **324**, L343 (1995).

47. L. A. Kibler, *Preparation and Characterization of Noble Metal Single Crystal Electrode Surfaces*, International Society of Electrochemistry, 2003.
48. N. M. Markovic and P. N. Ross, *Surf. Sci. Rep.* **45**, 121 (2002).
49. N. Markovic, H. Gasteiger, and P. N. Ross, *J. Electrochem. Soc.* **144**, 1591 (1997).
50. U. Frese, T. Iwasita, W. Schmickler, and U. Stimming, *J. Phys. Chem.* **89**, 1059 (1985).
51. U. Frese and U. Stimming, *J. Electroanal. Chem.* **198**, 409 (1986).
52. M. H. Dishner, M. M. Ivey, S. Gorer, J. C. Hemminger, and F. J. Feher, *J. Vac. Sci. Technol. A Vac. Surf. Films* **16**, 3295 (1998).
53. W. Haiss, D. Lackey, J. K. Sass, and K. H. Besocke, *J. Chem. Phys.* **95**, 2193 (1991).
54. D. M. Kolb, *Progress Surf. Sci.* **51**, 109 (1996).
55. O. M. Magnussen, J. Hagebock, J. Hotlos, and R. J. Behm, *Faraday Discuss.* **94**, 329 (1992).
56. G. J. Edens, X. P. Gao, and M. J. Weaver, *J. Electroanal. Chem.* **375**, 357 (1994).
57. T. Dretschkow and T. Wandlowski, *Ber. Buns. Gesell. Phys. Chem. Chem. Phys.* **101**, 749 (1997).
58. X. P. Gao, A. Hamelin, and M. J. Weaver, *Phys. Rev. B* **44**, 10983 (1991).
59. X. P. Gao, A. Hamelin, and M. J. Weaver, *Phys. Rev. Lett.* **67**, 618 (1991).
60. O. M. Magnussen, J. Hotlos, R. J. Behm, N. Batina, and D. M. Kolb, *Surf. Sci.* **296**, 310 (1993).
61. O. M. Magnussen, J. Wiechers, and R. J. Behm, *Surf. Sci.* **289**, 139 (1993).
62. A. Hamelin, *J. Electroanal. Chem.* **407**, 1 (1996).
63. M. Sturmat, R. Koch, and K. H. Rieder, *Phys. Rev. Lett.* **77**, 5071 (1996).
64. A. S. Dakkouri, *Solid State Ion.* **94**, 99 (1997).
65. G. J. Brug, M. Sluytersrehbach, J. H. Sluyters, and A. Hamelin, *J. Electroanal. Chem.* **181**, 245 (1984).
66. A. Hamelin and M. J. Weaver, *J. Electroanal. Chem.* **223**, 171 (1987).
67. A. Hamelin, L. Stoicoviciu, S. C. Chang, and M. J. Weaver, *J. Electroanal. Chem.* **307**, 183 (1991).
68. S. Trasatti, *Russ. J. Electrochem.* **41**, 1255 (2005).
69. M. Baldauf and D. M. Kolb, *Electrochim. Acta* **38**, 2145 (1993).
70. L. A. Kibler, M. Kleinert, R. Randler, and D. M. Kolb, *Surf. Sci.* **443**, 19 (1999).
71. L. A. Kibler, M. Kleinert, and D. M. Kolb, *Surf. Sci.* **461**, 155 (2000).
72. H. Naohara, S. Ye, and K. Uosaki, *Electrochim. Acta* **45**, 3305 (2000).
73. B. Alvarez, V. Climent, A. Rodes, and J. M. Feliu, *Phys. Chem. Chem. Phys.* **3**, 3269 (2001).
74. H. Naohara, S. Ye, and K. Uosaki, *J. Electroanal. Chem.* **500**, 435 (2001).
75. L. A. Kibler, M. Kleinert, V. Lazarescu, and D. M. Kolb, *Surf. Sci.* **498**, 175 (2002).
76. R. Hoyer, L. A. Kibler, and D. M. Kolb, *Electrochim. Acta* **49**, 63 (2003).
77. L. A. Kibler, A. M. El-Aziz, and D. M. Kolb, *J. Mol. Catal. A Chem.* **199**, 57 (2003).
78. M. Arenz, V. Stamenkovic, P. N. Ross, and N. M. Markovic, *Surf. Sci.* **573**, 57 (2004).
79. A. M. El-Aziz, R. Hoyer, L. A. Kibler, and D. M. Kolb, *Electrochim. Acta* **51**, 2518 (2006).
80. S. Pandelov and U. Stimming, *Electrochim. Acta* **52**, 5548 (2007).
81. Y. Pluntke, L. A. Kibler, and D. M. Kolb, *Phys. Chem. Chem. Phys.* **10**, 3684 (2008).

82. H. Wolfschmidt, R. Bussar, and U. Stimming, *J. Phys. Condens. Matt.* **20**, 374127 (2008).
83. B. Hammer and J. K. Nørskov, *Surf. Sci.* **343**, 211 (1995).
84. B. Hammer, Y. Morikawa, and J. K. Nørskov, *Phys. Rev. Lett.* **76**, 2141 (1996).
85. M. Mavrikakis, B. Hammer, and J. K. Nørskov, *Phys. Rev. Lett.* **81**, 2819 (1998).
86. B. Hammer and J. K. Nørskov, in *Chemisorption and Reactivity on Supported Cluster and Thin Films*, R. M. Lambert, G. Pacchiani, eds., Kluwer, Netherlands, 1997.
87. J. K. Nørskov, T. Bligaard, A. Logadottir, J. R. Kitchin, J. G. Chen, S. Pandelov, and U. Stimming, *J. Electrochem. Soc.* **152**, J23 (2005).
88. V. Pallassana, M. Neurock, L. B. Hansen, B. Hammer, and J. K. Nørskov, *Phys. Rev. B* **60**, 6146 (1999).
89. V. Pallassana, M. Neurock, L. B. Hansen, and J. K. Nørskov, *J. Chem. Phys.* **112**, 5435 (2000).
90. B. Hammer and J. K. Nørskov, *Adv. Catal.* **45**, 71 (2000).
91. A. Roudgar and A. Gross, *Phys. Rev. B* **67**, 033409 (2003).
92. A. Roudgar and A. Gross, *J. Electroanal. Chem.* **548**, 121 (2003).
93. M. Eikerling, J. Meier, and U. Stimming, *Z. Phys. Chem.* **217**, 395 (2003).
94. J. Greeley, T. F. Jaramillo, J. Bonde, I. B. Chorkendorff, and J. K. Nørskov, *Nature Mater* **5**, 909 (2006).
95. J. Greeley, J. K. Nørskov, L. A. Kibler, A. M. El-Aziz, and D. M. Kolb, *ChemPhysChem* **7**, 1032 (2006).
96. J. Greeley and J. K. Nørskov, *Surf. Sci.* **601**, 1590 (2007).
97. E. Santos, K. Potting, and W. Schmickler, *Faraday Discuss* **140**, 209 (2008).
98. E. Santos and W. Schmickler, *Angew. Chem. Int. Ed.* **46**, 8262 (2007).
99. E. Santos and W. Schmickler, *Chem. Phys.* **332**, 39 (2007).
100. E. Santos and W. Schmickler, *Electroch. Acta* **53**, 6149 (2008).
101. J. M. Ogden, M. M. Steinbugler, and T. G. Kreutz, *J. Power Sources* **79**, 143 (1999).
102. L. Carrette, K. A. Friedrich, and U. Stimming, *ChemPhysChem* **1**, 162 (2000).
103. N. M. Markovic and P. N. Ross, *Surf. Sci. Rep.* **45**, 121 (2002).
104. M. Z. Jacobson, W. G. Colella, and D. M. Golden, *Science* **308**, 1901 (2005).
105. J. O. M. Bockris, *J. Serb. Chem. Soc.* **70**, 475 (2005).
106. B. E. Conway and J. O. Bockris, *J. Chem. Phys.* **26**, 532 (1957).
107. R. Parsons, *Trans. Faraday Soc.* **54**, 1053 (1958).
108. H. Gerischer, *Angew. Chem. Int. Ed.* **70**, 107 (1958).
109. S. Trasatti, *J. Electroanal. Chem.* **39**, 163 (1972).
110. P. Sabatier, *Ber. Dtsch. Chem. Gesells.* **44**, 1984 (1911).
111. B. Hammer and J. K. Nørskov, *Surf. Sci.* **343**, 211 (1995).
112. A. Ruban, B. Hammer, P. Stoltze, H. L. Skriver, and J. K. Nørskov, *J. Mol. Catal. A Chem.* **115**, 421 (1997).
113. A. Roudgar and A. Gross, *Surf. Sci.* **597**, 42 (2005).
114. D. Menzel, *Science* **295**, 58 (2002).
115. P. J. Feibelman, *Science* **295**, 99 (2002).
116. P. J. Feibelman, *Phys. Rev. Lett.* **90**, 2003.

117. B. E. Conway, E. M. Beatty, and P. A. D. DeMaine, *Electrochim. Acta* **7**, 39 (1962).
118. J. Meier, K. A. Friedrich, and U. Stimming, *Faraday Discuss.* **121**, 365 (2002).
119. L. A. Kibler, M. Kleinert, R. Randler, and D. M. Kolb, *Surf. Sci.* **443**, 19 (1999).
120. M. Baldauf and D. M. Kolb, *J. Phys. Chem.* **100**, 11375 (1996).
121. M. Baldauf and D. M. Kolb, *Electrochim. Acta* **38**, 2145 (1993).
122. R. Hoyer, L. A. Kibler, and D. M. Kolb, *Electrochim. Acta* **49**, 63 (2003).
123. H. Naohara, S. Ye, and K. Uosaki, *J. Phys. Chem. B* **102**, 4366 (1998).
124. R. Hoyer, L. A. Kibler, and D. M. Kolb, *Surf. Sci.* **562**, 275 (2004).
125. S. R. Brankovic, J. McBreen, and R. R. Adzic, *Surf. Sci.* **479**, L363 (2001).
126. J. Tang, M. Petri, L. A. Kibler, and D. M. Kolb, *Electrochim. Acta* **51**, 125 (2005).
127. L. A. Kibler, A. M. El-Aziz, R. Hoyer, and D. M. Kolb, *Angew. Chem. Int. Ed.* **44**, 2080 (2005).
128. L. A. Kibler, A. M. El-Aziz, and D. M. Kolb, *J. Mol. Catal. A Chem.* **199**, 57 (2003).
129. B. Hammer and J. K. Nørskov, *Surf. Sci.* **343**, 211 (1995).
130. A. Ruban, B. Hammer, P. Stoltze, H. L. Skriver, and J. K. Nørskov, *J. Mol. Catal. A Chem.* **115**, 421 (1997).
131. J. Meier, PhD thesis, Technische Univ. München, 2003.
132. J. Meier, K. A. Friedrich, and U. Stimming, *Faraday Discuss.* **121**, 365 (2002).
133. J. Meier, J. Schiotz, P. Liu, J. K. Nørskov, and U. Stimming, *Chem. Phys. Lett.* **390**, 440 (2004).
134. S. Pandelov and U. Stimming, *Electrochim. Acta* **52**, 5548 (2007).
135. L. A. Kibler, *ChemPhysChem* **7**, 985 (2006).
136. J. Greeley, T. F. Jaramillo, J. Bonde, I. B. Chorkendorff, and J. K. Nørskov, *Nature Mater.* **5**, 909 (2006).
137. B. Hammer and J. K. Nørskov, *Adv. Catal.* **45**, 71 (2000).
138. B. Hammer and J. K. Nørskov, *Surf. Sci.* **343**, 211 (1995).
139. J. K. Nørskov, *Reac. Kinet. Dev. Catal. Process.* **122**, 3 (1999).
140. J. K. Nørskov, T. Bligaard, A. Logadottir, J. R. Kitchin, J. G. Chen, S. Pandelov, and U. Stimming, *J. Electrochem. Soc.* **152**, J23 (2005).
141. H. Wolfschmidt, R. Bussar, and U. Stimming, *J. Phys. Condens. Matt.* **20**, 374127 (2008).
142. N. M. Markovic and P. N. Ross, *Surf. Sci. Rep.* **45**, 121 (2002).
143. N. M. Markovic, C. A. Lucas, V. Climent, V. Stamenkovic, and P. N. Ross, *Surf. Sci.* **465**, 103 (2000).
144. F. Hernandez and H. Baltruschat, *J. Solid State Electrochem.* **11**, 877 (2007).
145. F. Hernandez and H. Baltruschat, *Langmuir* **22**, 4877 (2006).
146. J. Steidtner, F. Hernandez, and H. Baltruschat, *J. Phys. Chem. C* **111**, 12320 (2007).
147. M. Eikerling, J. Meier, and U. Stimming, *Z. Phys. Chem.* **217**, 395 (2003).
148. M. Mavrikakis, B. Hammer, and J. K. Nørskov, *Phys. Rev. Lett.* **81**, 2819 (1998).
149. A. Roudgar and A. Gross, *Phys. Rev. B* **67**, 033409 (2003).
150. A. Roudgar and A. Gross, *J. Electroanal. Chem.* **548**, 121 (2003).
151. J. Greeley, J. K. Nørskov, and M. Mavrikakis, *Annu. Rev. Phys. Chem.* **53**, 319 (2002).
152. J. R. Kitchin, J. K. Nørskov, M. A. Barteau, and J. G. Chen, *Phys. Rev. Lett.* **93**, 156801 (2004).

153. A. Roudgar and A. Gross, *J. Electroanal. Chem.* **548**, 121 (2003).
154. A. V. Ruban, H. L. Skriver, and J. K. Nørskov, *Phys. Rev. B* **59**, 15990 (1999).
155. C. Camus, diploma thesis, Technische Univ. München (Garching), 2004.
156. T. Brülle and U. Stimming, *J. Electroanal. Chem.* **636**, 10 (2009).
157. L. A. Kibler, *ChemPhysChem* **7**, 985 (2006).
158. L. A. Kibler, A. M. El-Aziz, and D. M. Kolb, *J. Mol. Catal. A Chem.* **199**, 57 (2003).
159. N. M. Markovic, C. A. Lucas, V. Climent, V. Stamenkovic, and P. N. Ross, *Surf. Sci.* **465**, 103 (2000).
160. N. M. Markovic and P. N. Ross, *Surf. Sci. Rep.* **45**, 121 (2002).
161. S. Pandelov and U. Stimming, *Electrochim. Acta* **52**, 5548 (2007).
162. H. Wolfschmidt, R. Bussar, and U. Stimming, *J. Phys. Condens. Matt.* **20**, 374127 (2008).
163. M. Baldauf and D. M. Kolb, *Electrochim. Acta* **38**, 2145 (1993).
164. J. Meier, K. A. Friedrich, and U. Stimming, *Faraday Discuss.* **121**, 365 (2002).
165. J. Meier, J. Schiotz, P. Liu, J. K. Norskov, and U. Stimming, *Chem. Phys. Lett.* **390**, 440 (2004).
166. M. Petri and D. M. Kolb, *Phys. Chem. Chem. Phys.* **4**, 1211 (2002).
167. Z. X. Xie and D. M. Kolb, *J. Electroanal. Chem.* **481**, 177 (2000).
168. S. G. Garcia, D. R. Salinas, C. E. Mayer, W. J. Lorenz, and G. Staikov, *Electrochim. Acta* **48**, 1279 (2003).
169. R. Schuster, V. Kirchner, P. Allongue, and G. Ertl, *Science* **289**, 98 (2000).
170. V. Kirchner, L. Cagnon, R. Schuster, and G. Ertl, *Appl. Phys. Lett.* **79**, 1721 (2001).
171. W. Schindler, D. Hofmann, and J. Kirschner, *J. Appl. Phys.* **87**, 7007 (2000).
172. R. T. Potzschke, G. Staikov, W. J. Lorenz, and W. Wiesbeck, *J. Electrochem. Soc.* **146**, 141 (1999).
173. W. Schindler, D. Hofmann, and J. Kirschner, *J. Electrochem. Soc.* **148**, C124(2001).
174. A. J. Bard and M. V. Mirkin, *Scanning Electrochemical Microscopy*, Wiley, New York, 2001.
175. N. Baltes, L. Thouin, C. Amatore, and J. Heinze, *Angew. Chem. Int. Ed.* **43**, 1431 (2004).
176. G. E. Engelmann, J. C. Ziegler, and D. M. Kolb, *Surf. Sci.* **401**, L420(1998).
177. M. G. Del Popolo, E. P. M. Leiva, H. Kleine, J. Meier, U. Stimming, M. Mariscal, and W. Schmickler, *Electrochim. Acta* **48**, 1287 (2003).
178. D. M. Kolb, R. Ullmann, and T. Will, *Science* **275**, 1097 (1997).
179. D. M. Kolb, R. Ullmann, and J. C. Ziegler, *Electrochim. Acta* **43**, 2751 (1998).
180. M. Del Popolo, E. Leiva, H. Kleine, J. Meier, U. Stimming, M. Mariscal, and W. Schmickler, *Appl. Phys. Lett.* **81**, 2635 (2002).
181. D. M. Kolb and F. C. Simeone, *Electrochim. Acta* **50**, 2989 (2005).
182. D. M. Kolb, G. E. Engelmann, and J. C. Ziegler, *Angew. Chem. Int. Ed.* **39**, 1123 (2000).
183. G. E. Engelman, J. C. Ziegler, and D. M. Kolb, *J. Electrochem. Soc.* **145**, 2970 (1998).
184. D. M. Kolb, G. E. Engelmann, and J. C. Ziegler, *Solid State Ion.* **131**, 69 (2000).
185. D. M. Kolb, J. C. Ziegler, and G. E. Engelmann, *Abstr. Papers Am. Chem. Soc* **219**, U347 (2000).
186. R. Ullmann, T. Will, and D. M. Kolb, *Chem. Phys. Lett.* **209**, 238 (1993).

187. R. Ullmann, T. Will, and D. M. Kolb, *Ber. Buns. Gesell. Phys. Chem. Chem. Phys.* **99**, 1414 (1995).
188. W. Schindler, P. Hugelmann, A. Hugelmann, and F. Kartner, *J. Electroanal. Chem.* **522**, 49 (2002).
189. M. Eikerling, J. Meier, and U. Stimming, *Z. Phys. Chem.* **217**, 395 (2003).
190. B. Hammer and J. K. Nørskov, *Adv. Catal.* **45**, 71 (2000).
191. A. Ruban, B. Hammer, P. Stoltze, H. L. Skriver, and J. K. Nørskov, *J. Mol. Catal. A Chem.* **115**, 421 (1997).
192. O. Savadogo, K. Amuzgar, and D. L. Piron, *Int. J. Hydrogen Energy* **15**, 783 (1990).
193. H. Naohara, S. Ye, and K. Uosaki, *J. Phys. Chem. B* **102**, 4366 (1998).
194. H. A. Gasteiger, N. M. Markovic, and P. N. Ross, *J. Phys. Chem.* **99**, 16757 (1995).
195. K. Sasaki, Y. Mo, J. X. Wang, M. Balasubramanian, F. Uribe, J. McBreen, and R. R. Adzic, *Electrochim. Acta* **48**, 3841 (2003).
196. S. L. Chen and A. Kucernak, *J. Phys. Chem. B* **108**, 13984 (2004).
197. M. Arenz, V. Stamenkovic, B. B. Blizanac, K. J. Mayrhofer, N. M. Markovic, and P. N. Ross, *J. Catal.* **232**, 402 (2005).
198. Y. Gimeno, A. H. Creus, P. Carro, S. Gonzalez, R. C. Salvarezza, and A. J. Arvia, *J. Phys. Chem. B* **106**, 4232 (2002).
199. G. J. Lu and G. Zangari, *J. Phys. Chem. B* **109**, 7998 (2005).
200. T. Brülle and U. Stimming, in preparation.

CHAPTER 2

# Comparison of Electrocatalysis and Bioelectrocatalysis of Hydrogen and Oxygen Redox Reactions

MARC T. M. KOPER and HENDRIK A. HEERING
Leiden Institute of Chemistry, Leiden University, Leiden, The Netherlands

This chapter compares the mechanisms of electrocatalytic hydrogen oxidation and evolution and oxygen reduction and evolution reactions, as catalyzed by metals, metal oxides, and enzymes. Our discussion stresses the multistep character of these reactions, the concomitant thermodynamic restrictions applicable to these steps, and the role of the catalyst in stabilizing the reaction intermediates. Despite the very different structures of the active sites of the various catalysts for these reactions, and the different concepts and ideas about their catalytic action, we find that there are important similarities between the ways electrocatalytic and bioelectrocatalytic systems work, and that in many cases synthetic (human-made) and natural catalysts face similar bottlenecks, and have provided similar solutions. Some of the existing misunderstandings may be resolved by realizing the importance of comparing these systems at the same thermodynamic driving force, as these may be vary significantly in different studies ("driving force gap"), pointing toward the usefulness of a technique such as protein film voltammetry.

## 2.1 INTRODUCTION

Fuel cells feature multielectron transfer reactions at both the cathode and the anode. The development of efficient catalysts for these reactions is one of the keys to the introduction of fuel cells in society [1], as well as to the large-scale application of water electrolysis for the production of hydrogen, more specifically, the reverse of the hydrogen–oxygen fuel cell. In the continued search for better, cheaper, more efficient,

*Fuel Cell Science: Theory, Fundamentals, and Biocatalysis*,
Edited by Andrzej Wieckowski and Jens K. Nørskov

and more stable electrocatalysts, it is increasingly recognized that the catalysts that nature has devised, namely, enzymes, to catalyze exactly the same redox reactions, may serve as an important inspiration for or even application to the development of better and/or new fuel cells and fuel cell catalysts.

The aim of this chapter is to compare the electrocatalysis and the bioelectrocatalysis of the hydrogen and oxygen redox reactions. Our incentive for writing this chapter was a mutual desire to bridge the gaps between different languages and concepts in both communities (MTMK being a "classical" electrochemist and HAH being a "bio-" electrochemist) in a single chapter, such that it would be understandable to both. Although some parts of the chapter may appear as rather trivial to a reader from a particular community, by reading each other's contributions, we both became aware of some important similarities and differences that we did not explicitly identify before. We hope that our report of this exercise, as summarized in the present chapter, will be equally useful for other researchers in this field.

Key elements in our comparison will be the identification of the potential-determining and rate-determining steps, the role of the catalyst in steering and accelerating the reaction by stabilizing certain intermediates, to a significant extent based on recent density functional theory (DFT) calculations, the experimental (spectrosopic) evidence for the alleged intermediates, and an evaluation and comparison of catalytic activities between different types of catalysts. We first discuss the hydrogen evolution reaction (HER) and hydrogen oxidation reaction (HOR), and summarize the mechanisms suggested for these reactions in both the electrocatalysis and the bioelectrocatalysis literature. The next two sections will do the same for the oxygen reduction reaction (ORR) and the oxygen evolution reaction (OER). Much of our discussion focuses on the fact that both the HER/HOR and the OER/ORR are multielectron, multiproton transfer reactions involving bond breaking. It is unlikely if not inconceivable that these reactions would happen in a single step. Certain restrictions applicable to multistep multielectron transfer reactions are imposed by thermodynamics, and these restrictions have important consequences for catalysis. Moreover, these restrictions are often at the basis of concepts that are, strictly speaking, outside the realm of thermodynamics, such as the Sabatier principle and the resulting volcano relationships. Much of our current understanding of the thermodynamics or energetics of the multistep mechanism for these reactions is based on DFT calculations, and therefore these studies play a central role in our discussion. Many of these theoretical studies also illustrate and highlight some of the main differences in redox catalysis by metals, oxides, and enzymes. The chapter ends with a few concluding remarks summarizing our main findings.

## 2.2 HYDROGEN OXIDATION AND EVOLUTION REACTION

### 2.2.1 Mechanism of HER/HOR on Metal Electrodes

#### *2.2.1.1 General Considerations*

The hydrogen oxidation/evolution reaction is given by the following redox reaction

$$H_2 \leftrightarrows 2H^+ + 2e^- \qquad E^\circ = 0\,V \tag{2.1}$$

with its well-known standard equilibrium potential of 0 (zero volt). Accepting the multistep character of this reaction, a mechanism must be proposed for this reaction, preferably consisting of a sequence of elementary reaction steps. We consider here the following generally accepted mechanism for the hydrogen oxidation/evolution reaction on metals [2], consisting of the following three elementary steps

$$H_2 + 2* \leftrightarrows 2H* \tag{2.2a}$$

$$H_2 + * \leftrightarrows H* + H^+ + e^- \tag{2.2b}$$

$$H* \leftrightarrows H^+ + * + e^- \tag{2.2c}$$

where we recognize the Tafel, Heyrovsky, and Volmer steps, respectively. The symbol * (asterisk) denotes a free adsorption site on the metal electrode surface. The H* intermediate is a hydrogen species chemisorbed on such a site on the surface. Note that in reactions (2.2b) and (2.2c), the proton and electron transfers are assumed to take place simultaneously. This concerted proton–electron transfer is well accepted in the electrochemical literature, and supported by detailed ab initio DFT calculations [3,4]. We emphasize this fact here, as this is an important difference with some of the suggested mechanisms for catalysis of the HER/HOR by hydrogenases, to be discussed in Section 2.2.2.

This multistep mechanism needs to satisfy thermodynamic constraints in the sense that the combined equilibrium constants need to conform to the Nernst equation of the overall reaction. In terms familiar to electrochemists, if HER/HOR were to follow a Heyrovsky–Volmer mechanism, then the sum of the standard equilibrium potentials of the Heyrovsky and Volmer reaction, divided by the total number of electrons in the reaction, would have to equal the standard equilibrium potential of the overall reaction [5]:

$$\frac{E^{\circ}_{(2.2b)} + E^{\circ}_{(2.2c)}}{2} = 0 \tag{2.3}$$

One can calculate the equilibrium potential of a redox reaction involving an adsorbed or chemisorbed state by defining the concomitant standard state. Consider the Volmer reaction (2.2c), for which the equilibrium condition is

$$\Delta G(H*) = \Delta G(H^+ + e^-) \tag{2.4}$$

Following Nørskov et al. [6], the term on the right-hand side (RHS) is 0 at $E = 0$ [vs. normal hydrogen electrode (NHE)] by definition, and $-e_0E$ at any other potential $E$, assuming that when the potential is changed only the energy of the electrons changes, and nothing else. This is not necessarily a trivial assumption, but it may be tested by DFT calculations. It has indeed been found that the binding energy of H* does not depend strongly on potential, or on an applied electric field [7–10]. This implies that the static dipole moment of the H* bond to the surface is close to zero, and that the electrosorption valence corresponding to reaction (2.2c) is close to 1 [10]. Therefore, the standard equilibrium potential for the Volmer reaction follows as

$$E^{\circ}_{(2.2c)} = -\Delta G^{\circ}(\mathrm{H}^*)/e_0 \tag{2.5}$$

where $\Delta G^{\circ}(\mathrm{H}^*)$ is now the adsorption energy of H* at some suitably defined standard state '0' [e.g., $\frac{1}{4}$ coverage on a (111) surface], with respect to the energy of $H_2$, which is equal to zero also by virtue of the above definition. If $\Delta G^{\circ}(\mathrm{H}^*) < 0$, that is, if $H_2$ dissociation onto the catalyst surface or catalyst site yields a thermodynamically more favorable situation, then $E^{\circ}_{(2.2c)} > 0$. A well-known example of this is hydrogen underpotential deposition (upd) on platinum and other transition metal electrodes. Note, however, that in terms of the overall reaction, such a situation implies that

$$E^{\circ}_{(2.2b)} = -E^{\circ}_{(2.2c)} < 0 \tag{2.6}$$

so that the Heyrovsky reaction would have a standard equilibrium potential (defined with the same standard state) below 0 V. Neglecting kinetic factors, this analysis would suggest that if in the HER, H* formation would be a thermodynamically favorable reaction at $E = 0$, molecular hydrogen formation as in the Heyrovsky reaction must be thermodynamically unfavorable. Following Nørskov et al. [6], one could argue that in such a case, the minimal "thermodynamic overvoltage" for the overall reaction, $\eta_T$, would be equal to the standard equilibrium potential of the thermodynamically least favorable reaction, namely, $\eta_T = E^{\circ}_{(2.2c)} - E^{\circ}_{(2.1)}$. This "thermodynamically least favorable reaction step" may be termed the *potential-determining step* (not to be confused with the rate-determining step, although often they may correspond to the same reaction step). Since the lowest possible thermodynamic overvoltage is zero, it follows that, on the basis of these thermodynamic considerations, we are looking for a catalyst with

$$\eta_T = E^{\circ}_{(2.2b)} = E^{\circ}_{(2.2c)} = \Delta G^{\circ}(\mathrm{H}^*)/e_0 = 0 \tag{2.7}$$

Any catalyst with $\Delta G^{\circ}(\mathrm{H}^*) \neq 0$ will have a $\eta_T < 0$, as either (1) the Volmer reaction has a standard equilibrium potential more negative than 0 [$\Delta G^{\circ}(\mathrm{H}^*) > 0$] or (2) the Heyrovsky reaction has a standard equilibrium potential more negative than 0 [$\Delta G^{\circ}(\mathrm{H}^*) < 0$]. Figure 2.1 plots $\eta_T$ as a function of $\Delta G^{\circ}(\mathrm{H}^*)$, and illustrates how the optimal catalyst is a compromise between strong and weak binding of H*, as expressed by the classical Sabatier principle. Figure 2.1 can therefore be considered as an example of a volcano relationship, in which the catalytic activity for a reaction is plotted as a function of some property of the catalyst, in this case the catalyst's ability to bind an intermediate. Such a property is also known as a *descriptor*, as it gives a general but essential description of the catalyst properties. Note how both concepts—the Sabatier principle and the volcano relationship—straightfowardly follow from thermodynamics combined with some very mild kinetic assumptions (one of the

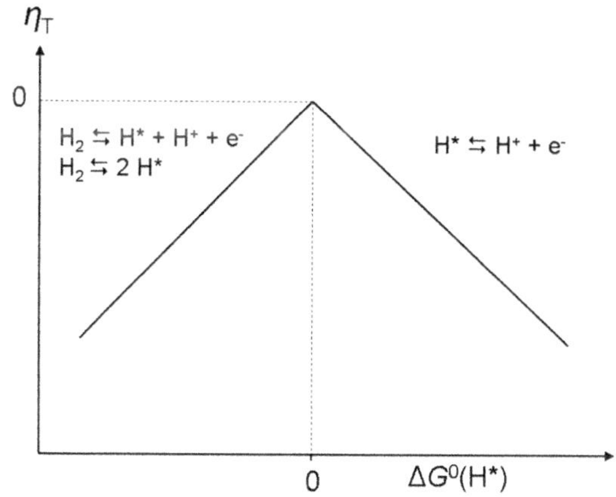


**FIGURE 2.1** "Thermodynamic" volcano plot for the hydrogen evolution reaction: thermodynamic overpotential (standard equilibrium potential of the potential-determining step) versus the free binding energy of hydrogen on the electrode surface[11,12]. On the right of the apex, the Volmer reaction is potential-determining; on the left of the apex, either the Heyrovsky or the Tafel reaction is potential-determining.

assumptions that we have made here is that of a relation between activation energy and thermodynamic driving force, i.e., the Brønsted–Evans–Polanyi postulate). Figure 2.1 is essentially a thermodynamic version of the kinetic volcano plot [i.e., log $i_0$ vs. $\Delta G^\circ$ (H*), where $i_0$ is the standard exchange current density] originally suggested by Parsons [11], and more recently revived on the basis of DFT-calculated values for $\Delta G^\circ$ (H*) by Nørskov et al. [12].

The same analysis may be applied to a Tafel–Volmer mechanism for HER/HOR. The only difference now is that reaction (2.2a) does not have a standard equilibrium potential, because it is not an electrochemical reaction. However, combined with reaction (2.2c), it must still satisfy the overall thermodynamic constraints of reaction (2.2). Defining the standard equilibrium constant of reaction (2.2a) as

$$K_{(2.2a)} = \exp[-\Delta G^\circ(\mathrm{H}^*)/RT] \tag{2.8}$$

the Tafel–Volmer mechanism has to obey the following relation:

$$E^\circ_{(2.2c)} - \Delta G^\circ(\mathrm{H}^*)/e_0 = 0 \tag{2.9}$$

In case $E^\circ_{(2.2c)} < 0$, the thermodynamic overvoltage is $\eta_T = E^\circ_{(2.2c)}$, as in the Heyrovsky–Volmer mechanism. In case $E^\circ_{(2.2c)} > 0$, the Tafel reaction becomes endothermic [Eq. (2.9)], and therefore rate-determining. The most straightforward definition of a *thermodynamic* overvoltage would then follow from Equation (2.9), leading to $\eta_T = -E^\circ_{(2.2c)}$. As a result, we see that the Heyrovsky–Volmer and the Tafel–Volmer mechanisms lead to "thermodynamic volcano curves" that are identical, and that more mechanistic assumptions are needed to distinguish between these two mechanisms. Nevertheless, this simple analysis lucidly illustrates the thermodynamic constraints applicable to multistep multielectron transfer reactions, and how catalysis may influence the thermodynamic overvoltage by optimizing the binding energy of the key intermediate, in this case H*.

One important consequence of a volcano-type relationship such as that illustrated in Figure 2.1 is that the potential-determining step is different on each side of the maximum. *Potential-determining* means that this step determines the (over) potential at which a certain current density may be reached. For $\Delta G°(H^*) > 0$, the Volmer reaction is potential-determining; for $\Delta G°(H^*) < 0$, either the Heyrovsky or Tafel reaction is potential-determining. It is important to point out the difference between the potential-determining and rate-determining steps. Potential-determining steps are not always rate-determining at the relevant potentials, as potential-determining steps suffer from an unfavorable thermodynamics, but not necessarily from an unfavorable kinetics at the potential at which the current finally starts flowing.

In electrochemical experiments, rate-determining steps are usually inferred from so-called Tafel slopes, which essentially express the overall potential dependence of the reaction rate (i.e., the current) in a certain potential range. Tafel slopes do not really have an equivalent in enzyme catalysis as there is no simple way to control the thermodynamic driving force (i.e., the potential) in such systems, unless these enzymes are immobilized on electrode surfaces in a technique called *protein film voltammetry* [13]. If the Volmer reaction is rate-determining and its rate obeys the Butler–Volmer model, then the potential dependence is equivalent to a Tafel slope of $\ln 10 \times RT/(\alpha FE) \sim 120$ mV/dec (millivolts per decade) if $a \sim 0.5$. If, on the other hand, the Heyrovsky reaction is rate-determining, and the preceding Volmer reaction in *quasiequilibrium*, the potential dependence of the reaction is equivalent to a Tafel slope of $\ln 10 \times RT/[(1 + a)FE] \sim 40$ mV/dec if $a \sim 0.5$. Finally, if the Tafel reaction is rate determining, the Tafel slope will be $\ln 10 \times RT/[2FE] \sim 30$ mV/dec. In these predictions of the Tafel slopes, saturation effects on the surface are not included.

The idea that the HER/HOR would follow a volcano-type dependence on the free adsorption energy of hydrogen dates back to the early work of Parsons [11] and Gerischer [14], later updated by Trasatti [15]. The idea was revived in 2005 by Nørskov et al. [12], who showed that $\Delta G°(H^*)$ may be calculated from first principles by using DFT in combination with some mild assumptions on entropic contributions. Moreover, trends in $\Delta G°(H^*)$ among different metals conform to the so-called *d*-band model proposed by Hammer and Nørskov [16–18]. This model claims a close one-to-one correspondence between the strength of a surface bond (of H, but also of other adsorbates such as O, OH, and CO) and the location of the center of the *d* band of the metal surface. The energy of the *d* band may be "engineered" by changing the properties of the metal surface, for instance, by (surface) alloying and overlayer formation. The *d*-band center also influences the activation energy for dissociative adsorption, primarily through the Brønsted–Evans–Polanyi relationship [18], which claims a linear correspondence between reaction energy and activation energy. The role of the *d* band in bond-breaking electron transfer reactions was recently modeled by Santos and Schmickler [19], essentially confirming the important role of the *d* band in HER/HOR.

#### 2.2.1.2 HER/HOR at Platinum Electrodes

As platinum is one of the best and most widely studied catalysts for HER/HOR, as well as the one used in fuel cells, it is useful to summarize here some of the pertinent experimental results. Platinum is generally agreed to be a very active catalyst for HER/HOR, and considered to be located near the apex of the volcano. Standard exchange current densities are not easy to obtain because in general it is difficult to correct accurately for mass transport limitations. Values ranging from $10^{-3}$ to 1 A/cm$^2$ have been reported for platinum [2,20–22], which would roughly correspond to a turnover rate of 0.1–100 s$^{-1}$ per Pt active site, at the standard equilibrium potential. The Pt "active site" is not well known, and there has been quite some controversy as to whether the HER/HOR is structure-sensitive on platinum. In contrast to early studies [23], the Markovic [2,20] and Conway [21] groups have found that the Pt(110) surface is the most active in catalyzing HER/HOR, typically an order of magnitude more active than the other low-index surfaces.

Markovic et al. [20] have measured Tafel slopes for both HER and the HOR on Pt single-crystal electrodes in acidic media. Tafel slopes also appear to depend on the crystal plane; the most active Pt(110) surface exhibits the 30 mV/dec Tafel slope well known from polycrystalline Pt [21,24]. This Tafel slope corresponds to a Tafel–Volmer mechanism, namely, a combination of reactions (2.2a) and (2.2c).

The HER/HOR reaction kinetics is first-order in the proton concentration over a wide pH range (0–14), meaning that overpotentials are similar on the reversible hydrogen electrode (RHE) scale, in full agreement with the idea that proton and electron transfer are coupled over the entire pH range. However, there is a nontrivial pH effect in the sense that in alkaline media, the rate of the reaction tends be about an order of magnitude lower than acidic media, at the same overpotential [25]. Conway et al. [25] have explained this effect by a more difficult abstraction of H from $H_2O$ than from $H_3O^+$. Pt(110) is still the most active surface, also in alkaline media, but the Tafel slope seems to be different, closer to 60 mV/dec, although no clear straight line was obtained [25].

An important discussion in the surface electrochemistry literature concerns the nature of surface-adsorbed hydrogen that acts as a precursor for hydrogen evolution in the HER, or for proton formation in the HOR [2,26]. It is well known that at potentials above the reversible hydrogen potential, from ~ 0.05 to 0.3 V versus RHE (i.e., where no hydrogen evolution takes place), adsorbed hydrogen is formed on platinum. This form of adsorbed hydrogen is referred to as *underpotential deposited* (upd) hydrogen [26]. The surface coverage of this $H_{upd}$ follows a Frumkin isotherm on Pt(111) and Pt(100) [26–28], with weak repulsive interactions between the adsorbed hydrogens. The coverages of $H_{upd}$ in acidic and alkaline media on Pt(111) are almost identical at the same overpotential, but again its kinetics of formation is much slower in alkaline media [29]. The adsorbed hydrogen that participates in the HER at potentials just below 0 V, is usually termed *overpotential deposited* (opd) hydrogen. The exact relation between $H_{upd}$ and $H_{opd}$ has led to ample discussion and speculation; for an illustrative account, see the paper by Conway and Jerkiewicz [30]. Most probably, $H_{upd}$ and $H_{opd}$ should be distinguished primarily on the basis of the different overall hydrogen coverages applicable to both states, with a typical $H_{upd}$ coverage being lower than 0.5–1.0, and a typical $H_{opd}$ coverage being close to or higher than 1 (i.e., including the $H_{upd}$ coverage, implying more than 1 H per surface Pt atom), suggesting that $H_{upd}$ and $H_{opd}$ reside at different sites on the surface. The different coverages lead to different lateral interactions, different energetics and site preferences, and therefore different activities toward $H_2$ evolution. There have been a number of attempts to probe the adsorbed hydrogen on platinum using vibrational spectroscopy. A surface-enhanced infrared (IR) study on a platinum film electrode by Kunimatsu et al. [31] confirmed earlier results by absorption–reflection IR [32] that, for potentials lower than $\sim 0.1\ V_{RHE}$, atop-bonded Pt–H is formed. The signal of this band correlated with the hydrogen evolution current, and therefore this species was identified as the intermediate in the HER. Kunimatsu et al. suggest that $H_{upd}$ adsorbs below 0.3 $V_{RHE}$ in hollow sites, although this state has not been resolved spectroscopically, and that, in addition to $H_{upd}$, $H_{opd}$ adsorbs below 0.1 $V_{RHE}$ at atop sites, and serves as the reaction intermediate.

#### 2.2.1.3 Experimental Evidence for Volcanic Behavior on Strained Monolayer and Bimetallic Systems

The evidence for a volcano-type relationship has been collected by Trasatti [15], and more recently by Nørskov and Stimming and coworkers [12]. One problem with such a correlation is that it is based partly on including metals that under experimental conditions are most likely covered by an oxide [33]. More recent studies by Kibler et al. [34,35] have employed the idea of using palladium monolayers on different substrates [from Pd on Re(0001) to Pd on Au(111)] to tune the reactivity toward hydrogen adsorption and hydrogen evolution. Since the potential of the $H_{upd}$ peak observed in voltammetry is directly related to the adsorption energy of hydrogen (or more accurately, of $H_{upd}$), the authors could relate the peak potential to the theoretically predicted *d*-band shift with respect to Pd(111) [34]. The good correlation found suggested that this monolayer approach is a viable and controlled way to tune reactivity. In a subsequent study, the same systems were studied for their HER activity [35]. Figure 2.2 shows the plot of the standard exchange current density versus the free energy of adsorption of hydrogen, clearly in agreement with the predictions of the Sabatier principle and the volcano plot. Note that from these results, an optimum standard exchange current density of $\sim 0.01$–$0.1\ A/cm^2$ is predicted.

Clearly, a catalyst consisting of Pd, Ru, and Pt is hardly an economical alternative for Pt, which is already quite a good catalyst for HER/HOR. Therefore, Greeley et al. [36] have carried out an extensive computational study of bimetallic surfaces, screening for specific combinations with a $\Delta G^{\circ}(H^*)$ close to zero, with a negative formation energy to ensure stability of the material, and preferably consisting of cheaper elements. The counterintuitive combination of Pt and Bi emerged as one of the possibly attractive materials, with a $\Delta G^{\circ}(H^*)$ closer to zero than Pt. An appropriately prepared Pt–Bi surface alloy was subsequently found to indeed have a higher activity for HER than does Pt, showing the promise and usefulness of this computational screening strategy.

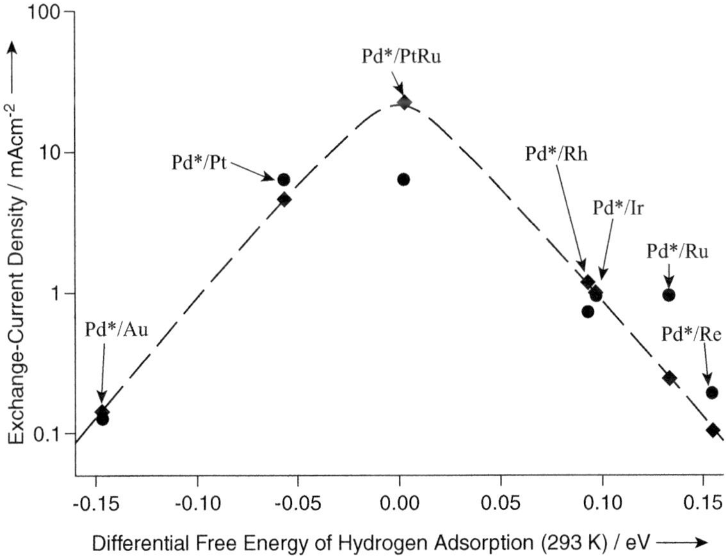


**FIGURE 2.2** Hydrogen evolution exchange current density as a function of the theoretical differential free energy of hydrogen adsorption, $\Delta G^{\circ}(H^*)$. The dashed line shows the functional form of the theoretical exchange current density, and the black diamonds correspond to the predictions for specific overlayers. The black circles represent experimental measurements (the measured exchange current density determines the ordinate for these points, and the abscissa is specified by the theoretical $\Delta G^{\circ}(H^*)$ value). Pd*/X denotes a pure overlayer of Pd on a substrate of metal X.

## 2.2.2 Mechanism of HER/HOR at Hydrogenases

Hydrogenases have been demonstrated to be effective catalysts in model biofuel cell (biologically based fuel cell) devices [37,38]. With protein film voltammetry, these enzymes exhibit maximum (overvoltage) turnover rates of $\sim 10^4/s^{-1}$ [39]. Three classes of hydrogenases are found in nature, and are named according to the metal ions constituting the active site [37,40,41]. [FeFe] hydrogenases contain an H cluster in the active site, consisting of a diiron site coupled to a [4Fe4S] cluster (see Fig. 2.3). [NiFe] hydrogenases contain a bimetallic active site consisting of a Ni and an Fe (Fig. 2.3). The [NiFeSe] hydrogenases form a subset of [NiFe] hydrogenases, in which the cysteine ligating the Ni, and acting as a general catalytic base involved in hydride formation, is replaced by selenocysteine. Finally, [Fe] hydrogenases contain a mononuclear center as active site.

Usually three additional iron–sulfur clusters are found in both [FeFe] and [NiFe] hydrogenases. These auxiliary redox sites are "lined up" to relay electrons between the buried active site and an electron donor/acceptor, or even an electrode surface. The [4Fe4S] cluster that is connected directly to the [FeFe] active site can also be considered an integral part of the catalytic site. Because the edge-to-edge distance between the [FeS] clusters in hydrogenases are typically around 1 nm, electron tunneling between clusters is not likely to limit the turnover rate [42].

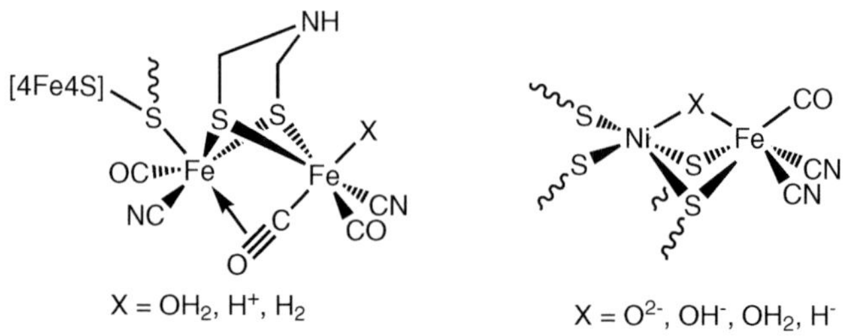


**FIGURE 2.3** Structures of the [FeFe] and [NiFe] hydrogenase active sites. (From Ref. 44.)

Nature's choice for Fe and Ni is interesting, since on the basis of the volcano plot, these metals (in bulk form) would be only second choice after transition metals such as Pt, which are not bioavailable. Remarkably, although not evolutionary related, all hydrogenases contain unusual CO ligands, and all except [Fe] hydrogenases also contain cyanide. The common motif in all three classes is an $Fe(CO)_xRS^-$ unit. Armstrong and Fontecilla-Camps have proposed that the $\pi$-acceptor nature of the CO ligands stabilize low-spin, low-valence iron, resulting in hydride binding and catalytic properties similar to second- and third-row transition metals such as platinum [43].

All hydrogenases catalyze the reversible hydrogen oxidation, and also catalyze single and double exchange of $D_2$ with bulk water protons. The [FeFe] hydrogenases are most active for hydrogen formation, while [NiFe] hydrogenases are poised for hydrogen oxidation. Mechanistic information has been obtained from enzyme kinetics, spectroscopy, X-ray structural data, and DFT calculations. A general feature is that hydrogen binding or dissociation, proton exchange, and electron transfer are separate steps. A second distinct difference with the metal surface-catalyzed reaction is the heterolytic splitting of hydrogen. A proton is abstracted by a nearby base, and hydride is stabilized by the metal ions rather than H*, although electron and spin densities are substantially delocalized. This is somewhat similar to the Heyrovsky reaction proposed for the mechanism on metals [reaction 2.2b]. As mentioned, on a platinum electrode the homolytic splitting [Tafel reaction (2.2a)] seems to be the preferred pathway.

The mechanism of HER/HOR by [FeFe] hydrogenases, as suggested by DFT calculations and experimental data, can be summarized as follows [45]:

$$[\mathrm{Fe^{II}Fe^{I}}]^{-1} + \mathrm{H}^+ \rightarrow [\mathrm{Fe^{II}Fe^{I}H}] \tag{2.10a}$$

$$[\mathrm{Fe^{II}Fe^{I}H}] + \mathrm{e}^- \rightarrow [\mathrm{Fe^{I}Fe^{I}H}]^{-1} \tag{2.10b}$$

$$[\mathrm{Fe^{I}Fe^{I}H}]^{-1} + \mathrm{H}^+ \rightarrow [\mathrm{Fe^{I}Fe^{I}HH}] \tag{2.10c}$$

$$[\mathrm{Fe^{I}Fe^{I}HH}] + \mathrm{e}^- \rightarrow [\mathrm{Fe^{I}Fe^{0}HH}]^{-1} \rightarrow [\mathrm{Fe^{II}Fe^{I}}\eta\text{-}\mathrm{H_2}]^{-1} \tag{2.10d}$$

$$[\mathrm{Fe^{II}Fe^{I}}\eta\text{-}\mathrm{H_2}]^{-1} \rightarrow [\mathrm{Fe^{II}Fe^{I}}]^{-1} + \mathrm{H_2} \tag{2.10e}$$

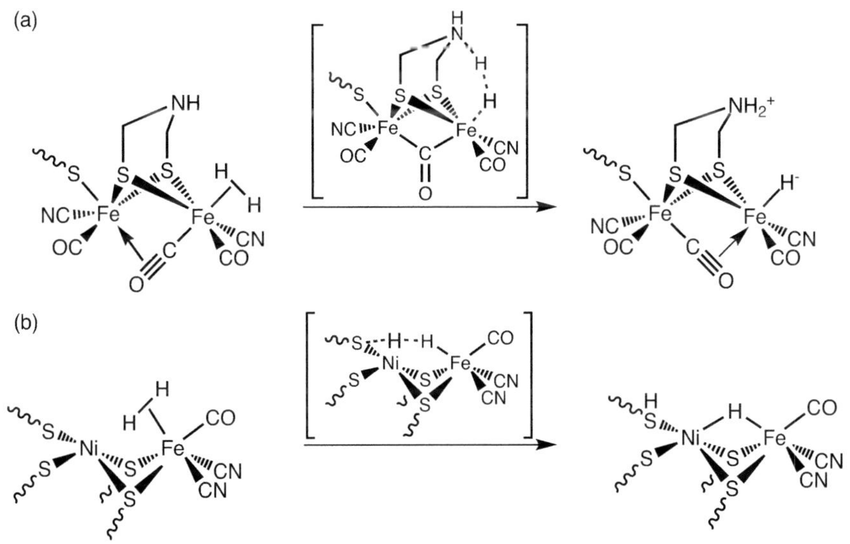


**FIGURE 2.4** Heterolytic splitting of hydrogen by [FeFe] and [NiFe] hydrogenases. (From Ref. 44.)

The electrons are rapidly transferred from the reductant (or the electrode) to the active site via the chain of [FeS] clusters. From electrochemical perspective, these [FeS] clusters thus are intramolecular mediators. Although Fe–H–Fe bridging intermediates have been proposed, the most likely mechanism involves one-Fe-bound H species and side-on-bound $H_2$. The secondary amine in the bridging ligand and $Fe^I$ appear to be equally potent as catalytic base. The iron most distant from the nearby [FeS] cluster, called the "$Fe_d$ site," which shuttles between the formal 1 + and neutral state in reactions (2.10a)–(2.10d), appears as the actual binding site for H and $H_2$ (see Fig. 2.4a). By taking together the proton and electron transfer steps in reactions (2.10a)–(2.10d), we can compare the DFT calculations performed by Sbraccia et al. [45] to the model presented in Section 2.2.1. For reactions (2.10a) and (2.10b) together

$$[Fe^{II}Fe^{I}]^{-1} + H^{+} + e^{-} \rightarrow [Fe^{I}Fe^{I}H]^{-1} \tag{2.10f}$$

the DFT-calculated reaction energy is −15.98 eV. Compared to a $H_2$ molecule, with a calculated total energy of 31.70 eV, the reaction energy is therefore −0.13 eV. For reactions (2.10c) and (2.10d) together

$$[Fe^{I}Fe^{I}H]^{-1} + H^{+} + e^{-} \rightarrow [Fe^{I}Fe^{0}HH]^{-1} \rightarrow [Fe^{II}Fe^{I}\eta{-}H_2]^{-1} \tag{2.10g}$$

the DFT-calculated reaction energy is −15.43 eV, implying a reaction energy of + 0.42 eV. For the final $H_2$ release [reaction (2.10e), the reaction energy is −0.29 eV (with a calculated activation energy of 0.07 eV). Note that the three calculated reaction energies now add up to 0, implying that we are considering the reaction at thermoneutral conditions. Although this is an analysis based on total energies, as calculated by DFT, and not on free energies, these numbers still suggest that (1) also for hydrogenases, the key to an efficient catalyst is to have an active site that binds H just as strong as $H_2$ (see also Ref. 46), and (2) that the addition of the second hydrogen [reaction (2.10g)] is the thermodymically uphill and hence potential-determining step, if we neglect zero-point energy contributions and entropy.

The mechanisms suggested for the HOR by [NiFe] and [NiFeSe] can be summarized as follow [47,48]:

$$[Ni^{II}/Fe^{II}] + H_2 \rightarrow [Ni^{II}/H_2/Fe^{II}] \tag{2.11a}$$

$$[Ni^{II}/H_2/Fe^{II}] \rightarrow [H^+/Ni^{II}/H^-/Fe^{II}] \tag{2.11b}$$

$$[H^+/Ni^{II}/H^-/Fe^{II}] \rightarrow [Ni^{III}/H^-/Fe^{II}] + H^+ + e^- \tag{2.11c}$$

$$[Ni^{III}/H^-/Fe^{II}] \rightarrow [H^+/Ni^{I}/Fe^{II}] \rightarrow [Ni^{II}/Fe^{II}] + H^+ + e^- \tag{2.11d}$$

One of the suggested molecular structures of the $[H^+/Ni^{II}/H^-/Fe^{II}]$ state is shown in Figure 2.4b. After the formation of this state, the released electrons are first transferred to the three iron–sulfur clusters in the enzyme, and then to an electron acceptor or an electrode. The consensus arising from DFT calculations on simplified models is that $H_2$ initially binds to the non-redox-active $Fe^{II}$, followed by heterolytic cleavage resulting in a bridging hydride and protonated S or Se ligand (Fig. 2.4). However, new calculations that also include a more substantial part of the peptide in the second coordination sphere indicate that $H_2$ initially binds to Ni, which is more in line with structural and spectroscopic data. Moreover, an alternative cycle is suggested in which hydrogen is activated via oxidative addition instead of heterolytic cleavage: the $[H^+/Ni^{I}/Fe^{II}]$ intermediate generated from the first hydrogen may bind a second $H_2$, resulting in $[H^+/H^-{-}Ni^{III}{-}H^-{-}Fe^{II}]$, with Ni(III) and two hydrides.

The common feature in the catalytic cycles of [NiFe] and [FeFe] hydrogenases is a side-on metal-bound $H_2$, and two proton and electron release/uptake steps. Formally, the one-electron redox reactions are metal-based, and catalysis involves two-electron proton reduction or hydride oxidation. [Fe] hydrogenase also forms a side-on bound $Fe{-}H_2$, which is deprotonated, and the hydride is then transferred to the organic cosubstrate methenyltetrahydromethanopterin [49,50] (compare to [NiFe] hydrogenase, where oxidation of hydride occurs via formal hydride transfer to the base and electronredistribution to form $Ni^{I}$). DFT calculations of the associated activation energies show that the rate-limiting step is hydride formation, [reaction (2.11b)], even though it is stabilized by metal ions and a catalytic base, and free hydride is already energetically favored over free H at physiological pH [with $E°(H^+/H) = -2.29$ V and $E°(H^+/H^-) = -1.26$ V in water [44,51,52] $\Delta G = 442$ kJ/mol for homolytic splitting of hydrogen, while deprotonation of hydrogen requires $\Delta G = 203$ kJ/mol at pH 0 and 162 kJ/mol at pH 7].

Using protein film voltammetry, Leger et al. have found that the maximum HOR current is generated by *Allochromatium vinosum* [NiFe] hydrogenase adsorbed on edge plane graphite essentially independent of pH, while the limiting HER current is pH-dependent and can be modeled with three protonation sites coupled to two one-electron reduction steps. This also results in one-electron potentials that depend on pH in a nonmonotonous fashion. Above pH 7, the first reduction step becomes independent of pH (vs. NHE), resulting in an increasing HOR overpotential at more alkaline pH, in particular at 1 bar hydrogen [53]. *Desulfovibrio desulfuricans* [FeFe] hydrogenase has an inverted pH dependence of the HOR rate, with a higher maximum activity at pH 5.9 than at pH 8.4 [54]. In addition, hydroxide acts as competitive inhibitor for [NiFe] hydrogenase, substituting for hydride in the oxidized ($Ni^{III}$) active site. In protein film voltammetry, this results in a time-dependent reversible inactivation at high HOR overpotentials [38,55]. These observations illustrate the more complex and subtler sensitivity of hydrogenases to pH compared to the situation with metal electrodes. Enzymes have been evolutionally optimized to function under certain restricted pH conditions, and at driving forces adapted to their role and place in the metabolic energy scheme. It would have been naive to assume that they would exhibit similar kinetics and overpotentials under very different conditions.

The Armstrong group compared electrocatalytic hydrogen oxidation by platinum and by hydrogenases. They show that the Levich plots at 0.24 V versus NHE, 1 bar $H_2$, 45°C, and pH 7 with a monolayer of *A. vinosum* [NiFe] hydrogenase on graphite, with platinized graphite, and with platinized gold are indistinguishable (Fig. 2.5). However, at lower overpotentials, the platinized electrodes are more active [56], because the hydrogenase electrode displays an increasing activity with overpotential due to poor electronic contact between the electrode and a subpopulation of the enzyme molecules [57]. Karyakin and coworkers compared the electrocatalytic activity of a 40 pmol/cm$^2$ of *Desulfomicrobium baculatum*[NiFeSe] hydrogenase, immobilized on a polymerized mediator electrode with 0.5 mg/cm$^2$ Pt–Vulcan, both at pH 7 and 0.2 V overpotential. The [NiFeSe] hydrogenase electrode and Pt are equally active up to 30% partial $H_2$ pressure, but Pt–Vulcan is 25% more active at 100% hydrogen [58].

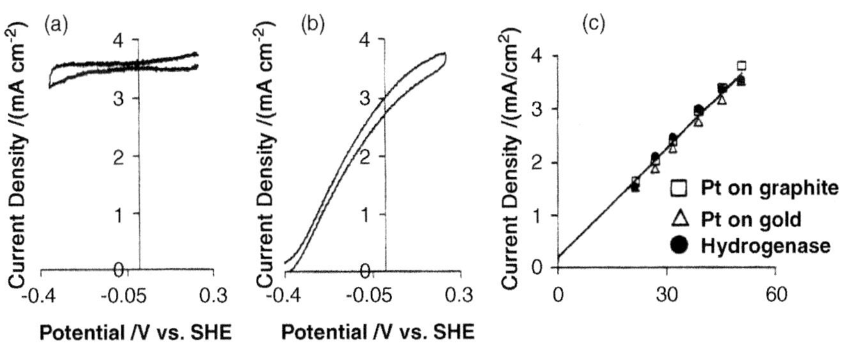


**FIGURE 2.5** Hydrogen oxidation currents on platinized gold and from hydrogenase on graphite at pH 7, 45°C, rotating at 2500 rpm, under 1 atm hydrogen, scan rate 1 V/s: (a) platinized gold; (b) hydrogenase; (c) Levich plots for hydrogenase (1 V/s) and platinum (chronoamperometry) at +0.242 V versus SHE. (Adapted from Ref. 156.)

## 2.3 OXYGEN REDUCTION REACTION

### 2.3.1 Mechanism of ORR at Metal Electrodes

There is very extensive literature on ORR as this reaction is generally recognized as one of the main bottlenecks in developing efficient low-temperature fuel cells. We will focus here on some of the more recent mechanistic advances; extensive reviews on ORR can be found in various papers and monographs [1,2,59,60]. The overall ORR reaction is

$$O_2 + 4H^+ + 4e^- \leftrightarrows 2H_2O \qquad E^\circ = 1.23\,V(\text{vs. RHE}) \tag{2.12}$$

It is now generally believed that the first step in ORR is the concerted proton–electron transfer to form a surface-adsorbed superoxide intermediate:

$$O_2 + H^+ + e^- \leftrightarrows OOH^* \tag{2.13a}$$

There is limited spectroscopic evidence for the OOH* intermediate, but it has been observed in DFT calculations of $O_2$ adsorbed on gold and platinum surfaces in the presence of water [61,62]. Also, the direct dissociation of $O_2$ into two atomic oxygens has been computed to have an activation energy of $\sim 0.5$ eV (in the absence of water), making this a less likely step [63]. The OOH* would be the precursor for O–O bond breaking to finally reduce to water [this occurs barrierless in DFT calculations on Pt(100) and Pt(110) [64]], following the pathway

$$OOH^* \leftrightarrows O^* + OH^* \tag{2.13b}$$

$$O^* + H^+ + e^- \leftrightarrows OH^* \tag{2.13c}$$

or the combined reaction

$$OOH^* + H^+ + e^- \leftrightarrows 2OH^* \tag{2.13d}$$

and finally the reduction of surface-bonded OH*:

$$OH^* + H^+ + e^- \leftrightarrows H_2O \tag{2.13e}$$

Alternatively, OOH* may react with another proton to give hydrogen peroxide:

$$OOH^* + H^+ + e^- \leftrightarrows H_2O_2 \tag{2.13f}$$

$H_2O_2$ may also serve as the precursor to O–O bond breaking.

A minimal mechanism for ORR now consists of at least four elementary steps. Consider, for instance an OER mechanism consisting of reaction (2.13a), (2.13b) + (2.13c) = (2.13d), and (2.13e). Each of these steps has a corresponding standard equilibrium potential, the values of which may be estimated from first-principles calculations in conjunction with a suitable choice of the standard state. A relation similar to Equation (2.3) applies:

$$\frac{E^{\circ}_{(2.13a)} + E^{\circ}_{(2.13d)} + 2E^{\circ}_{(2.13e)}}{4} = E^{\circ} = 1.23 \tag{2.14}$$

The *thermodynamic overvoltage* is again defined by the thermodymically least favorable reaction of the four elementary steps. Instead of a single intermediate, as in the HER/HOR reaction, we have now two or three, OOH*, OH*, and O*. However, Nørskov et al. [65,66] have shown that the binding energies of these three species, at least on metal surfaces, are strongly correlated. From detailed DFT calculations, it was found that $\Delta G^{\circ}(\text{OOH*}) \sim 0.5 \times \Delta G^{\circ}(\text{O*}) + \text{constant}$, and $\Delta G^{\circ}(\text{OH*}) \sim 0.5 \times \Delta G^{\circ}(\text{O*}) + \text{constant}$ [66]. Therefore, $\Delta G^{\circ}(\text{O*})$ will serve as a good descriptor, and the equilibrium potentials of reactions (2.13a), (2.13d), and (2.13e) may all be written as a function of this single descriptor (again assuming that the binding energy itself does not depend on potential) [6,67]. This leads to a plot of the thermodynamic overpotential $\eta_T$ as a function of $\Delta G^{\circ}(\text{O*})$, as illustrated in the "thermodynamic volcano plot" of Figure 2.6. Figure 2.6 illustrates that this plot again has two main lines, the location of the crossing point depending on exactly how each equilibrium potential depends on $\Delta G^{\circ}(\text{O*})$. If oxygen binding

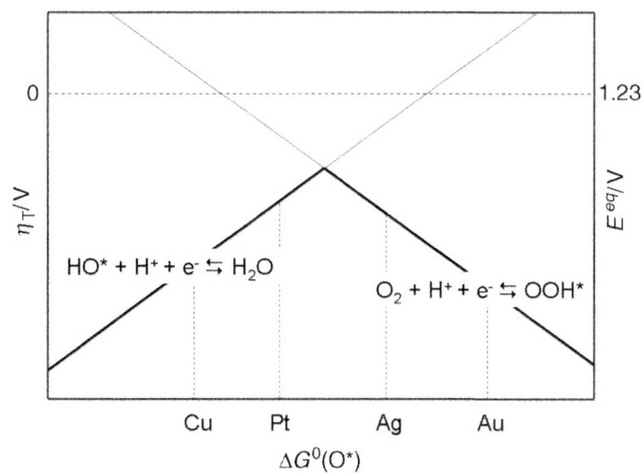


**FIGURE 2.6** Plot of the equilibrium potentials of reactions (2.13a) and (2.13e) as a function of the free binding energy of oxygen. Reaction (2.13d) has a much higher equilibrium potential and is therefore never the potential-determining reaction step. The relative positions of four metals, Cu, Pt, Ag, and Au, are indicated in the figure. The bold line gives the volcano plot of the thermodynamic overpotentials. The maximum lies close to 0.9 V. (Adapted from Ref. 67.)

is weak (i.e., on the RHS of the volcano in Fig. 2.6), calculations suggest that reaction (2.13a) is potential-determining. If oxygen binding is strong [i.e., on the left-hand side (LHS) of the volcano in Fig. 2.6], one may expect reaction (2.13e), that is, the reduction of surface-bonded OH*, to be potential-determining. Note that the crossing point now does *not* correspond to a situation of $\eta_T = 0$, in contrast to the HER/HOR situation. The reason for this lies in the fact that the binding energies of O*, OH*, and OOH* are all correlated through the relations derived by Nørskov et al. [65,66], and can therefore not vary independently. This restriction renders a situation of zero thermodynamic overpotential impossible. If $\eta_T = 0$ for the ORR should ever be reached, it will have to be with a class of materials that does not satisfy Nørskov's relations, or satisfies similar relationships but with a different functional form.

In the more detailed calculations of Nørskov et al., the predicted volcano plot is as shown in Figure 2.7, and compared to experimental data points obtained by Stamenkovic et al. [68]. It is observed that alloying Pt with less noble elements such as Fe, Co, and Ni weakens the bonding of oxygen to the surface, and thereby shifts the overpotential to the other leg of the volcano. These alloys all have a lower overpotential than does Pt, in good agreement with the predictions of the theory. The weaker bonding

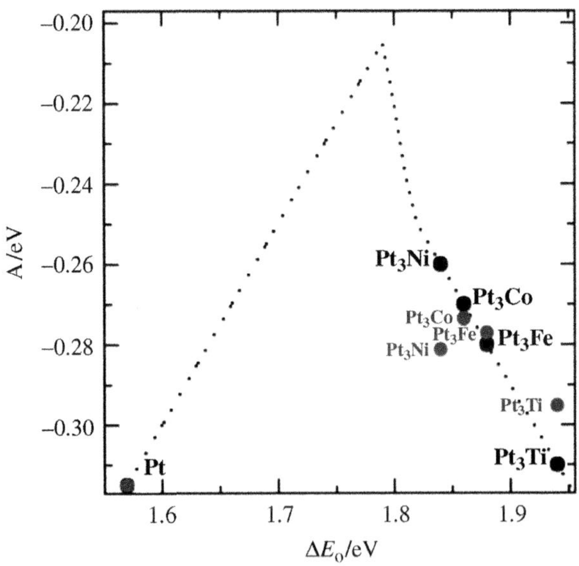


**FIGURE 2.7** Model of the activity at an electrode potential of 0.9 $V_{RHE}$ shown (dotted line) as a function of the adsorption energy of oxygen $\Delta E_O$. On the left side of the maximum in the volcano, the rate is limited by removal of adsorbed O and OH species; immediately to the right of the maximum, the rate is limited by $O_2$ dissociation; and on the extreme right, the limiting step is protonation of adsorbed $O_2$. Also shown in red are the measured activities relative to that of Pt. The predictions of the model for $Pt_3M$ (M = Ni, Co, Fe, Ti) alloys are shown relative to the predictions for Pt at the DFT-calculated values of $\Delta E_O$. (See the original reference [68] for details.)

of oxygen by adding a less noble second element to platinum is caused by a downshift of the *d* band of the Pt, due to the interaction with elements such as Ni, Fe, and Co.

The idea of weakening the bond of oxygen, or more specifically OH, to the platinum surface, by alloying or otherwise modifying the electronic properties of platinum, so as to shift the activity toward the apex of the volcano in Figure 2.6, has proved to be an extremely useful strategy in preparing a catalyst that is more active than platinum itself. A prominent success in this respect is the work on single-crystal PtNi alloys by Stamenkovic and Markovic [69]. These authors showed that by adding Ni to Pt, the binding of OH to the surface is indeed shifted to higher potentials, that is, becomes weaker. Their results are illustrated in Figure 2.8 for the specific case of the ORR on a single-crystalline $Pt_3Ni(111)$ surface in perchloric acid solution. The blank voltammetry in panel (a) shows that, compared to the Pt(111) surface, on the $Pt_3Ni(111)$ surface the hydrogen $H_{upd}$ region has shifted to lower potentials and the $OH_{ads}$ region has shifted to higher potentials. This implies a weaker bonding of $OH_{ads}$ to the $Pt_3Ni(111)$ surface, caused by a lowering of the *d* band with respect to pure Pt(111). Interestingly, using *low-energy ion scattering* (LEIS) and in situ surface *X-ray scattering*, Stamenkovic et al. also showed that the top layer of their $Pt_3Ni(111)$ surface consists of pure Pt, whereas the second layer is enriched in Ni ($\sim$ 50%). Figure 2.8c compares the ORR on $Pt_3Ni(111)$ to that on Pt(111), illustrating the almost 0.1 V gain in the reduction of the overpotential compared to Pt(111) (which is the most active Pt surface in perchloric acid). As illustrated in Figure 2.7, other platinum alloys, such as PtFe and PtCo, have also been found to have a higher activity for the ORR than pure Pt by weakening the bonding of OH to the surface [70]. Similar strategies have been employed by Adzic and coworkers by using mixed platinum overlayers on palladium [60,71].

The potential for oxygen reduction shifts the expected 60 mV/pH unit on the NHE scale, equivalent to a constant overpotential regardless of pH. This illustrates the intimate coupling between electron and proton transfer on metal electrodes. This is not to say that there are no differences between acidic and alkaline media [2]. Although the overpotentials in both media are similar, Pt(110) appears as the most active surface in sulfuric acid, whereas Pt(111) is the most active surface in perchloric acid and in alkaline media. This probably relates to the fact that Pt(111) interacts most weakly with $OH^-$ in solution. There is only relatively scant spectroscopic evidence for surface-adsorbed (su)peroxide intermediate during ORR on platinum (but see the more recent ATR-SEIRAS study by Shao et al. [72]). However, hydrogen peroxide may be a prominent product of the ORR on platinum, but only at high overpotentials in the potential region of $H_{upd}$ formation [2]. This suggests that coadsorption of other species may inhibit O–O bond splitting. Adsorbing halide ions have a very similar effect, enhancing the formation of hydrogen peroxide during ORR on Pt [2]. Interestingly, a metal such as Au is also active for oxygen reduction, especially in alkaline media, but it typically performs the two-electron reduction to hydrogen peroxide [73]. Only on a Au(100) surface in alkaline media, the ORR follows the four-electron pathway [73]. The exact reason for this remarkable structural effect has remained elusive [61,74].

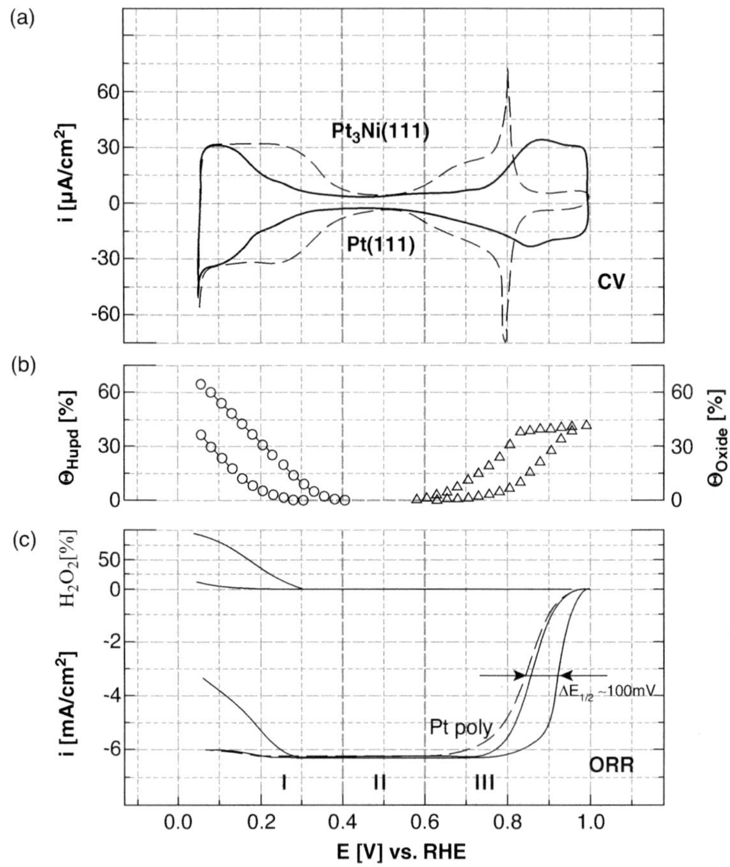


**FIGURE 2.8** In situ characterization of the $Pt_3Ni(111)$ surface in $HClO_4$ (0.1 M) at 333 K. (a) Cyclic voltammetry in designated potential region (red curve) as compared to the voltammetry obtained from Pt(111) surface (blue curve); (b) surface coverage of $H_{upd}$ and oxide (OH) calculated from cyclic voltammograms of $Pt_3Ni(111)$ (red curve) and Pt(111) (blue curve); (c) corresponding polarization curves for ORR obtained from rotating-ring–disk electrode (RRDE) measurements. Upper curve with green scale refer to hydrogen peroxide production in designated potential region, and lower curves give ORR currents measured on $Pt_3Ni(111)$ (red curve), Pt(111) (blue curve), and polycrystalline Pt (gray curve) surfaces. The arrows show the positive potential shift of 100 mV in electrode half-potential ($\Delta E_{1/2}$) between ORR polarization curves measured on Pt-poly and $Pt_3Ni(111)$ surfaces. Numerals I, II, and III represent potential region of $H_{upd}$ adsorption/desorption processes, double-layer region, and region of $OH_{ad}$ layer formation, respectively.

An interesting consequence of the model presented above is that whereas it seems firmly established that reaction (2.13e), that is, the reduction of OH* or, more generally, of Pt–surface oxygen species, is potential-determining, this reaction is not generally considered to be the rate-determining step on platinum in the potential region where a significant current flows (say, 0.9 $V_{RHE}$). The Tafel slope of 120 mV/

dec normally observed for Pt in acidic media [2,59,60] implies that reaction (2.13a) is the rate-determining step. A possible resolution to this apparent paradox lies in the realization that at the relevant potentials, reaction (2.13a) takes place on a surface partially covered by OH* or surface oxides. To describe this situation, Markovic and Ross [2] proposed the following equation for the current density:

$$j = -nFkc_{O_2}(1-\theta_{ad})^x \exp\left(-\frac{\alpha FE}{RT}\right)\exp\left(-\frac{r\theta_{ad}}{RT}\right) \qquad (2.15)$$

In this equation, the $(1-\theta_{ad})^x$ expresses the inhibition by adsorbates, primarily surface-bonded OH and surface oxides but also anions, through a site blocking effect; the first exponential term expresses the intrinsic Butler–Volmer-type potential dependence of the rate-determining step [i.e., reaction (2.13a)], and the second exponential term expresses any effects of the adsorbates on the activation energy of reaction (2.13a). The $(1-\theta_{ad})$ term in this equation is where the potential-determining step comes into play, as current will flow only if this term in nonzero. Note that this implies that the potential-determining step does not have to be the rate-determining step, as the $(1-\theta_{ad})$ term may be governed by an equilibrium process with an unfavorable equilibrium potential (i.e., $OH_{ad}$ formation/reduction). Stamenkovic et al. [69] have attributed the high activity of their $Pt_3Ni(111)$ surface to the favorable effect the weaker OH binding has on this $(1-\theta_{ad})$ term; specifically, it decreases site blocking. Wang et al. [75] have used an equation very similar to Eq. (2.15) to fit the detailed polarization curve for ORR on a Pt(111) electrode in perchloric acid, and found very good agreement between experiment and theory.

### 2.3.2 Mechanism of ORR by Enzymes

There are several types of enzymes that couple oxygen reduction to the oxidation or oxygenation of substrates, employing either an organic cofactor such as flavin, or metal ions. Here we discuss the catalytic mechanism of laccase and of cytochrome *c* oxidase, two enzymes that have been studied in detail, and catalyze the reduction of molecular oxygen to water, accepting four electrons in vitro from a range of donors or an electrode. Both types of oxidases have been coupled to cathodes in model biofuel cell devices [38,76]. Laccases are multicopper oxidases, containing a trinuclear Cu center as active site, in addition to a mononuclear auxiliary site. Cytochrome *c* oxidase is a heme–copper oxidase, containing an iron (in a heme) and a copper ion in the active site, and two auxiliary redox cofactors (a low-spin heme and a binuclear Cu site).

Both laccases and cytochrome *c* oxidase catalyze the four-electron reduction of molecular oxygen to water, and operate via a mechanism in which the active site is first fully reduced, and only then binds oxygen. When the reductant is present in excess (i.e, reduction is carried out at a high thermodynamic driving force or overpotential), the enzymes are in fact fully "charged" with four electrons before $O_2$ binds. The reduced active-site metal ions are poised for activation of the O–O bond and fast multielectron transfer. The resulting oxidized metal ions act as Lewis acids that stabilize the formed anionic products, postponing the requirement for protons. From a catalytic perspective, all that remains is product (water) release, which is promoted by subsequent internal electron and proton redistribution, proton uptake, and rereduction of the enzyme. Thus, molecular oxygen is fully reduced to water without releasing reactive intermediates.

#### *2.3.2.1 Laccase Mechanism*

Laccases are multicopper oxidases that catalyze the reduction of oxygen to water, using a wide range of organic and inorganic electron donors [77–80]. Fungal laccases are often, but not exclusively, involved in lignin degradation (wood rot), while laccases of plant origin are often involved in lignin synthesis. Laccases contain a trinuclear Cu center as active site (combining a type 2 mononuclear site, bound to the peptide by two histidine imidazoles, and a type 3 binuclear site, coordinated by 2 × 3 imidazoles) where molecular oxygen is reduced. It also harbors a mononuclear type 1 Cu(I/II) auxiliary site at 1.2 nm distance from the active site, involved in one-by-one electron uptake and transfer from reducing substrates.

Electrons from reducing substrates enter via the type 1 Cu site, and oxygen binds to the fully reduced trinuclear Cu site ($3Cu^{I}$) (see Fig. 2.9). Usually, the type 1 site is also reduced, in particular with two-electron donor substrates. In the rate-determining first step, a fully deprotonated peroxide intermediate is bound by one Cu(I) and two Cu(II). The subsequent formation of $O^{2-}$ and $OH^{-}$, stabilized by bridging between the cupric ions, is fast. However, when the type 1 copper is removed (by replacing the Cu by Hg, or in a mutant lacking one of the ligands), this step becomes the rate-limiting step, and a peroxi species accumulates. Reduction of the peroxy state (normally by type 1 $Cu^{I}$) first results in reduction of the type 2 Cu, weaking the bond to peroxide. The O–O bond

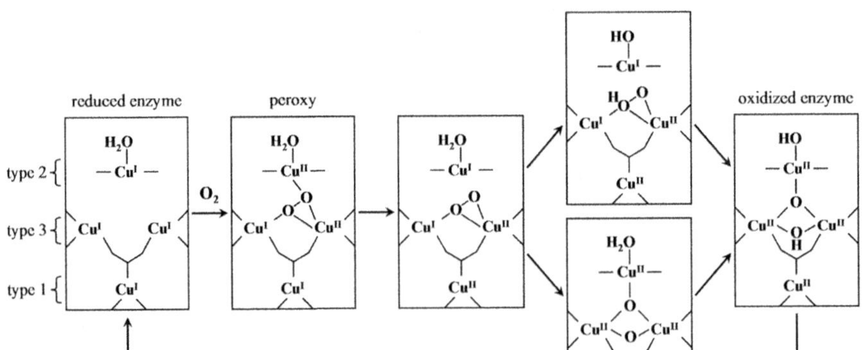


**FIGURE 2.9** Proposed intermediates during oxygen reduction by laccase, based on spectroscopic data and calculations. The oxidized enzyme intermediate, known as the *native intermediate*, differs from the isolated "resting" state in which the type 2 Cu is not connected with the type 3 ions. This resting state is formed by slow water release, and is only slowly reduced [77,80].

cleavage is catalyzed by proton transfer from a nearby carboxylic acid, H-bonded via a water molecule to the peroxide. On the basis of the calculated potential energy surface, Solomon concludes that at low pH, a proton is transferred before the transition state, while at high pH, protonation occurs after bond breaking [80]. Both pathways involve simultaneous and very fast transfer of two electrons, in which the Cu ions donate electrons of opposite spin into the peroxide $\sigma^*$, and stabilize it as Lewis acids.

Subsequently, the all-cupric enzyme is rereduced and four protons are taken up, resulting in dissociation of the water molecules. The fast reduction of peroxide implies that no peroxide is released and oxygen is always fully reduced to water. The ORR activity of laccases is optimal at acid pH. The activity rapidly decreases with increasing pH due to binding of hydroxide to the trinuclear site. This stabilizes the oxidized active site, thus inhibiting full reduction and oxygen binding. Effectively this means that product release becomes rate-limiting under alkaline conditions.

Density functional theory calculations by Anderson et al. [81] concluded that the low overpotential of laccase for the ORR reaction, as compared to platinum, may be explained by the high potential at which OH* may be reduced from the Cu sites to make them available for $O_2$ binding ("reduced enzyme" in Fig. 2.9). According to Anderson's calculations, this potential may lie some 0.2 V closer to the equilibrium potential than on Pt, but the OH* reduction to water is still the potential-determining reaction, similar to platinum.

#### 2.3.2.2 Cytochrome c Oxidase Mechanism

Respiratory oxygen reductases (terminal oxidases) are membrane-bound enzymes that can be divided into two subclasses: the *bd*-type quinol oxidases and the heme copper oxidases [82,83]. The active site of cytochrome *bd* quinol oxidase is a binuclear reaction center consisting of two high-spin heme irons, while a third, low-spin heme is believed to be involved in quinol oxidation [82]. Heme copper oxidases form a very diverse group of evolutionary related enzymes, and have been subject of detailed mechanistic studies. The common feature is a binuclear active site consisting of a heme-iron and a Cu ion coordinated by three histidine (His) imidazoles, in addition to a six-coordinate low-spin heme at close proximity to the active-site high-spin heme. Heme copper oxidases translocate protons across the membrane in addition to the four protons required to form water. The resulting proton electrochemical gradient stores the released redox energy and is used to drive ATP synthesis [84]. With the exception of quinol oxidases in this subclass, all heme copper oxidases also contain one or more additional redox cofactors that are involved in electron relay and storage [83].

Mitochondrial cytochrome *c* oxidase (CcO, also termed *respiratory complex IV*), and the closely related CcO from the soil bacterium *Paracoccus denitrificans* are heme–copper oxidases that have been studies in great detail. The active site is formed by heme $a_3$ and a $Cu_B$ site, coordinated by three His side chains. The auxiliary heme, a 6c heme *a*, is in rapid redox equilibrium with a binuclear $Cu_A$ mixed-valence site that accepts one electron from cytochrome *c* [83,85].

The mechanism for ORR by cytochrome *c* oxidase is schematically summarized in Figure 2.10. Electrons from cytochrome *c* are first transferred to the binuclear mixed-valence $Cu_A$, and then via heme *a* to the active site. Oxygen only binds to the fully

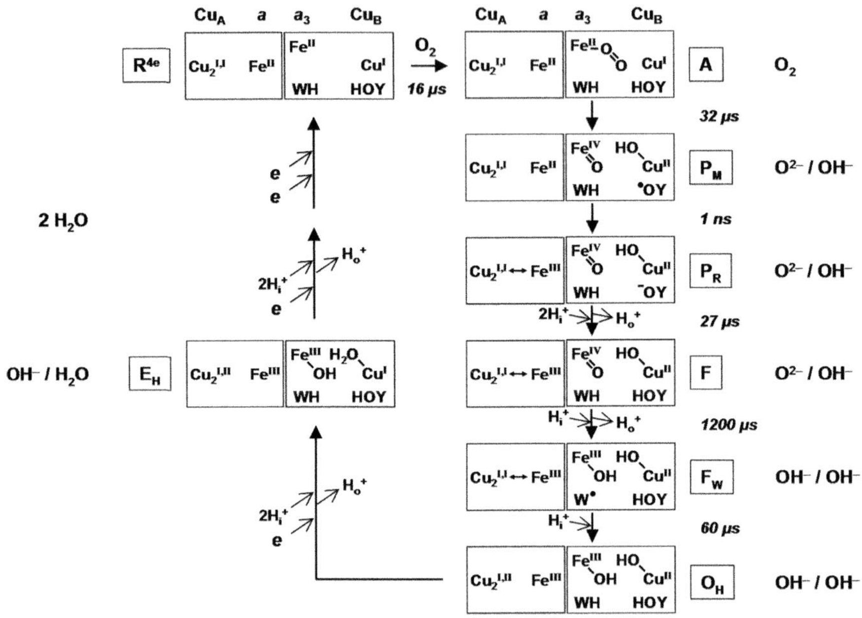


**FIGURE 2.10** Proposed oxygen reduction mechanism for cytochrome *c* oxydase [83,85]. The measured half-life times, the commonly used names of the observed intermediates, and the formal charge and protonation states of the bound oxygen species are indicated. For the $Cu_A$ site, the shorthand notation indicates the valences of both atoms (although $Cu_2^{I,II}$ is in fact a delocalized mixed-valence state). HOY is the tyrosine covalently bound to a $Cu_B$ imidazole ligand, and WH a tryptophan residue. $H_i^+$ is a proton from the inside, $H_o^+$ is a proton delivered to the outside.

reduced $Fe^{II}/Cu^{I}$ active site. Under conditions where cytochrome *c* is not limiting, $Cu_A$ and heme *a* are also reduced at this stage. In the first observed, rapidly formed intermediate after oxygen binding, four electrons have already been transferred and the O–O bond has been broken, resulting in an oxoferryl heme $a_3$, and $Cu_B$(II) hydroxide. This process may precede intramolecular electron transfer from the auxiliary heme *a* and $Cu_A$ sites, and thus only involves the ($Fe^{II}/Cu^{I}$) active site, creating two "holes"; the third electron generates Fe(IV), while the forth electron and the proton are derived from a Tyr (tyrosine), covalently bound to one of the His ligands of the Cu, under formation of an extremely shortlived neutral radical. It is possible that this intermediate, which is stabilized when starting from the two-electron reduced enzyme, is entirely bypassed when starting from the fully reduced enzyme. In contrast to laccase, no peroxo- or other partially reduced intermediates have ever been detected in heme copper oxidases, although the experimentally manifest very fast 4e reduction of iron-bound molecular oxygen to oxoferryl and cupric hydroxide might suggest the involvement of a transition state (or very shortlived intermediate) in which $O_2$ is activated by binding to both Fe and Cu. In fact, Noodleman and coworkers [86] did include formal Fe(III)–O–O–Cu(II) intermediates in their DFT calculations prior to oxoferryl formation.

Subsequently, the two electrons stored in heme *a* and $Cu_A$ are transferred to the active site and two protons are taken up. The released energy is conserved by pumping two protons across the membrane. The thus formed fully oxidized (ferric hemes, mixed-valence $Cu_A$, cupric $Cu_B$) enzyme intermediate is then reduced again by four cytochrome *c*, and two more protons are taken up to form water. In addition, two more protons are translocated over the membrane during this reductive part of the catalytic cycle. The DFT calculations by Noodleman [86] as well as those by Siegbahn and Blomberg [87] indicate that this part of the mechanism, the rereduction of the $O_H$ state to the R reduced state, where the enzyme is ready to accept another oxygen molecule, is the thermodynamically uphill part of the reaction cycle. Kjaergaard et al. [88] have compared DFT calculations of the energetics for ORR on CcO to Pt(111) and $Pt_3Ni$ (111), and found that the calculations by Kaukonen [89] and by Blomberg and Siegbahn [90] predict smaller thermodynamic overpotentials than those that have been calculated for the Pt-based ORR catalysts. According to the calculations by Kaukonen, the potential-determining step is either the reduction of O* or OH*, both with standard potentials close to 1.04 $V_{NHE}$, whereas the calculation by Blomberg and Siegbahn suggests that the reduction OH* is potential-determining, with a standard potential of 0.86 $V_{RHE}$. Both calculations suggest that CcO should be a better catalyst than either Pt or $Pt_3Ni$, which is not quite in agreement with the experimental literature employing protein film voltammetry (see Section 2.3.2.3). Nonetheless, these results suggest that the reduction of the oxygen dissociation products (O* or OH*) is the potential-determining step in the ORR mechanism on CcO, very similar to what has been concluded for platinum and laccase.

It can be noted that the initially deprotonated oxidized metal oxo sites temporarily store the energy, which is released by the subsequent protonation and reduction steps, and is used by heme copper oxidases to drive the translocation of up to four additional protons across the membrane. In fact, the oxidized state during turnover differs from the "resting" oxidized enzyme: the former seems to have stored some of the energy to drive subsequent proton pumping ("$O_H$" denotes oxidized, high-energy), while the latter does not pump protons on reduction [83]. The Tyr (tyrosine) and Trp (tryptophan) radical intermediates are suggested to be involved in proton pumping by translating electronic free energy into protonic free energy (redox-state-dependent p*K*), but other sites such as the imidazole ligand to $Cu_B$ could also accomplish this [86,91].

### 2.3.2.3 Comparing ORR by Laccase and Cytochrome c Oxidase

Despite the initial formation of relatively high-energy intermediates in CcO, the actual oxygen reduction is by no means rate-limiting. Rather, the reaction driving the second proton translocation is substantially slower. Nevertheless, CcO has a turnover rate comparable to most laccases, for which the first two-electron reduction of oxygen to a peroxi intermediate is rate-limiting (at least at acid pH). This illustrates the extreme oxygen reduction efficiency of heme copper oxidases, albeit at a rather high driving force, because the cofactor potentials are found between 0.22 and 0.42 V versus NHE [92,93]. For laccases, the overpotential may be somewhat smaller. The reduction potentials of the primary electron acceptor type 1 and the binuclear type 3 sites seem to be correlated, and range from 0.43 to 0.79 V versus NHE, while the type 2 site has a potential of ~0.4 V [77].

Heller and coworkers demonstrated that when a laccase containing a type 1 Cu with $E^\circ$~0.8 V is coimmobilized in a redox–hydrogel containing high-potential Os complexes as local electron shuttle sites, the overpotential for oxygen reduction becomes very small, and even better than for ORR on platinum [94]. However, care should be taken not to increase the mediator potential too far. Although the catalytic half-wave potential is close to the mediator potential, inversing the driving force for electron transfer from the Os mediator sites to the type 1 Cu site will ultimately result in lowering the maximum current at high driving force (amplitude of the sigmoid) due to rate-limiting electron transfer from the mediator to the enzyme. In fuel cell applications an optimum power output can be expected when the mediator potential is tuned such that the electron transfer rate becomes comparable to the maximum catalytic turnover rate.

A similar argument may hold when the auxiliary redox cofactors in enzymes are acting as mediators, although very rapid intramolecular electron transfer may result in a half-wave potential that is not correlated with the mediator site potential (i.e., the turnover rate no longer titrates with the redox state of the mediator) [95]. The Armstrong group achieved oxygen reduction at very low overpotential with laccase on graphite without mediators (Fig. 2.11). The electrode was modified with chrysene groups that bind to laccase near the type 1 Cu site, orienting the enzyme for optimal

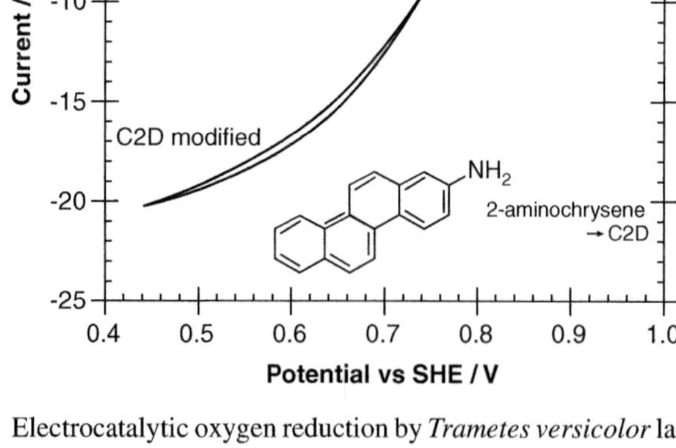


**FIGURE 2.11** Electrocatalytic oxygen reduction by *Trametes versicolor* laccase on graphite with and without electrode surface modification by chrysene-2-diazonium (C2D). C2D binds lacase near the type 1 Cu site. *Conditions:* 1 bar $O_2$, rotating at 3000 rpm, 25°C, scan rate 5 mV/s, in 200 mM sodium citrate pH 4.0 ($E^{\circ\prime}$ for $O_2/H_2O$ is 0.99 $V_{SHE}$, $E^{\circ\prime}$ of type 1 Cu is 0.78 $V_{SHE}$). (Adapted from Ref. 96.)

electron transfer. The half wave potential is close to the reduction potential of the type 1 Cu site, indicating that it acts as mediator [96].

In the full catalytic cycles of laccase at pH > 5, and of CcO, the rate-limiting step is located in the rereduction half-cycle, where proton uptake and product release take place. This agrees with the surface catalysis prediction that strong oxygen binding to the active site leads to rate-limiting or potential-determining product release. In laccase the release of product is indeed rate-limiting, unless it is promoted by rather low pH. CcO is a special enzyme that conserves energy by pumping protons, and the rate limiting step is associated with this "work." This implies that CcO is not an optimal oxygen reduction catalyst per se. It can be argued that its main physiological role is in fact not oxygen reduction, but the translocation of protons from a pH ~7 bath to a low-pH bath (or, more accurately, against the proton electrochemical gradient), driven by the potential difference between oxygen and cytochrome *c*. Even if the proton pumping is uncoupled (i.e., when the enzyme is solubilized by surfactant, and no lipid bilayer membrane is present), our more recent direct, unmediated protein film voltammetry experiments (Fig. 2.12) show that the overpotential does not decrease, but remains very similar to that observed in the presence of cytochrome *c* as mediator. This is to be expected, because the redox potentials of $Cu_A$, heme *a*, and the (water- or hydroxide-bound) active-site metal ions are optimized to accept electrons from cytochrome *c* ($E^{\circ\prime} = 0.26$ V at pH 7).

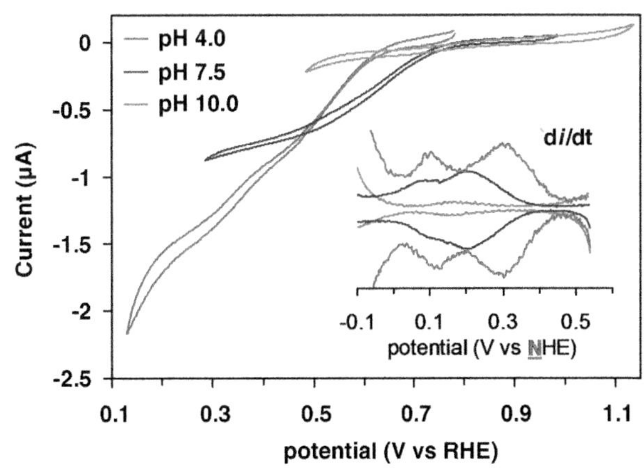


**FIGURE 2.12** Electrocatalytic oxygen reduction by a submonolayer (~10%) of *Paracoccus denitrificans* cytochrome *c* oxidase immobilized on a stationary gold electrode under air. Conditions: 50 mM buffer (citrate pH 4.0, phosphate pH 7.5, or Caps pH 10.0), scan rate 10 mV/s, 25°C, electrode area 3 mm$^2$. The current is plotted versus the reversible hydrogen electrode, the derivatives (inset) versus the normal hydrogen electrode to illustrate the moderate pH dependence of the half-wave potentials. (F. G. M. Wiertz, O.-M. Richter, B. Ludwig, and H. A. Heering, unpublished results.)

#### *2.3.2.4 Oxygen Reduction and Protonation Are Not Intimately Coupled*

Both laccase and CcO essentially operate via a mechanism in which oxygen is activated by binding to prereduced metal ions, and is efficiently split via $2 \times 2e$ or $1 \times 4e$ steps, involving formally only a single proton from an acid group in the protein. During the reductive part of the catalytic cycle the reduced enzyme is regenerated and protons are taken up to form water:

$$\text{Enzyme}(4e) + O_2 \rightarrow \text{enzyme}/O^{2-}/OH^- \rightarrow \text{enzyme}(4e) + 2H_2O \qquad (2.16)$$

Oxygen reduction and protonation are thus separated. Depending on the sequence of events during enzyme reduction and water release, and the (often not or only weakly proton-coupled) reduction potential of the primary cofactor that communicates with the electrode, the pH dependence of the catalytic potential (half-wave potential) typically is substantially smaller than that of the NHE. Notably, the reduction potentials of the type 1 Cu site in laccase [97] and of the $Cu_A$ site in CcO [92] are constant (not proton-coupled) over at least six and three pH units, respectively. For this reason, it is more convenient to plot the potentials versus the NHE, as illustrated in Figure 2.12. Consequently, the required overpotential for oxygen reduction is pH-dependent and more favorable at higher pH (even comparable to Pt at pH 10). However, both laccase and CcO are rapidly inactivated under alkaline conditions due to hydroxide inhibition and/or slow proton uptake.

## 2.4 OXYGEN EVOLUTION REACTION

### 2.4.1 Oxygen Evolution on Metal Oxide Electrodes

Oxygen evolution on metal electrodes occurs at potentials higher than 1.23 $V_{NHE}$, a potential at which all metal surfaces in an aqueous electrolyte (apart, perhaps, from gold) will be covered by a (thin) oxide layer. Therefore, practical catalysts for the OER are oxide electrodes [24,98]:

$$2H_2O \leftrightarrows O_2 + 4H^+ + 4e^- \qquad E^\circ = 1.23\text{V(vs. NHE)} \qquad (2.17)$$

Mechanisms suggested in the literature involve steps similar to those considered for ORR, only in the reverse direction. Water first dissociates on the metal oxide, forming surface-bonded OH*:

$$H_2O \leftrightarrows OH^* + H^+ + e^- \qquad (2.18a)$$

This OH* is usually considered to be further oxidized to O*

$$OH^* \leftrightarrows O^* + H^+ + e^- \qquad (2.18b)$$

followed by the formation of a superoxide intermediate

$$H_2O + O^* \leftrightarrows OOH^* + H^+ + e^- \tag{2.18c}$$

which is the precursor to $O_2$ evolution:

$$OOH^* \leftrightarrows O_2 + H^+ + e^- \tag{2.18d}$$

In the older OER literature one will also find the recombination of two O* on the surface to $O_2$ as one of the relevant reactions.

The mechanism given above, or slight modifications thereof, has been suggested for the two most active oxides for the OER, $RuO_2$, and $IrO_2$. Rossmeisl et al. have recently carried out DFT calculations of the intermediates represented in reactions (2.18a)–(2.18d) to analyze the OER mechanism on various metals [65] as well as on a model $RuO_2$(110) surface [99]. The sum of the (standard) equilibrium potentials of these four steps should again add up to $4 \times 1.23$ V [see Eqs. (2.3) and (2.13)]. As with ORR, the ideal catalyst would correspond to the one in which each step has an equilibrium potential equal to 1.23 V. The values calculated for $RuO_2$ by Rossmeisl et al. [99] are 1.32 V [reaction (2.18a)], 1.39 V [reaction (2.18b)], 1.60 V [reaction (2.18c)], and 0.65 V [reaction (2.18d)], suggesting that the formation of the OOH* intermediate is the potential-determining step on $RuO_2$(110). Again, a correlation between the binding energies of O*, OH*, and OOH* exists, although the exact functional form is on oxides differs from that on metals [99]. Therefore, equilibrium potentials may be calculated as a function of $\Delta G^\circ(O^*)$, where this quantity now refers to binding to an oxide, not a metal. Figure 2.13 shows the thus calculated thermodynamic overpotential as a function of $\Delta G^\circ(O^*)$, leading again to the familiar

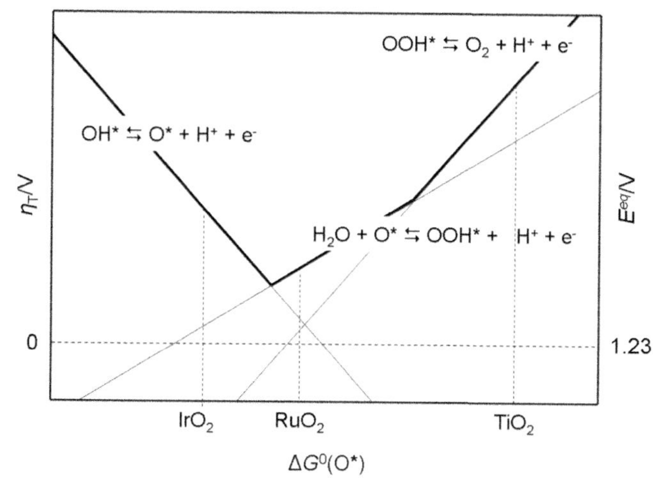


**FIGURE 2.13** Equilibrium potentials of reactions (2.18b)–(2.18d) versus the free binding energy of oxygen on metal oxides $\Delta G^\circ(O^*)$. The relative positions of $IrO_2$, $RuO_2$, and $TiO_2$ are indicated in the figure. (Adapted from Ref. 99.)

volcano-type relations. In the analysis by Rossmeisl et al., both $RuO_2$ and $IrO_2$ lie close to the top of the volcano, with $RuO_2$ binding oxygen a little too weakly and $IrO_2$ binding oxygen a little too strongly. This mechanistic model must be a rough representation of reality, as in fact rather little is known about the real nature of the active site and the exact influence of the catalyst structure in the OER on oxides. On the other hand, the analysis above is a first step toward an atomic-level understanding of the OER on oxides, and provides a useful conceptual framework for thinking about alternative catalysts. The model has also been extended to predict photoelectrocatalytic activity at a $TiO_2$ photoanode [100].

Experimentally, the OER at $RuO_2$ is observed to start at ~1.45 V in acidic media, with a Tafel slope of ~40 mV/dec of polycrystalline $RuO_2$ [98,101] (a Tafel slope of 60 mV/dec has been observed on single-crystal $RuO_2$ [102]). On $IrO_2$, the overpotential is slightly higher but the Tafel slope is also close to 40 mV/dec [103]. In most of the electrochemical literature, this Tafel slope is interpreted as the formation of O* from OH* being rate-determining. The actual observation of intermediates in the OER has proved extremely difficult, but isotope-labeling tactics combined with online electrochemical mass spectrometry detecting the evolved oxygen has shown that the oxide electrode is actually a partner in the reaction, as the oxygen from the electrode ends up in the oxygen gas generated during the reaction [104,105].

On both $RuO_2$ and $IrO_2$, the overpotential does not shift significantly with pH [98,106]. On the NHE scale, this is equivalent to a shift of 60 mV/pH in oxidation potential, implying again an intimate (Nernstian) coupling between proton and electron transfer in the reaction kinetics. Structural effects in the OER have been observed both for $RuO_2$ and $IrO_2$ nanoparticles [106–109]. Murray and coworkers [106] established an interesting linear relationship between the charge under the $Ir^{5+}/Ir^{4+}$ redox wave observed before the onset of the OER and the OER current at moderate overpotential, implying that their nanoparticle films remain active even if they consist of multiple layers of nanoparticles. Moreover, they observed the potential of the OER onset to coincide with the oxidation of the oxide electrode from the $Ir^{5+}$ state to the $Ir^{6+}$ state.

### 2.4.2 Oxygen Evolution by the Oxygen Evolving Complex of Photosystem II

Photosystem II is the only known natural system capable of oxidizing water to molecular oxygen. The active site where water is oxidized is the oxygen evolving complex (OEC), containing four manganese and one calcium ion. The OEC is oxidized in four one-electron steps, each driven by light-induced charge separation, forming a highly oxidizing $P_{680}{}^{+}$ radical ($E$~1.3 V). A Tyr residue ($Y_Z$) acts as mediator site, forming to a tyrosyl radical intermediate that is the electron acceptor from the OEC. The experimental results can be summarized by the Kok cycle (Fig. 2.14), in which each photon drives a one-electron oxidation of the OEC, generating the increasingly oxidized storage states $S_0$ to $S_4$. The most reduced state $S_0$ that is formed under steady-state turnover conditions has oxidation state [3 Mn(III), Mn(IV)], and is regenerated from $S_4$ by water uptake (one water is actually already

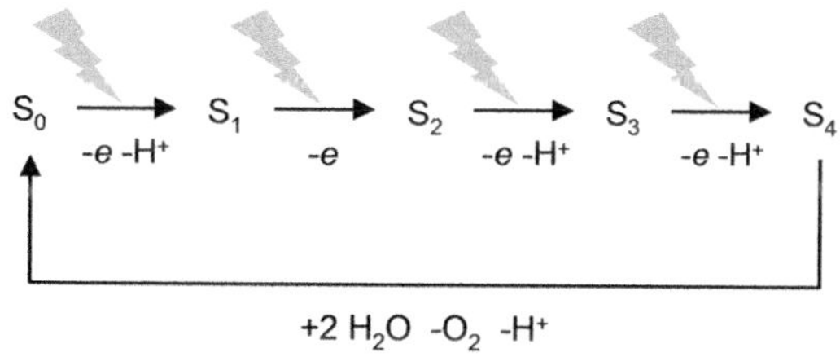


**FIGURE 2.14** The catalytic Kok cycle of the oxygen evolving complex in photosystem II [110,111].

taken up during one of the oxidation steps) and oxygen evolution. Although the redox cycling starts at $S_0$, state $S_1$ is the dark resting state [110,111]. All steps in the cycle, except $S_1$ to $S_2$, also involve removal of a proton [112].

With a potential of the $P_{680}{}^+$ radical of ~1.3 V and $E°(O_2/H_2O) = 0.82$ V at pH 7, the total driving force for water oxidation by photosystem II is +0.48 V, which is similar to the 0.56 V total driving force used by cytochrome *c* oxidase to reduce oxygen with cytochrome *c* ($E°$=0.26 V). However, photosystem II does not appear to couple this with equal efficiently to the generation of a proton electrochemical gradient [113]; while CcO removes eight protons from the inside (bacterial cytoplasm, mitochondrial matrix) and adds four electrons to the outside, photosystem II removes four protons from the inside (stroma) for the reduction of the electron acceptor plastoquinone to plastoquinol, and adds the four protons from water to the outside (lumen) via proton exit channels [114].

Two X-ray structures are available, but unfortunately at insufficient resolution (3–3.5 Å) to resolve the mechanistically important structural details of the OEC [115]. The structures of the OEC in each redox state have therefore been optimized by combining DFT calculations with information derived from the large database of spectroscopic and enzymatic results. [110,111,113,114,116–118]. Two types of DFT studies have been used to predict the geometric and electronic properties: Siegbahn and coworkers [113,116,117] used a cluster model with DFT employing the so-called B3LYP functional, and Batista and coworkers [114,118] used an embedded quantum mechanics/molecular mechanics (QM/MM) approach. The consensus structure is a (distorted) cubic arrangement of Ca and three Mn, connected via μ-oxo bridges (Fig. 2.15). The fourth Mn, addressed as "dangling Mn," is outside the cube, but nevertheless strongly bound to it via one or two oxo bridges. The complex is bound to the peptide via six carboxylates and an imidazole.

Both DFT models propose that further oxidation of the all-Mn(IV) $S_3$ state to $S_4$ generates an oxyl radical $O^{\bullet-}$ bound to the dangling Mn(IV) (Fig. 2.12). Subsequently, the O–O bond is formed by nucleophilic attack of a second, fully reduced oxygen atom bound to $Ca^{2+}$, and dioxygen is evolved after transfer of in total three electrons to the fully oxidized OEC. The derived catalytic mechanisms are summarized in Figures 2.16 and 2.17, focusing on the redox states of the Mn and the substrate.

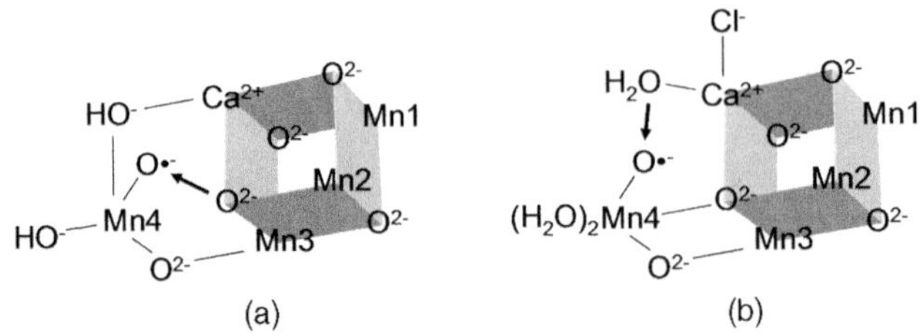


**FIGURE 2.15** Schematic cartoons of the fully oxidized oxygen evolving complex (state $S_4$) containing all Mn(IV) and an oxyl radical, according to Siegbahn and coworkers (a) [117], and Batista and coworkers (b) [114]. The geometries are simplified, and the protein-derived ligands are omitted to emphasize the cubic arrangement with a "dangling" manganese Mn4. The arrows indicate the nucleophilic attack that results in O–O bond formation.

The DFT models differ in the details of the structures and ligand arrangement, the structural rearrangements during the catalytic cycle, the state at which one of the substrate waters is bound to the cluster, and the nature of the second oxygen atom (either a μ-$O^{2-}$- or $Ca^{2+}$-bound water). The Siegbahn model is mostly deprotonated compared to the Batista model (see Fig. 2.15). Because water binds only when forming $S_3$ in this model, this leads to a pure electron transfer from $S_1$ to $S_2$, while several protons and coordinating water are present in the structures proposed by Batista (their $S_0$ state including ligands has charge +1). Both models include structural rearrangements involving a second Mn4–O–Mn3 bridge. In the Siegbahn model, opening this bridge is required for subsequent O–O formation, while in the Batista model it plays

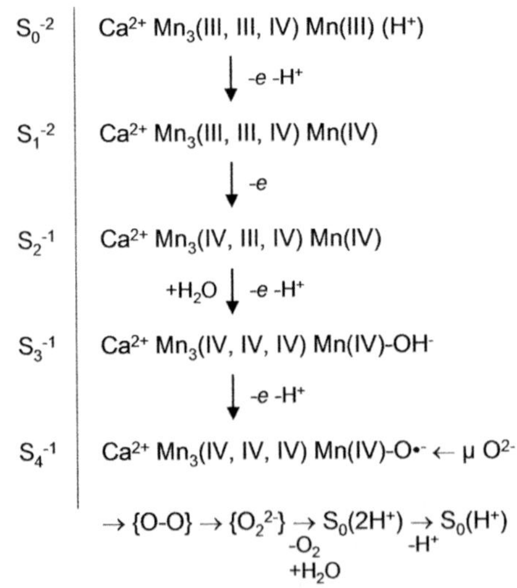


**FIGURE 2.16** Summary of the mechanism proposed by Siegbahn and coworkers [117]. The valences are given for Mn1, Mn2, Mn3, and Mn4, respectively. The reacting $O^{2-}$ is the μ-oxo bridge between Ca, Mn2, and Mn3. The net charges of the OEC plus ligands are given as superscripts.

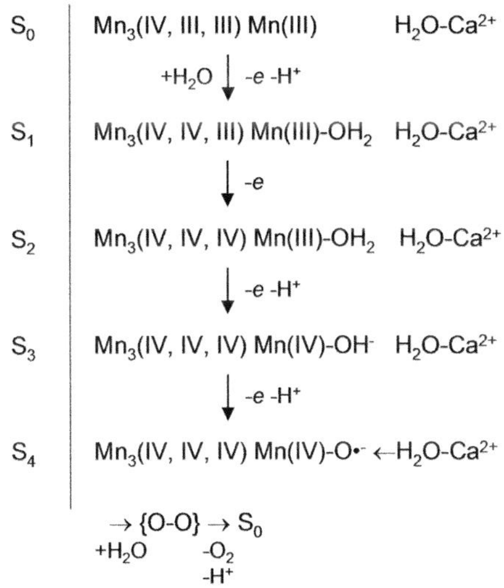


**FIGURE 2.17** Summary of the mechanism proposed by Batista and coworkers [114] The valences are given for Mn1, Mn2, Mn3, and Mn4, respectively. A water molecule from the second coordination sphere of Ca replaces the substrate water in concert with {O—O} formation.

an important role in proton binding and H-bonding to activate substrate water. Siegbahn did not include chloride in the full model [113,117], but Batista argues that experimentally a chloride is required to go beyond $S_2$. In the lowest energy configuration, $Cl^-$ was found to ligate Ca near $Y_Z$ in accordance with spectroscopic data.[114]

Other mechanisms have also been proposed that incorporate elements given in Figures 2.13 and 2.14, although oxidation of a μ-oxo bridge from $S_2$ to $S_3$ is also suggested instead of metal-based oxidation [111]. Messinger also proposed an oxyl radical reacting with a μ-oxo bridge [119]. Meyer and coworkers [120] propose a mechanism for the OEC that is inspired by that of the Ru(III)—O—Ru(III) "blue dimer," emphasizing the importance of proton-coupled electron transfer during the Kok cycle. The O—O bond is proposed to form on the dangling Mn by redox nucleophilic attack of Ca-bound hydroxide on $Mn_3$(IV,IV,IV)/Mn(IV) = O in the $S_3$ state resulting in $Mn_3$(IV,IV,III)/Mn(III)—OOH, from which $S_0$ is restored by the fourth oxidation and deprotonation, followed by oxygen release and water uptake. Pecoraro and coworkers [121] similarly proposed a reaction between Ca-bound water and Mn(IV) = O.

Rossmeisl et al. [122] have compared the thermodynamics of the Kok cycle for the OER by the OEC, using Siegbahn's DFT calculations for two suggested literature structures of the Mn cluster, with that calculated for $RuO_2$ (see Section 2.4.1), assuming the same reaction steps as suggested in Section 2.4.1 [i.e., reactions (2.18a)–(2.18d)]. In agreement with the suggested potential-determining step on $RuO_2$, reaction (2.18c) was also concluded to be potential-determining on the structure suggested by Loll et al. [115c], whereas reaction (2.18d) was found to be the potential-determining step on the structure suggested by Ferreira et al. [115a]. The calculated thermodynamic overvoltages were smaller on the OEC than on $RuO_2$, 0.32 and 0.21 V on the OEC versus 0.37 V on $RuO_2$, confirming the high efficiency of the OEC. Rossmeisl et al. attribute the lower thermodynamic overvoltage of the enzyme to the larger flexibility of the oxidation site and geometry, which cannot be matched by the rigid oxide surface. We might add to this that the $RuO_2$ catalyst, on the other hand, has a much longer lifetime [98].

## 2.5 CONCLUDING REMARKS

In this chapter, we have presented a brief comparative overview of the reaction mechanisms of hydrogen reduction and evolution, and oxygen reduction and evolution by both synthetic (human-made) (metals, metal oxides) and natural catalysts (enzymes), emphasizing the multistep character of these important reactions and role of catalysis and catalytic intermediates in accelerating their rates. On the basis of this comparison, we may draw a few general conclusions.

First, an important consideration when comparing electrocatalysis and bioelectrocatalysis lies in the experimental approach. In particular, for any kind of meaningful comparison, it is imperative that the reaction kinetics be compared at equal thermodynamic driving force. While studying a reaction at varying thermodynamic driving force is natural to surface electrocatalysis as the electrode potential is one of the key experimental parameters, classical biochemical methods for studying enzyme kinetics often start far from equilibrium, and are carried out in a homogeneous solution. Some of the ideas and catalytic concepts presented are therefore not straightforward to compare. Notably, much of the mechanistic information on enzymes is derived from pre-steady-state kinetics, in which a fully or partially “charged” (reduced/oxidized) enzyme reacts with substrate. This is to some extent equivalent to electrocatalysis at high driving force or high overpotential. We refer to this important issue as the “driving force gap,” and we believe that this gap lies at the basis of some of the misunderstandings that recur when enzymologists and surface electrochemists discuss what appears to be the same reaction, but studied under very different (thermodynamic) condtions. This driving force gap should also be considered when enzymologists and surface electrochemists use the term *efficient*. When used in combination with enzymeatic activity, *efficient* typically means fast, selective, with no noticeable production of intermediates. In surface electrocatalysis, *efficient* typically means a high activity close to the thermodynamic potential.

Protein film voltammetry, or electroenzymology, is therefore an obviously more suitable way to generate data that can be compared meaningfully to surface electrocatalysis. We have illustrated our discussion with a few examples of this comparison from the literature in this chapter. Many of these comparisons have focused on comparing overpotentials, but we believe that there are more detailed approaches possible that in many cases still need to be explored more fully. Studying the potential dependence of catalytic currents in terms of Tafel plots is a widespread and useful practice in surface electrocatalysis, but is much more problematic in protein film voltammetry as the potential dependent binding and orientation of the immobilized enzymes will need to be deconvolved from the overall current in order to obtain the intrinsic potential (driving force) dependence of the catalytic reaction. Therefore, understanding and controlling the potential dependent state of the surface-adsorbed protein will remain a crucial issue in the further development of protein film voltammetry. Second, a prominent feature often observed in protein electrochemistry, is the occurrence of "boosts" and "switchoffs" at high driving force (Fig. 2.12 shows an example of a *boost*). While this is a rather common aspect of surface electrocatalysis, often related to the formation of inhibiting surface oxides, it remains a very special feature in biocatalysis, and has given new insight in enzyme catalysis [13]. Comparing these more qualitative aspects of electrocatalysis at varying driving force has not been discussed in any detail here, but will become important as more detailed comparisons become commonplace. Finally, surface electrocatalysis has succesfully incorporated many different surface-sensitive spectroscopic techniques in order to resolve the relationship between structure, reactivity, and the chemical nature of reaction intermediates. This approach is still in its infancy in protein film voltammetry, in no small part due to greater degree of complexity of these interfaces, but there is no question in our minds that coupling protein film voltammetry to techniques such as infrared spectroscopy, surface-enhanced (resonance) Raman spectroscopy, online electrochemical mass spectrometry, and other nonelectrochemical techniques with molecular sensitivity, will be an important future avenue for this field of scientific research.

Returning to catalysis, the general conclusion from classical biochemistry is that multicentered enzymes tend to first accumulate all the electrons or holes before the active site binds the substrate, activates the bond, and promotes concerted multi-electron transfer. For a surface electrochemist, this conclusion should be understood on the basis of the abovementioned driving force gap, namely, that it implies studying the reaction at a high overpotential. Auxiliary sites act as intramolecular electron sources or sinks. In protein film voltammetry, these auxiliary sites usually operate as intramolecular mediators. Often (but not exclusively) the substrate has the highest affinity for the fully "charged" active site. As a first approximation, the auxiliary sites will titrate with the applied potential, while the active-site redox state depends on both the redox state of the auxiliary sites and substrate availability. At driving forces close to zero, the enzyme will be only partially, but dynamically, charged. This implies that the rate-limiting step will depend on both the electrode potential and substrate availability, and different intermediates will accumulate depending on conditions. A boost or attenuation of the activity may occur at high driving force, such as due to trapping the enzyme in a redox state with a higher or lower affinity for substrate, by selecting for a faster or slower branch in the mechanism due to an altered order of electrochemical and chemical steps, or indeed by fully "charging" the auxiliary sites, boosting electron transfer to/from the active site. Apart from acting as primary electron source or sink, active-site metal ions are also strong Lewis acids, favoring deprotonated intermediates. Product release is achieved by subsequent proton and electron transfer steps. The redox cycles of the auxiliary sites often do not involve protons, which may result in substantially uncoupled proton and electron transfer and a weak pH-dependence of the catalytic potential (vs. NHE). In surface electrochemistry, such uncoupled proton–electron transfer leading to deviations from Nernstian pH dependences would be considered anomalous.

Another important aspect, not addressed in detail here, is that the catalytic sites are embedded in a highly specific and adaptive peptide matrix that controls the dielectric and electronic properties and environment of the cofactor, selectively controls access of substrates and inhibitors, accumulates protons (proton channels, acid/base groups), electrons (auxiliary cofactors, aromatic amino acids, possibly specific tunneling pathways), and gaseous substrate (hydrophobic channels and cavities). The peptide is not a rigid scaffold, but may display redox-state-dependent fluctuations and (even large) movements. In the OEC, the local protein environment is an active part of the catalytic site, while in CcO the peptide is intimately involved in the coupling between the exothermic redox reaction and proton translocation. For reliable DFT calculations, a substantial part of at least the second coordination sphere (i.e., the peptide matrix) is therefore required, either directly in the QM core, or by applying QM/MM methods. Similarly, many of the current DFT calculations for metallic and oxidic systems have been based on a simplified description of the active site and its electrolyte surroundings. While such a description may be satisfactory for a qualitative approach, it is less ideal when detailed insight into the mechanism is desired. We should also not forget some of the more fundamental limitations of DFT, related primarily to the accurate description of electronic correlation, which plays an important role in oxides and oxygen. However, as new methods and functionals are being developed, and larger computational cells can be used, we may expect more accurate calculations that will continue to contribute significantly to the understanding of HER/HOR and ORR/OER.

In spite of the differences, in both the experimental approaches and the detailed mechanisms of catalysis, we believe that the comparison presented in this chapter provides a basis for the statement that synthetic and natural catalysts face the same catalytic bottlenecks in both the HER/HOR and the ORR/OER reactions, and have in essence been forced to opt for similar solutions largely imposed by thermodynamic restrictions. For an efficient HER/HOR catalyst, it is imperative to have a catalytically active site, whether this is a metal surface or a hydrogenase, that binds the atomic hydrogen (protonic, neutral, or hydride) just as strong as a hydrogen molecule. Only in this way will it be possible to reach a state of zero thermodynamic overvoltage. Very similar restrictions apply to ORR/OER, with the added complication that in this reaction more than one intermediate needs to be considered (i.e., OOH, OH, and O). This higher number of intermediates, and the relations that exist between their binding characteristics, renders it more difficult to reach a situation of zero thermodynamic overvoltage (i.e., to "line up" the equilibrium potentials of the individual steps in the mechanism). This observation provides a natural explanation for the fact that it is rather easy to catalyze the HER/HOR reaction, but much more difficult to catalyze OER/ORR near its thermodynamic potential. Interestingly, the best catalysts for ORR engineered by humans and nature, namely, $Pt_3Ni(111)$ and laccase, both seem to have the same potential-determining step: the reduction of OH to water so as to make the catalyst available for oxygen binding. Laccase seems to do this slightly more efficiently than $Pt_3Ni(111)$, at the expense of having a more complicated molecular structure and a much larger "footprint." For the OER, the existing analyses would also suggest that the oxygen evolving complex is more efficient than $RuO_2$, although both catalysts face similar potential determining steps in the sense that removing the strongly bound oxygen intermediate from the catalyst limits the overall efficiency.

From a chemist's perspective, the challenge lies in designing molecular catalysts that mimic the catalytic efficiency of enzymes and the atom efficiency of synthetic (human-made) catalysts. Molecular catalysis of HER/HOR and ORR/OER have been very active fields of research in inorganic chemistry [120,123–125], and we believe that many of the principles outlined in this chapter will also apply to these systems.

## REFERENCES


1. M. T. M. Koper, (Ed.), *Fuel Cell Catalysis: A Surface Science Approach*, Wiley-VCH, Hoboken, NJ, 2009.
2. N. M. Markovic and P. N. Ross Jr., *Surf. Sci. Rep.* **45**, 117 (2002).
3. M. Otani, I. Hamada, O. Sugino, Y. Morikawa, Y. Okamoto, and T. Ikeshoji, *J. Phys. Soc. Jpn.* **77**, 024802 (2008).
4. E. Skúlason, G. S. Karlberg, J. Rossmeisl, T. Bligaard, J. Greeley, H. Jonsson, and J. K. Nørskov, *Phys. Chem. Chem. Phys.* **9**, 3241–3250 (2007).
5. A. J. Bard and L. R. Faulkner, *Electrochemical Methods: Fundamentals and Applications*, Wiley, 2001.
6. J. K. Nørskov, J. Rossmeisl, A. Logadottir, L. Lundqvist, J. R. Kitchin, T. Bligaard, and H. Jónsson, *J. Phys. Chem. B* **108**, 17886 (2004).
7. M. T. M. Koper and R. A. van Santen, *J. Electroanal. Chem.* **472**, 126 (1999).
8. I. Hamada and Y. Morikawa, *J. Phys. Chem. C* **112**, 10889 (2008).
9. J. Rossmeisl, J. K. Nørskov, C. D. Taylor, M. J. Janik, and M. Neurock, *J. Phys. Chem. B* **110**, 21833 (2006).
10. M. T. M. Koper, *Faraday Discuss.* **140**, 11 (2009).
11. R. Parsons, *Trans. Faraday Soc.* **34**, 1053 (1958).
12. J. K. Nørskov, T. Bligaard, A. Logadottir, J. R. Kitchin, J. G. Chen, S. Pandelov, and U. Stimming, *J. Electrochem. Soc.* **152**, J23 (2005).
13. (a) C. Leger and P. Bertrand, *Chem. Rev.* **108**, 2379 (2008); (b) J. Hirst, *Biochim. Biophys. Acta* (1757) **225**, (2006); (c) F. A. Armstrong, *Curr. Opin. Chem. Biol.* **9**, 110 (2005); (d) K. A. Vincent, F. A. Armstrong, *Inorg. Chem.* **44**, 798 (2005); (e) L. J. C. Jeuken, *Biochim. Biophys. Acta* **1604**, 67 (2003); (f) F. A. Armstrong, R. Camba, H. A. Heering et al., *Discuss.* **116**, 191 (2000); (g) F. A. Armstrong, H. A. Heering, J. Hirst *J. Chem. Soc. Rev.* 169 (1997).
14. H. Gerischer, *Bull. Soc. Chim. Belg.* **67**, 506 (1958).
15. S. Trasatti, *J. Electroanal. Chem.* **39**, 163 (1977).
16. B. Hammer and J. K. Nørskov, *Nature* **376**, 238 (1995).
17. B. Hammer and J. K. Nørskov, *Adv. Catal.* **45**, 71 (2000).

18. J. K. Nørskov, T. Bligaard, B. Hvolbaek, F. Abild-Petersen, I. Chorkendorff, and C. H. Christensen, *Chem. Soc. Rev.* **37**, 2163 (2008).
19. E. Santos and W. Schmickler, *Angew. Chem. Int. Ed.* **46**, 8262 (2007).
20. N. M. Markovic, B. N. Grgur, and P. N. Ross, *J. Phys. Chem. B* **101**, 5405 (1997).
21. J. Barber, S. Morin, and B. E. Conway, *J. Electroanal. Chem.* **446**, 125 (1998).
22. O. A. Petrii and G. A. Tsirlina, *Electrochim. Acta* **39**, 1739 (1994).
23. R. Gomez, A. Fernandez-Vega, J. M. Feliu, and A. Aldaz, *J. Phys. Chem.* **97**, 4769 (1993).
24. E. Guerrini and S. Trasatti, Electrocatalysis in water electrolysis, in *Catalysis for Sustainable Energy Production*, P. Barbaro and C. Bianchini, eds., Wiley-VCH, Weinheim, 2009, p. 235.
25. J. H. Barber and B. E. Conway, *J. Electroanal. Chem.* **461**, 80 (1999).
26. G. Jerkiewicz, *Prog. Surf. Sci.* **57**, 137 (1998).
27. M. T. M. Koper and J. J. Lukkien, *J. Electroanal. Chem.* **485**, 161 (2000).
28. N. Garcia-Araez, J. J. Lukkien, M. T. M. Koper and J. M. Feliu, *J. Electroanal. Chem.* **588**, 1 (2006).
29. K. J. P. Schouten, M. J. T. C. van der Niet, and M. T. M. Koper, *Phys. Chem. Chem. Phys.*, submitted (2010).
30. B. E. Conway and G. Jerkiewicz, *Electrochim. Acta* **45**, 4075 (2000).
31. K. Kunimatsu, T. Senzaki, G. Samjeské, M. Tsushima and M. Osawa, *Electrochim. Acta* **52**, 5715 (2007).
32. R. J. Nichols and A. Bewick, *J. Electroanal. Chem.* **243**, 445 (1988).
33. W. Schmickler and S. Trasatti, *J. Electrochem. Soc.* **153**, L31 (2006).
34. L. A. Kibler, A. M. El-Aziz, R. Hoyer, and D. M. Kolb, *Angew. Chem. Int. Ed.* **44**, 2080 (2005).
35. J. Greeley, J. K. Nørskov, L. A. Kibler, A. M. El-Aziz, and D. M. Kolb, *Chem. Phys. Chem* **7**, 1032 (2006).
36. J. Greeley, T. Jaramillo, J. Bonde, I. Chorkendorff, and J. K. Nørskov, *Nature Mater.* **5**, 909 (2006).
37. F. A. Armstrong, N. A. Belsey, J. A. Cracknell, G. Goldet, A. Parkin, E. Reisner, K. A. Vincent and A. F. Wait, *Chem. Soc. Rev.* **38**, 36–51 (2009).
38. J. A. Cracknell, K. A. Vincent, and F. A. Armstrong, *Chem. Rev.* **108**, 2439–2461 (2008).
39. H. R. Pershad, J. L. C. Duff, H. A. Heering, E. C. Duin, S. P. J. Albracht, and F. A. Armstrong, *Biochemistry* **38**, 8992–8999 (1999).
40. J. C. Fontecilla-Camps, A. Volbeda, C. Cavazza, and Y. Nicolet, *Chem. Rev.* **107**, 4273–4303 (2007).
41. P. M. Vignais and B. Billoud, *Chem. Rev.* **107**, 4206–4272 (2007).
42. C. C. Page, C. C. Moser, and P. L. Dutton, *Curr. Opin. Chem. Biol.* **7**, 551–556 (2003).
43. F. A. Armstrong and J. C. Fontecilla-Camps, *Science* **321**, 498–499 (2008).
44. V. Artero and M. Fontecave, *Coord. Chem. Rev.* **249**, 1518–1535 (2005).
45. C. Sbraccia, F. Zipoli, R. Car, M. H. Cohen, G. C. Dismukes, and A. Selloni, *J. Phys. Chem. B* **112**, 13381–13390 (2008).
46. B. Hinneman, P. G. Moses, J. Bonde, K. P. Jørgensen, J. H. Nielsen, S. Horch, I. Chorkendorff, and J. K. Nørskov, *J. Am. Chem. Soc.* **127**, 5308 (2005).
47. H. Wu and M. B. Hall, *C. R. Chim.* **11**, 790–804 (2008).

48. S. O. N. Lill and P. E. M. Siegbahn, *Biochemistry* **48**, 1056–1066 (2009).
49. S. Shima, O. Pilak, S. Vogt, M. Schick, M. S. Stagni, W. Meyer-Klaucke, E. Warkentin, R. K. Thauer, and U. Ermler, *Science* **321**, 572–575 (2008).
50. S. Vogt, E. J. Lyon, S. Shima, and R. K. Thauer, *J. Biol Inorg. Chem.* **13**, 97–106 (2008).
51. D. D. M. Wayner and V. D. Parker, *Acc. Chem. Res.* **26**, 287 (1993).
52. V. D. Parker, *J. Am. Chem. Soc.* **114**, 7458 (1992).
53. C. Léger, A. K. Jones, W. Roseboom, S. P. J. Albracht, and F. A. Armstrong, *Biochemistry* **41**, 15736–15746 (2002).
54. A. Parkin, C. Cavazza, J. C. Fontecilla-Camps, and F. A. Armstrong, *J. Am. Chem. Soc.* **128**, 16808–16815 (2006).
55. A. K. Jones, S. E. Lamle, H. R. Pershad, K. A. Vincent, S. P. J. Albracht, and F. A. Armstrong, *J. Am. Chem. Soc.* **125**, 8505–8514 (2003).
56. A. K. Jones, E. Sillery, S. P. J. Albracht, and F. A. Armstrong, *Chem. Commun.* 866–867 (2002).
57. C. Léger, A. K. Jones, S. P. J. Albracht, and F. A. Armstrong, *J. Phys. Chem. B* **106**, 13058–13063 (2002).
58. A. A. Karyakin, S. V. Morozov, O. G. Voronin, N. A. Zorin, E. E. Karyakina, V. N. Fateyev, and S. Cosnier, *Angew. Chem. -Int. Ed.* **46**, 7244–7246 (2007).
59. S. Gottesfeld, in *Fuel Cell Catalysis: A Surface Science Approach*, M. T. M. Koper, ed., Wiley-VCH, Hoboken, NJ, 2009, p. 1.
60. Y. Xu, M. Shao, M. Mavrikakis, and R. R. Adzic, in *Fuel Cell Catalysis: A Surface Science Approach*, M. T. M. Koper, ed., Wiley-VCH, Hoboken, NJ, 2009, p. 271.
61. P. Vassilev and M. T. M. Koper, *J. Phys. Chem. C* **111**, 2607 (2007).
62. M. J. Janik, C. D. Taylor, and M. Neurock, *J. Electrochem. Soc.* **156**, B126 (2009).
63. M. P. Hyman and J. W. Medlin, *J. Phys. Chem. B* **109**, 6304 (2005).
64. A. Panchenko, M. T. M. Koper, T. E. Shubina, S. J. Mitchell, and E. Roduner, *J. Electrochem. Soc.* **151**, A2016 (2004).
65. J. Rossmeisl, A. Logadottir, and J. K. Nørskov, *Chem. Phys.* **319**, 178 (2005).
66. F. Abild-Petersen, J. Greeley, F. Studt, P. G. Moses, J. Rossmeisl, T. Munter, T. Bligaard, and J. K. Nørskov, *Phys. Rev. Lett.* **99**, 016105 (2007).
67. J. Rossmeisl, J. Greeley, and G. S. Karlberg, in *in Fuel Cell Catalysis: A Surface Science Approach*, M. T. M. Koper, ed., Wiley-VCH, Hoboken, NJ, 2009, p. 57.
68. V. Stamenkovic, B. S. Mun, K. J. J. Mayrhofer, P. N. Ross, N. M. Markovic, J. Rossmeisl, J. Greeley, and J. K. Nørskov, *Angew. Chem. Int. Ed.* **45**, 2897 (2006).
69. V. Stamenkovic, B. Fowler, B. S. Mun, G. Wang, P. N. Ross, C. A. Lucas, and N. M. Markovic, *Science* **315**, 493 (2007).
70. V. Stamenkovic, B. S. Mun, M. Arenz, K. J. J. Mayrhofer, C. A. Lucas, G. F. Wang, P. N. Ross, and N. M. Markovic, *Nature Mater.* **6**, 241 (2007).
71. J. Zhang, M. B. Vukmirovic, Y. Xu, M. Mavrikakis, and R. R. Adzic, *Angew. Chem. Int. Ed.* **44**, 2132 (2005).
72. M. H. Shao, P. Liu, and R. R. Adzic, *J. Am. Chem. Soc.* **128**, 7408 (2006).
73. R. R. Adzic, in *Electrocatalysis*, J. Lipkowski and P. N. Ross, eds., Wiley-VCH, New York, 1999.
74. J. Kim and A. A. Gewirth, *J. Phys. Chem. B* **110**, 2565 (2006).

75. J. A. Wang, N. M. Markovic, and R. R. Adzic, *J. Phys. Chem. B* **108**, 4127 (2004).
76. (a) I. Willner, Y. M. Yan, B. Willner, and R. Tel-Vered, *Fuel Cells* **9**, 7–24 (2009); (b) S. C. Barton, J. Gallaway, P. Atanassov, *Chem. Rev.* **104**, 4867–4886 (2004); (c) K. A. Vincent, J. A. Cracknell, J. R. Clark et al., *Chem. Commun.* 5033–5035 (2006); (d) K. A. Vincent, J. A. Cracknell, O. Lenz et al., *Proc. Natl. Acad. Sci. USA* **102**, 16951–16954 (2005); (e) S. Shleev, J. Tkac, A. Christenson, *Biosens. Bioelectron.* **20**, 2517–2554 (2005).
77. V. Morozova, G. P. Shumakovich, M. A. Gorbacheva, S. V. Shleev, and A. I. Yaropolov, *Biochemis.* (Moscow) **72**, 1136–1150 (2007).
78. P. Baldrian, *FEMS Microbiol. Rev.* **30**, 215–242 (2006).
79. E. I. Solomon, U. M. Sundaram, and T. E. Machonkin, *Chem. Rev.* **96**, 2563 (1996).
80. E. I. Solomon, A. J. Augustine, and J. Yoon, *Dalton Trans.* 3921–3932 (2008).
81. E. Vayner, H. Schweiger, and A. B. Anderson, *J. Electroanal. Chem.* **607**, 90–100 (2007).
82. (a) M. Tsubaki, H. Hori, and T. Mogi, *J. Inorg. Biochem.* **82**, 19–25 (2000); (b) S. Junemann, *Biochim. Biophys. Acta* **1321**, 107–275 (1997).
83. (a) I. Belevich and M. I. Verkhovsky, *Antioxid. Redox Signal.* **10**, 1–29 (2008); (b) M. Wikstrom, M. I. Verkhovsky, *Biochim. Biophys. Acta* **1767** 1200–1214 (2007); (c) M. Wikstrom, M. I. Verkhovsky, *Biochim. Biophys. Acta* **1757**, 1047 (2006); (d) K. Faxen, G. Gilderson, P. Adelroth, P. Brzezinski, *Nature* **437**, 286 (2005); (e) S. Papa, N. Capitanio, G. Capitanio, *Biochim. Biophys. Acta* **1655**, 353 (2004); (f) O. M. H. Richter, B. Ludwig, *Rev. Physiol. Biochem. Pharmacol.* **147**, 47–74 (2003); (g) M. Ruitenberg, A. Kannt, E. Bamberg, K. Fendler, H. Michel, *Nature* **417** 99 (2002); (h) M. M. Pereira, M. Santana, M. Teixeira, *Biochim. Biophys. Acta* **1505** 185–208 (2001); (i) S. Han, S. Takahashi, D. L. J. Rousseau, *Biol. Chem.* **275**, 1910 (2000); (j) R. B. Gennis, *Proc. Natl. Acad. Sci. USA* **95**, 12747 (1998).
84. P. Mitchell, Nobel Prize lecture (Chemistry, 1978), *Science* **206**, 1148 (1979). Nature 191, (1961) 144.
85. F. G. M. Wiertz, O. M. H. Richter, B. Ludwig, and S. de Vries, *J. Biol. Chem.* **282**, 31580 (2008).
86. J. A. Fee, D. A. Case, and L. J. Noodleman, *Am. Chem. Soc.* **130**, 15002–15021 (2008).
87. P. E. M. Siegbahn and M. R. A. Blomberg, *J. Phys. Chem. A* **112**, 12772–12780 (2008).
88. C. H. Kjaergaard, J. Rossmeisl, and J. K. Nørskov, *Inorg Chem.* (in press).
89. M. Kaukonen, *J. Phys. Chem. B* **111**, 12543 (2007).
90. M. R. A. Blomberg, and P. E. J. Siegbahn, *Comput. Chem.* **27**, 1373 (2006).
91. (a) S. de Vries, *Biochim. Biophys. Acta* **1777**, 925 (2008); (b) V. L. I. Kaila, M. P. Johansson, D. Sundhold, L. Laakkonen, M. Wikstrom, *Biochim. Biophys. Acta* **1787**, 221 (2009).
92. E. A. Gorbikova, K. Vuorilehto, M. Wikstrom, and M. I. Verkhovsky, *Biochemistry* **45**, 5641–5649 (2006).
93. P. Hellwig, S. Grzybek, J. Behr, B. Ludwig, H. Michel, and W. Mantele, *Biochemistry* **38**, 1685–1694 (1999).
94. (a) S. C. Barton, H. H. Kim, G. Binyamin, Y. C. Zhang, and A. Heller, *J. Am. Chem. Soc.* **123**, 5802–5803 (2001); (b) S. C. Barton, M. Pickard, R. Vazquez-Duhalt, A. Heller, *Biosens. Bioelectron.* **17**, 1071–1074 (2002); (c) V. Soukharev, N. Mano, A. Heller, *J. Am. Chem. Soc.* **126**, 8368–8369 (2004); (d) N. Mano, V. Soukharev, A. Heller, *J. Phys. Chem. B* **110**, 11180–11187 (2006).

95. (a) H. A. Heering, F. G. M. Wiertz, C. Dekker, and S. de Vries, *J.Am. Chem. Soc.* **126**, 11103–11112 (2004); (b) C. Léger, F. Lederer, B. Guigliarelli, P. Bertrand, *J. Am. Chem. Soc.* **128**, 180–187 (2006); (c) C. Léger, S. J. Elliott, K. R Hoke et al., *Biochemistry* **42** 8653–8662 (2003).

96. C. F. Blanford, C. E. Foster, R. S. Heath, and F. A. Armstrong, *Faraday Discuss.* **140**, 319–335 (2008).

97. F. Xu, *J. Biol. Chem.* **272**, 924–928 (1997).

98. S. Trasatti, *Electrochim. Acta* **29**, 1503 (1984).

99. J. Rossmeisl, Z. -W. Qu, H. Zhu, G. -J. Kroes, and J. K. Nørskov, *J. Electroanal. Chem.* **607**, 83 (2007).

100. A. Valdes, Z. -W. Qu, G. J. Kroes, J. Rossmeisl, and J. K. Nørskov, *J. Phys. Chem. C* **112**, 9872 (2008).

101. G. Lodi, E. Sivieri, A. De Battisti, and S. Trasatti, *J. Appl. Electrochem.* **8**, 135 (1978).

102. P. Castelli, S. Trasatti, F. H. Pollak, and W. E. O'Grady, *J. Electroanal. Chem.* **210**, 189 (1986).

103. A. Marshall, M. Tsypkin, B. Børresen, G. Hagen, and R. Tunold, *J. New Mater. Electrochem. Syst.* **7**, 197 (2004).

104. M. Wohlfahrt-Mehrens and J. Heitbaum, *J. Electroanal. Chem.* **237**, 251 (1987).

105. S. Fierro, T. Nagel, H. Baltruschat, and C. Comninellis, *Electrochem. Commun.* **9**, 1969 (2007).

106. T. Nakagawa, C. A. Beasley, and R. W. Murray, *J. Phys. Chem. C.* (in press).

107. M. Yagi, E. Tomita, S. Sakita, T. Kuwabara, and K. Nagai, *J. Phys. Chem. B* **109**, 21489 (2005).

108. T. Kuwabara, E. Tomita, S. Sakita, D. Hasegawa, K. Sone, and M. Yagi, *J. Phys. Chem. C* **112**, 3774 (2008).

109. J. Jirkovsky, M. Makarova, and P. Krtil, *Electrochem. Commun.* **8**, 1417 (2006).

110. (a) B. Conlan, *Photosynth. Res.* **98**, 687–700 (2008); (b) I. L. McConnell, *Photosynth. Res.* **98**, 261–276 (2008); (c) F. A. Armstrong, *Phil. Trans. R. Soc. B* **363**, 1263–1270 (2008).

111. G. Renger and T. Renger, *Photosynth. Res.* **98**, 53–80 (2008).

112. H. Suzuki, M. Sugiura, and T. J. Noguchi, *Am. Chem. Soc.* **131**, 7849–7857 (2009).

113. P. E. M. Siegbahn, *Phil. Trans. R. Soc. B* **363**, 1221–1228 (2008).

114. E. M. Sproviero, J. P. McEvoy, J. A. Gascon, G. W. Brudvig, and V. S. Batista, *Photosynth. Res.* **97**, 91–114 (2008).

115. (a) K. N. Ferreira, T. M. Iverson, K. Maghlaoui, J. Barber, and S. Iwata, *Science* **303** 1831–1838 (2004); (b) J. Biesiadka, B. Loll, J. Kern, K. D. Irrgang, A. Zouni, *Phys. Chem. Chem. Pys.* **6**, 4733–4736 (2004); (c) B. Loll, J. Kern, W. Saenger, A. Zouni, J. Biesiadka, *Nature* **438**, 1040–1044 (2005).

116. P. E. M. Siegbahn and F. Himo, *J. Biol. Inorg. Chem.* **14**, 643–651 (2009).

117. P. E. M. Siegbahn, *Chem. Eur. J.* **14**, 8290 (2008).

118. E. M. Sproviero, J. A. Gascon, J. P. McEvoy, G. W. Brudvig, and V. S. J. Batista, *Am. Chem. Soc.* **130**, 3428–3442 (2008).

119. J. Messinger, *Phys. Chem. Chem. Phys.* **6**, 4764–4771 (2004).

120. F. Liu, J. J. Concepcion, J. W. Jurss, T. Cardolaccia, J. L. Templeton, and T. J. Meyer, *Inorg. Chem.* **47**, 1727–1752 (2008).

121. V. L. Pecoraro, M. J. Baldwin, M. T. Caudle, W. Y. Hsieh, and N. A. Law, *Pure Appl. Chem.* **70**, 925–929 (1998).

122. J. Rossmeisel, K. Dimitrievski, P. Siegbahn, and J. K. Nørskov, *J. Phys. Chem. C* **111**, 18821–18823 (2007).

123. V. Artero and M. Fontecave, *Coord. Chem. Rev.* **240**, 1518–1535 (2005).

124. J. M. Savéant, *Chem. Rev.* **108**, 234–237 (2008).

125. R. Boulatov, in *Fuel Cell Catalysis: A Surface Science Approach*, M. T. M. Koper, ed., Wiley-VCH, Hoboken, NJ, 2009, p. 637.

CHAPTER 3

# Design of Palladium-Based Alloy Electrocatalysts for Hydrogen Oxidation Reaction in Fuel Cells

SUNG JONG YOO

Fuel Cell Center, Korea Institute of Science and Technology, Seoul, Korea

YUNG-EUN SUNG

World Class University Program of Chemical Convergence for Energy and Environment, School of Chemical and Biological Engineering, Seoul National University, Seoul, Korea

Fuel cells, which directly convert chemical energy into electrical energy, are promising and worthy candidates for high-efficiency and low-emission-energy conversion devices in mobile, stationary, and portable application sectors and are expected to be one of the major clean energy sources in the near future. Despite important advancements in fuel cell investigation, there is the major impediment to the large-scale production and commercialization of such applications, as follows: the decreased catalytic activity by long-term operation, the high cost the electrocatalyst (Pt) required, and the deficiency of kinetic mechanisms on the electrocatalytic surface, which are the urgent issues to be addressed since they determine the efficiency and the cost of this energy source. In this study, a new approach was developed for designing electrocatalysts for hydrogen oxidation reaction in fuel cells. These electrocatalysts consist of Pd (adsorbed spontaneously $H_2$ gas)-based alloy materials. The Pd-based alloy electrodes were fabricated by dual-gun radiofrequency magnetron sputtering, and the surface of these electrodes was very smooth.

It was found that the electronic properties of Pd-based electrodes could be fine-tuned by the electronic and strain effects introduced by the surface composition and the lateral effects of the hydrogen atoms. The role of alloy was found reflected in a volcano plot of the activity for the hydrogen oxidation reaction as a function of their measured *d*-band centers or *d*-band vacancies. The activity of Pd-based electrocatalyst for

*Fuel Cell Science: Theory, Fundamentals, and Biocatalysis*,
Edited by Andrzej Wieckowski and Jens K. Nørskov

hydrogen oxidation reaction was up to 5 times (Pt–Pd) or 1.5 times (Ir–Pd) higher than that for the pure Pt electrocatalyst. The activity is enhanced mainly by increased bond strength of Pt–H (or Ir–H) and the d-band vacancy of Pt (or Ir) induced by $PdH_x$. Fuel cell tests showed a very good performance of the Pt–Pd electrocatalysts. Our results demonstrated a viable way of designing the electrocatalysts that could successfully alleviate two issues facing commercialization of the fuel cell—the cost of electrocatalysts and their efficiency.

## 3.1 INTRODUCTION

The electrocatalytic properties of bimetallic surfaces on single-crystal metal surface have been extensively studied in ultra-high-vacuum systems [1–4]. In many cases, the formation of the bimetallic surfaces significantly changed the geometric and electronic properties of the primary metal, and pronounced differences were observed in the electrocatalytic reactivity of some transition metal monolayers on various primary metal substrates [1].

Nørskov advanced a new theory to elucidate simply the adsorption properties of metal atoms and small molecules on transition metal surfaces [5]. According to this theory, instead of the density of states at Fermi level, the *d*-band center (average energy of the whole *d* band) is used to elucidate the catalytic reactivity of the reaction sites on the metal surface. Density function theory (DFT) calculations showed that the binding energies and catalytic reactivity of small atom or molecule adsorption on strained surfaces correlate well with the position of the *d*-band center of surface metal atoms. The surface bond lengths changes in orbital overlap as a result of the strain effect and modifies the electronic structure of the metal. According to this argument, when the surface atoms are subjected to tensile strain, the *d*-orbital overlap is decreased, resulting in a sharpening of the *d* band and an upshift of the *d*-band center, which indicates stronger adsorption energy. On the other hand, when the surface atoms are under compressive strain, the *d*-orbital overlaps are increased, resulting in a broadening of the *d* band and a lowering of its average energy. Consequently the presence of other metals around a metal atom changes its electronic environment and modifies its electronic structure and its chemical properties. These results show the theoretical possibility to design the outstanding electrocatalyst and agree with data from many experimental researches [6–8].

Mukerjee et al. [9,10] have proposed that the bimetallic catalyst particles exist as an alloy with a Pt skin where the electronic properties of surface Pt atoms are modified by the subsurface alloy, reducing the formation of surface Pt–$OH_{ads}$, which has been found on each of the low-index Pt surfaces [11]. Existence of this Pt skin has been theoretically verified by calculation of segregation energies [12], where several alloys of fourth-row transition metals with precious metals, including Pd, are expected to have the same structure. Balbuena et al. [13] suggested that alloys of Pt with Co and Cr are optimal for the dissociative adsorption of molecular oxygen, leading to enhanced catalytic activity. Also, Pt alloys with limited amounts of a nonnoble transition metal may show an increase in performance by tuning the interatomic distance as well as increasing the electron density [14] at the catalytic sites. This agrees well with the experimental work of both Mukerjee et al. [9,10] and Stamenkovic et al. [15], who have shown that Pt alloys with Cr and Co have the highest activity and the activity of the $Pt_3Co$ catalysts was at least twice that of pure Pt. While it is clear that this approach has produced a greater understanding of the oxygen reduction reaction (ORR) from a fundamental perspective as well as a handful of materials that have improved electrocatalytic behavior and are less expensive than pure platinum catalysts, the costs associated with these materials remain prohibitive for a commercial-scale fuel cell technology. More recently, Stimming et al. found the enhancement of the hydrogen evolution reaction (HER) on Pd/Au(111) electrode surfaces [16]. This allowed us to determine critical ensembles for H adsorption and evolution on Pd/Au (111) electrode surfaces, which show distinct differences in their chemical properties as compared with bulk Au(111) or Pd monolayer-covered Au(111) electrodes. The enhanced hydrogen evolution was explained by the coverage effect and proton spillover effect. Marković et al. first reported an enhanced hydrogen oxidation reaction (HOR)/HER at an epitaxial Pd layer on Pt(111) electrode by elucidating the energetic effect and surface defects [17].

Therefore, next-generation HOR/ORR electrocatalysts should either consist of moderate to low-cost materials that exhibit a catalytic activity at least equal to that of platinum or "zero cost" materials [18]. More recently, several researchers have identified several alternative electrocatalysts in various forms including metal alloys [19,20], metal oxides [21–23] and both organic and inorganic and metallic complexes [24,25]. However, these anodes and cathodes generally exhibit a lower activity than Pt. Three groups have identified a family of Pd-based alloy electrocatalysts that have shown great promise for the HOR and ORR [26–28]. A significant amount of work remains to completely characterize these materials, including clarifying the role of composition on the catalytic activity, testing performance stability and elucidation of the mechanism.

In this chapter, the kinetics of the HOR were studied on the Pd-based electrocatalysts (Pd–Pt, Pd–Ir). The correlation of the volcano types were achieved for the first time between the catalytic activity (represented by exchange current density) and the *d*-band center measured by high-resolution X-ray photoemission spectroscopy (HR-XPS) or *d*-band vacancy measured by in situ X-ray absorption near-edge spectroscopy (XANES). The most promising electrocatalyst-based Pd was affected by Pd spontaneously absorbed hydrogen.

## 3.2 PROMOTIONAL EFFECT OF Pd ON HYDROGEN OXIDATION REACTION AT A Pt–Pd ALLOY ELECTRODE

Polymer electrolyte fuel cells (PEFCs), which directly convert chemical energy into electrical energy, are promising and worthy candidates for high-efficiency and low-emission energy conversion devices in mobile, stationary, and portable application sectors [29–31]. Despite important advancement in PEFCs investigation, there is the major impediment to the large-scale production and commercialization of such applications, as follows: the decreased catalytic activity by long-term operation and the high cost of the solid polymer electrolyte (Nafion) and of the electrocatalyst (Pt) required [32,33]. Unequal to the solid polymer electrolyte, the market prices of noble metal electrocatalysts for mass production will be limited when the exclusive use of electrocatalysts is caused by the expansion of the PEFC market. Thus, the reduction of electrocatalysts loading without sacrificing performance or the use of alternative bimetallic Pt-based electrocatalysts with the more improving performance is required [34,35].

The electrochemical oxidation of molecular hydrogen has been the subject of extensive investigations over the twentieth century. This is largely because the hydrogen oxidation reaction is important to energy conversion, specifically in the area of fuel cells where it has practical utility as an anode reaction [36,37]. Much effort has been devoted to understanding the kinetics and mechanism of this reaction in an attempt to search for low-cost and more efficient electrocatalysts for hydrogen oxidation, which would have widespread application in fuel cells and many other electrochemical applications. For instance, Pt has received the greatest attention as a result of the high catalytic hydrogen oxidation rate that can be obtained at an unoxidized Pt surface. The other noble metals of the Pt family, which include Rh, Pd, Ir, and Ru, have been studied to a lesser degree but are frequently used as catalyst components [38]. Hydrogen oxidation at electrodes requires a catalytic surface on which H atoms are adsorbed and is observed only with metals such as Pt, Pd, Rh, Ru, and Ir. On metal surfaces, it is now generally accepted that hydrogen oxidation at room temperature is a fast process that proceeds by initial adsorption of molecular hydrogen, which involves either slow dissociation of hydrogen molecules into atoms or dissociation into the ion and atom followed by a charge transfer step. The overall process can be divided into five steps [39]:

1. The transfer of the $H_2$ molecules to the platinum electrode via gas diffusion
2. The adsorption of hydrogen molecules from the electrolyte onto the electrode surface through the electrolytical double layer
3. The dissociative adsorption of hydrogen on platinum surface via Tafel reaction (or recombination reaction)

$$\mathrm{Pt{-}Pt} + \mathrm{H_2} \leftrightarrow \mathrm{Pt{-}H} + \mathrm{Pt{-}H} \tag{3.1}$$

4. The ionization of H atom giving one electron to the electrode and a hydronium ion the solution, leaving an empty platinum site through the Volmer reaction (or charge transfer reaction)

$$\mathrm{Pt{-}H} + \mathrm{H_2O} \leftrightarrow \mathrm{Pt} + \mathrm{H_3O^+} + \mathrm{e^-} \tag{3.2}$$

or by direct reaction of hydrogen molecules through Heyrovsky reaction (or ion-plus-atom reaction)

$$\mathrm{Pt} + \mathrm{H_2} + \mathrm{H_2O} \leftrightarrow \mathrm{Pt{-}H} + \mathrm{H_3O^+} + \mathrm{e^-} \tag{3.3}$$

5. The transport of the hydronium ions from the phase boundary to the electrolyte solution phase

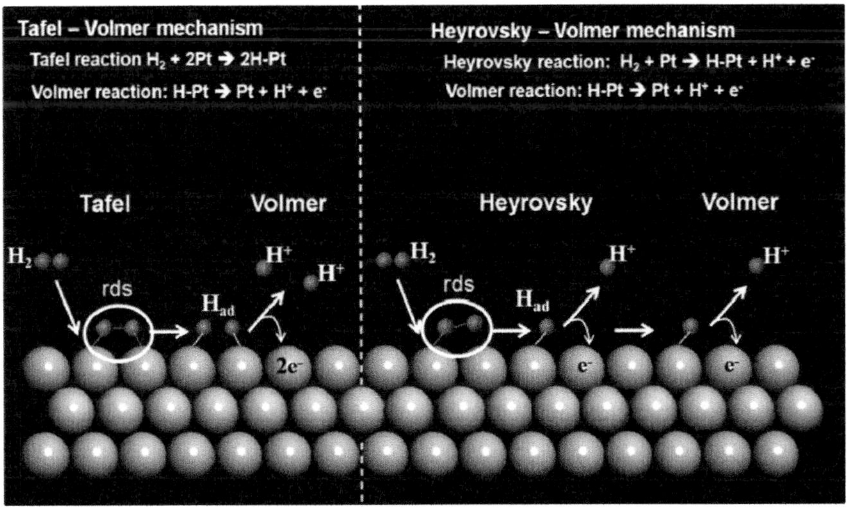


**FIGURE 3.1** Elementary reaction steps for HOR.

Therefore, the overall charge reaction (or oxidation) can be expressed as

$$H_2 + 2H_2O \leftrightarrow 2H_3O^+ + 2e^- \tag{3.4}$$

where Pt–H represents atomic hydrogen that is chemisorbed to a metal surface. Reactions (3.1)–(3.3) are historically referred to as the *Tafel–Volmer* [(3.1) and (3.2)] and *Heyrovsky–Volmer* [(3.2) and (3.3)] sequences, respectively (see Fig. 3.1), where the Tafel and Heyrovsky steps are generally believed to be the rate-determining steps (RDSs). From a steady-state current–potential curve, a plot of log $i$ versus overpotential leads to a Tafel plot from which a slope of $(1 - a)nF/RT$ and a value of the transfer coefficient $a$ are generally calculated. The exchange current is obtained from an extrapolation to an intercept of log $i_o$ [40]. In a number of previous studies of platinum concerning the hydrogen oxidation reaction, a Tafel slope of 30 mV was found, lending support to a Tafel–Volmer mechanism, although this generated a great deal of controversy that still continues.

There are more than enough signs that Pd has many of the desired electrocatalytic properties with respect to HOR/HER, since the electron transfer from the Pd surface into the antibonding orbital of the hydrogen molecule play a important role to break the hydrogen bonds and this process is associated the lower activation energy [41,42]. It was proposed that the major factor of catalytic activity was the electronic structure of transition metal [43]. More recent studies have been concerned with the correlation between electronic structure and catalytic activity of Pd [44]. The adsorption energy of hydrogen is in proportion to hydrogen–metal bond strength and relative level of $d$-band center [45]. It is reasonable to expect improved activity of the Pd from promoter or Pd-based binary electrocatalyst system [46].

In this chapter, we investigated the hydrogen oxidation at PtPd alloy electrodes with varying composition in comparison with pure Pt by using an in situ X-ray absorption near-edge structure (XANES) system for gas-phase experiments to analyze the electronic states. We have focused on the electronic structures of the PtPd alloys in the hydrogen gas ambient for the HOR. We have found a correlation between the *d*-band vacancy of Pt induced by alloying Pt–$PdH_x$ and the HOR activity. It is clearly indicated that the improved HOR activity at Pt–$PdH_x$ alloy electrodes is brought about by a stronger H-bond strength than at pure Pt electrode due to the modification of the electronic structure.

### 3.2.1 Electrochemical Characterization Using a Rotating Disk Electrode

The HOR on the Pt-based electrocatalysts in the aqueous electrolyte is faster than other electrochemical reactions [47]. Therefore, the concentrations of $H_2$ gas dissolved in the aqueous electrolyte are exhausted near the electrode, which means that the reaction is no longer kinetically controlled. For this reason, the rotating-disk electrode (RDE) technique has conventionally been used to investigate the kinetics of reactions such as the electrocatalysis of the HOR in order to maintain steady-state conditions. We studied the kinetics of the HOR on Pt–Pd alloy electrodes with various compositions and compared the kinetics of Pt–Pd alloys and pure Pt. The polarization curves for the RDEs, $Pt_{100}Pd_0$, $Pt_{91}Pd_9$, $Pt_{83}Pd_{17}$, $Pt_{72}Pd_{28}$, $Pt_{60}Pd_{40}$, and $Pt_{48}Pd_{52}$, in sulfuric acid saturated with $H_2$ gas at 276 K, obtained using a sweep rate of 20 mV/s in the range from 0.0 to 0.3 V at different rotational speeds from 1000 to 3000 rpm are displayed in Figure 3.2. The current density, with respect to the activation and diffusion, is described by the following equation:

$$1/i = 1/i_{\mathrm{k}} + 1/i_{l,c} \tag{3.5}$$

Here, $i_k$ is the kinetic current and $i_{l,c}$ is the current of diffusion in the electrolyte. The diffusion current, that is, the current that is determined by the rate of supply of dissolved hydrogen to the surface, was calculated with the *Levich* equation [48]:

$$i_{l,c} = 0.62nFAD_0^{2/3}\omega^{1/2}\nu^{-1/6}C_0^* = BC_0^*\omega^{1/2} \tag{3.6}$$

Here, $n$ is the number of electrons, $F$ is Faraday's constant, $D_0$ is the diffusion coefficient for hydrogen in the electrolyte, $\nu$ is the kinematic viscosity of the electrolyte, $C_0^*$ is the hydrogen concentration in the electrolyte, and $\omega$ is the electrode rotation rate. $C_0^*$, $\nu$, and $D_0$ calculations were based on Henry's law [49], the method reported in Reference 34, and the Stokes–Einstein equation [5,50], respectively. The values of $C_0^*$, $\nu$, and $D_0$, calculated as different temperatures, are summarized in Table 3.1. Then, given (3.6), formula (3.5) assumes the form of the following equation:

$$1/i = 1/i_k + 1/i_{l,c} = 1/i_k + 1/(0.62nFAD_0^{2/3}\omega^{1/2}\nu^{-1/6}C_0^*) = 1/i_{\mathrm{k}} + 1/BC_0^*\omega^{1/2} \tag{3.7}$$

where the calculated value of $BC_0^*$ is 4.72 mA $cm^{-2}$ $rpm^{-0.5}$ for a two-electron exchange reaction ($n = 2$) at 276 K. From the slopes of the *Koutecky–Levich* plots on a

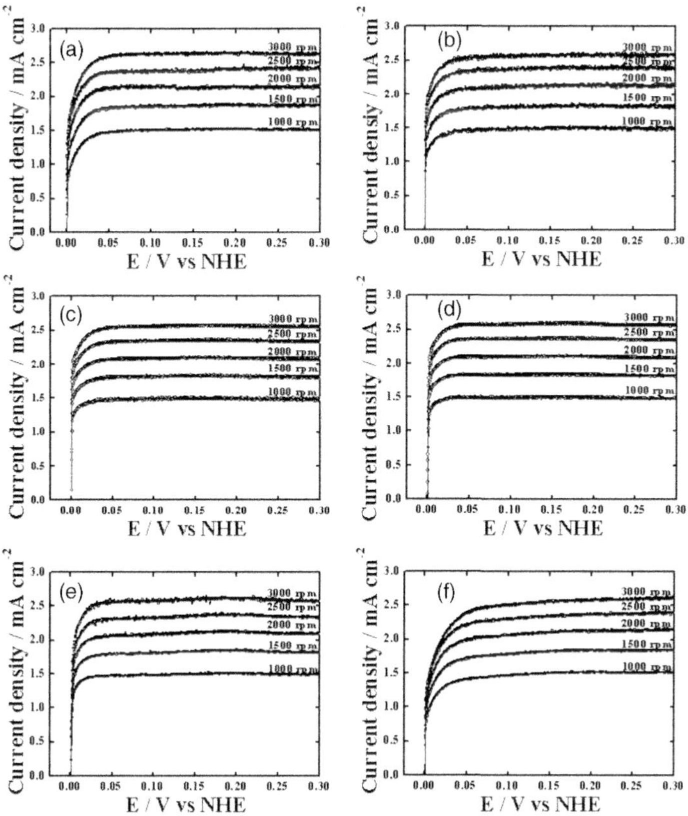


**FIGURE 3.2** Hydrogen oxidation current densities during positive-going sweeps (20 mV/s) on (a) $Pt_{100}Pd_0$, (b) $Pt_{91}Pd_9$, (c) $Pt_{83}Pd_{17}$, (d) $Pt_{72}Pd_{28}$, (e) $Pt_{60}Pd_{40}$, and (f) $Pt_{48}Pd_{52}$ RDEs in 0.5 M $H_2SO_4$ saturated with $H_2$ at 276 K [46].

pure Pt electrode and Pt–Pd alloy electrodes, the constant $BC_0^*$ values of 4.80 (+1.71% error) ~ 4.73 (+0.22% error) mA $cm^{-2}$ $rpm^{-0.5}$ were obtained, which agree relatively well with the calculated value for the two-electron oxidation of $H_2$ (see Table 3.2).

**TABLE 3.1 Errors between Experimental and Theoretical Values of $BC_0$ at 276 K**

| Electrode | $BC_0$ ($10^{-2}$ mA $cm^{-2}$ $rpm^{-0.5}$) | Error(%) |
|---|---|---|
| $Pt_{100}Pd_0$ | 4.80 | +1.71 |
| $Pt_{91}Pd_9$ | 4.73 | +0.22 |
| $Pt_{83}Pd_{17}$ | 4.75 | +0.65 |
| $Pt_{72}Pd_{28}$ | 4.78 | +1.28 |
| $Pt_{60}Pd_{40}$ | 4.75 | +0.65 |
| $Pt_{48}Pd_{52}$ | 4.76 | +0.86 |

*Source:* Reference 46.

**TABLE 3.2 Hydrogen Concentration ($C_{0^*}$), Kinematic Viscosity (ν),[a] Diffusion Coefficient (D),[b] and Levich Constant (*B*) over Full Range of Temperatures Employed in This Study**

| $T$ (K) | $C_0$ ($10^{-3}$ M) | ν ($10^{-2}$ cm$^2$ s$^{-1}$) | $D$ ($10^{-5}$ cm$^2$ s$^{-1}$) | $BC_0$ ($10^{-2}$ mA cm$^{-2}$ rpm$^{-0.5}$) |
|---|---|---|---|---|
| 276 | 8.80 | 1.60 | 1.83 | 4.719394 |
| 283 | 8.38 | 1.33 | 2.27 | 5.341602 |
| 293 | 7.82 | 0.97 | 3.23 | 6.648876 |
| 303 | 7.28 | 0.65 | 4.99 | 8.822094 |
| 313 | 6.72 | 0.39 | 8.62 | 12.79132 |
| 333 | 5.43 | 0.08 | 48.3 | 42.93944 |

[a] P. W. Atkins, *Physical Chemistry*, 5th ed., Oxford Univ. Press, 1994; E. W. Washburn, ed., *International Crirical Tables of Numerical Data, Physics, Chemistry and Technology*, McGraw-Hill, New York, 1928, Vol. 3.

[b] Dη/T = constant.

[c] *Source:* Reference 46.

### 3.2.2 Kinetics of Hydrogen Oxidation Reaction on Pt–Pd Alloy Electrodes

The mass-transfer-corrected Tafel plots of the Pt–Pd alloy electrodes are shown in Figure 3.3. The slope of the Tafel plot is useful to obtain mechanistic information about HOR [51,52]. The HOR mechanism on a polycrystalline Pt electrode in aqueous electrolyte is considered. This is usually assumed to proceed by the initial adsorption of hydrogen. This involves the electrochemical adsorption step, in which hydrogen molecule is dissociated [rate-determining step (RDS)], followed by the

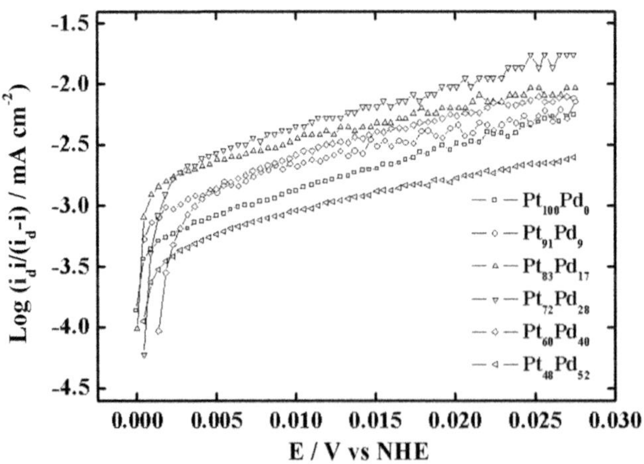


**FIGURE 3.3** Mass-transfer-corrected Tafel plots for hydrogen oxidation on Pt–Pd alloy electrodes with various compositions in a solution of 0.5 M $H_2SO_4$. These plots were obtained from Figure 3.2 [46].

**TABLE 3.3 Tafel Slopes as a Function of Degree of Alloying of Pt–Pd at 276 K**

| Electrode | Tafel Slope/(mV/dec) |
|---|---|
| $Pt_{100}Pd_0$ | 24 |
| $Pt_{91}Pd_9$ | 31 |
| $Pt_{83}Pd_{17}$ | 32 |
| $Pt_{72}Pd_{28}$ | 27 |
| $Pt_{60}Pd_{40}$ | 22 |
| $Pt_{48}Pd_{52}$ | 30 |

*Source:* Reference 46.

adsorbed hydrogen atom discharge step (fast charge transfer step), as described in Equations (3.1)–(3.3).

In the potential range of $0.05\,V > E > 0.2\,V$, the Tafel slopes for the Pt–Pd alloy electrodes were ~30 mV/decade, as shown in Table 3.3, which is coincident with that (28 mV/dec) of Pt reported by Markovic et al. [53]. From these results it is reasonable to propose that HOR takes place on Pt–Pd via the Tafel–Volmer (or Heyrovsky–Volmer) mechanism. So far, although there remain a number of open questions with regard to RDS of the electrochemical oxidation of hydrogen in the electrolyte solution, this indicates that the RDS of the Pt–Pd alloy electrodes is the same as that of a pure Pt electrode.

Figure 3.4 shows the Arrhenius plots, $\log(i_0)$ versus $1/T$, for the Pt–Pd alloy electrodes. These plots yield straight lines where the slope is equal to $-E_a/(2.3R)$. The HOR activation energy on $Pt_{72}Pd_{28}$ was much higher than that on Pt. A significant increase in the kinetic activation energy of the HOR was also observed when the temperature was increased, and the lowest activation energy (9.90 kJ/mol) was

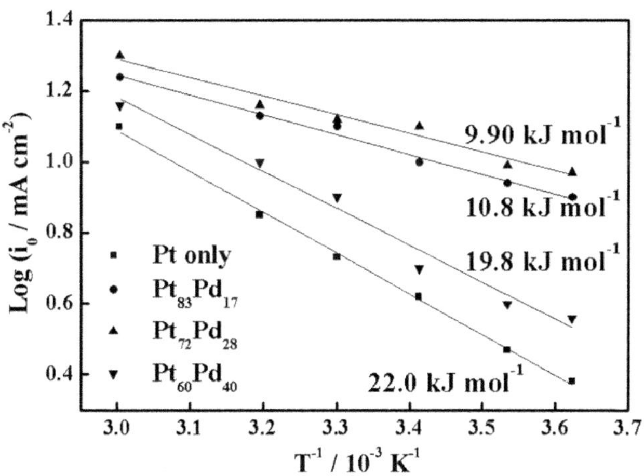


**FIGURE 3.4** Arrhenius plots of the exchange current densities ($i_0$) for the Pt–Pd alloy electrodes with various compositions. The activation energy was calculated from the slope $= -E_a/(2.3R)$ [46].

observed at the $Pt_{72}Pd_{28}$ electrode along with a much smaller temperature effect. The $Pt_{60}Pd_{40}$ and Pt electrodes exhibited the highest activation energies of 19.8 and 22.0 kJ/mol along with much larger temperature effects than those of $Pt_{72}Pd_{28}$ and $Pt_{83}Pd_{17}$, respectively. The activation energies of the $Pt_{60}Pd_{40}$ and Pt electrodes are nearly twice that of $Pt_{72}Pd_{28}$, so it is clear that the activation energy plays an important role in HOR performance on these materials.

### 3.2.3 A Volcano Plot for HOR on Pt–Pd Alloy Electrode (*d*-Band Center–Activity Correlation)

The volcano curve trends in the relationship between the exchange current density and the degree of alloying, as shown in Figure 3.5, indicate that two factors (electronic effect and lattice effect) may underlie the alloying effect. First, the mechanism for the hydrogen intercalation on a Pd electrode in aqueous electrolyte is considered. Hydrogen is a chemically (or electrochemically) reactive gas that adsorbs dissociatively on the Pd surface. Because of its small size of hydrogen and its lower cohesive energy on Pd than on Pt [Pt (564 kJ/mol) > Pd (376 kJ/mol)] [54], hydrogen, it can intercalate on a Pd surface, and this can lead to strong perturbations of the electronic structure of the Pd surface, resulting only in the relaxation of the surface lattice. The absorption of hydrogen on a Pd electrode also leads to modification of the electronic structure of Pd, as follows. Hydrogen induces electron states below the Fermi level and the *d*-band width is reduced, and the *d* band decreases relative to the Fermi level to an extent that increases with increasing H concentration, which results in a drastic reduction of the density of states (DOS) near the Fermi level [55–57]. Therefore, Pd

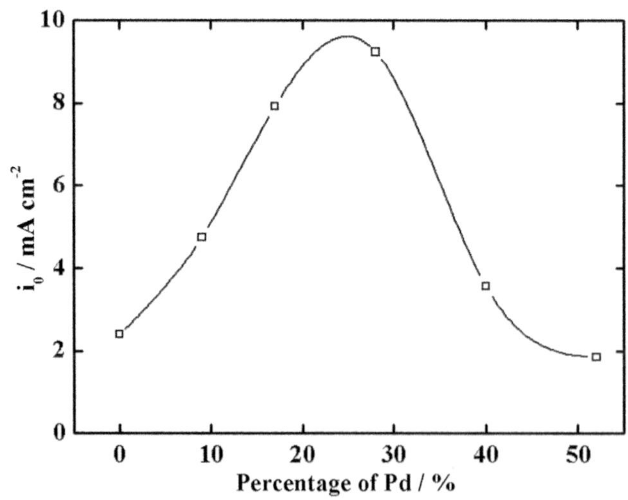


**FIGURE 3.5** Relationship between the exchange current density and the degree of alloying of Pt–Pd. The exchange current density was measured from the Tafel plots obtained in a solution of 0.5 M $H_2SO_4$, and the activation energies were calculated from Arrhenius plots (see Fig. 3.4) [46].

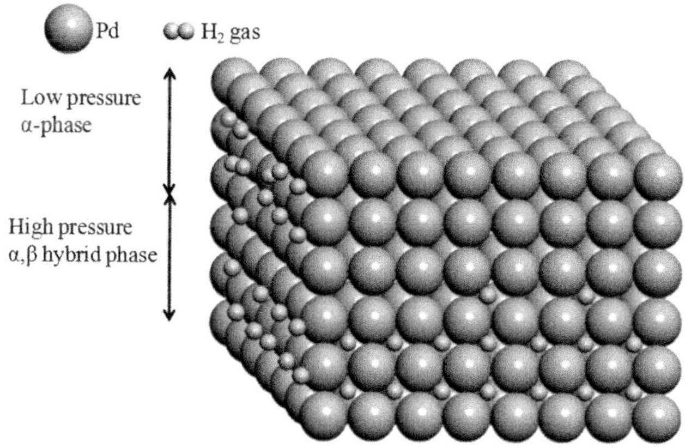


**FIGURE 3.6** Schematic model of metal structure with H atoms in the spaces between the Pd metal atoms [46].

undergoes surface reconstruction, which leads to the absorption of hydrogen into its bulk structure. Figure 3.6 schematically represents the reconstruction of the Pd surface and bulk; that is, the transition state is a mixture of α phase and β phases. The α phase is formed at a low hydrogen gas pressure, and the β phase is formed at a high hydrogen gas pressure at potentials near the equilibrium potential by an exothermic reaction. The lattice expansion of $Pt_{72}Pd_{28}H_x$ was observed by measuring the XRD patterns with and without hydrogen, as shown in Figure 3.7. Also, the presence of hydrogen intercalation on the $Pt_{72}Pd_{28}$ electrode is observed in Figure 3.8. The in situ visual observation of the hydrogen intercalation phenomenon using the gasochromic mechanism was performed using a He–Ne laser (633 nm) as the light source and a power meter for the transmittance measurement during the alternate exposure to hydrogen,

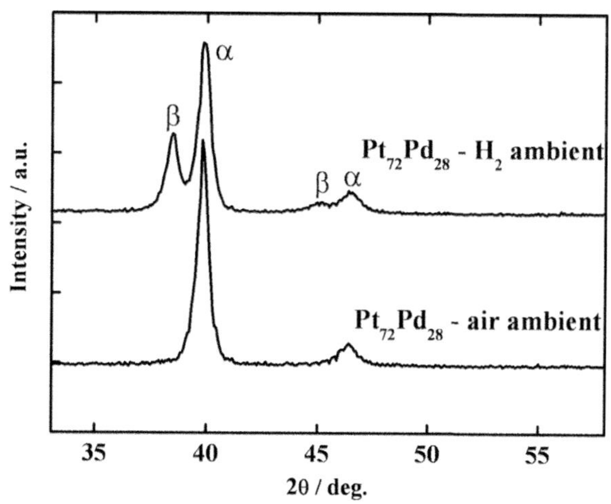


**FIGURE 3.7** XRD data for $Pt_{72}Pd_{28}$ before and after hydrogen charging [46].

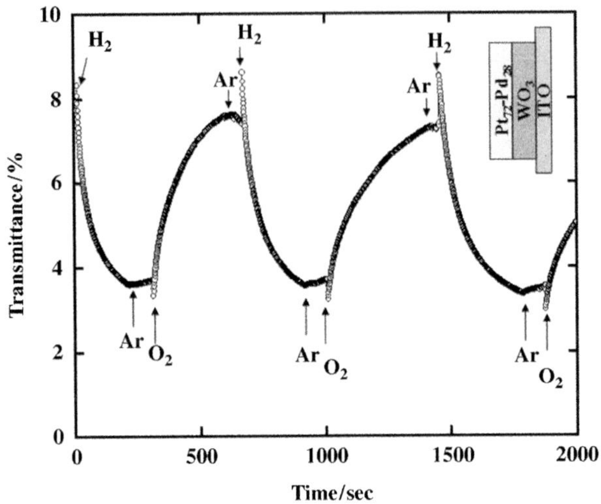


**FIGURE 3.8** Modulation of the optical transmittance of a $Pt_{72}Pd_{28}$ electrode alternately exposed to hydrogen and oxygen [46].

oxygen, and argon (see Fig. 3.9). The gasochromic mechanism can be written in a simplified form as [58]

$$xH^+ + xe^- + WO_3 \text{ (bleached state)} = H_xWO_3 \text{ (colored state)} \qquad (3.8)$$

A thin layer of Pt–Pd deposited on top of $WO_3$ acts as a catalyst to dissociate hydrogen molecules into protons and electrons. The protons and electrons diffuse into the $WO_3$ layer. The double injection of protons and electrons into the $WO_3$ layer causes a change in the optical transmittance through small-polaron transitions exactly like the electrochromic process [59]. This model has been widely accepted to explain gasochromic coloration.

Bennet and Gelatt observed a reduction in the number of unoccupied Pd 4*d* states on the Pd absorbed hydrogen and the narrowing of the *d*-band width caused by the lattice expansion and modification of the electronic structure [60,61]. This result

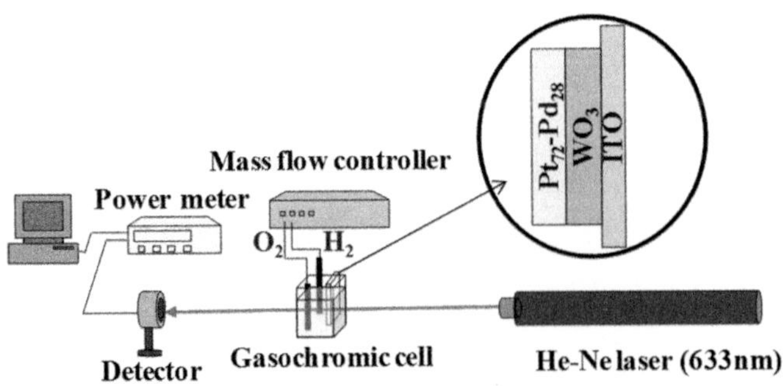


**FIGURE 3.9** Experimental setup for hydrogen intercalation on a $Pt_{72}Pd_{28}$ electrode [59].

shows that the relationship between the HOR and the electronic structure of Pt is affected by $PdH_x$, assuming that the HOR occurs mainly on the Pt surface, since the exchange current density of pure Pt is higher than that of pure Pd for the HOR in the low-overpotential region.

We measured the Pt $L_{III}$ and $L_{II}$ edges using an in situ XANES cell for gas-phase experiments to confirm that the relationship between HOR and the electronic structure of Pt is affected by the value of hydrogen in $PdH_x$. One significant aspect of XANES analysis is that it can provide important information on the Pt $d$-band vacancy. The $d$-band vacancy is derived from the analysis of the Pt $L_{III}$ and $L_{II}$ white lines. The $L_{II}$ and $L_{III}$ edges are due to the excitation of the $2p_{1/2}$ and $2p_{3/2}$ electrons, respectively. These electrons can undergo transitions to empty states in the vicinity of the Fermi level. Since the dipole selection rules in the XANES region restrict the transition to $L = 1$ and $J = 0,1$ (where $L$ and $J$ are the orbital angular and the total angular quantum number, respectively), the transitions to the $d$ orbitals are strongly favored [62]. For Pt, it has been shown that the final states with $J = \frac{5}{2}$ contribute 14 times more than do those with $J = \frac{3}{2}$. The $L_{III}$ transition ($2p_{3/2}$ to $5d_{5/2}$) is thus more highly favored [63], by the selection rules, than the $L_{II}$ transition ($2p_{3/2}$ to $5d_{5/2}$). The intensity of the $L_{III}$ peaks and to a lesser extent the $L_{II}$ peaks increases with increasing Pt $d$-band vacancy.

In general, for hydrogen dissociation, the strong binding of hydrogen would suggest a low hydrogen activation energy barrier [64–66]. When hydrogen is adsorbed onto a Pt metal surface, the $s$ (bonding) electrons of hydrogen form a coordinate covalent bond, which is shared with the Pt $d$-band vacancy. At the same time, the Pt $d$-band orbitals overlap with the empty hydrogen $s^*$ (antibonding) orbital to form backbonding, which strengthens the adsorptive bond between Pt and the head hydrogen atom and weakens the bond between the two hydrogen atoms. Then, how is the $d$-band vacancy related to the adsorptive bond strength of hydrogen? The large vacancy of the Pt $d$ band means that the DOS of Pt is pushed from below the Fermi level to just above the metal Pt valence band, which narrows the $d$-band width and shifts the $d$-band center upward in energy toward the Fermi level, as shown in Figure 3.10. Greeley et al. showed the correlation of the $d$-band states ($d$-band center) and the free energy of hydrogen adsorption (adsorptive bond strength of hydrogen), and concluded that the upshift of the $d$-band center leads to a reduction of the hydrogen absorption energy [67,68]. Therefore, this supposes that the larger vacancy of the Pt $d$ band results in stronger adsorptive bond strength of hydrogen.

Figure 3.11 shows the in situ XANES spectra with or without hydrogen on the Pt–Pd alloy electrodes, which were used for calculating the number of $d$-band vacancies from the Pt $L_{III}$ and $L_{II}$-edge spectra. The peak intensity of the Pt $L_{III}$ and $L_{II}$-edge XANES spectra with hydrogen was shown to be more sensitive to the unoccupied $d$ state than that of the spectra without hydrogen. Since the electronegativity of Pt (2.28) is roughly the same as that of Pd (2.20), no significant charge transfer is to be expected; hence, the change in the number of $d$-band vacancies of Pt without hydrogen should be small. On the other hand, after hydrogen charging, the DOS of Pt was significantly modified by the Pd adsorbed hydrogen.

Figure 3.12 shows the result of calculating the $d$-band vacancy of Pt as a function of the degree of alloying. From the difference in the areas under the Pt $L_{III}$ and $L_{II}$

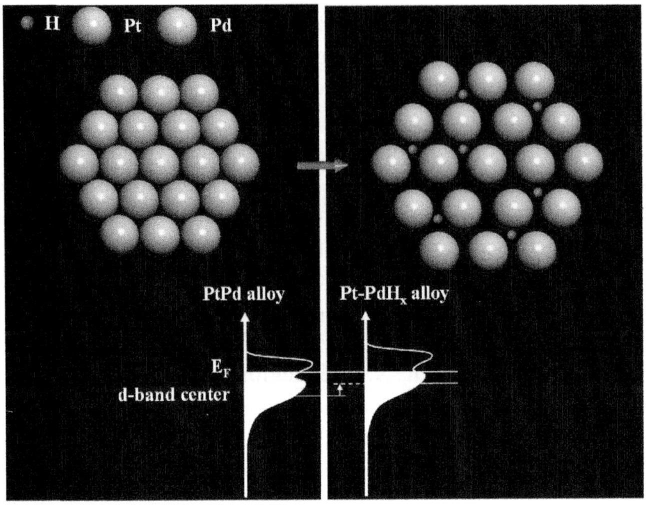


**FIGURE 3.10** Schematic explanation of the Pt–PdH$_x$ alloying effect on the *d*-band state of Pt [46].

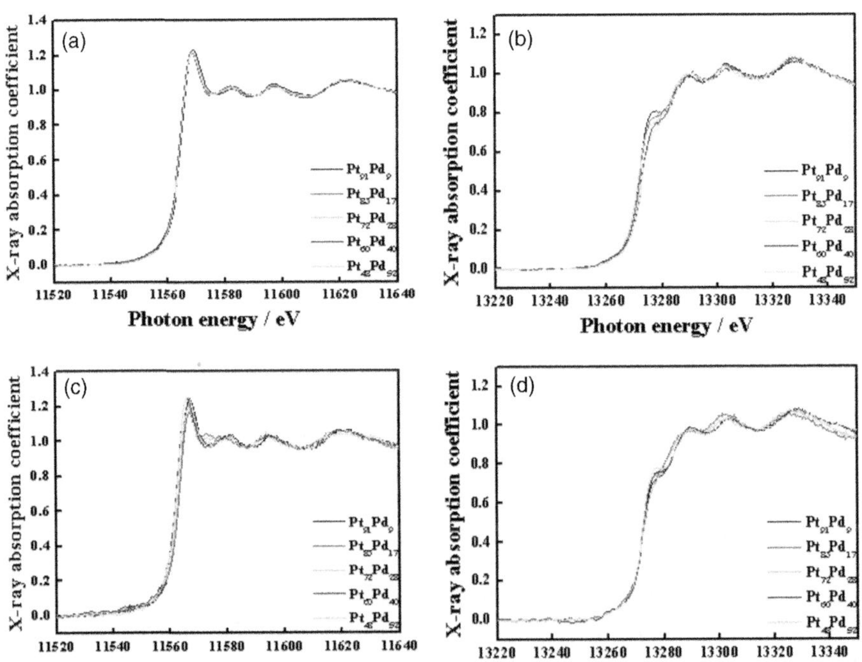


**FIGURE 3.11** (a) Pt $L_{III}$ and (b) Pt $L_{II}$ XANES spectra measured in air ambient and (c) Pt $L_{III}$ and (d) Pt $L_{II}$ XANES spectra measured after hydrogen charging ($\alpha$ or $\beta$ hydride) on Pt–Pd alloy electrodes with various compositions [46].

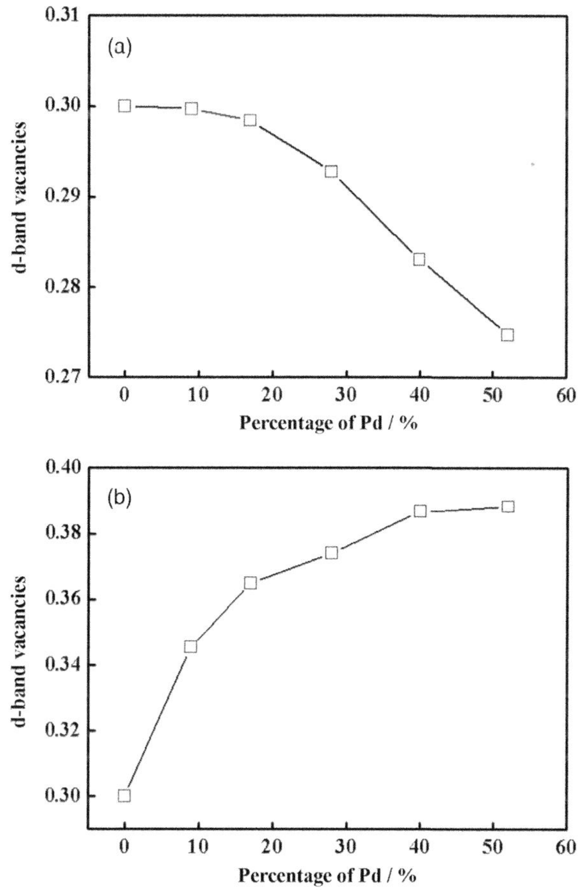


**FIGURE 3.12** (a) *d*-band vacancy from XANES data measured in air ambient and (b) *d*-band vacancy from XANES data measured after hydrogen charging [46].

absorption edges between the sample (pure Pt and Pt–Pd alloys) and a Pt reference foil, the fractional change in the number of *d*-band vacancies relative to the reference material ($f_d$) was estimated [69]. This technique therefore constitutes a new way of examining the role of the alloying element in Pt alloys in the *d*-band vacancy of Pt and hence is used in this investigation to elucidate the differences in the degree of electrocatalysis on the Pt versus Pt–Pd alloy electrocatalysts. The number of *d*-band vacancies of Pt in air ambient decreases as the percentage of Pd increases (Fig. 3.12a), and the number of *d*-band vacancies of Pt in hydrogen ambient increases as the percentage of Pd increases (Fig. 3.12b). This means that the *d*-band width of Pt is increased by the *d–d* hybridization of Pt and Pd in air ambient and that the *d*-band center of Pt is downshifted. In contrast, the *d*-band width of Pt is decreased by the hydrogenation of Pd and the *d*-band center of Pt is upshifted. These concepts are consistent with those of Nørskov et al. who also elucidated the coupling between the *d*-band width and *d*-band center for near-surface alloys [70–72].

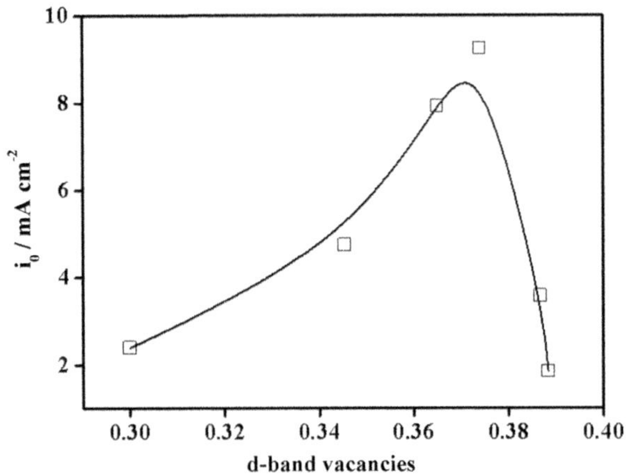


**FIGURE 3.13** Correlation between exchange current density and *d*-band vacancies from the in situ XANES data measured after hydrogen charging [46].

The measurement of the *d*-band vacancies allows us to directly correlate the catalytic activity for HOR. As shown in Figure 3.13, the *d*-band vacancies of Pt versus the exchange current density exhibits a volcano curve. This result means that HOR activity is given by the strength of the hydrogen–Pt bond interaction, which depends on the position of the Pt *d*-band states relative to the Fermi level. However, if the number of *d*-band vacancies of Pt becomes too high, the hydrogen oxidation currents decrease, owing to the stronger adsorptive bond strength of hydrogen, which increases the desorption energy barrier in the charge transfer step (Volmer reaction). This situation leads to a maximum in the catalytic activity when the ability of the Pt–Pd alloy to bind hydrogen is neither too weak nor too strong.

### 3.2.4 Fuel Cell Test of Pt–Pd Alloy Electrocatalysts

#### *3.2.4.1 Studies of Nanosized Electrocatalysts Fabricated Using a New Sputter Technique*

The utilization of Pt electrocatalysts plays an important role in practical application of PEFCs. Several articles devoted to the study of minimizing Pt loading have been published since 1989. One approach is to form catalyst layers by using direct sputter deposition. Srinivasan et al. deposited a Pt layer on the gas diffusion layer (GDL) using a sputtering technique. In this system, cell performance at a loading of 0.45 mg Pt/cm$^2$ was equivalent to that of conventional methods at a loading of 4.0 mg Pt/cm$^2$ [73]. Hirano et al. deposited a Pt layer (0.1 mg/cm$^2$), as a cathode electrode, using a sputter technique on an uncatalyzed GDL. This modification resulted in enhanced cell performance [74]. O'Hayre compared the power density of a commercial membrane electrode assembly (MEA) loaded with 0.4 mg Pt/cm$^2$ to that of a MEA with 0.04 mg Pt/cm$^2$ (15 nm) fabricated using the sputtering technique. The results showed that three-fifth of the maximum power density was produced with one-tenth the catalyst loading [75]. However, conventional sputtered Pt catalyst layers suffered from the deposition of Pt as a conformal film on the substrate [76]. Owing to this compact and dense structure, the Pt catalyst layer was not porous, and displayed a small electrochemically active surface (EAS) area.

More recently, we have reported that a high-pressure (>100 mTorr) sputtering technique can produce nanoscale particles, which form in the vapor as a result of the high pressure [77,78]. The sputtered atoms undergo many collisions in the vapor at these high pressures, subsequently nucleate, and grow in the vapor by rapid condensation. It is expected that the Pt catalyst layer formed using the high-pressure sputtering technique leads to an improvement in the Pt microstructure, such that the Pt catalysts layer is more porous, allowing enhanced Pt utilization.

This study investigated Pt nanocatalysts deposited using a high-pressure radiofrequency (RF) magnetron sputtering technique with a mixture of gaseous Ar and He for use in PEFCs. Nanoscale Pt particles were achieved using a sputtering pressure (200 mTorr, Ar/He = 1) that was higher than that used in normal thin-film processing (1–10 mTorr, Ar/He = 1). Figure 3.14a shows a typical SEM image of a dense Pt thin film produced on the GDL substrate by sputtering at 11 mTorr pressure with

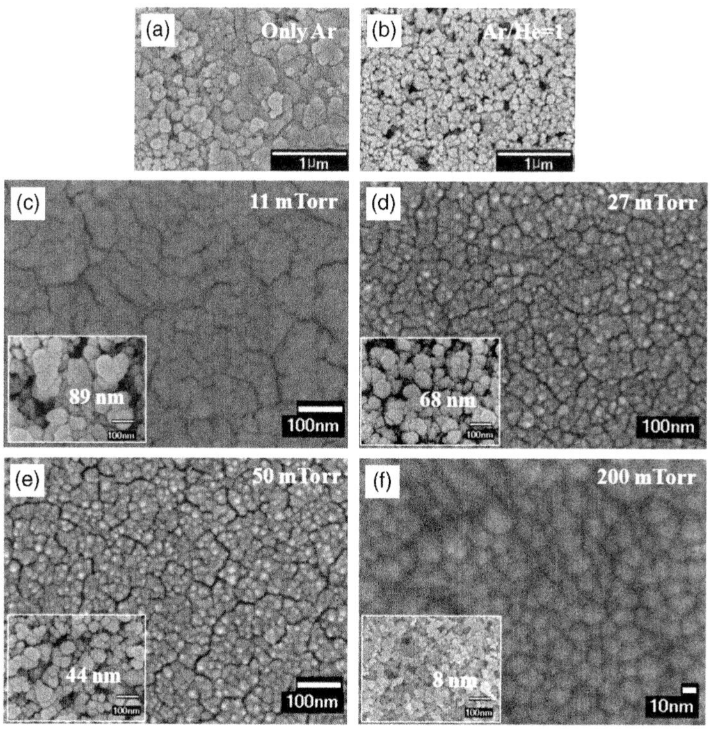


**FIGURE 3.14** SEM images of Pt catalyst layers deposited on the GDL substrate using sputtering technique with (a) only Ar (11 mTorr) and (b) both Ar and He (5.5 mTorr of Ar, total pressure 11 mTorr); SEM images of Pt catalyst layers deposited on the Si(100) substrate using different sputtering pressures with a mixture of gaseous Ar and He (Ar/He = 1): (c) 11 mTorr, (d) 27 mTorr, (e) 50 mTorr, and (f) 200 mTorr [31].

Ar (no He) gas. This image indicates that condensation of Pt particles induced by Ar gas results in the formation of a dense thin film. Figure 3.14b presents a SEM image of Pt nanoparticles on the GDL substrate that resulted when He (5.5 mTorr) gas was added to the Ar (5.5 mTorr) gas. The average Pt nanoparticle diameter was about 95 nm, and the size distribution was broad. As is evident in Figure 3.14, Ar and He gas pressure critically affect Pt nanoparticle production. More recent papers have reported that Ar gas is responsible for the sputtering process, while Pt particle condensation is effectively induced by He gas [79]. In other words, the sputtering process induced by Ar gas creates the Pt atomic species, which, in turn, form Pt nanoparticles on induction by He gas. Figure 3.14c–f shows SEM images of the variation in Pt nanoparticle size induced by the different sputtering pressures (Ar/He $= 1$) on the Si(100) substrates. Pt nanoparticle size increased at lower sputtering pressures, indicating the growth of nanoparticle nucleation. Figure 3.14c shows the formation of a dense film, which indicates that sputtering occurred below the supersaturated vapor pressure. Figure 3.14f shows the result of depositing Pt nanoparticles at a pressure of 200 mTorr, that is, above the supersaturated vapor pressure. The narrowness of the distribution is consistent with an average Pt nanoparticle size of $8.0 \pm 2$ nm. This result indicates that the Ar and He gas pressures determine the size range of the Pt particles produced by sputtering and subsequent condensation. Therefore, careful optimization of the gas pressure provides the versatility to produce, from a source of Pt atoms, a range of particle sizes, from a few atoms to a few thousand atoms per nanoparticle, depending on the sputtering pressure. In general, it has been reported that the growth of nanoparticles occurs in two stages [80]. Energetic atoms sputtered from the target are cooled by He gas, leading to the nucleation of nanoscale particles. For the nucleation of Pt atoms, a three-body collision between two sputtered atoms and a cooling He atom is essential, as it removes excess kinetic energy from the sputtered atoms. The growth of particles also occurs via two-body collisions, such as cluster–cluster collisions [81]. The two- and three-body collisions occur more frequently at high supersaturated vapor pressure compared with normal gas pressures. Therefore, well-defined nanoscale structures can be fabricated by controlling the sputtering pressure. SEM images of Pt deposited on GDL substrate using different sputtering pressures are shown in the Figure 3.14 inset, which also shows the dense Pt thin-film layer produced at 11 mTorr pressure. At 200 mTorr pressure, only Pt nanoparticle formation, but not film formation, was observed. The average size of the Pt nanoparticles decreased with increasing pressure and the size distribution narrowed.

The size of Pt nanoparticles on the GDL substrate produced at 200 mTorr pressure was confirmed by the XRD patterns shown in Figure 3.15a, and the average size of the Pt particles was about 8 nm, as calculated using the Debye–Scherrer equation [82]. Figure 3.15b illustrates the size distribution of the Pt nanoparticles as a function of the sputtering pressure. One hundred particles were measured to study the size distribution of the Pt nanoparticles. The SEM images (Fig. 3.14) showed that the diameters of Pt nanoparticles varied from ~89 to ~8 nm as the sputtering pressure was decreased. The kinetic energy with which the atoms and particles arrive at the GDL substrate determines the size of the resulting nanoparticles. The kinetic energy of the particles depends on the sputtering conditions.

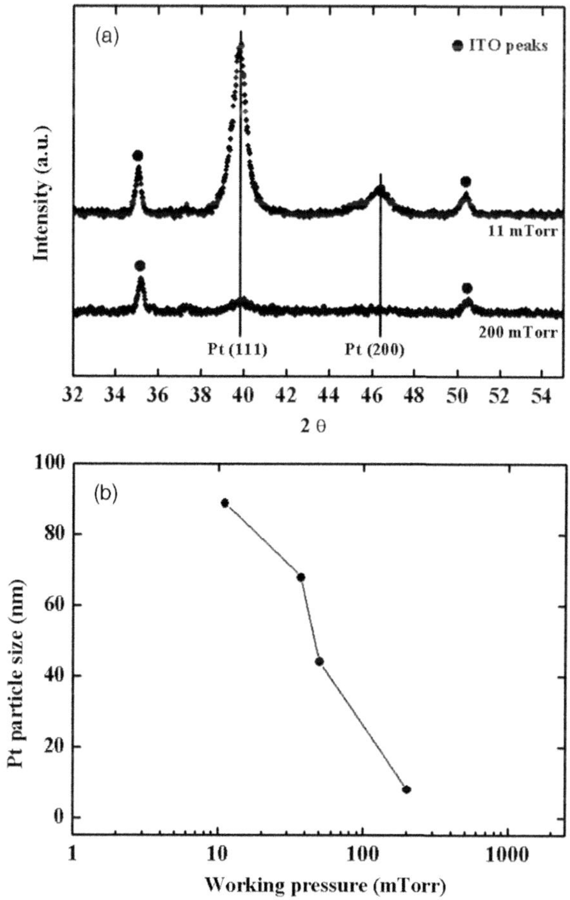


**FIGURE 3.15** (a) XRD pattern for different sputtering pressures and (b) average particle sizes of Pt nanocatalysts as a function of the sputtering pressure with a mixture of gaseous Ar and He (Ar/He = 1) [31].

On the basis of these results, it was concluded that particle size is a function of sputtering pressure. At very low pressures, the motion of the sputtered atoms is essentially that of an elastic collision, and a continuous film is formed on the GDL substrate. At higher gas pressures, the sputtered atoms undergo collision-induced aggregation. Therefore, the atoms sputtered at high pressure arrive at the GDL substrate not only in the form of nanoscale particles but also with lower kinetic energies. A nanoscale particle is thereby promoted, particularly if energetic atoms sputtered from the target are cooled by the He gas.

Figure 3.16 compares the electrochemical performances of PEFC single cells made using Pt nanocatalysts formed at four different sputtering pressures: 11, 27, 50, and 200 mTorr. In all cases, equivalent amounts of Pt (0.02 mg/cm$^2$) were added to the anode and cathode. Fuel cell performance curves were measured after allowing for a

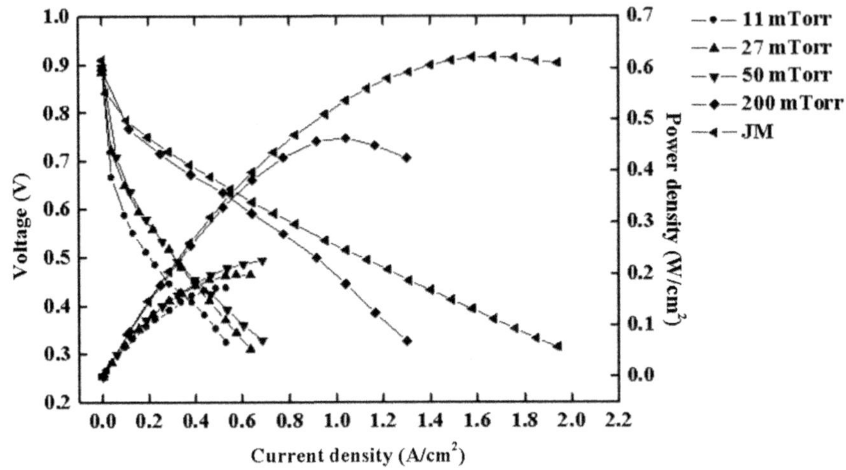


**FIGURE 3.16** Polarization curves and power density curves of MEAs with Johnson Matthey (JM) catalyst (-◄--) and Pt nanocatalyst layers formed at different sputtering pressures of 11 mTorr (-●-), 27 mTorr (-▲-), 50 mTorr (-▼-), and 200 mTorr (-■-) with a mixture of gaseous Ar and He mixture (Ar/He = 1) [31].

steady reading at each point during the current scan, as controlled by the fuel cell test station. The open-circuit voltage of all samples sputtered at 11 ~ 200 mTorr pressure showed an average open-circuit voltage of ~0.9 V, which is close to that of commercial Pt catalysts. From Figure 3.16, the results of the fuel cell performance of the Pt catalyst layers sputtered at 11, 27, and 50 mTorr show that the maximum power density per unit area increased slightly with an increase in the sputtering pressure. In particular, the fuel cell performance of Pt nanocatalysts (total loadings of 0.04 mg/cm$^2$ per MEA) sputtered at 200 mTorr pressure showed a maximum power density of 420 mW/cm$^2$, which was comparable to the 610 mW/cm$^2$ for the JM catalyst (40 wt% Pt/C, total loadings of 0.6 mg/cm$^2$ per MEA). However, the efficiency per total weight (10.5 W/mg) of the sputtered Pt nanocatalysts was 10-fold greater than that of the JM catalyst (1 W/mg). This finding indicates that the Pt nanocatalyst layer electrode fabricated using a high-pressure sputtering technique in a mixture of gaseous Ar and He gas has an improved the utility of Pt catalyst with low catalyst loading.

#### *3.2.4.2 Fuel Cell Test of Nanosized Pt–Pd Alloy Electrocatalysts Fabricated Using a Sputter Technique*

The size of the $Pt_{72}Pd_{28}$ alloy nanoparticles also was determined by means of the TEM images shown in Figure 3.17a. The dark points correspond to the $Pt_{72}Pd_{28}$ alloy nanoparticles (~8.5 nm), and the crystalline plane of the $Pt_{72}Pd_{28}$ alloy nanophase representing the *d* spacing of the (111) plane ($d_{111}$ ~ 0.226 nm) can be observed; this result was in agreement with the value based on Vegard's law ($d_{PtPd} = Xd_{pt} + (1 - X)d_{Pd}$, where $X$ is composition of Pt in the Pt–Pd electrode and $d_{PtPd}$, $d_{pt}$, $d_{Pd}$ is $d$ spacing of the (111) plane of Pt–Pd, Pt, and Pd, respectively). The formation of

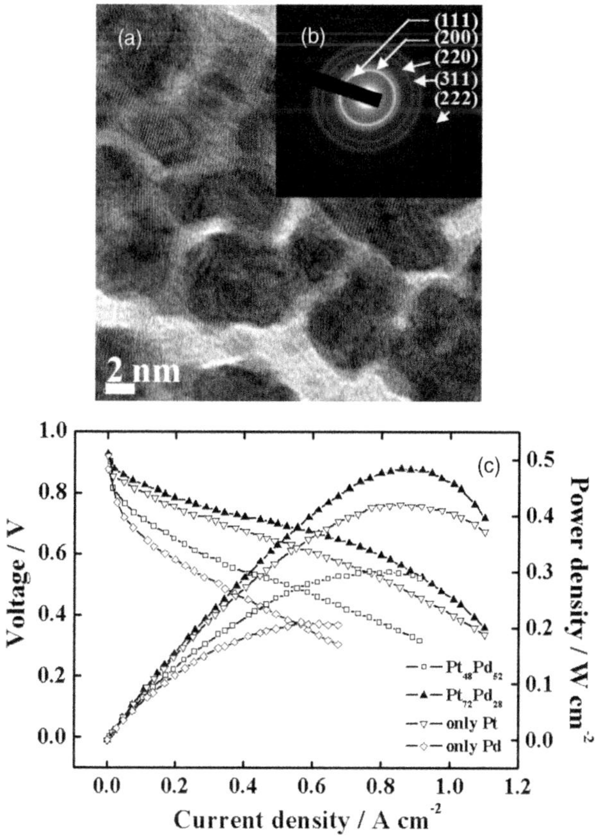


**FIGURE 3.17** (a) High-resolution TEM image; (b) TED pattern of $Pt_{72}$–$Pd_{28}$ nanocatalysts deposited at a sputtering pressure of 200 mTorr (Ar/He = 1); (c) polarization curves and power density curves of MEAs with only Pd nanocatalyst layers (-◇-), only Pt nanocatalyst layers (-▽-), $Pt_{72}Pd_{28}$ nanocatalyst layers (-▲-), and $Pt_{72}Pd_{28}$ nanocatalyst layers (-□-), formed at different sputtering pressures of 200 mTorr with a mixture of gaseous Ar and He mixture (Ar/He = 1).

polycrystalline $Pt_{72}Pd_{28}$ also was confirmed by TED analysis of the (111), (200), (220), (311), and (222) planes.

On the basis of this new sputtering technique, we carried out electrochemical performances of PEFC single cells $Pt_{72}Pd_{28}$, $Pt_{48}Pd_{58}$, pure Pt, and pure Pd electrocatalyst. In all cases, equivalent amounts of Pt (0.02 mg/cm$^2$) were added to the cathode and equivalent amounts of $Pt_x$–$Pd_y$ (0.02 mg/cm$^2$), Pt (0.02 mg/cm$^2$), and Pd (0.02 mg/cm$^2$) were added to the anode. Fuel cell performance curves were measured after allowing for a steady reading at each point during the current scan, as controlled by the fuel cell test station. The open-circuit voltage of all samples sputtered showed an average open-circuit voltage of ~0.9 V, which is close to that of commercial Pt catalysts. In particular, the fuel cell performance of $Pt_{72}Pd_{28}$ nanocatalysts (total loadings of

0.04 mg/cm$^2$ per MEA) sputtered at 200 mTorr pressure showed a maximum power density of 480 mW/cm$^2$, which was comparable to the 420 mW/cm$^2$ for the pure Pt catalyst (total loadings of 0.04 mg/cm$^2$ per MEA). This finding indicates that the $Pt_{72}Pd_{28}$ nanocatalyst layer electrode fabricated using a high-pressure sputtering technique in a mixture of gaseous Ar and He gas has an improved electrochemical surface reaction efficiency, which is consistent with results presented in the previous section.

### 3.2.5 Summary

In conclusion, we found that significant electrocatalysis occurred for the HOR at the Pt–Pd alloy electrodes, to an extent that depends on the alloy composition, and revealed the principal role of the electronic effect of Pt–Pd alloys in the HOR reaction. We demonstrated that the relationship between the exchange current density and the degree of Pt–Pd alloying exhibits a volcano curve trend and that, at the same temperature, the maximum exchange current density for the hydrogen oxidation is obtained at $Pt_{72}Pd_{28}$, which is similar to the trend of the activation energies. We also showed that there is a relationship between the number of *d*-band vacancies of Pt on the Pt–$PdH_x$ alloy electrodes as a function of the degree of alloying and HOR activity. Our results imply that the high performance of Pt–$PdH_x$ catalyst in HOR relies heavily on the electronic structure of Pt caused Pd spontaneously adsorbed hydrogen gas in the electrocatalytic process. The single-cell performance of the $Pt_{72}Pd_{28}$ nanocatalyst layer electrode fabricated using a high-pressure sputtering technique in a mixture of gaseous Ar and He gas was improved, due to the improvement of the electrochemical surface reaction efficiency.

## 3.3 HYDROGEN OXIDATION REACTION AT A Ir–Pd ALLOY ELECTRODE

Iridium was extensively studied as Pt-free electrocatalyst since the Pt-like catalytic behavior and electronic structure had been reported [83–85]. Although the Ir has the advantage of low-cost materials intrinsic activity was insufficient for independent use. In this study, we investigated the hydrogen oxidation at Ir–Pd alloy electrodes with varying composition in comparison with pure Ir by using an in situ XANES system for gas-phase experiments to analyze the electronic states. We have studied the electronic structures of the Ir–Pd alloys in the hydrogen gas ambient for HOR and found a correlation between the *d*-band vacancy of Ir induced by alloying Ir–$PdH_x$ and the HOR activity.

### 3.3.1 Composition Analysis of Ir–Pd Alloy Electrodes

Iridium–palladium alloy electrodes were grown using an RF magnetron cosputtering system consisted of a dual-sputtering gun. Si(100) and glassy carbon were used as substrates to characterize the structural and electrochemical properties, respectively.

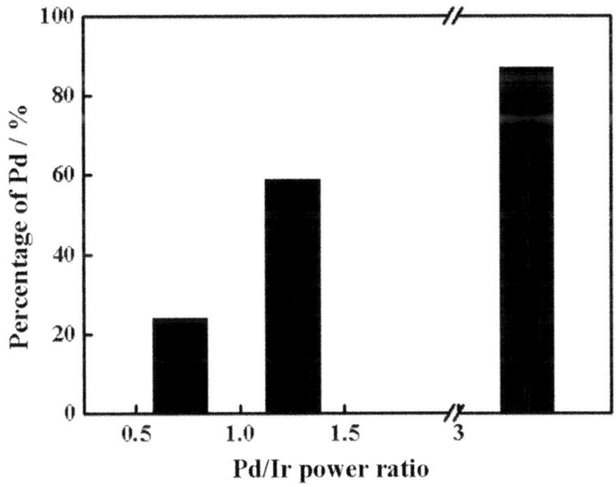


**FIGURE 3.18** Percentage of Pd as a function of Pd/Ir power ratio.

Cosputtering was performed under inert Ar gas at a flow rate of 26 standard cubic centimeters per minute (sccm) at room temperature for 10 min, which produced films about 100 nm thick. In order to fabricate the Ir–Pd thin-film electrodes with various compositions, the guns with the metal targets in the RF magnetron sputtering system were controlled as a function of the RF power. With increasing RF power of the Ir target gun at a fixed RF power of the Pd gun, the Pd concentration was varied from 0% to 87%. A pure Pt electrode was also produced. In order to identify the elemental composition in the surface region and analyze the interface composition, X-ray photoelectron spectrometric (XPS) images of Ir–Pd electrodes as a function of the ratio of the Pd power to Ir power were obtained, as shown in Figure 3.18, where we confirmed that the sputtering yield of Pd was larger than that of Ir. The XPS analysis is the most widely used analytical technique for obtaining the chemical composition of a solid surface. The main advantage of the analysis is its high sensitivity in the 0.5–2 nm region near the surface. In addition, it can be combined with ion-beam sputtering to remove the surface of metal film and to continue to monitor the chemical composition of the remaining surface as this surface moves into the sample. The well-defined alloy thin-film electrodes play a significant role in the development of chemical and electronic models for electrocatalysis in polycrystalline samples.

### 3.3.2 Kinetics of HOR on Ir–Pd Electrodes at 273 K

Hydrogen oxidation reactions on Ir or Pd electrodes in acid electrolytes are among the fastest known electrochemical reactions, and it is experimentally very difficult to measure anything except diffusion polarization. It is our contention that, in order to illuminate the fine details of the hydrogen reaction and to establish the role of Ir–Pd electrode surface geometry on the kinetics of the hydrogen reaction, the HOR reaction rate should be slowed down, that is, examined at the lowest temperature possible (273 K) in dilute solutions (0.5 M $H_2SO_4$). This will, in turn, enable us to extract true

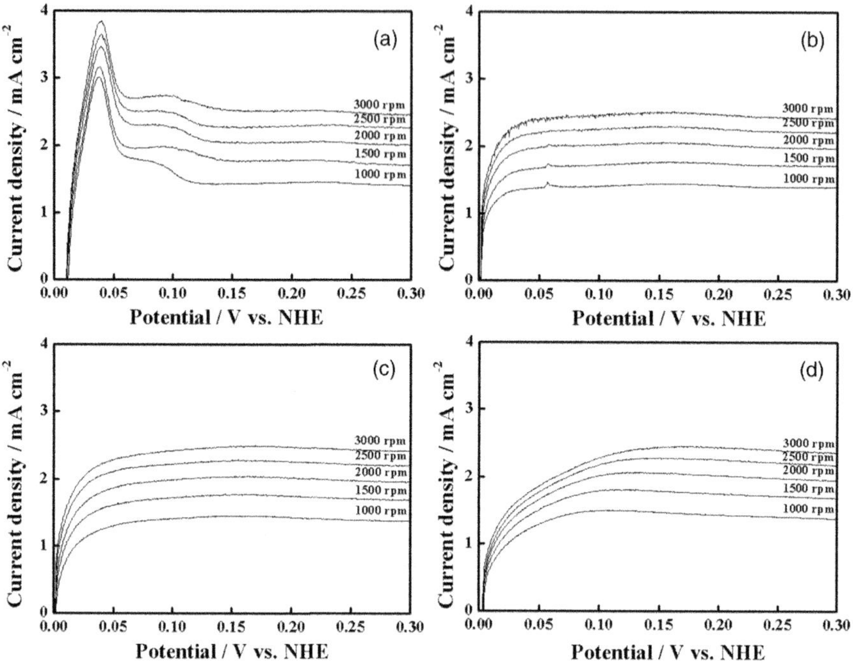


**FIGURE 3.19** Hydrogen oxidation current densities during positive-going sweeps (20 mV/s) on (a) $Pd_{87}Ir_{13}$, (b) $Pd_{59}Ir_{41}$, (c) $Pd_{24}Ir_{76}$, and (d) $Pd_0Ir_{100}$ RDEs in 0.5 M $H_2SO_4$ saturated with $H_2$ at 273 K.

kinetic rates by use of the mathematically well-established correlation for the hydrodynamics of the RDE assembly. Although a complete set of the polarization curves were recorded in the temperature range 273–333 K (at intervals of 10 K), only representative curves are shown and discussed here.

Polarization curves for HOR at 1 atm of $H_2$ recorded on rotating Ir–Pd disk electrodes in 0.5 M $H_2SO_4$ at 273 K are shown in Figure 3.19. A close inspection of this figure reveals that at low overpotentials there is little or no rotation dependence of the hydrogen oxidation currents. Figure 3.19 also shows that the anodic current rises sharply from the origin and reaches the limiting current at overpotentials of about 30 mV. This indicates that the HOR on Ir–Pd is kinetically controlled in a very narrow potential range and that hydrogen oxidation current densities $i$ on Ir–Pd electrodes are determined mainly by diffusion overpotential $\eta_{l,c}$, following closely the overpotential–current density relation [86]

$$\eta_{l,c} = -RT/2F \ln (1-i/i_{l,c}) \tag{3.9}$$

where $i_{l,c}$ represents the measured diffusion-limited current at any given rotation rate. Thus, to obtain kinetic data for this surface, we must concentrate on analysis of the small-polarization-potential region, as described below. Above 0.2 V on all three

**TABLE 3.4 Errors between Experimental and Theoretical Values of $BC_0^*$ at 273 K**

| Electrode | $BC_0^*$ ($10^{-2}$ mA cm$^{-2}$ rpm$^{-0.5}$) | Error (%) |
|---|---|---|
| $Pd_{87}Ir_{13}$ | 4.601 | +0.62 |
| $Pd_{59}Ir_{41}$ | 4.518 | −1.19 |
| $Pd_{24}Ir_{76}$ | 4.517 | −1.22 |
| $Pd_0Ir_{100}$ | 4.516 | −1.24 |

surfaces, there appears to be well-defined diffusion-limiting currents $i_{l,c}$ [87]. The calculated value of $BC_0^*$ at 273 K is $4.57 \times 10^{-2}$ (mA/cm$^2$)rpm$^{-1/2}$. The experimental values for $BC_0^*$ at 273 K at various rotation rates are recorded in Table 3.4. The experimental value of $BC_0^*$ on the Ir–Pd electrode is $4.6 \times 10^{-2}$ (mA/cm$^2$)rpm$^{-1/2}$ at 273 K, which are in reasonable agreement with the theoretical value at 273 K.

On the other hand, the polarization curves for $Pd_{83}Ir_{17}$ reveal the absorption of large amounts of hydrogen into the bulk lattice of Pd up to about +0.12 V. There is no rotation rate dependence, implying entirely kinetic resistance of HOR. Sweeping the electrode potential more positively, HOR is first under mixed kinetic–diffusion control, and above +0.3 V well-defined diffusion-limiting currents are observed. The anodic polarization curves for Ir–Pd electrodes (Fig. 3.19) show that the of HOR kinetics, initially under activation control, $0 < E < 0.03$ V, is followed by a region of mixed kinetic–diffusion control up to ~0.2 V, and above 0.2 V true diffusion-limiting currents are recorded. Qualitatively, these electrodes have a significantly higher activity than does $Pd_{83}Ir_{17}$. Figure 3.19 also indicates that the region relative to the HOR reaction is under mixed diffusion–kinetic control over a wider range of potentials and provides a better opportunity for detailed kinetic analysis.

### 3.3.3 Tafel Plots on Ir–Pd Electrodes at 273 K

It is well known that hydrogen absorption of Pd near 0 V [88,89]. Because the absorbed hydrogen in the Pd can be oxidized with dissolved hydrogen in electrolyte, it was not accurate measuring the HOR current using the potential sweep method [35]. To avoid the effect of oxidation of absorbed hydrogen, chronoamperometry was employed to measure HOR current. In the chronoamperometry, oxidation current was stabilized after 120 s and measurements were conducted at 120 s. A Tafel plot is a useful way to obtain mechanistic and kinetic information for electrode reactions, and it is used intensively for the HOR. As shown in the Figure 3.19a, the $Pd_{87}Ir_{13}$ electrode is largely absorbed hydrogen in the Pd (β phase), which could not be exactly measured by the HOR current–potential sweep method. So we conducted a chronoamperometry measurement for 120 s (Fig. 3.20).

Figure 3.21 shows Tafel plots of mass-transport-corrected currents for the HOR at Ir–Pd electrodes in 0.5 M $H_2SO_4$ at 273 K. The Tafel slopes are determined at low overpotentials, where measured current densities are essentially due to a mixed diffusion–kinetic resistance for the HOR. For only Ir electrode, we fitted the curves with one slope, ~32 mV/dec with an exchange current density of $\sim 0.57 \times 10^{-3}$ A/cm.

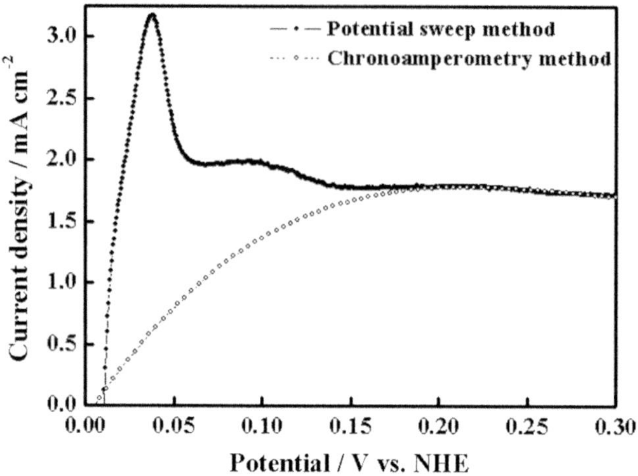


**FIGURE 3.20** HOR polarization curves obtained by linear potential sweep voltammetry (-●-) and chronoamperometry method (-○-) in steady state of HOR current density on $Pd_{87}Ir_{13}$ electrode.

On $Pd_{83}Ir_{17}$, HOR appears to have the highest Tafel slope, ~48 mV/dec with a HOR exchange current density of $\sim 0.11 \times 10^{-3}$ A/cm$^2$. The $Pd_{59}Ir_{41}$ electrode has the lowest Tafel slope of ~28 mV/dec observed; the HOR exchange current density is $\sim 3.5 \times 10^{-3}$ A/cm$^2$. On polycrystalline Pt, the most widely accepted mechanism in acid medium is the chemical adsorption step (Tafel) or electrochemical adsorption step (Heyrovsky), followed by the adsorbed hydrogen atom discharge step (Volmer). On polycrystalline Pt in a strongly acid medium the Tafel–Volmer mechanism has been proposed, with Tafel as the rate-determining step (RDS) [90]. Marković and

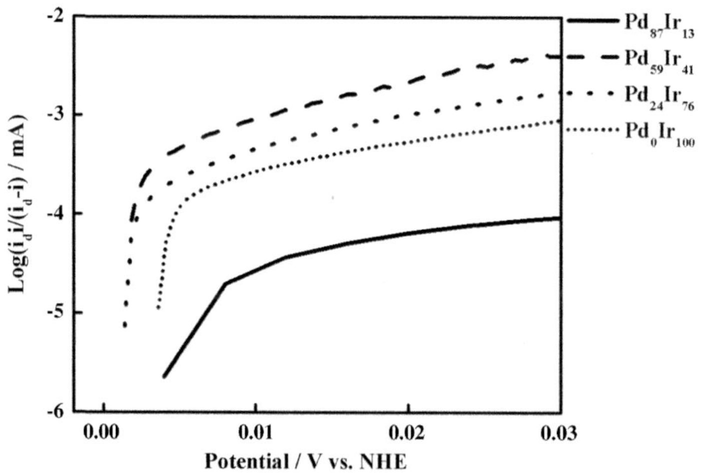


**FIGURE 3.21** Mass-transfer-corrected Tafel plots for hydrogen oxidation on Ir–Pd alloy electrodes with various compositions in a solution of 0.5 M $H_2SO_4$. These plots were obtained from Figure 3.19.

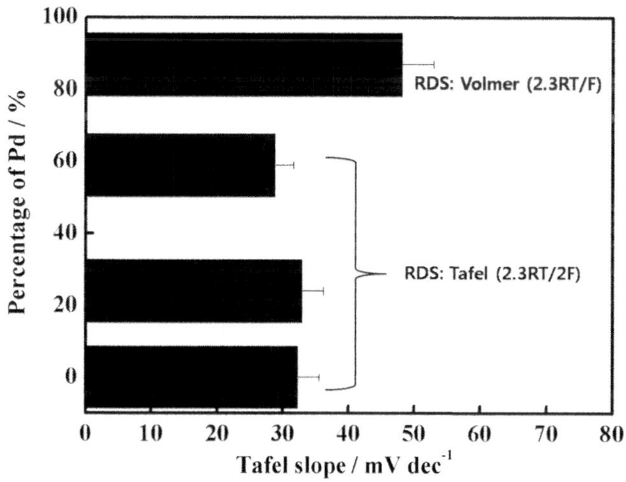


**FIGURE 3.22** Values of Tafel slope as a function of percentage of Pd.

coworkers [91] studied HOR on Pt (110) in 0.05 M $H_2SO_4$ at 303 K, and found a Tafel coefficient of 28 mV/dec. In the present work, a Tafel plot for HOR on Ir–Pd electrodes (except $Pd_{87}Ir_{13}$), corrected for mass transport influence, produced a value of Tafel slope, ~30 mV/dec in similar experimental conditions (see Fig. 3.22). From these results it is reasonable to assume that HOR takes place on Ir via the Tafel–Volmer mechanism with Tafel as RDS. On the other hand, the Tafel slope for the $Pd_{87}Ir_{13}$ electrode is ~58 mV/dec, which is reasonable to propose that the HOR takes place on Pd via the Tafel–Volmer mechanism with the Volmer step as RDS.

### 3.3.4 Exchange Current Densities and Activation Energies on Ir–Pd Electrodes

In the previous sections, we have shown that the kinetics of the hydrogen reaction on Ir–Pd electrodes is difficult to measure in the Tafel region because of its high exchange current density. However, the same property makes it possible to measure the exchange current density by conducting measurements in the vicinity of the reversible potential, often referred to as the *small-polarization region*. In our work, the polarization curves were analyzed in the small-polarization region ( +30 mV).

Figure 3.21 show that a linear relationship between the logarithm values of the current density and the overpotential is valid and also shows the magnification of currents in the vicinity of the small-polarization potential. At the same temperature (273 K), the exchange current densities increase in the order $Pd_{59}Ir_{41} > Pd_{24}Ir_{76} > Pd_0Ir_{100} > Pd_{87}Ir_{13}$, which gives the order of absolute kinetic activities of Ir–Pd electrodes. The values of the exchange current densities versus percentage of Pd determined from the data in Figure 3.21 are shown in Figure 3.23a. This volcano-type curve indirectly reveals that the Ir/Pd ratio would affect the electronic structure (valence band, ligand effect, lattice strain effect, etc.) of the electrocatalyst, which may mainly play a role in the electrocatalytic reaction.

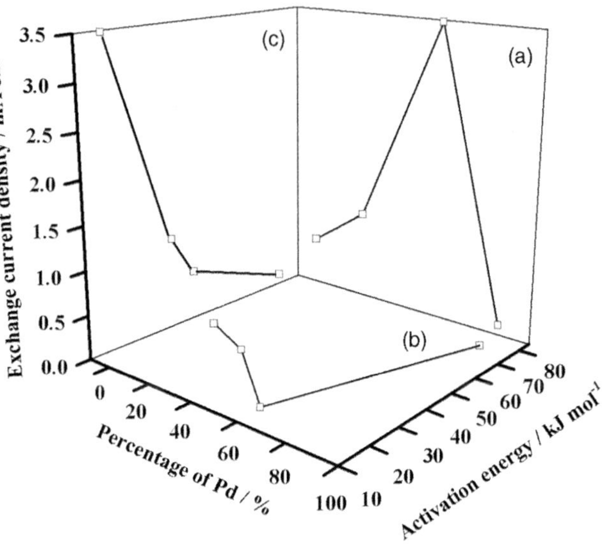


**FIGURE 3.23** (a) Relationship between exchange current density and degree of of Ir–Pd alloying. Exchange current density was measured using Tafel plots obtained in a solution of 0.5 M $H_2SO_4$, and activation energies were calculated from Arrhenius plots (see Fig. 3.21). (b) Arrhenius plots of exchange current densities ($i_0$) for Ir–Pd alloy electrodes with various compositions. The activation energy was calculated from the slope $= -E_a/(2.3R)$. (c) Exchange current density ($i_0$) versus activation energy on Ir–Pd electrodes.

At low overpotentials, where all electrodes have systematic log $i$ versus $E$ relationships, it is important to note that analysis of the polarization curves recorded for HOR revealed that the Tafel slopes were directly proportional to the temperature. It is quite clear that the HOR kinetics increase significantly when the temperature is increased even in this relatively narrow range of absolute temperature. Using the values of $i_0$ determined on Ir–Pd electrodes over the temperature range 273–333 K, we have determined the apparent activation energies for the hydrogen reaction on Ir–Pd electrodes in 0.5 M $H_2SO_4$ from the Arrhenius relation

$$d \log i_0/d(1/T) = -\Delta H^\circ/(2.3R) \tag{3.10}$$

where $\Delta H^\circ$ is the apparent enthalpy of activation at the reversible potential (hereafter referred to as *activation energies*) for the hydrogen reaction on Ir–Pd electrodes. The slopes obtained from Arrhenius plots for all Ir–Pd electrodes correspond to the activation energies: 76 kJ/mol for $Pd_{87}Ir_{13}$, 44 kJ/mol for only Ir, 37 kJ/mol for $Pd_{24}Ir_{76}$, and 18 kJ/mol for $Pd_{59}Ir_{41}$. As shown the Figure 3.23c, the fact that the most active surface also has the lowest activation energy is a classical result for electrocatalytic reactions and indicates that the energy of adsorption of the reactive intermediate plays a dominant role in the kinetics [92].

### 3.3.5 *d*-Band Center on Ir–Pd Electrodes

The electronic structures of Ir–Pd electrodes are investigated with high-resolution photoelectron spectroscopy (HR-XPS). The position of the center of the valence band is given by the equation

$$\frac{\int N(\mathrm{e})e\,de}{\int N(\mathrm{e})de} \qquad (3.11)$$

where *N*(*e*) is the DOS or, in our case, the XPS intensity after background subtraction. To obtain the *d*-band centers, we carried out the integration up to 10.0 eV binding energy (BE) with respect to $e_F$. The *d*-band center for the $Pd_{87}Ir_{13}$ electrode (the highest Pd content among all samples) is $2.45 \pm 0.08$ and is in good agreement with the value of 2.39 eV obtained from the gradient-corrected DFT periodic slab calculations by Pallassana et al. [93] on Pd(111). This demonstrates the validity of our method. The observed valence band spectra of the surface of Ir–Pd electrodes clearly revealed a reduction in the density of states near the Fermi level as an increase in Pd in Ir–Pd alloy and shifting of the *d*-band center to higher binding energy. As shown in Figure 3.24a, comparison of the *d*-band center of Ir and Pd reveals a rough linear relation between them, as a function of the Pd percentage, because the binding energy of the *d*-band center of Ir has a higher than that of Pd. Mun et al. reported a trend in the *d*-band center shift with changing Pd–Re surface composition [94]. Clearly, the direction of the *d*-band shift is consistent with Pd–Re surface composition.

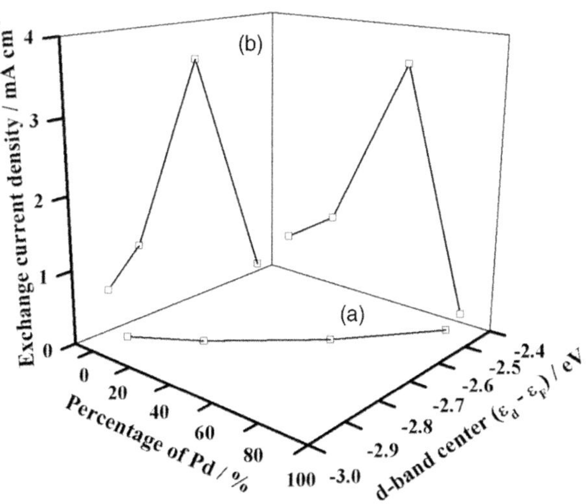


**FIGURE 3.24** (a) Linear relationship between the *d*-band center and the percentage of Pd; (b) exchange current density ($i_0$) for HOR on Ir–Pd electrode in a 0.5 M $H_2SO_4$ solution as a function of measured *d*-band centers ($e_d - e_F$; relative to the Fermi level). The *d*-band centers were measured by HR-XPS.

Plotting the kinetics currents obtained from *Koutecky–Levich plots* for hydrogen oxidation on Ir–Pd electrodes as a function of the measured *d*-band centers using synchrotron-based HR-XPS ($e_d$) (Fig. 3.24b) produces a volcano-type curve, with $Pd_{59}Ir_{41}$ possessing the maximum activity. The finding of this volcano-type dependence of Ir–Pd alloy electrodes is general. For a number of catalytic reactions, it has been shown that the activity of catalysts exhibits volcano-type dependence on some variable and that the best of the catalyst strikes a balance between two competing influences [49].

To ascertain the competing influence in the present case, an understanding the HOR mechanism is necessary, which, however, is still the subject of extensive discussion. As discussed in Section 3.2, evidence suggests that the dissociation of $H_2$ molecules into atoms or the ion–atom recombination step appears to be the rate-determining step on pure platinum. This step involves the breaking of the H−H bond. As Nørskov et al. demonstrated, a more reactive surface, such as one characterized by a higher-lying $e_d$, tends to bind adsorbates more strongly, thereby enhancing the kinetics of dissociation reactions producing these adsorbates [49,53,56]. On the other hand, a surface with a lower-lying $e_d$ tends to bind adsorbates more weakly and facilitates the formation of bonds among them. Thus it may be expected that the most activity Ir–Pd electrode should have a $e_d$ with an intermediate value.

The $Pd_{87}Ir_{13}$ electrode lies at the high-$e_d$ end of the plot and binds hydrogen much more strongly than does $Pd_{59}Ir_{41}$, where $Pd_{24}Ir_{76}$ and $Pd_0Ir_{100}$ lie at the other end and bind oxygen considerably more weakly than does $Pd_{59}Ir_{41}$. If $Pd_{59}Ir_{41}$ is taken to be close to the optimum balance as described above, then the position of the $e_d$ would suggest that $Pd_{24}Ir_{76}$, and $Pd_0Ir_{100}$ are less active for hydrogen oxidation because breaking the H−H bond is more difficult on these surfaces than on $Pd_{59}Ir_{41}$.

### 3.3.6 In Situ XANES Data of Ir–Pd Electrode-Induced Hydrogen

We measured Ir $L_{III}$ and $L_{II}$ edges using a specially designed in situ XANES cell for gas-phase experiments to confirm the relationship between HOR and the electronic structure of Ir affected by $PdH_x$ in Ir–Pd electrodes. One significant aspect of XANES analysis is that it can provide important information on the Ir *d*-band vacancy. When $H_2$ is absorbed onto an Ir metal surface, *s* electrons of $H_2$ form a coordinate covalent bond, shared with a vacant Ir *d* band. At the same time, Ir *d*-band orbitals overlap with the empty $H_2s^*$ (antibonding) orbital to form the backbonding. This strengthens the adsorptive bond between Ir and the head H atom and weakens the bond between H and the H atom in $H_2$, which is the major driving force in producing an adsorptive bond. On the other hand, this would be favorable for the hydrogen oxidation reaction, that is, Tafel reaction. Figure 3.25 shows the normalized X-ray absorption intensity near Ir $L_{III}$ and $L_{II}$ edges. As shown in Section 3.2, in metallic Pt a white line is observed at the $L_{III}$ edge but is almost absent at the $L_{II}$ edge, while for a Ir a white line is observed at both the $L_{III}$ and $L_{II}$ X-ray absorption edges. That is, for a free Pt atom the empty states in the 5*d* shell have a J value of only $\frac{5}{2}$, but the empty states of a Ir atom have two *J* values: $\frac{5}{2}$ and $\frac{3}{2}$. Figure 3.25 shows the result of calculating the *d*-band vacancy of Ir as the degree of alloying. From the difference in areas under the Ir $L_{III}$ and $L_{II}$ absorption edges between

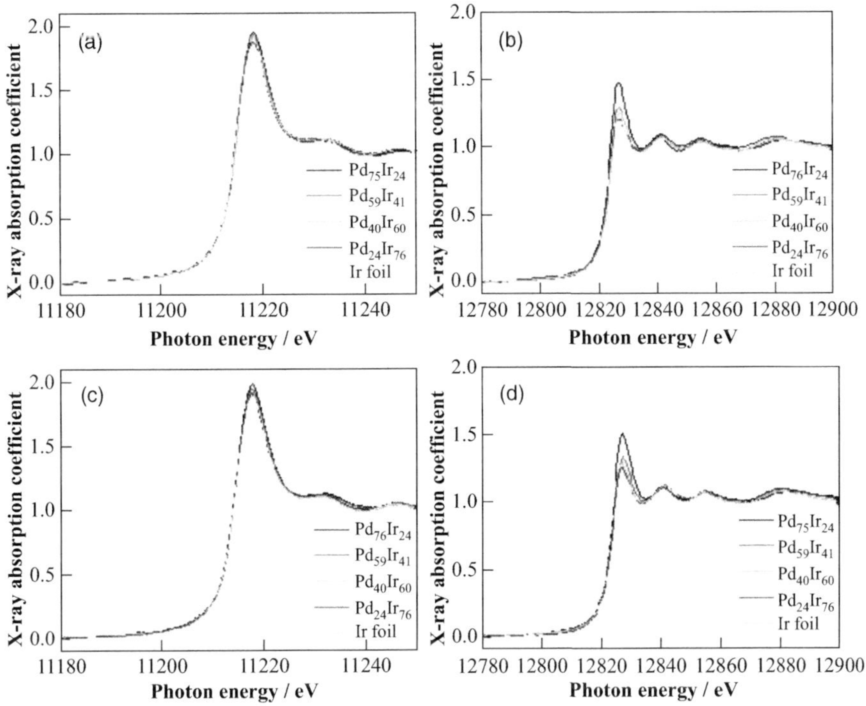


**FIGURE 3.25** (a) Ir $L_{III}$ and (b) Ir $L_{II}$ XANES spectra measured in air ambient and (c) Ir $L_{III}$ and (d) Ir $L_{II}$ XANES spectra measured after hydrogen charging ($\alpha$ or $\beta$ hydride) on Ir–Pd alloy electrodes of various compositions.

the Ir–Pd electrode samples and a Ir reference foil, the fractional change in the number of *d*-band vacancy relative to the reference material ($f_d$) was estimated [54]. This technique therefore constitutes a new way of examining the role of the alloying element in Ir alloys in terms of the *d*-band occupancy of Ir and hence is used in this investigation to elucidate the differences in electrocatalysis on the Ir–Pd electrode. As shown in Figure 3.26a, the *d*-band vacancy of Ir in air ambient is increased as the percentage of Pd increases. This is similar to the results of the *d*-band center of Ir–Pd electrodes measured by using HR-XPS (see Fig. 3.24). On the other hand, the *d*-band vacancy of Ir in $H_2$ ambient is increased more than in air ambient as the percentage of Pd increases. This means that the DOS bandwidth of Ir is decreased by the *d–d* hybridization of Ir and Pd in air ambient and the *d*-band center of Ir is upshifted. Compared with air and $H_2$ ambient, the DOS bandwidth of Ir sharply decreases by the hydrogenation of Pd, and the *d*-band center of Ir is further upshifted. These concepts are consistent with those of Nørskov et al. [49,53,56], who also elucidate the coupling between *d*-band width and *d*-band center for the near-surface alloys.

We can assume that the HOR reaction occurs mainly on the Ir surface, because Appleby et.al. reported that the exchange current density of pure Ir is higher than that of pure Pd for HOR in the low-overpotential region [95]. This allows us to directly

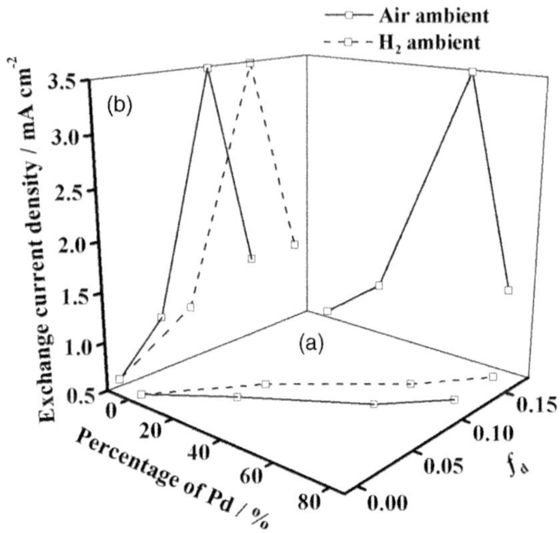


**FIGURE 3.26** (a) Fractional change in the number of *d*-band vacancies relative to the reference material ($f_d$) from XANES data measured in (—) air ambient and (- - -) $H_2$ ambient; (b) correlation between exchange current density and *d*-band vacancies from the in situ XANES data measured after hydrogen charging.

correlate the catalytic activity for HOR with the *d*-band vacancy of the electrocatalyst. As shown in Fig. 3.26b, the *d*-band vacancy of Ir versus the exchange current density exhibits the volcano curve. This result means that HOR activity is given by the strength of the H–Ir bond interaction, which depends on the position of the Ir *d*-band states relevant to the Fermi level. If the *d*-band vacancy of Ir becomes excessive, the hydrogen oxidation currents decrease, owing to a higher binding of adsorbates, which, in turn, increases the desorption energy barrier in the charge transfer step (Volmer reaction). This situation leads to a maximum in catalytic activity when the ability of the Ir–Pd alloy to bind adsorbates is neither too weak nor too strong.

## 3.4 SUMMARY

In summary, HOR electrocatalytic activity of Ir–Pd alloy electrodes showed a volcano-type dependence on *d*-band center or *d*-band vacancy, and the dependence on alloy composition revealed the principal role of the electronic effect of Ir–Pd alloys in HOR. For the first time, we also showed that there is a relationship between the number of *d*-band vacancies of Ir on Ir–$PdH_x$ alloy electrodes as a function of HOR activity. Our results imply that the high activity of the Ir-$PdH_x$ catalyst in the HOR relies heavily on the electronic structure of Ir, which is made favorable by Pd spontaneously adsorbed hydrogen gas in the electrocatalytic process. More important and from a practical perspective, we showed that it is possible to design a HOR electrocatalyst that does not contain platinum at all but can surpass the activity of pure platinum. Therefore, the

Ir–Pd alloy electrocatalyst represents a viable way to avoid using platinum and the associated cost of fuel cell electrode. The results also point to the important possibility of fine-tuning the electrocatalytic activity of transition metal.

## ACKNOWLEDGEMENT

This research was supported by WCU (World Class University) program through the National Research Foundation of Korea funded by the Ministry of Education, Science and Technology (R31-10013).

## REFERENCES

1. J. A. Rodriguez, *Surf. Sci.* **24**, 225 (1996).
2. M. Baldauf and D. M. Kolb, *J. Phys. Chem.* **1000**, 11375 (1996).
3. H. Naohara, S. Ye, and K. Uosaki, *Electrochim. Acta* **45**, 3305 (2000).
4. T. J. Schmidt, V. Stamenkovic, M. Arenz, N. M. Marković, and P. N. Ross, *Electrochim. Acta* **47**, 3765 (2002).
5. B. Hammer and J. K. Nørskov, *Adv. Catal.* **45**, 71 (2000).
6. Y. Xu, A. V. Ruban, and M. Mavrikakis, *J. Am. Chem. Soc.* **126**, 4714 (2004).
7. J. A. Rodriguez and D. W. Goodman, *Science* **257**, 897 (1992).
8. F. Buatier de Mongeot, M. Scherer, B. Gleich, E. Kopatzki, and R. J. Behm, *Surf. Sci.* **411**, 249 (1998).
9. S. Mukerjee, S. Srinivasan, M. P. Soriaga, and J. McBreen, *J. Electrochem. Soc.* **142**, 1409 (1995).
10. S. Mukerjee and S. Srinivasan, *J. Electroanal. Chem.* **357**, 201 (1993).
11. N. M. Marković and P. N. Ross, in *Interfacial Electrochemistry: Theory, Experiment and Applications*, A. Wieckowski, ed., Marcel Dekker, New York, 1999.
12. E. Christoffersen, P. Liu, A. Ruban, H. L. Skriver, and J. K. Nørskov, *J. Catal.* **199**, 123 (2001).
13. P. B. Balbuena, D. Altomare, L. A. Agapita, and J. M. Seminario, *J. Phys. Chem. B* **107**, 13671 (2003).
14. J. M. Seminario, L. A. Agapito, L. Yan, and P. B. Balbuena, *Chem. Phys. Lett.* **410**, 275 (2005).
15. V. Stamenkovic, T. J. Schmidt, P. N. Ross, and N. M. Marković, *J. Phys. Chem. B* **106**, 11970 (2002).
16. S. Pandelov and U. Stimming, *Electrochim. Acta* **52**, 5548 (2007).
17. N. M. Marković, C. A. Lucas, V. Climent, V. Stamenkovic, and P. N. Ross, *Surf. Sci.* **465**, 103 (2004).
18. H. A. Gasteiger, S. S. Kocha, B. Sompalli, and F. T. Wagner, *Appl. Catal. B* **56**, 9 (2005).
19. S. Ye and A. K. Vijh, *Electrochem. Commun.* **27**, 272 (2003).
20. V. R. Stamenkovic, B. Fowler, B. S. Mun, G. Wang, P. N. Ross, C. A. Lucas, and N. M. Marković, *Science* **315**, 493 (2007).

21. J. M. Zen and C. B. Wang, *J. Electrochem. Soc.* **141**, L51 (1994).
22. S. J. Yoo, T.-Y. Jeon, K.-S. Lee, K.-W Park, and Y.-E. Sung, *Chem. Comm.* **46**, 794 (2010).
23. A. Kowal, M. Li, M. Shao, K. Sasaki, M. B. Vukmirovic, J. Zhang, N. S. Marinkovic, P. Liu, A. I. Frenkel and R. R. Adzic, *Nature* **8**, 325 (2009).
24. K. Sawai and N. J. Suzuki, *J. Electrochem. Soc.* **151**, A682 (2004).
25. R. Bashyam and P. Zelenay, *Nature* **443**, 63 (2006).
26. W. E. Mustain, K. Kepler, and J. Prakash, *Electrochem. Commun.* **8**, 406 (2006).
27. B. Lim, M. Jiang, P. H. C. Camargo, E. C. Cho, J. Tao, X. Lu, Y. Zhu, and Y. Xia, *Science* **324**, 1302 (2009).
28. Y.-H. Cho, B. Choi, Y.-H. Cho, H.-S. Park, and Y.-E. Sung, *Electrochem. Commun* **9**, 37 (2007).
29. T. E. Springer, T. A. Zawodzinski, and S. Gottesfeld, *J. Electrochem. Soc.* **138**, 2334 (1991).
30. C. K. Dyer, *J. Power Sources* **106**, 31 (2002).
31. S. J. Yoo, Y.-H. Cho, H.-S. Park, J. K. Lee, and Y.-E. Sung, *J. Power Source* **178**, 547 (2008).
32. S. Litster and G. McLean, *J. Power Sources* **130**, 61 (2004).
33. G. J. K. Acres, J. C. Frost, G. A. Hards, R. J. Potter, T. R. Ralph, D. Thompsett, G. T. Burstein, and G. J. Hutchings, *Catal. Today* **38**, 393 (1997).
34. M. S. Wilson, J. A. Valerio, and S. Gottesfeld, *Electrochim. Acta* **40**, 353 (1995).
35. M. Watanabe and S. Motoo, *J. Electroanal. Chem.* **60**, 275 (1975).
36. L. A. Kibler, *ChemPhysChem* **7**, 985 (2006).
37. J. Lipkowski and P. N. Ross, *Electrocatalysis: Frontiers of Electrochemistry*, Wiley-VCH, New York, 1998.
38. B. E. Conway and B. V. Tilak, *Electrochim. Acta* **47**, 3571 (2002).
39. I. Fishtik, C. A. Callaghan, J. D. Fehribach, and R. Datta, *J. Electroanal. Chem.* **576**, 57 (2005).
40. D. T. Sawyer and J. L. Roberts, *Experimental Electrochemistry for Chemistry*, Wiley, New York, 1974.
41. F. A. Lewis, *The Palladium Hydrogen System*, Academic Press, New York, 1967.
42. C. T. Chan and S. G. Louie, *Phys. Rev. B* **27**, 3325 (1983).
43. F. Maroun, F. Ozanam, O. M. Magnussen, and R. J. Behm, *Science* **293**, 1811 (2001).
44. L. A. Kibler, A. M. El-Aziz, R. Hoyer, and D. M. Kolb, *Angew. Chem., Int. Ed.* **44**, 2080 (2005).
45. J. R. Kitchin, J. K. Nørskov, M. A. Barteau, and J. G. Chen, *J. Chem. Phys.* **120**, 10240 (2004).
46. S. J. Yoo, H.-Y. Park, T.-Y. Jeon, I.-S. Park, Y.-H. Cho, and Y.-E. Sung, *Angew. Chem., Int. Ed.* **120**, 9447 (2008).
47. H. A. Gasteiger, N. M. Marković, and P. N. Ross, *J. Phys. Chem.* **99**, 8290 (1995).
48. A. J. Bard and L. R. Faulkner, in *Electrochemical Methods*, Wiley, New York, 2001.
49. R. Sander, *Compilation of Henry's Law Constants for Inorganic and Organic Species of Potential Importance in Environmental Chemistry*, version 3, Mainz, Germany, 1999 (http: //www.henrys- law.org).
50. P. W. Atkins, *Physical Chemistry*, 6th ed., Oxford Univ. Press, Oxford, UK, 1998.
51. R. M. Q. Mello and E. A. Ticianelli, *Electrochim. Acta* **42**, 1031 (1997).

52. N. M. Marković and P. N. Ross, *Surf. Sci. Rep.* **45**, 117 (2002).
53. N. M. Marković, B. N. Grgur, and P. N. Ross, *J. Phys. Chem. B* **101**, 5405 (1997).
54. D. R. Lide, ed., *CRC Handbook of Chemistry and Physics*, 85th ed., CRC Press, Boca Raton, FL, 2004.
55. P. A. Bennett and J. C. Fuggle, *Phys. Rev. B* **26**, 6030 (1982).
56. S. Mizusaki, T. Miyatake, N. Sato, I. Yamamoto, M. Yamaguchi, M. Itou, and Y. Sakurai, *Phys. Rev. B* **73**, 113101 (2006).
57. J. H. Weaver, *Phys. Rev. B* **11**, 1416 (1975).
58. B. W. Faughnan, R. S. Crandall, and P. M. Heyman, *RCA Rev.* **36**, 177 (1975).
59. S. J. Yoo, Y. H. Jung, H. G. Choi, D. K. Kim, J. W. Lim, and Y.-E. Sung, *Appl. Phys. Lett.* **90**, 173126–1 (2007).
60. C. D. Gelatt Jr. H. Ehrenreich, and J. A. Weiss, *Phys. Rev. B* **17**, 1940 (1978).
61. P. A. Bennett and J. C. Fuggle, *Phys. Rev. B* **26**, 6030 (1982).
62. L. M. Falicov, W. Hanke, and M. P. Maple, *Valence Fluctuation in Solids*, North-Holland, Amsterdam, 1981.
63. A. E. Russell, S. Maniguet, R. J. Mathew, J. Yao, M. A. Roberts, and D. Thompsett, *J. Power Sources* **96**, 226 (2001).
64. J. K. Nørskov, T. Bligaarda, A. Logadottira, S. Bahna, L. B. Hansena, M. Bollingera, H. Bengaarda, B. Hammerb, Z. Sljivancaninb, M. Mavrikakisc, Y. Xuc, S. Dahld, and C. J. H. Jacobsend, *J. Catal.* **209**, 275 (2002).
65. Y. Xu, A. V. Ruban, and M. Mavrikakis, *J. Am. Chem. Soc.* **126**, 4717 (2004).
66. A. Michaelides, Z.-P. Liu, C. J. Zhang, A. Alavi, D. A. King, and P. Hu, *J. Am. Chem. Soc.* **125**, 3704 (2003).
67. J. Greeley and M. Mavrikakis, *Nature Mater.* **3**, 810 (2004).
68. J. Greeley, J. K. Nørskov, L. A. Kibler, A. M. El-Aziz, and D. M. Kolb, *ChemPhysChem* **7**, 1032 (2006).
69. A. N. Mansour, J. W. Cook, and D. E. Sayers, *J. Phys. Chem.* **88**, 2330 (1984).
70. B. E. Conway, E. M. Beatty, and P. A. D. DeMaine, *Electrochim. Acta* **7**, 39 (1962).
71. J. K. Nørskov, T. Bligaard, A. Logadottir, J. R. Kitchin, J. G. Chen, S. Pandelov, and U. Stimming, *J. Electrochem. Soc.* **152**, J23 (2005).
72. T. Bligaard and J. K. Nørskov, *Electrochim. Acta* **52**, 5512 (2007).
73. S. Srinivasan, D. J. Manko, H. Koch, M. A. Enayetullah, and A. J. Appleby, *J. Power Sources* **29**, 367 (1990).
74. S. Hirano, J. Kim, and S. Srinivasan, *Electrochim. Acta* **42**, 1587 (1997).
75. R. O'Hayre, S.-J. Lee, S.-W. Cha, and F. B. Prinz, *J. Power Sources* **109**, 483 (2002).
76. C. A. Cavalca, J. H. Arps, and M. Murthy, US Patent 6,300,000 (2001).
77. H. Hahn and R. S. Averback, *J. Appl. Phys.* **67**, 1113 (1990).
78. G. M. Chow, R. L. Holtz, A. Pattnaik, A. S. Edelstein, T. E. Schlesinger, and R. C. Cammarata, *Appl. Phys. Lett.* **56**, 1853 (1990).
79. C. G. Granqvist and R. A. Buhrman, *J. Appl. Phys.* **47**, 2200 (1976).
80. S. Yatsuya, T. Kamakura, K. Yamauchi, and K. Mihama, *Jpn. J. Appl. Phys.* **25**, 42 (1986).
81. J. M. Soler and N. Garcia, *Phys. Rev. Lett.* **49**, 1857 (1982).
82. V. Radmilvic, H. A. Gasteiger, and P. N. Ross, *J. Catal.* **154**, 98 (1995).

83. R. Woods, in *Electroanalytical Chemistry*, A. J. Bard, ed., Marcel Dekker, New York, 1976.
84. B. E. Conway, *Electrodes of Conductive Metallic Oxides, Part B*, S. Trasatti, ed., Elsevier, Amsterdam, 1981.
85. S. Chen and A. Kucernak, *J. Phys. Chem. B* **108**, 13984 (2004).
86. K. J. Vetter, in *Electrochemical Kinetics*, S. Bruckensein and B. Howard, eds., Academic Press, New York, 1967.
87. H. A. Gasteiger, N. M. Marković, and P. N. Ross, *J. Phys Chem.* **99**, 3411 (1995).
88. C. Gabrielli, P. P. Grand, A. Lasia, and H. Perrot, *J. Electrochem. Soc.* **151**, A1937 (2004).
89. C. Gabrielli, P. P. Grand, A. Lasia, and H. Perrot, *J. Electrochem. Soc.* **151**, A1925 (2004).
90. W. Vogel, L. Lundquist, P. Ross, and P. Stonehart, *Electrochim. Acta* **20**, 79 (1975).
91. R. F. Mann, J. C. Amphletta, B. A. Peppleya, and C. P. Thurgooda, *J. Power Sources* **163**, 679 (2007).
92. R. Parsons, *Trans. Faraday Soc.* **54**, 1053 (1958).
93. V. Pallassana, M. Neurock, L. B. Hansen, B. Hammer, and J. K. Nørskov, *Phys. Rev. B* **60**, 6140 (1999).
94. B. S. Mun, C. Lee, V. Stamenkovic, N. M. Marković, and P. N. Ross, *Phys. Rev. B* **71**(1), 115420 (2005).
95. A. J. Appleby, H. Kita, M. Chemla, and G. Bronoel, *Encyclopedia of the Electrochemistry of the Elements*, IXa, Marcel Dekker, New York, 1973.

CHAPTER 4

# Mechanism of an Enhanced Oxygen Reduction Reaction at Platinum-Based Electrocatalysts: Identification and Quantification of Oxygen Species Adsorbed on Electrodes by X-Ray Photoelectron Spectroscopy

MITSURU WAKISAKA and MASAHIRO WATANABE
Fuel Cell Nanomaterials Center, University of Yamanashi, Kofu, Japan

HIROYUKI UCHIDA
Clean Energy Research Center, University of Yamanashi, Kofu, Japan

In this chapter, a novel XPS study of the enhanced ORR mechanism of Pt-based alloy electrodes is presented in detail. The methodology employed for the identification and quantification of oxygen-containing species adsorbed on the Pt-based electrodes by XPS is described. The assignments of oxygen species are extensively discussed by utilizing a well-defined Pt(111) single-crystal electrode. Four types of oxygen species were distinguished, with binding energies at 529.6, 530.5, 531.1, and 532.6 eV; the first two were assigned to adsorbed atomic oxygen and hydroxyl, while the latter two were assigned to the bilayer water molecules. The coverage of each oxygen species was found to vary with the electrode potential and strongly depend on the Pt electrode types. The coverages of oxygen species in $O_2$-saturated solution were also found to differ from those in $N_2$-purged solution. Specifically, the coverage of atomic oxygen increased in the potential region where ORR took place. As compared to pure Pt, the enhanced ORR mechanism at the Pt-based alloy electrodes is discussed, on the basis of the coverage changes of atomic oxygen during ORR as well as apparent rate constants for the ORR evaluated by hydrodynamic voltammograms using channel flow double electrodes.

*Fuel Cell Science: Theory, Fundamentals, and Biocatalysis*,
Edited by Andrzej Wieckowski and Jens K. Nørskov

## 4.1 INTRODUCTION

Active catalysts for oxygen reduction reaction (ORR) are important for the development of efficient polymer electrolyte fuel cells [1]. Enhanced ORR activities of Pt alloyed with nonprecious metals such as Fe, Co, and Ni have been reported in acid electrolyte solutions [2–12]. In terms of the design of high-performance cathode catalysts, it is essential to analyze how such an enhanced ORR activity appears at these alloys. The ORR mechanism at several Pt alloys has been investigated intensively and discussed in terms of various factors, including structural effects such as changes in short-range atomic order, particle size effects, electronic effects involving increased Pt *d*-band vacancies at Pt skin layers, and inhibition of Pt–OH formation [2–12]. Mukerjee and coworkers have reported changes in lattice constants and Pt *d*-band vacancies at alloy electrodes as compared to pure Pt in their studies using X-ray diffraction (XRD) and X-ray absorption spectroscopy (XAS) [6,7]. We also have found core-level shifts of Pt 4*d* and 4*f* at the Pt skin layer on alloys by X-ray photoelectron spectroscopy (XPS), which correlate with the 5*d*-band center shifts [2–4,13].

Theoretical calculations have been employed for elucidating the enhanced ORR mechanism at Pt-based alloy electrodes [14–18]. Nørskov and coworkers have calculated the effects of alloying of Pt on the electronic structures and chemisorption energies of oxygen species by means of density functional theory (DFT) [17,18]. They concluded that the volcano dependence of ORR activity at the Pt-based alloy electrodes could be explained by the adsorption energy of atomic oxygen on the surface, which was strongly related to the *d*-band center of the alloys [18]. However, those theoretical studies have not considered the surface oxidation state of the electrode surface as well as the presence of water molecules under the influence of an electric field (electrode potential). It is still essential to establish the nature and coverages of the oxygen-containing species on the electrocatalyst surface during ORR in order to advance the understanding of the enhanced ORR mechanism. In situ extended X-ray absorption fine structure (EXAFS) has been utilized to analyze the oxygen species on Pt-based electrodes [6,7,19]. Changes in the coordination number of oxygen around Pt and the Pt–O bond length were reported in these papers [6,7,19]. However, it is difficult to distinguish the various types of possible adsorbed species, such as oxygen atoms ($O_{ads}$), hydroxyl groups ($OH_{ads}$), and water molecules ($H_2O_{ads}$), because of the insufficient resolution.

In contrast, XPS has distinct advantages in surface sensitivity and energy resolution, with the ability to detect small binding energy (BE) shifts [20]. In this chapter, a novel XPS study of the enhanced ORR mechanism at Pt-based alloy electrodes is presented in detail. The methodology employed for the identification and quantification of oxygen-containing species adsorbed on the Pt-based electrodes by XPS is described in Section 4.2. The assignments of oxygen species, utilizing a well-defined Pt(111) single-crystal electrode, are intensively discussed in Section 4.3. The surface oxidation process and intermediates of ORR at the Pt-based electrodes are demonstrated in Sections 4.3 and 4.4, respectively, on the basis of changes in the surface composition of oxygen species. Then, in Section 4.5, the enhanced ORR mechanism at Pt-based alloy electrodes is discussed, compared to that at pure Pt (polycrystalline), on the basis of coverage changes of oxygen species at the alloy electrodes as well as apparent rate constants for the ORR evaluated by hydrodynamic voltammograms using channel flow double electrodes (CFDEs) [21,22].

## 4.2 METHODOLOGY

### 4.2.1 EC-XPS Apparatus

A closed ultra-high-vacuum (UHV) system combined with an electrochemical cell was employed to characterize electrode surfaces without exposing them to air [13,23,24]. Combined electrochemical–UHV systems were originally developed by Hubbard and coworkers [25], and a sample transfer system and electrochemical (EC) chamber employed, as illustrated in Figure 4.1, were originally developed by Itaya and coworkers [26]. The methodology of the combined UHV systems has been reviewed in the literature [1,27]. The XPS chamber was equipped with a hemispherical electron spectrometer, an X-ray source, a monochromator, and an ion gun for surface cleaning [13]. The photoelectron takeoff angle and analysis area of the specimen were 45° and $0.8 \times 2$ mm, respectively. The electron spectrometer was calibrated by using Au $4f_{7/2}$ (83.98), Ag $3d_{5/2}$ (368.26), and Cu $2p_{3/2}$ (932.67 eV) lines before and after each experiment [28]. The EC chamber was backfilled with $N_2$ (purified by a catalyst filter) or $O_2$, and evacuated rapidly by two sorption pumps and a cryopump. A Neoflon-based electrochemical cell (EC cell), with a nozzle-like Luggin capillary in the center, was placed in a separate chamber, which was connected with a reservoir of 0.1 M HF electrolyte solution through Teflon tubing, as shown in Figure 4.1b [13]. The shape of the EC cell allows us to exchange solutions saturated with $N_2$ or $O_2$ rapidly, especially near the test electrode surface described below, under potential control. A Pt foil was placed in the EC cell as a counterelectrode (CE). A reversible hydrogen electrode (RHE) was used to control the electrode potential, which was placed out of the chamber and connected to the EC cell with a Teflon tube filled with 0.1 M HF solution [23,24].

### 4.2.2 Test Electrodes

Platinum-based alloy or pure Pt thin films (thickness 80 nm) were prepared on Au polycrystalline disks (diameter 13 mm; thickness 1 mm) by DC magnetron sputtering a Pt target together with or without a second metal target (Fe or Co) in an Ar plasma at room temperature. The composition of the Pt alloys was controlled by varying the ratio of the sputtering currents for the Pt and the second metal, and was analyzed by XPS. The grazing incidence XRD patterns of the alloys indicated the formation of solid solutions with face-centered cubic (fcc) phase. After the introduction into the UHV system, clean surfaces of the test electrodes were obtained by scanning $Ar^+$ ion bombardment at 1 keV (0.68 μA) for 10 min without annealing.

The alloy electrodes were electrochemically stabilized by the formation of a Pt skin layer during several repetitive potential sweeps in the HF solution: dissolution of alloy

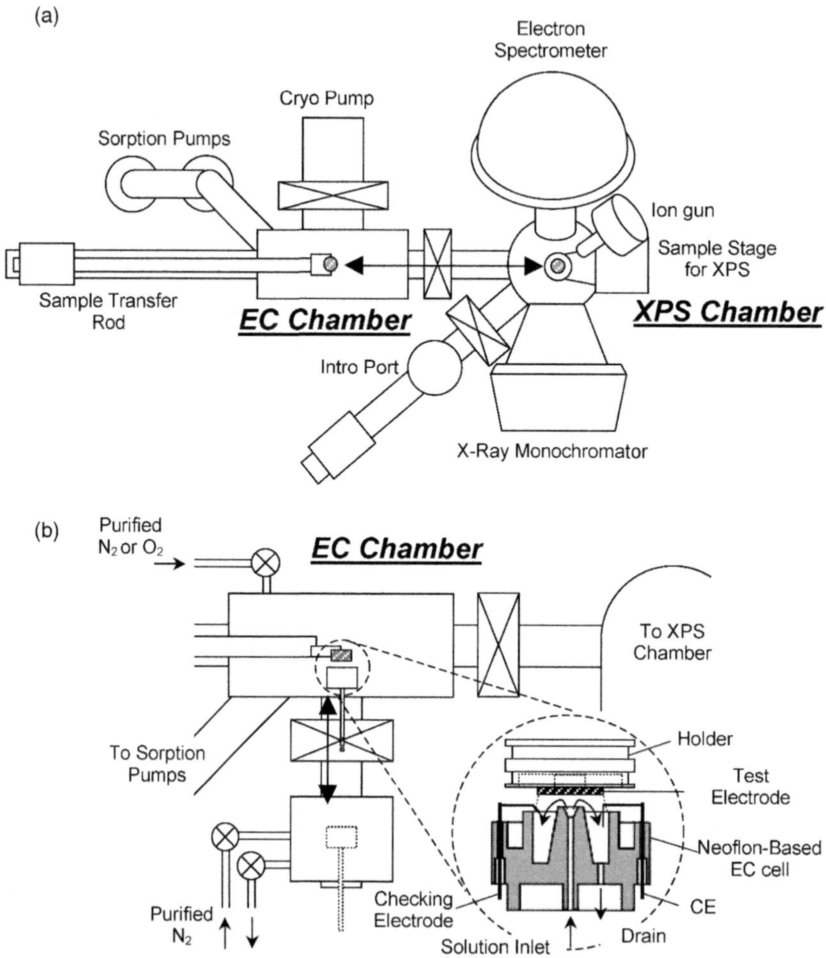


**FIGURE 4.1** Schematic views of (a) the entire UHV apparatus (topview) and (b) the EC chamber (sideview). The configuration of the EC cell and sample electrode is illustrated in the dashed-line circle.

components followed by a rearrangement of the surface [3,29,30]. The skin formation has been confirmed by electrochemical scanning tunneling microscopy [29] and quartz crystal microbalance studies [30]. The electrochemical stabilization of the alloy electrodes was performed in the XPS apparatus by the following procedure: (1) immersion of the clean Pt-based alloy electrode into the electrolyte solution saturated with $N_2$ under potential control at 75 mV and holding the electrode potential for 10 min, (2) repeated potential cycling between 75 and 1000 mV at a sweep rate of 100 mV/s until the cyclic voltammogram (CV) showed a steady state, and (3) maintaining the electrode potential at 75 mV again for 40 min to complete the rearrangement of the surface [29]. Figure 4.2 shows core-level (CL) spectra (Pt 4*f*) for the alloy

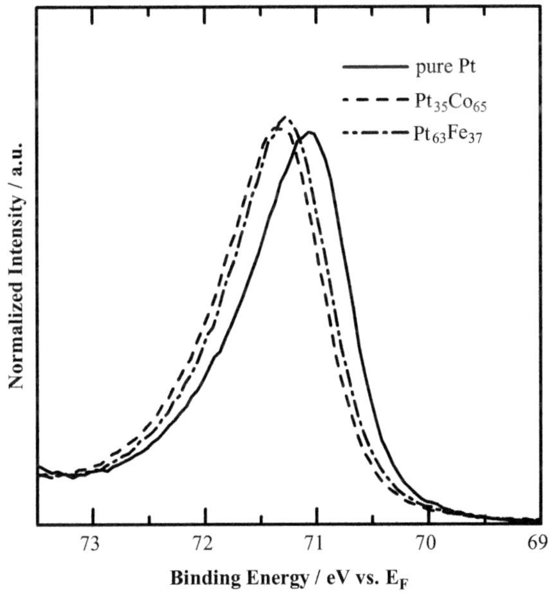


**FIGURE 4.2** Area-normalized core-level (CL) spectra of $Pt4f_{7/2}$ for pure Pt (solid line), Pt skin/$Pt_{35}Co_{65}$ (dashed line), and Pt skin/$Pt_{63}Fe_{37}$ (dashed–dotted line) after the electrochemical stabilization; the emersion potential was 1.00 V versus RHE.

electrodes after the electrochemical stabilization, normalized to total Pt 4*f* photoelectron intensity (area). The Pt 4*f* CLs for both Pt–Fe and Pt–Co electrodes are clearly observed to shift to higher binding energy with respect to pure Pt, clearly indicating the modified electronic structure of the Pt skin layer by the underlying alloys.

A well-defined Pt(111) single crystal (diameter 10 mm; thickness 0.6 mm) was also employed in order to confirm the assignments of the oxygen species based on binding energy shifts of O 1*s*. The Pt(111) surface was freshly prepared by annealing in a hydrogen flame and subsequently quenching with pure water before the XPS measurements [31].

### 4.2.3 Emersion Process

To obtain O 1*s* spectra without any interference, 0.1 M HF electrolyte solution was employed in this work, because HF contains neither oxygen nor a specifically adsorbing anion [23,24]. Each test electrode was polarized at a given potential for 5 min in $N_2$-purged solution, or for 15 min in an $O_2$-saturated solution. Then, the electrode was emersed from the solution under potential control and promptly transferred to a UHV chamber evacuated by a combination of pumps. The electrode surface was spontaneously cooled to a temperature below 220 K, due to the latent heat of evaporation of the electrolyte droplet remaining on the electrode surface during the transfer to UHV (*freeze–evacuation*) [24]. Figure 4.3 shows the temperature change

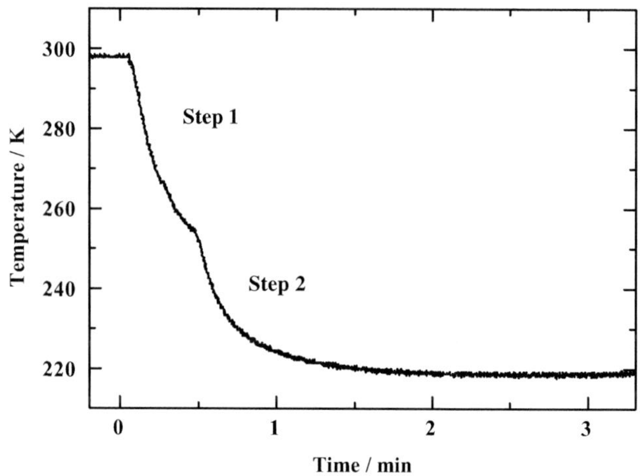


**FIGURE 4.3** Temperature changes of a test Pt electrode (diameter 13 mm; thickness 1 mm) during the evacuation process, measured by an attached thermocouple (chromel–alumel).

of a test Pt electrode (diameter 13 mm; thickness 1 mm) during the evacuation process, measured by an attached thermocouple. The temperature of the test electrode was found to decrease in two steps. In the first step, the electrode was cooled down to ~255 K following evaporation of the electrolyte droplet. At the same time, the electrolyte droplet was frozen on the electrode surface. In the next step, the electrode was further cooled down by sublimation of the frozen electrolyte on the surface in vacuum. At such low temperatures, bilayer water molecules can partially remain on the Pt surface, even in UHV, whereas the bulk water must be completely desorbed during the evacuation process. The XPS measurement was carried out within 6 min after the electrode emersion.

### 4.2.4 Deconvolution of O 1*s* Spectra

Oxygen 1*s* spectra obtained in the XPS measurements were deconvoluted into several asymmetric Lorentzian–Gaussian peaks with a linear background [20,23,24]. Peak parameters such as the full width at half-maximum (FWHM), tail scale (TS), and Gaussian-to-Lorentzian ratio (G/L), were referred to those for a submonolayer of atomic oxygen ($O_{ad}$) formed on a polycrystalline Pt surface by $O_2$ exposure in vacuum [32]. The curve fittings were performed for all spectra with the FWHM, TS, and G/L parameters fixed as constants while allowing the peak energies and areas to vary [23,24]. Figure 4.4 shows typical results of the deconvolution of the O 1*s* spectra for pure Pt (polycrystalline) after emersion from 0.1 M HF solution saturated with $N_2$ at 0.40 and 1.00 V [23]. The O 1*s* spectrum at 0.40 V was deconvoluted into two components, 532.4 and 531.1 eV, which can be assigned to two types of bilayer water molecules, because no oxidation of the Pt surface occurs at this low electrode potential. The O 1*s* spectrum at 1.00 V, where the electrode surface was oxidized,

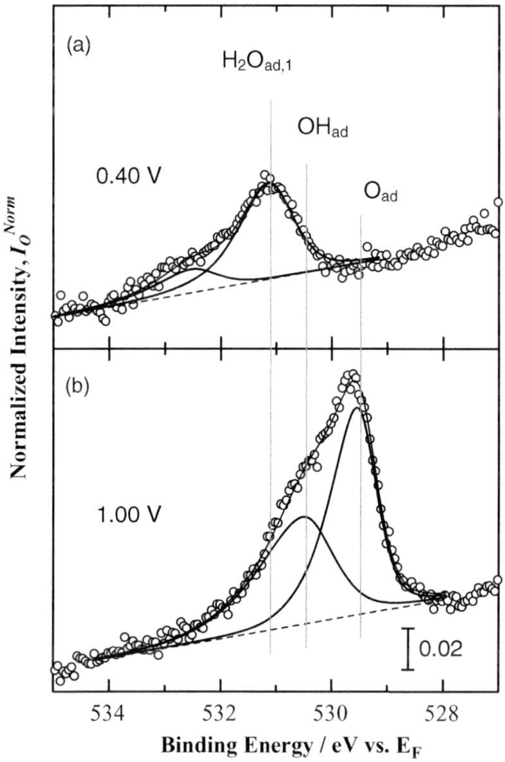


**FIGURE 4.4** Deconvolution of O 1*s* spectra for polycrystalline Pt at 0.4 and 1.00 V in 0.1 M HF solution saturated with $N_2$; the spectral intensities have been normalized to that of Pt. Open circles and bold lines indicate experimental data points and deconvoluted components, respectively. Dashed lines indicate the linear backgrounds, and thin solid lines show the resulting calculated fits to the experimental points.

was also deconvoluted into two components, 530.5 and 529.5 eV, assigned to hydroxyl ($OH_{ad}$) and atomic oxygen ($O_{ad}$), respectively. The assignments of oxygen-containing species, utilizing a well-defined Pt(111) single crystal, are extensively discussed in Section 4.3.

### 4.2.5 Coverage Calculation

To evaluate the changes in the amounts of adsorbed oxygen species, the intensities of the O 1*s* spectra at pure Pt and Pt-based alloy electrodes were normalized in area as follows. In the case of pure Pt, the intensity for O 1*s* was normalized by that of Pt 4*f* [23]

$$I_{\mathrm{O}}^{\mathrm{norm}} = \frac{I_{\mathrm{O}}/S_{\mathrm{O}}}{I_{\mathrm{Pt}}/S_{\mathrm{Pt}}} \tag{4.1}$$

where $I_O^{norm}$, $I_i$, and $S_i$ are the normalized O 1*s* intensity, measured photoelectron intensity, and sensitivity factor of element *i* ($i = O$ or Pt), respectively. We used the values of $S_O$ and $S_{Pt}$ provided by PHI [28]. In the case of the Pt skin/alloy electrodes, we calculated the $I_O^{norm}$ values using the following equation, taking into account the contribution of nonprecious metal 2*p* photoelectrons from the underlying alloy [23]:

$$I_O^{norm} = \frac{I_O/S_O}{I_{Pt}/S_{Pt} + I_{Me}/S_{Me}} \tag{4.2}$$

On the basis of the normalized photoelectron intensity $I_O^{norm}$ of each component, we calculated a fractional coverage $\theta$ of each species (atomic ratio of oxygen in the species to Pt). Because the photoelectron intensity of the metal substrate was only slightly attenuated by the submonolayer of oxygen species, the following simple relation can be utilized [20]

$$\theta = M_{O/Pt} \cdot I_O^{norm} \tag{4.3}$$

where $M_{O/Pt}$ is a monolayer matrix factor [20], 2.4, obtained from an $O_2$ exposure of $1 \times 10^8$ L to a polycrystalline Pt surface in vacuum at room temperature, yielding $\theta = 0.4$ [32]. In the present study, the photoelectron takeoff angle was set at 45° in the "magic angle" range, in which the substrate roughness does not interfere with the overlayer calculation [33–35].

## 4.3 ASSIGNMENT OF OXYGEN SPECIES (SURFACE OXIDATION PROCESS)

### 4.3.1 Pt(111) Single-Crystal and Polycrystalline Pt

Figure 4.5 shows two sets of O 1*s* spectra for Pt(111) single-crystal and polycrystalline electrodes emersed from 0.1 M HF solution saturated with $N_2$ at electrode potentials *E* from 0.4 to 1.1 V versus RHE [23,24]. The intensities and shapes of the O 1*s* spectra were found to change with the potential. The changes can be attributed to the changes in the surface state, namely, coverage and composition of oxygen-containing species adsorbed on the Pt electrode accompanied with the surface oxidation. Examples of the O 1*s* deconvolution at the Pt(111) single-crystal electrode are shown in Figure 4.6.

On the basis of the deconvolutions, four oxygen species with discrete BEs were distinguished, for potentials $E < 1.1$ V. Figure 4.7c shows the BEs of the deconvoluted oxygen species on Pt(111) as a function of the electrode potential [24]. Surprisingly, the same four oxygen species were distinguished at both the Pt(111) single crystal and polycrystalline Pt. The near identity of the BEs on both surfaces (see Fig. 4.7d) suggests that the chemical state of each oxygen species might be independent of the crystal face exposed. The BEs of the four oxygen species were found to be nearly constant at 529.6, 530.5, 531.1, and 532.6 eV, with variations of ±0.2 eV. The species at 529.6 and 530.5 eV can be assigned to $OH_{ad}$ and $O_{ad}$, in agreement with the

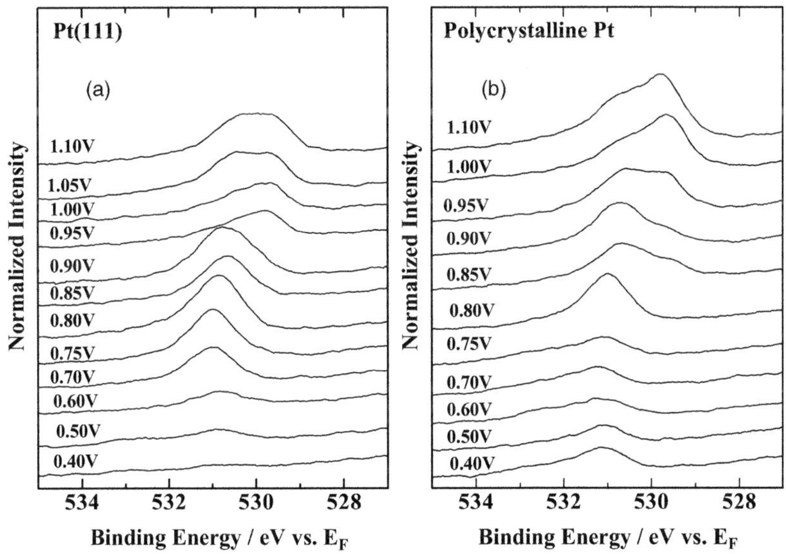


**FIGURE 4.5** Normalized O 1*s* spectra for (a) Pt(111) single-crystal and (b) polycrystalline Pt electrodes after emersion from 0.1 M HF solution purged with $N_2$ at various electrode potentials from 0.40 to 1.10 V.

assignments of these species on Pt(111) formed in UHV [36–38]. The species at 532.6 eV can be assigned to $H_2O_{ad}$, also in good agreement with the value found for Pt(111) in UHV [36,39–41]. The remaining species exhibited a BE of 531.1 eV, close to that for $OH_{ad}$. However, because this species was observed on polycrystalline Pt, even at $E = 0.4$ V, where no surface oxidation had occurred (see Figs. 4.4 and 4.7b), we assign this species to a second type of $H_2O_{ad}$, with a chemical state different from that for the one observed at 532.6 eV. The BEs of the respective oxygen species were found to be independent of the type of electrode, namely, single- or polycrystalline. Therefore, the two types of $H_2O_{ad}$ were not ascribed to different crystal facets. Also, the second type of $H_2O_{ad}$ was not attributed to a solvated water molecule interacting with $F^-$ or $HF_2^-$, because survey spectra for both Pt(111) and polycrystalline Pt revealed no F signal at any electrode potential [24].

It has been well established that water molecules form an icelike bilayer structure on the Pt(111) surface at low temperatures in UHV [39]. Ogasawara et al. have analyzed this bilayer structure by using XPS and evaluated the distance between the two water layers from the O 1*s* BE split, on the basis of a final-state hole-screening theory [42]. The theory explains the lower O 1*s* BE of the first-layer water molecule and predicts a short distance between the molecule and the Pt surface due to the image charge. Thus, the $H_2O_{ad}$ observed herein at 531.1 eV can be assigned to the first-layer water, which is bound directly to the Pt surface through an O lone pair, while the feature at 532.6 eV can be assigned to the second-layer water, which interacts with the first layer via hydrogen bonding [24]. The proposed bilayer model is illustrated in

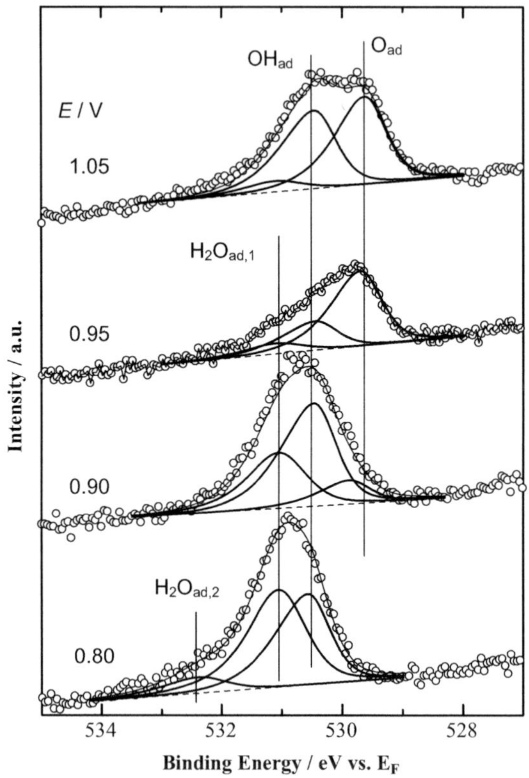


**FIGURE 4.6** Deconvolution of O 1*s* spectra for the Pt(111) single-crystal electrode at various potentials. Open circles and bold lines indicate experimental data and the resulting deconvoluted peaks of the components, respectively. Dashed lines indicate the linear backgrounds, and thin solid lines indicate the calculated fits.

Figure 4.8a [24]. According to this model, the theoretical coverage of $H_2O_{ad}$ is 0.67 at Pt(111). Hereafter, the two water species at 531.1 and 532.6 eV are denoted by $H_2O_{ad,1}$ and $H_2O_{ad,2}$, respectively. The BE of $H_2O_{ad,1}$ formed in HF solution was significantly lower than that reported for $H_2O_{ad}$ formed in UHV, indicating a shorter length of the bond to Pt, probably due to an electric field effect (i.e., a lowered Pt Fermi level) corresponding to higher electrode potentials. Toney et al. have reported such an effect on the bond distance between water and the Ag(111) electrode surface in their X-ray scattering study [43]. They observed that the average Ag–O distance decreased by 1 Å at potentials above that of zero charge (PZC) in NaF solution [43]. A theoretical study has reported that a change by 0.96 Å in the Pt–water distance yields a shift of 1.9 eV in BE [42]. Therefore, the low BE for $H_2O_{ad,1}$, 531.1 eV, is quite reasonable for $E >$ PZC (~0.3 V vs. RHE [44]). The intensity of $H_2O_{ad,2}$ was fairly small over the whole potential region, as compared to $H_2O_{ad,1}$. This might be due to the desorption of $H_2O_{ad,2}$ weakly bound to the first layer during the electrode transfer to UHV. In

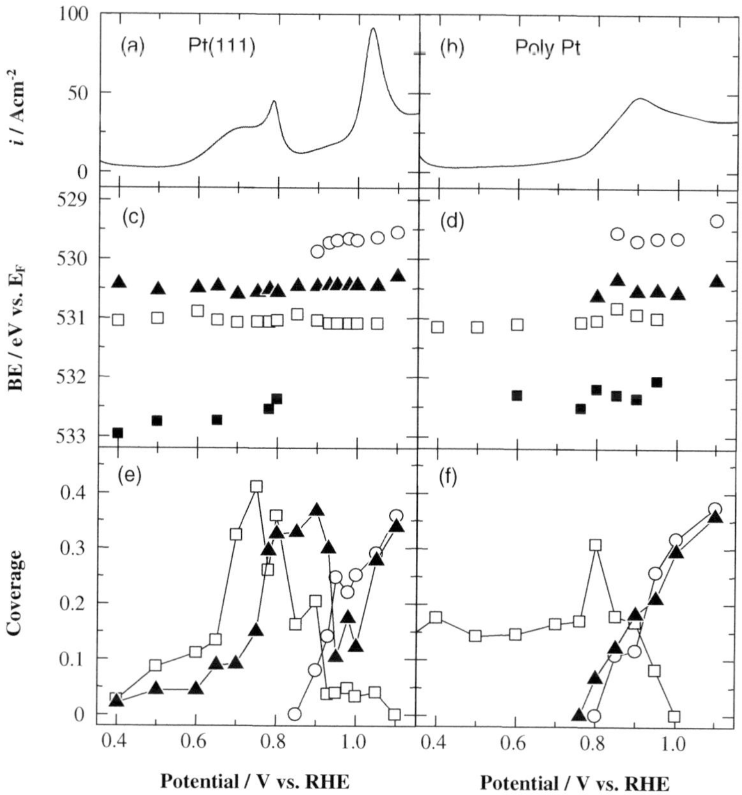


**FIGURE 4.7** Comparisons of the properties at Pt(111) single-crystal and polycrystalline Pt electrodes in 0.1 M HF solution purged with $N_2$: voltammograms (50 mV/s) at (a) Pt(111) and (b) polycrystalline Pt; deconvoluted BEs for different oxygen species at (c) Pt(111) and (d) polycrystalline Pt; and coverage changes of each oxygen species as a function of electrode potential at (e) Pt(111) and (f) polycrystalline Pt; symbols: (○) $O_{ad}$, (▲)$OH_{ad}$, (□) $H_2O_{ad,1}$, and (■) $H_2O_{ad,2}$. (Reproduced from Ref. 24.)

contrast, the strongly bound $H_2O_{ad,1}$ remained on the electrode surface during the XPS measurement, probably due to assistance by the freeze–evacuation process.

It is possible to argue that the species at 530.5 and 531.1 eV could be assigned to $OH_{ad}$ on different facets, rather than the latter being assigned to $H_2O_{ad,1}$, or that the species at 530.5 and 529.6 eV could be assigned to $OH_{ad}$ on different facets, rather than the latter being assigned to $O_{ad}$. A productive approach to resolve such questions is to confirm the agreement between spectroscopically estimated and electrochemically estimated values for the number of electrons per Pt site [$N$(e)/$N$(Pt)] for the adsorption of oxygen-containing species. The θ values for the adsorbates proposed here were determined from the photoelectron intensities. The adsorption steps involving each species, $H_2O_{ad,1}$, $OH_{ad}$, and $O_{ad}$, can be associated with electron transfer numbers ($z$) of 0, 1, and 2, respectively. Therefore, e/Pt values can be calculated for the adsorption

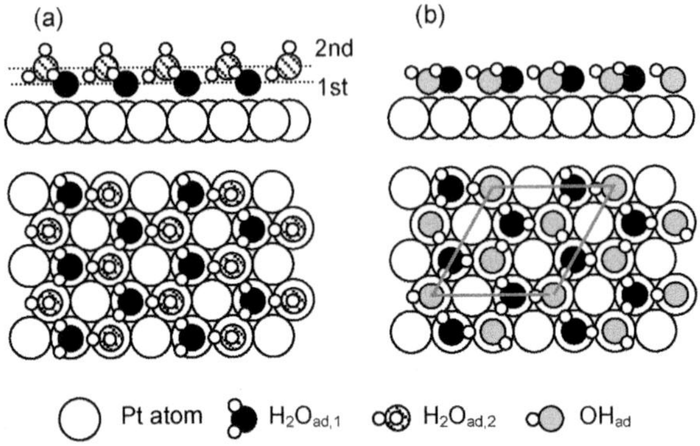


**FIGURE 4.8** Structural models for (a) the water bilayer on Pt(111) and (b) the mixed $OH_{ad}$/$H_2O_{ad,1}$ layer, denoted as Pt(111)–(3 × 3)–3(OH + $H_2O$).

of each of these species at each potential, based on the corresponding θ values. For example, for any given potential, we obtain

$$N(\mathrm{e})/N(\mathrm{Pt}) = \theta[\mathrm{OH_{ad}}] + 2\theta[\mathrm{O_{ad}}] \tag{4.4}$$

Four different assignment schemes could be postulated for both Pt(111) and polycrystalline Pt, as presented in Table 4.1, including the assignment mentioned above, which is designated as model 1 [23,24]. Alternatively, on the basis of the integrated charge for the CV, we have calculated the corresponding e/Pt values. Both results—those based on the XPS results and the CV results—are shown in Figure 4.9 for (a) Pt(111) and (b) polycrystalline Pt [23,24]. If the species at 531.1 eV were assigned to $OH_{ads}$ rather than $H_2O_{ads,1}$, the electron/surface Pt site (e/Pt) value based on the coverage must deviate from that independently measured from the voltammogram. Model 1, however, gives the best fit (the smallest $\sum\chi^2$) to the experimental values over the whole potential region for both electrode types. Hence, the component at 531.1 eV should be assigned to $H_2O_{ads,1}$ ($z = 0$), the component at 530.5 eV to $OH_{ads}$ ($z = 1$), and the component at 529.6 eV to $O_{ads}$ ($z = 2$) [23,24].

**TABLE 4.1 Proposed Models for Assignment of the Three Deconvoluted Oxygen 1*s* Species**

| | Oxygen-Containing Species[a] | | |
|---|---|---|---|
| Model | 531.1 eV | 530.5 eV | 529.6 eV |
| 1(●) | $H_2O_{ad}[0]$ | $OH_{ad}[1]$ | $O_{ad}[2]$ |
| 2(◇) | $H_2O_{ad}[0]$ | $OH_{ad}[1]$ | $OH_{ad}[1]$ |
| 3(▲) | $OH_{ad}[1]$ | $OH_{ad}[1]$ | $O_{ad}[2]$ |
| 4(□) | $OH_{ad}[1]$ | $O_{ad}[2]$ | $O_{ad}[2]$ |

[a] The number of electrons transferred (or valence) is shown in brackets.

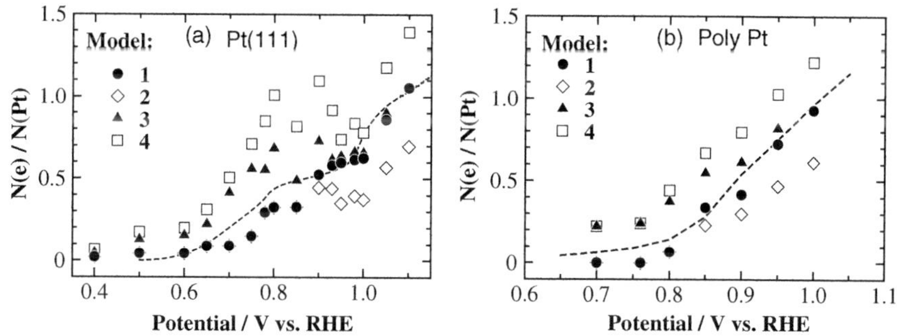


**FIGURE 4.9** Calculated N(e)/N(Pt) (transferred electron number/Pt site) for (a) Pt(111) and (b) polycrystalline Pt, using XPS-derived values for θ, with *N*(e) calculated for the four models, allocating the *z* values to each species shown in Table 4.1. The dashed line indicates the *N*(e)/*N*(Pt) curve based on the charge density for the surface oxidation at Pt(111) in the CV measurement.

Figure 4.7e,f show the fractional coverages of the oxygen species, $O_{ad}$, $OH_{ad}$, and $H_2O_{ad,1}$ (atomic ratio of each oxygen species to Pt), as a function of electrode potential at Pt(111) and polycrystalline Pt, respectively. On Pt(111), $OH_{ad}$ appeared at $E > 0.6$ V, accompanying the surface oxidation current in the "butterfly" region (see Fig. 4.7a) [31,45]. More recently, Rai et al. simulated the formation of $OH_{ads}$ in the butterfly region by density functional theory (DFT) [46]. However, to our knowledge, our very recently (2009) reported XPS result was the first direct experimental evidence [24] that the anodic current in this region is due to $OH_{ads}$ formation as the first step of Pt surface oxidation, via the following scheme:

$$H_2O_{ad} \rightarrow OH_{ad} + H^+ + e^- \tag{4.5}$$

At $E = 0.80$ V, the coverage θ of $OH_{ads}$ was found to reach 0.36 (86 μC cm$^{-2}$), in good agreement with the 92 μC/cm$^2$ charge density measured electrochemically from 0.50 to 0.80 V (excluding the sharp spike; see below). It should be noted that $\theta[H_2O_{ad,1}]$ also increased in the butterfly region. This increase can be explained on the basis of hydrogen bond formation between $H_2O_{ad,1}$ and the $OH_{ad}$ appearing in this region. It has been reported in a thermal desorption study that the water molecule was more strongly adsorbed on Pt surfaces covered with $OH_{ad}$ [36]. The increment in $H_2O_{ad,1}$ by the appearance of $OH_{ad}$ was found also for the polycrystalline Pt at 0.80 V, compared with that at 0.70 V (see Fig. 4.7f) [24].

Remarkably, when $\theta[OH_{ad}]$ and $\theta[H_2O_{ad,1}]$ became nearly equal on Pt(111) at $E = 0.80$ V, the sum of these coverages reached a maximum of 0.68, coinciding within experimental error with the reported coverage of bilayer water molecules [39,42]. This agreement suggests that the honeycomb framework of the bilayer remained during the initial surface oxidation process (0.50–0.80 V). A structural model for the mixed $OH_{ad}/H_2O_{ad,1}$ layer at 0.80 V is illustrated in Figure 4.8b [24]. This structure, with $\theta[OH_{ad}] = \theta[H_2O_{ad,1}]$, should be the most stable among the possible mixed

$OH_{ad}/H_2O_{ad}$ layers, because the hydrogen bonding network (HBN) is complete, as reported in a theoretical study [47]. The decrease in $\theta[H_2O_{ad,1}]$ at $E > 0.80$ V suggests that the HBN was disturbed, and $H_2O_{ad,1}$ became unstable. It should be noted that the sharp spike in the voltammogram around 0.80 V, similar to that observed in sulfuric acid solution [45], might reflect a structural change of this adlayer relating to the HBN. Berná et al. suggested in a discussion of their CV data that the sharp spike around 0.8 V in perchloric acid solution could be ascribed to the formation of OH originating from a water molecule interacting with the perchlorate anion [48]. In the present XPS research, however, such solvation water interacting with $F^-$ or $HF_2^-$ was never observed [24].

At $E = 0.90$ V, $\theta[O_{ad}]$ commenced to increase, whereas $\theta[OH_{ad}]$ decreased on Pt(111). This suggests that the $O_{ad}$ originated from an oxidation of $OH_{ad}$ via the following scheme:

$$OH_{ad} \rightarrow O_{ad} + H^+ + e^- \tag{4.6}$$

At $E > 1.00$ V, both $\theta[O_{ad}]$ and $\theta[OH_{ad}]$ were found to increase, accompanying the large anodic peak (see Fig. 4.7a). In contrast to the Pt(111), $\theta[O_{ad}]$ and $\theta[OH_{ad}]$ on polycrystalline Pt commenced to increase almost in the same manner at $E = 0.8$ V. It should be pointed out that the $O_{ad}$ forms on polycrystalline Pt at almost the same potential as Pt(111), whereas the $OH_{ad}$ appears on Pt(111) at lower potential than on polycrystalline Pt, suggesting that the formation of $OH_{ad}$ is very sensitive to crystal face structures such as terraces, steps, kinks, or grain size, while that of $O_{ad}$ is less dependent on such structure [24].

### 4.3.2 Pt Skin Layer on Alloy Electrode

Figure 4.10 shows normalized CVs, based on the real Pt surface area, at pure Pt (polycrystalline) and two Pt skin/Pt-based alloy electrodes (Pt–Fe and Pt–Co) in 0.1 M HF solution purged with $N_2$. Strictly speaking, the onset of oxide formation at the Pt skin/alloy electrodes during the positive-going scan was found to shift to a potential slightly higher than that at pure Pt. However, the electrical charge for reduction of the oxide formed, $\Delta Q_{OX}$, at the Pt skin/alloy electrodes up to 1.00 V were not changed significantly compared to that at pure Pt, suggesting almost the same surface oxidation state on both electrodes at 1.00 V.

Figure 4.11a shows normalized O 1*s* spectra for pure Pt and Pt skin/alloy electrodes after emersion at various potentials *E* from $N_2$-purged 0.1 M HF solution. At $E = 1.00$ V, the pure Pt and two Pt skin/alloy electrodes exhibited almost the same O 1*s* spectra, indicating that the surface oxidation state became almost the same, which is consistent with the identical $\Delta Q_{OX}$ at these electrodes as stated above. In contrast, at $E < 1.00$ V, the O 1*s* spectra were quite different in shape between pure Pt and the Pt skin/alloys. However, the deconvolution of these peaks clearly indicates that the O 1s spectra at the Pt skin/alloy electrodes consisted of the same four components as those for pure Pt. Figure 4.12 shows typical results of the O 1*s* deconvolution for the Pt skin/Pt–Fe alloy electrode [23]. Figure 4.13 shows the peak position of each species as

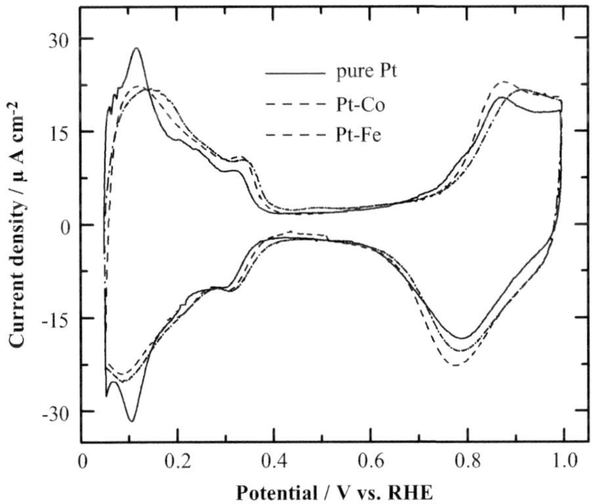


**FIGURE 4.10** Cyclic voltammograms at pure Pt (solid line), Pt skin/$Pt_{35}Co_{65}$ (dashed line), and Pt skin/$Pt_{63}Fe_{37}$ (dashed–dotted line) electrodes in $N_2$-purged 0.1 M HF solution after the electrochemical stabilization (see text). The scan rate was 20 mV/s; the CVs are normalized to the real Pt surface area.

a function of the potential at both pure Pt and Pt skin/alloy electrodes. It is clearly seen in Figure 4.13 that each peak position, or binding energy, is almost constant (within an error of ±0.2 eV) at potentials between 0.7 and 1.0 V. The difference in shape of the O 1*s* spectra between pure Pt and Pt skin/alloy electrodes can be explained mainly by the

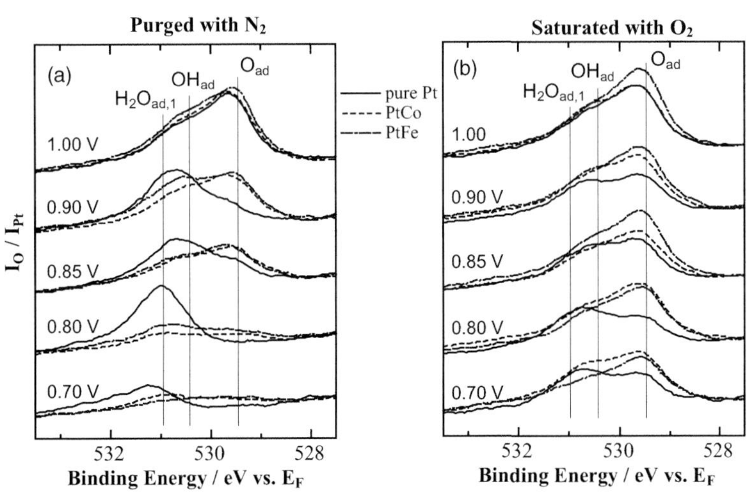


**FIGURE 4.11** Normalized XP spectra of O 1*s* for pure Pt (solid lines), Pt skin/$Pt_{35}Co_{65}$ (dashed lines), and $Pt_{63}Fe_{37}$ (dashed–dotted lines) electrodes after the emersion from 0.1 M HF solutions: (a) purged with $N_2$ and (b) with $O_2$.

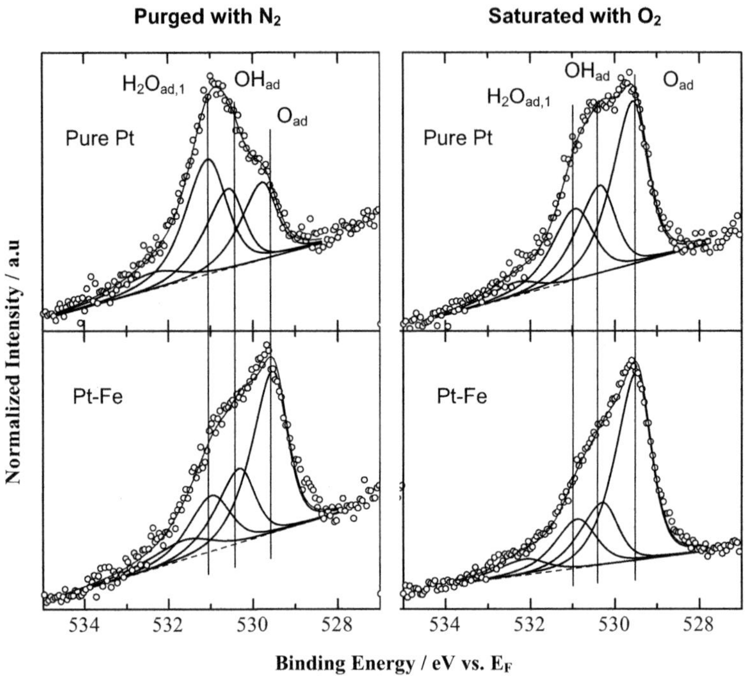


**FIGURE 4.12** Typical results of the deconvolution of O 1*s* spectra at 0.85 V for pure Pt and Pt skin/$Pt_{63}Fe_{37}$ electrodes.

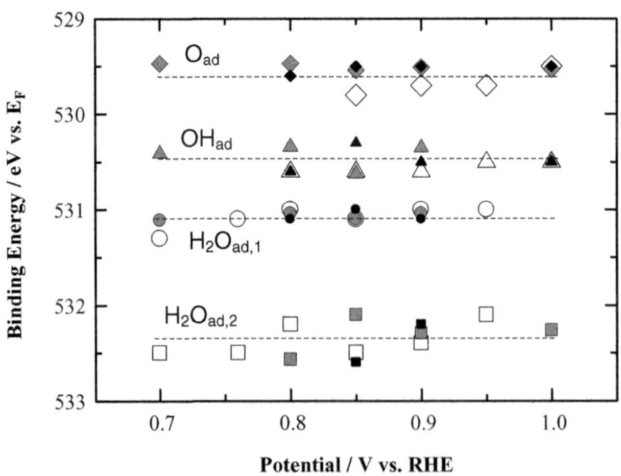


**FIGURE 4.13** XPS (BE) peak positions of each species as a function of the electrode potential at pure Pt, Pt skin/$Pt_{35}Co_{65}$, and Pt skin/$Pt_{63}Fe_{37}$ electrodes after emersion from 0.1 M HF solution purged with $N_2$, indicated by open, gray, and solid symbols, respectively.

difference in coverage of $H_2O_{ad,1}$ as shown in Figure 4.11a, suggesting a weaker bonding of $H_2O_{ad,1}$ to the Pt skin layer than that to pure Pt.

## 4.4 INTERMEDIATES OF ORR

After emersion from $O_2$-saturated solution, dramatic changes were observed in the O 1*s* spectra compared to those for the $N_2$-saturated solution, as shown in Figure 4.11b. However, the observed spectra were also deconvoluted into the same four components (see Fig. 4.12) [23]. In the potential region in which ORR proceeds ($E < 0.90$ V), the intensity of $O_{ad}$ increased markedly at both Pt skin/alloy and pure Pt, compared with those in Figure 4.11a. Such an increase in the oxygen species is certainly brought about by an adsorption of oxygen dissolved in the solution and/or intermediates during ORR.

At this point, we focus on the coverage $\theta$ of each oxygen-containing species quantitatively. In the potential range between 0.70 and 1.00 V, no $H_2O_2$ was detected by the present XPS measurement [23], which is consistent with our previous results with electrochemical measurements, in which $H_2O_2$ was not detected at a pure Pt electrode and only a 1% yield was detected at a Pt–Fe alloy electrode [22]. Therefore, we can consider the $O_{ad}$, $OH_{ad}$, and $H_2O_{ad,1}$ as the dominant species that are presumably involved in the ORR mechanism. Figure 4.14 shows $\theta$ values of these species as a function of each electrode potential $E$ at pure Pt and two Pt skin/alloy electrodes (Pt–Fe and Pt–Co) [49]. Compared to the $N_2$-purged solution, the $\theta[O_{ad}]$ values in $O_2$-saturated solution clearly increased on pure Pt at $E < 0.90$ V, where ORR occurs. The $\theta[OH_{ad}]$ values also increased in the region from 0.80 to 0.70 V, but $\theta[H_2O_{ad}]$ decreased slightly in the region from 0.90 to 0.80 V, perhaps due to displacement by $O_{ad}$. As seen in the voltammograms in Figure 4.14a, the ORR currents in the region from 1.00 to 0.75 V on both Pt skin/alloy electrodes were appreciably higher than that at pure Pt, indicating the enhanced ORR activity of the Pt skin/alloys. It is striking that the $\theta[O_{ad}]$ values for the Pt skin/alloy electrodes in $O_2$-saturated solution were approximately a factor of 2 larger than that at pure Pt in the potential region from 0.90 to 0.70 V.

## 4.5 MECHANISM OF ENHANCED ORR AT Pt-BASED ELECTRODES

From hydrodynamic voltammograms using CFDE [21,22], apparent rate constants $k_{app}$ for ORR at pure Pt and Pt alloys (Pt–Fe, Pt–Co, and Pt–Ni) were evaluated at a constant overpotential in 0.1 M $HClO_4$. For Arrhenius plots of the $k_{app}$ at 20–50 °C as shown in Figure 4.15, the value of $k_{app}$ on the $Pt_{54}Fe_{46}$ alloy was found to be larger than that on pure Pt by a factor of 4, but the apparent activation energy $e_a$ on both electrodes was almost identical (~40 kJ mol$^{-1}$ at 0.76 V) [22]. On the basis of such results, we previously proposed that the high ORR activities at the alloys could be ascribed to a large preexponential factor $Z$ in the following Arrhenius equation [22]:

$$k_{app} = Z \exp(-e_a/RT) \tag{4.7}$$

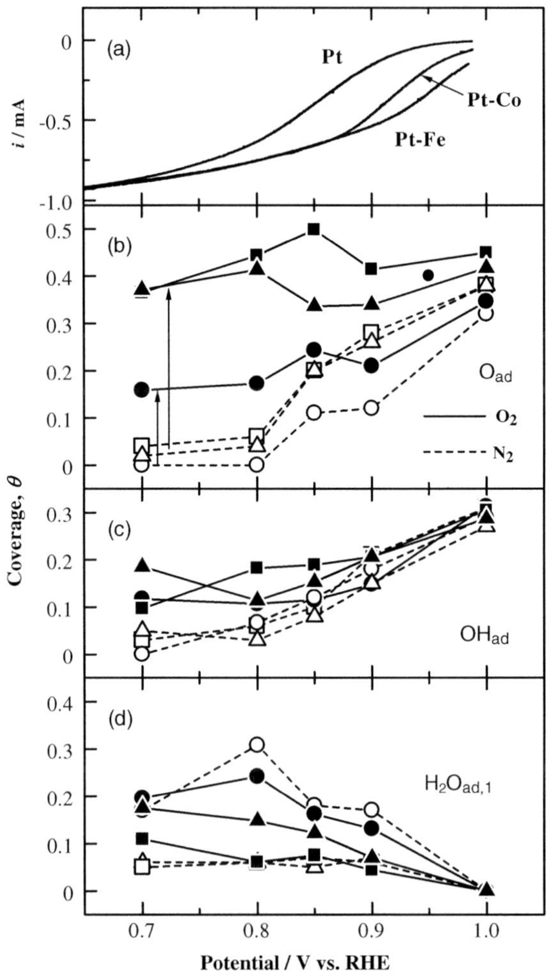


**FIGURE 4.14** (a) Linear sweep (2 mV/s, negative-going) voltammograms for ORR at pure Pt, Pt skin/$Pt_{35}Co_{65}$, and Pt skin/$Pt_{63}Fe_{37}$ electrodes in $O_2$-saturated 0.1 M HF solution. Coverages of oxygen species: (b) $O_{ad}$, (c) $OH_{ad}$, and (d) $H_2O_{ad,1}$ on the pure Pt (circles), Pt skin/$Pt_{35}Co_{65}$ (triangles) and Pt skin/$Pt_{63}Fe_{37}$ (squares) electrodes in $N_2$-(dashed lines) and $O_2$-saturated (solid lines) solutions.

As shown in Figure 4.14a, similar enhancements were observed in 0.1 M HF solution at the Pt skin layers on the Pt–Fe and Pt–Co alloy surfaces in the present work. The enrichment of $O_{ad}$ at the Pt skin/alloy electrodes observed in the EC-XPS measurements should be responsible for the increase in the $Z$ value, if $O_{ad}$ is included in the rate-determining step (RDS). It was also found that the Tafel lines for the alloys shifted to more positive potentials, with slopes similar to that at pure Pt: approximately −120 mV/dec in the high-current region and about −75 mV/dec in the low-current

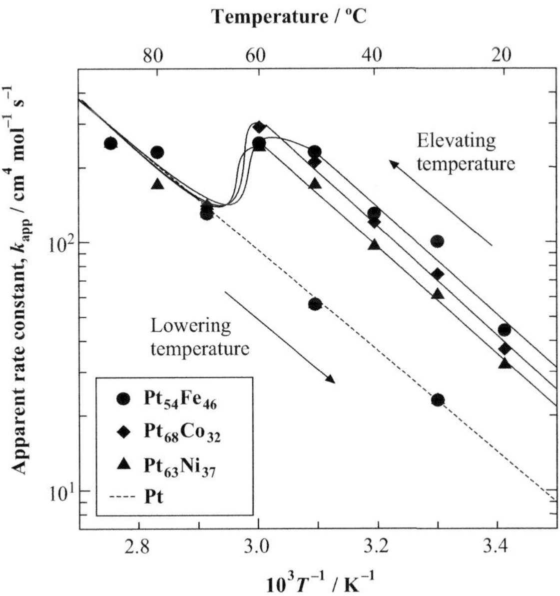


**FIGURE 4.15** Arrhenius plots of apparent rate constants $k_{app}$ for ORR at pure Pt, $Pt_{54}Fe_{46}$, and $Pt_{68}Co_{32}$ electrodes at $-0.525$ V versus $E^\circ$ [22].

region. This suggested that the increase in the ORR activities at the Pt skin layers over the alloys is dominantly brought about by promoting the same RDS as that for pure Pt, for example, the first one-electron transfer step [4,22].

$$x\,\mathrm{Pt} + \mathrm{O_2} + \mathrm{H^+} + \mathrm{e^-} \rightarrow \mathrm{Pt}_x\mathrm{-O_2^*-H} \tag{4.8}$$

where $Pt_x$-$O_2$*-H is the activated complex, which has not been identified thus far.

Here, based on the present EC-XPS results, we propose that $O_{ad}$ is one of the limiting reactants and that the RDS is the elementary step (4.10), followed by reaction (4.11):

$$\mathrm{O_2} \rightarrow 2\,\mathrm{O_{ad}} \tag{4.9}$$

$$\mathrm{O_{ad}} + \mathrm{H^+} + \mathrm{e^-} \rightarrow \mathrm{OH_{ad}} \tag{4.10}$$

$$\mathrm{OH_{ad}} + \mathrm{H^+} + \mathrm{e^-} \rightarrow \mathrm{H_2O} \tag{4.11}$$

Compared with pure Pt, the most prominent point for the Pt skin/alloy emersed from $O_2$-saturated solution is the large increase in $\theta[O_{ad}]$ around the ORR onset potential (1.00 V), indicating that the initial reaction (4.9) is relatively fast compared with the subsequent elementary steps (4.10) and/or (4.11). Hence, step (4.10) is certainly promoted on the alloy, presumably due to the large increase in $\theta[O_{ad}]$. It is a very important conclusion that the reaction rate of (4.11) is not so fast, based on the fact that

$OH_{ad}$ still remains on the surface, even in the limiting current region (0.70 V) for both electrodes.

Another essential feature of the Pt skin layer over the alloy is a considerable decrease in $\theta[H_2O_{ad}]$ with respect to that at pure Pt for both $N_2$- and $O_2$-saturated solutions. This suggests that the dissociative adsorption of the $O_2$ molecule may not be obstructed by the adsorbed water at the Pt skin. Such a lower affinity to water is consistent with that proposed earlier by Mukerjee and coworkers [7]. However, an inhibition of ORR due to "OH poisoning" [7,9] at the alloy electrodes was never observed in the present research. Because we confirmed the positive core-level shift in Pt 4*f* at the Pt skin in Section 4.2.2 [13], these differences of oxygen-containing species between pure Pt and Pt-based alloys can be roughly ascribed to modified electronic properties of the Pt skin layer by the underlying alloy [2–13]. Theoretical calculations using density functional theory (DFT) have been reported in order to explain the enhanced ORR activity at Pt alloys [18]. These calculations have shown that Pt alloys with Ni, Co, or Fe have smaller bonding energies to $O_{ad}$ and $OH_{ad}$ than does pure Pt, suggesting the inhibition of $O_{ad}$ and $OH_{ad}$ formation at the alloy electrodes [18]. However, it should be emphasized that such an inhibition of $OH_{ad}$ formation was definitely not observed in the present study, whereas an increase of $\theta[O_{ad}]$ was observed for Pt skin/alloy electrodes. In the theoretical studies [17,18], we note that the effect of the electronic states on the rate of step (4.10) has not been evaluated, for either pure metals or alloys, in the presence of water molecules under the influence of an electric field. As described in Section 4.2.2, the positive CL shift for Pt 4*f* can be explained by an upshift of the Fermi level, corresponding to a downshift of the *d*-band center. The upshift of the Fermi level (or corresponding *d*-band center shift) may cause an increase of backdonation from the *d* band of Pt to the $\pi^*$ orbital of the $O_2$ molecule, promoting the $O_2$ dissociation, as proposed in our previous study [4].

## 4.6 SUMMARY

In this chapter, a novel XPS study of the enhanced ORR mechanism at Pt-based alloy electrodes is presented in detail. The oxygen species on Pt(111) single-crystal, polycrystalline Pt and Pt skin/Pt alloy electrodes in $N_2$- and $O_2$-saturated 0.1 M HF solution have been positively identified by XPS. Four types of oxygen species were distinguished, with BEs at 529.6, 530.5, 531.1, and 532.6 eV; the first two were assigned to $O_{ad}$ and $OH_{ad}$, while the latter two were assigned to the bilayer water molecules, $H_2O_{ad,1}$ and $H_2O_{ad,2}$. The BEs of the respective oxygen species were found to be nearly constant over the whole potential region and independent of the Pt electrode types. In contrast, the quantitative coverage of each species varied considerably and was found to be dependent on the electrode potential, crystal face, and electronic properties. Furthermore, we have for the first time demonstrated clear differences in the surface oxidation processes for Pt(111) and polycrystalline Pt electrodes. The present XPS results provide new insights into the ORR mechanism at Pt-based electrodes and important clues for the design of new, higher-activity electrocatalysts.

## REFERENCES


1. W. Vielstich, A. Lamm, and H. A. Gasteiger, eds., *Handbook of Fuel Cell: Fundamentals, Technology and Applications*, Wiley, New York, 2003.
2. T. Toda, H. Igarashi, and M. Watanabe, *J. Electrochem. Soc.* **145**, 4185 (1998).
3. T. Toda, H. Igarashi, and M. Watanabe, *J. Electroanal. Chem.* **460**, 258 (1999).
4. T. Toda, H. Igarashi, H. Uchida, and M. Watanabe, *J. Electrochem. Soc.* **146**, 3750 (1999).
5. S. Mukerjee and S. Srinivasan, *J. Electroanal. Chem.* **357**, 201 (1993).
6. S. Mukerjee, S. Srinivasan, M. P. Soriaga, and J. McBreen, *J. Electrochem. Soc.* **142**, 1409 (1995).
7. V. S. Murthi, R. C. Urian, and S. Mukerjee, *J. Phys. Chem. B* **108**, 11011 (2004).
8. V. Stamenkovic, T. J. Schmidt, P. N. Ross, and N. M. Markovic, *J. Phys. Chem. B* **106**, 11970 (2002).
9. U. A. Paulus, A. Wokan, G. Scherer, T. J. Schmidt, V. Stamenkovic, N. M. Markovic, and P. M. Ross, *J. Phys. Chem. B* **106**, 4181 (2002).
10. M. Min, J. Cho, K. Cho, and H. Kim, *Electrochim. Acta* **45**, 4211 (2000).
11. L. Xiong and A. Manthiram, *J. Electrochem. Soc.* **152**, A697 (2005).
12. V. R. Stamenkovic, B. S. Mun, M. Arenz, K. J. J. Mayrhofer, C. A. Lucas, G. Wang, P. N. Ross, and N. M. Markovic, *Nature Mater.* **6**, 241 (2007).
13. M. Wakisaka, S. Mitsui, Y. Hirose, K. Kawashima, H. Uchida, and M. Watanabe, *J. Phys. Chem. B* **110**, 23489 (2006).
14. A. Anderson and T. V. Albu, *J. Electrochem. Soc.* **147**, 4229 (2000).
15. J. Roques and A. B. Anderson, *J. Electrochem. Soc.* **151**, E85 (2004).
16. Y. Xu, A. V. Ruban, and M. Mavrikakis, *J. Am. Chem. Soc.* **126**, 4717 (2004).
17. J. K. Nørskov, J. Rossmeisl, A. Logadottir, L. Lindqvist, J. R. Kitchin, T. Bligaard, and H. Jónsson, *J. Phys. Chem. B* **108**, 17886 (2004).
18. V. Stamenkovic, B. S. Mun, K. J. J. Mayrhofer, P. N. Ross, N. M. Markovic, J. Rossmeisl, J. Greeley, and J. K. Nørskov, *Angew. Chem. Int. Ed.* **45**, 2897 (2006).
19. M. Tada, S. Murata, T. Asakoka, K. Hiroshima, K. Okumura, H. Tanida, T. Uruga, H. Nakanishi, S. Matsumoto, Y. Inada, M. Nomura, and Y. Iwasawa, *Angew. Chem. Int. Ed.* **46**, 4310 (2007).
20. D. Briggs and M. P. Seah, eds., *Practical Surface Analysis*, 2nd ed., Vol. 1, *Auger and X-Ray Photoelectron Spectroscopy*; Wiley, New York, 1990.
21. N. Wakabayashi, M. Takeichi, H. Uchida, and M. Watanabe, *J. Electroanal. Chem.* **574**, 339 (2005).
22. N. Wakabayashi, M. Takeichi, H. Uchida, and M. Watanabe, *J. Phys. Chem. B* **109**, 5836 (2005).
23. M. Wakisaka, H. Suzuki, M. Mitsui, H. Uchida, and M. Watanabe, *J. Phys. Chem. C* **112**, 2750 (2008).
24. M. Wakisaka, H. Suzuki, M. Mitsui, H. Uchida, and M. Watanabe, *Langmuir* **25**, 1897 (2009).
25. R. M. Ishikawa and A. T. Hubbard, *J. Electroanal. Chem.* **69**, 317 (1976).
26. T. Yamada, N. Batina, and K. Itaya, *J. Phys. Chem.* **99**, 8817 (1995).

27. W. Chrzanowski and A. Wieckowski, in *Interfacial Electrochemistry: Theory, Experiment, and Application*, Marcel Dekker, New York, 1999, p. 937.
28. J. Chastain and R. C. King Jr., eds., *Handbook of X-Ray Photoelectron Spectroscopy*, Physical Electronics, Inc., USA, 1995.
29. L. J. Wan, T. Moriyama, M. Ito, H. Uchida, and M. Watanabe, *Chem. Commun.* 58 (2002).
30. H. Uchida, H. Ozuka, and M. Watanabe, *Electrochim. Acta* **47**, 3629 (2002).
31. J. Clavilier, K. E. Achi, and A. Rodes, *J. Electroanal. Chem.* **272**, 253 (1989).
32. P. R. Norton, *Surf. Sci.* **47**, 98 (1975).
33. W. S. M. Werner, *Surf. Interface Anal.* **23**, 696 (1995).
34. P. L. J. Gunter, O. L.J. Gijzeman, and J. W. Niemantsverdriet, *Appl. Surf. Sci.* **115**, 342 (1997).
35. P. Kappen, K. Reihs, C. Seidel, M. Voetz, and H. Fuchs, *Surf. Sci.* **465**, 40 (2000).
36. G. B. Fisher and J. L. Gland, *Surf. Sci.* **94**, 446 (1980).
37. M. Kinne, T. Fuhrmann, J. F. Zhu, B. Trankenschuh, R. Denecke, and H. P. Steinruck, *Langmuir* **20**, 1819 (2004).
38. C. Puglia, A. Nilsson, B. Hernnäs, O. Karis, P. Bennich, and N. Mårtensson, *Surf. Sci.* **342**, 119 (1995).
39. M. A. Henderson, *Surf. Sci. Rep.* **46**, 1 (2002).
40. W. Ranke, *Surf. Sci.* **209**, 57 (1989).
41. M. Kiskinova, G. Pirug, and H. P. Bonzel, *Surf. Sci.* **150**, 319 (1985).
42. H. Ogasawara, B. Brena, D. Nordlund, M. Nyberg, A. Pelmenschikov, L. G.M. Pettersson, and A. Nilsson, *Phys. Rev. Lett.* **89**, 276102 (2002).
43. M. F. Toney, J. N. Howard, J. Richer, G. L. Borges, J. G. Gordon, O. R. Melroy, D. G. Wiesler, D. Yee, and L. B. Sorensen, *Surf. Sci.* **335**, 326 (1995).
44. M. J. Weaver, *Langmuir* **14**, 3932 (1998).
45. F. T. Wagner and P. N. Ross, *J. Electroanal. Chem.* **250**, 301 (1988).
46. V. Rai, M. Aryanpour, and H. Pitsch, *J. Phys. Chem. C* **112**, 9760 (2008).
47. C. Clay, S. Haq, and A. Hodgson, *Phys. Rev. Lett.* **92**, 046102 (2004).
48. A. Berná, V. Climent, and J. M. Feliu, *Electrochem. Commun.* **9**, 2789 (2007).
49. M. Watanabe, M. Wakisaka, H. Yano, and H. Uchida, in *ECS Transactions, Proton Exchange Membrane Fuel Cells 8*, The Electrochemical Society, NJ, 2008, Vol. 16, p. 199.

CHAPTER 5

# Biocathodes for Dioxygen Reduction in Biofuel Cells

RENATA BILEWICZ

Faculty of Chemistry, University of Warsaw, Warsaw, Poland

MARCIN OPALLO

Institute of Physical Chemistry, Polish Academy of Sciences, Warsaw, Poland

This chapter reviews the development of biocathodes having potential application for biofuel cells. We have restricted this area to electrodes for dioxygen reduction because of the ubiquitous nature of this oxidant. The enzymes used for this reaction—laccase, bilirubin oxidase, and copper efflux oxidase—belong to the multicopper oxidase family. They catalyze four-electron dioxygen reduction to water without generating oxygen intermediates. Numerous methods of their immobilization on the electrode surfaces, including adsorption, covalent bonding, polymer encapsulation, or sol–gel processes, are discussed. The biocathodes can be classified in two groups, one of them involving mediated and the other, mediatorless electrocatalysis. The electrodes belonging to the former group are modified not only with the enzyme but also with the mediator shuttling the electrons between the protein and the conductive substrate. Osmium polymers and 2,2′-azino-bis(3-ethylbenzthiazoline-6-sulfonate) are among the most common mediators for laccase or bilirubin oxidase. Biocathodes exhibiting mediatorless electron transfer are often modified with conductive nanoparticles: gold or carbon, or carbon nanotubes enabling facile heterogeneous electron transfer and increasing the electrode working surface. Although the typical lifetime of the biocathode ranges from a few hours to a few days, much more recently some examples of electrodes working for weeks or even months have been reported.

*Fuel Cell Science: Theory, Fundamentals, and Biocatalysis*,
Edited by Andrzej Wieckowski and Jens K. Nørskov

## 5.1 INTRODUCTION

Since the first report on dioxygen–glucose biofuel cell working at neutral pH [1] construction of efficient enzyme modified electrodes suitable as the cathodes in fuel cells have become a clear and challenging research area. As the result, since the late 1990s a significant number of papers describing such electrodes with potential application in biofuel cells or examples of the whole devices have appeared in the literature. Although these studies do not go much beyond the evidence of concept stage, there are some areas in which possible application of such devices can be envisaged. One of the key points of this research is the construction of enzyme-modified biocathode capable of dioxygen reduction to water.

Contrary to the number of conventional low-temperature fuel cells, the number of possible biofuels depends only on the existence of an appropriate enzyme or microorganism catalyzing its oxidation. On the other hand, dioxygen remains almost the only oxidant for biofuel cells. This is because of ubiquitous nature of this molecule. Although there are many areas where biofuel cells can be applied, the use of their miniaturized version in living organisms is strongly advocated [2]. Fortunately, the dioxygen concentration in fluid presented there, like blood in a mammalian body, is sufficient to supply a biocathode of a low-power (microwatt to milliwatt range) fuel cell. This is the reason why so many research groups are looking for biocathodes (and bioanodes) applicable in membraneless biofuel cell and working in physiological conditions. The conventional, precious-metal-modified cathodes cannot be applied in such conditions, because they usually exhibit maximum activity in extreme pH conditions. Additionally, the overpotential of dioxygen reduction at these electrodes is large in neutral pH. Therefore more biocompatible and cheaper materials would be favored. Moreover, immobilization of the enzyme (and mediator) on the electrode allows for advocated miniaturization of the device [2].

At present enzyme-modified carbon cathodes prove to be a good choice for membraneless biofuel cells [2–5]. The main advantages of the use of enzymes as catalysts at cathodes in fuel cells [5] are (1) overpotential often close to zero, (2) infinitely renewable excellent specificity, (3) ability to scavenge fuel and oxidant from ambient environment, (4) biodegradability of devices, and last but not least (5) no requirement for using noble metals as catalysts and electrode materials.

The main obstacles to overcome are (1) low current density per catalyst volume, (2) poor stability and restricted temperature range, (3) difficulty in achieving good electronic coupling to the electrode, (4) poor stability and vulnerability to various solution components and temperature, and (5) long time and high cost of isolation and purification. The appropriate design of the biocathode may help overcome the problems (1–3) and partially problem (4). The solution of other problems requires the help of genetic engineers and biochemists.

The decrease in overpotential of dioxygen reduction is crucial to achieve large biofuel cell voltage and is realized by appropriate selection of the enzyme, typically from the oxidoreductase family and its modification. The electrode material has to be selected to achieve efficient electron exchange with a high-potential redox center of the enzyme usually by good electronic coupling. For mediated catalysis the same factor and proximity of the redox potentials has to be taken into account while selecting mediator shuttling electrons between the enzyme and the electrode. On the other hand, the electrode support and/or material encapsulating the enzyme is crucial for the lifetime of the biocatalytic layer. It is clear that many factors have to be taken considered in biocathode design. Moreover, the role of every component is complex—it affects different electrode features.

So far a significant number of reports have been limited to demonstration of bioelectrocatalytic dioxygen reduction at a given electrode. Although the problem is very complex, the efforts leading to improvement in the performance of biocathode are visible. In the early 1990s research concentrated on redox polymer film-modified electrodes providing a good electronic coupling between mediator and enzyme. However, more recently a large family of nanoparticulate and other materials were tested to find favorable conditions for mediatorless bioelectrocatalysis.

In this chapter reviews the efforts to construct an efficient enzyme-modified cathode for the biofuel cell. First, the properties of the enzymes applied for this purpose are presented. This is followed by a survey of the methods of their immobilization on the electrode. Next the performance of biocathodes exhibiting mediatorless and mediated electrocatalysis is reconsidered. Finally, our reflections on future directions of this research are presented.

## 5.2 ENZYMES FOR DIOXYGEN REDUCTION CATALYSIS

Multicopper oxidases—laccase, bilirubin oxidase, or copper efflux oxidase—catalyze four-electron reduction of dioxygen to water without generating oxygen intermediates when specific donors are available. Moreover, this reaction occurs at sufficiently high potential. This makes them interesting as catalysts for dioxygen reduction in biofuel cell applications [2,3,6–9]. A few examples of biocathodes modified by other enzymes such as horseradish peroxidase, microperoxidase-11, and cytochrome *c*/cytochrome oxidase systems and employed for bioelectrocatalytic dioxygen reduction can be found [10–13].

Laccases were the first copper oxidoreductases applied for electrocatalytic dioxygen reduction [14–16], and later they became the most popular biocathode element. This is not only due to specific properties of the enzyme but also because isolation and purification methods are well developed and described in the literature [17]. Also, they are relatively easy to handle and readily affordable. Laccases contain in the active center four copper atoms of the type $T_1$ (blue copper center), $T_2$ (normal copper center), and $T_3$ (two coupled Cu centers). $T_2$ and $T_3$ are close to each other (0.04 nm) and form the so-called copper cluster. The $T_1$ center is ~0.13 nm from the $T_2$/$T_3$ cluster. Crystallographic data of the enzyme–substrate complex shows that the substrate—an electron donor—binds to the negatively charged cavity close to the $T_1$ center, ~0.07 nm below the enzyme surface [18]. $T_1$ is the main electron acceptor, connected to the copper cluster through His–Cys–His tripeptide (Fig. 5.1).

The redox potential of the enzyme is crucial for its application as cathode catalyst in biofuel cells. For the majority of laccases the redox potential of the electron-accepting

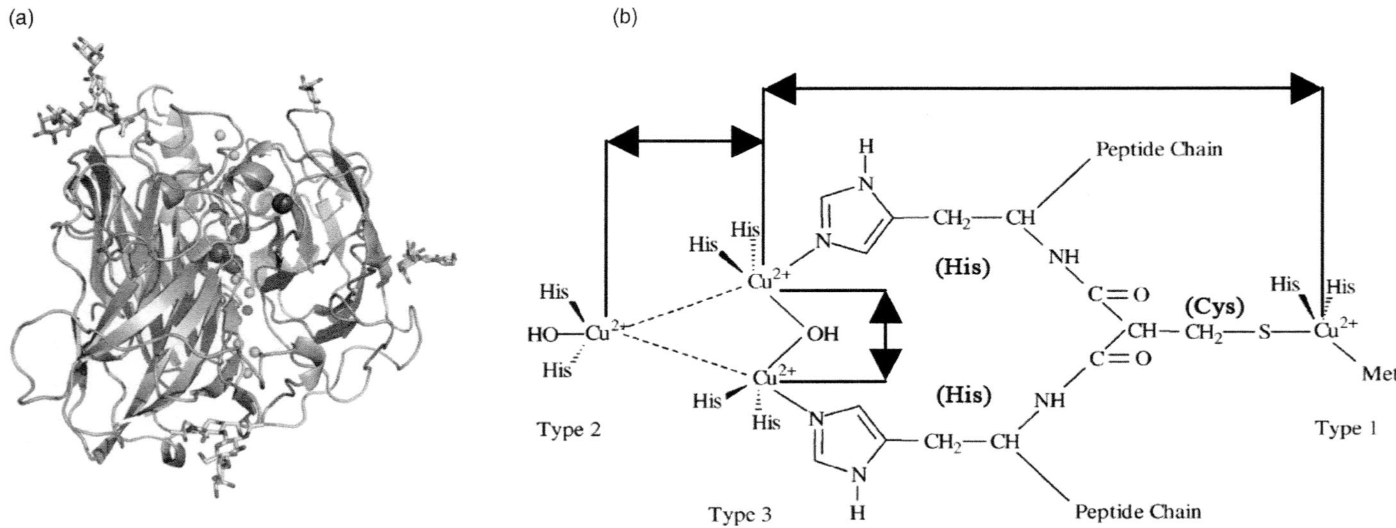


**FIGURE 5.1** (a) Three-dimensional structure of laccase from *Cerrena maxima*; (b) active center of laccase containing 4 copper atoms (types $T_1$, $T_2$, and $T_3$, from right to left) [19].

$T_1$ site is in the range 0.43–0.78 V versus NHE [20–27]. From this perspective, laccases can be divided into three groups: low-, medium-, and high- potential. The low-potential ones (isolated mainly from trees) are characterized by $T_1$ redox potential of ~0.43 V versus NHE [24,28]. The intermediate ones with $T_1$ redox potential in the range 0.47–0.71 V versus NHE are isolated from fungi [21,29]. For the high-potential laccases the value is ~0.78 V versus NHE [23,28]. Obviously, the latter group is most interesting for application in biocathodes. The redox potential of the $T_1$ center of laccase from *Cerrena unicolor* used by our groups is equal 0.75 V versus NHE [30–32].

Another classification of all blue multicopper oxidases is based on their primary structure [25]. The enzymes with low, intermediate, and high redox potentials contain methionine, leucine, and phenylalanine (Fig. 5.2), respectively, as an axial ligand to the $T_1$ center. The hydrophobic pocket formed next to the $T_1$ copper center (Fig. 5.2) is thought to be of crucial importance for the electronic coupling and thus direct electron transfer ability between the given laccase and the electrode surface [33,34].

The oxidation of simple substrates by laccase is described by the "ping-pong" mechanism [26,35–37]. The $T_1$ center is responsible for accepting electrons from the substrate and its transfer to the $T_2/T_3$ cluster binding the dioxygen molecule and reducing it to water. Although the number of mechanistic schemes already proposed agrees well with the currently available kinetic and structural data [26,38], both the mechanism of electron transfer and the mechanism of reduction of oxygen to water are still subjects of intensive studies.

The dependence of catalytic activity on the redox potential of the $T_1$ center suggests that the first electron transfer from the substrate to this copper ion is the rate-limiting

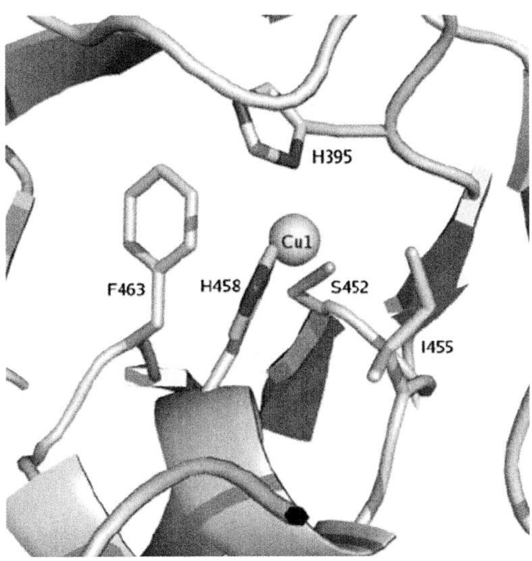


**FIGURE 5.2** The structure of mononuclear copper $T_1$ center of the laccase from *Cerrena maxima* with the hydrophobic pocket characteristic of high-redox-potential laccases [19].

step. The catalytic cycle starts from the formation of a native intermediate. The substrate is oxidized close to the $T_1$ center and the electron is transferred to the $T_2/T_3$ copper cluster, where dioxygen is reduced [26]. Slow disappearance of the half-product results in the formation of a fully oxidized form of laccase. Under these conditions the $T_1$ center can be continuously reduced by the substrate, but the electron transfer onto the copper cluster becomes too slow to be catalytically efficient.

This mechanism assumes that redox potential of laccase and possibly other copper oxidoreductases can increase through changes in the electron density distribution between the copper center and the neighboring amino acid [18]. Laccase and hemocyanine are examples of enzymes exhibiting large structural and electronic differences in space between their binuclear $T_3$ sites [19]. Interestingly, the binding of $O_2$ was found to be more favorable for laccase by ~9 kcal/mol [39]. For hemocyanine this reaction is driven by electrostatic destabilization of the deoxy state in which the two copper ions are at close distance and a facile two-electron reduction of $O_2$ occurs (Fig. 5.3). For laccase, greater Cu–Cu distance in the $T_3$ site and a low force constant in the deoxy form minimize the electrostatic repulsion of the two Cu ions [39]. The active-site flexibility results in intrinsic reactivity of the trinuclear Cu cluster leading to an intermediate required for the irreversible reduction of $O_2$ to $H_2O$. Indeed, for multicopper oxidoreductases the deoxy $T_3$ site does not react with dioxygen when the $T_2$ site is absent. The low activation barrier produced by the triangular structure provides good donor–acceptor molecular orbital overlap. The O–O bond cleavage in the trinuclear Cu cluster is achieved by the proton-assisted or proton-unassisted process. It was found that while the proton helps to stabilize the acceptor $O_2^{2-}$ $s^*$ orbital in the proton-assisted process for a good donor–acceptor molecular orbital overlap, the third oxidized Cu center in the trinuclear site plays the role of Lewis acid in the proton-unassisted O–O bond cleavage [40].

The distance between $T_1$ copper and one of the histidines in high-potential laccases is much larger than in the case of those of intermediate potential [41]. This may be due to hydrogen bonding formed between His479 and Ser113, present in all high-potential laccases. The serine is not found in intermediate potential laccases and thus is not very useful for biocathodes. Several other factors may affect the redox potential values of

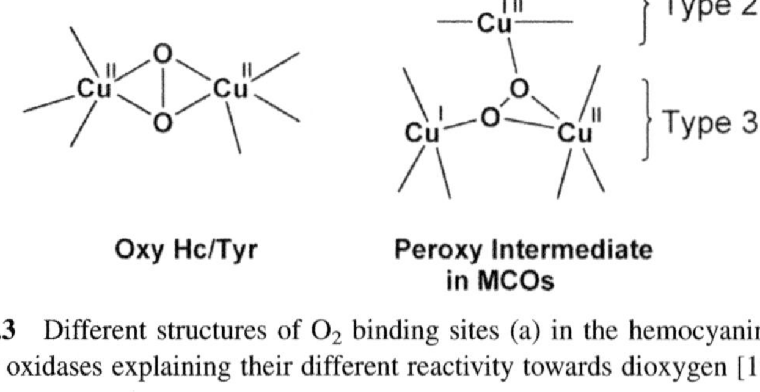


**FIGURE 5.3** Different structures of $O_2$ binding sites (a) in the hemocyanine and (b) the multicopper oxidases explaining their different reactivity towards dioxygen [19].

metalloproteins, for example, electrostatic and solvation effects [42,43], and the reasons for large differences between $T_1$ potentials for different laccases are not fully understood [44].

Bilirubin oxidases are another group of popular enzymes used in construction of biocathodes. Although they are much more difficult to obtain than laccase and very expensive, their popularity results from stronger activity at neutral pH and relative resistance to $Cl^-$ ions. These properties are superior for devices operating under physiological conditions [2,45,46]. The crystallographic structure of bilirubin oxidase is still unknown. Biochemical, spectral, and kinetic studies reveal that the $T_2/T_3$ cluster binds four histidine molecules while the $T_1$ center is surrounded by histidine, cysteine, and methionine [26,47]. The redox potential of the $T_1$ center in bilirubin oxidase isolated from *Myrothecium verrucaria* is 0.49 V versus NHE, which is a smaller value than that of high-potential laccase. It also catalyzes the oxidation of substrates other than laccase. These substrates usually contain tetrapyrrole units such as bilirubin and its oxidation to biliverdin, accompanied by reduction of dioxygen to water [26,48].

Another multicopper oxidoreductase—copper efflux oxidase (CueO) from *Escherichia coli* (Fig. 5.4)—has gained considerable interest because it enables direct electron transfer bioelectrocatalytic dioxygen reduction and is superior to other multicopper oxidases based on catalytic efficiency when adsorbed on carbon aerogel or Ketjenblack-modified carbon paper electrodes [49–52]. Unfortunately, the overpotential of this reaction is larger than that of laccase or bilirubin oxidase. This is because the $T_1$ site of copper efflux oxidase is located far from the surface of the enzyme covered with a large, methionine-rich α-helix region providing the fifth Cu

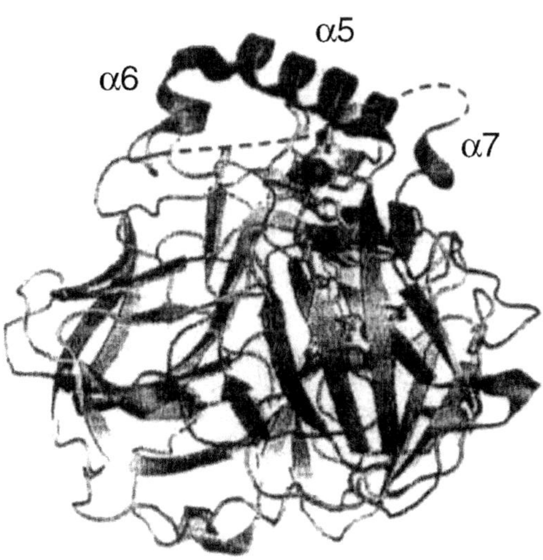


**FIGURE 5.4** The structure of copper efflux oxidase (to generate the Δα5–7 deletion mutant, the amino acid residues in the α-helical region from Pro357 to His406 (red ribbon region) were deleted and replaced by the dipeptide Gly–Gly) [53]. (See insert for color representation.)

binding site 0.75 nm from the $T_1$ site. The higher activity of the protein toward organic substrates can be increased by engineering the protein to delete part of the helical region amino acids without changing the $T_1$ redox potential.

A slight positive shift of the $T_1$ redox potential of copper efflux oxidase was achieved by site-directed mutagenesis [51]. This change results from mutation of methionine acting as the fourth ligand at axial position in the $T_1$ Cu and formation of additional hydrogen bonds with His443 coordinated to the same center. The protein engineering strategy can be a powerful tool in tuning the redox properties of enzymes used as electrocatalysts, provided that mutants are of sufficient stability, which is not always the case. Understanding of the structure–redox properties relationship is, however, of major importance for decreasing the oxygen reduction overpotential.

It is clear that both high-potential laccases and bilirubin oxidases are still strong candidates for modification of cathodes for biofuel cells. The use of copper efflux oxidases will be feasible only after its redox potential shift or with extremely low potential anodes.

## 5.3 IMMOBILIZATION OF THE ENZYME ON THE BIOCATHODE

Stable immobilization of the enzyme on the electrode is a prerequisite for construction of biocathode. One strategy involves immobilization of the protein next to or within the fraction or a few nanometers to the solid conductive surface (Fig. 5.5a). This arrangement is usually applied for systems exhibiting direct electron transfer between the enzyme and the electrode [5]. Obviously, direct contact between the electrode and the enzyme may affect protein structure and therefore its stability. The second approach is useful for mediated bioelectrocatalysis. It involves encapsulation of the protein into a thin film of material providing easy access to mediator from the solution or functional group shuttling the electrons between the enzyme and the electrode support (Fig. 5.5b) [54]. Obviously, both electrode arrangements have to provide easy access to the oxidant dissolved in the surrounding solution. There are also numerous

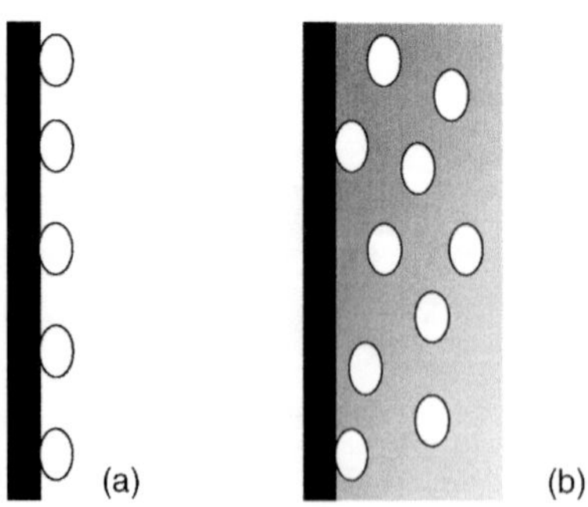


**FIGURE 5.5** Basic ways of immobilizing the enzyme on the electrode surface: (a) enzyme directly contacting the electrode support; (b) enzyme encapsulated in a thin film present on the electrode surface.

examples combining these two approaches, including immobilization of enzyme into a polymeric film containing conductive nanoparticles.

The first of the proposed strategies involves irreversible adsorption of the enzyme [30,33,50,52,55–85] or its covalent bonding to the electrode surface [30,33,34,84–92]. Two types of materials, carbon-based and bare and modified gold, were used to prepare biocathode with adsorbed laccase, bilirubin oxidase, copper efflux oxidase, microperoxidase-11, and Fe–Fe dehydrogenase. The first two enzymes were found to weakly adsorb on smooth carbon material—glassy carbon (GC) [56]. The other carbon-based materials for adsorption of these enzymes include graphite [30,55,76,77], carbon surface with high crystal graphite edge density [69], and edge plane of pyrolytic graphite [57,79,80]. A significant enhancement of laccase adsorption was found when pyrolytic graphite was covered by anthracene-based units immobilized by diazonium coupling [33].

The majority of reports on adsorption of enzymes exhibiting dioxygen reduction catalysis are concerned with different types of structured carbon-based materials. This is not surprising, because they have been known for a long time as suitable adsorbents for organic compounds. Their active electrode surface is substantially larger than those in solid graphite or glassy carbon. The strong adsorption of copper oxidoreductases has been observed to exhibit catalytic activity toward dioxygen on materials such as carbon black [78], finely divided colloidal graphite [59,65,74], acetylene carbon black [59,65,74,83], active carbon [64,67], porous carbon [75], ozone-treated carbon black [67], single-walled carbon nanotubes (SWCNTs) [93], carbon aerogel [58,68], graphite microparticles embedded in sol–gel processed silicate [94], Ketjenblack [58,66,95] multiwalled carbon nanotubes (MWCNTs) [85], lipid-modified MWCNTs [81] or polyaromatic compound–modified MWCNTs [84] adsorbed on GC, carbon felt [57], mesoporous carbon [52,82] and hydrophilic carbon nanoparticles [96], and pyrene sulfonate–modified SWCNTs [97] embedded in sol–gel processed silicate matrix. Biocathode obtained by microperoxidase-11 adsorption on CNTs immobilized on GC [70] or on Pt [71] has been used to investigate enzyme adsorption on carbon-based materials.

There are also composite materials, where adsorption of the enzyme on carbon-based component may play an important role in obtaining a stable biocathode. The stable electrode was obtained by incorporation of laccase into gel composed of CNTs and room-temperature ionic liquid (RTIL) [98–100]. It seems that adsorption of laccase on CNTs contributes to the stability of biocathode prepared by screen printing with ink having CNTs as one of the components [101]. Electrodes made of alginate carbon beads composite modified with laccase [102] and carbon paste consisting of paraffin oil, graphite powder, and bilirubin oxidase [103] belong to the same class.

Gold is another important material used as support for the adsorbed enzyme. Initially structured metal such as rough gold for adsorbed microperoxidase-11 [61,72,73] and nanoporous gold for adsorbed laccase [62] were used. For the latter enzyme, smooth gold and the same metal covered by 4-aminothiophenol were also found to be the suitable surface [63]. However, it was noted that only modification of the Au surface protects enzymatic activity [63]. Tethering laccase was also obtained on Au modified with dithiobis-*N*-succinimidyl propionate [104]; however, in this case electrostatic interaction between the film and the enzyme plays an important role.

The preparation of biocathode via covalent bonding of the enzyme to the electrode surface was rarely attempted because it complicates electrode preparation. Metal electrodes are favorite substrates for this, because of easy and controlled premodification of the surface providing a variety of functional groups. Laccase was immobilized onto the amine terminated thiol monolayers on gold surface by glutaraldehyde coupling [87]. The same enzyme was attached to a self-assembled monolayer of mercaptoundecanoic or mercaptopropionic acids [86]. Silanization of platinum surface with the aminated derivative can be effectively used for glutaraldehyde coupling of the same enzyme [91,105]. The peptide coupling with carboxylic groups introduced on the glassy carbon surface was found to be less effective [105]. Another approach involves covalent attachment of bilirubin oxidase to MWCNTs further immobilized on the electrode surface [84,85,92].

Other procedures involve more components and multistage procedures. They include grafting of laccase on the aminopolypyrrole film–modified electrode via the reaction with glutaraldehyde [106] or grafting of avidin to the same film followed by complexation of biotinylated laccase [106] (Fig. 5.6). The bindings of aromatic functionalities to the carbon surface (by diazonium coupling reaction) to hydrophobic residues close to the "blue" Cu site were utilized to obtain the laccase-modified biocathode [34]. Probably the most complicated procedure involves the adsorption of MWCNT on gold surface followed by adsorption of aromatic compounds having carboxylic functionalities on nanotube surfaces [84]. The carboxylic groups were further used for covalent coupling of bilirubin oxidase with *N*-hydroxysulfosuccinimide [84].

All methods of enzyme immobilization discussed above provide direct contact with the environment and conductive surface. However, the main obstacle in monolayer modification of the electrodes is the low density of catalytic sites because of the large

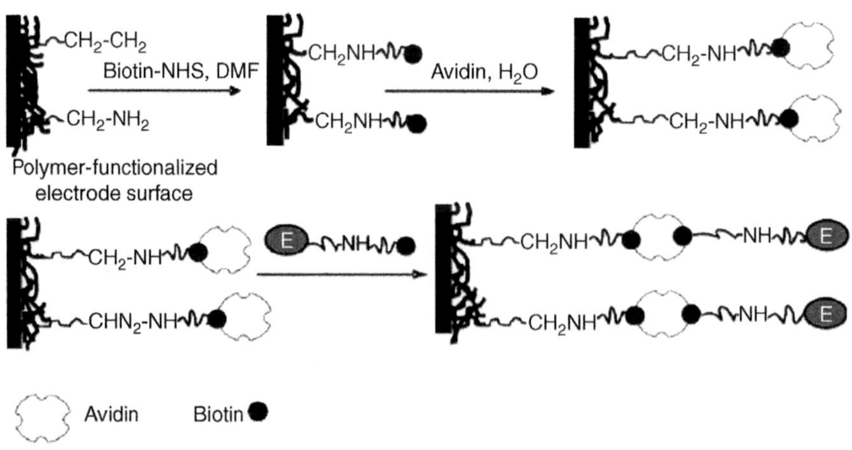


**FIGURE 5.6** Construction of the laccase (E) electrodes by molecular recognition [106].

size of the enzyme. Moreover, for systems, its orientation relative to the electrode surface is inappropriate for efficient communication with the $T_1$ site. The encapsulation of protein into a thin film covering the electrode surface is another strategy. Its most important function is to separate the enzyme from the outer environment, allowing for penetration by reaction substrates. Quite often the same material serves as a reservoir of redox-active groups serving as a mediator for electron exchange between the protein and the conductive material. Following this approach, the enzymes were most often encapsulated into a polymeric thin film [45,46,54,55,82,104,107–148]. The dense polymeric network does not allow for protein escape into the surrounding solution because of size matching.

Among polymers, ionomers were most frequently selected for enzyme encapsulation [45,46,55,82,104,107–127,133–136,138–142,144,147,149–151] because the pores of this material are large enough to enable penetration by the surrounding solution, thus increasing protein stability. Also protein electrostatic interaction with charged groups of ionomer may be helpful in preserving activity. Among ionomers, osmium, and ruthenium, polypyridine ionomers are very popular for laccase or bilirubin oxidase encapsulation [45,46,54,82,104,107–126,138,144,151]. These polymers are water-soluble hydrogels with a moderate degree of crosslinking [54]. They exhibit segmental mobility thereby do not significantly affect the structure of the encapsulated enzyme. Their backbones are terminated by ligands participating in formation of transition metal ion inert complexes or other charged groups (i.e., carboxylic). The concentration of the ionic pendant groups, the type of crosslinker, and degree of crosslinking affect hydration and swelling of the polymer, and thereby the encapsulated enzyme stability. The osmium polypyridine hydrogels are also unique in the sense that they consist of both cationic and anionic functional groups. This promotes formation of electrostatic adducts between the polymer and the enzyme [54]. These interactions have more recently been explored to form osmium polymer laccase electrode by layer-by-layer self-assembling on Au modified with 3-mercaptopropyl sulfonate [104]. The use of nonfunctionalized thermoresponsive poly (*N*-isopropylacrylamide) gel for stable laccase immobilization should be mentioned in this context [148].

Clearly, maximum stability of the electrode depends on the type of enzyme, namely, its isoelectric point [110]. The film made of osmium polypyridine ionomer increases laccase stability against methanol [125,144] because of good resistance against deswelling. It also allows for efficient catalysis of dioxygen reduction by laccase in physiological buffer despite its maximum activity [2,45,46,108,112,114,115,118,119,126,152]. This is probably the result of blocking access of chloride ions, decreasing activity of the enzyme, and higher local pH. The addition of the doped cubic-phase lyotropic liquid crystal to such a polymeric film severely impedes permeation of urate-damaging bilirubin oxidase [123].

Nafion is another ionomer used for laccase, horseradish, or cabbage peroxidase encapsulation [32,133,135,136,140]. This polymer exhibits blocking effects against solution components, decreasing enzymatic activity. In particular, it was shown that Nafion film treated with tetrabutylammonium bromide prevents bilirubin oxidase from interactions with ethanol present in the surrounding solution [140]. A similar blocking effect of methanol was observed for graphite electrode with adsorbed laccase and covered with Nafion [34]. The same ionomer was also applied as the top layer on osmium polypyridine ionomer mentioned above in order to increase the stability of encapsulated bilirubin oxidase in serum by exclusion of urate [112].

Two other ionomers bearing sulfonate ionomers—poly(aniline sulfonate) [127] and poly(styrene sulfonate) [134]—were also employed for biocathode preparation. Poly(anilinesulfonic acid) was used for electrostatic immobilization of laccase in multilayer films consisting also of cytochrome *c*. [127] Immobilization of bilirubin oxidase in poly(styrene sulfonate), polylysine, and CNT composite was reported [134]. In this case not only electrostatic interactions but also adsorption of enzyme on CNT may play an important role. Polylysine was used for the same enzyme electrostatic entrapment on different substrates [142,150], also in multilayer configuration [141]. The same polymer was used for layer-by-layer immobilization of laccase together with MWCNT on different substrates including three-dimensional macroporous gold [149,153]. Also in this approach both electrostatic interactions and adsorption were employed.

Out of the conducting polymers, only polypyrrole [128,129,131,132,143,154–156] and poly-1,2-phenylenediamine [157] were used for laccase or bilirubin oxidase encapsulation by electropolymerization. Polypyrrole enhances enzyme stability against methanol [129], and its reticulation by glutaraldehyde treatment improves its performance in the flow system [131].

Encapsulation of the enzyme by glutaraldehyde crosslinking, a well-known procedure developed to fix proteins, was used for laccase, bilirubin oxidase, microperoxidase-8, and cytochrome *c*/cytochrome oxidase adduct immobilization on the bare or modified electrode surface [158–168]. It was used for immobilization of native enzyme or together with bovine serum albumine [158,160,163–166]. The latter component was added to slow down enzyme deactivation in serum [158] or in the presence of ethanol [163]. Another crosslinking reagent—poly(ethylene glycol) diglycidyl ether—was applied for bilirubin oxidase encapsulation [159].

Inorganic polymer, namely, sol–gel processed silicate films, were used for laccase or bilirubin oxidase immobilization [96,139,169–176]. This is not surprising, because in the early 1990s it was found that encapsulation of enzyme in this material does not affect its activity [177,178]. The porous silicate matrix provides easy access to the dissolved substrate, and the pores are small enough to avoid leakage of the enzyme into solution. The film of another inorganic material redox active layered double hydroxide with ABTS counterions was used as support for laccase [160], but stable immobilization was achieved after glutaraldehyde crosslinking.

In addition to film, geometry of bioelectrocatalytic material was reported. Laccase was incorporated into crosslinked poly(ethylene imine) microcapsules, and this material was adsorbed on rough GC [137].

Biocompatible liquid crystalline phases were employed for encapsulation of laccase [32,123,179–184]. At appropriate composition lipid–water mixtures form cubic phases consisting of soft organic material penetrated by a network of water-filled channels (Fig. 5.7). Most of the cubic phases composed of membrane lipids were found to be bicontinuous, thus having regions that are continuous with respect to both

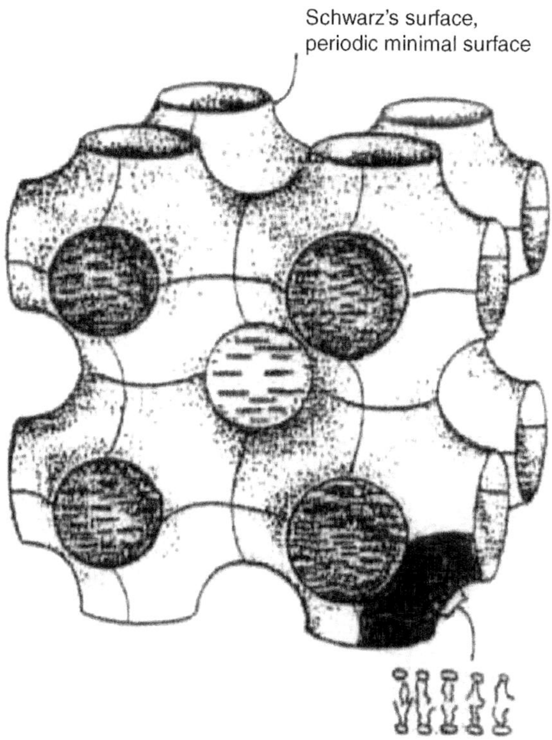


**FIGURE 5.7** Structure of the lipid liquid crystalline cubic phase [181].

polar (water) and nonpolar component. The abovementioned application of this class of materials is an attractive approach for biofuel cells and biosensor applications since they allow for efficient transport of both hydrophilic and hydrophobic substrates throughout the film (Fig. 5.7). The use other biocompatible material such as cytochrome *c* as structural film element obviously prevents laccase [127] or bilirubin oxidase [185] from denaturation in the film.

Last but not least, some composite and less conventional materials used for preparation of biocathodes should be mentioned. Horseradish peroxidase was incorporated into solid-carbon-based matrix for spray printing [13]. Similarly, laccase retained its activity in conducting ink prepared from carbon and carbon nantotubes after screen printed electrodes were produced [101]. Liquid-based materials were also used for enzyme immobilization. They include classic paraffin oil carbon paste for immobilization of bilirubin oxidase [103] and carbon nanotubes—ionic liquid gel for immobilization of laccase [98–100].

To summarize, one can say that numerous materials have already been tested for encapsulation of active enzymes for biocathodes. It seems that in most cases the material design was oriented toward high catalytic reactivity rather than prevention of interaction between the deactivating reagent and protein. Moreover, most investigations show that a given electrode works only on the voltammetric timescale.

Long-term experiments are scarce, and no systematic studies of the effect of immobilization on the enzyme stability could be found in the literature.

## 5.4 BIOCATHODES EMPLOYING MEDIATED ELECTROCATALYSIS

The majority of research is aimed at the construction of cathode for membraneless biofuel cell where immobilization of both the enzyme and the mediator is essential. The electrode material has to provide relatively high surface concentration of the molecules or redox functional groups exchanging electrons with protein and easy access for dioxygen dissolved in solution. For the two-compartment biofuel cell, the mediator can be water-soluble, and its easy access from catholyte to the immobilized enzyme has to be provided. For mediated catalysis, the orientation of the enzyme is less important because the electrons are exchanged with the electrode by the mediator.

Starting from the first report on mediated bioelectrocatalysis of dioxygen reduction to water [6], almost all biocathodes involve laccase or bilirubin oxidase from different sources. Two types od mediators dominate the research: $Os^{3+/2+}$ redox center stabilized by ligands attached to polymer backbone [186] and $ABTS^{2-}$ dianion [6].

The advantage of osmium polymer or copolymer film is that it serves as both electron relay and host for for laccase or bilirubin oxidase [45,46,55,82,104,107–126,130,144–147,151,186,187]. The redox centers are stabilized by a set of ligands containing heterocyclic nitrogen atoms having octahedral geometry (Fig. 5.8). They are selected such that redox potential of $Os^{3+/2+}$ redox couple can be optimized to be close to that of the active redox center of the enzyme [110,188]. The redox centers are tethered to polymer backbone with chains consisting of a few or more than 10 carbon atoms and heteroatoms such as nitrogen. Because of their flexibility, the distance between the redox centers in oxidized and reduced states is frequently shortened, allowing for efficient electron exchange on their collision [189]. Other hydrated anionic and cationic groups are appended to the polymeric backbone, and such hydrogel structure allows for motion of counterions within polymer network. More importantly, $Os^{3+/2+}$ redox centers can exchange electrons with the enzyme. As a result of such "wiring" of the enzyme to the electrode [116,118,126,188], the electron-hopping chain is formed between the electrode substrate and the enzyme. Enzyme leakage is avoided not only by encapsulation of the large enzyme by crosslinked polymer backbones but also by electrostatic attraction between charged (cationic) redox groups of polymer and (anionic) protein. The highly hydrophilic polymer matrix provides water-rich environment suitable for stable enzyme immobilization. It also allows for transport of dissolved dioxygen to the redox center of the enzyme and exhibits strong affinity to the product of the four electrons dioxygen reduction.

The $Os^{3+/2+}$ polymeric film with encapsulated blue copper oxidoreductase immersed in a simulated physiological solution was found to be superior to Pt electrode in acidic media in terms of overpotential (Fig. 5.9)[116].

Since 2001 the structure of these polymers has been continuously optimized to provide enhanced catalytic current density. Now it is clear that the magnitude of this parameter depends on osmium polymer structure and the matching of their redox

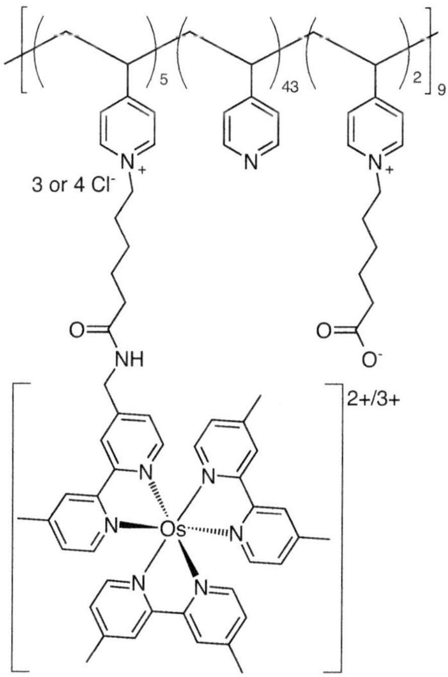


**FIGURE 5.8** An example of Os polymer used as electron relay and matrix for copper oxidoreductase encapsulation [54].

potential and the redox potential of the polymer. It is determined by the type of the ligands and can be optimized [110] (Fig. 5.10).

It was demonstrated that the bimolecular reaction rate constant exhibits sensitivity on the redox potential of the mediator (Fig. 5.10). In some conditions a linear

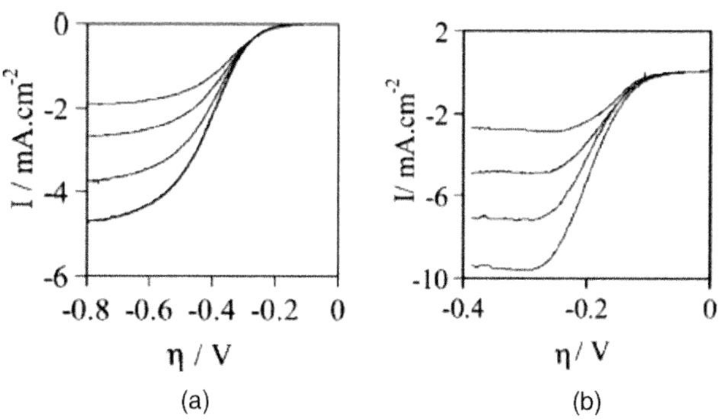


**FIGURE 5.9** Comparison of the dependence of current density on overvoltage on rotating-disk electrodes in dioxygenated solution at 37°C at scan rate 5 mV/s at different rotation rates: (a) Pt electrode in 0.5 mol/dm$^3$ $H_2SO_4$; (b) vitreous carbon modified by Os polymer with encapsulated laccase in saline phosphate buffer [116].

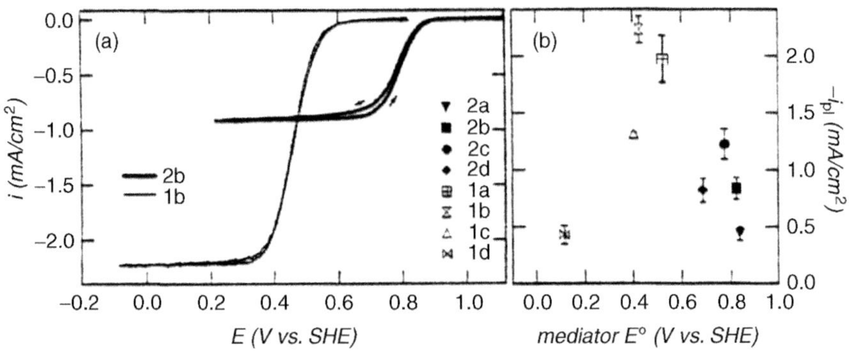


**FIGURE 5.10** (a) Polarization curves obtained with rotated glassy carbon disk electrode modified by two selected osmium polymers in 1 atm $O_2$-saturated citrate buffer at 900 rpm rotation, scan rate 1 mV/s (b) plateau current values for electrode modified with different osmium polymers versus their redox potentials [110].

dependence was found (kinetic control), whereas in others no dependence was observed (diffusional control) [110] (Fig. 5.11). The film composition, namely, the ratio of enzyme crosslinker and redox polymer, is also important for the catalytic activity of the electrode. It can be optimized with the help of scanning electrochemical microscopy screening [116,130].

An interesting example of osmium polymer brush—laccase based pH switchable bioelectrocatalytic electrode controlled by the enzyme-based logic gate—has been

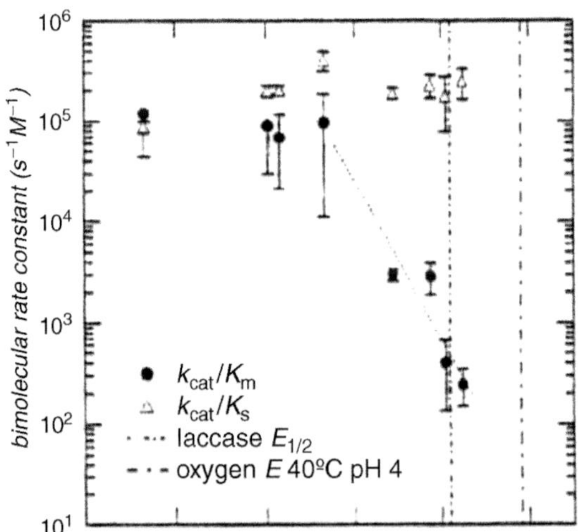


**FIGURE 5.11** The effect of osmium polymer mediator redox potential $E^{\circ}$ on laccase–mediator bimolecular rate constant $k_{cat}/K_m$ and laccase–oxygen rate constant $k_{cat}/K_s$. Vertical lines mark redox potentials of laccase and $O_2/H_2O$ at pH 4. Dotted line shows a linear free-energy region observed for high $E^{\circ}$ values [110].

presented [151]. The catalysis is switched off by pH increase produced by decomposition of urea to ammonia by urease. At higher pH the motion of polymeric chains is frozen and electron transport within polymer inhibited [151].

Insufficient matching of redox potentials of enzyme and $Ru^{3+/2+}$ polymers of a similar structure was found to be the reason of their poor performance as mediator in biocathode [107]. However, modification of high-potential laccase by ruthenium complexes was observed to increase the efficiency of bioelectrocatalytic dioxygen reduction at high potentials [190]. In 2008 the potentiality of perfluoroalkyl-modified 2,2-bipyridyl ruthenium complex in a Nafion membrane as an effective mediator was again suggested [191].

The novel concept based on coassembling two bimolecular building blocks seems to be promising for electrode construction [192]. One of them consists of triblock polypeptide with $Os^{3+/2+}$ bispyridine complex attached to the histidine unit. The other is a chimera of artificial α-helical leucine zipper and random coil domains fused to a polyphenol oxidase, small laccase. This material may be useful as a biocathode film but its utility to dioxygen electroreduction has not yet been demonstrated.

It should be emphasized that the comparison of catalytic efficiencies of dioxygen electroreduction from different sources is difficult, because usually the data reported are obtained with a rotating-disk electrode (RDE) at varying rotation speed. However, it is fair to say that a fraction of $mA/cm^2$ is easy to obtain with the electrodes modified with osmium polymer and laccase or bilirubin oxidase in quiet conditions. It is at least 10 times below the value predicted by the one-dimensional model based on macrohomogeneous and thin-film theories accounting for dual-substrate enzyme kinetics [193]. The same theory predicts another one order of magnitude increase for gas-phase diffusive or a convective, flow-through transport approach. Indeed, osmium polymer and laccase biocathode operating in contact with the gas phase exhibit higher current density of catalytic current than in aqueous solution [144,145].

The electrodes described above are ready to serve as cathode for the membraneless biofuel cell and related devices. This is not always the case for the biocathodes based on $ABTS^{2-}$ (Fig. 5.12) and copper oxidoreductases because of their water solubility.

This compound has been known for a long time as a good mediator for laccase [26]. The redox potential of $ABTS^{2-/\bullet-}$ redox couple is low enough to provide reasonable driving force for dioxygen electroreduction catalyzed by laccase [6,194] and bilirubin oxidase [195]. Although the formed anion radical slowly deactivates laccase [196–198], this mediator is frequently used as a "model" system.

Because of its water-solubility, $ABTS^{2-}$ is a good mediator for the cathode in the membrane biofuel cell. The bioelectrocatalysis of dioxygen electroreduction is observed for GC [56] or porous carbon [75] electrode modified by adsorbed laccase immersed in $ABTS^{2-}$ buffer solution. In the former case this mediator was found to be

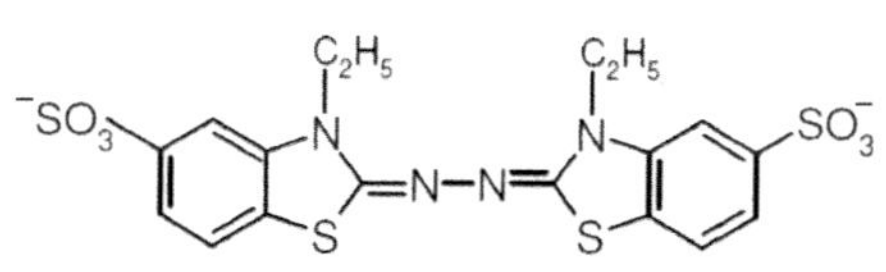


**FIGURE 5.12** 2,2′-Azino-bis(3-ethylbenzthiazoline-6-sulfonate) ($ABTS^{2-}$).

superior to other water-soluble mediators in terms of current density [56]. The catalytic effect is also observed with laccase encapsulated in porous material such as dialysis membrane [195], sol–gel processed silicate [91,139,169,173,176,199], lipid cubic phases [179,180,182], and poly(*N*-isopropylacrylamide) gel [148] or bilirubin oxidase encapsulated in the sol–gel processed film modified with CNTs [171,172]. The pores formed by the sol–gel processed polymer (silicate) or peptide assembly-–like cubic phases (Fig. 5.6) allow for efficient mediator and dioxygen transport. The electrode modified by laccase encapsulated nonporous material, such as in gel composed of MWCNTs and RTIL [200], also exhibits catalysis in these conditions because of exchange of RTIL anion with $ABTS^{2-}$ ionic. The cathodic unit of membrane biofuel cell filled with $ABTS^{2-}$ and laccase solution works well in quiet conditions [201]. The microfluidic biofuel cell [202] with the catholyte stream consisting of $ABTS^{2-}$ and laccase solution is the only example of a membraneless device with a dissolved mediator. This system seems to be promising because the continuous supply of the enzyme solves the problem of its instability while microfluidic geometry allows for membraneless contact between the catholyte and the anolyte. However, for the membraneless biofuel cell working in quiet conditions, immobilization of $ABTS^{2-}$ occurs [101,106,128,129,131–134,139,143,155,160,170,175,181,184,203].

Only one attempt at the synthesis and use of polymer with ABTS pendant groups for electrode modification was reported [203], but enzyme immobilization in this film was not reported. It was tested only in the presence of laccase in solution and the catalytic current density obtained was rather low.

Other attempts involve encapsulation of both the mediator and the enzyme. $ABTS^{2-}$ counterions were doped into conductive polymer (polypyrrole and its amino derivative) film together with laccase during the electropolymerization process [106,128,129,131,132,143]. In order to increase the active electrode surface, such a modified polypyrrole film was immobilized on the CNT-modified electrode [155]. Polypyrrole encapsulated $ABTS^{2-}$ was also used as a mediator for laccase or bilirubin oxidase grafted on the amino polypyrrole film [106,156]. Interestingly, laccase was found to catalyze oxidative polymerization of pyrrole in the presence of $ABTS^{2-}$ and dioxygen [204], but this material was never tested in an electrochemical cell.

The attraction between functional groups of ionomer film and $ABTS^{2-}$ [205] was explored for mediator immobilization [134,139]. The composite film consisting of ionomer–polyallylamine for $ABTS^{2-}$ immobilization and sol–gel processed silicate for laccase encapsulation was proposed as biocathode material [139]. It is significantly enhanced to one of the highest values reported for this mediator ($> 0.2\ mA/cm^2$) when enzyme and polymer immobilization are preceded by indium tin oxide nanoparticles [139] (Fig. 5.13a). In another study both bilirubin oxidase and mediator were encapsulated in the film of polylysine and bioelectrocatalysis was observed [134]. Electrostatic interactions were also employed in $ABTS^{2-}$ incorporation in inorganic material—double-layered hydroxide: $Zn_2Cr(OH)_6ABTS$. The film electrode consisting of this clay, laccase, and BSA, obtained by glutaraldehyde treatment, exhibited catalytic activity toward dioxygen.

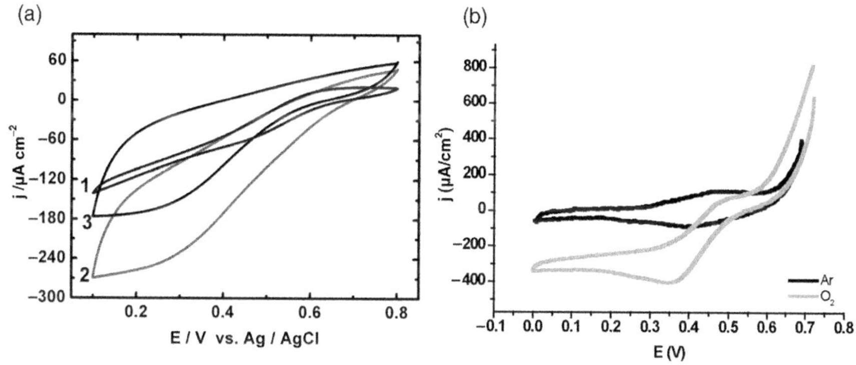


**FIGURE 5.13** Voltammograms of $ABTS^{2-}$ and laccase-modified electrodes in dioxygen saturated electrolyte solution: (a) silicate–polyallylamine film with ITO nanoparticles [139]; (b) monoolein liquid crystalline film with MWCNTs [181].

The liquid-crystalline-phase film was employed to host laccase and $ABTS^{2-}$ [181]. The glassy carbon electrode modified with laccase and the mediator embedded in the lyotropic liquid crystalline phase–monoolein liquid crystalline film exhibits high catalytic activity towards dioxygen as was exhibited by an even larger catalytic current for this mediator (> 0.3 mA/cm$^2$) (Fig. 5.13b) [184]. This electrode was successfully applied as a cathode in a biofuel cell based on cubic lipid phases modified electrodes [181].

Another method of preparation of $ABTS^{2-}$-modified film is based on direct immobilization on carbon-based materials. This can be done by mediator irreversible adsorption on porous carbon [128] or MWCNTs [133,170]. The screen-printed electrode produced from carbon and CNTs is an important example, because it retains activity for months [101]. Adsorption occurs due to the delocalized π-electron system of $ABTS^{2-}$ anions and graphene structure of CNTs (Fig. 5.14a) and the modified

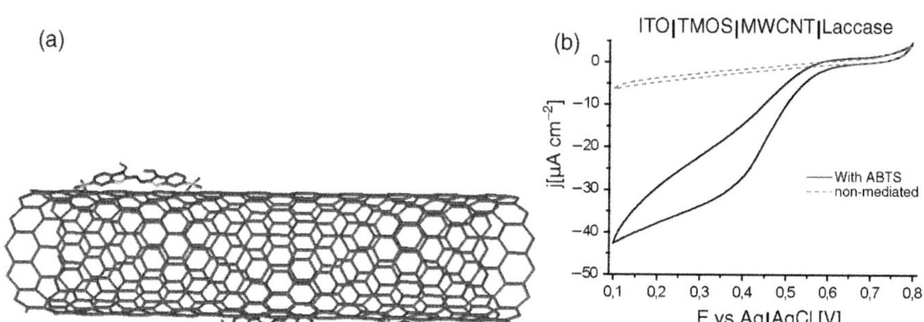


**FIGURE 5.14** A scheme of adsorbed $ABTS^{2-}$ on carbon nanotube (a) The voltammetric curve (scan rate 1 mV/s) was obtained with electrode modified by the sol–gel processed silicate film with encapsulated laccase in dioxygen saturated buffer solution; the solid curve was obtained after adsorption of $ABTS^{2-}$ [170].

CNTs become water-soluble. Two procedures were successfully tested: adsorption on MWCNTs before immobilization in a Nafion film [133] or adsorption on MWCNTs already immobilized in the sol–gel processed silicate film [170]. In both cases immobilization of the mediator is stable, it retains its electroactivity, and after immobilization of laccase the electrodes exhibit efficient dioxygen catalysis. Interestingly, the experiments with unmodified carbon nanotubes reveal a mediatorless component of bioelectrocatalysis [170] (Fig. 5.14b).

In order to avoid any form of desorption of the mediator, covalent immobilization of $ABTS^{2-}$ on SWCNTs surface was proposed [181,206–208]. This functionalization is based on the use of diazonium salts generated in situ. It enables selective stable attachement of the mediator, in both its reduced and oxidized forms, to the walls or ends of SWCNTs [206–208], which in turn affects the efficiency of the bioelectrocatalytic process. Functionalized SWCNTs are suitable for immobilization in Nafion film at the electrode surface [184,209]. This functionalization can also be carried out in situ, and CNTs can be derivatized at the ends/defect sites or on their side walls (Fig. 5.15) [207,208].

The population of the mediator at the electrode surface is significantly larger when derivatization of nanotubes is carried out at the walls (Fig. 5.15, SWCNT-ABTS side). Thus catalytic current densities higher than those on pristine nanotubes are obtained (Fig. 5.16). It is also very probable that, contrary to the latter case, the ABTS/SWCNT-

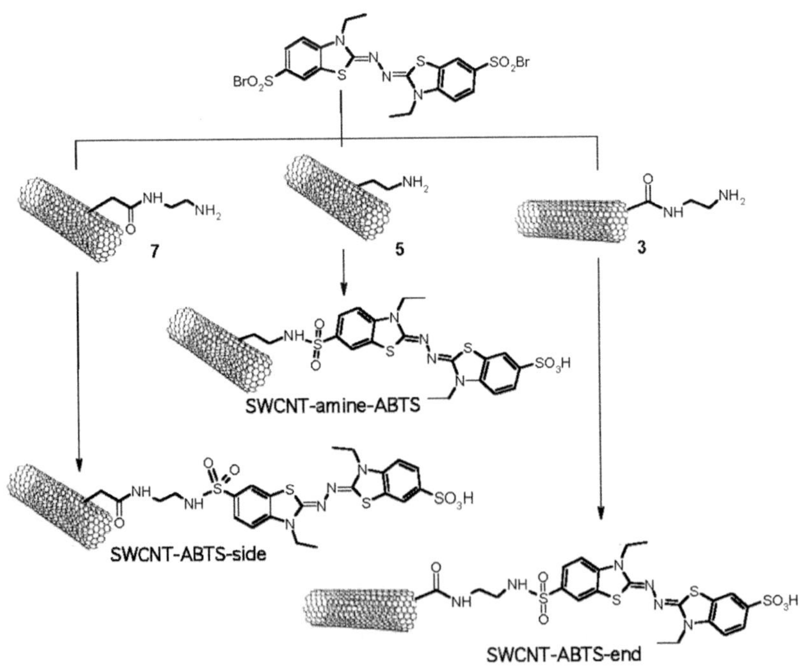


**FIGURE 5.15** A scheme of covalent derivatization of SWCNTs with ABTS [208].

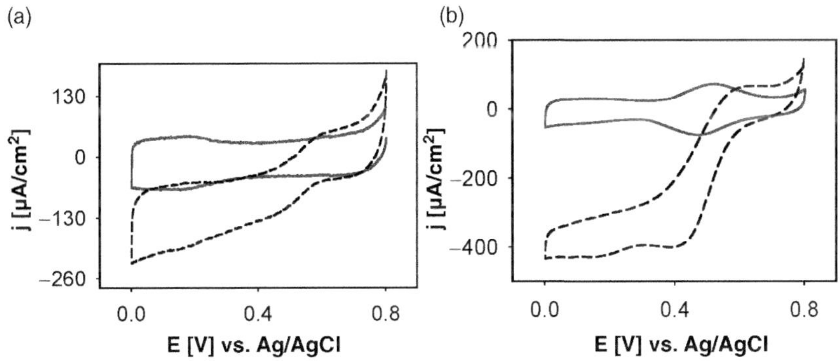


**FIGURE 5.16** Cyclic voltammograms obtained with electrodes modified by (a) pristine SWCNTs and (b) SWCNTs–amine–ABTS encapsulated in a laccase–Nafion polymer film. The electrode was immersed in (—) deoxygenated and (---) dioxygen saturated McIlvaine buffer (pH 5.2). Scan rate 1 mV/s.

modified electrode exhibits both mediated and nonmediated catalysis (see also Table 5.1).

Despite the smaller number of mediator moieties per mole of nanotubes at the electrode surface, the highest catalytic efficiency is achieved for SWCNTs-ABTS-end [184,208] (Table 5.1).

Again we note that there is the mediatorless contribution to the catalytic process. This is also suggested for the ABTS-modified MWCNTs film [170]. The unusual polarization behavior of the electrodeposited film of laccase and osmium redox copolymer on the SWCNT-modified electrode [146] also indicates such an effect. In the latter case the catalytic current is observed at potentials far more positive compared with the redox potential of the mediating osmium-based redox polymer, close to the potential of the $T_1$ site in *Trametes versicolor* laccase.

Immobilization of $ABTS^{2-}$ and the enzyme in separate materials being in contact with each other represents another approach [175]. The mediator is immobilized in bulk electrode material: hydrophobic sol–gel processed matrix containing graphite particles [210]. A thin hydrophilic silicate film with encapsulated laccase is deposited on the top of this composite (Fig. 5.17). High bioelectrocatalytic activity of this electrode toward dioxygen shows that there is good electronic contact between the mediator and laccase. Although one can argue that the solubility of $ABTS^{2-}$ in water helps in penetration of the hydrophilic film, a similar electrocatalytic efficiency is also observed with less water-soluble mediator syringaldazine [174]. In the last case the mediator redox potential is modulated by solution pH, because its electrode reaction involves protons.

Other mediators for laccase and bilirubin oxidase were tested in catalyzed dioxygen electroreduction. Multicyanometallates dissolved in solution or immobilized in cationic ionomer are suitable for this purpose [60,137,142,150,159,165]. The efficiency of dioxygen reduction was found to depend on the difference between the medatior and bilirubin oxidase redox potentials [142]. From this perspective, $W(CN)_8{}^{4-}$ is a better mediator than $Os(CN)_6{}^{4-}$ and $Mo(CN)_8{}^{4-}$. It was used for

**TABLE 5.1 Characteristics of ABTS Redox Processes and Oxygen Catalytic Current[a]**

| Electrode | $E_{pa}$ (V) | $E_{pa}-E_{pc}$ (V) | $Q$ (μC/cm$^2$) | $G_{ABTS}$ (nmol/cm$^2$) | $j_{cat}$ (A/cm$^2$) at +0.2 V | $j_{cat}/G_{ABTS}$ (μA/nmol) |
|---|---|---|---|---|---|---|
| GCE/SWCNT with laccase in Nafion film | | | | | 186.7 | |
| GCE/SWCNT–ABTS side with laccase in Nafion film | 0.534 | 0.050 | 6862.8 | 35.6 | 199.4 | 5.6 |
| GCE/SWCNT–ABTS end with laccase in Nafion film | 0.547 | 0.061 | 199.4 | 1.0 | 66.2 | 66.2 |
| GCE/SWCNT–amine–ABTS with laccase in Nafion film | 0.523 | 0.046 | 7062.2 | 36.6 | 421.5 | 11.5 |

[a] Per nanomole of nanotubes electroactive on the surface of the nanostructured carbon electrode. The electrode was immersed in dioxygen saturated McIlvaine buffer (pH 5.2). Scan rate 1 mV $s^{-1}$

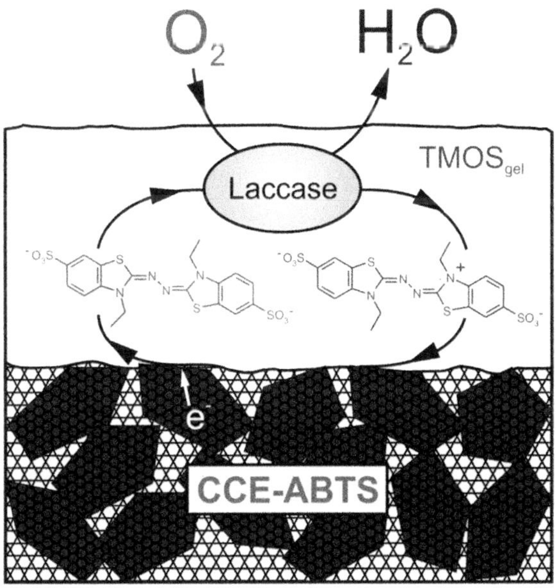


**FIGURE 5.17** A scheme of the structure and reaction mechanism at ABTS modified carbon ceramic electrode covered by a thin film of sol–gel processed hydrophilic silicate with encapsulated laccase.

construction of the efficient biocathode composed of carbon felt with immobilized bilirubin oxidase and polylysine with a nylon-covered membrane [142]. The same cationic polymer [150] or methylimidazolium units grafted on MWCNTs surface [165] were used for electrostatic entrapment of $Fe(CN)_6^{3-}$ together with bilirubin oxidase or laccase, respectively, and the catalysis was demonstrated. Pentacyanoferrate bound poly(1-vinylimidazole) was tested as a mediator for immobilized bilirubin oxidase [159]. The electrode covered by a film composed of this redox polymer, poly (ethyleneglycol diglycidyl ether) and entrapped enzyme, exhibits good electrocatalytic activity. A study of polyammonium cations as bilirubin oxidase stabilizer [60] is important in terms of biocathode construction. The results show that these cations, constituting quaternary ammonium nitrogen in the main chain and hydrophilic groups, such as hydroxyl and amide groups, may act as a protective colloid. The stabilizing effect of these polymers having $Fe(CN)_6^{3-}$ counterions on bilirubin oxidase immobilized on a membrane covered electrode was found [60]. In a more study $Mo(CN)_8^{4-}$ anions were used as mediators for laccase entrapped in microcapsules deposited on the electrode surface [137]. It is also important to say that buffer solution of $Fe(CN)_6^{3-}$ and bilirubin oxidase were used in cathodic compartment of membrane glucose–dioxygen biofuel cell developed for Sony Corp. [211,212].

A few mediators other than $ABTS^{2-}$ or metallocyanocomplexes were also tested for mediated bioelectrocatalysis of dioxygen reduction. Quinones or water-soluble ferrocene derivatives dissolved in aqueous solution were tested with laccase present in solution [194], covalently bonded to self-assembled monolayers of mercaptoundecanoic or mercaptopropionic acids [86], and encapsulated in monoolein cubic phases [179,182] or poly(*N*-isopropylacrylamide) gel [148]. The film, formed by gel comprising MWCNT and room-temperature ionic liquid, served as host for laccase or bilirubin oxidase and ferrocene mono- or dicarboxylic acids [98,99]. This electrode was tested as a biocathode in the biofuel cell powered by fruit juice. The electrode covered with the tetrabutylammonium bromide–treated Nafion film with immobilized $Ru(bipy)_3{}^{2+}$ and bilirubin oxidase exhibits catalytic activity in the presence of ethanol [140]. Finally, there is a single example of the electrode modified by laccase encapsulated in conducting polymer exhibiting efficient mediated catalysis of dioxygen reduction [157]. Unfortunately, poly-1,2-phenylene diamine redox properties shift the onset of catalytic current to potentials too low for the glucose–air biofuel cell, but it may be useful in hybrid devices [213].

Not only small redox polymers, conducting polymers organic, or inorganic molecules but also a small protein can act as electron mediators in bioelectrocatalytic dioxygen electroreduction. Cytochrome *c* dissolved in solution or adsorbed on mercaptoundecanol-modified Au electrode can act as a promoter for bilirubin oxidase dissolved in solution [185].

There are also a few reports of the application of enzymes other than copper oxidoreductases for modification of biocathodes exhibiting dioxygen reduction. One of them was cytochrome oxidase. This enzyme forms a supramolecular complex with thiolated cytochrome *c*, which is assembled on the Au electrode [12]. This architecture allows for efficient four-electron dioxygen electroreduction. A similar electrode structure was obtained by crosslinking [90]. The linkage of cytochrome *c*/cytochrome oxidase complex to $Cu^{2+}$ (polyacrylic acid) film allows one to obtain an electrochemically switchable cathode [89]. The effect of magnetic field on this biocathode was also studied [88].

Although horseradish peroxidase does not catalyze four-electron dioxygen electroreduction, this enzyme was used in combination with Co-protoporphyrin in chlorides solution [135,214]. Such a bifuctional catalyst deposited on the MWCNT-modified electrode surface turned out to be an effective biocathode. Dioxygen reduction is initiated on the redox center of the metal complex, and unwanted hydrogen peroxide is further reduced by horseradish peroxidase. The cabbage peroxidase is also effective in such a bifunctional system [135].

There are also examples of devices where hydrogen is produced on the cathode of the biofuel cell [57,215]. In one of them this electrode consists of carbon paper with *Thiocapsa roseopersicina* hydrogenase [57]. The cathodic reaction is possible owing to the mediator—methyl viologen dissolved in the cathodic compartment.

## 5.5 BIOCATHODES EMPLOYING MEDIATIORLESS ELECTROCATALYSIS

Electronic communication with the $T_1$ site of copper oxidases is crucial for efficient bioelectrocatalysis. For some enzymes and electrode materials, mediatorless electrocatalysis can be achieved, so the electrons can pass through the peripheral redox centers. Such electrodes are now favored since this approach eliminates problems with mediator stability, selectivity, or thermodynamic losses, due to the lack of mismatch between the enzyme and mediator formal potential, mediator price, and safety for the environment [68]. Not surprisingly, it is believed that direct electron-transfer-based biofuel cells would be easier to construct and more attractive for commercialization [5,25,161] despite the fact that the current and power densities are usually lower as compared with the mediated electrocatalytic systems. Additionally, the problem of proper orientation of the enzyme toward conductive substrate and the shortening of the distance between the active site ($T_1$ site of laccase or bilirubin oxidase) is crucial for efficient electron exchange [216,217].

Mediatorless bioelectrocatalysis of dioxygen reduction to water was reported more than two decades earlier [14] than the mediated one [6]. Tarasevich et al. [14,16,74] first reported that some *Trametes versicolor* laccase adsorbed on nanostructured carbon electrodes catalyzes four-electron electroreduction of dioxygen to water without the mediator with appreciable current density of 1 mA/cm$^2$. The superiority of mediatorless bioelectrocatalysis on carbon-based materials over noble metal catalyst was demonstrated in 2007 [68] (Fig. 5.18).

The overpotential of diffusion-controlled mediatorless dioxygen reduction at pH 7 using laccase (pH 5) or bilirubin oxidase (pH 7) adsorbed carbon aerogel electrode is

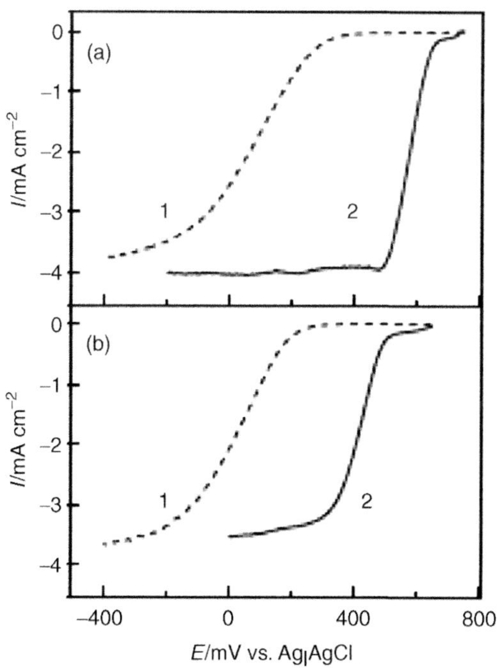


**FIGURE 5.18** Voltammograms of (a) Pt (1) and laccase-adsorbed carbon aerogel (2) electrode in pH 5 McIlvaine buffer and (b) Pt (1) and bilirubin oxidase adsorbed carbon aerogel (2) in pH 7 phosphate buffer at 25°C; the measurements were made at rotation rate of 1000 rpm and scan rate of 20 mV/s [68].

reduced by 0.55 and 0.4 V, respectively, in comparison with Pt electrode. The difference between the behavior of two biocatalysts results from the difference between their redox potentials equal to 0.12 V.

Such a decrease of overpotential is indispensable in order to obtain any voltage from the most popular dioxygen–glucose biofuel cell based on mediatorless bioelectrocatalysis. Significant improvement can be obtained with other sugars such as fructose or cellobiose as fuels. The replacement of this protein with fructose dehydrogenase increases the open-circuit voltage to 0.79 V and the power density to 850 $\mu W/cm^2$ under stirring conditions [58]. With cellobiose dehydrogenase, an open-circuit voltage of 0.73 V is obtained, but the power density is still rather low [55]. Unfortunately, the replacement of glucose-based anode will preclude the use of such a device in living organisms as envisioned by Heller [213]. The replacement of bioanode with a ionomer covered zinc anode may substantially increase the voltage of such a hybrid device [213]. Two examples with nanoparticulate biocathodes modified with laccase have been reported in the literature [96,101]. Both devices exhibit high open-circuit voltage close to 1.5 V. and 0.6–0.8 V under load. The stability of the device with screen-printed electrodes is remarkable [101].

Numerous studies show mediatorless catalysis with copper oxidoreductases on different types of carbon electrodes as glassy carbon, pyrolytic graphite or spectrographite [15,25,30,33,34,55,56,63,64,69,218–220] and boron doped diamond [32,183]. At short timescale, significant oxygen reduction currents were noted on laccase-modified carbon and graphite electrodes earlier [25,221,222]. The mechanism of this process involves donation of electrons to the $T_2/T_3$ cluster through the $T_1$ center on the cathode. The oxygen reduction wave appears at 0.775 V versus NHE and agrees well with the formal potential of $T_1$ site of the *Trametes versicolor* laccase. The reaction is controlled by catalysis rather than diffusion since the power output was not highly dependent on stirring. Anthracene-modified pyrolytic edge plane graphite substrate allowing fungal laccases to adsorb in a proper orientation represents a striking example of efficient mediatorless bioelectrocatalysis ($> 0.5\ mA/cm^2$) and remarkable biocathode stability [33,34] (Fig. 5.19). It seems that the affinity of anthracene groups with the hydrophobic pockets on the enzyme directed toward the $T_1$ site stabilizes *Pycnoporus cinnabarinus* laccase under resting conditions.

In terms of stability, the covalent binding of *Trametes hirsuta* laccase to electrodes functionalized with aminophenyl groups leads to better operational stability and resistance against chloride ions. This attachment is based on the formation of ester and/or amide bonds between the protein and hydroxyl and/or amino groups of the modified electrode [217] (Fig. 5.19).

The amino groups favor DET between $T_1$ and the electrode. At pH 5.1 catalytic current also reaches 0.5 $mA/cm^2$. It is much larger than that detected for laccase bound to the organothiol monolayer-modified electrodes, pointing to better-oriented biocatalyst molecules on the surface [194,219,223]. The current density decreases at high ionic strength [217], which shows that electrostatic interactions between positively charged attached amino groups and negatively charged regions of the protein surface play an important role in orienting laccase during immobilization (see Fig. 5.20). Indeed, negatively charged pockets due to carboxylic residues in the protein surround

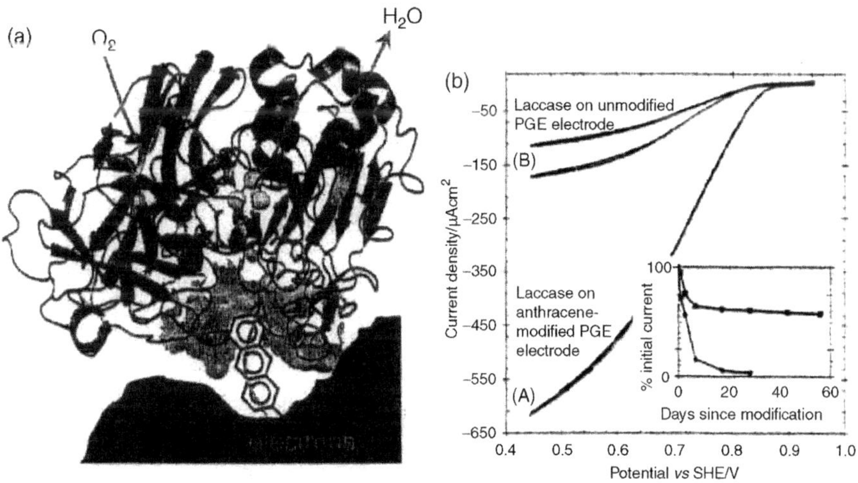


**FIGURE 5.19** 2-Aminoanthracene as the electron transfer unit to the $T_1$ center of laccase. The hydrophobic pocket interacting with anthracene is in blue B) [5]. (See insert for color representation.)

the $T_1$ site [224,225]. As much as 75% of initial catalytic current was retained after 10 days of continuous operation of the electrode [217]. Moreover, $Cl^-$ anions do not suppress the activity of this electrode as it is the case of $F^-$ anions. The fomer are bulkier and can approach only laccase $T_1$ site, whereas $F^-$ anions block oxygen by binding to the $T_2/T_3$ cluster. It should be emphasized that such a favorable situation is possible only in mediatorless dioxygen bioelectrocatalysis.

Catalytic activity was also observed for *Myrothecium verrucaria* bilirubin oxidase adsorbed at unmodified graphite [69]. The current density reaches an enzyme kinetic

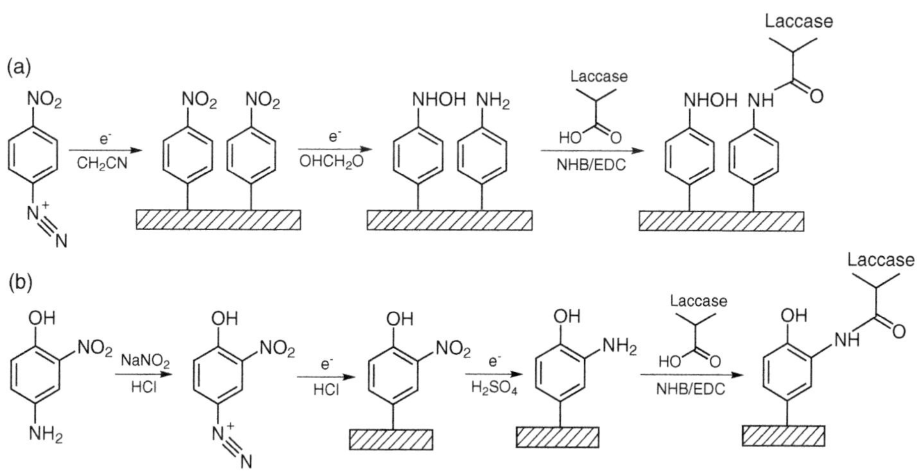


**FIGURE 5.20** A scheme of electrode modification with laccase via (a) aminophenyl groups and (b) 2-aminophenol groups [217].

limit of ~1 mA/cm$^2$ at the RDE, where larger densities are usually noted. Bilirubin oxidase can be assembled by means of alternate electrostatic adsorption with positively charged polyethylenimine to form a large three-dimensional multilayer structure, which promotes preservation of the enzyme catalytic activity [226]. The surface concentration of electroactive bilirubin oxidase molecules is $3.2 \times 10^{-10}$ mol/cm$^2$, which is 1–2% of the total immobilized enzyme. This experiment confirms that mediatorless electrical communication between redox centers of bilirubin oxidase and the electrode takes place mainly through direct physical contact with the electrode surface. For this electrode, efficient electroreduction of oxygen takes place at 0.46 V versus NHE close to the redox potential of the $T_1$ center of bilirubin oxidase [226].

The mediatorless catalysis with copper efflux oxidases (CueO) from *Eschericia coli* was observed only on nanostructured material [50–52]. The impressive large current density of 4 mA/cm$^2$ at RDE was obtained with the unfavorably high (0.5 V) overpotential. It can be decreased by creation of mutants [51]. As more recently reported, the current density of $O_2$ reduction reached values as high as 20 mA/cm$^2$ when CueO and an air diffusion biocathode were employed [49]. The onset of catalytic current was ~ 0.35 V, which is close to formal potential of the type 1 site in CueO (0.26 V) [50]. Usually the biocathodes rely on dioxygen dissolved in the electrolyte solution; thus the maximum current is limited by $O_2$ diffusion because of low solubility and small diffusion coefficient of oxygen. The $O_2$ diffusion depends on stirring or rotation rate, or flow rate of the electrolyte solution. The air-diffusion-type biocathode was constructed using nanostructured carbon particles, carbon paper, and a binder to glue the particles on paper (Fig. 5.21).

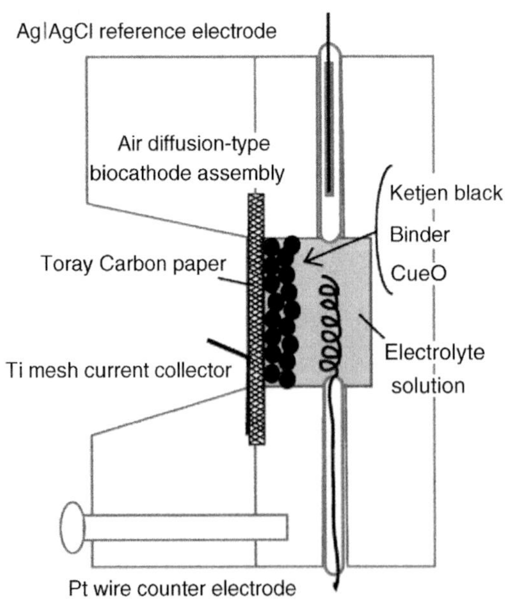


**FIGURE 5.21** The air diffusion CueO-modified electrode [49].

Following the earliest experiments with nanostructured electrodes modified by laccase [16], different, mainly carbon-based, nanomaterials are employed to improve the heterogeneous electron transfer kinetics on the enzyme-modified electrodes and to increase the catalytic current density. Enzymes adsorb on these materials or are covalently bonded (see section 5.3). Only ~10% of the immobilized enzyme molecules take part in the bioelectrocatalytic reaction because of the unfavorable orientation of the adsorbed laccase to the conducting surface [227]. This can be improved with nanostructured surfaces by allowing more laccase molecules to attain a favorable orientation and distance from the conducting material [32,68,96,209,228–232]. Moreover, the electrode is "extended into 3D space" [5], which allows dissolved dioxygen to reach a larger number of biocatalyst centers. This, in turn, leads to higher efficiency of bioelectrocatalysis. For example, mesoporous carbon structures such as the aerogel stabilize the enzyme electrode and allow one to reach $O_2$ reduction current of 10 mA/cm$^2$ at 8000 rpm (Fig. 5.16) [51,68] larger than for the mediated catalysis. Stable current responses were obtained in a flow cell (flow rate 200 mL/h) with the laccase adsorbed carbon aerogel electrode for as long as 10 days [68]. This is probably because nanosized elements of the electrode do not significantly affect the conformation of the enzyme molecule of similar size.

In mediatorless catalysis quite often the rotating-disk electrode (RDE) is used for bioelectrocatalysis study and this makes a comparison difficult. Also, one has to be careful with comparison of such data for mediatorless and mediated catalysis. The onset potential of mediatorless dioxygen reduction is higher for the electrode modified by laccase than that modified by bilirubin oxidase [68], reflecting the difference between the redox potentials of the $T_1$ site of these enzymes. This indicates that this center serves as an acceptor of electrons from the electrode and thus makes the laccase-modified electrode more promising for the biofuel cell. This process is predominantly diffusion-controlled. At the RDE the current density is linearly dependent on the square root of angular velocity. As was mentioned above, at a high rotation rate ($>$ 3000 rpm) [51,68] up to 10 mA/cm$^2$ can be achieved. At very high rotation rates the slope of dependence of the current on square root of angular velocity decreases because of fast mass transfer of dissolved dioxygen, which exceeds the kinetic limitation of the enzyme. This deviation occurs even at high overvoltage and is related to the interfacial electron transfer kinetic limitation. The latter effect is often the case for mediatorless reduction and is explained as being due to a disorder of enzyme molecules at the electrode surface and large tunneling distances [216,233].

A few approaches have been combined to construct the efficient biocathode for mediatorless dioxygen reduction [96]. They include the application of high-potential laccase, immobilization of this enzyme in sol–gel processed matrix [176], and the use of hydrophilic (phenyl sulfonated) carbon nanoparticles (CNPs) of diameter ~ 7.8 nm as the electroactive component [234]. *Cerrena unicolor* laccase is adsorbed on CNPs prior to their immobilization in a sol–gel processed silicate film. Although the redox process of the enzyme is not observed on this electrode, almost 0.1 mA/cm$^2$ catalytic current density was obtained [96]. Its maximum value was obtained at pH 4.8, which is slightly different from the maximum activity of the enzyme (pH = 5.3) [235]. This may be caused by a local pH shift due to larger concentration of hydrated protons in

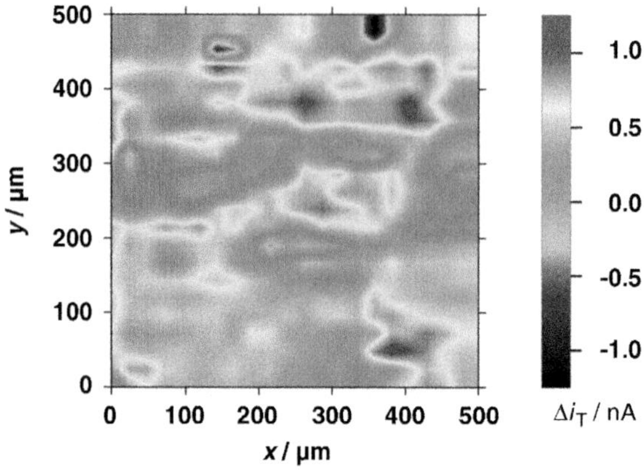


**FIGURE 5.22** Scanning electrochemical microscopy (redox competition mode) visualization of laccase distribution in a film consisting of carbon nanoparticles dispersed in sol–gel processed silicate immersed in 0.1 M McIlvaine buffer (pH 4.8) saturated with air [96]. (See insert for color representation.)

comparison with the bulk of solution. Under these conditions dioxygen electroreduction commences at a value close to the formal potential of the $T_1$ redox center of the laccase. The bioelectrocatalytic activity at neutral pH is decreased by about 30 times, which is explained by blocking the dioxygen access to the $T_2/T_3$ cluster by $OH^-$ anions [56].

Scanning electrochemical microscopy (SECM) (Fig. 5.22) and atomic force microscopy (AFM) experiments show that the presence of both components of the film prevents their aggregation earlier than was observed for laccase [173] and CNPs [234] immobilized in a silicate film. This is probably due to protein adsorption on CNPs providing favorable conditions for direct electron transfer and mediatorless bioelectrocatalysis [96], which was not observed for the same laccase immobilized in the sol–gel processed silicate film [176].

This observation suggests that laccase aggregation is responsible for the lack of mediatorless catalysis in bioelectrocatalytic dioxygen reduction observed in various matrices in the absence of nanoparticles. On the other hand, the adsorption of enzyme on carbon nanoparticulate material results in an increased electrode resistance [96]. Adsorption of enzyme may reduce the number of percolation pathways within the film and decrease the probability of electron transfer between conductive nanoparticles and with the solid electrode substrate. This distortion of the electronic contact between the particles within the composite material leads to the observed decrease of heterogeneous electron transfer rate constants [74,96,100]. The fractal and percolation properties of the active layer were modeled by Tarasevich et al. [74], and the fractal percolation limitation was suggested to hinder the increase of active-layer thickness. To escape the fractal percolation limitation, one should increase the concentration of carbon particles, which hence decrease the average fractal cluster size.

The sol–gel processed CNPs film electrode with adsorbed bilirubin oxidase is also effective for mediatorless dioxygen electrocatalysis [236]. This process is also

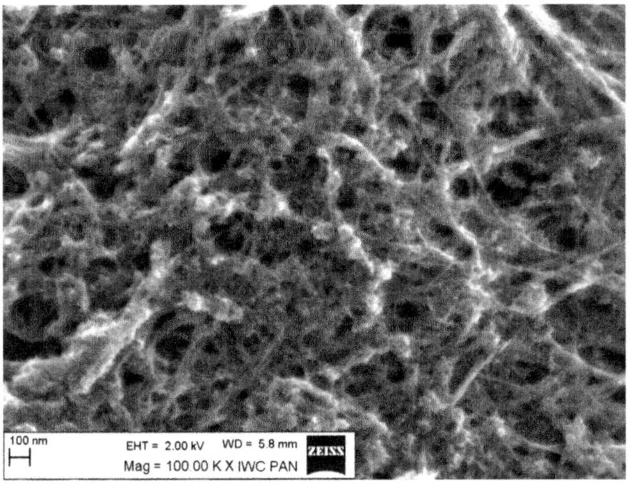


**FIGURE 5.23** Scanning electron microscopy of film electrode prepared from SWCNT modified with pyrene sulfonate and laccase encapsulated in sol–gel processed film [97].

observed for CCE modified by adsorbed bilirubin oxidase. The enrichment of CCE body with CNPs increases the catalytic current density in dioxygen saturated solution from 0.14 to 0.34 mA/cm$^2$ [236].

Carbon nanotubes are most the frequently used carbon-based conductive nanomaterial for preparation of biocathode exhibiting mediatorless catalysis. Their electronic properties are unique as they can be either metallic or semiconducting, depending on their structure [237]. Exhibiting good conductivity, they are obviously favored for biocathode construction. When immobilized on the electrode surface, they form an extended network and the electroactive surface is significantly increased (Fig. 5.23). Moreover, their diameter is comparable in size to laccase or bilirubin oxidase, and therefore they are favored for mediatorless bioelectrocatalysis.

The majority of reports have shown that chemical modification of the nanotube sidewall or terminus is generally needed to control dispersion and assembly of CNTs into usable devices [81,84,85,92,97,133,162,164,165,170,206,207,238–241]. Both laccase and bilirubin oxidase were immobilized on CNT film by physical adsorption, layer-by-layer deposition, or direct covalent attachment [84,85,92]. This approach leads to a higher content of biocatalyst addressable from the electrode, and electrically contacted and higher catalytic current density is observed.

Glassy carbon electrode covered by a SWCNT network encapsulated in a wide range of materials such as Nafion, lecithin, hydrophobin, or liquid crystalline lipid cubic phase together with laccase exhibits high mediatorless catalysis of dioxygen reduction [32,209]. The composition of the film enhances the stability of the enzyme and prevents it from leaking into solution. This electrode also exhibits high background current due to a large electroactive surface of SWCNTs. Therefore the ratio of the catalytic to background current has to be taken into account in data analysis (Table 5.2), which indicates Nafion as the favored matrix.

**TABLE 5.2 Characteristics of Catalytic Reduction of Oxygen at Different Modified Electrodes Containing ~ 40 μg Laccase in Matrix; McIlvaine Buffer Solution (pH 5.2)**

| Electrode | $E$ onset[a] (V) | $E_{1/2}$ (V) | $j$ (μA/cm$^2$) (at 200 mV) | $i_{cat}/i_{bcg}$ (at 200 mV) |
|---|---|---|---|---|
| BDD with laccase in cubic-phase film | 0.620 | 0.368 | 0.7 | 3.6 |
| GCE/nanotubes with adsorbed laccase | 0.620 | 0.441 | 58.2 | 3.7 |
| GCE/nanotubes with laccase on hydrophobin film | 0.599 | 0.465 | 17.2 | 6.6 |
| GCE/nanotubes with laccase in lecithin film | 0.611 | 0.494 | 40.6 | 5.0 |
| GCE/nanotubes with laccase in Nafion film | 0.611 | 0.448 | 110.0 | 10.6 |

[a] Potential of onset of the catalytic dioxygen reduction current.
[b] *Source:* Reference 32.

As proved by data in Table 5.1, carbon nanotubes, indeed, increase the electrode working area and improve the conductivity of the film. While a common strategy is to incubate the CNTs in enzyme solution [242,243], simple adsorption may not provide stable modification with the enzyme.

The appreciable current density can also be observed for SWCNTs and enzyme immobilized in the polymeric film. This is the case of the film based on glutaraldehyde cross linked laccase [164] or bilirubin oxidase [168] covering SWCNT-modified carbon fiber microelectrode. Interestingly, the catalytic current is also seen in nitrogen-saturated solution, indicating strong affinity of dioxygen to the electrode material. The current density of dioxygen reduction on the electrode modified by laccase and SWCNTs can be significantly increased by functionalization of nanotubes with pyrenesulfonate derivatives [97] (Figs. 5.23 and 5.24). This functionalization obviously increases SWCNT solubility and decreases aggregation when the film (in this case sol–gel processed silicate) is formed. The role of sulfonate groups in electrocatalysis is not clear; however, it should be noted that mediatorless catalysis of the same magnitude is observed for the phenylsulfonated carbon nanoparticle film [96]. The SECM experiments also suggest deaggregation of laccase in a silicate film [244].

Although nanoparticulate films already provide a three-dimensional structure of the electrode, the control of the increase in electrochemically active surface and the amount of encapsulated protein can be provided by building up a multilayered structure [127,149,153]. The significant increase in catalytic efficiency with a multilayer structured film consisting of MWCNTs and laccase with poly-L-lysine and branched polyethyleneimine as linkers represents an important example [149]

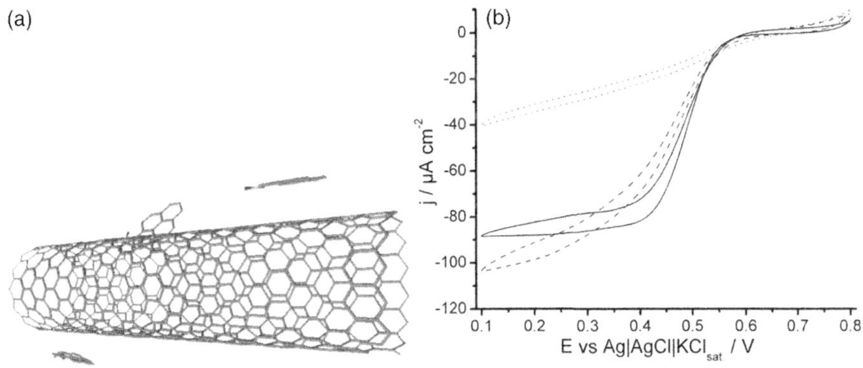


**FIGURE 5.24** Scheme of (a) SWCNT modified by pyrene sulfonate and cyclic voltammogram (scan rate 1 mV/s[1]) of the film electrodes consisting of laccase and SWCNT encapsulated in sol–gel processed silicate matrix obtained in dioxygen saturated McIlvaine buffer at pH 4.8; (b) SWCNTs unmodified (dotted lines) or modified by pyrene sulfonate (solid lines) and pyrene tetrasulfonate (dotted lines) [97].

(Figs. 5.25 and 5.26). This procedure allows one to obtain the substrate with extended contact with a large amount of biocatalyst being accessible by substrate of the electrode reaction.

Although carbon-based materials dominate the research on mediatorless biocathodes, direct electron transfer was reported for laccase at gold electrodes [25,62,63,74,86,92,127,153,221,223,245–247]. However, in these conditions bioelectrocatalytic reduction of dioxygen on gold occurs at potentials more than 0.2 V lower than that of the $T_1$ site, decreasing the applicability of gold electrodes in dioxygen–glucose biofuel cell. Moreover, $H_2O_2$ was detected among the products, thus pointing to a different mechanism of catalysis [63,223]. The mechanism was

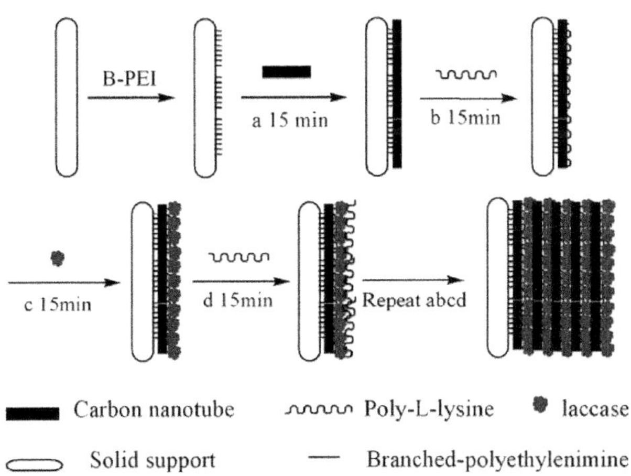


**FIGURE 5.25** A scheme of self-assembly of the multilayered laccase MWCNT electrode [149].

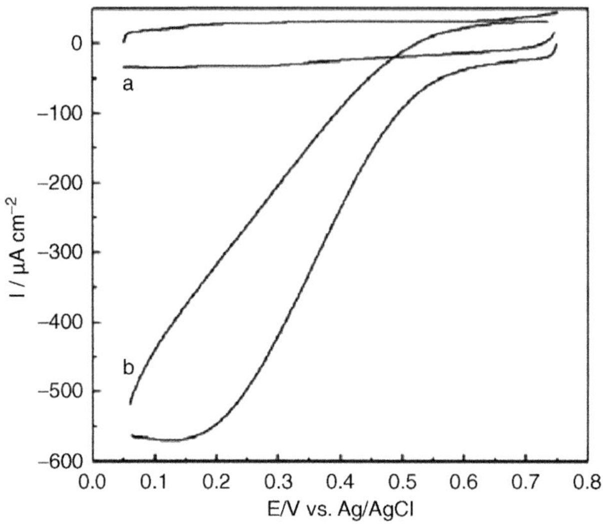


**FIGURE 5.26** Cyclic voltammogram (scan rate 50 mV/s) of the film electrodes prepared by the method presented in Fig. 5.17 under nitrogen (curve a) and air (curve b) saturated solution in PBS buffer at pH 6 [149].

explained by assuming that the $T_2$ Cu site was exchanging electrons with the electrode instead of $T_1$. Quite surprisingly, such a negative potential shift was also detected for laccase adsorbed on CCE composed of graphite microparticles [94].

Despite such disadvantages of gold as biocathode material, the efficient (800 μA/cm$^2$) biocathode was constructed of laccase and Au nanoparticles [153]. The

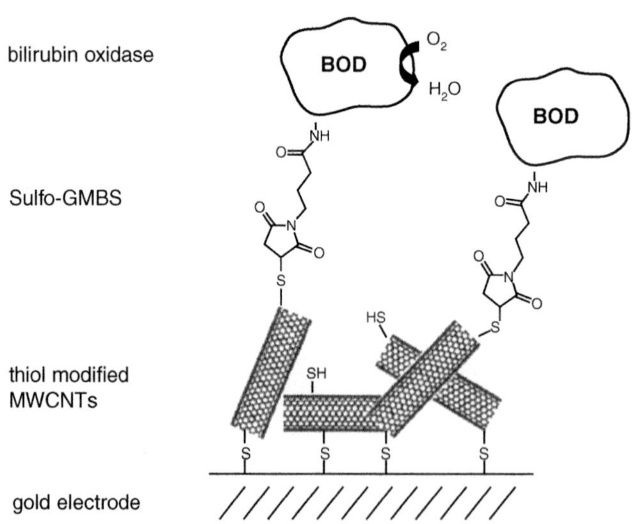


**FIGURE 5.27** A scheme of self-assembly of thiol functionalized and bilirubine oxidase appended MWCNTs on gold electrode [92].

macroporous structure was synthesized electrochemically by the inverted colloidal crystal template technique. This allows one to control the thickness of the film and the number of nanoparticles and protein molecules. Although the onset potential of catalytic current remained almost unchanged in comparison with the smooth electrode, it was applied to the dioxygen–glucose biofuel cell [153].

The most recent example of bioelectrocatalysis on gold electrodes employs the affinity of thiol-modified MWCNTs with covalently bonded bilirubin oxidase [92] (Fig. 5.27).

## 5.6 SUMMARY AND OUTLOOK

The design of a biocathode, as of any enzyme-modified electrode, is not an easy task because of several tough requirements to be fulfilled. They include selection of appropriate enzyme and (in some cases) the mediator to provide a high potential of oxidant bioelectrocatalytic reduction. Its high efficiency depends on the same factors and electrode construction. At the same time, a friendly environment for fragile protein immobilization has to be provided. This is crucial from the practical perspective because stable performance of the device is expected.

So far the development of biocathodes for biofuel cells has been almost restricted to electrodes for dioxygen reduction to water. At early stages only a few papers, mostly from the Tarasevich group [14–16], were devoted to mediatorless bioelectrocatalysis. Some results obtained with carbon blacks and laccase-modified electrodes are still competitive in terms of current density; however, their reproducibility was questioned [68]. Following the first paper reporting the application of $ABTS^{2-}$ as the laccase mediator [6], attention turned to mediated bioelectrocatalysis. The seminal papers from the Heller group introducing the concept of "wiring" the enzyme to redox polymer [118,186,187] shifted the research path to biocathodes modified by laccase or bilirubin oxidase encapsulated in osmium redox polymer, and these electrodes have been continuously developed. The inspiration by early Tarasevich papers combined with the advent of nanotechnology directed the research back to mediatorless electron transfer, in particular to construction of biocathodes modified by conductive nanomaterials and enzymes [229]. Simplification of the electrode construction was used as an argument for these electrodes. However, this was not always the case, and multicomponent biocathodes were proposed for mediatorless catalysis [32,96,149,153,248,249]. Certainly, higher contents of the enzyme could be accumulated with nanomaterials, and much larger current densities of oxygen reduction could be achieved.

A significant part of research on biocathodes is aimed at reducing dioxygen reduction overpotential. This is because of the final goal to produce the biofuel cell of as high a voltage as possible. From this perspective the laccase-modified electrode is superior, and for the most popular glucose–dioxygen biofuel cell, the value of open-circuit potential 0.78 V obtained with "wired" osmium polymers remains the limit at the moment [122]. The more recently proposed replacement of glucose oxidase by fructose dehydrogenase [58] and cellobiose dehydrogenase [55] allows one to obtain similar values (0.79 and 0.73 V, respectively) with no help of mediators. However, for the more recently proposed [213] and realized [96,101] hybrid zinc–dioxygen cell, it is not so important because of the low redox potential of the $Zn^{2+/0}$ couple. For these hybrid systems, efficient electrodes modified by enzymes with lower redox potentials, such as copper efflux oxidase, can be applied. The reduction potential depends on the redox potential of the $T_1$ enzyme active site exchanging electrons with the electrode support. The change of this parameter for a given enzyme is not easy. When protein is adsorbed on the electrode surface, this can be fine-tuned by appropriate selection of the electrode material in order to decrease the distance between the electron conductive surface and the enzyme redox site [25]. The different behavior of laccases at carbon and gold electrodes is an example of such differences [25,77]. The shift of the reduction potential due to replacement of solid metal by nanoparticulate material thus far has not been impressive but remains a possible direction [153]. The genetic modification of the enzyme, in particular site-directed mutagenesis, needed to change ligands at the $T_1$ copper center and move the potential of the $T_1$ site to more positive values [51]. Of course, the stability of mutants remains an important aspect.

Increase in catalytic efficiency is another goal of the research. It is crucial for biofuel cell design because in many devices the cathode is reported to be less efficient than the bioanode. It can be achieved by appropriate design of the electrode–mediator and mediator–enzyme contact or the use of nanomaterials for electrode modification. This is not only the case of mediatorless catalysis. The modification of nanomaterials such as carbon nanoparticles or CNTs by the mediator seems to be prospective. This requires the optimization process to provide the largest concentration of percolation paths and optimum conditions for (direct) electron exchange between conductive nanoparticles and the enzyme and efficient dioxygen transport. Fortunately, the use of nanomaterials increases biocathode durability and decreases the problems connected with enzyme aggregation [96,97,236] (see text below). Three-dimensional structures allowing one to increase the enzyme load can be prepared using CNTs or metallic nanoparticles [127,149,153]. This direction of research may result in a significant increase of efficiency for both mediated and mediatorless catalysis. However, this approach may create new problems connected with transport of dioxygen across the multilayered film pores, which can be solved applying the flow system [68,202].

Despite significant efforts and progress in biocathode and biofuel cell design and even the first attempts at commercialization of the device [211,212], this area of research is still at its infancy. Most of published research is oriented toward a more challenging goal: the electrode applicable in a membraneless device. However, the so called "biobattery generating electricity from sugar" [212] consists of two chambers separated by a membrane [211]. Both mediators and biocatalyst are dissolved in a catholyte and an anolyte, and the device is designed to power a Walkman (small portable audioplayer). This is rather far from the very exciting dream about the implantable biofuel cell. The majority of papers demonstrate a new design of biocathode, most often applicable in the membraneless biofuel cell, and some of them present improvement in terms of current density. This approach is good at the beginning but at some time has to be changed. The issue of electrode stability on timescale longer than a week is rather ignored with a few notable most recent exceptions where good performance of the device for several months was reported [33,101,156]. The use of a microfluidic device where biocatalyst is continuously supplied to the electrodes [202] may solve the problem of enzyme instability due to contact with the electrode support, but it requires continuous supply of the protein.

Therefore, from time to time, one has to ask a question: "Are we heading in a good direction?" This is because we are working with biomolecules unstable in terms of the lifetime of typical electric power supplies. We will gradually increase the efficiency of dioxygen reduction and shift its potential closer and closer to its thermodynamic value, but our efforts will be futile because of the inherent instability of the enzyme. It is known that appropriate encapsulation of the enzyme may increase its stability. However, most stable biocathodes were prepared in a simple way; laccase is adsorbed on the carefully designed graphite surface or nanomaterials [33,101]. These two examples may indicate that the mildest immobilization technique not separating the enzyme from the surrounding solution is superior in terms of stability. Although the first results are not very encouraging [51,53], one can envisage that genetic engineering will be as successful in creation of more stable copper oxidoreductases as was done with genetically modified cereal crops. This means that we have to be ready with new electrode designs for the stable, genetically modified enzymes.

Biomimetic catalysts for dioxygen electroreduction, like most popular iron porphyrin complexes, would be a convenient alternative [250–252]. Unfortunately, the chemical instability of electrodes modified by these complexes was observed and spurious intermediates were produced. Moreover, their catalytic efficiencies toward $O_2$ were too low for practical use in fuel cells. A more recent attempt at combinatorial biomimetics inspired by catalytic activity of copper oxidoreductases should be noted [253]. Atlhough the potential for dioxygen electroreduction obtained on electrodes modified with the simple copper(II) polyhistidine complex is too low to be acceptable for biocathodes in fuel cells, this direction of research may become fruitful in the future. There is also some progress in the development of other abiotic cathodes, but the results are still far from impressive in terms of both stability and current efficiency [254].

## REFERENCES


1. A. T. Yahiro, S. M. Lee, and D. O. Kimble, *Biochim. Biophys. Acta* **88**, 375 (1964).
2. A. Heller, *Phys. Chem. Chem. Phys.* **6**, 209 (2004).
3. S. C. Barton, J. Gallaway, and P. Atanassov, *Chem. Rev.* **104**, 4867 (2004).
4. B. Wang, *J. Power Sources* **152**, 1 (2005).
5. J. A. Cracknell, K. A. Vincent, and F. A. Armstrong, *Chem. Rev.* **108**, 2439 (2008).
6. G. T. R. Palmore and H. H. Kim, *J. Electroanal. Chem.* **464**, 110 (1999).
7. R. A. Bullen, T. C. Arnot, J. B. Lakeman, and F. C. Walsh, *Biosens. & Bioelectron.* **21**, 2015 (2006).
8. F. Davis and S. P. J. Higson, *Biosens. & Bioelectron.* **22**, 1224 (2007).
9. M. J. Cooney, V. Svoboda, C. Lau, and G. Martin, *Energy Environ. Sci.* **1**, 320 (2008).

10. H. A. O. Hill, N. J. Walton, and I. J. Higgins, *FEBS Lett.* **126**, 282 (1981).
11. H. A. O. Hill and N. J. Walton, *J. Am. Chem. Soc.* **104**, 6515 (1982).
12. E. Katz, I. Willner, and A. B. Kotlyar, *J. Electroanal. Chem.* **479**, 64 (1999).
13. A. Pizzariello, M. Stred'ansky, and S. Miertus, *Bioelectrochemistry* **56**, 99 (2002).
14. I. V. Berezin, V. A. Bogdanovskaya, S. D. Varfolomeev, M. R. Tarasevich, and A. I. Yaropolov, *Bioelectrocatalysis* **240**, 615 (1978).
15. C. W. Lee, H. B. Gray, and F. C. Anson, *J. Electroanal. Chem.* **172**, 289 (1984).
16. M. R. Tarasevich, A. I. Yaropolov, V. A. Bogdanovskaya, and S. D. Varfolomeev, *BIoelectrochem. Bioenerg.* **6**, 587 (1979).
17. J. Rogalski and A. Leonowicz, in *Conscise Encyclopedia of Bioresource Technology*, A. Pandey, ed., Haworth Press, Binghamton, NY, 2004.
18. K. Piontek, M. Antorini, and T. Choinowski, *J. Biol. Chem.* **277**, 37663 (2002).
19. A. V. Lyashenko, I. Bento, V. N. Zaitsev, N. E. Zhukhlistova, Y. N. Zhukova, A. G. Gabdoulkhakov, E. Y. Morgunova, W. Voelter, G. S. Kachalova, E. V. Stepanova, O. V. Koroleva, V. S. Lamzin, V. I. Tishkov, C. Betzel, P. F. Lindley, and A. M. Mikhailov, *J. Biol. Inorg. Chem.* **11**, 963 (2006).
20. H. P. Call and I. Mucke, *J. Biotechnol.* **53**, 163 (1997).
21. P. Schneider, M. B. Caspersen, K. Mondorf, T. Halkier, L. K. Skov, P. R. Ostergaard, K. M. Brown, S. H. Brown, and F. Xu, *Enzyme Microb. Technol.* **25**, 502 (1999).
22. F. Xu, A. E. Palmer, D. S. Yaver, R. M. Berka, G. A. Gambetta, S. H. Brown, and E. I. Solomon, *J. Biol. Chem.* **274**, 12372 (1999).
23. O. V. Koroleva, E. V. Stepanova, V. P. Gavrilova, V. I. Biniukov, and A. M. Pronin, *Biochemi (Moscow)* **66**, 960 (2001).
24. B. R. M. Reinhammar, *Biochim. Biophys. Acta* **275**, 245 (1972).
25. S. Shleev, J. Tkac, A. Christenson, T. Ruzgas, A. I. Yaropolov, J. W. Whittaker, and L. Gorton, *Biosens. & Bioelectron.* **20**, 2517 (2005).
26. E. I. Solomon, U. M. Sundaram, and T. E. Machonkin, *Chem. Rev.* **96**, 2563 (1996).
27. S. Shleev, A. Christenson, V. Serezhenkov, D. Burbaev, A. Yaropolov, L. Gorton, and T. Ruzgas, *Biochem. J.* **385**, 745 (2005).
28. B. R. M. Reinhammar and B. Vanngar, *Eur. J. Biochem.* **18**, 463 (1971).
29. A. Klonowska, C. Gaudin, A. Fournel, M. Asso, J. Le Petit, M. Giorgi, and T. Tron, *Eur. J. Biochem.* **269**, 6119 (2002).
30. S. Shleev, M. Klis, Y. Wang, J. Rogalski, R. Bilewicz, and L. Gorton, *Electroanalysis* **19**, 1039 (2007).
31. E. Nazaruk, A. Michota, J. Bukowska, S. Shleev, L. Gorton, and R. Bilewicz, *J. Biol. Inorg. Chem.* **12**, 335 (2007).
32. K. Stolarczyk, E. Nazaruk, J. Rogalski, and R. Bilewicz, *Electrochim. Acta* **53**, 3983 (2008).
33. C. F. Blanford, R. S. Heath, and F. A. Armstrong, *Chem. Commun.* **2007**, 1710 (2007).
34. C. F. Blanford, C. E. Foster, R. S. Heath, and F. A. Armstrong, *Faraday Discuss.* **140**, 319 (2008).
35. A. I. Yaropolov, O. V. Skorobogatko, S. S. Vartanov, and S. D. Varfolomeyev, *Biochem. Biotechnol.* **49**, 257 (1994).
36. L. C. Petersen and H. Degn, *Biochim. Biophys. Acta* **526**, 85 (2000).

37. A. I. Yaropolov, A. Naki, G. S. Pepanyan, E. L. Saenko, and S. D. Varfolomeyev, *Moscow Univ. Chem. Bull.* **27**, 93 (1986).

38. S. K. Lee, S. D. George, W. E. Antholine, B. Hedman, K. O. Hodgson, and E. I. Solomon, *J. Am. Chem. Soc.* **124**, 6180 (2002).

39. J. Yoon, S. Fuji, and E. I. Solomon, *Proc. Natl. Acad. Sci. USA* **106**, 6585 (2009).

40. J. Yoon, S. Fuji, and E. I. Solomon, *J. Am. Chem. Soc.* **129**, 13127 (2007).

41. V. Ducros, A. M. Brzozowski, K. S. Wilson, S. H. Brown, P. Ostergaard, P. Schneider, D. S. Yaver, A. H. Pedersen, and G. J. Davies, *Nature Struct. Biol.* **5**, 310 (1998).

42. P. J. Stephens, D. R. Jollie, and A. Warshel, *Chem. Rev.* **96**, 2491 (1996).

43. G. Battistuzzi, M. Borsari, L. Loschi, M. C. Menziani, F. De Rienzo, and M. Sola, *Biochemistry* **40**, 6422 (2001).

44. E. I. Solomon, R. K. Szilagyi, S. D. George, and L. Basumallick, *Chem. Rev.* **104**, 419 (2004).

45. S. Tsujimura, K. Kano, and T. Ikeda, *Electrochemistry* **70**, 940 (2002).

46. N. Mano, H. H. Kim, Y. C. Zhang, and A. Heller, *J. Am. Chem. Soc.* **124**, 6480 (2002).

47. L. H. Guo and H. A. O. Hill, *Adv. Inorg. Chem.* **36**, 341 (1991).

48. A. Shimizu, J. H. Kwon, T. Sasaki, T. Satoh, N. Sakurai, T. Sakurai, S. Yamaguchi, and T. Samejima, *Biochemistry* **38**, 3034 (1999).

49. R. Kontani, S. Tsujimura, and K. Kano, *Bioelectrochemistry* (in press).

50. Y. Miura, S. Tsujimura, Y. Kamitaka, S. Kurose, K. Kataoka, T. Sakurai, and K. Kano, *Chem. Lett.* **36**, 132 (2007).

51. Y. Miura, S. Tsujimura, S. Kurose, Y. Kamitaka, K. Kataoka, T. Sakurai, and K. Kano, *Fuel Cells* **9**, 70 (2009).

52. S. Tsujimura, Y. Miura, and K. Kano, *Electrochim. Acta* **53**, 5716 (2008).

53. K. Kataoka, H. Komori, Y. Ueki, Y. Konno, Y. Kamitaka, S. Kurose, S. Tsujimura, Y. Higuchi, K. Kano, D. Seo, and T. Sakurai, *J. Mol. Biol.* **373**, 141 (2007).

54. A. Heller, *Curr. Opin. Chem. Biol.* **10**, 664 (2006).

55. V. Coman, C. Vaz-Dominguez, R. Ludwig, W. Herreither, D. Haltrich, A. L. De Lacey, T. Ruzgas, L. Gorton, and S. Shleev, *Phys. Chem. Chem. Phys.* **10**, 6093 (2008).

56. C. Fernandez-Sanchez, T. Tzanov, G. M. Gubitz, and A. Cavaco-Paulo, *Bioelectrochemistry* **58**, 149 (2002).

57. M. Hambourger, M. Gervaldo, D. Svedruzic, P. W. King, D. Gust, M. Ghirardi, A. L. Moore, and T. A. Moore, *J. Am. Chem. Soc.* **130**, 2015 (2008).

58. Y. Kamitaka, S. Tsujimura, N. Setoyama, T. Kajino, and K. Kano, *Phys. Chem. Chem. Phys.* **9**, 1793 (2007).

59. A. V. Kapustin, M. R. Tarasevich, Y. G. Chirkov, and V. A. Bogdanovskaya, *Russ. J. Electrochem.* **40**, 909 (2004).

60. H. Katano, H. Tatsumi, T. Hibi, T. Ikeda, and T. Tsukatani, *Anal. Sci.* **24**, 1421 (2008).

61. E. Katz, B. Filanovsky, and I. Willner, *New J. Chem.* **23**, 481 (1999).

62. H. J. Qiu, C. X. Xu, X. R. Huang, Y. Ding, Y. B. Qu, and P. J. Gao, *J. Phys. Chem. C* **112**, 14781 (2008).

63. S. Shleev, M. Pita, A. I. Yaropolov, T. Ruzgas, and L. Gorton, *Electroanalysis* **18**, 1901 (2006).

64. D. M. Sun, C. X. Cai, W. Xing, and T. H. Lu, *Chin. Sci. Bull.* **49**, 2452 (2004).

65. M. R. Tarasevich, V. A. Bogdanovskaya, and A. V. Kapustin, *Electrochem. Commun.* **5**, 491 (2003).
66. M. Togo, A. Takamura, T. Asai, H. Kaji, and M. Nishizawa, *J. Power Sources* **178**, 53 (2008).
67. M. Tominaga, M. Otani, M. Kishikawa, and I. Taniguchi, *Chem. Lett.* **35**, 1174 (2006).
68. S. Tsujimura, Y. Kamitaka, and K. Kano, *Fuel Cells* **7**, 463 (2007).
69. S. Tsujimura, T. Nakagawa, K. Kano, and T. Ikeda, *Electrochemistry* **72**, 437 (2004).
70. M. K. Wang, Y. Shen, Y. Liu, T. Wang, F. Zhao, B. F. Liu, and S. J. Dong, *J. Electroanal. Chem.* **578**, 121 (2005).
71. M. K. Wang, F. Zhao, Y. Liu, and S. J. Dong, *Biosens. Bioelectron.* **21**, 159 (2005).
72. I. Willner, G. Arad, and E. Katz, *Bioelectrochem. Bioenerg.* **44**, 209 (1998).
73. I. Willner, E. Katz, F. Patolsky, and A. F. Buckmann, *J. Chem. Soc. Perkin Trans.* **2**, 1817 (1998).
74. M. R. Tarasevich, Y. G. Chirkov, V. A. Bogdanovskaya, and A. V. Kapustin, *Electrochim. Acta* **51**, 418 (2005).
75. Y. Liu, M. K. Wang, F. Zhao, B. F. Liu, and S. J. Dong, *Chem. Eur. J.* **11**, 4970 (2005).
76. A. I. Yaropolov, A. N. Kharybin, J. Emnéus, G. Marko-Varga, and L. Gorton, *Bioelectrochem. Bioenerg.* **70**, 199 (1996).
77. S. Shleev, Y. Wang, M. Gorbacheva, A. Christenson, D. Haltrich, R. Ludwig, T. Ritzgas, and L. Gorton, *Electroanalysis* **20**, 963 (2008).
78. M. R. Tarasevich, A. I. Yaropolov, V. A. Bogdanovskaya, and S. D. Varfolomeev, *Bioelectrochem. Bioenerg.* **6**, 393 (1979).
79. C. W. Lee, H. B. Gray, and F. C. J. E. C. Anson, *J. Electroanaly. Chem.* **172**, 289 (1984).
80. M. H. Thuesen, O. Farver, B. Reinhammar, and J. Ulstrup, *Acta Chem. Scandi.* **52**, 555 (1998).
81. Y. M. Yan, L. Su, and L. Q. Mao, *J. Nanosci. Nanotechnol.* **7**, 1625 (2007).
82. J. Gallaway, I. Wheeldon, R. Rincon, P. Atanassov, S. Banta, and S. C. Barton, *Biosens. Bioelectron.* **23**, 1229 (2008).
83. B. A. Kuznetsov, G. P. Shumakovich, O. V. Koroleva, and A. I. Yaropolov, *Biosens. Bioelectron.* **16**, 73 (2001).
84. G. Gobel and F. Lisdat, *Electrochem. Commun.* **10**, 1691 (2008).
85. M. C. Weigel, E. Tritscher, and F. Lisdat, *Electrochem. Commun.* **9**, 689 (2007).
86. M. Klis, E. Maicka, A. Michota, J. Bukowska, S. Sek, J. Rogalski, and R. Bilewicz, *Electrochim. Acta* **52**, 5591 (2007).
87. G. Gupta, V. Rajendran, and P. Atanassov, *Electroanalysis* **15**, 1577 (2003).
88. E. Katz, O. Lioubashevski, and I. Willner, *J. Am. Chem. Soc.* **127**, 3979 (2005).
89. E. Katz and I. Willner, *J. Am. Chem. Soc.* **125**, 6803 (2003).
90. V. Pardo-Yissar, E. Katz, I. Willner, A. B. Kotlyar, C. Sanders, and H. Lill, *Faraday Discuss.* **116** 119 (2000).
91. D. Quan, Y. Kim, and W. Shin, *J. Electroanal. Chem.* **561**, 181 (2004).
92. K. Schubert, G. Goebel, and F. Lisdat, *Electrochim. Acta* **54**, 3033 (2009).
93. Y. M. Yan, W. Zheng, L. Su, and L. Q. Mao, *Adv. Mater.* **18**, 2639 (2006).

94. B. Haghighi, A. Rahmati-Panah, S. Shleev, and L. Gorton, *Electroanalysis* **19**, 907 (2007).
95. Y. Kamitaka, S. Tsujimura, and K. Kano, *Chem. Lett.* **36**, 218 (2007).
96. K. Szot, W. Nogala, J. Niedziolka-Jonsson, M. Jonsson-Niedziolka, F. Marken, J. Rogalski, C. Nunez-Kirchner, G. Wittstock, and M. Opallo, *Electrochim. Acta* **54**, 4620 (2009).
97. M. Jonsson-Niedziolka, K. Szot, J. Rogalski, and M. Opallo, *Electrochem. Commun.* **11**, 1042 (2009).
98. Y. Liu and S. J. Dong, *Biosens. Bioelectron.* **23**, 593 (2007).
99. Y. Liu and S. J. Dong, *Electrochem. Commun.* **9**, 1423 (2007).
100. Y. Liu, L. Huang, and S. Dong, *Biosens. Bioelectron.* **23**, 35 (2007).
101. M. Smolander, H. Boer, M. Valkiainen, R. Roozeman, M. Bergelin, J. E. Eriksson, X. C. Zhang, A. Koivula, and L. Viikari, *Enzyme Microb. Technol.* **43**, 93 (2008).
102. Z. Khani, C. Jolivalt, M. Cretin, S. Tingry, and C. Innocent, *Biotechnol. Lett.* **28**, 1779 (2006).
103. N. Yuhashi, M. Tomiyama, J. Okuda, S. Igarashi, K. Ikebukuro, and K. Sode, *Biosens. Bioelectron.* **20**, 2145 (2005).
104. R. Szamocki, V. Flexer, L. Levin, F. Forchiasin, and E. J. Calvo, *Electrochim. Acta* **54**, 1970 (2009).
105. D. Quan and W. Shin, *Mater. Sci. Eng. C* **24**, 113 (2004).
106. G. Merle, L. Brunel, S. Tingry, M. Cretin, M. Rolland, K. Servat, C. Jolivalt, C. Innocent, and P. Seta, *Mater. Sci. Eng. C Biomimet. Supramol. Sys.* **28**, 932 (2008).
107. F. Barriere, Y. Ferry, D. Rochefort, and D. Leech, *Electrochem. Commun.* **6**, 237 (2004).
108. F. Barriere, P. Kavanagh, and D. Leech, *Electrochim. Acta* **51**, 5187 (2006).
109. T. Chen, S. C. Barton, G. Binyamin, Z. Q. Gao, Y. C. Zhang, H. H. Kim, and A. Heller, *J. Am. Chem. Soc.* **123**, 8630 (2001).
110. J. W. Gallaway and S. A. C. Barton, *J. Am. Chem. Soc.* **130**, 8527 (2008).
111. C. Kang, H. Shin, and A. Heller, *Bioelectrochemistry* **68**, 22 (2006).
112. C. Kang, H. Shin, Y. C. Zhang, and A. Heller, *Bioelectrochemistry* **65**, 83 (2004).
113. K. Karnicka, K. Eckhard, D. A. Guschin, L. Stoica, P. J. Kulesza, and W. Schuhmann, *Electrochem. Commun.* **9**, 1998 (2007).
114. P. Kavanagh, P. Jenkins, and D. Leech, *Electrochem. Commun.* **10**, 970 (2008).
115. H. H. Kim, N. Mano, X. C. Zhang, and A. Heller, *J. Electrochem. Soc.* **150**, A209 (2003).
116. N. Mano, J. L. Fernandez, Y. Kim, W. Shin, A. J. Bard, and A. Heller, *J. Am. Chem. Soc.* **125**, 15290 (2003).
117. N. Mano, *Chem. Commun.* 2221 (2008).
118. N. Mano, H. H. Kim, and A. Heller, *J. Phys. Chem. B* **106**, 8842 (2002).
119. N. Mano, F. Mao, and A. Heller, *J. Am. Chem. Soc.* **124**, 12962 (2002).
120. N. Mano, F. Mao, and A. Heller, *J. Am. Chem. Soc.* **125**, 6588 (2003).
121. N. Mano, F. Mao, and A. Heller, *Chembiochem* **5**, 1703 (2004).
122. N. Mano, F. Mao, W. Shin, T. Chen, and A. Heller, *Chem. Commun.* 518 (2003).
123. P. Rowinski, C. Kang, H. Shin, and A. Heller, *Anal. Chem.* **79**, 1173 (2007).
124. V. Soukharev, N. Mano, and A. Heller, *J. Am. Chem. Soc.* **126**, 8368 (2004).

125. Y. H. Sun and S. C. Barton, *J. Electroanal. Chem.* **590**, 57 (2006).
126. S. C. Barton, M. Pickard, R. Vazquez-Duhalt, and A. Heller, *Biosens. Bioelectron.* **17**, 1071 (2002).
127. T. Balkenhohl, S. Adelt, R. Dronov, and F. Lisdat, *Electrochem. Commun.* **10**, 914 (2008).
128. L. Brunel, J. Denele, K. Servat, K. B. Kokoh, C. Jolivalt, C. Innocent, M. Cretin, M. Rolland, and S. Tingry, *Electrochem. Commun.* **9**, 331 (2007).
129. J. F. Fei, H. K. Song, and G. T. R. Palmore, *Chem. Mater.* **19**, 1565 (2007).
130. J. L. Fernandez, N. Mano, A. Heller, and A. J. Bard, *Angew. Chem. Int. Ed.* **43**, 6355 (2004).
131. A. Habrioux, G. Merle, K. Servat, K. B. Kokoh, C. Innocent, M. Cretin, and S. Tingry, *J. Electroanal. Chem.* **622**, 97 (2008).
132. A. Habrioux, E. Sibert, K. Servat, W. Vogel, K. B. Kokoh, and N. Alonso-Vante, *J. Phys. Chem. B* **111**, 10329 (2007).
133. K. Karnicka, K. Miecznikowski, B. Kowalewska, M. Skunik, M. Opallo, J. Rogalski, W. Schuhmann, and P. J. Kulesza, *Anal. Chem.* **80**, 7643 (2008).
134. S. Komaba, T. Mitsuhashi, and S. Shraishi, *Electrochemistry* **76**, 619 (2008).
135. B. Kowalewska, M. Skunik, K. Karnicka, K. Miecmikowski, M. Chojak, G. Ginalska, A. Belcarz, and P. J. Kulesza, *Electrochim. Acta* **53**, 2408 (2008).
136. M. A. Osina, V. A. Bogdanovskaya, and B. N. Efremov, *Russ. J. Electrochem.* **38**, 1082 (2002).
137. D. Rochefort, L. Kouisni, and K. Gendron, *J. Electroanal. Chem.* **617**, 53 (2008).
138. H. Shin, C. Kang, and A. Heller, *Electroanalysis* **19**, 638 (2007).
139. K. Szot, J. Niedziolka, J. Rogalski, F. Marken, and M. Opallo, *J. Electroanal. Chem.* **612**, 1 (2008).
140. S. Topcagic and S. D. Minteer, *Electrochim. Acta* **51**, 2168 (2006).
141. S. Tsujimura, K. Kano, and T. Ikeda, *J. Electroanal. Chem.* **576**, 113 (2005).
142. S. Tsujimura, M. Kawaharada, T. Nakagawa, K. Kano, and T. Ikeda, *Electrochem. Commun.* **5**, 138 (2003).
143. K. Servat, S. Tingry, L. Brunel, S. Querelle, M. Cretin, C. Innocent, C. Jolivalt, and M. Rolland, *J. App. Electrochem.* **37**, 121 (2007).
144. N. S. Hudak and S. C. Barton, *J. Electrochem. Soc.* **152**, A876 (2005)
145. N. S. Hudak, J. W. Gallaway, and S. C. Barton, *J. Electrochem. Soc.* **156**, B9 (2009).
146. N. S. Hudak, J. W. Gallaway, and S. C. Barton, *J. Electroanal. Chem.* **629**, 57 (2009).
147. L. Stoica, N. Dimcheva, Y. Ackermann, K. Karnicka, D. A. Guschin, P. J. Kulesza, J. Rogalski, D. Haltrich, R. Ludwig, L. Gorton, and W. Schuhmann, *Fuel Cells* **9**, 53 (2009).
148. M. Klis, M. Karbarz, Z. Stojek, J. Rogalski, and R. Bilewicz, *J. Phys. Chem. B* **113**, 6062 (2009).
149. L. Deng, L. Shang, Y. Z. Wang, T. Wang, H. J. Chen, and S. J. Dong, *Electrochem. Commun.* **10**, 1012 (2008).
150. T. Nakagawa, S. Tsujimura, K. Kano, and T. Ikeda, *Chem. Lett.* **32**, 54 (2003).
151. L. Amir, T. K. Tam, M. Pita, M. M. Meijler, L. Alfonta, and E. Katz, *J. Am. Chem. Soc.* **131**, 826 (2009).
152. N. Mano and A. Heller, *J. Electrochem. Soc.* **150**, A1136 (2003).

153. L. Deng, F. Wang, H. J. Chen, L. Shang, L. Wang, T. Wang, and S. J. Dong, *Biosens. Bioelectron.* **24**, 329 (2008).

154. L. Brunel, S. Tingry, K. Servat, M. Cretin, C. Innocent, C. Jolivalt, and M. Rolland, *Desalination* **199**, 426 (2006).

155. Y. M. Tan, Q. J. Xie, J. H. Huang, W. S. Duan, M. Ma, and S. Z. Yao, *Electroanalysis* **20**, 1599 (2008).

156. G. Merle, A. Habrioux, K. Servat, M. Rolland, C. Innocent, K. B. Kokoh, and S. Tingry, *Electrochim. Acta* **54**, 2998 (2009).

157. B. Palys, A. Bokun, and J. Rogalski, *Electrochim. Acta* **52**, 7075 (2007).

158. F. Gao, Y. M. Yan, L. Su, L. Wang, and L. Q. Mao, *Electrochem. Commun.* **9**, 989 (2007).

159. K. Ishibashi, S. Tsujimura, and K. Kano, *Electrochemistry* **76**, 594 (2008).

160. C. Mousty, L. Vieille, and S. Cosnier, *Biosens. Bioelectron.* **22**, 1733 (2007).

161. A. Ramanavicius, A. Kausaite, and A. Ramanaviciene, *Biosens. Bioelectron.* **20**, 1962 (2005).

162. W. Zheng, Q. F. Li, L. Su, Y. M. Yan, J. Zhang, and L. Q. Mao, *Electroanalysis* **18**, 587 (2006).

163. Y. M. Yan, O. Yehezkeli, and I. Willner, *Chem. Eur. J.* **13**, 10168 (2007).

164. X. C. Li, H. J. Zhou, P. Yu, L. Su, T. Ohsaka, and L. Q. Mao, *Electrochem. Commun.* **10**, 851 (2008).

165. L. Xiang, Z. N. Zhang, P. Yu, J. Zhang, L. Su, T. Ohsaka, and L. Q. Mao, *Anal. Chem.* **80**, 6587 (2008).

166. W. Zheng, H. M. Zho, Y. F. Zheng, and N. Wang, *Chem. Phys. Lett.* **457**, 381 (2008).

167. E. Katz, A. F. Buckmann, and I. Willner, *J. Am. Chem. Soc.* **123**, 10752 (2001).

168. X. Li, L. Zhang, L. Su, T. Ohsaka, and L. Mao, *Fuel Cells* **9**, 85 (2009).

169. W. E. Farneth and M. B. D'Amore, *J. Electroanal. Chem.* **581**, 197 (2005).

170. M. Jonsson, K. Szot, J. Niedziolka, J. Rogalski, K. Karnicka, P. J. Kulesza, and M. Opallo, *J. Nanosci. Nanotechnol.* **9**, 2346 (2009).

171. J. Lim, N. Cirigliano, J. Wang, and B. Dunn, *Phys. Chem. Chem. Phys.* **9**, 1809 (2007).

172. J. Lim, P. Malati, F. Bonet, and B. Dunn, *J. Electrochem. Soc.* **154**, A140 (2007).

173. W. Nogala, M. Burchardt, M. Opallo, J. Rogalski, and G. Wittstock, *Bioelectrochemistry* **72**, 174 (2008).

174. W. Nogala, E. Rozniecka, J. Rogalski, and M. Opallo, *J. Electroanal. Chem.* **608**, 31 (2007).

175. W. Nogala, E. Rozniecka, I. Zawisza, J. Rogalski, and M. Opallo, *Electrochem. Commun.* **8**, 1850 (2006).

176. I. Zawisza, J. Rogalski, and M. Opallo, *J. Electroanal. Chem.* **588**, 244 (2006).

177. S. Braun, S. Rappoport, R. Zusman, D. Avnir, and M. Ottolenghi, *Mater. Lett.* **10**, 1 (1991).

178. D. Avnir, S. Braun, O. Lev, and M. Ottolenghi, *Chem. Mater.* **6**, 1605 (1994).

179. R. Bilewicz, P. Rowinski, and E. Rogalska, *Bioelectrochemistry* **66**, 3 (2005).

180. E. Nazaruk and R. Bilewicz, *Bioelectrochemistry* **71**, 8 (2007).

181. E. Nazaruk, S. Smolinski, M. Swatko-Ossor, G. Ginalska, J. Fiedurek, J. Rogalski, and R. Bilewicz, *J. Power Sources* **183**, 533 (2008).

182. P. Rowinski, R. Bilewicz, M. J. Stebe, and E. Rogalska, *Analy. Chem.* **76**, 283 (2004).

183. K. Stolarczyk, E. Nazaruk, J. Rogalski, and R. Bilewicz, *Electrochem. Commun.* **9**, 115 (2007).

184. R. Bilewicz, K. Stolarczyk, K. Sadowska, J. Rogalski, and J. F. Biernat, *ECS Trans.* **19**(6), 27–36 (2009).

185. R. Dronov, D. G. Kurth, F. W. Scheller, and F. Lisdat, *Electroanalysis* **19**, 1642 (2007).

186. S. C. Barton, H. H. Kim, G. Binyamin, Y. C. Zhang, and A. Heller, *J. Am. Chem. Soc.* **123**, 5802 (2001).

187. S. C. Barton, H. H. Kim, G. Binyamin, Y. C. Zhang, and A. Heller, *J. Phys. Chem. B* **105**, 11917 (2001).

188. N. Mano, V. Soukharev, and A. Heller, *J. Phys. Chem. B* **110**, 11180 (2006).

189. F. Mao, N. Mano, and A. Heller, *J. Am. Chem. Soc.* **125**, 4951 (2003).

190. T. V. Fedorova, A. S. Vilesov, S. A. Kurzeev, E. V. Stepanova, E. O. Landesman, and O. V. Koroleva, *Appl. Biochem. Microbiol.* **42**, 550 (2006).

191. P. A. Jelliss, S. D. Minteer, M. Patel, A. Siemiarczuk, M. Watt, and R. E. K. Winter, *J. Mater. Chem.* **18**, 2104 (2008).

192. I. R. Wheeldon, J. W. Gallaway, S. C. Barton, and S. Banta, *Proc. Natl. Acad. Sci. USA* **105**, 15275 (2008).

193. S. C. Barton, *Electrochim. Acta* **50**, 2145 (2005).

194. M. Klis, J. Rogalski, and R. Bilewicz, *Bioelectrochemistry* **71**, 2 (2007).

195. S. Tsujimura, B. Tatsumi, J. Ogawa, S. Shimizu, K. Kano, and T. Ikeda, *J. Electroanal. Chem.* **496**, 69 (2001).

196. J. Rogalski, A. L. Dawidowicz, and A. Leonowicz, *Acta Biotechnol.* **10**, 261 (1990).

197. K. J. Davies and M. E. Delisignore, *J. Biol. Chem.* **262**, 9908 (1987).

198. S. D. Maleknia, J. W. H. Wong, and K. M. Downard, *Photochem. Photobiol. Sci.* **3**, 741 (2004).

199. W. E. Farneth, B. A. Diner, T. D. Gierke, and M. B. D'Amore, *J. Electroanal. Chem.* **581**, 190 (2005).

200. Y. Liu and S. J. Dong, *Electroanalysis* **20**, 827 (2008).

201. A. P. Borole, S. LaBarge, and B. A. Spott, *J. Power Sources* **188**, 421 (2009).

202. A. Zebda, J. Renaud, M. Cretin, F. Pichot, C. Innocent, R. Ferrigno, and S. Tingry, *Electrochem. Commun.* **11**, 592 (2009).

203. J. F. Fei, A. Basu, F. T. Xue, and G. T. R. Palmore, *Org. Lett.* **8**, 3 (2006).

204. H. K. Song and G. T. R. Palmore, *J. Phys. Chem. B* **109**, 19278 (2005).

205. J. H. Thomas, J. M. Drake, J. R. Paddock, S. Conklin, J. Johnson, C. J. Seliskar, H. B. Halsall, and W. R. Heineman, *Electroanalysis* **16**, 547 (2004).

206. K. Sadowska, K. P. Roberts, R. Wiser, J. F. Biernat, E. Jablonowska, and R. Bilewicz, *Carbon* **47**, 1501 (2009).

207. K. Sadowska, E. Jablonowska, K. Stolarczyk, R. Wiser, R. Bilewicz, K. P. Roberts, and J. F. Biernat, *Polish J. Chem.* **82**, 1309 (2008).

208. K. Sadowska, K. Stolarczyk, J. F. Biernat, K. P. Roberts, J. Rogalski, and R. Bilewicz, *Carbon* (in press).

209. E. Nazaruk, K. Sadowska, K. Madrak, J. F. Biernat, J. Rogalski, and R. Bilewicz, *Electroanalysis* **21**, 507 (2009).

210. L. Rabinovich and O. Lev, *Electroanalysis* **13**, 265 (2001).

211. H. Sakai, T. Nakagawa, Y. Tokita, T. Hatazawa, T. Ikeda, S. Tsujimura, and K. Kano, *Energy Environ. Sci.* **2**, 133 (2009).

212. http://www.sony.net/SonyInfo/News/Press/20070807-074E/index.html, 2007.

213. A. Heller, *Anal. Bioanal. Chem.* **385**, 469 (2006).

214. P. J. Kulesza, R. Marassi, K. Karnicka, R. Wlodarczyk, K. Miecmikowskil, M. Skunik, B. Kowalewska, M. Chojak, B. Baranowska, A. Kolary-Zurowska, and G. Ginalska, *Rev. Adv. Mater. Sci.* **15**, 225 (2007).

215. Y. K. Oh, Y. J. Lee, E. H. Choi, and M. S. Kim, *Int. J. Hydrogen Energy* **33**, 5218 (2008).

216. C. Leger, A. K. Jones, S. P. J. Albracht, and F. A. Armstrong, *J. Phys. Chem. B* **106**, 13058 (2002).

217. C. Vaz-Dominguez, S. Campuzano, O. Rudiger, M. Pita, M. Gorbacheva, S. Shleev, V. M. Fernandez, and A. L. De Lacey, *Biosens. Bioelectron.* **24**, 531 (2008).

218. M. H. Thuesen, O. Farver, B. Reinhammar, and J. Ulstrup, *Acta Chem. Scand.* **52**, 555 (1998).

219. D. L. Johnson, J. L. Thompson, S. M. Brinkmann, K. A. Schuller, and L. L. Martin, *Biochemistry* **42**, 10229 (2003).

220. S. Shleev, O. Nikitina, A. Christenson, C. T. Reimann, A. I. Yaropolov, T. Ruzgas, and L. Gorton, *Bioorg. Chem.* **35**, 35 (2007).

221. A. I. Yaropolov, A. N. Kharybin, J. Emneus, G. Marko-Varga, and L. Gorton, *Bloelectrochem. Bioenerg.* **40**, 49 (1996).

222. S. Shleev, A. Jarosz-Wilkolazka, A. Khalunina, O. Morozova, A. Yaropolov, T. Ruzgas, and L. Gorton, *Bioelectrochemistry* **67**, 115 (2005).

223. M. Pita, S. Shleev, T. Ruzgas, V. M. Fernandez, A. I. Yaropolov, and L. Gorton, *Electrochem. Commun.* **8**, 747 (2006).

224. I. Bento, M. A. Carrondo, and P. F. Lindley, *J. Biol. Inorg. Chem.* **11**, 539 (2006).

225. A. V. Lyashenko, I. Bento, V. N. Zaitsev, N. E. Zhukhlistova, Y. N. Zhukova, A. G. Gabdoulkhakov, F. Y. Morgunova, W. Voelter, G. S. Kachalova, E. V. Stepanova, G. V. Koroleva, V. S. Lamzin, V. I. Tishkov, C. Betzel, P. F. Lindley, and A. M. Mikhailov, *J. Biol. Inorg. Chem.* **11**, 963 (2006).

226. D. Ivnitski, K. Artyushkova, and P. Atanassov, *Bioelectrochemistry* **74**, 101 (2008).

227. M. R. Tarasevich, V. A. Bogdanovskaya, and L. N. Kuznetsova, *Russ. J. Electrochem.* **37**, 833 (2001).

228. Y. Xiao, F. Patolsky, E. Katz, J. F. Hainfeld, and I. Willner, *Science* **299**, 1877 (2003).

229. J. Kim, H. F. Jia, and P. Wang, *Biotechnol. Adv.* **24**, 296 (2006).

230. B. Willner, E. Katz, and I. Willner, *Curr. Opin. Biotechnol.* **17**, 589 (2006).

231. I. Willner, R. Baron, and B. Willner, *Biosens. Bioelectron.* **22**, 1841 (2007).

232. M. Pumera, S. Sanchez, I. Ichinose, and J. Tang, *Sens. Actuat. B Chem.* **123**, 1195 (2007).

233. F. A. Armstrong, *Curr. Opin. Chem. Biol.* **9**, 110 (2005).

234. S. M. Macdonald, K. Szot, J. Niedziolka, F. Marken, and M. Opallo, *J. Solid State Electrochem.* **12**, 287 (2008).

235. M. Bradford, *Analy. Biochem.* **72**, 248 (1976).

236. W. Nogala, A. Celebanska, K. Szot, G. Wittstock, and M. Opallo, *Electrochim. Acta* (in press).

237. A. Krueger, in *Strained Hydrocarbons*, H. Dodziuk, ed., Wiley-VCH, Weinheim, 2009, p. 335.
238. S. Banerjee, T. Hemraj-Benny, and S. S. Wong, *Adv. Mater.* **17**, 17 (2005).
239. I. Willner, Y. M. Yan, B. Willner, and R. Tel-Vered, *Fuel Cells* **9**, 7 (2009).
240. H. Yan, H. J. Zhou, P. Yu, L. Su, and L. Q. Mao, *Adv. Mater.* **20**, 2899 (2008).
241. D. Wang, J. J. Rack, and L. W. Chen, *J. Nanosci. Nanotechnol.* **9**, 2310 (2009).
242. D. Ivnitski, K. Artyushkova, R. A. Rincon, P. Atanassov, H. R. Luckarift, and G. R. Johnson, *Small* **4**, 357 (2008).
243. D. Ivnitski, B. Branch, P. Atanassov, and C. Apblett, *Electrochem. Commun.* **8**, 1204 (2006).
244. W. Nogala, K. Szot, M. Burchardt, M. Jonsson-Niedziolka, J. Rogalski, G. Wittstock, and M. Opallo, *Biolectrochemistry* **79**, 101 (2010).
245. A. Christenson, N. Dimcheva, E. E. Ferapontova, L. Gorton, T. Ruzgas, L. Stoica, S. Shleev, A. L. Yaropolov, D. Haltrich, R. N. F. Thorneley, and S. D. Aust, *Electroanalysis* **16**, 1074 (2004).
246. R. Santucci, T. Ferri, L. Morpurgo, I. Savini, and L. Avigliano, *Biochem. J.* **332**, 611 (1998).
247. M. Tominaga, M. Ohtani, and I. Taniguchi, *Phys. Chem. Chem. Phys.* **10**, 6928 (2008).
248. Y. M. Tan, W. F. Deng, B. Ge, Q. J. Xie, J. H. Huang, and S. Z. Yao, *Biosens. Bioelectron.* **24**, 2225 (2009).
249. D. Chen, G. Wang, and J. Li, *J. Phys. Chem. C* **111**, 2351 (2007).
250. J. P. Collman, N. K. Devaraj, R. A. Decreau, Y. Yang, Y. L. Yan, W. Ebina, T. A. Eberspacher, and C. E. D. Chidsey, *Science* **315**, 1565 (2007).
251. J. P. Collman, L. Fu, P. C. Herrmann, and X. M. Zhang, *Science* **275**, 949 (1997).
252. J. P. Collman, P. Denisevich, and Y. Konai, *J. Am. Chem. Soc.* **103**, 6027 (1980).
253. Y. C. Weng, F. R. F. Fan, and A. J. Bard, *J. Am. Chem. Soc.* **127**, 17576 (2005).
254. S. Kerzenmacher, J. Ducree, R. Zengerle, and F. von Stetten, *J. Power Sources* **182**, 1 (2008).

CHAPTER 6

# Platinum Monolayer Electrocatalysts: Improving Structure and Activity

KOTARO SASAKI, MIOMIR B. VUKMIROVIC, JIA X. WANG, and RADOSLAV R. ADZIC

Brookhaven National Laboratory, Upton, New York

We introduce the concept of platinum monolayer electrocatalysts for the $O_2$ reduction reaction using the initial studies data that clarified the role of substrate-induced strain, position of the *d*-band center of the monolayer, and the Pt–substrate electronic interaction. Further, we review significant improvements in their activity and stability, describe the scaleup synthesis demonstrated in more recent work, and analyze a broad range of possibilities for further improvements. Examples include (1) fine-tuning of the Pt monolayer–support interaction using a Pd monolayer as an "interlayer" where Pd acts as a buffer between Pt and the support, which alleviates lattice-induced strain and curvature-induced strain and decreases the number of low coordination sites in nanoparticles, thus boosting activity for oxygen reduction; (2) using alloys and modified alloy surfaces as supports for a Pt monolayer that facilitates a weaker binding energy of OH and its easier removal from a Pt monolayer, thereby enhancing activity; (3) using ordered intermetallic compounds that have excellent thermodynamic stability as cores to support a Pt monolayer; and (4) controlling the core particle's composition, size, and shape. These methods of tailoring the structure and properties of a Pt monolayer can enhance Pt mass activity by about one order of magnitude, and produce a two- and three fold larger specific activity than do the standard Pt/C catalysts. Finally, we describe scaleup syntheses of three different electrocatalysts whose activity and durability observed in fuel cell tests verifies this uniquely effective method of using Pt to synthesize catalysts with the desired properties. The results confirm the great potential of Pt monolayer electrocatalysts for resolving problems of the availability of Pt in future catalysis and electrocatalysis and removing the main obstacles to the broad-scale application of fuel cell technology.

*Fuel Cell Science: Theory, Fundamentals, and Biocatalysis,*
Edited by Andrzej Wieckowski and Jens K. Nørskov

## 6.1 INTRODUCTION TO Pt MONOLAYER ELECTROCATALYSTS

The slow kinetics of oxygen reduction reaction (ORR) at fuel cell cathodes even on the best Pt catalysts contributes significantly to the loss of the cell voltage, thereby lowering the efficiency of energy conversion in this power source. Consequently, high Pt loading is required in cathode electrocatalysts. To resolve these impediments to commercialization of fuel cells, one of the main goals of current research in electrocatalysis aims to increase the activity of Pt for the ORR [1–3]. Addressing these issues, we have inaugurated Pt monolayers as electrocatalysts for the ORR and Pt submonolayer for the CO-tolerant electrocatalysts for the $H_2$ oxidation reaction (HOR) [4,5]. These electrocatalysts, consisting of a single Pt monolayer on suitable nanoparticle substrates, are characterized by an ultimately low Pt content and almost complete utilization of Pt, and exhibit support-induced tunable activity and good stability.

The activity of a Pt monolayer can be enhanced by selecting the proper nanoparticle substrate through structural and electronic effects [6,7]. These findings led to development of three types of Pt monolayer electrocatalysts for the ORR: (1) Pt monolayer on carbon-supported Pd, or Pd alloy nanoparticles [5]; (2) mixed Pt-transition metal monolayer on Pd nanoparticles [8]; and (3) Pt monolayer on core–shell nanoparticles consisting of a noble metal shell around the non-noble-metal core [9–11].

The role of substrate in determining the activity of a Pt monolayer was first established in the studies of ORR on Pt monolayers deposited on five different single-crystal surfaces and confirmed with nanoparticle supports [5]. The ORR kinetic data and the density functional theory calculations (DFT) of the binding energy of atomic oxygen on the six surfaces result in the volcano dependence of kinetic currents as a function of the *d*-band center energy $\varepsilon_d$ [12] having the $Pt_{ML}$/Pd(111) surface at the top of the curve. Its activity was higher than that of the Pt(111) surface. The positions of the $\varepsilon_d$ of the Pt monolayers do not correlate strictly with the amount of strain because the position of the $\varepsilon_d$ for these monolayers depends on both the strain (geometric effects) and the electronic interaction between the Pt monolayer and its substrate (ligand effect) [6,7].

Experimentally, the superior ORR catalytic activity of $Pt_{ML}$/Pd(111) was found to be associated with reduced OH coverage [5], as inferred from voltammetry and X-ray absorption near-edge structure spectroscopy (XANES) that identified a shift of OH adsorption, respectively, on $Pt_{ML}$/Pd(111), or $Pt_{ML}$/Pd/C to potentials more positive than those on Pt(111), or Pt/C [13]. This was attributed to an enhanced repulsive interaction of OH groups on the $Pt_{ML}$/Pd(111) compared to that on pure Pt(111), originating from the fact that Pd–OH forms at less positive potentials than does Pt–OH [12].

Synthesis of Pt monolayer electrocatalysts was accomplished by placing a monolayer of Pt atoms on metal nanoparticles using a galvanic displacement of a under-potentially deposited (upd) Cu monolayer by a Pt monolayer [14]. Pt monolayers deposited on carbon-supported Pd nanoparticles as electrocatalysts for ORR have a very high activity, very high Pt mass activity, and good long-term stability [15].

On the basis of the increased catalytic activity measured for the Pt monolayer on Pd(111), ascribed to the weaker binding energy of OH in relation to pure Pt(111), a new class of mixed monolayer (Pt + M) was designed. A metal M that adsorbs OH or O strongly further reduces OH coverage on Pt, due to the lateral repulsion between the OH adsorbed on Pt, and the OH or O adsorbed on a neighboring surface metal atom M. The concept of adsorbate–adsorbate repulsion was further verified by correlating the kinetic currents for five systems against the effective repulsion energy between two OHs(a) or an O(a) and an OH(a) calculated from first-principles methods [8].

The third class of electrocatalysts was designed to further lower the mass of noble metals by depositing Pt monolayers on non-noble-metal–noble metal core–shell nanoparticles [9–11]. The activity of a Pt monolayer is enhanced through structural and electronic effects [6,7]. The core–shell structure was synthesized by thermally induced segregation of the noble metals. Another method for fabricating the supporting core–shell nanoparticles employs a redox–transmetallation reaction, involving the oxidation of carbon-supported non-noble-metal nanoparticles by noble metal cations, yielding a noble metal shell around the non-noble-metal core. Figure 6.1a illustrates the high activity obtained with a Pt/AuNi/C electrocatalyst comprising a Pt monolayer on noble metal–non-noble-metal core–shell nanoparticle support. Its high activity is reflected in $E_{1/2} = 0.89$ V, and Pt mass activity of 0.9 A/mg$_{Pt}$ or five-fold increase in total noble metal mass activity compared with Pt/C. Figure 6.1b shows the elemental distribution of the three components of Pt/AuNi/C electrocatalyst obtained using scanning energy dispersive spectroscopy (EDS). A uniform distribution of Pt and Au over Ni core is observed. The peak for the Ni component in the middle of the particle indicates its core–shell structure. A Pt mass activity and total noble metal mass activity at 0.9V for Pt/C and Pt/AuNi$_5$/C electrocatalysts are shown in Figure 6.1b.

These results indicate that we have a class of highly promising catalysts that are characterized by a high activity of the ORR; almost complete Pt utilization, and good long-term performance stability. The properties of these electrocatalysts address the major problems of fuel cell electrocatalysis. That work described an analysis of Pt monolayer electrocatalysts including the correlation of their activity with the center of the Pt *d* band; the effects of the types of supports, namely, metal, alloy, core–shell, and oxide nanoparticles; and the corresponding mechanistic underpinnings afforded by electronic structure theory. A small contraction of a Pt monolayer, preferably (111)-oriented, and a shift of Pt-OH formation positively are conducive to high activity and reasonable durability. Below we describe new improvements of Pt monolayer electrocatalysts with particular emphasis on the effect of composition, shape, size, and surface structure of nanoparticle cores on their activity and stability.

## 6.2 IMPROVING ACTIVITY AND STABILITY OF Pt MONOLAYER ELECTROCATALYSTS

### 6.2.1 Effects of Pt Shell Thickness, Facet, and Particle Size on Catalytic Activity

To optimize core-shell nanoparticle catalysts information on the effects of the thickness of the Pt shell, lattice mismatch, and particle size on specific and mass

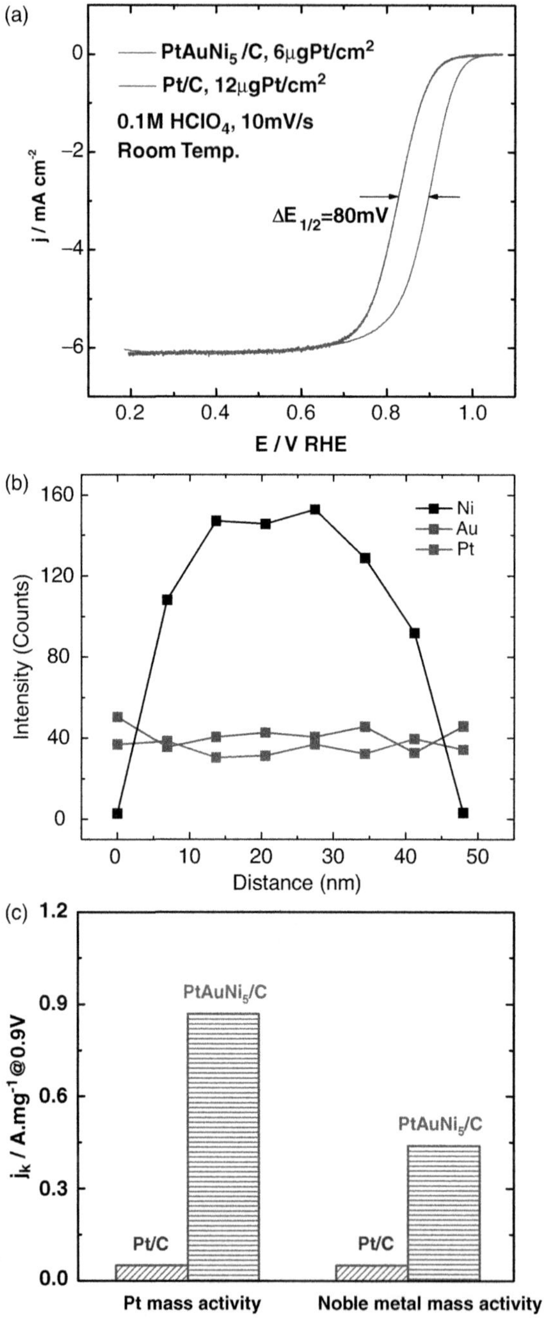


**FIGURE 6.1** Polarization curves for Pt monolayer on an Au/Ni/C core–shell nanoparticle. The curve for commercial Pt/C is given for comparison. Other parameters are indicated in the graph (a); distribution of components in a Pt/Ni/Au/C nanoparticle obtained by scanning probe energy dispersive spectroscopy (b); Pt mass activity and total noble metal mass activity at 0.9V for Pt/C and Pt/AuNi5/C electrocatalysts (c).

activities is needed. These were obtained from the changes in effective surface area and activity induced by stepwise Pt monolayer depositions on Pd and $Pd_3Co$ nanoparticles. The core-shell structure was characterized at the atomic level using Z-contrast scanning transmission electron microscopy (STEM) coupled with element-sensitive electron energy loss spectroscopy (EELS). Support for the origin of enhanced specific activity was obtained from DFT calculations using a nanoparticle model, revealing the effect of nanoscale-induced surface contraction on facet-dependent oxygen binding energy. The results suggest that a moderately compressed (111) facet is most conducive to oxygen reduction reaction on small nanoparticles and indicate the importance of concerted structure and component optimization for enhancing core–shell nanocatalys activity and durability.

To generate Pt shells with controlled morphology and thickness on a Pd core, *viz*, $Pd_CPt_n$ nanoparticles [where $n$ is the number of Pt layers in the shell on Pd cores ($Pd_C$)], we utilized a Cu upd-mediated electrodeposition method. Figure 6.2a shows

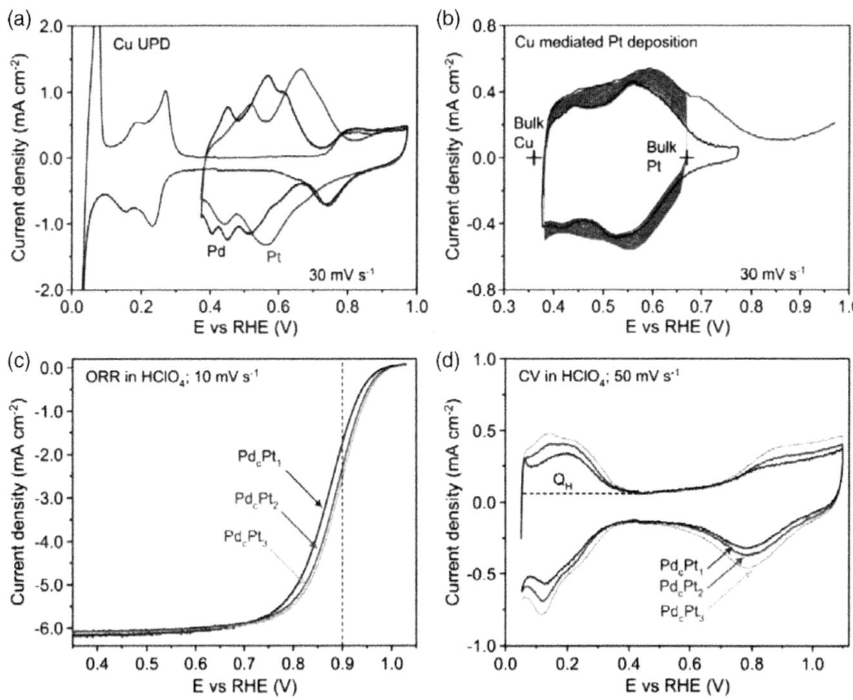


**FIGURE 6.2** (a) Voltammetry curves for Cu upd on Pd (blue) and Pt (red) nanoparticles in Ar-saturated solutions containing 50 mM $H_2SO_4$ and 50 mM $CuSO_4$, together with a baseline voltammetry curve for Pd in a solution without Cu (black); (b) voltammetry curves before (blue), during (green), and after (red) Pt monolayer deposition on Pd in the above solution with additional 0.1 mM $K_2PtCl_4$; (c) ORR polarization; (d) voltammetry curves measured for one- to three-monolayer-thick Pt shells on carbon-supported Pd nanoparticles (10 wt%, 7 μg/cm$^2$ Pd loading) using a rotating-disk electrode (1600 rpm) in oxygen- and argon-saturated 0.1 M $HClO_4$ solutions, respectively [19].

voltammetry curves for Cu upd on Pd and Pt nanoparticle catalysts, wherein a full monolayer of Cu is deposited and removed during potential cycles positive of the Cu bulk's deposition potential (0.36 V). An irreversible deposition of Pt under diffusion control takes place below its bulk deposition potential (0.67 V) in the solution containing 0.1 mM $K_2PtCl_4$ and 50 mM $Cu^{2+}$. With repeated potential cycles between 0.37 and 0.67 V, the Cu upd current gradually increases (green curves in Fig. 6.2b), reflecting the enlargement of the surface area as the size of the core–shell nanoparticles grows by irreversibly deposited Pt. The role of Cu upd and stripping cycle in this electrodeposition process is to lower the rate of Pt deposition and to enhance surface diffusion of Pt adatoms so that smooth layer forms [16,17]. At the end of Pt deposition, a linear potential sweep, up to 1 V, completely removes Cu (Fig. 6.2a,b). The number of potential cycles required to deposit a full monolayer of Pt was determined by comparing the rate of increase in measured surface area with the expected values based on the core particles' average diameter. Estimates of Pt loading obtained from averaged Cu upd stripping charges measured before and after depositing a monolayer yielded a Pt : Pd mass ratio within 15% of the values obtained from measurements of inductively coupled plasma atomic emission spectroscopy.

The increase in ORR activity with added Pt layers is nearly proportional to the increase in effective surface area measured by the integrated hydrogen desorption (Fig. 6.2c,d). Table 6.1 lists the specific and mass activities obtained from the experiments using 4 nm Pd and 4.6 nm $Pd_3Co$ cores with those for 3 nm Pt as the reference (2–3 nm Pt/C has the highest Pt mass activity). The enhancement factors up to 3 in specific activity, 9 in Pt mass activity, and 2 in the mass activity of the platinum group metals (PGM) illustrate the effectiveness of the core-shell approach in enhancing catalytic performance and reducing costs.

The atomic structures and component distributions of the $Pd_CPt_n$ nanoparticles were examined using an aberration-corrected STEM equipped with EELS. Figure 6.3,

**TABLE 6.1 ORR Activities at 0.9 V Measured at 10 mV/s[1] in Oxygen-Saturated $HClO_4$ Solutions as a Function of Pt Shell Thickness on Pd and Pd Alloy Cores[a]**

| Number of Pt Layers | Specific Activity ($mA/cm^2$) | Pt Mass (mA/μg) | PGM Mass (mA/μg) |
|---|---|---|---|
| | *4.0 nm Pd/C (10 wt%)* | | |
| 1 | 0.50 | 0.96 | 0.25 |
| 2 | 0.52 | 0.59 | 0.27 |
| 3 | 0.51 | 0.43 | 0.25 |
| | *4.6 nm $Pd_3Co$/C (14 wt%)* | | |
| 1 | 0.80 | 1.56 | 0.35 |
| 2 | 0.78 | 0.99 | 0.42 |
| 3 | 0.79 | 0.75 | 0.44 |
| | *3 nm Pt/C (20 wt%)* | | |
| 0 | 0.25 | 0.17 | 0.17 |

[a]Values for Pt are included as the reference.
[b]*Source:* Reference 19.

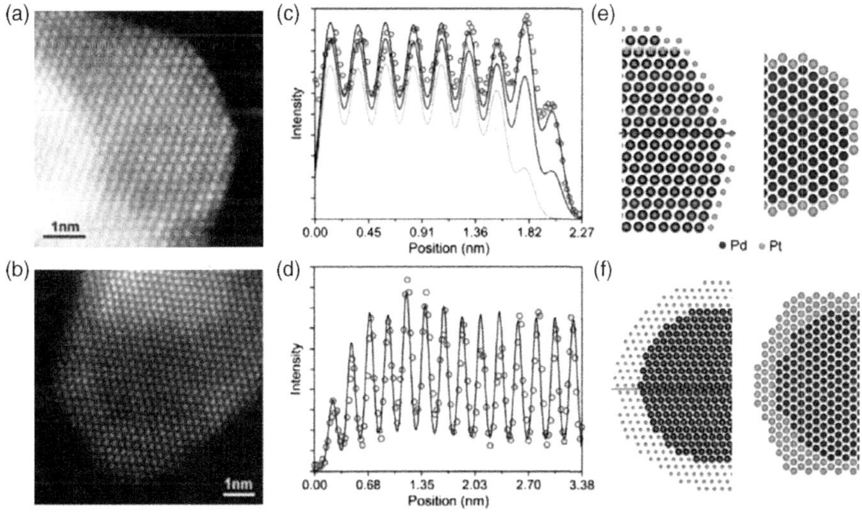


**FIGURE 6.3** (a,b) HAADF-STEM images of the Pd(core)–Pt(shell) nanoparticles obtained for the $Pd_CPt_1$ and $Pd_CPt_4$ samples, respectively. We note that the nanoparticles overlap with others at the left side in (a), and at the top side in (b). (c,d) Intensity profiles from the scan lines in (a) and (b) (open circles), and the best fits (black lines), based on the structure models shown in (e) and (f) for Pd particles with one and three to four Pt surface layers, respectively. The calculations are convoluted with a Gaussian point spread function. Two additional lines in (c) show the calculations made with all the Pt surface atoms being removed (green line), or replaced by Pd (blue line). (e) Projection of the one-Pt-layer core–shell structure model along the [101] direction (left), and the atomic arrangement (right) of the (1–1–1) plane at the position indicated by the line in the left. (f) Projection of the structure model with four Pt layers on the STEM image plane (left) and the arrangements of atoms in the vertical columns along the scan line (right) [19]. (See insert for color representation.)

panels (a) and (b) show atomically resolved high-angle annual dark field (HAADF)-STEM images for the $Pd_CPt_1$ and $Pd_CPt_4$ samples, respectively. The analysis of the *Z*-contrast HAADF intensity profiles along the scan lines from the center to the edge, using three-dimensional atomic structural models generated from the observed particles' sizes, shapes, and facets show nearly perfect fits to the data (black lines in Fig. 6.3c,d). Panels (e) and (f) illustrate the projections of the structure models in the STEM image planes, and the arrangements of atoms in the vertical columns along the scan lines. The models obtained from the best fits indicate that all and the only surface atoms are Pt for the $Pd_CPt_1$ sample (Fig. 6.3e), and, the Pt shell is three or four-layers thick for the $Pd_CPt_4$ sample (Fig. 6.3f), confirming the formation of well-defined Pt shells on Pd metal cores.

Huang et al. [18] reported a coordination-dependent surface atomic contraction on 3–5 nm Au nanoparticles. The out-of-plane contraction was largest for the atoms at edges, significant at the (100) facets, and smallest in the middle of the (111) terraces, resulting in a significant in-plane contraction at the (111) facets. In agreement with their experimental findings, our DFT calculations for a Pt nanoparticle (using the

$Pt_CPt_1$ model) demonstrate a considerable decrease in average in-plane spacing on the (111) terraces compared to that on the extended Pt(111) surface (−4.2%) [19]. This nanoscale-induced in-plane contraction affects the binding energy for oxygen (BE-O), and thus influences ORR activity [20–22]. The BE-O was found site-dependent, where the most stable configuration for O adsorption was the bridging adsorption at the edges with a BE-O of 4.50 eV, followed by bridging adsorption at the (100) terraces (4.16 eV), and then fcc adsorption at the (111) terraces (3.89 eV). The −0.27 eV difference in BE-O between the (111) and (100) facets on nanoparticles distinctly differs from that of 0.07 eV on extended surfaces. Since slight weakening of the oxygen binding with respect to that on the extended Pt(111) surface enhances ORR activity, the highly coordinated (111) facets are most conducive to the ORR on small nanoparticles.

For the $Pd_CPt_n$ particles, their size and the lattice mismatch (−0.9%) jointly determine the strain that affects the BE-O on the (111) facets. Since the intrinsic ORR activity is determined largely by the BE-O [20–22], this fact indicates that particle size affects not only the fraction of active sites at surface but also the turnover frequency (TOF) per active site. The product of the two determines the specific activity. A previous study of the ORR on transition metal alloys suggested that the best ORR electrocatalysts may bind oxygen more weakly, by about 0.2 eV [23], than Pt(111) does, that is, corresponding to 3.89 eV in our calculations (dashed line in Fig. 6.4c). This value is reached on 1.8-nm Pt particles solely by the nanoscale-induced

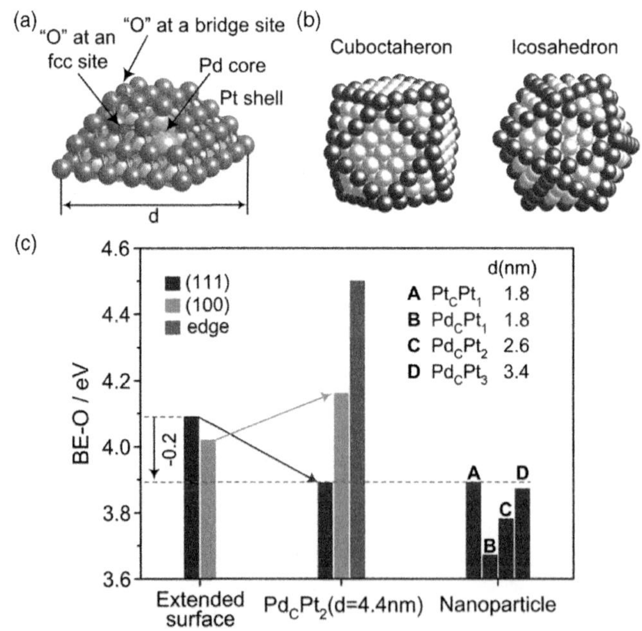


**FIGURE 6.4** (a) A semispherelike $Pd_CPt_n$ nanoparticle model used in the DFT calculations with a diameter of *d*; (b) two spherelike shapes with different fraction of atoms at facets; (c) binding energy for oxygen on different sites calculated using the slab model for extended surfaces (as a reference), and the nanoparticle model for various core–shell nanoparticles [19].

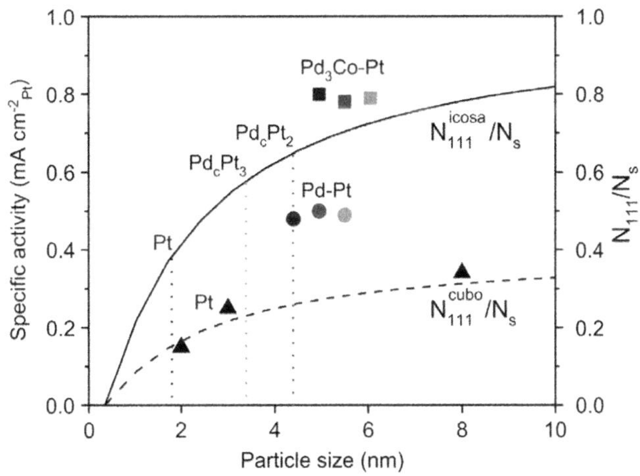


**FIGURE 6.5** ORR specific activities at 0.9 V measured at 10 mV/s (symbols, left axis) and surface fraction of atoms on the (111) facets (solid and dashed lines, right axis) as a function of particle size. The vertical dotted lines mark the sizes of three nanoparticle models in which the O binding energies are close to the optimal value of 3.89 eV. The blue, red, and green colors are used, respectively, for particles with one, two, and three Pt layers [19].

surface contraction. Larger particle sizes for optimal BE-O were found using the 3.4-nm $Pd_CPt_3$ and 4.4-nm $Pd_CPt_2$ models where reduced nanosize effect is compensated for by increased influence of Pd core.

Figure 6.5 shows the measured specific activities as a function of particle size, together with the calculated surface fraction of atoms on the (111) facets ($N_{111}/N_s$) of the icosahedral and cuboctahedral particles [24]. As the dotted vertical lines show, one advantage of the $Pd_CPt_n$ particles is that they have the optimal BE-O in the 3–6 nm size range where the $N_{111}/N_s$is significantly higher than at 1.8 nm. The large difference in $N_{111}/N_s$ between the two spherelike particle shapes suggests the potential of enhancing the ORR activity through shape- and morphology-controlled syntheses. Since low-coordination sites are prone to oxidation and dissolution, ensuring a smooth and (111)-orientation-rich Pt surface bodes well for simultaneously enhancing the ORR catalyst's activity and stability. These results point the way to tailoring a Pt shell's structure and properties by controlling core particle components, size, and shape for enhancing the activity and durability of the ORR electrocatalysts.

### 6.2.2 Fine-Tuning the Pt Monolayer–Support Interaction Using a Pd Interlayer

Given the great effect of the supporting nanoparticles on the activity of a Pt monolayer, the ORR catalyst limits the number of suitable materials. If the costs or stability of the nanoparticle material are suitable, but the effect on the Pt monolayer activity is adverse, a modification of Pt–support interaction can be effectuate using an "interlayer," such as a Pd monolayer. We discuss two systems—Ru(0001) and

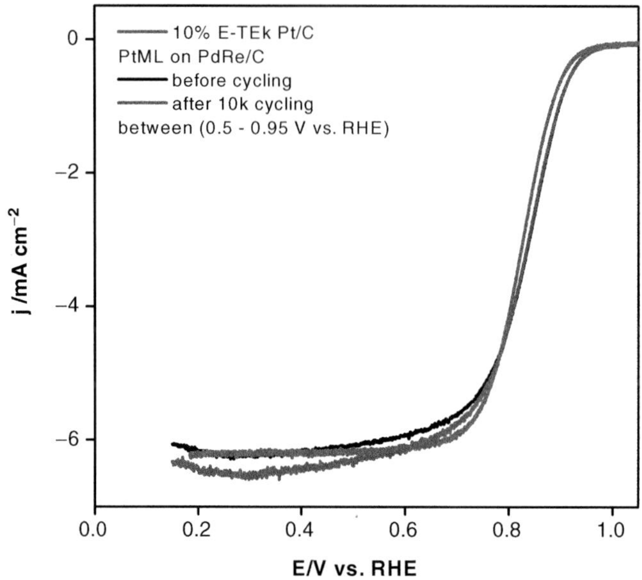


**FIGURE 6.6** Polarization curves for Pt monolayer on Pd/Re/C, by displacing the Cu monolayers with Pd and Pt; the curve for Pt/C by E-TEK and the curve after stability test involving 10,000 potential cycles from 0.5 to 0.95 V are given for comparison [25].

Re(0001)—that are both improved by a Pd interlayer. Pt monolayer on Re(0001) is compressed by 0.5% compared to Pt(111), while on Ru(0001) the compression is considerably larger amounting to 2.5%. The activity of $Pt_{ML}$ on these surfaces with a Pd interlayer is much larger than without it. For Ru(0001), the interlayer partially removes compressive strain caused by the large lattice constant mismatch between Pt and Ru. For Re(0001), the strain on a Pt monolayer is small, similar to that caused by Pd lattice. Thus, an electronic effect seems to be operative in the case of Re. This is quite feasible given the low activity of Re(0001) and its position on the volcano curve. Similar improvements were obtained with nanoparticle electrocatalysts [25]. Figure 6.6 displays the polarization curves for a Pt monolayer on Pd/Re/C. The curve for Pt/C and the curve after stability test involving 10,000 potential cycles ranging from 0.5 to 0.95 V are given for comparison. The activity of the Pt/Pd/Re/C electrocatalyst is similar to that of a commercial Pt/C, and considerably higher than that of Pt/Re/C [25]. The addition of the second Pd monolayer causes further increase in activity after correction for the increase in surface area.

The influence of the Pd interlayer on activity for ORR could be explained by the position of the $Pt_{ML}$ *d*-band center ($\varepsilon_d$) [6,7]. The Pt monolayer on Ru(0001) is compressed by 2.5% compared to Pt(111) causing a downshift in energy of $\varepsilon_d$. As a consequence, oxygen bind less strongly on that $Pt_{ML}$ than on Pt(111), leading to a slow step of breaking the O–O bond, which results in low activity for ORR compared to Pt(111) [12]. The $Pd_{ML}$ on Ru(0001) is compressed by 1.65% compared to Pd(111), while the Pt monolayer on Pd(111) is compressed by 0.85% compared to Pt(111).

Thus, introducing the Pd layer as a buffer between Pt and Ru will alleviate some of the compressive strain causing upshift in energy of $\varepsilon_d$, thus boosting activity for ORR. In addition, Pd interlayer decreases the number of low-coordination sites and decreases the curvature-induced strain; all of these effects increase the electrocatalyst's stability.

### 6.2.3 Modified Alloy Surfaces as Support for Pt Monolayers

The fine-tuning of the Pt monolayer–support interaction can be applied to alloy supports. There exists the great potential for designing electrocatalysts with alloy nanoparticles used as cores for Pt monolayer catalysts. An annealed $Pd_3Fe(111)$ single-crystal alloy support has a segregated Pd layer that was verified using low-energy ion scattering and low-energy electron diffraction techniques. This segregated Pd layer, with the same structure as Pd(111), is formed on the surface of high-temperature-annealed $Pd_3Fe(111)$. It has somewhat different electronic properties [i.e., $-0.25$ eV downshift of the *d*-band center compared to Pt(111)], and it is considerably more active than Pd(111); its ORR kinetics are comparable to those of a Pt(111) surface [26]. Figure 6.7a displays the polarization curves for the ORR on the annealed $Pd_3Fe(111)$ and the Pt monolayer covered $Pd_3Fe(111)$ surfaces and oxidation of $H_2O_2$ generated in the reaction, on disk and ring electrodes, respectively. For comparison the curves for the polished $Pd_3Fe(111)$ and a Pt(111) surfaces are also given. The enhanced catalytic activity of the segregated Pd layer compared to bulk Pd apparently reflects the modification of Pd surface's electronic properties by underlying Fe. The $Pd_3Fe(111)$ suffers a large loss in ORR activity when the subsurface Fe is depleted by potential cycling, that is, repeated excursions to high potentials in acid solutions.

Figure 6.7b shows their corresponding specific activities at 0.9 V. A Pt monolayer supported on Pd/annealed–$Pd_3Fe(111)$ shows the highest ORR kinetics among these three surfaces, by a factor of $\geq 2$ compared with the Pt(111) surface, and also demonstrates significantly increased ORR activity in comparison with $Pt_{ML}$/Pd(111) surface. An important property of the electrode's surface is the almost complete elimination of $H_2O_2$ formation. The maximum $H_2O_2$ amount calculated from the measured ring current is less than 1% at 0.08 V, suggesting a complete four-electron reduction process throughout this wide range of potential

The high activity of the Pd layer is ascribed to the modification of the Pd surface's electronic properties by subsurface Fe. This is in agreement with directly relating the position of the *d*-band center to adsorption energies according to the *d*-band center theory of Nørskov and coworkers [6,7]. Our DFT studies suggest that the observed enhancement of ORR activity originates mainly from the destabilization of OH binding, and the decreased Pt–OH coverage on the Pt/Pd/$Pd_3Fe(111)$ surface. The activity of $Pt_{ML}$/Pd(111) and Pt(111) is limited by OH removal, while the activity of $Pt_{ML}$/Pd/$Pd_3Fe(111)$ is limited by the O–O bond scission, which places these two surfaces on the two sides of the volcano plot. On the basis of DFT analysis, the binding energy of OH on $Pt_{ML}$/Pd(111) is weaker than that of OH on Pt(111), and OH removal is easier on $Pt_{ML}$/Pd(111) than it is on Pt(111); see Table 6.2. The combination of these two factors are responsible for the increased ORR activity of $Pt_{ML}$/Pd(111) compared

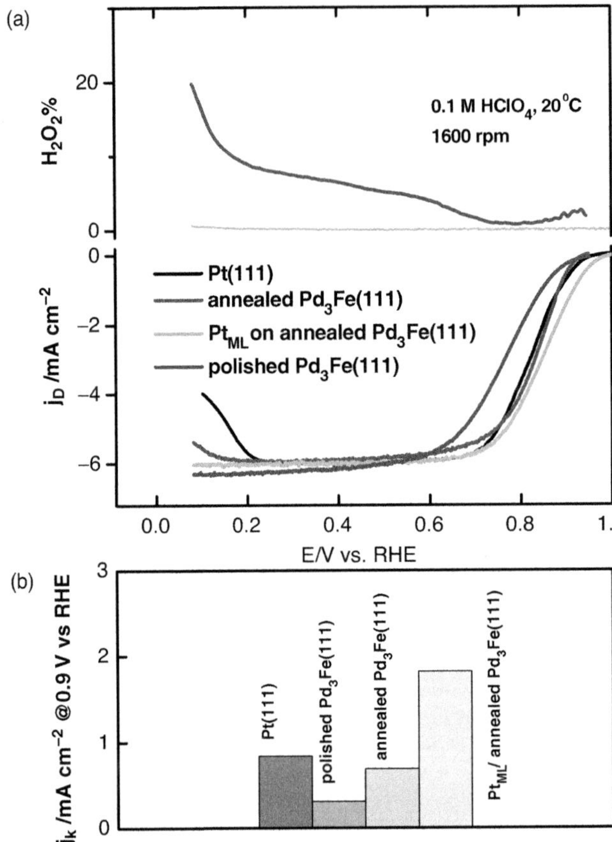


**FIGURE 6.7** Polarization curves of disk ($j_D$) (b), and fraction of $H_2O_2$ production (a) detected on ring (anodic sweep direction) for ORR on the annealed $Pd_3Fe(111)$ and the Pt monolayer covered $Pd_3Fe(111)$ surfaces in oxygen-saturated 0.1 M $HClO_4$ at room temperature; ring potential: 1.28 V; ring and disk areas are 0.126 and 0.307 $cm^2$, respectively; collection efficiency: $N = 0.245$, scan rate: 20 mV/s, 1600 rpm. For comparison the curves for the polished $Pd_3Fe(111)$ (blue) and a Pt(111) surfaces (black) are given. Panel (b) shows the bar plot for the comparison of ORR specific activities of the above surfaces at 0.9 $V_{RHE}$ [26].

**TABLE 6.2 Binding Energies of OH on Surfaces of Pt(111), $Pt_{ML}$/Pd(111), and $Pt_{ML}/Pd/Pd_3Fe(111)$[a]**

| Energy | Pt(111) | $Pt_{ML}$/Pd(111) | $Pt_{ML}/Pd/Pd_3Fe(111)$ |
|---|---|---|---|
| $BE_{OH}$(eV) | −2.09 | −2.07 | −1.93 |

[a] Note that the differences in BEs in Figure 6.3 and those in this table are due to the difference in calculation parameters used [26].

to Pt(111). As shown in Table 6.2, in $Pt_{ML}/Pd/Pd_3Fe(111)$, which is an appropriate model for the $Pt_{ML}$ on the Pd/annealed $Pd_3Fe(111)$ surface, the binding of OH is destabilized much more so than the destabilization between $Pt_{ML}/Pd(111)$ and Pt(111). This leads to $Pt_{ML}/Pd/Pd_3Fe(111)$ having more OH-free sites for $O_2$ adsorption and reactions than the $Pt_{ML}/Pd(111)$ and Pt(111) surfaces. In addition, annealing may have decreased the number of low-coordination surface sites.

### 6.2.4 Intermetallics as Supports for Pt Monolayer Electrocatalysts

Many intermetallic compound nanoparticles are stable and highly active catalysts for small organic molecule oxidation [27,28]. Given the higher thermodynamic stability of intermetallic compounds than alloys, using them as support for a Pt monolayer facilitated obtaining electrocatalysts with remarkable properties. The PtPb and PdFe intermetallics that use Pd affords an excellent support for a Pt monolayer, while the high activity of PdFe alloys [29,30] for the ORR, in addition offer the economic advantage of being a Pt-free system. Figure 6.8 shows a bar plot of the kinetic current densities at 0.9 V. The Pt/PtPb catalysts gave currents at 0.9 V as high as 2.7 $mA/cm^2$ (geometric surface area). The stability of the catalysts was assessed by applying potential steps between 0.6 and 1.13 V at 10-s intervals. The unsupported, Pt/PtPb and Pt/Pd catalysts showed no diminution in activity after 6000 cycles. However, a loss of activity was observed in the corresponding supported catalyst, suggesting that carbon corrosion or another problem related to the support was responsible for the fall in this capacity. After 16,000 cycles, the unsupported catalysts showed a small loss of activity.

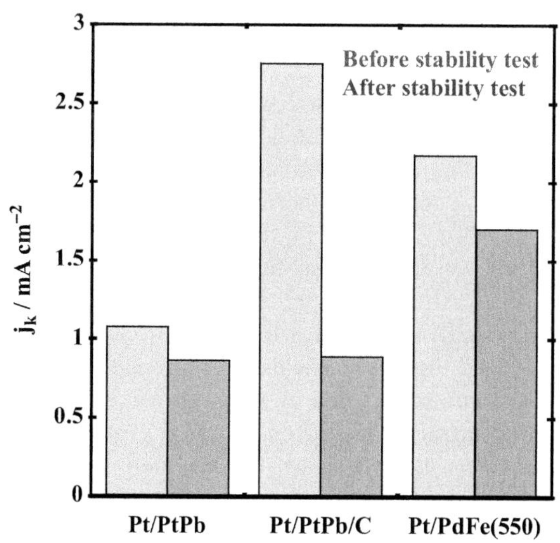


**FIGURE 6.8** Bar plot of the kinetic current density at 0.9 V for the electrocatalysts indicated in the previous graph both before and after stability tests (see text). The Pt/PtPb catalyst is the same from Table 6.1 with a lower initial PtPb loading [31].

In this new class of core–shell electrocatalysts for ORR comprising a Pt monolayer shell and ordered intermetallic compounds cores, we coupled highly stable, inexpensive intermetallics with a Pt monolayer (high activity, low metal content). Strategies suitable for selecting the types of intermetallic compounds as the support for a Pt monolayer are discussed further in the paper by Ghosh et al. [31].

## 6.3 SCALEUP SYNTHESIS, STRUCTURE, AND ACTIVITY

The results with Pt monolayer electrocatalysts may significantly impact the science of electrocatalysis and fuel cell technology, as they have demonstrated an exceptionally effective way of using Pt that can resolve problems of other approaches, including electrocatalysts' inadequate activity and high Pt content. Thus, a scaleup synthesis of these electrocatalysts needs to be addressed. At the beginning, the synthesis was carried out on a glassy carbon electrode (with diameter ~5 mm), which is impractical for producing larger quantities of the electrocatalyst. We have established a scaleup synthesis method using a new electrochemical cell that allows a synthesis of gram quantities of Pt monolayer electrocatalysts.

The cell for a scaleup synthesis has a main body of a Ti container that acts as a working electrode coated with $RuO_2$. It is corrosion-resistant, and the upd of Cu is not taking place on this surface. A Pt black counterelectrode (CE) lies in a compartment with a fritted glass to avoid a contact between CE and nanoparticles. A 2-g batch of carbon-supported Pd nanoparticles (30 wt%) was successfully covered by Cu and then replaced by a Pt monolayer. The average diameter of the Pd nanoparticles was 4.2 nm. The Cu upd process was carried out until the current became a steady value near zero. After the Cu upd monolayer formation, a $Pt^{2+}$ solution ($K_2PtCl_4$ in a 50 mM $H_2SO_4$ solution) in a separate reservoir was injected slowly into the cell while the solution was vigorously stirred, to allow Pt atoms to replace Cu monolayers on Pd surfaces galvanically. The resultant Pt/Pd/C electrocatalyst was washed repeatedly by filtering. All solutions were purged by Ar gas thoroughly. Cu-mediated layer-by-layer deposition of Pt monolayer on Pd nanoparticles has been carried out with this cell.

The morphology and structure of Pt/Pd/C electrocatalyst synthesized by the new scaleup methodology were examined by HAADF-STEM techniques. The Z-contrast images in Figure 6.9 show bright shells on relatively darker nanoparticles, which suggest the formation of a core–shell structure, namely, a Pt ($Z=78$) monolayer or bilayer (shell) on a Pd ($Z=46$) nanoparticle (core). Individual bright dots are observed around the nanoparticles in Figure 6.9b,c; they are presumably considered to be Pt atoms bombarded by the high-voltage electron beam during observation. HAADF images showed only partial coverage of shells on core nanoparticles, or gave no clear contrast of core–shell structures when samples were tilted from the zone axis; the image contrast becomes one predicted by atomic number variation (Z contrast). This is likely to be one of reasons why some pictures show incomplete Pt monolayers.

In situ EXAFS of the Pt/Pd/C electrocatalyst gave average information on atomic structures from thousands of nanoparticles [32]. Figure 6.10 shows in situ EXAFS of

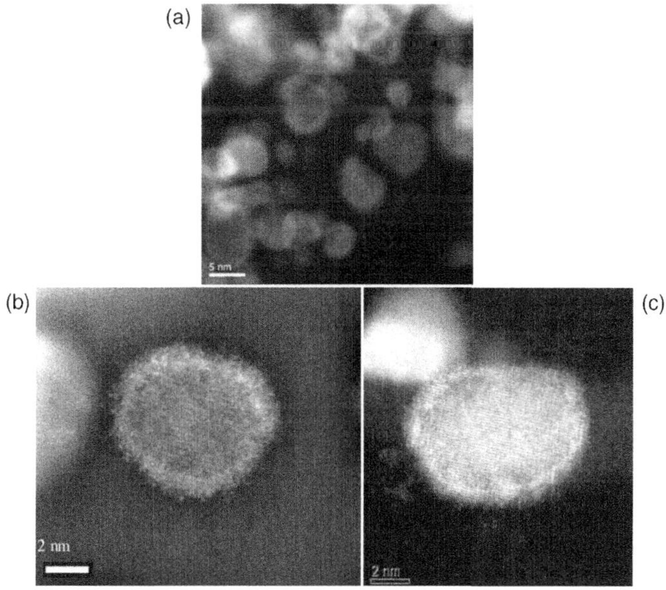


**FIGURE 6.9** HAADF, *Z*-contrast images of $Pt_{ML}$/Pd/C electrocatalyst synthesized in a 2-g batch using a new scaleup method; a Pt shell is clearly distinguished from a Pd core (higher brightness) [37].

Pt $L_3$ edge from the Pt/Pd/C electrocatalyst in 1 M $HClO_4$ at a potential of 0.41 V/RHE together with those from reference foil, respectively. The amplitude of oscillation from Pt deposits on Pd nanoparticles is much smaller than that of Pt foil, indicating that the size of Pt deposits is quite small [33]. The phase in oscillation of the Pt depositions is markedly shifted from that of Pt foil, indicating that the atomic structure of the Pt deposited on Pd nanoparticles is significantly different from 3D Pt bulk. On the other hand, the *k*-space spectrum from Pd nanoparticles indicates that there may be only the size effect between these Pd components.

The results of fitting are shown in Figure 6.11; the Pt–Pt coordination number (CN) obtained is $5.8 \pm 0.8$ and the Pt–Pd CN is $2.7 \pm 0.7$. The Pt–Pt and Pt–Pd bond lengths are determined to be $2.729 \pm 0.005$ Å and $2.724 \pm 0.007$ Å, respectively, which are smaller than that of pure Pt (2.775 Å). For a complete Pt monolayer on a Pd(111) surface, the Pt–Pt CN is 6 and the Pt–Pd CN is 3. These numbers may be slightly different for monolayers on nanoparticles [24,34]. Depending on the particle shape, there is an average surface CN of 4.8 and an average surface–core CN of 3.2 for a cubic octahedron nanoparticle with 4–5 nm diameter [35]. For icosahedron nanoparticles in (111) planes and edges, the average surface CN is 6 and an average surface–core CN is 2.7 for nanoparticles with 4–5 nm diameter. Further analysis indicates that Pt–Pt CN and Pt–Pd CN obtained verify the Pt monolayer formation on Pd nanoparticle surfaces.

We have demonstrated that a further enhancement in ORR activity can be achieved by replacing a portion of the Pt monolayer with other metal atoms that oxidize at lower

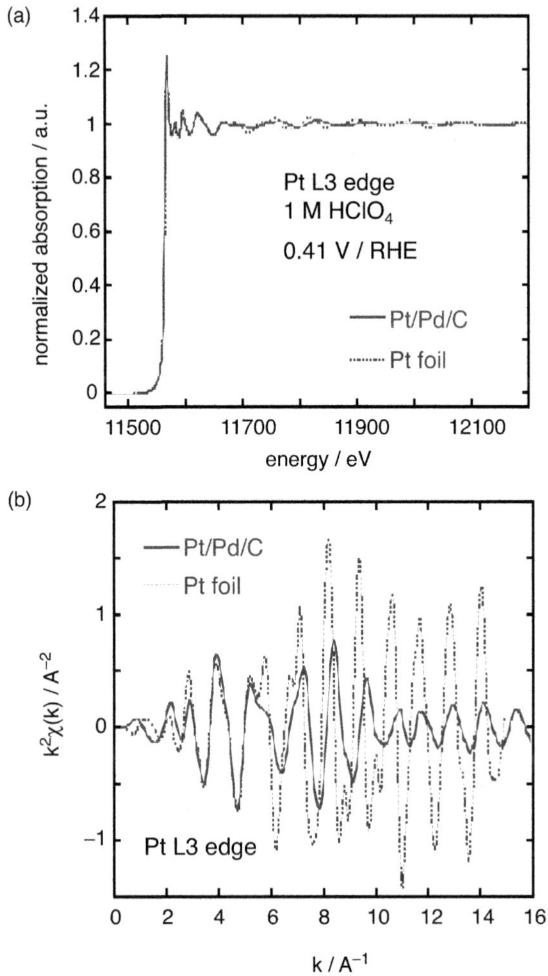


**FIGURE 6.10** (a) In situ EXAFS and (b) ($k^2$-weighted) $k$-space spectra of Pt $L_3$ edge obtained from $Pt_{ML}$/Pd/C electrocatalysts in 1 M $HClO_4$ at a potential of 0.41 V/RHE (solid lines), together with those from a Pt foil (dotted lines) [37].

potentials than Pt does (such as Ir, Os, and Re) [8]. A scaleup synthesis of a mixed $Pt_{0.8}Ir_{0.2}$ monolayer on Pd nanoparticle electrocatalyst was performed using the same electrochemical cell but by injecting a mixed Pt–Ir solution (instead of a Pt-only solution) after the upd Cu formation on the Pd nanoparticle surfaces. Figure 6.12 shows the polarization curve obtained from the mixed $Pt_{0.8}Ir_{0.2}$ monolayer on Pd nanoparticle electrocatalyst at a rotation rate of 1600 rpm, together with that from the Pt monolayer on Pd electrocatalyst obtained in a scaleup synthesis. The ORR activity of the mixed $Pt_{0.8}Ir_{0.2}$ monolayer on Pd nanoparticle electrocatalyst is higher than that of the Pt-only monolayer electrocatalyst; the half-wave potential is 0.902 V/RHE (the

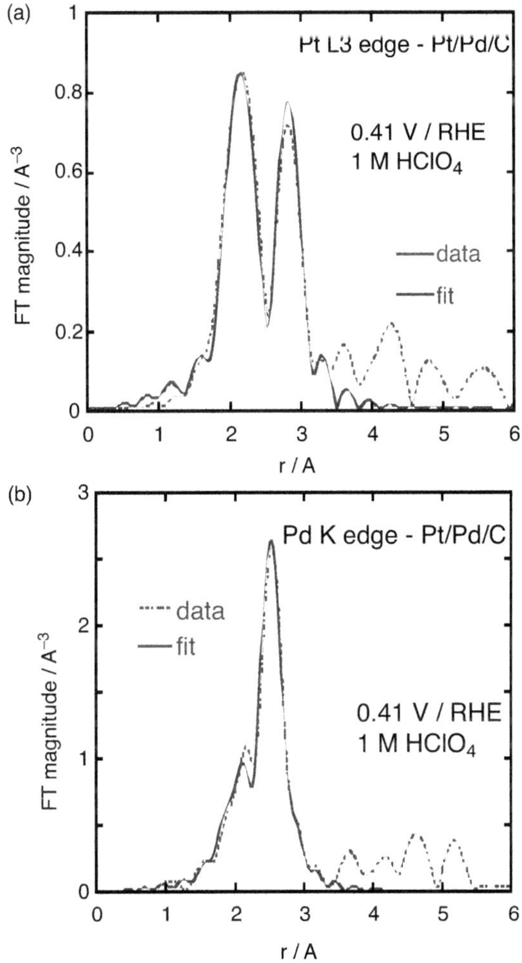


**FIGURE 6.11** Fourier transform magnitudes of the data (dotted red line) and first-shell fit (solid blue line) of (a) Pt $L_3$ and (b) Pd $K$ edges from $Pt_{ML}$/Pd/C electrocatalysts in 1 M $HClO_4$ at a potential of 0.41 V/RHE [37].

Pt loading was 9.7 nmol) and the Pt mass activity is 0.66 A/$mg_{Pt}$ at 0.9 V/RHE. As stated in Section 6.1, analyses strongly indicate that Ir by forming Ir-OH, prevents Pt-OH formation due to the lateral repulsion between the two oxidized adsorbates. This facilitates the enhanced ORR activity. In situ XANES verifies the reduced Pt-OH formation due to the presence of Ir [8].

Establishing the scaleup synthesis methodology allows us to prepare gram quantities of Pt monolayer electrocatalysts for real fuel cell tests and operations, which makes it possible to bridge the gap between nanoscience of electrocatalysis and fuel cell technology and offers an exceptionally effective way of minimizing the use of Pt.

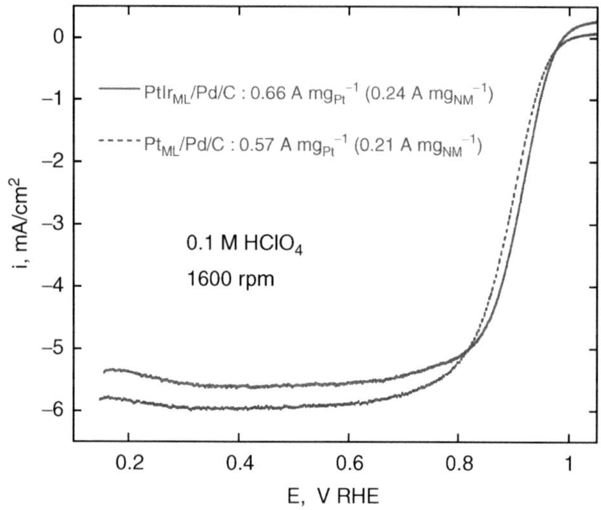


**FIGURE 6.12** Polarization curve for ORR on the mixed $Pt_{0.8}Ir_{0.2}$ monolayer on Pd nanoparticle electrocatalyst in 0.1 M $HClO_4$ at a rotation rate of 1600 rpm; also shown is that from the Pt monolayer on Pd electrocatalyst (Fig. 6.8) [electrolyte, $O_2$-saturated 0.1 M $HClO_4$; sweep rate, 10 mV/s; Pt, 9.7 nmol (9.7 μg/cm$^2$); ir, 2.0 nmol (1.9 μg/cm$^2$); Pd, 34 nmol (18.6 μg/cm$^2$)] [37].

## 6.4 FUEL CELL PERFORMANCE AND STABILITY

The ultimate test of Pt monolayer electrocatalysts clearly involves validating their performance in fuel cells. Several tests were carried out at Los Alamos National Laboratory. The total loss was approximately 0.140 V by the end of the testing (approximately 3000 h). The loss in activity is attributed to either dissolution of Pt at the operating potential or to the embedding of Pt atoms into the Pd substrate. In another long-term stability test with constant current (0.417 A/cm$^2$) during 450 h maintained an average voltage of 0.602 V. So, the results presented here clearly indicate that the Pt monolayer electrocatalysts are very promising for practical applications.

More recently, membrane electrode assembly (MEA: 25 cm$^2$) was fabricated by decaling $Pt_{ML}$/Pd/C and Pt/C as a cathode and an anode, respectively, on a Nafion NRE-211cs (DuPont Fuel Cells) membrane. Cell performance and durability of catalysts were tested by small single cell in accordance with US Department of Energy (DOE) protocol. Figure 6.13 shows the changes in (a) mass activity (MA) at 0.9 V and (b) electrochemical surface area (ECSA) during the potential modulations between 0.7 V and 0.9 V at 80°C. At the beginning of the test, $Pt_{ML}$/Pd/C showed more than 0.4 A/mg$_{Pt}$ at 0.9 V of MA and 118 m$^2$/g$_{Pt}$ of ECSA. These values are much better than those of reference Pt/C catalyst. After 30,000 potential cycles, MA and ECSA of $Pt_{ML}$/Pd/C decreased by 18% and 19%, respectively. These degradation ratios are less than the half of those of Pt/C and are better than the DOE target. A continuation of this test brought the number of cycles to 60,000 with a small additional loss of activity.

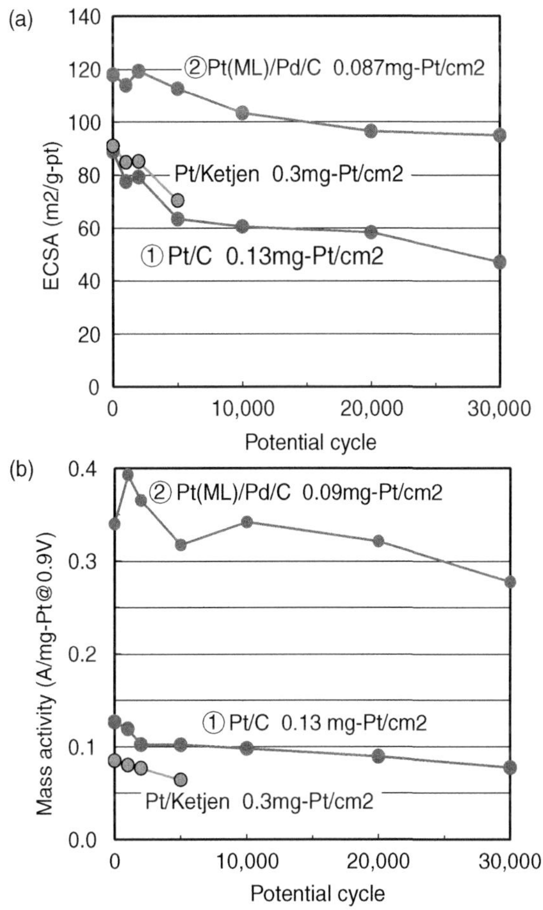


**FIGURE 6.13** Fuel cell test of the stability of Pt(ML)/Pd/C on potential cycling from in comparison with two commercial Pt/C electrocatalysts, showing changes in (a) mass activity and (b) electrochemical surface area during potential cycling between 0.7 and 0.9 V at 80°C [38].

Transportation applications are expected to be an especially important for fuel cells. A major obstacle to this application is Pt dissolution during the stop-and-go driving of an electric (electricity-powered) car that generates large electrode potential excursions. Thus, the abovementioned durability of Pt monolayer electrocatalysts indicates that we have a quite promising solution to the stability obstacle, in addition to performance and costs issues. The stability of Pt monolayer electrocatalysts on Pd and several other supports, although it may appear paradoxical, is better than that of carbon supported Pt nanoparticles. This is a consequence of a shift of Pt oxidation (Pt-OH formation) to potentials more positive than that of Pt nanoparticles caused by the electronic effects of the underlying Pd and reduced reactivity of Pt by small contraction caused by a Pd lattice as discussed above.

In addition to their intrinsic stability, Zhang et al. demonstrated that Pt and Pt monolayer oxygen reduction fuel cell electrocatalysts can be stabilized against dissolution under potential cycling regimes by modifying the Pt with small Au clusters [36].

## 6.5 SUMMARY

The results of our more recent studies of Pt monolayer electrocatalysts demonstrate significant improvements in their activity and stability, including the samples obtained in scaleup syntheses. We also demonstrate that there is a broad range of possibilities for further improving their activity and durability. Fine-tuning the Pt monolayer–support interaction using a Pd monolayer as an "interlayer" allows for increased activity for oxygen reduction. Using alloys and modified alloy surfaces as supports for a Pt monolayer facilitates both a weaker binding energy of OH and its removal from a Pt monolayer. Employing ordered intermetallic compounds with excellent thermodynamic stability as cores to support a Pt monolayer can also enhance activity. These methods of tailoring the structure and properties of a Pt monolayer can enhance the Pt mass activity by about one order of magnitude at 0.9 V, and produce a two- and threefold larger specific activity than the standard Pt/C catalysts. The results present ample illustration of the broad possibilities of the Pt monolayer–core–shell approach for designing the ORR electrocatalysts in accordance with predictions of initial studies [4,5]. They point the way to tailoring a Pt shell's structure and properties by controlling core particle composition, size, and shape for enhancing the activity and durability of ORR electrocatalysts. A smooth Pt monolayer deposit predominantly (111)-oriented is particularly suitable for simultaneously enhancing ORR catalyst activity and stability. Finally, we describe scaleup syntheses of three different electrocatalysts whose activity and durability observed in fuel cell tests verify this uniquely effective method of using Pt to synthesize catalysts with the desired properties. The results confirm the great potential of Pt monolayer electrocatalysts for resolving problems of availability of Pt in future catalysis and electrocatalysis and removing the main obstacles to the broad scale application of fuel cell technology.

## ACKNOWLEDGMENT

This work is supported by US Department of Energy, Divisions of Chemical and Material Sciences, under the Contract DE-AC02-98CH10886.

## REFERENCES

1. M. R. Tarasevich, A. Sadkowski, and E. Yeager, Comprehensive treatise of electrochemistry, in *Oxygen Electrochemistry*, B. E. Conway, et al., eds., Plenum Press, New York, 1983, p. 301.

2. R. R. Adzic, Recent advances in the kinetics of oxygen reduction, in *Electrocatalysis*, J. Lipkowski, P. N Ross, eds., Wiley-VCH, Weinheim, 1998, p. 197.
3. S. Gottesfeld and T. A. Zawodzinski, in *Advances in Electrochemical Science and Engineering*, D. M. Kolb, R. C. Alkire, eds., Wiley-VCH, Weinheim, 1997, p. 195.
4. R. R. Adzic, J. Zhang, K. Sasaki, M. B. Vukmirovic, M. Shao, J. X. Wang, A. U. Nilekar, M. Mavrikakis, J. A. Valerio, and F. Uribe, *Top. Catal.* **46**, 249 (2007).
5. J. Zhang, Y. Mo, M. B. Vukmirovic, R. Klie, K. Sasaki, and R. R. Adzic, *J. Phys. Chem. B* **108**, 10955 (2004).
6. B. Hammer and J. K. Nørskov, *Adv. Catal.* **450**, 71 (2000).
7. J. Greeley, J. K. Nørskov, and M. Mavrikakis, *Annual Review of Physical Chemistry* **53**, 319 (2002).
8. J. L. Zhang, M. B. Vukmirovic, K. Sasaki, A. U. Nilekar, M. Mavrikakis, and R. R. Adzic, *J. Am. Chem. Soc.* **127**, 12480 (2005).
9. J. Zhang, F. H. B. Lima, M. Shao, K. Sasaki, J. X. Wang, J. Hanson, and R. R. Adzic, *J. Phys. Chem.* **109**, 22701 (2005).
10. F. H. B. Lima, J. Zhang, M. Shao, K. Sasaki, M. B. Vukmirovic, E. A. Ticianelli, and R. R. Adzic, *J. Solid State Electrochem.* **12**, 399 (2008).
11. M. Shao, K. Sasaki, N. Marinkovic, L. Zhang, and R. R. Adzic, *Electrochem. Commun.* **9**, 2848 (2007).
12. J. Zhang, M. B. Vukmirovic, Y. Xu, M. Mavrikakis, and R. R. Adzic, *Angew. Chem. Int. Ed.* **44**, 2132 (2005).
13. J. Zhang, M. B. Vukmirovic, K. Sasaki, F. Uribe, and R. R. Adzic, *J. Serb. Chem. Soc.* **70**, 513 (2005).
14. S. R. Brankovic, J. X. Wang, and R. R. Adzic, *Surf. Sci.* **477**, L173 (2001).
15. K. Sasaki, J. X. Wang, M. Balasubramanian, J. McBreen, F. Uribe, and R. R. Adzic, *Electrochim. Acta* **49**, 3873 (2004).
16. K. Sieradzki, S. B. Brankovic, and N. Dimitrov, *Science* **284**, 138 (1999).
17. J. X. Wang, B. M. Ocko, and R. R. Adzic, *Surf. Sci.* **540**, 230 (2003).
18. W. J. Huang, R. Sun, J. Tao, L. D. Menard, R. G. Nuzzo, and J. M. Zuo, *Nature Materials* **7**, 308 (2008).
19. J. X. Wang, H. Inada, L. Wu, Y. Zhu, Y. M. Choi, P. Liu, W. -P. Zhou, and R. R. Adzic, *J. Am. Chem. Soc.* (in press).
20. J. K. Nørskov, J. Rossmeisl, A. Logadottir, L. Lindqvist, J. R. Kitchin, T. Bligaard, and H. Jonsson, *J. Phys. Chem. B* **108**, 17886 (2004).
21. A. U. Nilekar and M. Mavrikakis, *Surf. Sci.* **602**, L89 (2008).
22. V. Stamenkovic, B. S. Mun, K. J. J. Mayrhofer, P. N. Ross, N. M. Markovic, J. Rossmeisl, J. Greeley, and J. K. Nørskov, *Angew. Chem. Int. Ed.* **45**, 2897 (2006).
23. V. Stamenkovic, B. S. Mun, K. J. J. Mayrhofer, P. N. Ross, N. M. Markovic, J. Rossmeisl, J. Greeley, and J. K. Nørskov, *Angew. Chem. Int. Ed.* **118**, 2963 (2006).
24. H. G. Frische and R. E. Benfield, *Atoms Mol. Clust.* **D26**, S15 (1993).
25. M. B. Vukmirovic, W. -P. Zhou, W. Chen, K. Sasaki, J. Jiao, M. Mavrikakis, and R. R. Adzic, Abstract 843, 216th ECS Meeting in Vienna, Austria, 2009.
26. W. P. Zhou, X. Yang, M. B. Vukmirovic, E. K. Bruce, J. Jiao, M. Mavrikakis, and R.R. Adzic, *J. Am. Chem. Soc.* **131**, 12755 (2009).

27. C. Roychowdhury, F. Matsumoto, V. B. Zeldovich, S. C. Warren, P. F. Mutolo, M. Ballesteros, U. Wiesner, H. D. Abruna, and F. J. DiSalvo, *Chem. Mater.* **18**, 3365 (2006).

28. L. R. Alden, C. Roychowdhury, F. Matsumoto, D. K. Han, V. B. Zeldovich, H. D. Abruna, and F. J. DiSalvo, *Langmuir* **22**, 10465 (2006).

29. M. H. Shao, K. Sasaki, P. Liu, and R. R. Adzic, *Z. Phys. Chem. Int. J. Res. Phys. Chem. Chem. Phys.* **221**, 1175 (2007).

30. M. H. Shao, K. Sasaki, and R. R. Adzic, *J. Am. Chem. Soc.* **128**, 3526 (2006).

31. T. Ghosh, M. B. Vukmirovic, F. J. DiSalvo, and R. R. Adzic, *J. Am. Chem. Soc.* **132**, 906 (2009).

32. K. Sasaki and R. R. Adzic, *Synchrotron Radiat. News* **22**, 17 (2009).

33. A. Frenkel, C. W. Hills, and R. G. Nuzzo, *J. Phys. Chem. B* **105**, 12689 (2001).

34. J. McBreen and S. Mukerjee, In situ x-ray absorption studies of carbon-supported Pt and Pt alloy nanoparticles, in *Interfacial Electrochemistry: Theory, Experiment, and Applications*, A. Wieckowski, ed., Marcel Dekker, New York, 1999, p. 895.

35. A. Frenkel, *Z. Kristallogr.* **222**, 605 (2007).

36. J. Zhang, K. Sasaki, E. Sutter, and R. R. Adzic, *Science* **315**, 220 (2007).

37. K. Sasaki, J. X. Wang, H. Naohara, N. Marinkovic, K. More, H. Inada, and R. R. Adzic, *Electrochim. Acta* **55**, 2645 (2010).

38. H. Naohara, K. Sasaki, and R. R. Adzic, Abstract 980, 216th ECS Meeting in Vienna, Austria, 2009.

CHAPTER 7

# The Importance of Enzymes: Benchmarks for Electrocatalysts

FRASER A. ARMSTRONG
Department of Chemistry, Oxford University, Oxford, United Kingdom

## 7.1 INTRODUCTION

This chapter is not about fuel cells themselves but is intended as a summary of the various enzymes under investigation as fuel cell catalysts and the various ways to connect them electrically to an electrode. More recent reviews and other chapters in this book have dealt with the design, operation, and performance of biological fuel cells [1,2]. Quite a number of enzymes are now established to be highly efficient molecular electrocatalysts [3]. As discussed below, their role in technology is likely to be restricted to niche devices where power or stability would not be issues [4], but they also have an important indirect role in providing inspiration for future electrocatalyst design.

Studies by protein film electrochemistry now provide many examples of reactions catalyzed by enzymes that appear electrochemically reversible—these include the energy-relevant redox couples $H^+/H_2$, $NAD^+/NADH$, $CO_2/HCOOH$, and $CO_2/CO$, although so far not $O_2/H_2O$ [5–8]. Electrochemical measurements have an advantage over conventional kinetic studies in that efficiency and overpotential are clearly measurable. Although the $H^+/H_2$ couple is reversible at Pt electrodes, the other redox couples named above have so far been shown to be reversible only at enzyme-modified electrodes: even with the best synthetic catalysts so far investigated, substantial overpotentials are required to drive the reactions in either direction. Enzymes exist for a wide range of fuels, including alcohols, and for the common oxidants $O_2$ and $H_2O_2$. Enzymes are highly specific for their reactants, and provided $O_2$ or $H_2O_2$ does not interfere with the oxidation of fuels and there are no safety concerns, there is no

*Fuel Cell Science: Theory, Fundamentals, and Biocatalysis*,
Edited by Andrzej Wieckowski and Jens K. Nørskov

requirement to separate fuel and oxidant by a membrane. The simplification that this provides makes enzyme fuel cells suitable for miniaturization. Enzymes tend to be less prone to inhibition than synthetic electrocatalysts, and their turnover frequencies are usually far higher; for respiratory (energy-producing) enzymes or enzymes required to destroy harmful substances such as hydrogen peroxide, rates range from at least $10^2$ to upward of $10^4\ s^{-1}$. Second-order rate constants are often close to the diffusion limit, and this fact becomes even more evident when they are studied electrochemically, because hydrodynamic assistance, such as a rotating electrode, is usually required to counteract reagent depletion [9].

It is appropriate next to consider the disadvantages either real or perceived, of using enzymes as electrocatalysts. The first disadvantage is their size, because an enzyme makes a much larger footprint on an electrode support than the few atoms needed to define a conventional electrocatalyst. A typical relative molecular mass is upward of 40 kg (NADH dehydrogenase, an electrochemically reversible catalyst, has a molecular mass of nearly a million) compared to a few hundred for a conventional catalyst (Pt = 195). Radii of enzymes often exceed 50 Å, corresponding to a footprint of more than 7500 Å compared to around 6 Å for a Pt atom. As we will consider in more detail later, there are good reasons for enzymes being so large because it is important to "bury," at least partially, the active site at which the reaction occurs. However, burying the active site creates problems for electron transfer with an electrode. There are usually two extreme situations for an active site that is located deep within the enzyme: (1) both the electron acceptor and the donor are small molecules that react by bond formation (*inner-sphere* electron or atom transfer); and (2) there is a relay system (sometimes called a "wire") consisting of fast, one-electron redox sites extending out to reach the enzyme surface at a docking patch for the redox partner. To overcome situation 1, it is essential to either use a small electron transfer mediator that can penetrate the enzyme, or otherwise engineer a series of redox centers into the enzyme to produce an artificial wire. In situation 2, the problem is to ensure that electrical contact is made with the enzyme at the natural docking site where electrons enter or leave; this requires an electrode surface that favors a specific protein orientation, or embedding the enzyme in a 3D conducting matrix where numerous contact points are made. Transferring electrons over a long distance also incurs an overpotential penalty, whether the process is facilitated by a natural intramolecular relay (such as a series of FeS clusters) or a small redox mediator.

A second disadvantage of enzymes, at least at the present time, is that pure, isolated samples are expensive; however, costs are likely to fall—the opposite of precious metals. Producing a biological catalyst involves several steps with expensive consumables, and yields can be small (even a gram of material contains only a few moles or units of catalyst). A third disadvantage is that enzymes are unstable, although this needs qualifying further. A conventional surface electrocatalyst is also unstable (at the atomic level), but loss of atoms from the electrode surface is usually compensated for by underlying atoms—Pt atoms, for example, are constantly being dislodged from their electrode support. Enzymes are usually restricted to moderate temperatures and aqueous environments, and certainly not dry conditions, although there may be considerable scope for exploiting the more robust enzymes produced by extremophiles, in particular thermophilic organisms. In summary, enzymes are likely to be practical only in niche applications, particularly those where a long lifetime is not required and the fuel cell could be disposable.

All this may sound rather pessimistic, so we turn next to a sound scientific reason for studying redox enzymes as fuel cell catalysts, namely, the inspiration that they provide. The active sites of redox enzymes are abundant metals such as Fe and Cu or organic cofactors such as flavins or a special quinone cofactor like pyrroloquinoline quinone (PQQ). In nonbiological catalysis the redox reactions occurring in low-temperature fuel cells usually require Pt or other precious metals that will only continue to become more costly and scarce. The immediate environment of the active site and surrounding structure are crucial, and this importance underlies the reason why enzymes are so large. Evolution of enzymes has been driven by efficiency, and the active sites have become optimized at the first, second, and *n*th coordination spheres, with exquisite attention to supramolecular detail that is difficult to approach even with surface chemistry or coordination chemistry. With a simple molecular or surface catalyst, the region outside the immediate focus of attention is structurally unknown or ill-defined, and certainly poorly controlled because it comprises numerous solvent or gas molecules undergoing random motions and collisions. In contrast, for an enzyme whose structure has been determined by X-ray diffraction, the surrounding environment is defined to atomic resolution (revealing nonhydrogen atoms) many tens of angstroms out from the active site. The object of such a large complex structure is to hold the active site in place, with all components at optimal positions (perhaps resembling the transition state) and to control the entry and exit of reactants (including protons) and inhibitors. Rejection of water may be a key issue, as was discussed in a paper by the theoretical chemist Michael Dewar [10].

To compensate for the large footprint of enzymes (on a "per volume" basis, enzymatic activity can be very low), it is necessary to increase the surface area or dimensionality of the electrode. This has been achieved by embedding enzymes in porous conducting materials or in redox hydrogels containing a large number of redox mediators such as Os complexes (see later). In such a way very large enhancements in current density are possible provided substrate transport or electron transfer does not become rate-limiting. Ideally, one would like to decrease ("trim down") the size of enzymes to arrive at the minimum size that will carry out the required reaction.

Many enzymes that are currently being studied as fuel cell electrocatalysts are bound to a membrane in their normal function in the organism. The enzyme molecule that is actually used in electrochemical and fuel cell investigations is then a soluble subcomplex that has been separated from its membrane anchor protein (often the two parts—referred to as *membrane-intrinsic* and *membrane-extrinsic*—are not bound together very tightly). A membrane-extrinsic enzyme has evolved naturally to transfer electrons over a long distance to a redox partner that is present in a membrane. There are also strong similarities between the theoretical tools used to understand the electric nature of electrode surfaces and biological membranes.

The fuels that are oxidized by enzymes range from dissolved gases ($H_2$) to various nonvolatile water-soluble substances. Methanol and ethanol are products of microbial

**TABLE 7.1 Redox Properties of Some Oxidants and Fuels and the Enzymes that Catalyze Their Conversions**

| Fuel (Oxidant) | Reduction Potential (V vs. SHE at pH 7) | Suitable Enzymes |
|---|---|---|
| $H_2O_2$ (→ $H_2O$) | 1.35 | Peroxidase |
| $O_2$ (→ $H_2O$) | 0.82 | Blue Cu oxidases (laccases, bilirubin oxidases) |
| $CH_3OH$ (→ $CH_2O$) | −0.18 | Alcohol dehydrogenase, methanol dehydrogenase |
| $CH_2O$ (→ $CO_2$) | −0.40 | Formate dehydrogenase |
| $H_2$ (→ $H_+$) | −0.41 | Hydrogenases |
| Glucose (→ gluconolactone) | −0.45 | Glucose oxidase, glucose dehydrogenase, cellobiose dehydrogenase |
| CO (→ $CO_2$) | −0.51 | Carbon monoxide dehydrogenase |

metabolism, glucose is the "fuel" of animal tissue and the bloodstream, and fructose and higher-order saccharrides are abundant in the plant world. The oxidation product should ideally be water or a nontoxic gas such as $CO_2$ that escapes naturally. A nonvolatile product is difficult to remove, thus complicating the device and lowering efficiency. Properties of common fuels are listed in Table 7.1.

Next we examine some active sites of enzymes and try to understand how they perform so well with turnover frequencies often exceeding 1000 per second. We need to consider what they are required to do in the organism that expresses them. We then consider how they are attached and electrically connected to an electrode.

## 7.2 ACTIVE SITES AND OVERALL STRUCTURES OF FUEL CELL ENZYMES

We start with hydrogenases, mainly because of the inspiration that they provide for producing future "supercatalysts" for $H_2$ technologies. Like Pt, hydrogenases catalyze electrochemically reversible $H_+/H_2$ interconversion when immobilized on an electrode. Just a minuscule overpotential is required, in either direction, to oxidize $H_2$ or produce it from protons [11,12].

Hydrogenases are microbial metalloenzymes, and their high activity is attributed to a remarkable environment around the Fe (and Ni) atoms that has evolved over nearly three billion years to catalyze $H_2$ oxidation and $H_2$ production, and in the last two billion years to perform these functions even in the presence of $O_2$ [13]. The active sites of the two main classes, known as [NiFe] or [FeFe] hydrogenases, are shown schematically in Figure 7.1a and more exactly in Figure 7.1b. The entire enzyme molecules with long-range electron relays are shown in Figure 7.1c. The hydrogenases so far studied successfully by protein film electrochemistry and applied in fuel cell demonstrations fall into two categories; they are either (1) the soluble subcomplex of an otherwise membrane-bound complex or (2) naturally soluble enzymes that

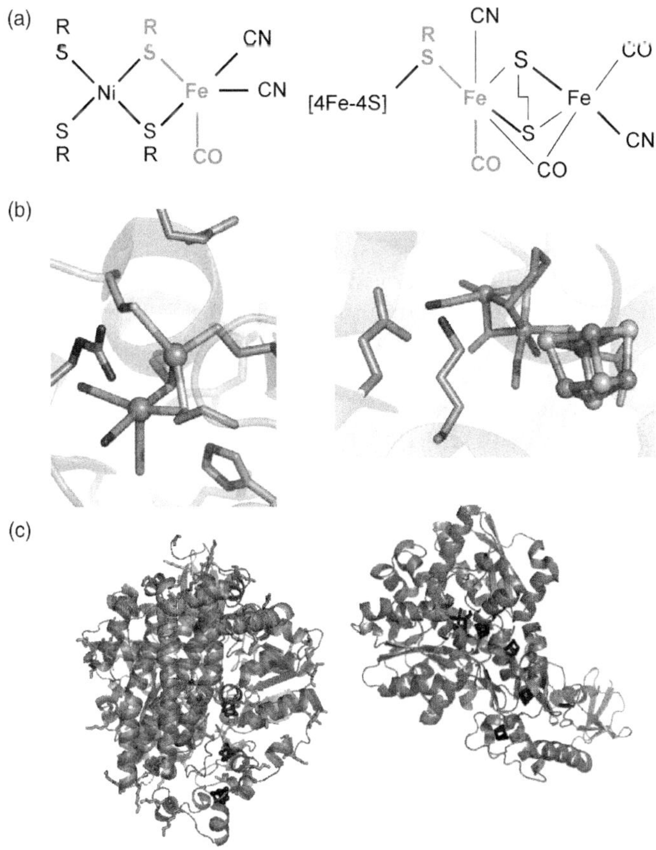


**FIGURE 7.1** Features of the structures of hydrogenases: (a) schematic structures of (left) [NiFe] and (right) [FeFe] active sites, emphasizing the RS–Fe–CO minimum structure lighter shade common to both centers; (b) close-up views of the active sites of the [NiFe] hydrogenases from *Desulfovibrio fructosovorans* (left) and the [FeFe] hydrogenase from *Clostridium pasteurianum* (right) showing sidechains of amino acids believed to be important; (c) overall structures of *D. fructosovorans* [NiFe] hydrogenase (left) and *Clostridium pasteurianum* [FeFe] hydrogenase (right) showing the long-range electron transfer relays formed by FeS clusters, leading into the buried active site.

exchange electrons with a protein partner. Hydrogenases that are particularly proficient at oxidizing $H_2$ consist of two subunits, one (the α subunit) housing the active site and the other (the β subunit) containing the FeS clusters of the long-range electron transfer relay. The two subunits are bound together very tightly, and the αβ enzyme remains intact under the harshest conditions. The overall molecular mass of a [NiFe] $H_2$-oxidizing enzyme is typically around 90–100 kg/mol. In the organism the αβ heterodimer is usually associated with a membrane-bound "anchor" protein so that the resulting electrons are supplied to a pool of quinol.

The minimal common component of the active site of all hydrogenases characterized so far is an Fe atom coordinated by one thiolate (RS) and one carbon monoxide. This motif is extended with further coordination—by cyanides in addition to carbonyl, by a second metal atom (Fe or Ni), and by more thiolate ligands. Somehow this molecular environment, far more resembling an organometallic compound than a Werner coordination complex, makes Fe (and Ni) behave as a third-row *d*-block (precious) metal catalyst. Unlike the mechanism usually considered for Pt, it is widely stated that hydrogenases catalyze the oxidation or production of $H_2$ by a *heterolytic* mechanism; in other words, $H_2$ is formed or broken up in a sequence that involves a hydrido species and a proton. Dihydrogen is trapped, either at the beginning or near the end of the reaction depending on the direction, by $H_2$–metal coordination analogous to that of the Dewar–Chatt–Duncanson model for alkene binding. A crucial point in understanding the chemistry of $H_2$ activation by metals came from experiments carried out by G. Kubas in the 1980s [14]. Both [NiFe] and [FeFe] hydrogenases have a 16-electron low-spin Fe(II) that is well suited for binding $H_2$ in the manner described by Kubas. Cleavage of the H–H bond bond is assisted by two factors: (1) coordination of $H^-$ (in [NiFe] hydrogenases this is probably the Ni that formally becomes Ni(III)] and (2) protonation of a base located very close by (perhaps one of the cysteine S ligands) noting that a proton cannot move far in a single step. The entire reaction occurs within a radius of about 5 Å, yet somehow the periphery (and beyond) has great importance. The fragile site must be protected against intruders, particularly $O_2$, CO, and $H_2O$—yet $H_2$, electrons, and protons must have free access. Electron transfer is mediated by a series of FeS clusters; in [NiFe] hydrogenases the first of these is about 12 Å from the Ni atom; in [FeFe] hydrogenases, the binuclear Fe site is directly attached to a [4Fe–4S] cluster that is part of the overall active site known as the "H cluster." Proton transfer over a longer range is mediated by protein bases such as the carboxylate sidechains of aspartate and glutamate, the terminal amine group of lysines, and the guanidinium group of arginine. Water molecules may also play a role but are not allowed to interfere with the active site. Could such unique arrangements be achieved in small molecular catalysts? This is a pressing issue that is under intense investigation by several groups worldwide.

The exact mechanisms of the two types of hydrogenase remain unclear; CO is a competitive inhibitor of $H_2$ oxidation and for the [FeFe] hydrogenases a CO adduct has been crystallographically defined, with CO bound to the distal Fe atom (that farthest from the [4Fe–4S] cluster) [15,16]. Therefore, this site is probably the one at which $H_2$ is coordinated. Unusual oxidation states are involved—Ni(III), Ni(I), and Fe(I) in the case of the [FeFe] hydrogenases; these are not normally encountered in coordination compounds and show again how the enzyme dictates unusual properties. A problem with hydrogenases is that they convert to inactive states at high potential [17]. Whereas reactions with CO are usually reversible, encounters with $O_2$ give rise to reactions that are not so reversible and sometimes result in destruction. Hydrogenases are being discovered that can operate in the presence of atmospheric $O_2$ levels and even in the presence of CO; these are mentioned later [18,19].

The internal electron relay means that electron mediators are not needed to overcome the barrier of a buried center, but since the active site is almost inaccessible to molecules larger than $O_2$ or $H_2S$, it is highly likely that mediators such as viologens would be excluded also. However, the electron relay reaches the protein surface in a comparatively small area close to the distal FeS cluster, and electron mediators could help to ensure that this zone remains electroactive even when not facing the electrode. Compared to most enzymes, hydrogenases depend on a relatively large number of genes for their expression, including those responsible for synthesising and transporting CO, $CN^-$, and Ni, all of which are rather toxic to an organism. This complexity makes the production of hydrogenases more demanding than that of other fuel-oxidizing enzymes.

Next to be addressed are the "blue" Cu oxidases, which catalyze the four-electron reduction of $O_2$ to water without release of hydrogen peroxide. Compared to the various anode reactions, the cathodic reaction is a much more focused area of enzyme fuel cell research, where $O_2$ is the oxidant of choice in almost all cases. Blue Cu oxidases are grouped into various types—laccases from plants and fungi, along with bilirubin oxidase, ascorbate oxidase, and ceruloplasmin; but it is the fungal laccases and bilirubin oxidase (the latter also produced by fungi) that are of interest for fuel cells. Laccases and bilirubin oxidases consist of a single peptide chain and tend to be extracellular proteins (they are secreted from a cell to break down compounds in the surrounding environment). They are usually glycosylated (they have carbohydrates (glycans) attached to the sidechains of certain amino acids). Their molecular masses are typically in the region of 40–50 kg/mol.

In conventional PEM fuel cells the $O_2$ reduction reaction at the Pt cathode is a major source of inefficiency because it requires a substantial overpotential—not surprisingly, a concerted four-electron/four-proton reaction is mechanistically very demanding. In the case of two 2-electron steps, an overpotential would arise because of the inherent instability of hydrogen peroxide: oxidation state diagrams for oxygen show that free peroxide (hydrogen peroxide or hydroperoxide anion) is inherently unstable (should disproportionate to $O_2$ and $H_2O$) regardless of pH. Therefore, as a limiting case, the reduction of $O_2$ should commence at the $O_2/H_2O_2$ potential, which is about 0.5 V more negative than the four-electron value. This expectation will be relaxed according to how favorably the catalyst binds peroxide, but any such stabilization effect should not be too great; otherwise completion of the reaction to produce $H_2O$ would be too sluggish. A similar argument applies for electrolytic production of $O_2$ where a similar overpotential requirement is encountered in the opposite direction. Four-electron oxidation of water during photosynthesis is achieved by a special [4Mn–4OCa] cluster contained in photosystem II. The energetic details are not established, but it is clear that a potential of 1–1.25 V is required to form $O_2$ from $H_2O$ at pH 7 ($E_{pH7} = 0.83$ V). Hence there is much interest in a simple Co phosphate compound that was reported in 2008 to be an excellent electrocatalyst for water oxidation [20].

The active site of blue Cu oxidases shown in Figure 7.2a,b consists of four Cu atoms, providing an immediately available reservoir of four electrons. Three of the Cu atoms are arranged in a site that rapidly binds $O_2$ and converts it to a peroxide that remains bound but is easily reduced further [21]. The fourth site, known as the "blue" Cu center, is located about 12 Å away and is linked to the trinuclear site by a

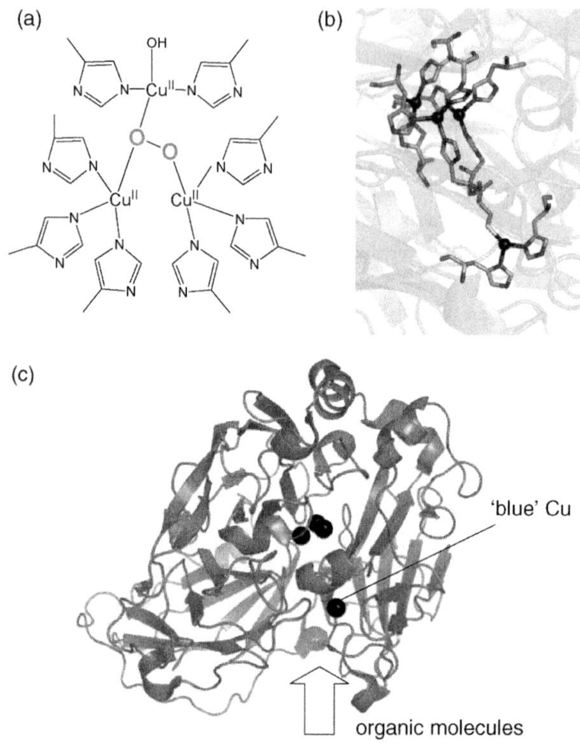


**FIGURE 7.2** Features of the structures of "blue" Cu oxidases: (a) schematic structure of the site at which $O_2$ is reduced; (b) close-up of the four Cu atoms that make up the active site and the electron relay (the "blue" Cu center); (c) the overall structure of laccase from *Trametes hirsuta* (pdb code 3FPX) showing the region for binding organic molecules and interacting with an electrode.

sequence of two amino acids. The blue Cu center, in which Cu is coordinated by two histidine imidazoles and a cysteine thiolate, is special as it has a particularly low reorganization energy for electron transfer [Cu(II)/Cu(I)], making it suitable for fast electron transfer with minimal driving force. In fungal laccases and bilirubin oxidase, the blue Cu has a particular high reduction potential (usually 0.6–0.8 V) and functions as an electron relay center between the $O_2$ reduction site and organic molecules (lignin degradation products, bilirubin, etc.) that bind nearby in a pocket and are oxidized to radicals. Experiments with laccase and bilirubin oxidase show that $O_2$ reduction commences at a potential that is significantly more positive than Pt at neutral pH [22]. The overpotential is comparable with more recent improvements such as $Pt_3Ni$ [23].

Blue Cu oxidases are quite stable enzymes that operate over pH values 4–8, with laccases being more active below pH 5 whereas bilirubin oxidase can operate well at slightly higher pH. The pH dependence is thought to arise from the mechanistic requirement for fast proton transfer to the trinuclear Cu site that involves a glutamate residue. Laccases are inhibited by halides and pseudohalides, particularly $F^-$ and $N_3^-$ (azide) and also by sulfide ions, although bilirubin oxidase appears to be less strongly inhibited. At least some blue Cu oxidases are not inhibited by methanol, even at very high concentrations, and this property may be useful in designing a methanol fuel cell that does not require a membrane [24].

Alcohols and carbohydrates are important biological fuels, and their oxidation is catalyzed by several enzymes; the important ones (for electrochemical applications) are those containing an organic redox cofactor, usually flavin adenine dinucleotide (FAD) or PQQ. Glucose oxidase (GOx) is well known to bioelectrochemists because it has achieved great importance as the catalyst in amperometric glucose sensors for diabetics. The protein environment surrounding the active-site FAD is highly evolved—outside their host apoenzymes, flavins show no specific catalytic activity. Flavins can be oxidized or reduced by two electrons in one-electron steps and then undergo discrete two-electron reactions by hydride/atom ($H^-$, "O") transfer. Glucose oxidase catalyzes a discrete two-electron transfer between two fairly small molecules, glucose and $O_2$; no long-range electron transfer is involved, and it was clear early on that direct electron exchange with an electrode might not be feasible. Glucose oxidase can be used in fuel cells if a suitable mediator is also present either in solution or as a redox hydrogel polymer [25,26]

To achieve high rates of electrocatalysis of glucose oxidation, enzymes containing natural electron relay systems have been sought. Glucose dehydrogenase offers an advantage over GOx because it does not use $O_2$ as an electron acceptor but instead uses a long-range electron transfer pathway via a *c*-type heme group. It is thus naturally equipped for direct electron exchange with an electrode, although the potential at which electrochemical exchange occurs is that of the heme group rather than the flavin, which operates closer to the glucose/gluconolactone potential. The intramolecular long-range electron transfer necessary to complete the reaction is thus an inherent source of overpotential [27]. Heme groups (Fe porphyrins) have more positive potentials than do the FeS clusters found in hydrogenases. Another naturally "wired" enzyme is cellobiose dehydrogenase (CDH), which is able to oxidize a range of disaccharides and oligosaccharides to their corresponding C1-lactones over a wide pH range. Cellobiose dehydrogenase (Fig. 7.4) is an extracellular flavocytochrome secreted by lignocellulose-degrading fungi, and it consists of two domains, one containing a flavin group and the other containing a heme group for mediating long-range electron transfer, similar to glucose dehydrogenase. The enzyme from *Phanerochaete chrysosporium* has been structurally defined by X-ray diffraction studies (for practical reasons carried out separately on the flavin and heme domains) [28]. The substrate-binding pocket differs from that of GOx in the identities and positions of groups interacting with and orienting the sugar molecule. The CDH from *Myriococcum thermophilum* (*Mt*) has a broader substrate range than *Pc* CDH and is able also to oxidize glucose and maltose. Thus *Mt* CDH has been investigated for its ability to function as a versatile fuel cell electrocatalyst [27].

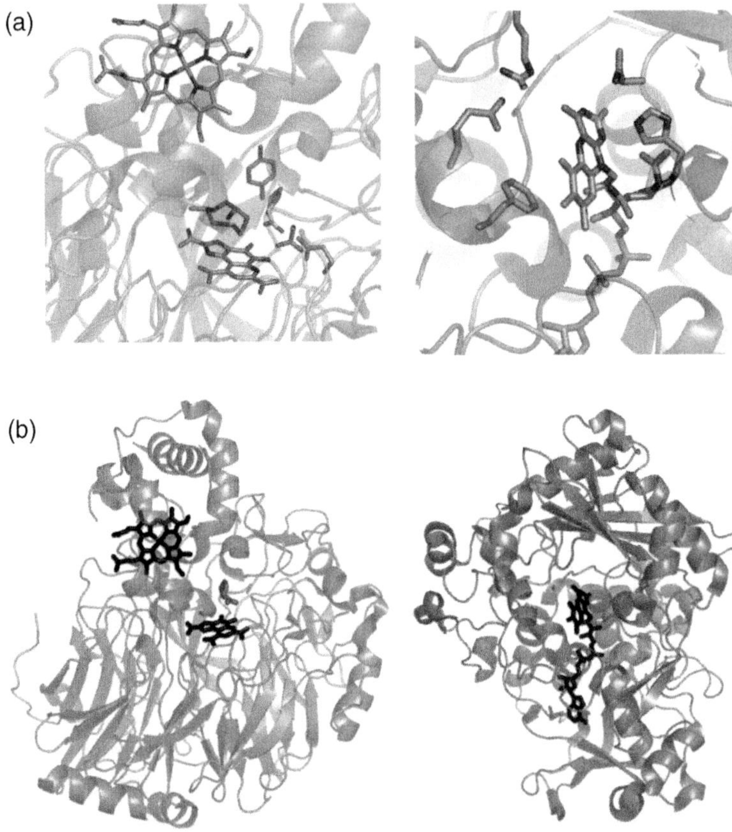


**FIGURE 7.3** Features of the structures of cellobiose oxidase and alcohol oxidase: (a) close-up views of the active sites of the PQQ-containing alcohol dehydrogenase from *Comamonas testosteroni* (left, pdb code 1KB0) and the flavin-containing cellobiose dehydrogenase from *Phanerochaete chrysosporium* (right, pdb code 1KDG); (b) overall structures of alcohol dehydrogenase (left) and cellobiose dehydrogenase (right). The full cellobiose dehydrogenase molecule contains a heme domain that is revealed in a separate structure deteremination.

Alcohol oxidation is usually carried out by Zn-containing enzymes that link the reaction directly to NAD reduction, and there is no long-range electron transfer activity to link into. Alternative alcohol dehydrogenases are known that contain a PQQ cofactor as the active center and a heme group to relay electrons to redox partners binding at the protein surface. The enzyme and its active-site region are shown in Figure 7.3. The small molecule in the active site is believed to be the product of alcohol oxidation [29].

An enzyme capable of oxidizing D-fructose to 5-keto-D-fructose is isolated from *Gluconobacter.* Fructose dehydrogenase (FDH) is a membrane-bound enzyme in vivo and contains a flavin cofactor at which fructose is oxidized and a heme group to relay electrons. Conceptually this is similar to cellobiose dehydrogenase and alcohol dehydrogenase just described, although fructose dehydrogenase has not yet been structurally characterized. The enzyme has a molecular mass of 140 kg/mol. In fuel cell experiments FDH has shown extraordinary performance in conjunction with bilirubin oxidase or laccase (see text below).

## 7.3 ATTACHING AND ELECTRICALLY CONNECTING ENZYMES TO ANODE AND CATHODE

Having identified enzymes for electrocatalysis on the basis of their inherent properties (substrate-binding specificity, turnover frequency, inherent overpotential requirement, electron transfer capability and stability), it is necessary to apply them in such a way as to achieve as high a current as possible with the smallest possible voltage losses. Recalling our earlier discussion of enzyme size, electrocatalytic current is linked directly to the turnover frequency of the electrocatalyst, and even with the most active of enzymes a monolayer on a flat surface will not be enough to give a useful current. A geometric (flat surface) monolayer of a large enzyme corresponds at most to about 5 pmol/cm$^2$, and even with a turnover frequency of 1000 s$^{-1}$, this will produce a current density of just 1 mA/cm$^2$. The highest current densities for an enzyme attached to a nominally flat electrode surface have been recorded for some hydrogenases (>1 mA/cm$^2$) and for a fungal laccase (> 0.6 mA/cm$^2$). The coverage of laccases and bilirubin oxidase on a PGE or other carbon electrode is enhanced and stabilized by derivatizing the electrode surface with polyaromatic functionalities using the diazonium coupling method introduced by Savéant and coworkers [30,31]. Laccase binds strongly at a carbon surface derivatized by the very hydrophobic species anthracene or chrysene, whereas bilirubin oxidase adsorbs strongly if the surface is derivatized with naphthyl carboxylate groups (Fig. 7.4). It is likely but not proven that these functionalities, perhaps present at a much lower density on naturally polished graphite, form strong interactions with the natural substrate binding regions close to the blue Cu center.

To increase current densities for enzyme electrodes much above 1 mA/cm$^2$, more effective tactics are required. Therefore efforts are directed at greatly increasing the actual number of enzyme molecules by creating the equivalent of a multilayer assembly extending *out* from the electrode surface. This can be achieved by either increasing the dimensionality of the true conducting electrode itself (extending it far into solution) or incorporating the enzyme into a dense polymer hydrogel containing a large number of immobilized redox-active molecules to mediate electron transfer reactions [32]. In either case the resulting "3D" electrode must accommodate fast electron transfer as well as rapid transport of ions and molecules.

In the first of these procedures a highly porous carbon electrode is made by pyrolyzing a carbonaceous gel onto a conducting support. The resulting *aerogel* electrode has an extremely large surface area and high mobility for incoming and outgoing species. Alternatively, an extremely rough carbon electrode is made by coating with a thick layer of conducting nanoparticles such as that available as the

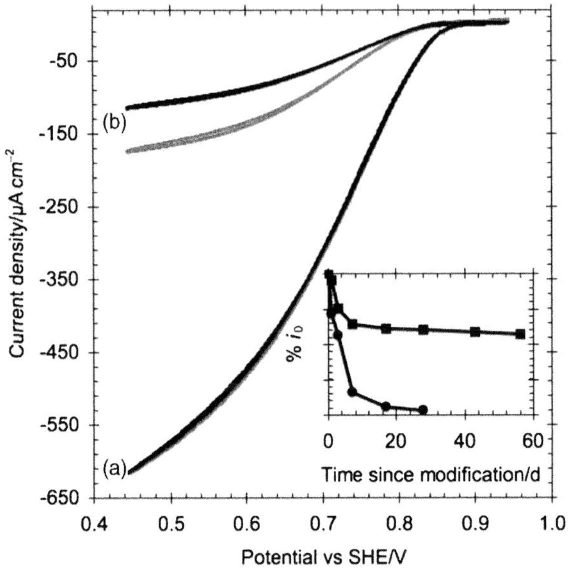


**FIGURE 7.4** Electrocatalytic activity of laccase from *Pycnoporus cinnabarinus* attached to (curve a) a PGE electrode that has been derivatized with anthracene functionalities by diazonium coupling and (curves b) an unmodified PGE electrode. The black curves show the catalytic waves immediately after spotting on laccase solution; the lighter shaded curves, the catalytic waves after cell solution was exchanged for fresh buffer, leaving no enzyme in solution (~30 min after the initial enzyme application). Inset shows the long-term change in catalytic current for a modified (■) and unmodified (●) electrode. The *y* axis shows the change in current density at 0.44 V versus SHE relative to the first wave with no enzyme in solution; scan rate = 5 mV/s, rotation rate = 2500 rpm. (Reproduced with permission of the Royal Society of Chemistry.)

commercial product known as *carbon black*. These enhancements to direct, unmediated electron transfer have been the subject of an important study by Kano and coworkers, who demonstrated a particularly high current density (> 4 mA/cm$^2$ at 25°C) for fructose oxidation by FDH at a carbon paper electrode modified by carbon black nanoparticles and 4 mA/cm$^2$ for a fungal laccase adsorbed on a carbon aerogel electrode (Fig. 7.5) [33]. Note, however, the high potential at which catalysis begins for fructose oxidation, the result of intramolecular electron transfer through the heme group, which has quite a high potential. The current densities translated into power densities of nearly 0.4 mW/cm$^2$ in a realistic fuel cell situation and over 0.8 mW/cm$^2$ was achieved on stirring.

Viewed from the eyes of a technology investor, the most persuasive experiments so far are those in which the enzymes are embedded in a crosslinked redox polymer hydrogel containing a high density of redox-active complexes covalently attached to the polymer backbone using long flexible linkers [26,32,34]. Heller, Leach, Calabrese-Barton, and others have carried out extensive investigations of a library of substitution-inert osmium complexes with pyridyl-*N* ligands that exhibit fast electron

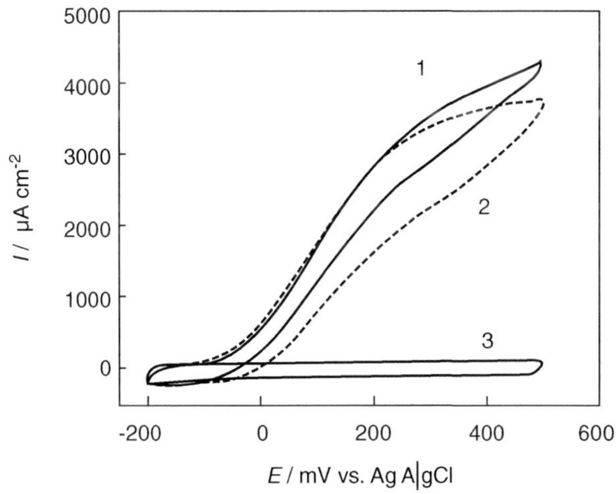


**FIGURE 7.5** The high current density displayed in cyclic voltammograms of the oxidation of fructose (0.2 M) by the heme-containing flavoenzyme fructose dehydrogenase from *Gluconobacter* adsorbed on a carbon-black-modified carbon paste electrode, pH 5.0, 20 mV/s. Curve 1 represents a stirred solution; curve 2, an unstirred solution; curve 3, a solution without fructose. (Reproduced from Ref. 33 with permission, Royal Society of Chemistry.)

exchange kinetics and can be designed to operate over a very large potential range. The concept has been applied to both anode and cathode, because the cell potential is determined mainly by the Os(III)/(II) potentials at each electrode, which can be varied over a very wide range by altering the ligands and substituents on the bipyridine rings. For example, the two complexes shown in Figure 7.6 vary in potential by more than 0.7 V [34]. The potential of the mediator should be matched as closely as possible to that of the enzyme—too much of a difference incurs a large voltage loss penalty, while too small a difference may give a smaller current. Any disadvantage of the voltage drop incurred by having an Os complex as a mediator is usually more than offset by the higher current density that becomes possible by ensuring that many more enzyme molecules are electrically connected to the electrode than would be the case if one relied instead on direct electron transfer to a monolayer or less of enzyme. For high-potential blue Cu oxidases, the potential at which $O_2$ reduction commences is linked strongly to the reduction potential of the Os complex. The voltage drop need not be problematic because a cathode featuring laccase or bilirubin oxidase incorporated within these hydrogels produces a large current density with an overpotential that is still superior to Pt under neutral pH conditions.

## 7.4 STABILITY ISSUES

Improving stability is illustrated in studies of a [NiFe] hydrogenase from *Desulfovibrio gigas*, in which the enzyme is covalently attached to a bed of carbon nanorods [35].

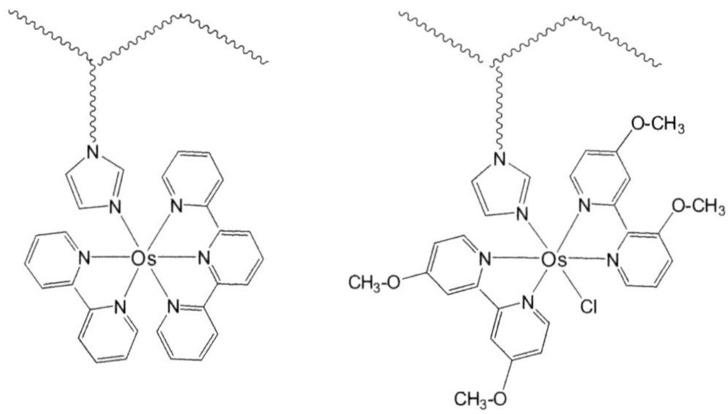

Polymer with $[Os(tpy)(bpy)(\underline{imid})]^{3+/2+}$

$E^o$ = 0.82 V vs SHE

Polymer with $[Os(dmo\text{-}bpy)_2(\underline{imid})Cl]^{2+/+}$

$E^o$ = 0.11 V vs SHE

**FIGURE 7.6** Substitution-inert Os complexes attached to polyvinylimidazole supports through extended flexible linkers. The two examples highlight the ability to alter the reduction potentials for the Os(III)/(II) couple across a very large range. In fuel cells these complexes are suited for mediating electron transfer at either cathode (high potential) or anode (low potential).

The nanorods were first derivatized with 4-aminophenyl groups by diazonium coupling with 4-nitrobenzenediazonium followed by reduction. The hydrogenase was then linked by carbodiimide coupling between the amino groups on the nanorods and carboxylate sidechains close to the distal [4Fe–4S] cluster. The electrochemistry is shown in Figure 7.7, where it is compared with the results achieved without covalent linking. Under anaerobic conditions the enzyme remains active on the electrode for periods of several weeks.

A serious problem with hydrogenases is their $O_2$ sensitivity, and for any plausible application, such as a self-powered $H_2$ sensor, it is necessary to identify and use hydrogenases that are $O_2$-tolerant. The term "$O_2$-tolerant" means that the hydrogenase can operate even in the presence of $O_2$, which is an inescapable requirement given that fuel cell membranes always have some gas permeability [36]. The reaction with $O_2$ is usually reversible for [NiFe] hydrogenases, which can form a hydroxide adduct, and irreversible for [FeFe] hydrogenases, which may be particularly susceptible to attack at the [4Fe–4S] subcluster. An enzyme that is not tolerant to $O_2$ is either permanently damaged or rapidly converted to a form that requires a long period of reductive activation to recover. An $O_2$-tolerant hydrogenase either does not react very rapidly with $O_2$ or recovers essentially immediately from $O_2$ attack [19]. Interestingly, $O_2$ tolerance has been demonstrated on the basis of the ability to use a hydrogenase in a membraneless fuel cell. A particular example is an experiment in which 3% $H_2$ in air is the mixed fuel and oxidant in a membraneless fuel cell powering a small electronic device (Fig. 7.8) [37]. An electronic clock may not demand much power, but the achievement is highly significant in terms of the hydrogenase being able to oxidize $H_2$ when there is a sevenfold excess of $O_2$.

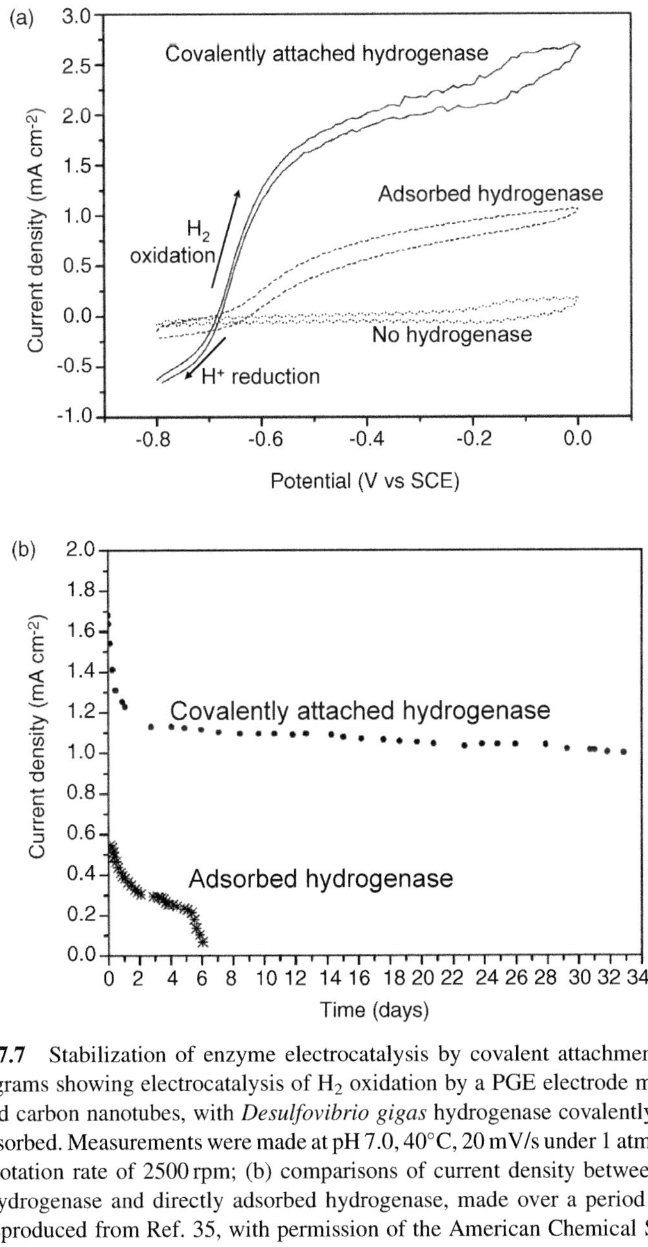


**FIGURE 7.7** Stabilization of enzyme electrocatalysis by covalent attachment: (a) cyclic voltammograms showing electrocatalysis of $H_2$ oxidation by a PGE electrode modified with multiwalled carbon nanotubes, with *Desulfovibrio gigas* hydrogenase covalently attached or directly adsorbed. Measurements were made at pH 7.0, 40°C, 20 mV/s under 1 atm $H_2$ and at an electrode rotation rate of 2500 rpm; (b) comparisons of current density between covalently attached hydrogenase and directly adsorbed hydrogenase, made over a period of a month. (Figures reproduced from Ref. 35, with permission of the American Chemical Society.)

## 7.5 CONCLUSIONS

It is difficult to predict how far fuel cells based on isolated enzymes will go in technology. The continuing search for niche applications is stimulating many new ideas, such as roles for enzyme fuel cells in self-powered implantable sensors or logic gates [38,39] The more fundamental studies are revealing interesting properties of enzymes as energy producers, relevant to physiology and biological energy production. Finally, knowledge that such important reactions as $H_2$ oxidation are

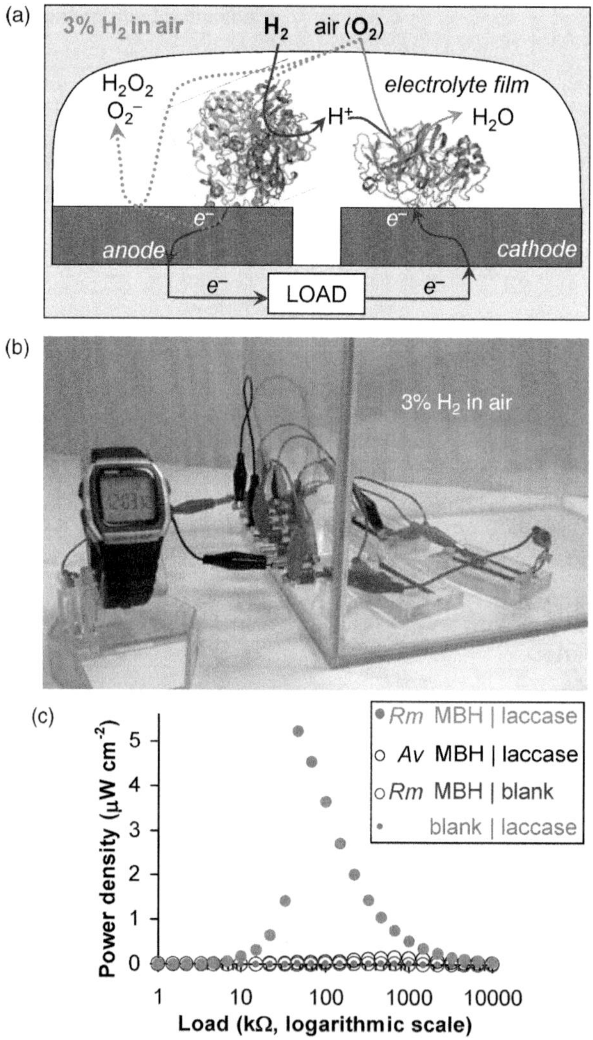


**FIGURE 7.8** A $H_2/O_2$ fuel cell based on an $O_2$-tolerant hydrogenase. This fuel cell runs on a mixture of 3% $H_2$ in air. The anode is a piece of graphite electrode modified by adsorption of the [NiFe] hydrogenase from *Ralstonia metallidurans* (*Rm*); the cathode is a piece of graphite modified by adsorption of laccase from *Trametes hirsuta*. (a) Schematic diagram; (b) actual device, with three fuel cells connected in series to provide 2.7 V; (c) power curve (power as a function of load), Av = *Allochromatium vinosum* [NiFe] hydrogenase, which is not $O_2$-tolerant; (d) voltammogram of *Rm* [NiFe] hydrogenase in the presence of $O_2$.

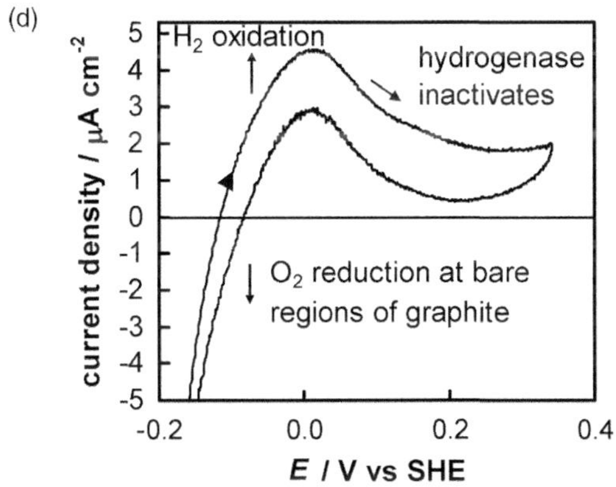


**FIGURE 7.8** *(Continued)*

electrochemically reversible at an Fe or Fe/Ni complex or that the $O_2$ overpotential is smaller at a Cu catalyst than Pt may eventually remove human dependence on precious metal catalysts. Many efforts are being made to mimic the action of enzymes; a recent example being a Cu catalyst for $O_2$ reduction [40]. It is hoped that all that will be required will be to reproduce enzyme function by creating the equivalent, exacting environments of active sites in inexpensive molecules, and combining these with materials that can enhance their density on an electrode.

## ACKNOWLEDGMENTS

Research on hydrogenases, blue copper oxidases, and enzyme fuel cells in the author's laboratory has been supported by the UK BBSRC and EPSRC.

## REFERENCES

1. I. Willner et al., Integrated enzyme-based biofuel cells—a review, *Fuel Cells* **9**(1); 7–24 (2009).
2. S. C. Barton, J. Gallaway, and P. Atanassov, Enzymatic biofuel cells for implantable and microscale devices, *Chem. Rev.* **104**(10); 4867–4886 (2004).
3. J. A. Cracknell, K. A. Vincent, and F. A. Armstrong, Enzymes as working or inspirational electrocatalysts for fuel cells and electrolysis, *Chem. Rev.* **108**; 2439–2461 (2008).
4. A. Heller et al., Miniature biological fuel cell that is operational under physiological conditions comprising enzymes, Patent Application WO WO, Therasense, Inc., (2003).
5. A. Parkin et al., Rapid and efficient electrocatalytic $CO_2$/CO interconversions by *Carboxydothermus hydrogenoformans* CO dehydrogenase I on an electrode, *J. Am. Chem. Soc.* **129**(34); 10328–10329 (2007).
6. T. Reda et al., Reversible interconversion of carbon dioxide and formate by an electroactive enzyme, *Proc. Natl. Acad. Sci. USA* **105**(31); 10654–10658 (2008).

7. Y. B. Zu, R. J. Shannon, and J. Hirst, Reversible, electrochemical interconversion of NADH and NAD(+) by the Catalytic (I lambda) subcomplex of mitochondrial NADH: Ubiquinone oxidoreductase (Complex I), *J. Am. Chem. Soc.* **125**(20); 6020–6021 (2003).
8. K. A. Vincent, A. Parkin, and F. A. Armstrong, Investigating and exploiting the electrocatalytic properties of hydrogenases, *Chem. Rev.* **107**(10); 4366–4413 (2007).
9. C. Léger and P. Bertrand, Direct electrochemistry of redox enzymes as a tool for mechanistic studies, *Chem. Rev.* **108**(7); 2379–2438 (2008).
10. M. J. S. Dewar, Alternative view of enzyme reactions, *Proc. Natl. Acad. Sci. USA* **82**(8); 2225 (1985).
11. M. Hambourger et al., [FeFe]-hydrogenase-catalysed $H_2$ production in a photoelectrochemical biofuel cell, *J. Am. Chem. Soc.* **130**, 2015–2022 (2008).
12. A. K. Jones et al., Direct comparison of the electrocatalytic oxidation of hydrogen by an enzyme and a platinum catalyst, *Chem. Commun.* 866–867 (2002).
13. J. C. Fontecilla-Camps et al., Structure/function relationships of [NiFe]- and [FeFe]-hydrogenases, *Chem. Rev.* **107**; 4273–4303 (2007).
14. G. J. Kubas, Fundamentals of $H_2$ binding and reactivity on transition metals underlying hydrogenase function and $H_2$ production and storage, *Chem. Rev.* **107**(10); 4152–4205 (2007).
15. B. J. Lemon and J. W. Peters, Photochemistry at the active site of the carbon monoxide inhibited form of the iron-only hydrogenase (CpI), *J. Am. Chem. Soc.* **122**(15); 3793–3794 (2000).
16. B. J. Lemon and J. W. Peters, Binding of exogenously added carbon monoxide at the active site of the iron-only hydrogenase (CpI) from *Clostridium pasteurianum*, *Biochemistry* **38**(40); 12969–12973 (1999).
17. A. K. Jones et al., Enzyme electrokinetics: Electrochemical studies of the anaerobic interconversions between active and inactive states of *Allochromatium vinosum* [NiFe]-hydrogenase, *J. Am. Chem. Soc.* **125**; 8505–8514 (2003).
18. K. A. Vincent et al., Electrocatalytic hydrogen oxidation by an enzyme at high carbon monoxide or oxygen levels, *Proc. Natl. Acad. Sci. USA* **102**; 16951–16954 (2005).
19. J. A. Cracknell et al., Enzymatic oxidation of $H_2$ in Atmospheric $O_2$: The electrochemistry of energy generation from trace $H_2$ by aerobic microorganisms, *J. Am. Chem. Soc.* **130**; 424–425 (2008).
20. M. W. Kanan and D. G. Nocera, In situ formation of an oxygen-evolving catalyst in neutral water containing phosphate and $Co^{2+}$, *Science* **321**(5892); 1072–1075 (2008).
21. E. I. Solomon et al., $O_2$ and $N_2O$ activation by bi-, tri-, and tetranuclear Cu clusters in biology, *Acc. Chem. Res.* **40**(7); 581–591 (2007).
22. N. Mano et al., Oxygen is electroreduced to water on a "wired" enzyme electrode at a lesser overpotential than on platinum, *J. Am. Chem. Soc.* **125**(50); 15290–15291 (2003).
23. V. R. Stamenkovic et al., Improved oxygen reduction activity on $Pt_3Ni(111)$ via increased surface site availability, *Science* **315**(5811); 493–497 (2007).
24. Y. Sun and S. C. Barton, Methanol tolerance of a mediated, biocatalytic oxygen cathode, *J. Electroanal. Chem.* **590**(1); 57–65 (2006).
25. T. Chen et al., A miniature biofuel cell, *J. Am. Chem. Soc.* **123**(35); 8630–8631 (2001).
26. P. Kavanagh et al., Performance of a glucose/$O_2$ enzymatic biofuel cell containing a mediated *Melanocarpus albomyces* laccase cathode in a physiological buffer, *Fuel Cells* **9**(1); 79–84 (2009).

27. F. Tasca et al., Highly efficient and versatile anodes for biofuel cells based on cellobiose dehydrogenase from *Myriococcum thermophilum*, *J. Phys. Chem. C* **112**(35); 13668–13673 (2008).
28. B. M. Hallberg et al., Crystal structure of the flavoprotein domain of the extracellular flavocytochrome cellobiose dehydrogenase, *J. Mol. Biol.* **315**(3); 421–434 (2002).
29. A. Oubrie et al., Crystal structure of quinohemoprotein alcohol dehydrogenase from *Comamonas testosteroni*—structural basis for substrate oxidation and electron transfer, *J. Biol. Chem.* **277**(5); 3727–3732 (2002).
30. C. F. Blanford et al., Efficient electrocatalytic oxygen reduction by the "blue" copper oxidase, laccase, directly attached to chemically modified carbons, *Faraday Discuss.* **140**; 319–335 (2008).
31. M. Delamar et al., Modification of carbon fiber surfaces by electrochemical reduction of aryl diazonium salts: Application to carbon epoxy composites, *Carbon* **35**(6); 801–807 (1997).
32. A. Heller, Electron-conducting redox hydrogels: Design, characteristics and synthesis, *Curr. Opin. Chem. Biol.* **10**; 664–672 (2006).
33. Y. Kamitaka et al., Fructose/dioxygen biofuel cell based on direct electron transfer-type bioelectrocatalysis, *Phys. Chem. Chem. Phys.* **9**(15); 1793–1801 (2007).
34. J. W. Gallaway and S. A. Calabrese Barton, Kinetics of redox polymer-mediated enzyme electrodes, *J. Am. Chem. Soc.* **130**(26); 8527–8536 (2008).
35. M. A. Alonso-Lomillo et al., Hydrogenase-coated carbon nanotubes for efficient $H_2$ oxidation, *Nano Lett.* **7**(6); 1603–1608 (2007).
36. M. Ludwig et al., Oxygen-tolerant $H_2$ oxidation by membrane-bound [NiFe]- hydrogenases of Ralstonia species: Coping with low-level $H_2$ in air, *J. Biol. Chem.* **284**; 465–477 (2009).
37. K. A. Vincent et al., Electricity from low-level $H_2$ in still air—an ultimate test for an oxygen tolerant hydrogenase, *Chem. Commun.* **2006** (48); 5033–5035 (2006).
38. A. Heller, Potentially implantable miniature batteries, *Anal. Bioanal. Chem.* **385**(3); 469–473 (2006).
39. L. Amir et al., Biofuel cell controlled by enzyme logic systems, *J. Am. Chem. Soc.* **131**(2); 826–832 (2009).
40. M. S. Thorum, J. Yadav, and A. A. Gewirth, Oxygen reduction activity of a copper complex of 3,5-diamino-1,2,4-triazole supported on carbon black, *Angew. Chem. Int. Ed.* **48**(1); 165–167 (2009).

## CHAPTER 8

# Approach to Microbial Fuel Cells and Their Applications

JUAN PABLO BUSALMEN

Laboratorio de Bioelectroquímica, INTEMA(CONICET), Mar del Plata, Argentina

ABRAHAM ESTEVE-NUÑEZ

Departamento de Química Analítica e Ingeniería Química, Universidad de Alcalá, Madrid, Spain

JUAN MIGUEL FELIU

Instituto de Electroquímica, Universidad de Alicante, Alicante, Spain

## 8.1 INTRODUCTION

As outlined in this book, an important group of research papers in electrochemistry deal with the investigation of adsorbed layers on electrode surfaces. As a first simple case, almost universally present in electrode–solution interfaces, anion-specific adsorption on metal electrodes, including single crystals, is a classical topic in textbooks [1]. The complexity of the problem increases when dealing with further adatom adsorption, in which hydrogen and underpotential deposition of foreign metals on higher-workfunction metallic substrates could be taken as examples, because in this case anion adsorption will also take place [2,3]. Increasing the complexity of the adsorbate, the adsorption of small molecules such as CO or NO, on well-characterized electrode surfaces has supplied a number of probe reactions whose quantitative characteristics, including molecular coverage, have to be consistently incorporated in the scenario defined for simpler systems [4,5]. Indeed, the reactions undergone by these adsorbed molecules are now considered as model surface reactions. Increasing the complexity of the adlayer, the behavior of larger molecules such as benzene and

*Fuel Cell Science: Theory, Fundamentals, and Biocatalysis*,
Edited by Andrzej Wieckowski and Jens K. Nørskov

its derivatives show interesting adsorption properties when adsorbed on surfaces at a monolayer level. This knowledge could be applied in molecular electronics [6] and serve as the basis for a more in-depth understanding of electropolymerization reactions [7].

Adsorbed dendrimers on electrode surfaces [8] could be consider next as species with a larger degree of chemical complexity, equivalent to the formation of self-assembled monolayers (SAMs). This latter case merits special attention, not only because of the charge transfer–distance effects [9] but also because they supply a soft environment to adsorb larger molecules such as enzymes [10]. In this respect, at increasing again the chemical complexity, the direct adsorption of these complex molecular entities on metallic electrodes becomes affected by the structural modifications that the adsorption step can cause on the catalytic activity for a particular reaction; the strong adsorbate–substrate interaction denaturalizes the activity of the catalyst. This is resolved in some cases by adsorbing these complex molecules over the adequate SAM, which will cause less damage on the three-dimensional structure of the relevant molecule [10].

Increasing the size of the adsorbates, we could include all the cases of nanoparticles attached on conducting substrates, as used in fuel cells. The size of nanoparticles spreads within 1–2 and 50–100 nm, thus enabling us to identify size effects as well as structural effects in electrocatalysis [11,12]. Although the chemical composition of nanoparticles usually is limited to one to three different elements [13], it should be noted that the surface structure of the nanoparticles also contributes to the reactivity.

The step beyond in structural complexity actually under the scope of electrochemists is adsorption of whole bacteria, a process that can be considered as the result of a molecular interaction between the relevant surface groups on this organism and the metal. Although this interaction is considered as electrostatic, specific surface bonding cannot be discarded a priori. Bacteria have some advantages over other large (1–3-μm) adsorbates because they are living systems that are stable for a long time and are able to maintain relevant electron transfer groups in its original environment, so denaturalization of the reactive groups is preserved, allowing the electrochemical analysis

In this chapter we address the problem of bacteria attached to electrodes as the origin of the so-called bacterial or microbial fuel cells. Although the language and literature sources are different from those used by the researchers in electrocatalysis, the underlying concepts are quite equivalent; we need to have some species adsorbed to and electrode in order to gain electrochemical information, and the electroactive groups should be close to the surface to be detected. Much information exists on phenomenological aspects, mainly for bacteria adsorbed on carbon substrates, but, in order to gain molecular information, such as that derived from spectroscopic data, more convenient electrode supports are required. Understanding of the electron transfer processes at the bacteria–electrode interface is still in its infancy and clearly represent a bottleneck for the possibility to use microbial fuel cells (MFCs) for practical applications in the future. Most of available information is clearly coming from biologists and engineers, while electrochemical work is still lacking. Consequently, the aim of this chapter is to attract the attention of those interested in bioelectrocatalysis on the brand-new field of bacterial electrochemistry.

## 8.2 MICROBIAL FUEL CELL FUNDAMENTALS

A *microbial fuel cell* (MFC) is a quite simple device. As shown in Figure 8.1, it comprises an anodic anaerobic compartment physically separated from the cathodic one by an ion exchange membrane. Usually carbon or graphite electrodes are located in both the anodic and the cathodic compartments, connected through an external circuitry that allows the electronic circulation and the control of usable electric current. At the anode, the "burning" or oxidation of an organic fuel produces electrons that reach the cathode to be used in reduction reactions, ideally the reduction of dissolved oxygen from a practical viewpoint. MFCs are not different from other fuel cells, except because *microbes*, rather than conventional chemical catalysts, facilitate the electrode reactions.

Bacteria have some minimal requirements to develop; in addition to several micronutrients, they require a carbon source, an electron donor, and an electron acceptor. All of these compounds are generally found in wastewaters, and this is the reason why MFCs find most popular application in the production of electricity from these synthetic (human-made) environments. During electricity production, dissolved organic material (DOM) in the wastewater is both the source of both carbon and the electron, while the electrode in the anodic chamber acts as the electron acceptor (see Section 8.3). Polarization of the anode during current harvesting usually selects an electroactive microbiological population from the indigenous microbiota of the waste, leading to the enrichment in the population of useful microorganisms. Owing to the metabolic diversity of the microbial world, many different oxidation processes can be

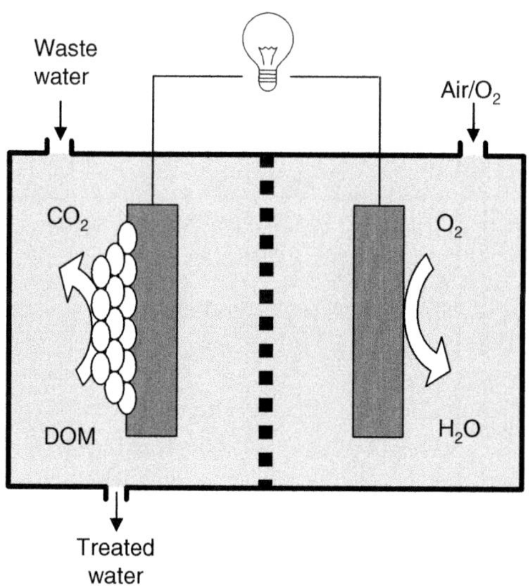


**FIGURE 8.1** Schematic representation of a microbial fuel cell producing electricity from wastewater. A microbial biofilm (white symbols) on the electrode in the anodic compartment catalyzes the oxidation of dissolved organic material (DOM) to $CO_2$ using the reduction of oxygen to water as the cathodic counterpart.

exploited to produce electricity (see Section 8.4), as the nature of the compound(s) to be oxidized is one of the factors in determining the composition of the final selected population. The metabolic routes available in that microbial population will, in turn, determine the degradation sequence of the compound(s) and the quality of intermediates and byproducts. Taking this into account, it seems necessary to know the identity of the different catalyst strains for a complete definition of a microbial fuel cell.

For research purposes a simplified version of a MFC is usually constructed in the laboratory by using a three-electrode H cell that contains the working electrode support (usually carbon) covered with a single bacterial strain (typically *Geobacter sp.*) that harvests energy from an artificial waste solution containing some organic material (acetate, fumarate, etc.) as the unique carbon and electron source to the working electrode. This simple device helps in recognizing a unique distinguishing feature of MFCs—the fuel is not directly oxidized ON the carbon electrode, but through complex biological machinery. In other words, there is no direct physical contact between the fuel and the electrode during the oxidation process, which highlights the importance of understanding biological mechanisms that participate in the catalytic process and the transport of electrons to the collecting electrode.

The basic sequence of events in the travel of electrons from acetate to the electrode can be summarized as follows (Fig. 8.2): (1) the biological oxidation of acetate to

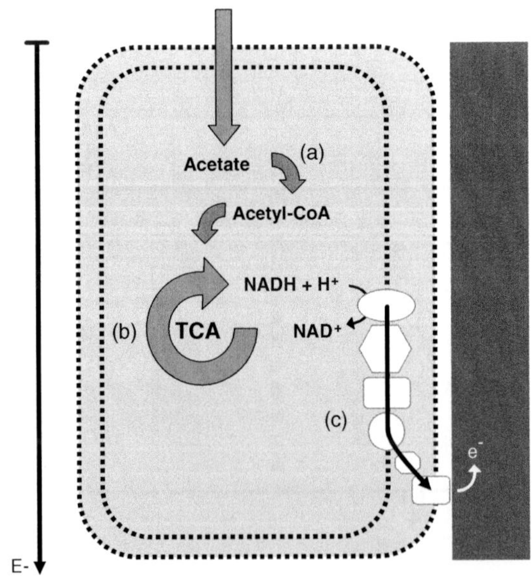


**FIGURE 8.2** Schematic representation of bacterial metabolic pathways through which electrons are channeled from acetate to the electrode. (a) production of acetyl-CoA; (b) tricarboxylic acid (TCA) circle; (c) respiratory electron transport chain. The arrow indicates the change in potential energy of transported electrons.

produce acetyl–coenzyme A; (2) the transport of electrons along the tricarboxylic acid cycle, in which the main electron-bearing product is reduced nicotinamide adenine dinucleotide ($NADH_2$); and (3) the stepped transport of electrons along the respiratory chain of bacteria that receive electrons from $NADH_2$ to transport them upstream to a final electron acceptor of higher potential. Biological processes coupled to this transport produce energy in the form of adenosine-5′-triphosphate (ATP) to support cell demands consuming a portion of the potential energy of traveling electrons (Fig. 8.2). At the end of the respiratory chain a final reductase enzyme interacts with the electrode to connect this kind of internal "biological circuit" to the electric circuit, transforming remaining potential energy in electrons that flow in the external circuit.

## 8.3 THE BACTERIA–ELECTRODE INTERFACE

### 8.3.1 Bacterial Adhesion to Electrodes

On approach to the surface of the conducting support, the adhesion behavior of a bacterial unit is dictated by poorly specified variables as the relative charge and hydrophobicity of the interacting surfaces (i.e., the carbon surface and the outermost bacterial membrane layer) [14]. The balance of these forces at the interface may determine adhesion (either reversible or irreversible) or repulsion. Also, specific interactions are known to occur that can sometimes overcome the unfavorable energetic balance of adhesion [15,16]. Specific interactions are determined mainly by the molecular composition of the bacterial surface, which, in turn, depends on the structure of the outermost bacterial envelope. Two large groups of bacteria are recognized on the basis of their response to the staining method developed by Hans C. Gram in 1884 [17]. Gram-positive bacteria bearing a cell wall of peptidoglycan surrounding the cytoplasmic membrane, and Gram-negative ones having a three-layer envelope composed by (inside to outside) the cytoplasmic membrane (CM), an intermembrane (or periplasmic) space, and the outer membrane (OM). Until now only Gram-negative bacteria have been shown to electrically interact at the electrode interface, although this possibility cannot be neglected for Gram-positive ones [18].

The outer membrane of Gram-negative bacteria is a bilipidic layer in which amphipatic lipids are located with their fatty acid chain directed to the interior of the layer and their polar heads facing either the periplasmic space or the bacterial cell exterior (i.e., the aqueous solution medium). The external face of the OM consists of lipopolysaccharides (LPSs) with a lipid A anchoring to the membrane and a hydrophilic carbohydrate domain that protrudes to the surrounding liquid [19]. Lipid A consists of fatty acids linked to a phosphorylated disaccharide (two glucosamine residues), while the carbohydrate side includes an oligosaccharide core usually substituted with phosphoric acid, ethanolamine, phosphoethanolamine, or acetyl groups (Fig. 8.3a). Finally there are O-specific polysaccharide chains that extend into the solution, consisting of repeated oligosaccharide units whose structure may significantly vary in different strains [19] (Fig. 8.3a).

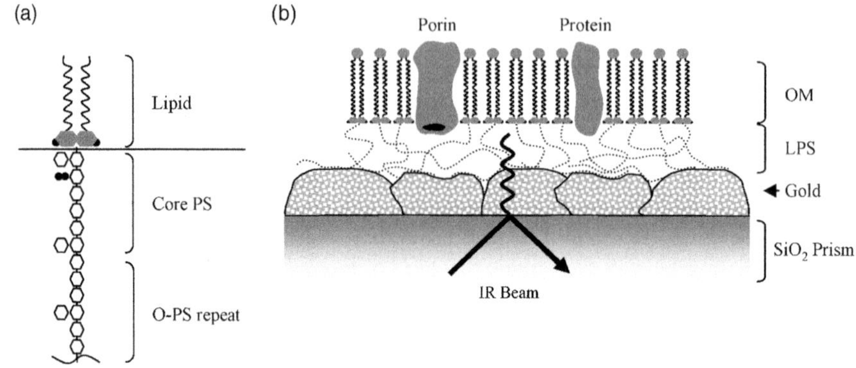


**FIGURE 8.3** (a) Schematic representation of a typical lipopolysaccharide molecule showing lipid A, the core polysaccharide (Core PS), and the oligopolysaccharide repeats (O-PS); (b) scheme of the bacteria–thin-film gold electrode interface in the ATR-SEIRAS configuration showing the interaction with the attenuated IR evanescent wave (OM, outer membrane; LPS, lipopolysaccharides; • phosphate groups.) (Reproduced with permission from Ref. 22.)

Charged groups in these structures are mainly carboxyl, phosphate, and amino groups that dissociate or protonate depending on pH, giving rise to a net electrostatic charge at the bacterial surface. With only a few exceptions, the surface of bacteria bear a negative charge, since the number of carboxyl and phosphate groups exceeds that of amino groups [14]. This is the case with *Shewanella* and *Geobacter* species, representatives of electrode-reducing organisms [20].

Adhesion to a conducting surface has been studied in detail for the case of *Pseudomonas fluorescens* (Gram-negative) adhering to a gold thin-film electrode. The study was performed at open-circuit potential as well as under polarization to show the influence of nonspecific electrostatic interactions [15]. In general, negatively charged bacteria readily adhered to positively charged electrodes and were repulsed from negatively charged ones, although irreversible adhesion was also sometimes observed at negative potentials in spite of the electrostatic repulsion. This can be justified by the specific bonding of bacterial surface molecules to Lewis sites at the gold surface.

Total internal reflection aqueous fluorescens (TIRAF) measurements indicate that a thin layer of water remains between the cells and the surface [21] during bacterial adhesion, but more sensitive ATR-SEIRAS measurements indicate that some water is displaced from the surface as a result of adsorption of contacting molecules [22]. These studies clearly point to the possible occurrence of specific interactions that can overcome an electrostatic barrier if present.

Contacting molecules have been identified by surface-enhanced IR absorption spectroscopy (SEIRAS) [22]. An schematic representation of the interface is presented in Figure 8.3b. In accordance with results reported by others for *P. aeruginosa* using conventional ATR on Ge or ZnSe crystals [23,24], the main features in the IR spectra of cells adhered to gold were those related to the presence of proteins (amide I

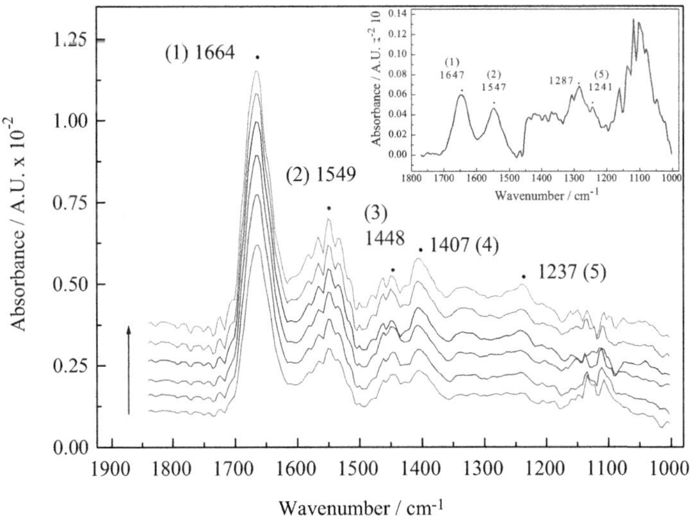


**FIGURE 8.4** Fingerprint region of ATR-SEIRAS spectra obtained after 6, 8, 10, 12, 15, and 20 min of exposure of a thin-film gold electrode to a concentrated suspension of *Pseudomonas fluorescens*. The arrow indicates the direction of time increment. Inset shows conventional ATR spectra of a concentrated suspension of *P. fluorescens* on $SiO_2$. Approximate positions corresponding to (1) amide I, (2) amide II, (3) $CH_2$, (4) $\nu_s COO^-$, and (5) amide III/P=O bands are indicated in the figure. Spectra were referred to those obtained in the absence of bacteria. (Reproduced with permission from Ref. 22.)

and II bands in Fig. 8.4); IR-absorbing vibrations at 1100–1200 $cm^{-1}$ were also observed to a minor extent, known to be related to the hydroxyl groups in carbohydrates [24].

Polarizing a conducting surface modifies its surface charge. Depending on the applied potential as relative to the potential of zero charge (PZC) of the material, polarization may prevent or favor the adsorption of ionic or dipolar elements in the solution, including bacteria [15,25]. Although this problem has not been investigated yet, it may be of importance in selecting the most appropriate external load value for MFC startup and operation. As higher the polarization potential of the anode increases, the electrostatic contribution to initial bacterial adhesion will change from repulsive at potential negative to the PZC, to attractive at more positive potentials. The values of the PZC of different metal surfaces that can be used as bacterial supports is well established from interfacial studies [26,27]. In the case of carbon, the existence of a well-defined differential capacitance minimum has been reported [28]. However, this does not correspond to the usual Gouy–Chapman minimum for the carbon–solution interface. The minimum is the result of the complexity of the carbon materials, which contributes with two other capacitive terms in the carbon side, the space charge and the charge in surface states [29] to the whole measurement. This consideration should be taken into account to understand bacterial adhesion from an electrostatic viewpoint.

Both *Shewanella* and *Geobacter* species have been found to be electronegative at circumneutral pH [20], suggesting a favorable adhesion balance with positively charged electrodes. In fact, the adhesion of *Shewanella* to hematite was shown to be governed by electrostatic interactions, as seems to be the case also for *G. sulfurreducens*, whose adhesion to graphite increases with the electrode potential [30]. Electronegativity of *Shewanella* species is assigned to abundant phosphorylation of the LPS core that, due to the screening by the external oligosaccharide chains (see Fig. 8.3 for reference), renders a relatively low surface charge. O-side chains and PS are also responsible for a net hydrophilic repulsion with hematite that are low enough to prevent adhesion [20].

As previously shown for *P. fluorescens*, the ATR spectra of *G. sulfurreducens* is dominated by amide bands, emphasizing the importance of protein molecules in bacterial adhesion to electrodes [31] (Fig. 8.5). Both *Shewanella* and *Geobacter* bear a great amount of proteins at the cell surface that would increase its hydrophobic character. Nevertheless, it has been argued that surface proteins in *Shewanella* may be

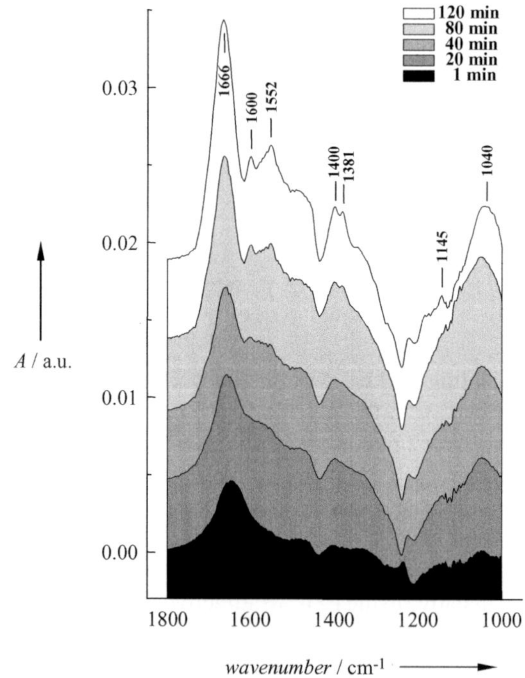


**FIGURE 8.5** Attenuated total reflection–SEIRAS spectra during the adsorption process of cells on a thin-film gold electrode. The electrode was initially polarized to 0.2 V in the acetate-containing electrolyte in a $N_2$ : $CO_2$ (80 : 20) atmosphere. After acquisition of the spectrum corresponding to the electrode–electrolyte interface, a bacterial suspension (109 cells $mL^{-1}$) was added to the cell and current produced by bacteria was continuously measured; single-beam spectra were recorded with a 5 min time interval after the addition of cells (au=arbitrary units). (Reproduced with permission from Ref. 56.)

exceptionally more hydrophilic than those of other Gram-negative bacteria [20]. The same can be true for *Geobacter* surface proteins, as suggested by their outside location. Indeed, hydrophilicity can be interpreted as a requirement to ensure the poor association with the outer membrane exhibited by some *Geobacter* proteins [32] that facilitate their physiological function as connectors to external solid electron acceptors. Of primary importance for the purposes of this chapter, the interfacial availability or demand of electrons at increasing imposed overpotentials may facilitate the heterogeneous charge transfer with electroactive compounds in the cell surface, giving rise to reduction or oxidation of cell components, respectively. These topics are discussed below.

### 8.3.2 Interfacial Charge Transfer

It has been recognized thus far that microbial metabolism has the potentiality for useful biotechnological applications and that electrical interfacing with a bacterial cell would open a variety of possibilities to take advantage of that potentiality. Consequently, the study of charge transfer across the bacteria–electrode interface has attracted the attention of pioneer researchers who followed this originally Utopian line of investigation. It was the case of Milton J. Allen and coworkers during the 1960s, whose creative work illustrates the effort made and provides very interesting guidelines for actual research [33–39].

A simple voltammetric analysis (Fig. 8.6) shows why dielectric properties of biological membranes have strongly limited practical opportunities for further developments for several decades [40]. No charge transfer processes other than those

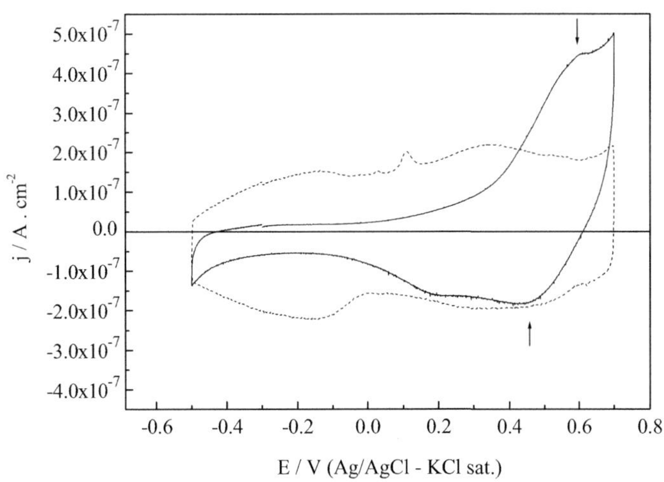


**FIGURE 8.6** Cyclic voltammograms on a polyoriented single-crystal gold electrode in 0.5% NaCl, pH 7, in the presence (—) and absence (- - -) of adsorbed bacterial cells. Adsorption was performed by exposing the electrode to a concentrated bacterial suspension during 10 min. Scan rate: 0.01 V/s. The arrows indicate the peak position for the bacterial redox processes. (Reproduced with permission from Ref. 22.)

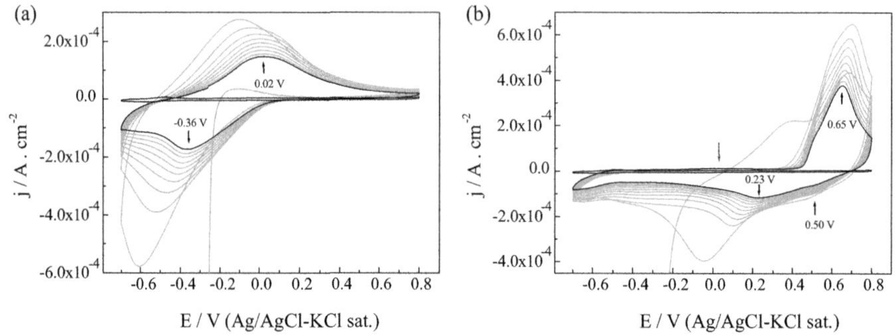


**FIGURE 8.7** Cyclic voltammograms obtained on graphite electrodes after 18 h exposure to a fumarate-limited culture of *G. sulfurreducens* under continuous polarization at (a) 0.1 V and (b) 0.6 V. The estimated peak potential of redox processes is indicated. The red arrow indicates the position of a very low-amplitude peak at 0.02 V. Gray lines: cycles 1–9. Blue line: cycle 10. Black line: cycle 10 of a control experiment performed at the open-circuit potential. (Reproduced with permission from Ref. 30.)

derived from the oxidation and denaturing of molecules in the cell surface at high potential [22] (Fig. 8.6) can be detected between typical bacterial cells and the electrode substrate.

Surprisingly, discoveries done more recently have opened the way for the expected developments and changed the paradigm in the field of microbial electrochemistry; electroactive microorganisms with the outstanding ability to oxidize organic compounds while exchanging electrons with external solid surfaces have been isolated and their properties investigated [41]. A number of studies have appeared presenting voltammograms as those shown in Figure 8.7, in which heterogeneous charge transfer to and from bacterial cells can be observed.

## 8.4 WHOLE-BACTERIUM CELL ELECTROCHEMISTRY

### 8.4.1 Bacterial Film Voltammetry (BFV)

At present, it is accepted that electron transport to the electrode is performed by redox proteins present in the external face of bacterial cells that are connected to the bacterial respiratory chain (where metabolic energy is produced) through a number of intermediate transporters [42]. Consequently, electrons collected by the electrode are those coming from central metabolic pathways (Fig. 8.2), and the measured current is directly related to the rate of respiration of the bacterial population.

Communication to the electrode may proceed by direct electron transfer (DET) or mediated electron transfer (MET) depending on the involved organism; excellent updated revisions on these mechanisms have appeared in the literature [43–45]. In this section we focus on direct electron transfer, although most of interpretations done here can be extended to the case of indirect electric communication.

Interfacial charge transfer can be studied with classical theoretical and experimental tools of electrochemistry, but the interpretation of results in the case of evaluating the whole bacteria–electrode interface is conditioned by the unique properties of the biological counterpart. In the following paragraphs we discuss some of the ways in which properties of bacterial cells and molecules can condition the electrochemical response.

Analysis of the bacteria–electrode interface has been approached using classical electrochemical tools as voltammetry, chronoamperometry, chronopotentiometry, and impedance spectroscopy and the knowledge of the process of electricity production has grown relatively rapidy since 2007 or so [30,46–54]. In a typical experiment, a biofilm is grown on the working electrode of a two-compartment electrochemical cell by exposing it to an appropriate culture media inoculated with the bacteria of interest. Importantly, the electrode must be polarized to a (preferentially) positive potential and no electron acceptors (including oxygen) have to be present in the solution side to favor the bacterial recognition of the charged electrode itself for that function. Compartments are usually separated by a cation exchange membrane for two purposes: (1) to allow ionic current flow between anode and cathode and (2) to prevent oxygen migration from the cathodic side. Polarization required for biofilm growth can be applied using a potentiostat, or simply connecting anode and cathode through an external resistor of low value (e.g., 500 Ω). In the former case the anodic potential can be suitably controlled against a reference electrode, while in the latter it will depend mainly on the resistor value, the kinetics of the cathodic reaction, and the biological response. Positive anodic potentials cannot be implemented using this latter method.

Carbon electrodes have been the most popular option for testing MFC's performance and applications [55], while gold electrodes have also been selected for fundamental studies [56]. It is important to recall that, although graphite is a convenient material for development of applications, it has a complex surface with a number of different chemical groups that may interfere in the electrochemical analysis. On the other hand, gold is very well characterized and presents a negligible electrochemical activity in the potential window of interest. Indeed, most of electrochemical and spectroelectrochemical tools can be performed with gold electrodes, allowing a much more detailed analysis of fundamental bacterial processes related to electricity production. In this respect, fundamental studies allowing the collection of data to understand the electrochemical behavior of bacteria would be more conveniently gained from gold substrates.

Duplication times of bacteria as *Geobacter* are in the range on tens of hours under suitable conditions. This long delay makes it possible to analyze processes related to changes in the initially adhered cells at relative short times of polarization [30]. It is also possible to perform these studies in a fully grown biofilm [46], provided that enough growth time is given.

Since the biofilm uses the electrode as the final electron acceptor, growth will be confined to a finite layer of cells with electrical contact to the surface. It is usually not a bacterial monolayer but a biofilm composed of bacterial clusters, channels, and voids, following the classical model of biofilm structure [57–59]. The biogenic current from electrode-reducing biofilms is directly related to its protein content, at least for biofilm thicknesses below ~50 μm [60], likely showing that the whole population is electrically wired with the electrode serving as the electron collector. Cells in intimate contact with the electrode are known to be wired to the electrode by external cytochromes [56], but the electrical communication of cells located at a long distance from the surface is not known yet. A role is presumed for special type IV pili to provide structural support and perhaps electrical conduction [60,61], but it is clear that much research is needed on these topics. The participation of external cytochromes in the cell-to-cell electrical communication is a much more attractive possibility, reinforced by known general features of multiheme cytochromes. New evidences has been provided indicating that OmcZ, a *c*-type external cytochrome with seven heme groups found in *Geobacter*, would be responsible for homogeneous electron transfer inside biofilms [61,62]. Although data on this protein are still scarce, PpcA, a triheme *Geobacter* cythochrome, has been found packaged in such a way that allows a close interaction between heme groups of adjacent molecules. Heme IV of one molecule seems to be able to transfer electrons to heme I in the adjacent one, lending structural support for an intermolecular electron transfer pathway [63].

Having a confined layer of active cells for the catalytic conversion of an organic substrate, such as acetate into electricity, resembles the analogous case of protein film voltammetry (PFV) [64] and offers an unprecedented opportunity to explore the catalytic properties of the layer, as well as the mechanism of the heterogeneous electron transfer. Differently from the situation of isolated proteins, where unfolding of the molecules after contacting the electrode is a major constraint [65], the electrochemical response of whole bacterial cells seems to be rather stable, probably benefiting from the preservation of electroactive molecules in their biological environment. Also, the continuous turnover of cell material gives a chance for the monitoring of bacterial response over the time required to perform the experiments (e.g., differential expression depending on polarization potential [30]).

Electrochemistry of living cells may differ from that of isolated molecules in a way that is not completely explained with the present knowledge of DET [47], but at least two practical situations with different voltammetric response can be distinguished: (1) the analysis is performed in the absence of any electron donor, so cells are starved and the electrochemical response is limited to surface redox processes; and (2) the analysis includes a source of electrons, and carbon and catalytic processes can be evaluated.

#### 8.4.1.1 Starved Bacterial Cells

Voltammetric signals of starved electrode-reducing bacteria typically consist of at least one pair of well-shaped oxidation and reduction peaks (Fig. 8.7a). In the absence of any electron flow from metabolic processes, these are the "fingerprint signals" of electroactive molecules at the bacterial cell surface, and their dependence on potential and time can be studied in detail. Importantly, their formal potential can be estimated as the average of peak potentials or through more precise chronopotentiometric measurements [30] and compared to those measured for isolated adsorbed proteins.

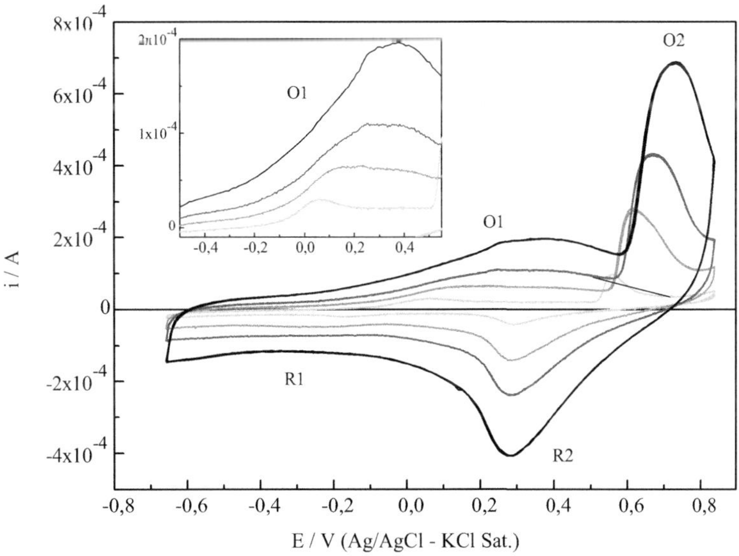


**FIGURE 8.8** Cyclic voltammograms of *G. sulfurreducens* whole cells on a thin-film gold electrode after 8 h exposure under continuous polarization at 0.6 V. Scan rate: (—) 0.005, (—) 0.010, (—) 0.050, and (—) 0.100 V/s. O—oxidation; R—reduction processes. Inset presents scale magnification to show the O1 process in detail. (Reproduced with permission from Ref. 66.)

Beyond the apparent simplicity of bacterial redox response, simple voltammetric analysis at increasing scan rates reveals that the main redox processes at the cell surface may include several steps with different degree of reversibility (Fig. 8.8). Such a response is presumed to be due to the cooperative and sequential electrochemical transition of several cytochromes (i.e., an enzymatic complex), or to the participation of multiheme cytochromes in the process [66].

As discussed previously [66,67], bacteria have not been evolutionary pressed to produce electricity, but to transfer electrons to insoluble acceptors, which in terms of heterogeneous electron transfer is almost the same. Perhaps the distinctive property of electrodes as substrates for bacterial respiration is that their potential can be varied in a continual way, aiming to fulfill the requirements of the running fuel cell process or the user demands. It has been demonstrated that potential changes trigger a bacterial response to adapt their electron transport mechanism optimizing the biological energy yield, producing wiring molecules with different formal potential [30,66]. This result clearly shows that the biological counterpart of a MFC plays a central role in determining the range of operative fuel cell potentials.

*Is there a biological limit for the fuel cell potential?* The composition of the cell surface is complex and includes a number of redox elements that, either participating or not in the respiratory process, can be accessible to the electrode giving rise to multiple signals; for example, 30 external cytochromes have been putatively located in the outer membrane of *Geobacter sulfurreducens*, while 18 have been described for *Shewanella oneidensis* evidencing the redox complexity of the bacterial surface. Redox pairs so far reported in *Geobacter* from electrochemical experiments are distributed in a wide potential window with half-cell potentials ranging from −0.51 [47] to −0.08 V [30] (vs. Ag/AgCl – KCl sat.) [or −0.22 V to 0.14 V vs. standard hydrogen electrode (SHE)] on carbon electrodes. From a biological perspective, to be involved in electricity generation, a redox pair must have a formal potential compatible with the production of metabolic energy. This means that a difference of at least 0.15 V should exist between the redox pair and that of the $NADH_2/NAD^+$ couple (−0.32 V SHE) to allow the synthesis of at least one ATP molecule (Fig. 8.9). The limit can be moved even higher if the observation that *Geobacter metallireducens* most likely conserves energy by the transference of electrons from $NADH_2$ to menaquinone (MQ; −0.07 V SHE) [68] is considered. Indeed, well-known external solid electron acceptors have a formal potential much more positive than MQ, likely indicating that initial steps in the upstream electron transport to insoluble acceptors may occur at the inner membrane, thus putting the limit above −0.07 V for the respiration of electrodes with enough energy yield to sustain growth. From the MFC perspective, provided that the structure of the initial

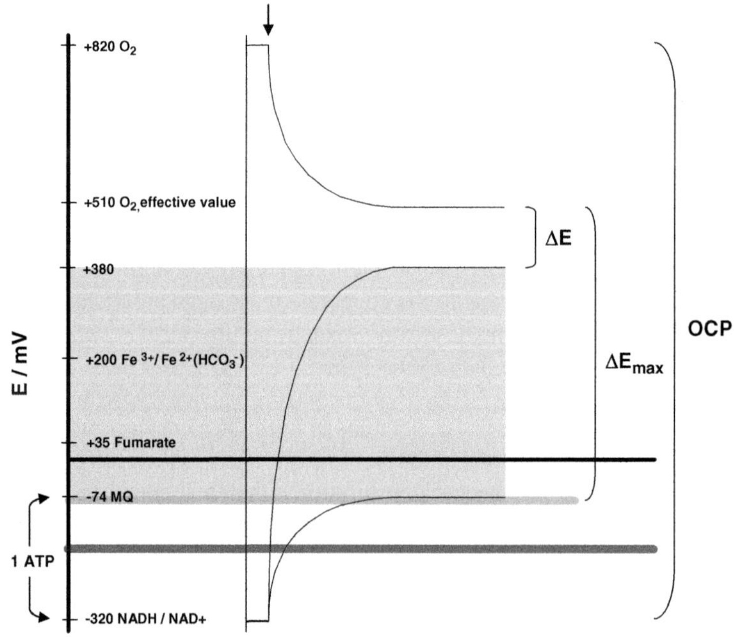


**FIGURE 8.9** Schematic representation showing the evolution of typical fuel cell potentials on connection. The dark gray line indicates the lower anode potential that allows the production of one ATP molecule. The dashed zone indicates the potential range that can be expected from a bacterial population whose lower limit locates positive to the menaquinone potential (light gray line), after the first conservation energy site. The arrow indicates the moment of cell connection.

steps in the electron transport chain of known bacteria is similar to those described above, anode potentials negative to −0.07 V SHE (−0.290 V Ag/AgCl) are not expected to be sustained by any bacterial population through DET. This is in good agreement with the experimental values usually found in practical MFCs [69]. Maximum redox potentials exhibited by bacterial populations, on the other hand, are around 0.38 V (SHE) (0.18 Ag/AgCl) [70,71]. Considering an effective potential of 0.51 V for the oxygen reduction reaction at pH 7 [51], the cell potential may vary from 0.58 to 0.13 V (Fig. 8.9).

#### *8.4.1.2 Metabolizing Bacteria*

In the presence of an electron donor, bacterial couples catalyzing current production from organic material can be identified by its electrocatalytic responses. The behavior parallels the case of determining catalytic sites in an adsorbed enzyme [64]. Provided that the source of electrons is present in the culture media, a continuous flow of "reduction power" (i.e., electrons) to the electrode reductases is ensured from the metabolic activity of cells. It produces the rereduction of electrode-wiring molecules immediately after the electrochemical oxidation occurs, yielding a catalytic turnover with a typical sigmoidal curve such as that shown in Figure 8.10a.

This behavior is similar to that observed in electrocatalysis for some particular cases. A clear example is the reduction of oxygen to peroxide on carbon electrodes modified by grafted quinones [72]. On modification of the quinone substituents, the redox potential of the surface couple shifts, following its particular HOMO-LUMO (highest occupied–lowest unoccupied molecular orbital) orbital energy. By selecting the appropriate quinone derivative, the reduction of oxygen can take place at more positive potentials on the modified electrodes [73]. Perhaps this strategy could be transferred to the bacterial layers, because conditioning at different potentials leads to the observation of different redox processes as noted previously (Fig. 8.7).

*Is there a biological limit for electricity production?* In the experiment shown in Figure 8.10a, the current becomes independent of potential at positive potential values, showing that the electron transfer to the electrode is fast enough as it is not rate-limiting. In protein film voltammetry (PFV), a limit like this is interpreted as the maximum turnover rate of the enzyme [74], as could be done here for an isolated electrode reductase. Nevertheless, the voltammetric signature of whole cells correspond to molecules that are (the final) part of a metabolic process that includes at least the steps described in Section 8.1 (Fig. 8.2).

Given that oxidizing power at the electrode surface exceeds the requirement of the biofilm and that electron tunneling is usually in the range of milliseconds across biological molecules [75], the limitation to current flow might be determined by the same control strategy that regulates bacterial respiration, namely, the rate of proton translocation across the inner cell membrane (during ATP synthesis), which, in turn, depends on the ATP/ADP ratio of bacterial cells [76]. On this regard, Izallalen et al. [77] have demonstrated that genetically modifying *G. sulfurreducens*, to add an extra ATP unit, increased the respiration rate, while lowering growth rate and cell yield. The phenotype was supported by increased levels of transcripts of genes involved in energy metabolism (the TCA cycle and the NADH deshydrogenase) and

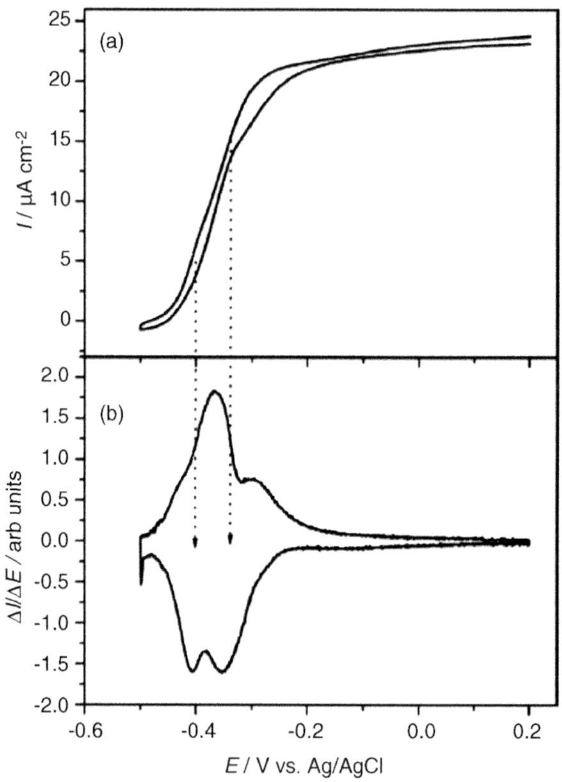


**FIGURE 8.10** (a) Cyclic voltammogram of a *G. sulfurreducens* biofilm-metabolizing acetate; (b) first derivative of the voltammetric curve over the potential. (Reproduced with permission from Ref. 47.)

reduction of electron acceptors. On the other hand, genes involved in cell growth, amino acid synthesis, and motility presented lowered transcript levels [77]. Surprisingly, the modified phenotype including the ATP drain did not produce increased current in a MFC as compared to the wild type, in spite of its increased respiration rate [77], although details of these experiments have not been published. Alternatively, much more recent data lead to the proposition that current limitation is at the rate of acetate uptake by the microorganisms [62]; nevertheless conclusive results have not been presented yet.

## 8.5 MFC APPLICATIONS

### 8.5.1 General Perspective

Microbial fuel cells are interesting bioelectrochemical devices for investigating the interaction between cells and insoluble electron acceptors such as electrodes. The system can mimic other environmentally relevant processes such as microbial redox interaction with iron oxides that have shaped the earth is crust for ages. The electrochemical analyses described above will contribute for sure to define new redox processes in the exciting bacteria–electrode interface, which still is not well understood by researchers. However, MFCs were initially conceived on the basis of the applied concept of generating power. So, the reader of this chapter might be interested in knowing the current scenario for MFC applications.

In contrast to classical fuel cells, one of the most remarkable features of MFCs is the wide diversity of fuels that can feed these bioelectrochemical devices. The fact that bacteria count on a vast number of metabolic pathways for biodegrading almost every organic compound in nature entices biotechnologists to play with the attractive concept of transforming organic material into electricity. However, the production of current should not be interpreted merely as a useful energy source, because it can also be used for another redox-based purpose, or even as an electroanalytical signal that correlates with the presence of a certain chemical, or, indeed, its concentration.

#### 8.5.1.1 MFCs and Wastewater

Although any kind of biodegradable substance could potentially be used for harvesting electricity, MFC researchers have focused mainly on those that constitute a waste whose treatment would require a high energy expenditure. Therefore, the main tested fuels for MFCs are the organic contents of wastewater. The energy stored in the chemical bonds of this heterogeneous mixture of compounds would be, if ideal transformation took place, released in the form of electricity enough for copying 9.3 times the energy invested in wastewater treatment through energy-intensive aeration-based processes [78]. So, the waste *would pay* for its own remediation.

Intensive investigation in designing devices for conversion of wastewater into electricity has been developed since 2004, and the power density has increased by several orders of magnitude [55]. A maximum power density of 6.9 $W/m^2$ of the anode area was achieved in 2008 [79] using a cathode that was 14 times larger than the anode. An unexpected result was that a well-known electrochemical procedure such as increasing the counterelectrode area to avoid current limitations took almost 10 years before it was used and reported in the MFC field. On the other hand, this reveals that the design and architecture of these bioelectrochemical devices can still be improved, and it seems reasonable to think that power production will increase in the near future.

Regarding cell design, different configurations, ranging from the flat plate [80], single chamber [81], upflow bed MFC [82], to stacked cassettes [83] have been tested while searching for improving the conversion process. Because of that effort, we are closer than ever to the target power of 1 $kW/m^3$ of wastewater established during the first International Meeting on the Microbial Fuel Cell (Pennsylvania, US, 2008) that would make this technology feasible for engineering applications. So far, this target value has been reached [84], although only on bench scale.

Most of the primary research performed thus for did not use real wastewater but synthetic mixtures or just acetate, which have been the subject of review [85]. On the other hand, more recent research includes studies with real wastewater from industrial sources as diverse as beer brewery [86] or paper-recycling plants [87], setting the trend for applications.

#### 8.5.1.2 MFCs and Vegetal Biomass

Another attractive concept is the use of vegetal biomass as a fuel for bioelectrogenic bacteria. In fact, solar energy has been invested by plants for making biopolymers such as cellulose or lignine, which are widely distributed along the earth is crust and are ubiquitous source of fuel for MFCs. In that sense, microorganisms from cow rumen have been shown to hydrolyze cellulose and produce a maximal power of 55 $mW/m^2$ [88]. In this study bacteria from the genus *Clostridium* were identified in the anode community, but no bioelectrogenesis was obtained when pure strains of cellulose-oxidizing bacteria such as *Clostridium cellulolyticum* were assayed. The complex structure and very low hydrolysis rate of cellulose presents challenges for using this material in an MFC. However, it was demonstrated that the addition of cellulolytic enzymes can help overcome this limitation, and power production as high as 102 $mW/m^2$ has been achieved [89]. Interestingly, a specially designed U-tube MFC has been used for the isolation of a *Enterobacter cloacae* strain that demonstrated for the first time that it is possible to generate electricity from cellulose using a single bacterial strain without exogenous mediators [90].

#### 8.5.1.3 MFCs and Inorganic Pollutants

Microbial oxidation of substrates in MFC is not merely an area for organic energy sources but also includes inorganic compounds such as sulfide [91–93], sulfate [94,95], and ammonium [95]. Sulfate removal seems a promising technology, based on coupling the biological sulfate reduction by *Desulfovibrio desulfuricans* with the in situ anodic oxidative depletion of sulfide, which precipitates as sulfur [94,95]. Another interesting research example, still in a preliminary stage, is ammonia oxidation. Ammonia is one of the most important inorganic compounds in wastewater or agricultural wastes, and its microbial oxidation coupled to generation of electricity has been reported by using a rotating-cathode MFC [96]. Although ammonium removal efficiency was close to 70%, the coulombic efficiency (CE) was as low as 0.34%, so the technology should be drastically improved before any reasonable use could be suggested.

### 8.5.2 Biocathodes: Electrodes as Electron Donors for Microorganisms

The reduction of oxygen at the cathode is one of the major bottlenecks of MFCs. However, oxygen reduction can be catalyzed by bacteria on the cathode (biocathode) of an MFC, allowing for bacterial growth using electrons coming from the anode [97]. Bacteria can obtain energy from this reaction because electrons enter the bacteria on the biocathode at a potential lower than that needed for oxygen reduction. Biocathodes have also been used for nitrate reduction [98] and hydrogen production [99]. Although most of MFC applications actually under research for bioremediation are based on anodic oxidation, the fact that bacteria can accept electrons from electrodes opens an unexplored and exciting bioelectrochemical field because certain pollutants require at least a partial reduction in order to minimize their toxicity. So, it is convenient to identify different microorganisms able to accept electrons from electrodes and transfer them to pollutants. In this context, there have been reports of the microbial reductive dechlorination of trichloroethene (TCE) by using a methylviologen-modified electrode polarized at −500 mV (SHE) [100], or tetrachloroethene (PCE) by using *Geobacter lovelyi* accepting electrons directly from an electrode polarized at −300 mV [101]. This concept of accepting electrons from a solid phase could also be applied in metal bioremediation for microbial uranium compounds reduction [102].

### 8.5.3 MFCs in the Environment: Remote Power Generation

Sedimentary microbial fuel cells (SMFCs) are electrochemical devices that harvest electrical energy from the potential differences across sediment–water interfaces in the environment. Burying an anode in the sediment and allocating a cathode in the overlying water enables electrical energy to flow through the external circuit. The first microorganisms with electrogenic capacity isolated from SMFC anodes were from the genus *Geobacter* [41]. Since then, different variations of environmental electrogenic devices have been described [103]. In contrast to the typical MFC, the natural fuel for SMFC is not simply organic matter since part of the current is generated by the oxidation of sulfide produced in situ by sulfate microbial reduction. Interestingly, another group of microorganisms can reoxidize sulfur to sulfate, leading to a cycle that, if properly balanced, would constitute a never-ending fuel source for these environmental fuel cells [104]. This technology cannot compete with the electrical grid in power, but actually satisfy the requirements for low power devices that work in remote places were the grid is not available. As an example, it can empower electrochemical sensors and small telemetry systems to transmit the data acquired [105], constituting an interesting and reasonable application that could be extremely useful for monitoring parameters characterizing water quality, such as temperature, pH, and concentrations of heavy metals in remote environments. Another interesting remote application is the use of sediment power for cathodic protection, a method that has been successfully tested at laboratory scale [106].

A novel SMFC application of generating electricity using rice crops in either the open field or the greenhouse has been reported. This is an interesting environment for implementing SMFCs because the usual practice of flooding the soil generates a suitable place to allocate the cathode (see Fig. 8.11). For a long time it was known that root exudates are rich in organic compounds and attract microorganisms that colonize the soil surrounding the root (rhizosphere). This organic matter can be oxidized by electrogenic bacteria from the sediment, transferring the electrons to graphite anodes placed in the root environment. It has already been observed that electricity generation was sunlight-dependent, thus exhibiting circadian oscillation. These results suggest that the rice–SMFC system can act as a photovoltaic cell in which plant photosynthesis is coupled to microbial conversion of organics to electricity (Fig. 8.1). Indeed, other authors suggest that variations in the cell potential ($E_{cell}$) associated with day–night cycles are due to perturbations in anodic environment since the fuel (root exudates) supply depends on plant photosynthesis. In contrast, we have more recently demonstrated that photosynthesis is involved in these $E_{cell}$ oscillations, but no necessarily

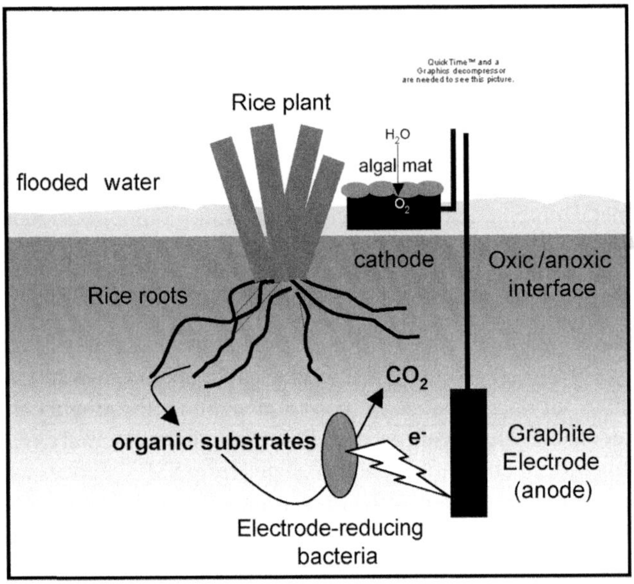


**FIGURE 8.11** Scheme of a sedimentary MFC fueled by root exudates from rice crop and alga. Oxygen is biologically supplied to the cathode by photosynthetic microbial mat.

related to plants but to algal and cyanobacteria populations that colonize aerobic cathode and supply oxygen for cathodic reduction. Variations in cathode potential perfectly match the $E_{cell}$ profile during day–night (24-h) periods (see Fig. 8.12).

Another important aspect of this environmental application is related to the emission of methane by the methanogenic microbial communities existing in the

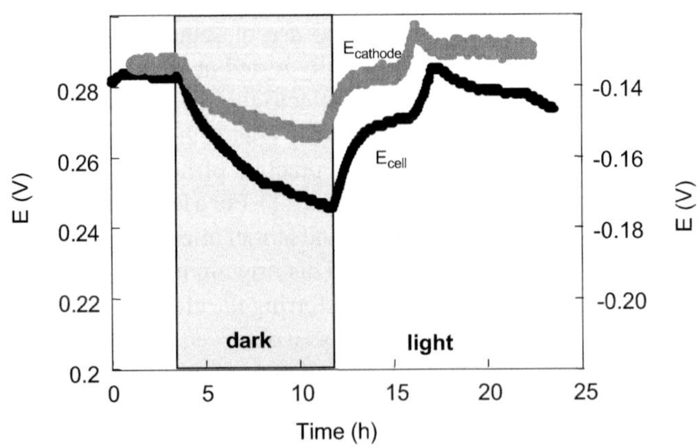


**FIGURE 8.12** Potential evolution during chronopotentiometric measurements performed on sedimentary MFC fueled by rice root exudates (black line) and on cathode (Ag/AgCl–KCl sat.) (gray line). The MFC was set up using a 560 Ω resistor, and day–night cycles were performed by artificial light in a greenhouse. (From Esteve-Núñez et al., personal communication.)

flooded rice soil. The emission of this greenhouse gas from this particular crop environment represents an enormous problem that contributes significantly to the global methane cycle (25% of the total annual methane emission into the atmosphere). The presence of anodes bearing electrode-reducing biofilms may change the fait of electrons from methane to electricity, replacing methane by $CO_2$ as the final product of microbial degradation of organic matter.

## 8.6 CONCLUDING REMARKS

Microbial fuel cells have the potentiality to offer bacteria the equivalent of aerobic respiration, because the final electron acceptor in the device is generally oxygen. Nevertheless, for cells to have a higher energetic yield they need to respire by using final reductases of high potential. Both the redox potential of a natural community in sedimentary fuel cells [107] and that of a pure culture of *Geobacter sulfurreducens* [30] have been shown to depend on the conditioning potential used to induce the growth of the population; in other words, the potential of the electron acceptor seems to determine the kind of external reductase used by the organisms. In *G. sulfurreducens* the expression of a potential-induced *c*-type cytochrome has been demonstrated. The same seems to occur with natural populations [107]. It once again emphasizes the distinctive feature of MFCs; they are "alive" systems with a dynamic response mediated by biological entities. To understand the complexity of such a response will surely demand a great research effort and warrant the discovery of fascinating mechanisms not ever suspected in bioelectrocatalisys.

Indeed, it should be stressed that the interaction of the attached bacteria with the electrode will depend on distance. Thus, the activity of the first monolayer of bacteria directly attached to the conducting material may be different from that of bacteria in the biofilm. Studies to identify these possibly different behaviors are required.

In this case, and because of the bacterial size, the number of species under study will be well below the usual figure of $10^{15}$ species/cm$^2$ in surface electrochemistry, but entities have a much more complex surface with local features depending on biological processes. Impedance and Raman and infrared spectroscopic techniques appear as powerful tools for the collection of required data. Potential perturbation techniques, namely, potentiodynamic versus potentiostatic, could also give complementary information, as the time window will enable the observation of different postkinetic reactions. Finally, electrochemical quartz crystal microbalance (EQCM) and scanning vibrating electrode (SVE) techniques still wait to be applied to explore the complexity of the electrogenic bacteria–electrode interface.

## REFERENCES

1. A. J. Bard and L. R. Faulkner, *Electrochemical Methods: Fundamental and Applications*, Wiley, Toronto, 1980.
2. D. A. Huckaby, M. D. Legault, and L. Blum, *J. Chem. Phys.* **109,** 3600 (1998).

3. J. M. Orts, R. Gomez, J. M. Feliu, A. Aldaz, and J. Clavilier, *J. Phys. Chem.* **100**, 2334 (1996).
4. R. Gómez, J. M. Feliu, A. Aldaz, and M. J. Weaver, *Surf. Sci.* **410**, 48 (1998).
5. M. Weaver, *Top. Catal.* **8**, 65 (1999).
6. M. Rodriguez-Lopez, A. Rodes, A. Berna, V. Climent, E. Herrero, P. Tunon, J. M. Feliu, A. Aldaz and A. Carrasquillo, *Langmuir* **24**, 3551 (2008).
7. M. F. Suarez-Herrera and J. M. Feliu, *J. Phys. Chem. B* **113**, 1899 (2009).
8. C. Amatore, Y. Bouret, E. Maisonhaute, J. I. Goldsmith, and H. D. Abruña, *ChemPhysChem* **2**, 130 (2001).
9. A. Kiani, M. A. Alpuche-Aviles, P. K. Eggers, M. Jones, J. J. Gooding, M. N. Paddon-Row, and A. J. Bard, *Langmuir* **24**, 2841 (2008).
10. F. A. Armstrong and G. S. Wilson, *Electrochim. Acta* **45**, 2623 (2000).
11. S. Guerin, B. E. Hayden, D. Pletcher, M. E. Rendall, J.-P. Suchsland, and L. J. Williams, *J. Combin. Chem.* **8**, 791 (2006).
12. J. Hernández, J. Solla Gullón, E. Herrero, J. M. Feliu, and A. Aldaz, *J. Nanosc. Nanotechnol.* **9**, 2256 (2009).
13. V. R. Stamenkovic, B. S. Mun, M. Arenz, K. J. J. Mayrhofer, C. A. Lucas, G. Wang, P. N. Ross, and N. M. Markovic, *Nat. Mater.* **6**, 241 (2007).
14. A. T. Poortinga, R. Bos, W. Norde, and H. J. Busscher, *Surf. Sci. Rep.* **47**, 3 (2002).
15. J. P. Busalmen and S. R. de Sanchez, *Appl. Environ. Microbiol.* **67**, 3188 (2001).
16. H. J. Busscher and A. H. Weerkamp, *FEMS Microbiol. Lett.* **46**, 165 (1987).
17. H. C. Gram, *Fortsch. Med.* **2**, 185 (1884).
18. H. L. Ehrlich, *Geobiology* **6**, 220 (2008).
19. S. N. Veremeichenko, M. A. Vodyanik, and G. M. Zdorovenko, *Appl. Biochem. Microbiol.* **41**, 365 (2005).
20. A. Korenevsky and T. J. Beveridge, *Microbiology* **153**, 1872 (2007).
21. L. V. Smith, L. K. Tamm and R. M. Ford, *Langmuir* **18**, 5247 (2002).
22. J. P. Busalmen, A. Berna, and J. M. Feliu, *Langmuir* **23**, 6459 (2007).
23. S. Y. Kang, P. J. Bremer, K. W. Kim, and A. J. McQuillan, *Langmuir* **22**, 286 (2006).
24. J. Wei, A. Saxena, B. Song, B. B. Ward, T. J. Beveridge, and S. C. B. Myneni, *Langmuir* **20**, 11433 (2004).
25. J. P. Busalmen and S. R. de Sanchez, *Appl. Environ. Microbiol.* **71**, 6235 (2005).
26. A. Hamelin, T. Vitanov, E. Sevastyanov, and A. Popov, *J. Electroanal. Chem.* **145**, 225 (1983).
27. S. Trasatti and E. Lust, eds., *Modern Aspects of Electrochemistry*, Vol. 33, Kluwer Academic/Plenum Publishers, New York, 1999.
28. J.-P. Randin and E. Yeager, *J. Electroanal. Chem.* **36**, 257 (1972).
29. H. Gerischer, R. McIntyre, D. Scherson, and W. Storck, *J. Phys. Chem.* **91**, 1930 (2002).
30. J. P. Busalmen, A. Esteve-Nuñez, and J. M. Feliu, *Environ. Sci. Technol.* **42**, 2445 (2008).
31. A. Barth, *Biochim. Biophys. Acta Bioenerg.* **1767**, 1073 (2007).
32. T. Mehta, M. V. Coppi, S. E. Childers, and D. R. Lovley, *Appl. Environ. Microbiol.* **71**, 8634 (2005).
33. M. J. Allen, R. J. Bowen, M. Nicholson, and B. M. Vasta, *Electrochim. Acta* **8**, 991 (1962).

34. M. J. Allen, *Electrochim. Acta* **9**, 1429 (1964).
35. M. J. Allen, *Bacteriol. Rev.* **30**, 80 (1966).
36. M. J. Allen, *Electrochim. Acta* **11**, 7 (1966).
37. M. J. Allen, *Electrochim. Acta* **11**, 15 (1966).
38. M. J. Allen, *Electrochim. Acta* **12**, 563 (1967).
39. M. J. Allen, *Electrochim. Acta* **15**, 1565 (1970).
40. R. Holzel, *Biochim. Biophys. Acta* **1450**, 53 (1999).
41. D. R. Bond, D. E. Holmes, L. M. Tender, and D. R. Lovley, *Science* **295**, 483 (2002).
42. D. R. Lovley, *Geobiology* **6**, 225 (2008).
43. T. J. DiChristina, J. K. Fredrickson, and J. M. Zachara, *Rev. Miner. Geochem.* **59**, 27 (2005).
44. J. A. Gralnick and D. K. Newman, *Mol. Microbiol.* **65**, 1 (2007).
45. L. Shi, T. C. Squier, J. M. Zachara and J. K. Fredrickson, *Mol. Microbiol.* **65**, 12 (2007).
46. E. Marsili, J. B. Rollefson, D. B. Baron, R. M. Hozalski, and D. R. Bond, *Appl. Environ. Microbiol.* **74**, 7329 (2008).
47. K. Fricke, F. Harnisch, and U. Schroder, *Energy Environ. Sci.* **1**, 144 (2008).
48. Z. He and F. Mansfeld, *Energy Environ. Sci.* **2**, 215 (2009).
49. E. Marsili, D. B. Baron, I. D. Shikhare, D. Coursolle, J. A. Gralnick, and D. R. Bond, *Proc. Nat. Acad. Sci. USA.* 0710525105 **105**(10), 3968–3973 (2008).
50. A. K. Manohar, O. Bretschger, K. H. Nealson, and F. Mansfeld, *Bioelectrochemistry* **72**, 149 (2008).
51. U. Schröder, *Phys. Chem. Chem. Phys.* **9**, 2619 (2007).
52. K. Rabaey, J. Rodriguez, L. L. Blackall, J. Keller, P. Gross, D. Batstone, W. Verstraete, and K. H. Nealson, *ISME J.* **1**, 9 (2007).
53. S. Ouitrakul, M. Sriyudthsak, S. Charojrochkul, and T. Kakizono, *Biosens. Bioelectron.* **23**, 721 (2007).
54. C. Dumas, A. Mollica, D. Feron, R. Basseguy, L. Etcheverry, and A. Bergel, *Electrochim. Acta* **53**, 468 (2007).
55. B. E. Logan, *Nature Rev. Microbial.* **7**, 375 (2009).
56. J. P. Busalmen, A. Esteve-Núñez, A. Berná, and J. M. Feliu, *Angew. Chem. Int. Ed.* **47**, 4874 (2008).
57. P. S. Stewart and M. J. Franklin, *Nature Rev. Microbiol.* **6**, 199 (2008).
58. G. O'Toole, H. B. Kaplan, and R. Kolter, *Annu. Rev. Microbiol.* **54**, 49 (2000).
59. J. W. Costerton, Z. Lewandowski, D. E. Caldwell, D. R. Korber, and H. M. Lappinscott, *Annu. Rev. Microbiol.* **49**, 711 (1995).
60. G. Reguera, K. P. Nevin, J. S. Nicoll, S. F. Covalla, T. L. Woodard, and D. R. Lovley, *Appl. Environ. Microbiol.* **72**, 7345 (2006).
61. K. P. Nevin, B.-C. Kim, R. H. Glaven, J. P. Johnson, T. L. Woodard, B. A. MethÃ©, R. J. DiDonato, Jr., S. F. Covalla, A. E. Franks, A. Liu, and D. R. Lovley, *PLoS ONE* **4**, e5628 (2009).
62. H. Richter, K. P. Nevin, H. Jia, D. A. Lowy, D. R. Lovley, and L. M. Tender, *Energy Environ. Sci.* **2**, 506 (2009).
63. C. G. Mowat and S. K. Chapman, *Dalton Trans.* **2005**, 3381 (2005).

64. F. A. Armstrong, *J. Chem. Soc. Dalton Trans.* **2002**, 661 (2002).
65. F. Armstrong, in *Bioelectrochemistry*, Vol. 9, G. Wilson, ed., Wiley-VCH Verlag GmbH, Weinheim, 2002, p. 11.
66. J. P. Busalmen, A. Esteve-Nuñez, A. Berná, and J. M. Feliu, *Bioelectrochemistry* (doi: 10. 1016/j.bioelechem.2009.04.011). (2009).
67. D. R. Lovley, *Nature Rev. Microbiol.* **4**, 497 (2006).
68. J. E. Champine, B. Underhill, J. M. Johnston, W. W. Lilly, and S. Goodwin, *Anaerobe* **6**, 187 (2000).
69. B. E. Logan, *Microbial Fuel Cells*, Wiley, Hoboken, NJ, 2008.
70. C. Dumas, R. Basseguy, and A. Bergel, *Electrochim. Acta* **53**, 3200 (2008).
71. K. Rabaey, N. Boon, S. D. Siciliano, M. Verhaege, and W. Verstraete, *Appl. Environ. Microbiol.* **70**, 5373 (2004).
72. K. Tammeveski, K. Kontturi, R. J. Nichols, R. J. Potter, and D. J. Schiffrin, *J. Electroanal. Chem.* **515**, 101 (2001).
73. F. Mirkhalaf, K. Tammeveski, and D. J. Schiffrin, *Phys. Chem. Chem. Phys.* **6**, 1321 (2004).
74. F. A. Armstrong, *Curr. Opin. Chem. Biol.* **9**, 110 (2005).
75. P. N. Barlett, in *Bioelectrochemistry*, P. N. Barlett, ed., Wiley, Chichester, UK, 2008, p. 1.
76. D. G. Nicholls and S. J. Ferguson, *Bioenergetics*, Academic Press, London, 1992.
77. M. Izallalen, R. Mahadevan, A. Burgard, B. Postier, R. Didonato Jr., J. Sun, C. H. Schilling, and D. R. Lovley, *Metab. Eng.* **10**, 267 (2008).
78. I. Shizas and D. M. Bagley, *J. Energy Eng.* **130**, 45 (2004).
79. Y. Fan, E. Sharbrough, and H. Liu, *Environ. Sci. Technol.* **42**, 8101 (2008).
80. B. Min and B. E. Logan, *Environ. Sci. Technol.* **38**, 5809 (2004).
81. S. Venkata Mohan, S. Veer Raghavulu, and P. N. Sarma, *Biosens. Bioelectron.* **23**, 1326 (2008).
82. Z. He, S. D. Minteer, and L. T. Angenent, *Environ. Sci. Technol.* **39**, 5262 (2005).
83. T. Shimoyama, S. Komukai, A. Yamazawa, Y. Ueno, B. E. Logan, and K. Watanabe, *Appl. Microbiol. Biotechnol.* **80**, 325 (2008).
84. Y. Fan, H. Hu, and H. Liu, *Environ. Sci. Technol.* **41**, 8154 (2007).
85. Z. Du, H. Li, and T. Gu, *Biotechnol. Adv.* **25**, 464 (2007).
86. Y. Feng, X. Wang, B. Logan, and H. Lee, *Appl. Microbiol. Biotechnol.* **78**, 873 (2008).
87. L. Huang, S. Cheng, F. Rezaei, and B. E. Logan, Reducing organic loads in wastewater effluents from paper recycling plants using microbial fuel cells. *Environ. Technol.* **30**(5), 499–504 (2009).
88. H. Rismani-Yazdi, A. D. Christy, B. A. Dehority, M. Morrison, Z. Yu, and O. H. Tuovinen, *Biotechnol. Bioeng.* **97**, 1398 (2007).
89. R. Farzaneh, L. R. Tom, and E. L. Bruce, *Biotechnol. Bioeng.* **101**, 1163 (2008).
90. F. Rezaei, D. Xing, R. Wagner, J. M. Regan, T. L. Richard, and B. E. Logan, *Appl. Environ. Microbiol.* **75**, 3673 (2009).
91. M. Sun, Z.-X. Mu, Y.-P. Chen, G.-P. Sheng, X.-W. Liu, Y.-Z. Chen, Y. Zhao, H.-L. Wang, H.-Q. Yu, L. Wei, and F. Ma, *Environ. Sci. Technol.* **43**, 3372 (2009).

92. B.-G. Zhang, S.-G. Zhou, H.-Z. Zhao, C.-H. Shi, L.-C. Kong, J.-J. Sun, Y. Yang, and J.-R. Ni, *Bioprocess Biosyst. Eng.* (in press)

93. K. Rabaey, K. VandeSompel, L. Maignien, N. Boon, P. Aelterman, P. Clauwaert, L. DeSchamphelaire, H. T. Pham, J. Vermeulen, M. Verhaege, P. Lens, and W. Verstraete, *Environ. Sci. Technol.* **40**, 5218 (2006).

94. F. Zhao, N. Rahunen, J. R. Varcoe, A. Chandra, C. Avignone-Rossa, A. E. Thumser, and R. C. T. Slade, *Environ. Sci. Technol.* **42**, 4971 (2008).

95. Z. He, J. Kan, Y. Wang, Y. Huang, F. Mansfeld, and K. H. Nealson, *Environ. Sci. Technol.* **43**, 3391 (2009).

96. F. Zhao, N. Rahunen, J. R. Varcoe, A. J. Roberts, C. Avignone-Rossa, A. E. Thumser, and R. C. T. Slade, *Biosens. Bioelectron.* **24**, 1931 (2009).

97. P. Clauwaert, D. van der Ha, N. Boon, K. Verbeken, M. Verhaege, K. Rabaey, and W. Verstraete, *Environ. Sci. Technol.* **41**, 7564 (2007).

98. P. Clauwaert, K. Rabaey, P. Aelterman, L. De Schamphelaire, T. H. Pham, P. Boeckx, N. Boon, and W. Verstraete, *Environ. Sci. Technol.* **41**, 3354 (2007).

99. R. A. Rozendal, A. W. Jeremiasse, H. V. M. Hamelers, and C. J. N. Buisman, *Environ. Sci. Technol.* **42**, 629 (2008).

100. F. Aulenta, A. Catervi, M. Majone, S. Panero, P. Reale, and S. Rossetti, *Environ. Sci. Technol.* **41**, 2554 (2007).

101. S. M. Strycharz, T. L. Woodard, J. P. Johnson, K. P. Nevin, R. A. Sanford, F. E. Loffler, and D. R. Lovley, *Appl. Environ. Microbiol.* **74**, 5943 (2008).

102. K. B. Gregory and D. R. Lovley, *Environ. Sci. Technol.* **39**, 8943 (2005).

103. L. M. Tender, C. E. Reimers, H. A. Stecher, D. E. Holmes, D. R. Bond, D. A. Lowy, K. Pilobello, S. J. Fertig, and D. R. Lovley, *Nature Biotechnol.* **20**, 821 (2002).

104. A. Shantaram, H. Beyenal, R. Raajan, A. Veluchamy, and Z. Lewandowski, *Environ. Sci. Technol.* **39**, 5037 (2005).

105. L. H. Orfei, S. Simison, and J. P. Busalmen, *Environ. Sci. Technol.* **40**, 6473 (2006).

106. W. Liesack, S. Schnell, and N. P. Revsbech, *FEMS Microbiol. Rev.* **24**, 625 (2000).

107. D. A. Finkelstein, L. M. Tender, and J. G. Zeikus, *Environ. Sci. Technol.* **40**, 6990 (2006).

CHAPTER 9

# Half-Cell Investigations of Cathode Catalysts for PEM Fuel Cells: From Model Systems to High-Surface-Area Catalysts

MATTHIAS ARENZ

Department of Chemistry, University of Copenhagen, Copenhagen, Denmark

NENAD M. MARKOVIC

Materials Science Division, Argonne National Laboratory, Argonne, Illinois

The aim of this chapter is to critically review more recent electrochemical investigations concerning PEM fuel cell catalysts utilizing model systems, and to link these observations to the behavior of industrial high-surface-area catalysts. Thereby we concentrate on investigations concerning general properties of cathode catalysts utilizing electrochemical half-cell measurements, including work in which the authors have been directly involved. Since many of the results reported in the literature regarding, for instance, the oxygen reduction reaction (ORR) mechanism or the influence of particle size on the ORR rate are conflicting, special emphasis is placed on the experimental procedures for obtaining these results, and possible reasons for the differences observed.

## 9.1 INTRODUCTION

Compared to other fields in physical chemistry, the development of electrochemistry in the first half of the twentieth century had been slow, mainly because of its empirical nature [1]. Since the 1980s, therefore, one of the main goals in modern

*Fuel Cell Science: Theory, Fundamentals, and Biocatalysis*,
Edited by Andrzej Wieckowski and Jens K. Nørskov

electrocatalysis was to create a link between the microscopic level of understanding the atomic and electronic structure of a catalyst, and the macroscopic reaction rate of a process, namely, the activity of the catalyst. For this kind of fundamental research the use of well-defined single crystals is essential, since investigations on high-surface-area materials are often too complex and thus difficult to interpret. Despite the advances in our understanding of such model systems, however, the transfer and application of this knowledge to commercially available technologies has proved to be difficult.

The aim of this chapter is to critically review more recent electrochemical (EC) investigations concerning PEM fuel cell catalysts utilizing model systems, and to link these observations to the behavior of industrial high-surface-area catalysts. Because of the sheer volume of work, we concentrate on investigations concerning general properties of cathode catalysts utilizing electrochemical half-cell measurements, including work in which the authors have been directly involved. Special emphasis is placed on the experimental procedures by which these results are obtained, and possible reasons for the differences observed.

## 9.2 EXPERIMENTAL CONSIDERATIONS

In a fuel cell the processes at the anode (hydrogen/methanol oxidation) and the cathode side (oxygen reduction) proceed at the same rate. The cell voltage, that is, the potential difference between the cathode and the anode, adjusts as a function of the current drawn from the system. The absolute potential value of the electrodes (vs. a reference electrode) is in general unknown.[1] For a given electrocatalyst the performance of a FC stack or a membrane electrode assembly (MEA) is a complex function of not only the operating conditions (temperature, gas pressure, etc.,) but also the preparation of membrane electrode assembly.

By comparison, in half-cell investigations the anode and the cathode reaction are investigated separately. By limiting the external influences on the performance, these studies enable one to focus on the function of the catalyst itself. With the help of a kinetic analysis, the true kinetic reaction rates of the processes can be extracted in a certain potential window. The main advantages of half-cell measurements compared to catalyst performance investigations in real fuel cells is the considerably reduced complexity of the system, as well as the possibility of investigating both model systems and high-surface-area catalysts under the same experimental conditions. Despite these advantages, the determination of reproducible kinetic reaction rates from electrochemical half-cell measurements is not as straightforward as it might seem. Even for simple systems like polycrystalline Pt there is a large discrepancy in the literature between the reported kinetic reaction rates of the ORR.

[1]In some cases a reference electrode is embedded in the MEA, which, however, can interfere with performance.

The most suitable methodology to investigate the ORR in electrochemical half-cells is the rotating (ring)-disk electrode (R(R)DE) [2], on which the focus of this chapter is placed. Alternatives are electrochemical flow cells [3,4] or microelectrode setups [5]. More recently the authors have devoted a considerable amount of time to improving measurement reproducibility of the ORR in a RDE configuration, and establishing benchmark activities. Therefore, before discussing more recent experimental results on cathode materials for low-temperature fuel cells, a description of the crucial experimental considerations is given along with a discussion of the pitfalls in determining reproducible values for the kinetic reaction rate of the oxygen reduction reaction. A complete description of an RDE electrochemical setup and the analysis of electrochemical measurements, however, are beyond the scope of this chapter. For this the interested reader is referred to the literature [2,6,7].

### 9.2.1 Potential Sources of Error in a RDE Electrochemical Setup

In order to determine reproducible kinetic reaction rates it is essential to eliminate a series of possible sources of error that are common to rotating-disk electrode measurements. From the experience of the authors the four major sources of error are those described in Sections 9.2.1.1–9.2.1.4.

#### *9.2.1.1 The Reference Electrode Potential*

In kinetic measurements small potential shifts in the polarization curves considerably influence the calculated kinetic currents. Typical reference electrodes such as the calomel or the silver/silver chloride electrode provide a stable potential that, however, is independent of the solution pH. In order to compare kinetic rates of pH-dependent processes as the ones found in a PEMFC, it is essential that the exact reference electrode potential with respect to the reversible hydrogen electrode (RHE) is known. A solution to this problem is best achieved by using, for instance, a calomel electrode and to calibrate its potential toward the RHE prior to each measurement. For this, polarization curves in a fully hydrogen saturated electrolyte are measured in an RDE setup using a platinum electrode. The RHE potential with respect to the deployed RE is then obtained where the measured current crosses the zero $y$ scale.

#### *9.2.1.2 Uncompensated Solution Resistance*

In order to provide a suitable electrolyte conductance, the electrolyte concentration in kinetic measurements is in general equal to or higher than 0.1 mol/L. Nevertheless, a rarely discussed source of error in kinetic measurements is the remaining solution resistance $R_u$ between the WE and RE. For minimizing $R_u$, the RE has to be positioned as close as possible to the WE without interfering with mass flow patterns in the cell, and usually a Luggin capillary is used for this. However, even employing a Luggin capillary, residual $R_u$ in the range between 10 and 20 Ω depending on the electrolyte cannot be avoided in a RDE setup as discussed in the mid-1960s by Newman [8,9]. The influence of $R_u$ on the rate can be exemplified by a simple "back of the envelope" calculation: Assuming a measured kinetic current for the ORR of only 1 mA, the abovementioned values of $R_u$ result in $iR$ drops of 10–20 mV. With a measured Tafel slope of 60 mV/dec, the *iR*-drop-uncorrected kinetic ORR currents are significantly underestimated. In order to obtain reproducible measurements of absolute kinetic rates, the residual $R_u$ must be compensated for. A suitable way to correct the uncompensated solution resistance during the measurement is to first determine the $R_u$ by applying an AC signal with a frequency of 5 kHz and 5 mV in amplitude, and then applying compensation by a positive feedback scheme [10].

#### 9.2.1.3 Traces of Impurities

In electrochemical half-cell measurements of fuel cell reactions, one objective is to realistically mimic the conditions of actual fuel cells under operation. In a PEMFC the electrolyte is a proton-conducting polymer, such as Nafion, which is considered as a nonadsorbing electrolyte. In order to mimic "Nafion behavior," perchloric acid solution has lately become the electrolyte of choice, since it is generally considered an almost non adsorbing electrolyte. However, more recently is has been shown that in fact even the "purest" available perchloric acid contains small amounts of impurities, such as chloride anions, which, indeed, can affect the catalytic activity of a cathode catalyst. Given that small amounts of impurities can have a huge effect on the determined kinetic reaction rate of the ORR [11], the comparison of a catalyst in different solutions is of fundamental interest. Alongside acid electrolyte, alkaline media such as potassium and sodium hydroxide solutions are also used as supporting electrolytes. An issue often neglected in alkaline measurements, however, is the possible influence of trace levels of heavy-metal cations. As shown in recently [12,13], as a result of the corrosion of glass in alkaline solution, lead as well as silicate is introduced into the system, which severely influences the measured kinetic rates. Not only can reduced reaction rates be observed (due to silica) but also enhanced reaction rates (due to lead) depending on the experimental protocol. As a consequence, kinetic studies in alkaline media should be performed only in nonglass electrochemical cells. An all-Teflon electrochemical cell designed for such measurements in alkaline electrolyte (which is suitable for measurements in acid solution as well) is shown in Figure 9.1.

#### 9.2.1.4 Electrode Preparation

Two types of electrodes are frequently used in electrocatalytic measurements of ORR, model bulk electrodes, and high-surface-area catalysts. Traditionally, bulk electrodes can be prepared ex situ in an ultra-high-vacuum (UHV) environment, or by heating in a hydrogen atmosphere. The UHV preparation of bulk electrodes has the advantage that in UHV there is a variety of electron/ion spectroscopies available to test the cleanliness of the electrode surface [Auger electron spectroscopy (AES)] as well as to determine both the surface structure [low-energy electron diffraction (LEED)] and composition [low-energy ion scattering (LEIS)]. This is particularly useful for bimetallic surfaces that cannot be prepared by the flame annealing technique, as developed by Clavilier [15,16]. In the latter technique the crystal is heated in a hydrogen/air or hydrogen/oxygen flame, thus removing organic and inorganic impurities from the surface. Furthermore, the high temperature allows ordering of the surface atoms, which is especially important for single crystals. More recently a third method, utilizing inductive heating [17,18], that is, sending an alternating current

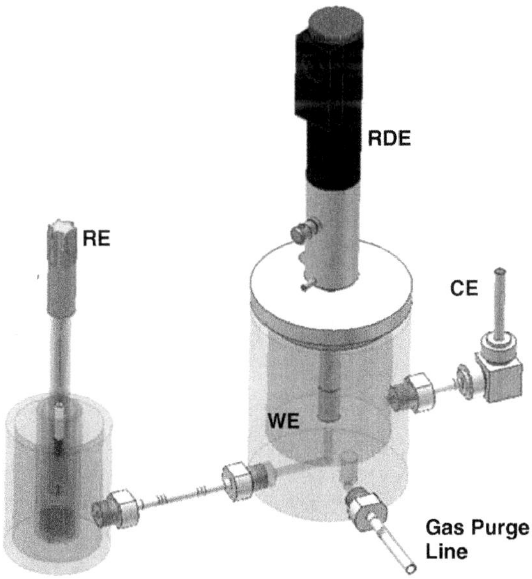


**FIGURE 9.1** Electrochemical setup with all-Teflon cell and rotating-disk electrode (RDE), working electrode (WE), counterelectrode (CE) site compartment, and reference electrode (RE) in site compartment with ion-conducting membrane separator [14].

through the induction coil outside the glass vessel in which the electrode is located, has been introduced. By this technique, which allows a more reproducible preparation of electrodes, the electrode is exposed during annealing to a defined gas atmosphere, its temperature is controlled in a reproducible manner, and after annealing the electrode surface can be subjected to a gas-purged electrolyte (e.g., hydrogen) without being exposed to oxygen. As shown by Kibler et al. [17] as well as Strmcnik et al. [19], all these parameters define the prehistory of the surface, which has a substantial effect on the surface morphology and thus catalytic measurements.

Compared to bulk electrodes, because of their high surface area, supported catalysts are less prone to contamination. High-surface-area catalysts are typically prepared by ultrasonically dispersing the catalyst powder in ultrapure water to give a desired concentration. Additives such as 2-propanol and Nafion [20] may also be added to the catalyst ink in order to enhance the dispersion. Afterward, the ink is pipetted onto a polished glassy carbon substrate and dried in an inert-gas stream. The verification of surface cleanliness, which has become standard practice in the surface science of single crystals, is fundamentally more difficult for supported particles [21]. Prepared catalyst films may contain impurities for which removal conditions are required that affect particle size or morphology of the catalyst. A typical cleaning procedure is the continuous potential cycling in argon-saturated solution up to a potential of at least 1000 $mV_{RHE}$ until a stable voltammogram is obtained. Although it is well known that potential cycling has a strong surface-structuring effect, in most studies little attention is paid to the impact of these treatment conditions on the properties of the catalyst nanoparticles [22].

## 9.3 OXYGEN REDUCTION REACTION

Oxygen reduction is a crucial reaction in terms of not only fuel cell technology but also corrosion science. In a PEMFC, the oxygen reduction reaction is the most sluggish reaction, and thus the cathode catalyst loading is the major contributing factor to overall catalyst costs [23]. To date no catalyst material has been found to exhibit any measurable reduction current close to the equilibrium potential of 1.23 $V_{RHE}$. Even for the most active catalyst materials (i.e., Pt group metals), significant reduction currents are observed only below 1.0 $V_{RHE}$.

It is beyond the scope of this chapter to give a complete overview of the work considering the reaction pathway of the oxygen reduction reaction. Despite the continuous efforts to unravel it, there seems to be still no consensus on the mechanism. Nevertheless, it is generally accepted that the four-electron process of the complete reduction of oxygen to water may include a number of elementary steps involving different reaction intermediates, and various reaction schemes based on experimental and theoretical work have been proposed (for an overview, see Refs. 24–26). A *simplified* version (shown in Fig. 9.2) based on previous schemes given by Bagotsky et al. [27] and Wroblowa et al. [28] is frequently used by the authors' groups to discuss the *most important* factors concerning the activity of a given electrocatalysts toward ORR [29–31].

According to the scheme in Figure 9.2, $O_2$ can be electrochemically reduced to water via two different pathways, either in the so-called direct pathway, without intermediate formation of $H_2O_{2,ad}$, or via $H_2O_{2,ad}$ in the series pathway. Possible side reactions include the catalytic decomposition of adsorbed peroxide or its desorption into the bulk of the solution. While low amounts of desorbed $H_2O_2$ detected in RRDE measurements are often taken as a support for a direct pathway, our studies [32–34] suggest that the series pathway via an $(H_2O_2)_{ad}$ intermediate is the dominant process on all Pt and Pt bimetallic catalysts. In these studies it was shown that on different Pt surfaces polarization curves of the reduction of peroxide, which is added to the oxygen free electrolyte, match those of the ORR on the same surface, indicating a close

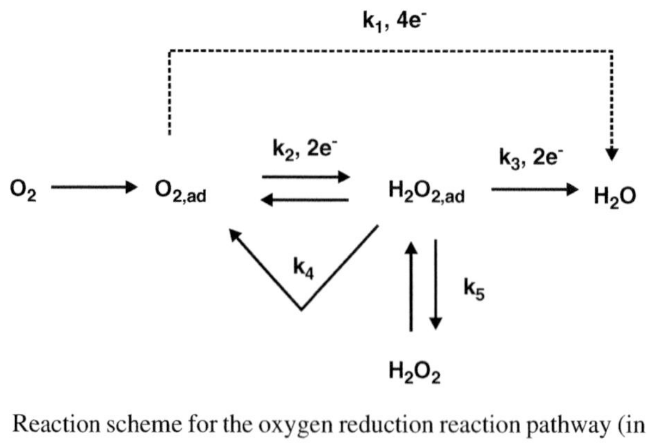


**FIGURE 9.2** Reaction scheme for the oxygen reduction reaction pathway (in acid solution).

relation between both reactions. Furthermore, the peroxide detected in RRDE measurements is significantly increased by adding strongly adsorbing anions such as bromide to the solution. Consequently, the ORR pathway was considered as a special case of the general mechanism where $k_1$ is essentially zero; thus, there is no splitting of the O–O bond before a peroxide species is formed. Further evidence for this assumption has been given by other authors showing that increasing hydrogen peroxide production is monitored in ORR investigation when decreasing the loading of a Pt/C catalyst in a TF-RRDE study or when increasing the particle distance of nanostructured Pt/GC electrodes in a double-disk thin-layer flow cell, respectively [35,36].

On the basis of the very simple reaction scheme (Fig. 9.2), the rate-determining step of ORR might be (no experimental evidence) the addition of the first electron to the oxygen molecule. The rate expression for the reduction current density $J$ is then given by [25,33,37]

$$J = nFkc_{O_2}(1-\theta_{ad})^x \exp(-\beta FE/RT)\exp(\gamma\Delta G_\theta)/RT) \tag{9.1}$$

where $n$ is the number of electrons, $k$ is the rate constant, $c_{O_2}$ is the concentration of $O_2$ in the solution, $\theta_{ad}$ is the total surface coverage by anions ($\theta_{anions}$) and $OH_{ad}$ ($\theta_{OHad}$), $x$ is either 1 or 2 depending on sites requirements of the adsorbates, $\beta$ and $\gamma$ are the symmetry factors (assumed to be $\frac{1}{2}$), and $\Delta G_\theta$ is the coverage-dependent Gibbs energy of adsorption of the reactive intermediates. In deriving Equation (9.1), it is assumed that the reactive intermediates, $(O_2^-)_{ad}$ and $(HO_2^-)_{ad}$, are adsorbed at low coverage, that is, are not a significant part of $\theta_{ad}$.

The most important insight comprised in this rate expression is that ORR does not proceed on a bare, but on a modified, catalyst surface. The kinetics are determined by the potential-dependent number of free platinum sites available for the adsorption of $O_2$ [$1 - \theta_{ad}$ term in Eq. (9.1)] and/or by the Gibbs energy of adsorption of the reaction intermediates, which may depend on $\theta_{OH,ad}$ or spectator species $\theta_{anions}$ as well. Hence, in order to evaluate the effects of the potential or oxygen pressure on the measured ORR kinetics, one has to be aware that the catalyst surface is not a static bare metal surface [38].

## 9.3.1 Kinetic Analysis of RDE Measurements

### *9.3.1.1 Turnover Frequency (TOF)*

The current density in Equation (9.1) is directly related to the turnover frequency (TOF), which is commonly used in heterogeneous catalysis [39]. The TOF is defined as the reaction rate per surface site in unit time. In electrocatalysis the concept of TOF is especially useful for comparison of Pt-based and Pt-free catalysts. The TOF can be determined from the faradaic current $i_f$ by the following equation

$$\mathrm{TOF} = \frac{i_f}{neF \times N} \tag{9.2}$$

where $e$ is the elementary charge, $n$ is the number of transferred electrons for the reaction of interest, $F$ is the Faraday constant, and $N$ is the number of electrochemical active sites determined by an adequate in situ EC technique (see text below).

As shown below, TOF can easily be related to the specific activity of platinum by multiplying the TOF value by the normalization constant for the electrochemical surface area and the number of transferred electrons in that reaction.

#### 9.3.1.2 Activity Determination

The intrinsic catalytic activity of an electrochemical reaction is defined by the exchange current density $J_0$. In the case of the ORR, however, the observed reaction rate close to reversible potential $E_0$ is essentially zero. Therefore, the oxygen reduction currents can be measured only at high overpotentials, and the catalytic activity of a cathode catalyst is commonly expressed by the kinetic current densities $J_k$ at a certain electrode potential.

The kinetic current density $J_k$ of the reaction can be obtained from RDE measurements of both extended and high-surface-area catalysts by correcting the observed disk current densities in the polarization curves for mass transport limitations:

$$\frac{1}{J} = \frac{1}{J_k} + \frac{1}{J_D} \Leftrightarrow J_k = \frac{J \times J_D}{J_D - J} \tag{9.3}$$

The potential window where $J_k$ can be extracted is limited to the potential region of mixed kinetic-diffusion controlled currents, ideally at current densities $J < J_D/2$. For high-surface-area catalysts the current densities (per geometric area) depend on the loading of the glassy carbon disk with the catalyst. Thus the potential region accessible to a kinetic analysis can be tuned to some extent by varying the catalyst loading [40,41]. Two critical points must be taken into account: (1) the GC electrode should be completely covered by the catalyst film and (2) the kinetic current density should be depend linearly on the catalyst loading.

While for bulk electrodes only surface-area-specific activities can be obtained, the catalytic activity of a high-surface-area electrocatalyst toward ORR can be expressed by two equally important characteristics: (1) the mass specific activity (in short, mass activity) $J_{k:m}$

$$J_{k:m} = \frac{J_k}{s_L} \tag{9.4}$$

where $s_L$ is the catalyst loading per geometric surface area, and (2) the surface-area-specific activity (specific activity) $J_{k:SA}$

$$J_{k:SA} = \frac{J_k}{R_f} \tag{9.5}$$

where $R_f$ is the roughness factor, defined as the ratio of the electrochemically active (ECA) surface area and the geometric surface area of the glassy carbon (GC) disk.

For a complete characterization of a high surface area catalyst both values should be determined.

#### 9.3.1.3 Mass Activity

The mass activity is primarily of economic importance, since it directly indicates the cost of energy conversion related to the deployed amount of catalyst. It is expressed in terms of $A/mg_{Pt}$ (or $A/mg_{precious\ metal}$ for alloy catalysts). Its determination is in principle rather straightforward. If the amount of deployed catalyst is known, the mass activity can be directly calculated from the total kinetic current in a Tafel plot. Care, however, should be taken to ensure that all the catalyst is, indeed, available for the reaction, which is not always the case. The complete accessibility of the catalyst can be probed by ensuring that the total kinetic current depends linearly on the catalyst loading. Only then can an unambiguous mass activity be obtained for the respective catalyst. Otherwise, part of the catalyst surface is inaccessible for the reaction, and thus the mass activity is underestimated. Furthermore, if too thick catalyst films are employed, resistance to oxygen diffusion through the catalyst film may occur. In early work on high-surface-area catalyst (HSAC) measurements via RDE, this effect was taken into account by introducing the so-called effectiveness factor. It turned out, however, that even then the analysis of these measurements is rather complicated if thick catalyst films are used. A breakthrough, therefore was the application of very thin catalyst films in RDE measurements by Schmidt et al. [42], the thin-film RDE methodology, which avoided diffusion resistance.

#### 9.3.1.4 Specific Activity

In comparison to the mass activity, from a fundamental point of view the specific activity is the more meaningful property of a catalyst: (1) it is directly related to the turnover frequency, and (2) only the specific activity allows for a comparison of model systems, such as bulk electrodes, with industrial high-surface-area catalysts. The specific activity for the ORR is expressed in terms of $\mu A/cm^2_{Pt}$.

The ORR specific activity is less affected by an incomplete utilization of the catalyst than the mass activity. However, for its calculation the determination ECA surface area is necessary. For Pt-based catalysts, the ECA surface area is determined by either the stripping charge of underpotential deposited hydrogen ($H_{upd}$) or oxidation of an adsorbed CO monolayer (CO stripping) [12]. For nonplatinum metal catalysts, these rather straightforward methods are seldom feasible. A range of alternative methods have been reviewed by Trasatti and Petrii [43].

A weakness of most methodologies for the ECA determination via the charge related to a certain adsorption/desorption or oxidation processes is that the equilibrium coverage of the molecule or atom on the respective polycrystalline material has to be known. Not only has each polycrystalline material an intrinsic surface roughness; the surface coverage can depend on both geometric and electronic properties of the surface (in general, both properties cannot be separated). A prominent example is the different $H_{upd}$ or CO stripping charges on the low-index planes of Pt single crystals [44,45]. Furthermore, the $H_{upd}$ charge of a Pt alloy surface can also be drastically different from that of the corresponding pure Pt surface, as seen in the case of the voltammogram of $Pt_3Ni(111)$ compared to Pt (111) [46]. In addition to these intrinsic difficulties, all currents that do not originate from the surface stripping process, such as double-layer charging, must be properly subtracted.

In the experience of the authors for Pt-based catalysts CO, stripping is the most accurate method for determining the ECA. Carbon monoxide adsorbs quite strongly on Pt at a rather constant equilibrium coverage. By comparison, as shown by Strmcnik et al. [45] on Pt single crystals, the $H_{upd}$ coverage on a Pt catalyst might not always be completed before the onset of hydrogen evolution.

#### *9.3.1.5 Guidelines for Determining the Activity of Cathode Catalysts via RDE Measurements*

The following guidelines are recommended for the characterization of various kinds of catalysts using the thin-film RDE methodology [10,41]:

1. The number of electrochemical active sites must be determined. The value obtained should be independent of the applied methodology.
2. Capacitive currents must be subtracted for the evaluation of electrocatalytic activity.
3. Compensation schemes for the solution resistance ($R_u$) must be implemented.
4. The diffusion limit current of ORR should be within 10% of the theoretical value.
5. The current at the potential of interest must be within 10% and 80% of the diffusion limit current, ideally below the half-wave potential.
6. The electrode thickness for the thin-layer method should not exceed 0.1 μm.
7. Reaction rate and ECA must depend linearly on loading.
8. Normalized activities, namely, mass and specific activity, should be expressed in a fuel cell relevant potential window versus the reversible hydrogen electrode.

Guidelines 1–5 are applicable to all activity measurement, while guidelines 6–8 are specific to nanosized catalysts. Measurement of the ECA sites and the subtraction of capacitive currents enable adequate evaluation and normalization of the activity. Compensation of the solution resistance ensures a correct potential determination and application, whilst reaching the expected diffusion limited current density indicates complete coverage of the GC with the catalyst. The potential window of measurable kinetic currents can be shifted by changing the catalyst loading, thereby avoiding extrapolation errors. The range of useful loadings has an upper limit, since the catalyst layer must be kept thin in order to avoid any diffusion resistance. At too high a catalyst loading, the usage is below 100% as shown by a deviation from the linear relationship between overall reaction rate and catalyst loading. Finally, the data should be presented in a manner that enables a straightforward comparison; not only to other RDE measurements, but also to flow cell and fuel cell data.

#### 9.3.1.6 Relating RDE Studies to PEMFC Measurements

Since the late 1990s there has been major disagreement between PEMFC measurements and RDE studies. In a PEMFC, the activity is determined by Tafel plots, specifically

$$\log(J_k) = \frac{E - E_0}{b} \tag{9.6}$$

where $E$ is the electrode potential, $E_0$ is the reversible potential for the oxygen reaction, and $b$ is the Tafel slope. As shown in a study by Neyerlin et al. [47], if one corrects for all resistances, a Tafel slope of 60 mV/dec is obtained in a PEMFC over the whole accessible potential range. Many older RDE studies, however, report two different Tafel slopes for the ORR on Pt depending on the applied potential. At low potentials, where the surface is assumed to be free from oxides, slopes of 120 mV/dec are extracted, whereas at higher potentials a value 60 mV/dec is determined. This finding is often interpreted such that 120 mV/dec is the intrinsic value for ORR that should be related to the ORR rate-determining step [48]. The Tafel slope obtained at potentials higher than 60 mV/dec is considered as an apparent slope due the potential dependent change of the oxide coverage on the catalyst surface. Considering all the guidelines for RDE measurements, given above, however, results in fairly good agreement with PEMFC studies are obtained as demonstrated in Figure 9.3. In this limited potential window a Tafel slope of about $60 \pm 5$ mV/dec (not a completely uniform slope) is obtained for both scan directions, that is, for the positive-going scan, which starts with

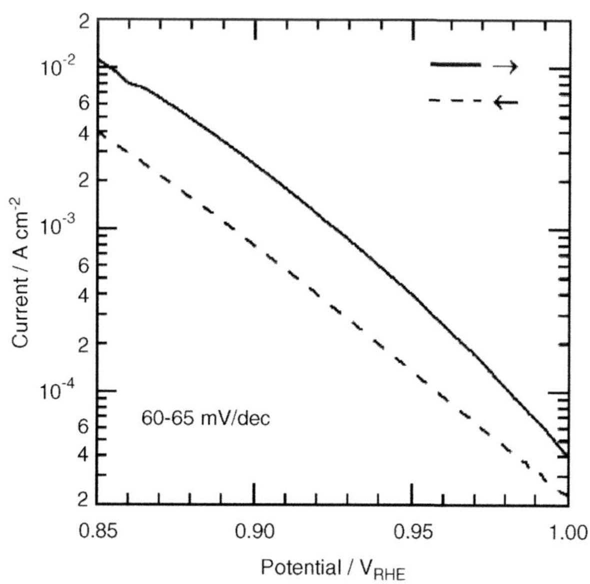


**FIGURE 9.3** Tafel plot of ORR on polycrysalline Pt in 0.1 M $HClO_4$ at room temperature with correction of solution resistance $R_u$. The scan direction is indicated by the arrows.

a reduced surface and for the negative-going scan, which starts with an oxide-covered surface.

With the longstanding discussion of the "right" Tafel slope, it seems instructive to discuss the experimental parameters affecting the measured, apparent Tafel slope. As pointed out by Gasteiger et al., the analysis of RDE curves close to the diffusion-limiting currents, where higher Tafel slopes are found, is problematic. The intrinsic uncertainty in this potential region is highest as the denominator in expression (9.3) approaches zero. An additional factor that is often neglected is the influence of the solution resistance $R_u$ on the Tafel slopes [49]. Even with the use of a Luggin capillary, without proper compensation of the residual resistance (which in older studies is rarely the case), the curvature in the Tafel slope toward lower potentials and thus higher currents is artificially increased. A claim of an intrinsic Tafel slope of 120 mV/dec therefore seems arbitrary as its determination is hardly certain.

### 9.3.2 Structure Sensitivity of ORR on Platinum

In electrocatalytic measurements, the electrode preparation and the potential prehistory play an important role in evaluating structure–function relationships. Early studies that reported insensitivity of ORR to the Pt surface structure [50] can be clearly related to preparation difficulties of well-ordered Pt single-crystal surfaces. A breakthrough in analysis of the structure sensitivity of Pt therefore was the relatively straightforward preparation of Pt single crystals by flame annealing and its clean insertion into an RDE setup [29,51,52]. As discussed in Section 9.2, the preparation of well-defined surfaces was further improved, with the introduction of inductive annealing [17,18].

Following the straightforward preparation of Pt single crystals, it could be demonstrated that ORR on Pt is a structure-sensitive reaction. This structure sensitivity arises not only from changes in the adsorption energies of reactive intermediates but also, more significantly, from the structure sensitivity adsorption of anions contained in the supporting electrolyte. Studies in different electrolytes revealed that adsorbed anions, such as sulfate [52], hydroxyl [29], and halides [33], are not intermediates in the ORR, and instead act as spectator species that impede the kinetics of the reaction. Therefore, the structure sensitivity of ORR on Pt is most pronounced in electrolytes with strongly adsorbing anions [44]. The activation energy for ORR on Pt, however, seems to be affected by neither the surface structure nor the electrolyte. In acid electrolyte a value of about 42 kJ/mol was determined for Pt single crystals [32], and corresponding values of 37–45 kJ/mol were found in alkaline solution [53].

A comparison of the cyclic voltammograms and ORR polarization curves obtained for Pt(111), Pt(100), and Pt(110), in both perchloric acid solution and sulfuric acid solution, is shown in Figure 9.4. From the cyclic voltammograms it is seen that adsorption/desorption processes mostly shift toward more negative potentials with increasing temperature, except in the case of bisulfate adsorption on a (111) surface. For smooth bulk electrodes, the activity is directly linked to the half-wave potential, that is, the potential where in the polarization curves the observed reduction current reaches half the value of the diffusion-limited current. Therefore, from Figure 9.4

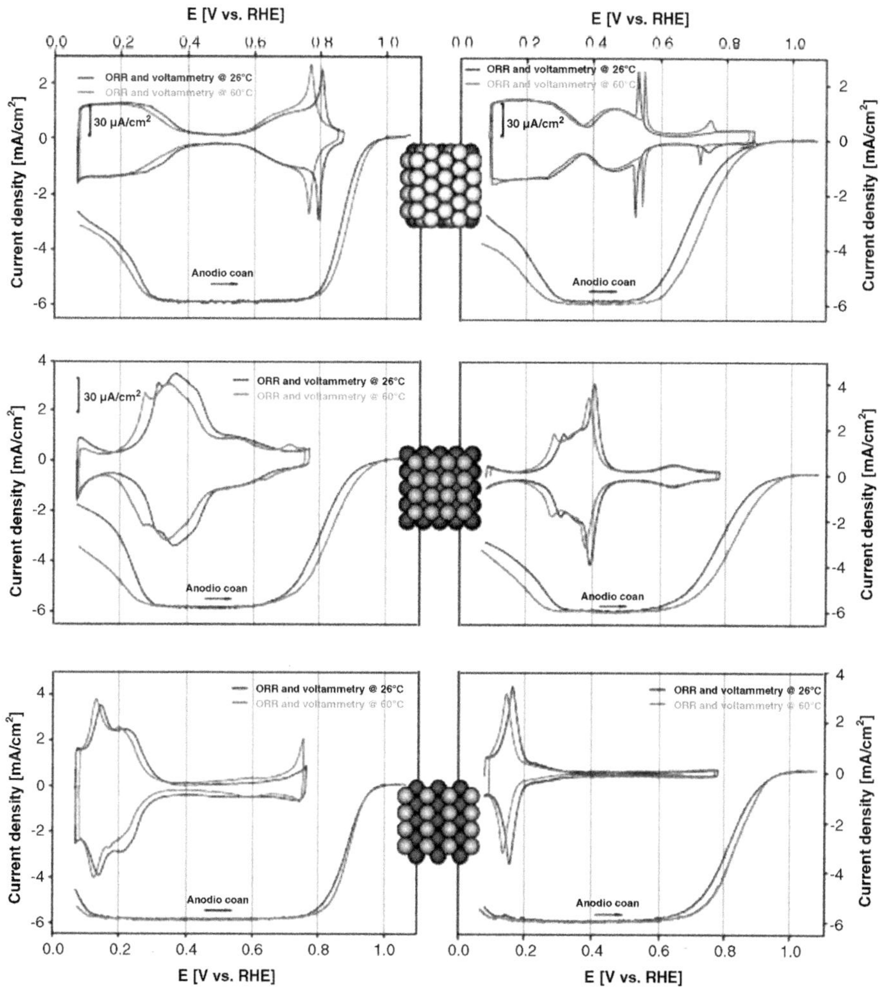


**FIGURE 9.4** ORR polarization curves and cyclic voltammetry of Pt(*hkl*) in 0.1 M $HClO_4$ (left side) and 0.05 M $H_2SO_4$ (right side). The sweep rate was 50 mV/s and the temperature 298 K (blue curves) and 333 K (red curves), respectively. (Figure taken with permission from Ref. 54.)

it can be seen that changes in activity manifest in small shifts of the polarization curves demanding a high accuracy in the measurements.

#### *9.3.2.1 Activity Benchmarking*

As Gasteiger et al. pointed out in a 2005 review [23], well-defined activity benchmarks and measurement methodologies for state-of-the-art Pt/C catalysts are required in order to effectively develop improved ORR catalysts for PEMFCs. In practice, newly proposed catalysts are compared to existing Pt/C catalysts or polycrystalline Pt. The reported activities of these standards, however, often vary by an order of magnitude in

**TABLE 9.1 Benchmark Activities for ORR Expressed as Specific Current Density**

| | 0.1 M $HClO_4$ | | 0.05 M $H_2SO_4$ | |
|---|---|---|---|---|
| Surface | $J_k$ at 0.9 V, 298 K ($\mu A/cm^2_{Pt}$) | $J_k$ at 0.9 V, 333 K ($\mu A/cm^2_{Pt}$) | $J_k$ at 0.85 V, 298 K ($\mu A/cm^2_{Pt}$) | $J_k$ at 0.85 V, 333 K ($\mu A/cm^2_{Pt}$) |
| Pt(111) | 1500 | 2800 | 220 | 500 |
| Pt(100) | 520 | 1110 | 1090 | 3000 |
| Pt(110) | 4200 | 4600 | 2200 | 3600 |
| Pt-poly | 2100 | 2600 | 1760 | 3200 |

*Source:* Data taken from Reference 54.

both rotating-disk electrode (RDE) experiments and in PEMFC testing [23]. As a consequence, proposed new catalysts often show performance inferior to that of existing state-of-the-art catalysts. In order to avoid such inhibiting effects in the effort to develop improved catalysts, standard procedures and benchmarks would enhance the comparability of results obtained by different research groups. While guidelines for catalyst testing in MEAs can be found in the paper by Gasteiger et al. [23], the authors extended the RDE guidelines, enabling the efficient comparison of Pt-based and Pt-free catalysts.

Table 9.1 lists benchmark activities for ORR on well-defined Pt single crystals and polycrystalline Pt. The activities were determined from the polarization curves shown in Figure 9.4 using Equation (9.3). The bulk crystals were prepared by inductive heating and measured with compensated solution resistance. The standard deviation of the measurements was typically 5% or less, proving the high reproducibility of sample preparation, transfer, and measurement. The potentials for the benchmark values were selected such that the measured reduction currents are close to the kinetically controlled potential region, but the contribution of pseudocapacitve currents is still negligible.

The order of activity in perchloric acid at room temperature is Pt(110) > Pt-poly ~ Pt(111) > Pt(100). At 333 K a twofold increase is observed at Pt(100) and Pt(111), whereas the kinetic current does not increase much on Pt(110) and Pt-poly. The influence of temperature on catalytic activity can be understood in terms of two effects on the Pt model catalysts: (1) it promotes the kinetics of the ORR and (2) it influences the adsorption of spectator species (at least for $OH_{ad}$ as seen from cyclic voltammograms in Fig. 9.4). The combination of these two effects determines the relative increase of the kinetic current with temperature. In sulfuric acid, the order of activity at room temperature is Pt(110) ~ Pt-poly > Pt(100) > Pt(111). A twofold increase in activity is observed at 333 K.

#### 9.3.2.2 *Particle Size Effect*

The effect of particle size on the electrocatalytic activity of Pt toward ORR has been subject of many controversial studies. While the majority of investigations conclude that there are reduced TOFs on small nanoparticles [23,24,40,55–62], there are also reports of size-independent TOFs [63–68]. The question surrounding the particle size effect is, whether, and at which particle size the mass activity gain achieved by using highly dispersed Pt (small particle size) is outweighed by a loss in specific activity. This makes understanding of the particle size effect and its underlying factors a crucial issue for conceptual strategies in cathode catalyst development.

The earliest studies concerning the influence of particle size on the specific activity (or TOF) date back to work on phosphoric acid fuel cells [55,56]. Bregoli was one of the first to indicate that efforts to increase the active surface area of a catalyst by reducing the particle size of Pt result in a loss in the specific activity for the oxygen reduction reaction. Sattler et al. found a similar but more pronounced behavior reporting a drastic drop in specific activity for Pt particles below 2 nm in diameter, with a projected near-zero activity for a monoatomically dispersed state [57]. At the same time a relatively constant specific activity was reported for "larger" particles (i.e., >3.5 nm). As a consequence, the technologically relevant mass activity peaked at a size of 3.5 nm. By utilizing TEM, the authors correlated the observed behavior with changes in the morphology of the particles with size. They argued that the relative amount of surface atoms in different crystal facets changes with size and that for small particles the (111) face dominates the surface while they are rounded off by the formation of (110) facets. It was proposed that the (111) face and rough (110) facets can be considered to bond oxygen in a less active form, leading to lower activity of small particles. On the basis of experimental studies by Sattler et al. [57] and Peuckert et al. [69], Kinoshita presented an interpretation of particle size effect based on the mass-averaged distribution (MAD) and surface-averaged distribution (SAD) of atoms on the (111) and (100) crystal faces and on the edge and corner sites of ideal cubo-octahedral particles as a function of their size. In this popular model [24,58] it was assumed the reactivity of small Pt particles for oxygen reduction is directly related to MAD of the different faces in the case of mass activity and SAD for the specific activity. It was shown that the reported maximum in mass activity at a particle size of ~3.5 nm correlates with the maximum MAD of the (100) and the (111) face of cubooctahedral particles, with the latter 7 times larger than the MAD (100) value. Thus both faces were identified as possible candidates for being active sites for the oxygen reduction. The corresponding increase of SAD (111) and (100) as a function of particle size explains the increase in specific activity with the size of highly dispersed Pt particles. Newer studies (see above), however, have revealed that in acid electrolyte the low-index Pt surface Pt(110) is the most active, while the order of activity of Pt(111) and Pt(100) strongly depends on the adsorption strength of the anion in solution.

More recent models explaining the particle size effect concentrate on the size-dependent electronic properties of nanoparticles and their effect on the adsorption strength of the reactants and oxygenated spectator species. A stronger adsorption of oxygen atoms on smaller particles is also well documented in the gas phase [70], where it has been found that oxygen atoms appear to dissolve quite easily into small metal particles, so that a clear distinction between chemisorption and bulk oxidation cannot always be drawn. As a consequence, in heterogeneous catalysis the decrease in the rate of oxidation reaction of organic molecules with decreasing particle size is usually attributed to a stronger adsorption of oxygen atoms on smaller particles. In accordance with gas-phase catalysis, Takasu et al. [60] attributed the particle size effect to stronger binding of reactants, in particular molecular oxygen, to the catalyst surface. This was derived from XPS measurements of the core-level binding energy of electrochemically aged Pt particles on a glassy carbon electrode. A similar conclusion was drawn from in situ XAS studies by Mukerjee et al. [61], namely, the adsorption of $OH_{ad}$, among other molecules, is stronger on smaller particles. Since the ORR does not proceed on bare Pt, but on surfaces modified by the electrolyte, the particle size effect is explained by the influence of electronic effects on the adsorption and desorption potential of oxygenated spectator species. The large fraction of low-coordination surface atoms on nanoparticles shifts the adsorption/desorption of OH or anions toward lower potentials. Indications that the adsorption strength of $OH_{ad}$ species in an electrochemical environment, indeed, changes with Pt particle size can be found by using a number of different methodologies. Comparing, for example, the cyclic voltammetry Pt catalysts with different mean particles sizes, it is observed that the features commonly related to the adsorption and desorption of OH shift toward lower potentials with decreasing size [23,40]. A similar effect is revealed by investigating the oxidation of CO molecules in solution (CO bulk oxidation). Adsorbed CO molecules strongly inhibit this reaction at low potentials and act as a surface poison. Only at higher potentials where OH species adsorb, the reaction can proceed. In polarization curves, therefore, the onset of CO bulk oxidation in the positive-going curve as well as its suppression in the negative curve indicate the potential-dependent surface coverage of oxygenated species. While in the positive-going curve no unambiguous differences in the onset of CO bulk oxidation with particle size are found (note that the reaction starts with a CO precovered surface), in the negative-going curve the suppression potential shifts with smaller sizes toward lower potentials, indicating higher adsorption strengths for $OH_{ad}$. However, it should not be ignored that CO stripping experiments show a different behavior, which is not in perfect agreement with a higher oxophilicity of small particles. In these measurements it is found that smaller particles are less active for CO monolayer oxidation than are larger particles, indicating that particle edges are not the active sites for this reaction and thus for OH adsorption. An explanation of this counterintuitive behavior of small particles was given by proposing to a reduced CO mobility [71,72] and a higher defect density on small particles, respectively [73,74].

A methodology that can be used as a sensitive in situ tool for the electronic state of a catalyst is determination of the so-called potential of zero total charge (PZTC) by CO displacement. In such measurements an influence of the particle size on the PZTC was found. It is observed that the PZTC shifts to more negative potentials for smaller particles in accordance with a higher oxophilicity.

Figure 9.5 summarizes the findings of a study from the authors concerning the particle size effect of Pt in perchloric acid solution. It can be concluded that the PZTC shifts approximately 35 mV in the negative direction by decreasing particle size from 30 nm down to 1 nm (Fig. 9.5a), while at the same time the reaction rate of the CO bulk oxidation as well as the ORR (Fig. 9.5d) inversely shift with the particle size, indicating an increased adsorption strength of oxygenated species for smaller particles. On the CO bulk oxidation, this property has a beneficial effect, while at the same time it is detrimental for the ORR, as $OH_{ad}$ actively blocks the active sites

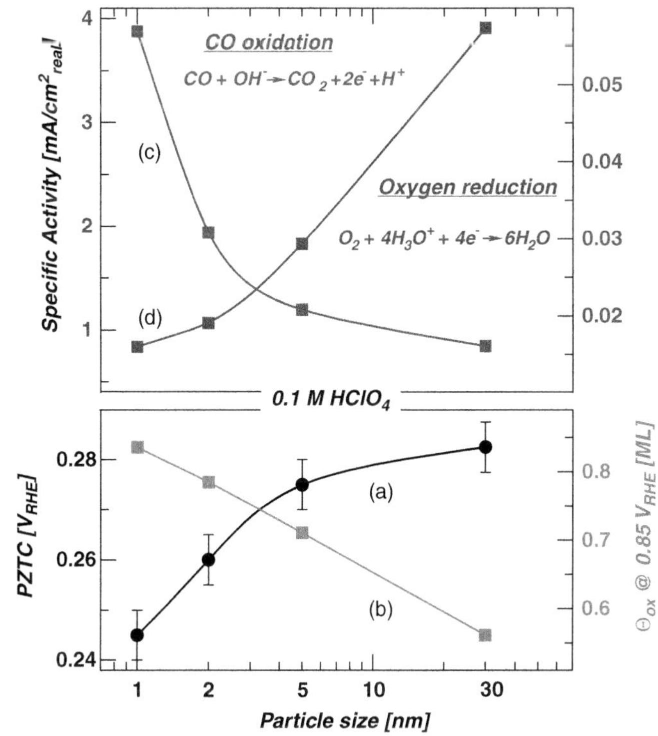


**FIGURE 9.5** Summary of the particle size dependent PZTC (curve a) and the oxide coverage (curve b) obtained from CO displacement and cyclic voltammetry, respectively, as well as the negative going curve of CO bulk oxidation (curve c) and the specific activity of the ORR (curve d). The measurements showcase the adsorption of OH and its impact on the electrochemistry at the Pt electrode–solution interface. (Figure taken with permission from Ref. 40.)

required for the adsorption of $O_2$ and/or the splitting of the O–O bond. Despite the advances in reporting and explaining the Pt particle size effect for the ORR, there are still some open questions. In the study by Takasu et al. as well as in more recent studies the most drastic difference in specific activity is observed by comparing Pt nanoparticles in the size range of 5 nm with Pt black particles in the size range of ~40 nm or to polycrystalline Pt [23,40,60]. These findings cannot be explained by the purely geometric model of Pt nanoparticles and are counterintuitive to an explanation based on electronic properties, as the largest changes are expected at sizes below 5 nm. Furthermore, the reported net reaction rates under comparable conditions differ widely. For example, in the study by Takasu et al. [60], specific activities for 5-nm Pt particles of $\sim 10\,\mu A/cm^2_{Pt}$ at $0.9\,V_{RHE}$ and room temperature are compared to $200\,\mu A/cm^2_{Pt}$ in the study by Mayrhofer et al. [40].

Watanabe and coworkers proposed a completely different explanation for the observed particle size dependence of the reaction rate [64,66,68]. According to the territory theory, a mutual influence on the diffusion or some other parameters exist when Pt particles are close together. As a consequence, not all of the Pt surface area is usable for the reaction. The particle size effect is thereafter not truly dependent on crystallite size, but rather on interparticle distances, with a critical distance of 20 nm above which the specific activity was constant. While an increase in peroxide production and thus incomplete reduction of oxygen with increasing particle distance has been reported by Schneider et al. [36], the territory hypothesis was questioned in an extensive effort by Giordano et al. [56], who could not find conclusive evidence for the significance of interparticle distance on the electrocatalytic activity. Furthermore, from the measurement of the zero point of charge of carbon-supported Pt, it was concluded that the interaction between the electrocatalyst and the carbon support is negligible, and does not influence the oxygen reduction.

#### 9.3.2.3 Influence of Surface Defects on ORR

An issue closely related to the particle size effect is elucidation of the active site for the oxygen reduction reaction. As the surface structure influences not only the adsorption of the reactants and intermediates but also, perhaps even more importantly, the adsorption of spectator species, the inherent structure sensitivity of the reaction itself is difficult to access. One possible approach is to study the influence of in situ surface treatment on the surface structure and its reactivity toward certain reactions. Several studies by the authors concern the effects of potential cycling on restructuring of the catalyst surface. While the phenomenon termed *electrochemical annealing* [75,76] is well known for a variety of different metals, for example, for Au and Cu crystals [17,77–79] in halide-containing electrolytes and Ru and Pt crystals in CO-saturated solution [74,80–84], its influence on reactivity has been investigated less. In a study concerning the particle size effect of Pt with regard to CO oxidation, it was observed that potential cycling in CO-saturated solution (i.e., CO annealing) decreases the reactivity of the respective surface for CO stripping [73,74]. It was proposed that CO annealing leads to the removal of surface defect sites [74] controlling the reactivity for CO oxidation. This hypothesis was recently substantiated in a combined experimental–theoretical study showing that during CO annealing of Pt(111), adislands are removed from the surface resulting in the formation of large regular terraces [83]. However, CO annealing (limited to $\sim$0.75 $V_{RHE}$) neither influences particle size nor induces any particle agglomeration that could affect the catalytic properties of the catalyst. While CO annealing influences CO oxidation, no influence of the treatment conditions on the oxygen reduction reaction was observed following the CO adlayer oxidation. Thus it was concluded that both reactions are governed by different active sites [22].

### 9.3.3 Bimetallic Catalysts for ORR

The concept of bimetallic catalysts is not new and originates from gas-phase catalysis, where many examples of superior properties as compared to the pure metals are reported (see Refs. 39 and 85–87 and sources cited and therein). Bimetallic catalysts may consist of alloyed metals, but different types of bimetallic model systems such as thin metal films deposited on a foreign substrate [88–90] have been extensively investigated as well. The observed changes in reactivity of bimetallic systems can be attributed to electronic, structural (e.g., surface strain), and/or ensemble effects. Electronic effects are closely related to structural effects and are in most cases difficult to separate. They originate in the fact that the electronic structure of the reactant or intermediate is modified by the different surroundings that affect the reactivity of the surface. The relation between the electronic properties of a catalyst and the rate of reaction often manifests itself in "volcano-type" curves, with the platinum group metals at or near the top of the "volcano plot" for many reactions. For the ORR, the "volcano" relationship appears to be valid for bimetallic alloys as well as for pure metals, showing variations in reaction rates by five or six orders of magnitude [91,92]. Besides the electronic effects in most bimetallic systems with respect to the pure metal, the distribution and types of reaction sites (ensembles) is altered as well, which can have an important effect on the measured reactivity and selectivity of a catalyzed surface reaction [93].

Thus, the effect of a bimetallic surface on a catalytic reaction can be manifold, including changing the reaction pathway, modifying the stability of intermediates, and a better suppression of competing and undesired reaction pathways. An enhanced reactivity may be due to a lowering of a barrier for the rate-limiting step, but it could just as well be due to a destabilization of adsorbed reaction products blocking sites [86]. The optimization of an electrocatalyst must, therefore, accommodate many contributing factors, including rate dependence on the geometry of the surface atomic structure.

#### *9.3.3.1 Pt Alloy Model Catalysts*

Several studies have been carried out to elucidate the role of alloying for the electrocatalytic activity of Pt with respect to the ORR (see Ref. 94 and sources cited therein). As the reported reaction rates differ broadly even for pure Pt high-surface-area catalyst, it is difficult to unambiguously correlate the different origins of improved catalytic performance on industrial bimetallic catalysts. One of the main inhibiting factors of understanding bimetallic catalysts is, in general, the unknown surface composition of the prepared catalyst particles, which can be quite different from its overall composition. Even more importantly, the particle composition is not necessarily stable in a reactive environment. As a consequence, studies have been performed on model systems in which the initial state can be characterized in UHV with surface science tools. The most promising candidates for PEMFC cathode catalysts are alloys of Pt with Ni, Co, or Fe [95–100].

The influence of the preparation procedure on bulk alloy systems was demonstrated in several studies by Stamenkovic et al. utilizing surface-sensitive analytical UHV techniques [99–103]. The composition of the topmost surface layer was monitored with low-energy ion scattering (LEIS), whereas Auger electron spectroscopy revealed changes in composition of the surface region of ~5 to ~10 Å deep. For a variety of alloys of Pt with different transition metals at a ratio of 3 : 1, two different surface compositions can be obtained by preparation in UHV: the so-called Pt-skin structure with only Pt surface atoms and a surface composition that is equal to the nominal composition of the alloy. Pt skin structures are produced by inducing Pt surface segregation, that is, an exchange of Pt and the less noble second component in the first layer, due to annealing the alloy to 1000 K. The strong enrichment of Pt at the surface is counterbalanced by the depletion of Pt in the first two or three layers, resulting in a concentration profile that oscillates around the bulk value [104–107].

In contrast, the nominal surface composition is obtained by light sputtering of the surface with Ar ions. Bulk alloys with nominal surface composition, however, are not stable in an acid solution. As shown by subsequent retransfer into UHV, in acid solution the nonnoble component of the surface is leached out into the electrolyte. Therefore, surfaces prepared by light sputtering were coined as "skeleton structures." In Figure 9.6 the improved catalytic properties of these bimetallic alloys are

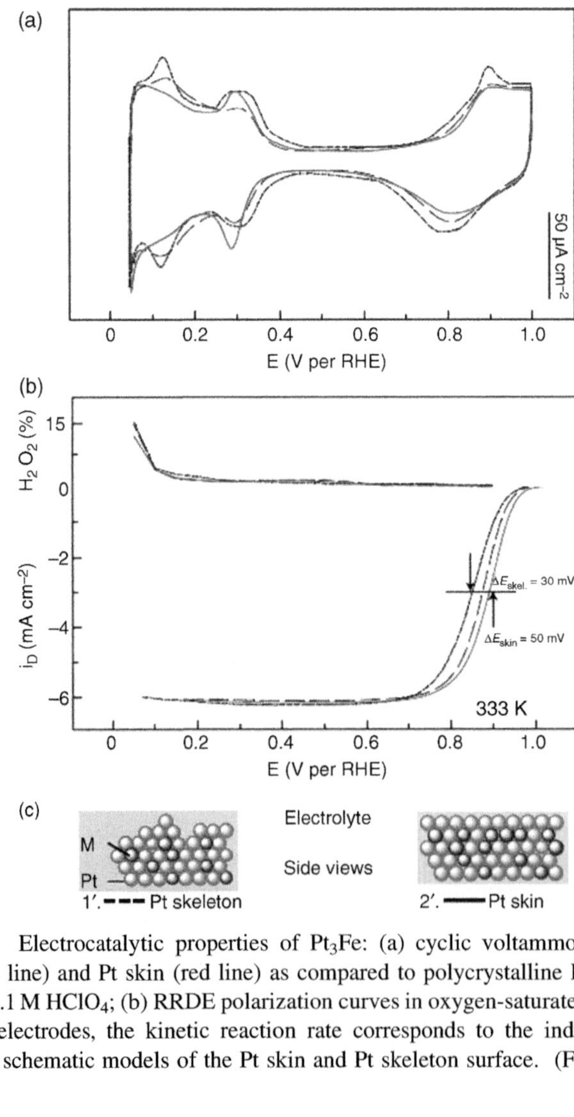


**FIGURE 9.6** Electrocatalytic properties of $Pt_3Fe$: (a) cyclic voltammograms of the Pt skeleton (blue line) and Pt skin (red line) as compared to polycrystalline Pt (black line) in Ar-saturated 0.1 M $HClO_4$; (b) RRDE polarization curves in oxygen-saturated electrolyte (for smooth bulk electrodes, the kinetic reaction rate corresponds to the indicated half-wave potential); (c) schematic models of the Pt skin and Pt skeleton surface. (Figure taken from Ref. 103.)

demonstrated for $Pt_3Fe$. In comparison to polycrystalline Pt, the Pt skin and corresponding Pt skeleton structure exhibit improved reaction rates toward the ORR. That Pt skin is more active than the skeleton surface, is an observation that holds true for most investigated alloys. For the majority of polycrystalline transition metal Pt alloys, the ORR activity compared to pure polycrystalline Pt is increased by a factor of 3–5 [103]. Single crystals, meanwhile, can exhibit even more astonishing properties. A highlight of the model singe crystal systems investigated so far is certainly the $Pt_3Ni$ (111) surface, which has been shown to exhibit an improvement by a factor of 10 compared to Pt(111) [103].

The improved catalytic properties of alloy model catalysts can be explained on the basis of a combination of theoretical calculations and high-resolution ultraviolet photon spectroscopy (UPS) by a slight modification of the electronic structure of the Pt surface atoms [101]. As the interaction energy of the oxygen reactant and possible intermediates is not accessible experimentally, the variation in surface electronic structure induced by alloying was determined by measuring the *d*-band centers of the different surfaces. An alternative to the *d*-band center is the determination of core-level shift by X-ray photoelectron spectroscopy [97]. Both values are correlated; that is, the *d* states and the core levels shift accordingly [108]. Plotting the variations in *d*-band centers (or the surface core-level shifts) of a variety of alloys against their catalytic activity for the ORR in acid solution results in a classical volcano-shaped dependence. In agreement with the experimental considerations described in the previous section, such behavior can be interpreted such that for metals that bind oxygen too strongly the rate is limited by the removal of adsorbed O and OH species. In contrast, for metal surfaces that bind oxygen too weakly the rate is limited by the dissociation of $O_2$. This phenomenological picture of the operation of alloy catalysts is substantiated by density functional theory (DFT) calculations. By determining the free energies of the possible intermediates of the ORR as a function of cell potential, a kinetic model was developed that gives the rate of the ORR at a given potential as a function of a single parameter characterizing the catalyst surface, the oxygen chemisorption energy [101]. Thus the description of the complex oxygen reduction reaction was considerably simplified by utilizing a single *descriptor* for the reaction. In agreement with the experimental findings, the theoretical model predicts a volcano-shaped dependence of the reaction rate on the oxygen chemisorption energy, which is related to the center of the *d* electrons. Thus the theoretical work not only offers insight into the underlying reasons for observed trends in catalytic activity but can also be utilized in screening a large variety of different materials for their catalytic properties. It therefore becomes an increasingly important tool in the search for novel catalyst materials.

#### 9.3.3.2 Concepts for Improved High-Surface-Area Alloy Catalysts

Inspired by the impressive activity gains obtained for the ORR on bulk model systems, the transfer of knowledge for the development of applied cathode high-surface-area catalysts has been in the focus of the latest research efforts. Here core–shell-type nanoparticles similar to the Pt skin and Pt skeleton structures, respectively, are highlighted. These catalyst particles contain a less expensive, nonnoble (or alloy) core surrounded by a noble metal shell. In the ideal case the catalytically active metal, namely Pt, is located only on the surface on which reactions take place while the reaction rate does not suffer from the cheap inactive core material. Thus an optimal utilization of the expensive active material would be guaranteed.

For the synthesis of such improved high-surface-area cathode catalysts, different approaches are pursued. Core–shell structures can be obtained by high-temperature annealing [109], chemical leaching of the nonnoble material [68,96,110], or electrochemical deposition techniques [111,112]. The practical applications of these concepts differ. While some of the preparation routes, such as the Pt monolayer deposition by galvanostatic displacement of copper [111], are difficult to apply on a large scale, these studies nevertheless give important conclusions about the properties of nanoparticle catalysts.

An experimental approach to obtaining core–shell particles that may be potentially upscaled is the electrochemical leaching of alloy catalyst, that is, dealloyed high-surface-area catalysts. As demonstrated by Strasser and coworkers starting with homogeneous alloy particles with a high content of a nonnoble component, skeleton structures are obtained in a electrochemical leaching process. A system especially suitable for this procedure is Pt–Cu. Both materials crystallize in a face-centered cubic (FCC) structure and form a complete solid solution. The catalyst synthesis starts with the impregnation of a high-surface-area carbon support with a Pt–Cu alloy of high copper content, such as $Pt_1Cu_3$. As the copper atoms are considerably smaller than the Pt atoms, the distance between two atoms in this alloy is significantly lower than in the pure metal. In the next step the copper is selectively dissolved. Because of the quite different Nernst potentials of dissolution of Pt and Cu this can easily be done by a treatment in hot acid without potential control. An alternative route is to apply an electrochemical potential positive of the Cu dissolution potential but still negative of Pt corrosion. In the latter case the advantage is that the speed of the dealloying process can be adjusted by the potential. However the leached copper atoms have a detrimental effect on the conductivity of the membrane in a fuel cell. Therefore in a fuel cell after dealloying, the membrane has to be cleansed from the Cu ions by a chemical treatment with an inorganic acid before being operational [110]. This process is schematically shown in Figure 9.7 for a base metal rich Pt–Cu–Co catalyst.

In either preparation, the dealloying process forces the remaining Pt atoms to reorganize. The goal is to maintain a core of the copper-rich phase to obtain a gradient in the composition in the two phases. Reported are structures where the total Pt content from 25 atom % in the precursor changed after dealloying to about 80% [113]. In discussing the ORR activity of these dealloyed high-surface-area catalysts (HSACs), one has to distinguish between the mass and the specific activity (see text above). In both cases the improvements depend on the annealing temperature as well as the composition of the precursor. For the $Pt_1Cu_3$ precursor the specific activity improves by increasing the alloying temperature from 600°C to 800°C. Increasing the temperature further to 950°C gives only a slightly better performance. The mass activity, however, is similar for the first two annealing temperatures, 600°C

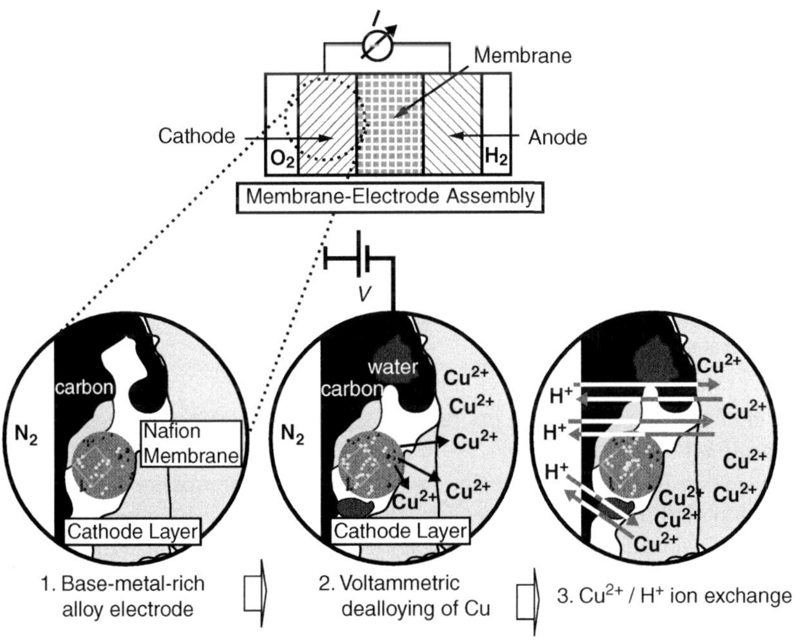


**FIGURE 9.7** Three-step procedure for voltammetric dealloying of base-metal-rich Pt–Cu–Co nanoparticle catalysts inside fuel cell electrode layers. The top frame illustrates a membrane–electrode assembly. After the in situ formation of the active catalyst phase, the leached Cu ions inside the fuel-cell membrane are removed by a chemical ion exchange. (Figure taken from Ref. 110.) Srivastava et al. Angew. Chem. Int. Ed. 46, 8988 (2007) Copyright Wiley-VCH Verlag GmbH & Co. KGaA. Reproduced with permission.

and 800°C, while it considerably drops by annealing to 950°C [110]. As the average particle size increases by high-temperature annealing, the drop in mass activities is a simple consequence of a reduced surface-to-bulk ratio of the Pt atoms.

Mayrhofer et al. [114] presented a preparation procedure for core–shell-type catalysts by which the dissolution of the innoble component can be avoided. In this study an alkaline environment was utilized for the in situ observation of surface segregation processes on the basis that alloy catalysts are stable in such electrolyte. In contrast to acid solution, where in cyclic voltammetry only the Pt surface is stable, in 0.1 M KOH, untreated Pt alloy precursors can be compared with leached catalysts. For the $Pt_3Co$ system it was demonstrated that Co surface atoms exhibit a distinctive feature in the cyclic voltammogram (CV) and that the structure of alloy catalyst can be changed utilizing electrochemical treatments. Pt-skin-type particles can be generated without the loss of material or particle growth by CO annealing. A scheme of the obtained particle structure is shown in Figure 9.8.

With the CO annealing procedure an improvement in catalytic activity similar to that for the polycrystalline bulk alloy model systems could be observed on $Pt_3Co$ HSAC (see Table 9.2). In alkaline solution the activity normalized to the mass of Pt of the untreated $Pt_3Co$ catalyst is comparable to that of a Pt HSAC catalyst, whereas after a CO-annealing treatment it increases significantly by a factor of 2.5–3. Interestingly,

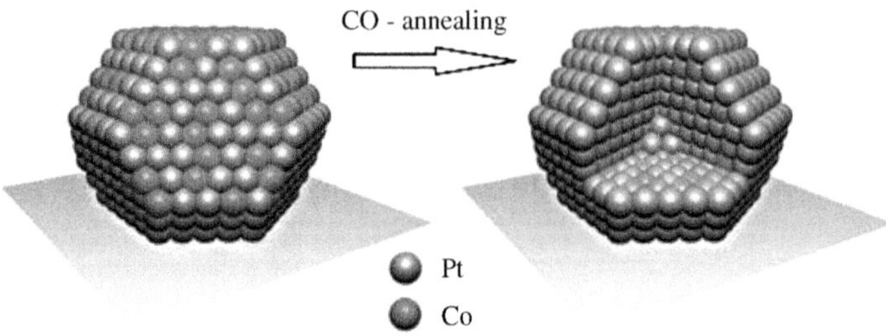


**FIGURE 9.8** Scheme of induced surface segregation. (Taken from Ref. 114.) (See insert for color representation.)

the effect of CO annealing in the gas phase is similar to the effect of electrochemical CO annealing [114].

In acid electrolyte the CO-annealed $Pt_3Co$ catalyst shows superior activity as well compared to plain Pt HSAC the specific activity increased by a factor of 3. Furthermore, the CO-annealed $Pt_3Co$ nanoparticles are also more active than the respective chemical-leached sample. This observation indicates that by leaching, skeleton-type particles are obtained whereas the CO annealing procedure induces an effect similar to that of thermal annealing, that is, forming a Pt shell around an alloy core without any leaching of Co into the electrolyte. In contrast to thermal annealing of particles, however, CO annealing is better suited to nanocatalysts, since the particle size remains unaltered and no active-surface-area loss occurs.

Not only from a fundamental perspective is it important to obtain a closer insight into the mechanisms causing surface segregation processes; it is well known [115] and has been utilized for the UHV preparation of the alloy model catalysts that, depending on the heat of segregation and the surface mixing energy, the composition of the surface of a bimetallic system can be very different from that of the bulk. The fact that temperature annealing of polycrystalline Pt alloys leads to the formation of Pt skin structures indicates that it is thermodynamically favorable for the Pt to segregate to the surface while the less-noble component, such as cobalt, displaces to the core. Nerlov et al. showed for the Ni/Cu(100) system, that in addition to the surface energy in the gas phase one has to consider the chemical potential of the components in segregation processes as well. Therefore, the bonding of an adsorbate to the surface

**TABLE 9.2 Comparison of the ORR Mass and Specific Activities at 0.90 $V_{RHE}$ in 0.1 M KOH ($mA/\mu g_{Pt}$) and 0.1 M $HClO_4$ ($mA/cm^2_{real}$), respectively**

| ORR | Pt (5 nm) | $Pt_3Co^{a,b}$ | $Pt_3Co^{c}$ | $Pt_3Co^{d}$ |
|---|---|---|---|---|
| 0.1 M KOH | 0.12 | 0.13 | 0.29 | 0.36 |
| 0.1 M $HClO_4$ | 0.45 | 0.86 | 1.15 | — |

[a] Untreated catalyst (KOH).
[b] Leached catalyst ($HClO_4$).
[c] In situ CO-annealed.
[d] Gas-phase CO-annealed.

might be a substantial driving force. This was demonstrated by Tao et al. [116] who analyzed the structural and chemical changes of Rh–Pd and Pt–Pd nanoparticles in response to reactive environments at ambient pressures. Although a large segregation effect due to the environment was observed for Rh–Pd, the effect was only small for Pt–Pd. In the former case, in a reducing environment a core–shell particle with a Pd shell developed, whereas the reverse process occurred in an oxidizing environment. The explanation for this effect considered the surface energies of the metals and the metal oxides, respectively. As the surface energy of Pd is less than that of Rh, but the surface energy of Pd oxide is higher than that of Rh oxide, the latter are assumed to be present in an oxidizing environment where the reverse surface segregation effect is observed. As demonstrated, the effect of CO annealing on nanoparticles is similar to the effect of thermal annealing of bulk alloys, and the same apparently energetically favored Pt-skin structure arises. This indicates that in a reduced state the Pt skin is the thermodynamically most stable structure and that carbon monoxide acts as a reducing agent. Furthermore, CO may also adsorb selectively on Pt atoms, thus favoring a Pt skin. In the case of cycling to higher potentials, however, cobalt surface atoms seem to be energetically preferred. At these potentials oxygen species are available to the surface and thus induce an enrichment of CO atoms at the surface because of their higher oxophilic character compared to Pt. Consequently, the observed surface segregation in $Pt_3CO$ nanoparticles by electrochemical treatments might be triggered by the oxidizing or reducing character of the treatment. As the activation barrier for segregation has to be surpassed, the treatments might also influence the mobility of the surface atoms.

The observed segregation phenomena triggered by the reactive environment indicate that a great challenge for the practical use of the presented tailor-made nanoparticle catalysts is the long-term maintenance of their preferred atomic structure and composition under realistic fuel cell conditions over an extended period in time.

## 9.4 ELECTROCHEMICAL CORROSION OF NANOPARTICLES

With more recent advances in the performance of PEMFC catalysts, improvement in the catalyst durability during operation remains a key challenge for developing PEMFCs with acceptable lifetimes in transportation and stationary sectors [117]. In order to determine whether current carbon-supported platinum (or platinum alloy) catalysts will meet long-term performance targets, PEMFC life tests under standard conditions as well as accelerated lifetime tests are performed (see Ref. 118 and other sources cited therein). Whereas under standard conditions the measurements often can last thousands of hours in accelerated tests the degradation occurs much faster because of the application of harsh, artificial conditions. Since this is less time-consuming, fuel cell accelerated life tests have become popular. However, the appropriateness of a given accelerated condition for predicting a general durability is not necessarily clear and the development of standard protocols for accelerated fuel cell tests is pursued by several research groups and organizations [118].

The decrease in performance of a PEMFC is closely related to its loss in electrochemically active surface area at the cathode, and the degradation is often

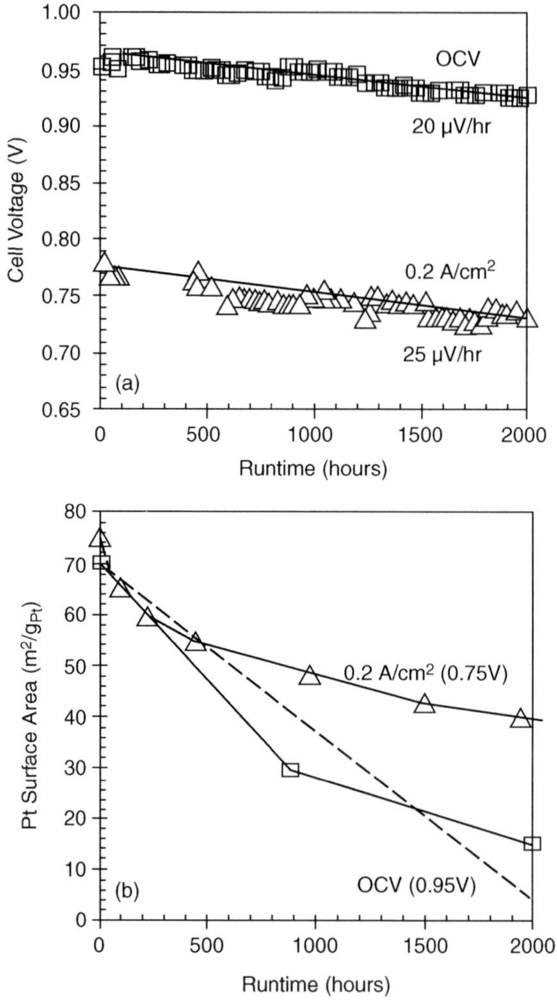


**FIGURE 9.9** Cell votage (a) and platinum surface area (b) as a function of shortstack runtime. Dashed line in (b) corresponds to platinum to calculated platinum surface area loss based on the measured $Pt^{2+}$ equilibrium concentration in sulfuric acid at 0.95 V and 353 K. (Reproduced from Ref. [117] by permission of The Electrochemical Society.)

studied by determining the ECA loss as a function of time. Figure 9.9 shows an example in which the observed loss in cell potential is coupled to the loss in ECA measured in situ in a PEMFC by cyclic voltammetry. The increase in cell potential considerably enhances the degradation rate under steady-state conditions. At the same time it was observed that equilibrium concentrations of dissolved platinum species in hot sulfuric acid increase with the applied potential as well. It was concluded that the loss of surface area is related to formation of soluble Pt species such as $Pt^{2+}$ [117].

By comparison, voltage cycling considerably accelerates the platinum surface area loss. The information on the surface area loss alone is not sufficient for understanding the underlying degradation mechanisms. Therefore, such studies are often coupled to an analysis of the fuel cell MEA at the end of the test, that is, a postmortem analysis. Utilizing X-ray diffraction (XRD) and TEM, the growth in average crystallite size of the catalyst or complete size distributions after the tests can be evaluated. Owing to the destructive nature of these techniques, however, only a single measurement can be conducted on a certain sample, in contrast to, for example, in situ ultra-high-vacuum TEM studies of catalysts [119, 120].

Despite a considerable number of more recent degradation studies, there is no general consensus on the main degradation mechanisms and how they are affected by the operation conditions. In the literature several theories are proposed in order to explain the loss in the active surface area of catalysts. The four primary mechanisms believed to be of relevance to low-temperature fuel cell catalysts are [117,118,121]: (1) *Ostwald-ripening*—metal ions dissolve from smaller particles, diffuse, and redeposit onto larger particles, resulting in reduced metal surface area via a minimization in surface energy; (2) *reprecipitation*—Pt dissolves into the ionomer phase within the cathode and then precipitates again as newly formed Pt particles via the reduction of hydrogen; (3) *particle coalescence*—Pt particles that are in close proximity sinter together to form larger particles; and (4) *corrosion of the carbon support* that anchors the Pt particles and provides electrical contact. These particle growth mechanisms and their rates may vary as a function of electrode potential, cell voltage cycling conditions, current density, particle size and shape, the hydration state of the membrane, and operating conditions. One inhibiting factor in gaining further knowledge is the lack of appropriate methods that enable a more efficient analysis of degradation processes in easy and fast experiments, such as those utilizing electrochemical half-cells.

### 9.4.1 Identical Location Transmission Electron Microscopy (IL-TEM)

In conventional experimental procedures, combining RDE with TEM or XRD measurements the catalyst is first subjected to an electrochemical measurement in an RDE setup and afterwards the catalyst film is removed from the RDE tip and suspended on a TEM grid for analysis. The results are then compared to the pristine catalyst powder. The drawback of such a method is that catalyst agglomerations observed by TEM cannot be unambiguously related to the effect of the electrochemical treatment, but could as well be due to an improper dispersion of the catalyst before drying. Consequently all such studies remain qualitative at best.

Realizing the need for better investigative techniques, the authors have developed a novel, nondestructive method based on a combination of electrochemical half-cell measurements and TEM investigations. In contrast to previous studies the catalyst film dried on a gold finder grid is used directly as working electrode in electrochemical half-cell measurements. This enables the nondestructive investigation of *identical locations* on a catalyst before and after electrochemical treatments (IL–TEM), and statistics of the same particles before and after a certain electrochemical treatment therefore becomes possible. The initial state of the catalyst is evaluated by choosing several areas where only very thin quasi-two-dimensional catalyst layers without agglomerates are observed. Afterwards, the TEM grid with the catalyst film is transferred into an electrochemical cell, contacted by a Pt wire, and subjected to a specific treatment. After the electrochemical treatment, the TEM grid is then carefully dried and transferred back into the electron microscope. By means of an alphabetic index, identical quadrants on the grid are retrieved at low magnification, and the location of the chosen areas on the finder grid can be easily be traced back by the very specific shape of the carbon support.

The methodology is demonstrated in Figure 9.10, where a standard carbon-supported Pt catalyst is subjected to an accelerated aging test of potential cycling between 0.4 V and 1.4 $V_{RHE}$ at room temperature with 1.0 V/s for 2 and 4 h, respectively. Because of the distinctive shape of the carbon support, it is seen that the same location was relocated after the treatments. Despite the transfer and harsh conditions imposed, the carbon support did not deform or corrode to a significant degree. In an analysis of the diameter of several particles before and after the

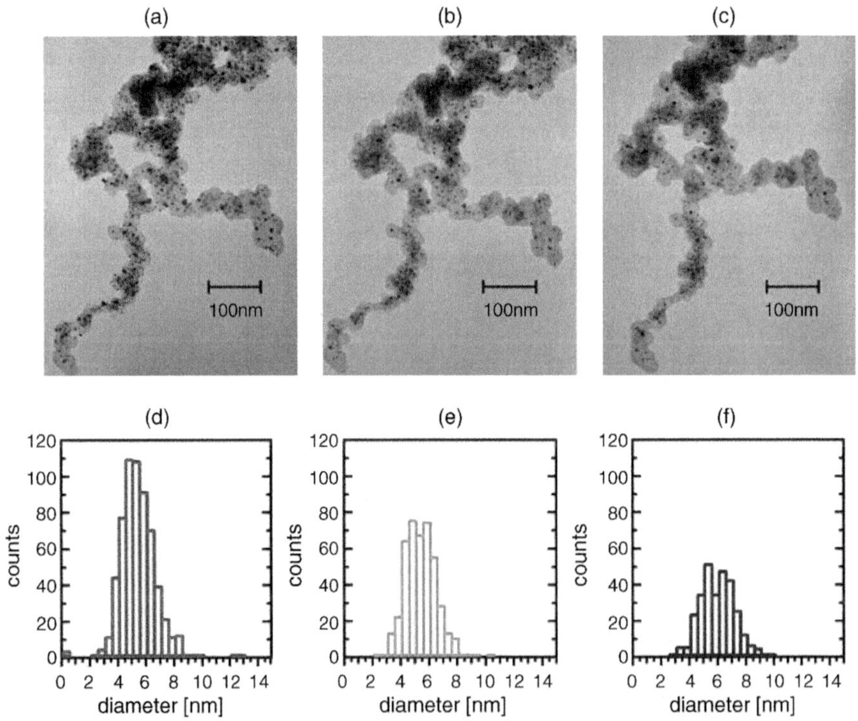


**FIGURE 9.10** TEM micrographs recorded at an identical location of a carbon-supported Pt catalyst. Catalyst before electrochemical treatment, with corresponding size distribution derived from several micrographs (a,d), catalyst after 2 h of acceleratied aging treatment (b,e), and a total of 4 h of treatment (c, f). Only identical locations of unambiguous, nonoverlapping catalyst areas were analyzed. (Figure adapted from Ref. 122.)

**TABLE 9.3 ECA Loss versus Change in Particle Size, Due to Accelerated Aging Test[a]**

| Sample | IL-TEM | EC | XRD | ICP-OES |
|---|---|---|---|---|
| Pristine sample | 100% | 100% | — | 21 (34) ppb |
| After 2 h | 71% | 69% | — | 22 (23) ppb |
| After 4 h | 47% | 50% | — | — |
| Pristine sample | 4.9 nm | — | 4.8 nm | — |
| After 2 h | 5.0 nm | — | — | — |
| After 4 h | 5.6 nm | — | 5.3 nm | — |

[a]Results obtained from particle analysis via IL-TEM, electrochemical CO stripping measurements in TF-RDE measurements (EC), and XRD. In the rightmost column the concentration of Pt in the electrolyte determined by ICP-OES is indicated together with the values calculated from the ECA loss (in parentheses).
*Source:* Reference 122.

treatments (Fig. 9.10d–f), it is apparent that the accelerated degradation treatment did not lead to a significant particle growth, rather than to a detachment of the Pt nanoparticles from the carbon support. Visually, almost no sign of either particle agglomeration or Ostwald ripening could be discerned (under these operating conditions), although some particles were found to move as a whole on the carbon substrate.

Such an analysis would not be feasible without the significant advantage of IL-TEM when compared to conventional TEM investigations, since not only are the size distributions before and after treatment comparable to one another; the absolute number of nanoparticles can also be determined directly. The fact that the local analysis of the catalyst is representative for the macroscopic catalyst is demonstrated in Table 9.3, where measurements of the ECA loss in RDE measurements under identical conditions are compared to the IL-TEM results.

The results from electrochemical ECA measurements are in full agreement with those from the IL-TEM analysis. The study provided clear evidence that under the applied accelerated aging conditions, the ECA loss was primarily due to particle detachment from the carbon support. Only minor Ostwald ripening and no particle agglomeration played a role. While the latter is certainly also affected by the density of the catalyst layer, the lack of Ostwald ripening is puzzling and needs further investigation. The potential of the methodology in discerning different corrosion mechanisms as a function of the applied conditions is clearly demonstrated.

## ACKNOWLEDGMENT

The authors would like to acknowledge longstanding colleagues and collaborators who played in integral part in the presented work: Phil Ross, Hubert Gasteiger, Vojislav Stamenkovic, Dusan Strmncnik, Karl Mayrhofer, Gustav Wiberg, Sean Ashton, Katrin Hartl, and Viktorija Juhart. MA acknowledges support by the German DFG through the Emmy-Noether project ARE852/1-1. NMM acknowledges the support by the Director, Office of Science, Office of Basic Energy Sciences, Division of Materials Sciences, U.S. Department of Energy under Contract No. DE-AC02-06CH11357.

## REFERENCES


1. E. Yeager, *J. Electrochem. Soc.* **113**, C153 (1966).
2. V. G. Levich, *Physicochemical Hydrodynamics*, Prentice-Hall, Englewood Cliffs, NJ, 1962.
3. Z. Jusys, J. Kaiser, and R. J. Behm, *Electrochim. Acta* **49**, 1297 (2004).
4. L. Colmenares, E. Guerrini, Z. Jusys, K. S. Nagabhushana, E. Dinjus, S. Behrens, W. Habicht, H. Bonnemann, and R. J. Behm, *J. Appl. Electrochem.* **37**, 1413 (2007).
5. D. Pletcher and S. Sotiropoulos, *J. Electrochem. Chem.* **356**, 109 (1993).
6. A. J. Bard and L. R. Faulkner, *Electrochemical Methods: Fundamentals and Applications*, Wiley, New York, 1980.
7. E. Gileadi, *Electrode Kinetics for Chemists, Chemical Engineers, and Materials Scientists*, Wiley-VCH, New York, 1993.
8. J. Newman, *J. Electrochem. Soc.* **113**, 1235 (1966).
9. J. Newman, *J. Electrochem. Soc.* **113**, 501 (1966).
10. G. K. H. Wiberg, K. J. J. Mayrhofer, and M. Arenz, Fuel Cells 2010, in print.
11. M. Arenz, V. Stamenkovic, T. J. Schmidt, K. Wandelt, P. N. Ross, and N. M. Markovic, *Surf. Sci.* **523**, 199 (2003).
12. K. J. J. Mayrhofer, G. K. H. Wiberg, and M. Arenz, *J. Electrochem. Soc.* **155**, P1 (2008).
13. K. J. J. Mayrhofer, A. S. Crampton, G. K. H. Wiberg, and M. Arenz, *J. Electrochem. Soc.* **155**, P78 (2008).
14. K. J. J. Mayrhofer, S. J. Ashton, J. Kreuzer, and M. Arenz, *Int. J. Electrochem. Sci.* **4**, 1 (2009).
15. J. Clavilier, R. Faure, G. Guinet, and R. Durand, *J. Electroanal. Chem.* **107**, 205 (1980).
16. J. Clavilier, D. Armand, and B. L. Wu, *J. Electrochem. Chem.* **135**, 159 (1982).
17. L. A. Kibler, A. Cuesta, M. Kleinert, and D. M. Kolb, *J. Electroanal. Chem.* **484**, 73 (2000).
18. M. Schweizer and D. M. Kolb, *Surf. Sci.* **544**, 93 (2003).
19. D. S. Strmcnik, D. V. Tripkovic, D. van der Vliet, K. C. Chang, V. Komanicky, H. You, G. Karapetrov, J. Greeley, V. R. Stamenkovic, and N. M. Markovic, *J. Am. Chem. Soc.* **130**, 15332 (2008).
20. H. Schulenburg, E. Muller, G. Khelashvili, T. Roser, H. Bonnemann, A. Wokaun, and G. G. Scherer, *J. Phys. Chem. C* **113**, 4069 (2009).
21. P. L. J. Gunter, J. W. Niemantsverdriet, F. H. Ribeiro, and G. A. Somorjai, *Catal. Rev. Sci. Eng.* **39**, 77 (1997).
22. K. J. J. Mayrhofer, M. Hanzlik, and M. Arenz, *Electrochim. Acta* **54**, 5018 (2009).
23. H. A. Gasteiger, S. S. Kocha, B. Sompalli, and F. T. Wagner, *Appl. Catal. B Environ.* **56**, 9 (2005).
24. K. Kinoshita, *Electrochemical Oxygen Technology*, Wiley, New York, 1992.

25. M. R. Tarasevich, A. Sadkowski, and E. Yeager, in *Comprehensive Treatise in Electrochemistry*, Vol 7, B. E. Conway, J. O. M. Bockris, and E. Yeager, et al., eds., Plenum Press, New York, 1983.
26. R. R. Adzic, in *Electrocatalysis*, J. Lipkowski and P. N. Ross, eds., Wiley-VCH, New York, 1998.
27. V. S. Bagotskii, M. R. Tarasevich, and Filinovskij, *Élektrokhimiya* **5**, 1218 (1969).
28. H. Wroblowa, Y. C. Pan, and J. Razumney, *J. Electroanal. Chem.* **69**, 195 (1976).
29. N. M. Markovic, H. A. Gasteiger, and P. N. Ross, *J. Phys. Chem.* **99**, 3411 (1995).
30. N. M. Markovic and P. N. Ross, *Electrochim. Acta* **45**, 4101 (2000).
31. M. Arenz, T. J. Schmidt, K. Wandelt, P. N. Ross, and N. M. Markovic, *J. Phys. Chem. B* **107**, 9813 (2003).
32. B. N. Grgur, N. M. Markovic, and P. N. Ross, *Can. J. Chem. Rev.* **75**, 1465 (1997).
33. N. M. Markovic, H. A. Gasteiger, B. N. Grgur, and P. N. Ross, *J. Electroanal. Chem.* **467**, 157 (1999).
34. V. Stamenkovic, N. M. Markovic, and P. N. Ross, *J. Electroanal. Chem.* **500**, 44 (2001).
35. M. Inaba, H. Yamada, J. Tokunaga, and A. Tasaka, *Electrochem. Solid State Lett.* **7**, A474 (2004).
36. A. Schneider, L. Colmenares, Y. E. Seidel, Z. Jusys, B. Wickman, B. Kasemo, and R. J. Behm, *Phys. Chem. Chem. Phys.* **10**, 1931 (2008).
37. F. Uribe, M. S. Wilson, T. Springer, and S. Gottesfeld, *Proc. Workshop on Structural Effects in Electrocatalysis and Oxygen Electrochemistry*, PV92-11, 1992, p. 494.
38. S. Gottesfeld, *ECS* **6**, 51 (2008).
39. G. A. Somorjai, *Introduction to Surface Chemistry and Catalysis*, Wiley, New York, 1993.
40. K. J. J. Mayrhofer, B. B. Blizanac, M. Arenz, V. R. Stamenkovic, P. N. Ross, and N. M. Markovic, *J. Phys. Chem. B* **109**, 14433 (2005).
41. K. J. J. Mayrhofer, D. Strmcnik, B. B. Blizanac, V. Stamenkovic, M. Arenz, and N. M. Markovic, *Electrochim. Acta* **53**, 3181 (2008).
42. T. J. Schmidt, H. A. Gasteiger, G. D. Stab, P. M. Urban, D. M. Kolb, and R. J. Behm, *J. Electrochem. Soc.* **145**, 2354 (1998).
43. S. Trasatti and O. A. Petrii, *Pure Appl. Chem.* **63**(5),711–734 (1991).
44. N. M. Markovic and P. N. Ross, *Surf. Sci. Rep.* **45**, 121 (2002).
45. D. Strmcnik, D. Tripkovic, D. van der Vliet, V. Stamenkovic, and N. M. Markovic, *Electrochem. Commun.* **10**, 1602 (2008).
46. V. R. Stamenkovic, B. Fowler, B. S. Mun, G. Wang, P. N. Ross, C. A. Lucas, and N. M. Markovic, *Science* **315**, 493 (2007).
47. K. C. Neyerlin, W. B. Gu, J. Jorne, and H. A. Gasteiger, *J. Electrochem. Soc.* **153**, A1955 (2006).
48. E. Yeager, *Electrochim. Acta* **29**, 1527 (1984).
49. S. J. Clouser, J. C. Huang, and E. Yeager, *J. Appl. Electrochem.* **23**, 597 (1993).
50. P. N. Ross, *J. Electrochem. Soc.* **126**, 78 (1979).
51. N. M. Markovic, R. R. Adzic, B. D. Cahan, and E. B. Yeager, *J. Electroanal. Chem.* **377**, 249 (1994).
52. N. M. Markovic, H. A. Gasteiger, and N. Philip, *J. Phys. Chem.* **100**, 6715 (1996).

53. T. J. Schmidt, V. Stamenkovic, P. N. Ross, and N. M. Markovic, *Phys. Chem. Chem. Phys.* **5**, 400 (2003).
54. D. Strmcnik, PhD thesis, Univ. Ljubljana, 2007.
55. L. J. Bregoli, *Electrochim. Acta* **23**, 489 (1978).
56. N. Giordano, E. Passalacqua, L. Pino, A. S. Arico, V. Antonucci, M. Vivaldi, and K. Kinoshita, *Electrochim. Acta* **36**, 1979 (1991).
57. M. L. Sattler and P. N. Ross, *Ultramicroscopy* **20**, 21 (1986).
58. K. Kinoshita, *J. Electrochem. Soc.* **137**, 845 (1990).
59. S. Mukerjee, *J. Appl. Electrochem.* **20**, 537 (1990).
60. Y. Takasu, N. Ohashi, X. G. Zhang, Y. Murakami, H. Minagawa, S. Sato, and K. Yahikozawa, *Electrochim. Acta* **41**, 2595 (1996).
61. S. Mukerjee and J. McBreen, *J. Electroanal. Chem.* **448**, 163 (1998).
62. F. Maillard, M. Martin, F. Gloaguen, and J. M. Leger, *Electrochim. Acta* **47**, 3431 (2002).
63. J. Bett and J. Lundquis, *Electrochim. Acta* **18**, 343 (1973).
64. M. Watanabe, H. Sei, and P. Stonehart, *J. Electroanal. Chem.* **261**, 375 (1989).
65. M. Watanabe, S. Saegusa, and P. Stonehart, *Chem. Lett.* 1487 (1988).
66. M. Watanabe, H. Sei, and P. Stonehart, *J. Electrochem. Soc.* **135**, C157 (1988).
67. P. Stonehart, *Ber. Bun. Gesell. Phys. Chem. Chem. Phys.* **94**, 913 (1990).
68. H. Yano, J. Inukai, H. Uchida, M. Watanabe, P. K. Babu, T. Kobayashi, J. H. Chung, E. Oldfield, and A. Wieckowski, *Phys. Chem. Chem. Phys.* **8**, 4932 (2006).
69. M. Peuckert, T. Yoneda, R. A. D. Betta, and M. Boudart, *J. Electrochem. Soc.* **133**, 944 (1986).
70. F. V. Hanson and M. Boudart, *J. Catal.* **53**, 56 (1978).
71. F. Maillard, M. Eikerling, O. V. Cherstiouk, S. Schreier, E. Savinova, and U. Stimming, *Faraday Discussi.* **125**, 357 (2004).
72. B. Andreaus, F. Maillard, J. Kocylo, E. R. Savinova, and M. Eikerling, *J. Phys. Chem. B* **110**, 21028 (2006).
73. K. J. J. Mayrhofer, M. Arenz, B. B. Blizanac, V. Stamenkovic, P. N. Ross, and N. M. Markovic, *Electrochim. Acta* **50**, 5144 (2005).
74. M. Arenz, K. J. J. Mayrhofer, V. Stamenkovic, B. B. Blizanac, T. Tomoyuki, P. N. Ross, and N. M. Markovic, *J. Am. Chem. Soc.* **127**, 6819 (2005).
75. J. A. Last and M. D. Ward, *Adv. Mater.* **8**, 730 (1996).
76. G. T. Burstein, I. M. Hutchings, and K. Sasaki, *Nature* **407**, 885 (2000).
77. S. Huemann, N. T. M. Hai, P. Broekmann, K. Wandelt, H. Zajonz, H. Dosch, and F. Renner, *J. Phys. Chem. B* **110**, 24955 (2006).
78. M. Giesen, G. Beltramo, S. Dieluweit, J. Muller, H. Ibach, and W. Schmickler, *Surf. Sci.* **595**, 127 (2005).
79. M. Lennartz, P. Broekmann, M. Arenz, C. Stuhlmann, and K. Wandelt, *Surf. Sci.* **442**, 215 (1999).
80. V. Komanicky, A. Menzel, K. C. Chang, and H. You, *J. Phys. Chem. B* **109**, 23543 (2005).
81. D. S. Strmcnik, P. Rebec, M. Gaberscek, D. Tripkovic, V. Stamenkovic, C. Lucas, and N. M. Markovic, *J. Phys. Chem. C* **111**, 18672 (2007).
82. M. S. Zei and G. Ertl, *Phys. Chem. Chem. Phys.* **2**, 3855 (2000).

83. D. S. Strmcnik, D. V. Tripkovic, D. van der Vliet, K.-C. Chang, V. Komanicky, H. You, G. Karapetrov, J. P. Greeley, V. R. Stamenkovic, and N. M. Markovic, *J. Am. Chem. Soc.* **130**, 15332 (2008).
84. V. Komanicky, A. Menzel, and H. You, *J. Phys. Chem. B* **109**, 23550 (2005).
85. J. A. Rodriguez, *Surf. Sci. Rep.* **24**, 223 (1996).
86. J. H. Larsen and I. Chorkendorff, *Surf. Sci. Rep.* **35**, 163 (1999).
87. J. H. Sinfelt, *Surf. Sci.* **500**, 923–946 (2002).
88. J. A. Rodriguez, R. A. Campbell, and D. W. Goodman, *J. Phys. Chem.* **94**, 6936 (1990).
89. J. A. Rodriguez, R. A. Campbell, and D. W. Goodman, *J. Vac. Sci. Technol. A Vac. Surf. Films* **9**, 1698 (1991).
90. J. A. Rodriguez, R. A. Campbell, and D. W. Goodman, *Surf. Sci.* **309**, 377 (1994).
91. A. J. Appleby, *Catal. Rev.* **4**, 221 (1970).
92. A. J. Appleby, *Surf. Sci.* **27**, 225 (1971).
93. E. Iglesia and M. Boudart, *J. Phys. Chem.* **95**, 7011 (2002).
94. D. Thompsett, W. Vielstich, H. Gasteiger, and A. Lamm, Handbook of Fuel Cells—Fundamentals, Technology and Application, Vol. 3. Wiley, Chichester, UK, 2003, p. 467.
95. S. Mukerjee, S. Srinivasan, M. P. Soriaga, and J. McBreen, *J. Electrochem. Soc.* **142**, 1409 (1995).
96. T. Toda, H. Igarashi, and M. Watanabe, *J. Electroanal. Chem.* **460**, 258 (1999).
97. T. Toda, H. Igarashi, H. Uchida, and M. Watanabe, *J. Electrochem. Soc.* **146**, 3750 (1999).
98. T. Toda, H. Igarashi, and M. Watanabe, *J. Electrochem. Soc.* **145**, 4185 (1998).
99. V. Stamenkovic, T. J. Schmidt, P. N. Ross, and N. M. Markovic, *J. Electroanal. Chem.* **554**, 191 (2003).
100. V. Stamenkovic, T. J. Schmidt, P. N. Ross, and N. M. Markovic, *J. Phys. Chem. B* **106**, 11970 (2002).
101. V. Stamenkovic, B. S. Mun, K. J. J. Mayrhofer, P. N. Ross, N. M. Markovic, J. Rossmeisl, J. Greeley, and J. K. Norskov, *Angew. Chem. Int. Ed.* **45**, 2897 (2006).
102. V. R. Stamenkovic, B. S. Mun, K. J. J. Mayrhofer, P. N. Ross, and N. M. Markovic, *J. Am. Chem. Soc.* **128**, 8813 (2006).
103. V. R. Stamenkovic, B. S. Mun, M. Arenz, K. J. J. Mayrhofer, C. A. Lucas, G. F. Wang, P. N. Ross, and N. M. Markovic, *Nature Mater.* **6**, 241 (2007).
104. Y. Gauthier, Y. Joly, R. Baudoing, and J. Rundgren, *Phys. Rev. B* **31**, 6216–6218 (1985).
105. M. Lundberg, *Phys. Rev. B* **36**, 4692–4699 (1987).
106. U. Bardi, A. Atrei, E. Zanazzi, G. Rovida, and P. N. Ross, *Vacuum* **41**, 437–440 (1990).
107. Y. Gauthier, *Surf. Rev. Lett.* **3**, 1663–1689 (1996).
108. B. Hammer, Y. Morikawa, and J. K. Norskov, *Phys. Rev. Lett.* **76**, 2141–2144 (1996).
109. S. Koh, J. Leisch, M. F. Toney, and P. Strasser, *J. Phys. Chem. C* **111**, 3744 (2007).
110. R. Srivastava, P. Mani, N. Hahn, and P. Strasser, *Angew. Chem. Int. Ed.* **46**, 8988 (2007).
111. J. Zhang, F. H. B. Lima, M. H. Shao, K. Sasaki, J. X. Wang, J. Hanson, and R. R. Adzic, *J. Phys. Chem. B Lett.* **109**, 22701 (2005).
112. S. Papadimitriou, A. Tegou, E. Pavlidou, S. Armyanov, E. Valova, G. Kokkinidis, and S. Sotiropoulos, *Electrochim. Acta* **53**, 6559 (2008).
113. S. Koh and P. Strasser, *J. Am. Chem. Soc.* **129**, 12624 (2007).

114. K. J. J. Mayrhofer, V. Juhart, K. Hartl, M. Hanzlik, and M. Arenz, *Angew. Chem. Int. Ed.* **48**, 3529 (2009).

115. A. Christensen, A. V. Ruban, P. Stoltze, K. W. Jacobsen, H. L. Skriver, J. K. Nørskov, and F. Besenbacher, *Phys. Rev. B* **56**, 5822 (1997).

116. F. Tao, M. E. Grass, Y. Zhang, D. R. Butcher, J. R. Renzas, Z. Liu, J. Y. Chung, B. S. Mun, M. Salmeron, and G. A. Somorjai, *Science* **322**, 932–934 (2008).

117. P. J. Ferreira, G. J. la O, Y. Shao-Horn, D. Morgan, R. Makharia, S. Kocha, and H. A. Gasteiger, *J. Electrochem. Soc.* **152**, A2256 (2005).

118. R. Borup, J. Meyers, B. Pivovar, Y. S. Kim, R. Mukundan, N. Garland, D. Myers, M. Wilson, F. Garzon, D. Wood, P. Zelenay, K. More, K. Stroh, T. Zawodzinski, J. Boncella, J. E. McGrath, M. Inaba, K. Miyatake, M. Hori, K. Ota, Z. Ogumi, S. Miyata, A. Nishikata, Z. Siroma, Y. Uchimoto, K. Yasuda, K. i. Kimijima, and N. Iwashita, *Chem. Rev.* **107**, 3904 (2007).

119. J. G. Lee, J. Lee, T. Tanaka, and H. Mori, *Phys. Rev. Lett.* **96**, 075504 (2006).

120. T. Akita, M. Okumura, K. Tanaka, M. Kohyama, and M. Haruta, *J. Mater. Sci.* **40**, 3101 (2005).

121. R. L. Borup, J. R. Davey, F. H. Garzon, D. L. Wood, and M. A. Inbody, *J. Power Sources* **163**, 76 (2006).

122. K. J. J. Mayrhofer, J. C. Meier, S. J. Ashton, G. K. H. Wiberg, F. Kraus, M. Hanzlik, and M. Arenz, *Electrochem. Communi.* **10**, 1144 (2008).

CHAPTER 10

# Nanoscale Phenomena in Catalyst Layers for PEM Fuel Cells: From Fundamental Physics to Benign Design

KAREN CHAN, ATA ROUDGAR, LIYA WANG, and MICHAEL EIKERLING

Department of Chemistry, Simon Fraser University, Burnaby, British Columbia, Canada

Critical progress in power density, cost reduction, and durability of polymer electrolyte fuel cells hinges on advances in catalyst layer design. The objective is to maximize the rate of desired electrochemical reactions, in particular the oxygen reduction reaction (ORR), at ultralow content of platinum. This task entails a hierarchy of structural effects on transport and interfacial processes, with major scientific challenges arising at nano- and mesoscales. Physical theory and molecular modeling, described in this chapter, establish the basis for the systematic structural optimization of Pt-based catalyst layers. Independent studies of CO electrooxidation and oxygen adsorption on Pt nanoparticles both suggest that nanoparticle reactivity depends on peculiar size effects and high site selectivity of elementary surface processes. Results from ab initio calculations and thermochemical modeling of elementary steps of ORR on a $Pt_4$ cluster exhibit significant differences from slab calculations and highlight the importance of active sites to the reaction mechanism. A kinetic model of hydrogen evolution on supported catalyst nanoparticle arrays shows geometric substrate effects on reactivity due to adsorbate spillover. In Section 10.4.3, we outline a 2D nanopore model that accounts for the effects of nanoporosity and pore surface charge on the effectiveness factor of Pt utilization.

*Fuel Cell Science: Theory, Fundamentals, and Biocatalysis*,
Edited by Andrzej Wieckowski and Jens K. Nørskov

## 10.1 INTRODUCTION

General principles of the structure and operation of catalyst layers are discussed in this introductory section.

### 10.1.1 Perspectives of Polymer Electrolyte Fuel Cells (PEFCs)

The looming depletion of fossil fuels and the adverse environmental effects of their current use are among humankind's greatest challenges [1]. It is, thus, a matter of utmost urgency to develop new energy sources, energy conversion technologies, and energy supply infrastructures. Over the past decades, research worldwide has explored polymer electrolyte fuel cells (PEFCs) as the next generation of energy converters for vehicles and portable devices [2,3]. PEFCs convert hydrogen and oxygen into water. The reaction enthalpy released in the process generates a flow of electric charge (electricity), which provides power to electrical devices that are inserted between the fuel cell electrodes, as well as heat. PEFCs exhibit unrivalled thermodynamic efficiencies, high energy densities, and ideal compatibility with hydrogen. Despite the abundance of promising demonstrations of vehicles and PEFC-powered devices, formidable technological hurdles to wide-scale commercialization persist. New generations of fuel cells have to surpass established energy conversion technologies in power density, operational flexibility, stability, and cost. To give an idea of the magnitude of this challenge: currently used PEFC stacks in prototype cars are about 100 times more expensive than conventional combustion engines (~$30/kW), albeit still distinctly inferior in durability [4].

In the late 1990s, it was a common view among stakeholders in industry and academia that required progress could be achieved with engineering-type optimization, relying on existing materials and components. Nowadays, it is widely recognized that progress in PEFC technology hinges on breakthroughs in design, fabrication, and implementation of innovative materials. Specific targets for improvement involve (1) increasing rates of current conversion at electrodes, (2) minimizing parasitic voltage losses due to transport of protons and electrons in conduction media and of reactants and water in porous diffusion media, (3) providing uniform distributions of reaction rates, and (4) balancing water and heat fluxes. Overcoming these multidisciplinary challenges will require a combination of contributions from fundamental physics and chemistry, diagnostics, materials science, and engineering.

### 10.1.2 Challenges and Opportunities in Catalyst-Layer Research

In operational PEFCs, all components must cooperate in order to optimize the complex interplay of transport and reaction. This optimization involves several tens of parameters. The catalyst layers (CLs) represent the major competitive ground in view of this challenge. We will focus on the cathode catalyst layer (CCL), since fuel cell research since the late 1990s has revealed that drastic improvements in power density, durability, and cost are impossible without advances in CCL design.

Nanoparticles of Pt or Pt alloys exhibit the highest electrocatalytic activity for PEFC reactions. In existing CLs, Pt particles with sizes of 2–5 nm are dispersed on porous carbon substrates [5,6]. Interpenetrating networks of C/Pt, ionomer, and pores ensure good transport of electrons, protons, chemical reactants ($H_2$, methanol, $O_2$, etc.), and product water to or from the hydrated, electrochemically active Pt surface sites. Optimized CL structures enable high conversion rates of reactants into electrical current. Owing to the formation of water and heat in the electrochemical reaction, water transport, vaporization exchange, and heat exchange, are also vital functions of the CCL.

Electrochemical reactions in CCLs occur only at Pt nanoparticles, at which protons, electrons, and oxygen molecules meet. A basic requirement is that the Pt surface be wetted by liquid water. Major shortcomings of the pertinent composite design are (1) statistical limitations of Pt utilization due to the random three-phase morphology and (2) highly nonuniform reaction rate distributions due to the high thickness ($\sim$10 μm). These conditions cause inefficient utilization of Pt and problematic voltage losses; those due to oxygen reduction in the CCL ($\sim$400 mV) diminish cell efficiency by 30–40%. A 10-fold increase in the surface area of Pt reduces losses by 60–120 mV. Pt is, however, an expensive and limited resource. A simple calculation reveals that replacement of combustion engines in a considerable portion of existing vehicles by fuel cell drive systems at no significant loss in power would severely deplete the known reserves of Pt.

The most critical hurdle to be cleared en route to PEFC commercialization is, therefore, to raise the ratio of the current conversion rate to the Pt loading by a factor of 10–100. Assuming that water-based polymer electrolyte membranes, namely. Nafion and the like, and Pt remain the best materials, such progress has to be realized at $T < 90\,^\circ$C and with Pt-based catalysts. Major efforts in research, thus, focus on maximizing activity and utilization of Pt. A $>10\times$ improvement in the rate of oxygen reduction compared to current fuel cell catalysts has been reported by a group of researchers at Argonne and Lawrence Berkeley National Laboratories for a nanosegregated $Pt_3Ni$ alloy, due to the so-called Pt skin effect [7–9]. Independently, structural models of catalyst-layer operation suggest that, with the present CL design, less than 10% of the Pt is effectively utilized for reactions [10]. Despite ideal conditions, under which these results were obtained, they nurture expectations that reducing the Pt loading to $< 10\%$ of the present amount at no loss in performance is a credible goal.

The optimization target for electrode design is the (mass-specific) effectiveness factor of Pt utilization. This property is the integral of the spatially varying faradaic current density over the active Pt surface area, divided by the ideal current generated if each Pt atom were equally active and exposed to equal, externally provided concentrations of protons and reactants (oxygen) as well as to the same electrode potential. The effectiveness factor, defined in this way, constitutes a yardstick for comparing different designs of catalyst layers. It accounts for the spatial variations in the statistical utilization of Pt due to the heterogeneous electrode structure as well as for the effects of uneven reaction rate distributions caused by insufficient mass and charge transport [10–14].

Structural studies employ experiments and molecular modeling to understand the self-organization phenomena that occur in the CL formation process [5,15]. Structural organization processes of particular interest are the formation of agglomerates of carbon/Pt, the structure and distribution of ionomer domains relative to these agglomerates, the porous morphology, and the heterogeneous wettability of the composite medium. These details determine transport and interfacial properties. A major challenge for structural studies is to rationalize the nature of active interfaces. Pt|vapour interfaces are disconnected from the proton-supplying network; they are thus inactive, which implies that water in CCLs is produced in liquid form. This conclusion is vital to understanding the water management issues arising in CCL [11]. The relative contributions to electrocatalytic activity from Pt–ionomer and Pt|water interfaces depend on carbon/Pt-ionomer phase segregation, porous structure, and wetting properties of pores. A larger fraction of hydrophilic micropores will increase the contribution to the Faradaic current of Pt|water interfaces relative to that of Pt–ionomer interfaces [11,16].

Often, claims that only Pt in contact with the ionomer phase is electrocatalytically active are made on the premise that electroneutrality should apply at all scales. However, charge separation in the double-layer region of a metal–electrolyte interface is a fundamental electrochemical phenomenon. This phenomenon forms the basis of a model of reactivity in water-filled agglomerates of catalyst layers. It lies at the heart of the operation of ionomer-free ultrathin catalyst layers (UTCLs). We address this in detail in Section 10.4.

The nature and abundance of electrocatalytically active catalyst surfaces is related to the vital but largely unsettled issue of agglomeration of carbon/Pt particles and their spontaneous phase segregation from ionomer molecules. The majority of data supporting the existence of C/Pt agglomerates in CL are based on TEM micrographs and porosimetry studies [17–19]. Controversy evolves around size and composition of agglomerates. Commonly considered views are (1) uniform mixtures of ionomer and carbon/Pt particles in agglomerates, (2) a thin uniform film of ionomer surrounding agglomerates of C/Pt, or (3) random uncorrelated percolating phases of C/Pt and ionomer.

Molecular modeling studies at Canada's National Research Council (NRC) Institute for Fuel Cell Innovation in Vancouver, conducted in collaboration with Nissan Motor Inc., employ coarse-grained molecular dynamics calculations to unravel the structural picture of the catalyst layer. Results suggest that the ionomer phase forms a thin adhesive film or a fibrous network attached to surfaces of micro-to-mesoporous Pt/C agglomerates, whereas pores inside agglomerates are ionomer-free [5,20]. The thickness of the ionomer film or fibers is $<10$ nm. Water hydrates only the exterior ionomer surface, while their hydrophobic interior expels water. The orientation of ionized ionomer sidechains relative to carbon/Pt agglomerates depends on the wetting properties of the carbon. These insights, scrutinized by Brunauer–Emmett–Teller (BET) analysis of nitrogen adsorption data, water sorption studies, and electrochemical measurements, furnish a distinctly redefined picture of catalyst-layer morphology. Results will be published soon.

The role of the bimodal porous morphology, heterogeneous wettability, and water accumulation in pores has more recently moved into the focus of modeling studies.

A large fraction of small hydrophilic pores with sizes of 5–10 nm inside carbon/Pt agglomerates is advantageous in that it provides a large surface area for deposition of the catalyst. These pores should be hydrophilic to maximize the wetted surface fraction of Pt. Larger secondary pores with sizes of 10–40 nm between agglomerates are vital for the gaseous transport of reactants. These pores should be hydrophobic. As emphasized in References 11 and 16, understanding the interplay between composition, porous structure, wetting properties, and operating conditions is crucial for determining the distribution, fluxes, and transformation of water in CCL. These phenomena lead to a transition from a partially saturated to a fully flooded state of the CCL, which is accompanied by a significant drop in fuel cell voltage. The transition region in steady-state voltage vs. current density response curves could exhibit bistable behavior. The occurrence of the bistability phenomenon implies that at a given current density two values of the voltage response are obtained, which correspond to the coexistence of partially saturated and fully flooded states.

The exchange current density per unit geometric surface area of the electrode, $j^0$ (mA/cm$^2$), is the key physical property of a CL, representing its ability to convert chemical flux of reactants into electrical current. The trouble is that $j^0$ is very small because of the sluggish kinetics of ORR, even if Pt is used as the catalyst [2,21–25]. Enhancing $j^0$ by simply increasing the Pt loading, $m_{Pt}$ (mg/cm$^2$), is prohibitive because of cost. The following expression elucidates this dilemma [11],

$$j^0 = 2 \cdot 10^3 [m_{Pt}] \, j^{0*} \varepsilon^{S/V} \Gamma_a g(S_r) f(X_{PtC}, X_{el}) / X_{PtC}. \quad (10.1)$$

In order to minimize parasitic voltage losses in the cell, $j^0$ should be maximal, while $m_{Pt}$ should be minimal to minimize costs. All remaining factors in Equation (10.1) should, therefore, be maximal. The factor $2 \cdot 10^3$ is the ideal Pt surface area that corresponds to spreading 1 mg of Pt as a monoatomic layer in Pt(111) configuration; this value is a theoretical baseline for total active surface area at the atomistic scale.

The intrinsic exchange current density of the catalyst, $j^{0*}$ (mA/cm$^2$), is the target of fundamental studies in electrocatalysis [6–9]. Other factors in (10.1) account for statistical limitations of catalyst utilization that occur on a hierarchy of scales. $\varepsilon^{S/V}$ is the surface-to-volume atom ratio of Pt nanoparticles, which scales roughly as $1/r$ with the particle radius, as we will discuss in Section 10.2.1. Additional statistical effects at the nanoparticle scale include the presence of active sites on the catalyst surface, discussed in Section 10.2.3 [26]. The effectiveness factor of catalyst utilization at the agglomerate level $\Gamma_a$ is determined mainly by electrostatic phenomena, as explored in Section 10.4.1 [27]. The factor $g(S_r)$ is the wetted fraction of pores, where $S_r$ is the liquid water saturation; it accounts for the fact that only water-wetted Pt is accessible for protons. The structure factor $f(X_{PtC}, X_{el})/X_{PtC}$, discussed in Reference 28, represents the statistical fraction of active Pt particles; it is a function of the volumetric amounts of solid carbon/Pt phase (volume fraction $X_{PtC}$) and ionomer phase ($X_{el}$). The form of the function $f(X_{PtC}, X_{el})$ depends on the micromorphology of the layer. If carbon/Pt and ionomer form uncorrelated random percolating phases, $f(X_{PtC}, X_{el})/X_{PtC}$ corresponds to the fraction of Pt particles at the triple phase boundary. In this case, the form of this function could be derived from the theory of active bonds in three-phase composite media; the maximum fraction of active

Pt particles would correspond to $f(X_{PtC}, X_{el})/X_{PtC} \simeq 0.4$. If the ionomer forms a uniform thin film that uniformly covers the surfaces of carbon/Pt agglomerates, as suggested by more recent molecular modeling studies [5], we would find $f(X_{PtC}, X_{el})/X_{PtC} \simeq 1$. Equation (10.1) provides a quantitative means for evaluating structural designs of CLs.

The challenge in catalyst-layer research is thus to maximize the mass-specific exchange current density, $j^0/m_{Pt} = i_m^* \cdot \Gamma_s$, which mandates improvements in mass-specific activity, $i_m^*$ (A/mg), and statistical utilization of Pt, $\Gamma_s$. Catalyst-layer performance is furthermore affected by insufficient transport of protons, oxygen, and water that result in nonuniform reaction rate distributions and inefficient catalyst utilization under operation [11,13]. The interplay of the exchange current density $j^0$ with any of the transport parameters of gaseous reactants, protons, and electrons could be used to define a corresponding *reaction penetration depth* $\delta$. The criterion for uniform reaction rate distributions at the macroscopic scale is that $\delta$ for each transport process be comparable to or larger than the thickness of the CL (i.e., $\delta \geq L$). This requirement implies that thickness must be adjusted well to the transport properties of the composite medium [11,13,14].

The foregoing discussion suggests that improvements in structure and function of catalyst layers should be pursued in three areas: (1) nanoparticle electrocatalysis, (2) statistical utilization of the catalyst, and (3) mixed transport in composite media. These areas encompass a hierarchy of kinetic and transport processes across many length scales. The main structural effects include atom arrangement on the catalyst surface, nanoparticle size and shape, nanoparticle distribution on the substrate, and random heterogeneous morphology of the layer. In view of the complex challenges, approaches in CL design demand concerted efforts in modeling, experiment, materials design, and fuel cell testing.

### 10.1.3 Approaches in Catalyst-Layer Design

A key parameter for the classification of CL design is the thickness $L$. For conventional CLs ($L \simeq 5-20$ μm) modeling and experiment have explored the effects of composition, porous morphology, and thickness on performance [11,13,14,28–30,32]. It is well corroborated that this design underutilizes Pt because of partial wetting of Pt, percolation effects, and transport limitations; it does not leave much room for improvement of $j^0/m_{Pt}$. Pertinent accounts of the status of physical modeling of conventional three-phase catalyst layers can be found in References 12–14.

A number of studies raise expectations that improvements in CL design should be addressed by introducing ultrathin catalyst layers (UTCLs, $L \sim 100$ nm), which are ionomer-free and nanoporous to provide a high electrochemically active surface area. UTCL need to be designed for operation at full liquid water saturation, since water in substrate pores serves as the medium for proton transport and enhances catalyst utilization. The reduced thickness renders oxygen transport limitations negligible. A successful example is CLs fabricated at 3 M, which utilize a substrate of highly oriented organic whiskers, onto which a continuous layer of Pt is sputter-deposited [9]. Other demonstrations employ porous metal substrates [8,10]. Local electrochemical reaction conditions in UTCLs are poorly understood; hitherto, systematic theoretical studies of their structure and physicochemical properties have not been pursued.

### 10.1.4 Catalyst-Layer Modeling at Nanoscale and Mesoscale

Major challenges for the design and operation of catalyst layers, in particular of nanostructured ultrathin layers, arise at the nanoscale and mesoscale. They include nanoparticle reactivity and dissolution kinetics, effects of the substrate on reactivity and stability, effects of nanoporosity, and effects of surface heterogeneity and roughness. Approaches to address these challenges are the main focus of this contribution. The objective of the modeling of nano-to-mesoscale phenomena is to relate electrochemical properties and performance of catalyst layers to the structure of the nanoparticle–substrate system and to the local reaction conditions in nanopores.

The majority of theoretical works on fuel cell catalyst layers have employed continuum-type, macrohomogeneous models [11–14,28–30]. These models represent the nanocomposite catalyst layer as a homogenous medium; relevant physical variables and parameters that characterize composition, microstructure, species transport, and interfacial processes in catalyst layers are averaged over representative elementary volumes (REVs), using concepts from the theory of random heterogeneous media [31]. The characteristic size of the REVs has to be large compared to that of spatial heterogeneities at nano-to-mesoscales, and small compared to the size of the catalyst layer. The latter restriction ensures that macrohomogeneous approaches could still provide spatial resolution in the distributions, at least in one dimension, of physical properties and variables in catalyst layers. The required spatially varying functions are the electrical potential of metal and electrolyte phases, pressures and concentrations of oxygen, protons and water, rates of electrochemical reaction and vaporization exchange, and fluxes of oxygen, protons, liquid water, and water vapor, which are consumed or produced during catalyst layer operation. These spatial distributions can be related to voltage–current response curves, effectiveness of Pt utilization, and water handling, which are all indicators of global performance. Relations among composition, effective properties, and performance resulting from macrohomogeneous models have been used extensively to diagnose the causes of performance losses or failure in catalyst layers [10–14,16,28,30,32]. Moreover, they have been exploited with some success for the systematic optimization of thickness and composition of conventional catalyst layers [13,14,19,28]. The main limitation of this approach is the neglect or the approximations of structural organization and correlation effects at the nanometer scale [5]. For the design of nanostructured catalyst layers with enhanced performance-to-cost ratio, the structural details that macrohomogeneous approaches neglect are essential.

The diametric modeling strategy is to employ fully self-consistent ab initio computer simulations that explicitly account for molecular degrees of freedom and electronic structure of the catalyst and substrate material. The downside of these first–principles approaches is their enormous computational complexity, which limits their applicability to idealized, highly ordered structures comprised of $\sim$100 atoms or less, which may poorly resemble the disordered, multicomponent structures in fuel cell catalyst layers. In addition, there is a gap between the characteristic timescales of fuel cell reactions and those for which the evolution of molecular systems can be followed by ab initio approaches (i.e., $\sim$ 100 ps). These structure and timescale gaps restrict ab initio studies to a highly simplified subset of relevant electrochemical and transport processes. Examples of such molecular scale studies can be found in the literature [33–38].

Theoretical studies employ a hierarchy of approaches to bridge the gaps existing between structure and elementary processes at the atomistic or molecular level and complex structures and processes at scales that are relevant for the operation of materials in fuel cells. Pertinent tools include ab initio quantum mechanical approaches, atomistic and coarse-grained molecular dynamics, Monte Carlo methods, kinetic modeling, and continuum approaches. In the following sections, we discuss applications of these tools to challenges in the operation of catalyst layers for PEFCs at the nano-to-mesoscale.

## 10.2 MODELING OF NANOPARTICLE STABILITY AND REACTIVITY

### 10.2.1 Size Effects in Nanoparticle Electrocatalysis

The electrical, magnetic, optical, and catalytic properties of transition metal nanoparticles differ strongly from those of the bulk phase. The peculiar effects of nanoparticle size, shape, and electronic structure are of interest from the perspectives of both fundamental research and applications in nanotechnology [39,40]. Interesting phenomena that occur at particle sizes below $\sim$2 nm arise from the confinement of (quasi)free electrons and the increasingly discrete electronic structure [41]. In electrocatalysis, the primary interests are to understand (1) the classical effects due to atom arrangement and (2) the heterogeneous electronic structure at the nanoparticle surface, which controls adsorption and charge transfer phenomena.

A key to achieving the targeted improvements in electrocatalytic activity in fuel cell electrodes lies in understanding how the structure of the catalyst determines the rates of relevant surface processes [3,12,26,42–46]. On nanoparticles, the proportions of the different surface sites (facets, edges, and corners) and the surface electronic structure are both closely related to particle size [42,47–51]. These particle size effects exhibit varying trends for different reactions [52–58]. Oxygen reduction on Pt nanoparticles shows a maximum in mass-specific activity for a particle diameter of $\sim$3 nm [59,60]; the mass activity for hydrogen reduction on Pd particles increases monotonically with decreasing Pd cluster size [61]; the $CO_{ad}$ electrooxidation reaction rate on Pt nanoparticles decreases when the particle size decreases to $<$ 5 nm due a reduction in the surface diffusivity of $CO_{ad}$ [12,26].

Studies of stable conformations of supported nanoparticles and of the elementary processes on their surfaces demand density functional theory (DFT) calculations and kinetic modeling of reactivities using Monte Carlo simulations or mean-field approaches [12,26]. Because of the complex nature of the structural effects, only few theoretical studies have been successful in establishing relations between structure and electrocatalytic reactivity. The widely known $d$-band model of Hammer and Nørskov [62,63] relates trends in chemisorption energies for various adsorbates on transition metal surfaces to the position of the $d$-band center, the first moment of the density of states from the Fermi level. A shift of the $d$-band center toward the Fermi level due to $d$-band narrowing leads to an increase in adsorption energy. Systematic DFT calculations and experiments on a series of polycrystalline alloy films of the type $Pt_3M$ (M = Ni, Co, Fe, Ti) have confirmed predicted correlations between the position of the $d$-band center, chemisorption energies, and electrode activities [64]. This success of the $d$-band model has catalyzed efforts in DFT-based, high-throughput combinatorial screening schemes for identifying highly active electrocatalyst materials [65,66].

However, because of the heterogeneous surface morphology of nanoparticles, the applicability of insights from studies at well-defined extended surfaces to small Pt nanoparticles or clusters is limited. In systems of supported catalyst nanoparticles, quantum confinement effects, irregular surface structure, and substrate effects have to be accounted for [67–71]. Claus and Hofmeister [72] suggested that the electrocatalytic activity of nanoparticles depends on the local environment of surface sites, namely, the coordination geometry and electronic structure.

Experimentally observable, facet-type Pt nanoparticles with regular crystalline structure and sizes <2 nm are shown in Figure 10.1 [73–77]. For large, roughly spherical particles, the surface-to-volume atom ratio, $\varepsilon^{S/V} = N_S/N$, introduced in Equation (10.1), is expected to be proportional to $N^{-1/3}$, where $N_S$ is the number of surface atoms and $N$ is the total number of atoms. We have evaluated this relation for particles with atom numbers $N < 200$ for the different shapes depicted in Figure 10.1. The relation between $\varepsilon^{S/V} = N_S/N$ and $N$ is shown in Figure 10.2. The discrete values at particle sizes that are permitted for the given shapes are in good agreement with the relation

$$\varepsilon^{S/V} = N_S/N = A + \alpha N^{-(1/3)} \tag{10.2}$$

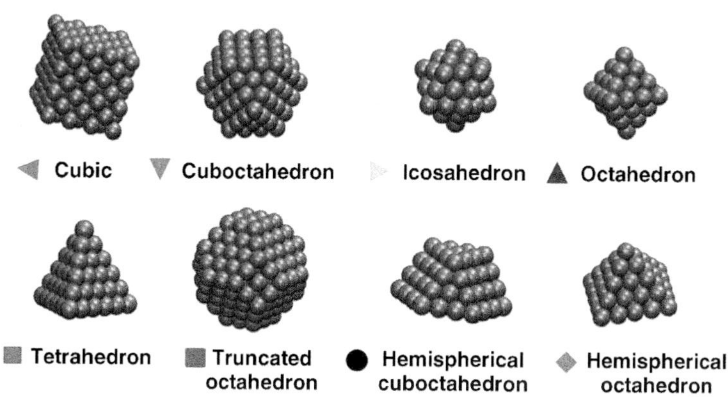


**FIGURE 10.1** Stable conformations of facet-type Pt nanoparticles. (Reprinted with permission from *J. Phys. Chem. C*, 2009, 113(42), 17989–17996. Copyright 2009 American Chemical Society.)

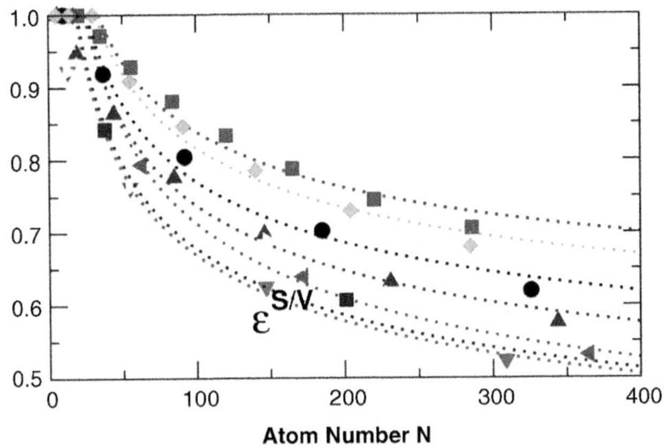


**FIGURE 10.2** Surface-to-volume atom ratio $\varepsilon^{S/V}$ over total number of atoms $N$ for various particle shapes (symbols as introduced in Fig. 10.1). Dotted lines indicate the scaling with $N^{-1/3}$. (Reprinted with permission from *J. Phys. Chem. C*, 2009, 113(42), 17989–17996. Copyright 2009 American Chemical Society.)

where $A$ and $\alpha$ depend on particle shape. Equation (10.2) is represented by the dotted lines in Figure 10.2. The tetrahedral structure, enclosed by four Pt(111) facets, corresponds to the largest $\varepsilon^{S/V}$. Maximum (tetrahedral) and minimum (cubocahedral) $\varepsilon^{S/V}$ values from the particle shapes explored differ by 15–20%. In general, more nonspherical shapes of nanoparticles correspond to larger values of $\varepsilon^{S/V}$. The variation in $\varepsilon^{S/V}$ with particle shape suggests that shape control could be a suitable means of enhancing the mass activity of Pt [76,77].

The nanocrystalline particles in Figure 10.1 are enclosed by (111) and (100) facets. Surface atoms on distinct facets, corner atoms, and edge atoms possess varying coordination numbers. The proportion of undercoordinated corner and edge atoms increases strongly with decreasing particle sizes below 2–3 nm. This renders the nanoparticles physically unstable, causing the strong size dependence of particle dissolution rates. The stability of nanoparticles can be studied by calculating their cohesive energies using DFT.

As a result of the geometric and electronic heterogeneity of the particle surface, the activities of surface reactions exhibit significant spatial fluctuations at the atomistic scale. Another important influence on the electronic structure and thus on surface processes is lattice contraction, which arises from the undercoordination of surface atoms. It is a foremost challenge for theory and modeling to understand the relations between nanoparticle size and shape and their stability and electrocatalytic activity.

### 10.2.2 Cohesive Energies of Pt Nanoparticles

We have performed ab initio calculations based on DFT with generalized-gradient approximation (GGA) using the Vienna Ab initio Software Package (VASP) to explore the stability of Pt nanoparticles. The ionic cores are represented by projected augmented waves (PAWs). The Kohn–Sham one-electron wavefunctions are expanded in a plane-wave basis set with kinetic energy cutoff of 400 eV. We calculated the average cohesive energies of Pt nanoparticles by dividing the total energy of the optimized particle conformation by the atom number $N$,

$$E_{\mathrm{coh}} = E_{\mathrm{tot}}/N - E_{\mathrm{tot}}^{\mathrm{gas}}, \tag{10.3}$$

where $E_{\mathrm{tot}}^{\mathrm{gas}}$ is the total energy of one Pt atom in the gas phase.

For sufficiently large, spherical particles, the Gibbs–Thompson equation relates the particle Gibbs energy to particle size. The chemical potential per metal atom in a particle of radius $r$, relative to the chemical potential $\mu(\infty)$ in the bulk metal, is

$$\mu(r) = \mu(\infty) + \frac{2\gamma}{\rho\, r} \quad \text{or} \quad |\mu(r)| = |\mu(\infty)| - \frac{2\gamma}{\rho\, r} \tag{10.4}$$

since $\mu(r), \mu(\infty) < 0$, where $\gamma$ is the surface tension and $\rho$ is the atomic density of the bulk metal [78,79]. Assuming, for simplicity, that the atomic density and the surface tension are independent of particle size, we can transform this into a function of $N$

$$|\mu(N)| = |\mu(\infty)| - 2\left(\frac{4\pi}{3}\right)^{1/3} \frac{\gamma}{\rho^{2/3}} N^{-1/3} \tag{10.5}$$

Entropy contributions to Gibbs energy due to the vibrations of atoms in the nanoparticle are small and vary very little with particle size. Thus, we expect the relation $|E_{\mathrm{coh}}(N)| \approx |\mu(N)|$ to be valid in the range of particle sizes in which the spherical particle approximation applies.

In a plot of $|E_{\mathrm{coh}}|$ versus $N^{-1/3}$, the intersection with the vertical ordinate ($N \to \infty$) corresponds to the bulk value of the chemical potential of the metal, $|\mu(\infty)|$, and the slope gives $\gamma/\rho^{2/3}$. Lin et al. [80]. tested the relation between $|E_{\mathrm{coh}}|$ and $N^{-1/3}$ for Pt nanoparticles with fewer than 25 atoms. Although their data could be represented reasonably well by a linear relation, the particle sizes considered were too small for a meaningful extrapolation to the bulk limit. Figure 10.3a depicts our results for $|E_{\mathrm{coh}}|$ vs. $N^{-1/3}$ for particles with up to 92 atoms and with all the morphologies shown in Figure 10.1, together with data from the work of Lin et al. We can distinguish two regions with a transition at $N \approx 9$,

$$|E_{\mathrm{coh}}(N)| = \begin{cases} 6.64\ \mathrm{eV} - 6.61\ \mathrm{eV} \cdot N^{-1/3} & \text{for} \quad N \le 9 \\ 5.83\ \mathrm{eV} - 4.93\ \mathrm{eV} \cdot N^{-1/3} & \text{for} \quad N > 9. \end{cases} \tag{10.6}$$

For $N > 9$, $|E_{\mathrm{coh}}|$ scales linearly with the relative proportion of surface atoms, $N^{-1/3}$. Extrapolation of Equation (10.6) to $\vec{N} \to \infty$ provides an estimate of the bulk cohesive energy, $|E_{\mathrm{coh}}(\vec{N} \to \infty)| = 5.83\ \mathrm{eV}$, which is in excellent agreement with the experimental value for the cohesive energy of bulk

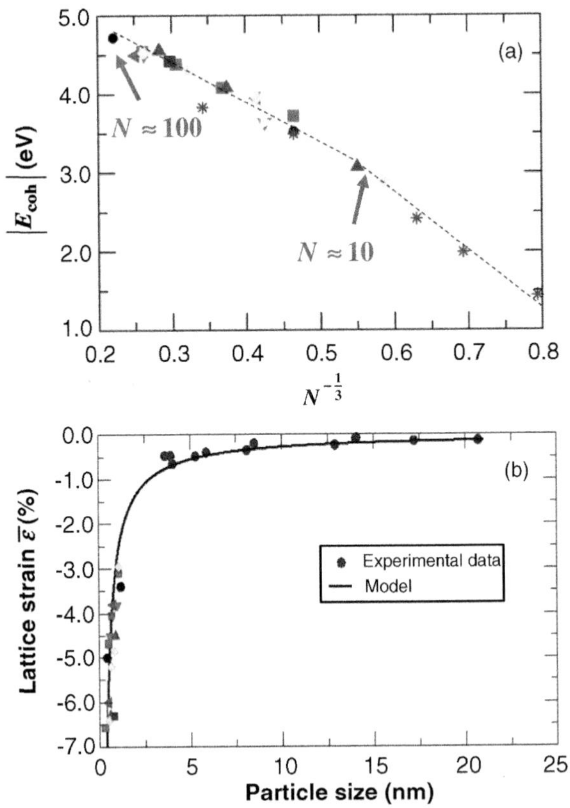


**FIGURE 10.3** (a) Average cohesive energy per atom vs. $N^{-1/3}$ for Pt nanoparticles with up to 100 atoms and various shapes shown in Figure 10.1. The symbol * indicates data points taken from Ref. 80. (b) DFT calculation results of average lattice strain in nanoparticles as a function of particle size in comparison with experimental data [87] and a thermodynamic model [88] of particle formation. (Reprinted with permission from *J. Phys. Chem. C*, 2009, 113 (42), 17989–17996. Copyright 2009 American Chemical Society.)

Pt, $|\mu_{exp}(\infty)| = 5.85$ eV [81]. The purported scaling of cohesive energy with the specific surface area extends to clusters with as few as nine atoms, corresponding to cluster sizes of $\sim 7$ Å. Using the slope of the graph and the bulk density of Pt, $\rho = 6.62 \cdot 10^{22}$ cm$^3$, we obtained an estimate of the surface tension, $\gamma_{np} = 3.8$ J/m$^2$, which agrees reasonably well with the experimental value for the surface tension of Pt in vacuo, $\gamma_{Pt} \approx 3.2$ J/m$^{-2}$ [82]. These findings suggest that metallic bonding, mediated by delocalized electrons, persists down to these small metal clusters. For $N \leq 9$, a linear trend could be discerned from the data of Lin et al. [80] to still provide a reasonable linear fit, but it is evident that the bonding mechanism deviates from metallic bonding.

No systematic trend of particle shape on $|E_{coh}|$ is discernible in Figure 10.3a. At a given $N$, variations in $|E_{coh}|$ for different shapes are $< 0.07$ eV. This relative shape invariance indicates that, in large particle ensembles, one should expect a statistical distribution of shapes; moreover, it suggests that small excitations could suffice to incur shape fluctuations in any given narrow interval of $N$ [83].

The variation of $|E_{\mathrm{coh}}|$ with $N$ implies a general drift in statistical particle size distributions toward larger sizes, namely particle dissolution and growth. This is a kinetic phenomenon that is driven by the difference in surface energies between small and large particles. The dependence of the particle dissolution rate on particle radius is given by

$$\nu(r,t) \propto \exp\left(\frac{2\beta_{\mathrm{Ost}}\gamma_{\mathrm{np}}\bar{V}_{\mathrm{np}}}{RT}\frac{1}{r}\right) \qquad (10.7)$$

where $\beta_{\mathrm{Ost}}$ is the Ostwald transfer coefficient, $\bar{V}_{\mathrm{np}}$ the molar volume of the particle, and $R$ the gas constant. According to Equation (10.7) smaller particles are thermodynamically less stable and so dissolve at a higher rate. The change in the slope of $|E_{\mathrm{coh}}|$ versus $N^{-1/3}$ at $N < 9$ suggests that the smallest clusters dissolve at a significantly enhanced rate. This effect could be related to the transition from predominantly metallic-to-covalent bonding.

The cohesive energies of small metal nanoclusters and nanoparticles is strongly dependent on coordination number and interatomic distances [84]. The shift in interatomic distance from bulk values is quantified by the lattice strain $\varepsilon = (a_{\mathrm{np}} - a_{\mathrm{cryst}})/a_{\mathrm{cryst}}$, where $a_{\mathrm{np}}$ and $a_{\mathrm{cryst}}$ are the lattice constants of the nanoparticle and the extended crystal configuration, respectively [85,86]. The dependence of $\varepsilon$ on particle size and shape arises from nanoscale confinement and undercoordination of surface atoms. Both the coordination number and the lattice strain in a nanoparticle exhibit spatial fluctuations. However, in order to identify general trends of variations in particle size and shape, we evaluate these effects using particle-averaged coordination number $\bar{z}$ and lattice strain $\bar{\varepsilon}$.

Figure 10.3b shows experimental data [87] of $\bar{\varepsilon}$ as a function of particle size in the range of 3–30 nm. The solid line is the lattice strain obtained from a simple thermodynamic model that assumes the particles to be spherical [88]. Predictions of the thermodynamic model are in very good agreement with experimental data from Reference 87. In the range of particle sizes $< 3$ nm, the lattice contraction strongly increases with decreasing size. Lattice strains obtained from our DFT calculations for particles with sizes $< 1.5$ nm are in good agreement with the thermodynamic model.

The increasing undercoordination of surface atoms with decreasing particle size reduces the average bond strength; this effect competes with the impact of the lattice contraction, which enhances the average bond strength [89]. Figure 10.4 shows $|E_{\mathrm{coh}}|$ as a function of $\bar{z}$ and $\bar{\varepsilon}$,

$$|E_{\mathrm{coh}}| = g(\bar{z}, \bar{\varepsilon}) \qquad (10.8)$$

$|E_{\mathrm{coh}}|$ increases with increasing $\bar{z}$ and decreasing $\bar{\varepsilon}$. The effect of undercoordination supersedes the effect of lattice contraction; the net effect of reducing the particle size is a destabilization of the particle. For $-7\% < \bar{\varepsilon} \leq -6\%$, corresponding to a size range

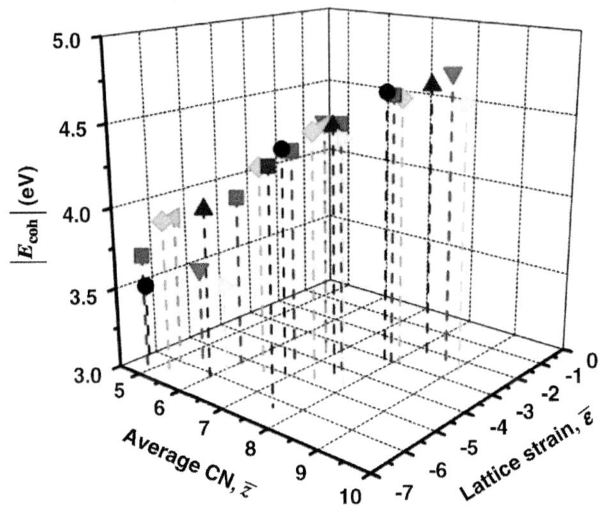


**FIGURE 10.4** Absolute value of cohesive energy $|E_{\text{coh}}|$ as a function of average coordination number $\bar{z}$ and lattice strain $\bar{\varepsilon}$ (%) for Pt nanoparticles in the size range from 1 to 2 nm with the shapes shown in Figure 10.1. (Reprinted with permission from *J. Phys. Chem. C*, 2009, 113 (42), 17989–17996. Copyright 2009 American Chemical Society.)

of 0.5–1 nm, we find a wide variation of $\bar{z}$, $5 < \bar{z} < 8$. In this range, $|E_{\text{coh}}|$ is strongly dependent on particle morphology. For $\bar{\varepsilon} > -6\%$, $|E_{\text{coh}}|$ decreases monotonically with decreasing $\bar{z}$.

### 10.2.3 Modeling of Particle Size Effects in $CO_{ad}$ Electrooxidation

The net rate of a desired electrochemical reaction is the result of a complex interplay of elementary surface processes. These include molecular adsorption, charge transfer, surface diffusion, and recombination of adsorbed intermediates, and desorption of products. The effect of particle morphology on the overall reactivity arises from the structural sensitivity of the elementary steps [47,48,51,90,91].

Electrooxidation of $CO_{ad}$ is a widely studied model reaction in electrochemistry, because the influence of the particle size and surface morphology on the catalytic activity is particularly pronounced for reactions involving $CO_{ad}$. $CO_{ad}$ electrooxidation is also relevant in PEFCs, where anode poisoning by strongly adsorbed $CO_{ad}$ occurs. The latter problem arises if the hydrogen gas supplied at the fuel cell anode contains traces of CO as a byproduct of the re-forming of hydrocarbons. In direct methanol fuel cells (DMFCs), $CO_{ad}$ can be formed as a strongly adsorbed intermediate of methanol oxidation. For $CO_{ad}$ electrooxidation, the maximum oxidation rate found in potential step measurements decreased by a factor of ~5 when the nanoparticle size dropped from > 5 nm to 2 nm [44,92–95].

The main steps in for CO electrooxidation according to the Langmuir–Hinshelwood reaction scheme [96] are

$$\text{Step 1:} \quad CO + nPt \leftrightarrow nPt{-}CO_{ad} \tag{10.9}$$

$$\text{Step 2:} \quad H_2O + \nu Pt \leftrightarrow \nu Pt{-}OH_{ad} + H^+ + e^- \tag{10.10}$$

$$\text{Step 3:} \quad nPt\text{-}CO_{ad} + \nu Pt\text{-}OH_{ad} \rightarrow CO_2 + H^+ + e^- + (\nu + n)Pt \tag{10.11}$$

When only a fraction of surface sites constitutes reaction centers, surface diffusion of reactants to these sites becomes a pivotal process. Trapping of $CO_{ad}$ at low-coordination sites and stronger binding on crystal facets could be the cause of reduced $CO_{ad}$ mobility on nanoparticles. NMR experiments at the solid–gas interface indicate increasing activation energies for surface diffusion with decreasing particle sizes, from 0.27 eV for 10-nm particles to 0.44 eV for 1.2-nm particles [97]. This corresponds to a lowering of the surface mobility on the small particles by a factor $\sim 10^3$. Using NMR spectroscopy, Kobayashi et al. found that the tracer diffusion coefficient of adsorbed $CO_{ad}$ at electrolyte-immersed Pt black particles (5.5 nm), $D_{CO} = 3.6 \cdot 10^{-13}\ cm^2/s^1$, is considerably lower than the value for single crystalline surfaces [98].

The theoretical model presented in References 26 and 99 employed the active-site concept to investigate reaction rate distributions on heterogeneous surfaces of Pt nanoparticles. The distinction of active and inactive sites is evident on bimetallic catalysts, which utilize the bifunctional mechanism [100]. For PtRu catalysts, the Ru surface atoms constitute the active sites, which promote $OH_{ad}$ formation via water splitting, whereas the Pt surface atoms represent the inactive sites, which act as a reservoir for adsorbed $CO_{ad}$. The distinct components of the bimetallic catalyst thus promote different reactions.

In the model, the heterogeneous 3D nanoparticle surface was mapped onto a regular hexagonal 2D array with the two surface states, as shown in Figure 10.5. Despite its simplicity, the model is designed with realistic nanoparticle surfaces in mind. The total number of surface sites is $N_s = 3n(n-1) + 1$, where $n$ is the edge length of the hexagon. $N_s$ was adjusted to match the total number of surface atoms on a *cuboctahedral* particle. For particles of diameter 3 nm we used, for example, $N = 397$. The particle surface was partitioned into electrocatalytically active sites as the exclusive sites, at which $OH_{ad}$ can be formed, and inactive sites, which act as reservoir for $CO_{ad}$. The structural parameters are the fraction of active sites $\xi_{tot}$ and the numbers of nearest neighbors (NN) of an active site with other active sites $z_{aa}$ or with inactive sites $z_{an}$. These NN numbers represent the degree of active site clustering on the surface. The active sites are not a priori assigned to specific surface sites, such as edge or corner sites, of the real 3D nanoparticle. At present, more complex structural models with specific active sites are not warranted since thus far the potential energy landscapes of nanoparticle crystallites have not been revealed by first-principles calculations, nor is the detailed surface structure of nanoparticles known from experimental studies.

The inclusion of finite surface mobility of adsorbed $CO_{ad}$ to active sites leads to a complex interplay of reactant surface mobility and on-site reactivity. The solution of the general model that includes details of the heterogeneous surface structure and the effects of finite CO surface mobility requires stochastic methods. The kinetic Monte Carlo (kMC) method is a versatile technique for the study of surface reactions.

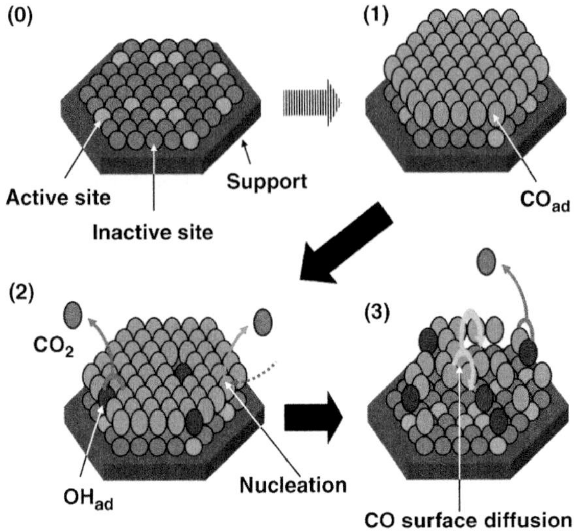


**FIGURE 10.5** Active-site model of nanoparticle surface adopted in kinetic model of nanoparticle reactivity. Scheme (0) is a possible configuration of the plain catalyst surface with active and inactive sites. The surface structure can be generated randomly or deliberately in order to mimic typical distributions of active sites. Schemes (1)–(3) illustrate adsorbate structures and processes on the surface at distinct stages of $CO_{ad}$ monolayer oxidation. A $CO_{ad}$ monolayer applied onto the surface, shown in (1), starts to oxidize through formation of $OH_{ad}$ on active sites, as depicted in (2). At later stages, indicated in (3), surface diffusion of $CO_{ad}$ to active sites will become an important process in the oxidation reaction.

Instead of solving a coupled system of deterministic reaction–diffusion equations, the kMC approach is based on a stochastic formulation. It enables the incorporation of structural details at an atomistic level. Moreover, kMC simulations proceed in real time and permit the straightforward inclusion of time-dependent transition rates. Several research groups have applied this approach to the study of $CO_{ad}$ electro-oxidation, focusing mostly on homogeneous surfaces [101–104]. For the simulation of chronoamperometric transients at nanoparticles, we have implemented a variable-timestep algorithm [105,106].

In the case of rapid $CO_{ad}$ surface diffusion, a mean-field (MF) model can be employed, which represents the surface state in terms of effective surface coverages of $CO_{ad}$ on inactive sites $\theta_{CO}$, $CO_{ad}$-free active sites $\theta_\xi$, and $OH_{ad}$ on active sites $\theta_{OH}$. The surface coverage ranges are $0 \leq \theta co \leq 1$, normalized to the saturation coverage on inactive sites, as well as $0 \leq \theta_{OH} \leq 1$ and $0 \leq \theta_\xi \leq 1$, both normalized to the saturation coverage on active sites. The average NN numbers $z_{an}$ and $z_{aa}$ are retained in the mean-field formulation to account for the degree of active-site clustering.

On nanoparticles, the exact fraction of active sites and the kinetics of their formation are unknown. In view of this unclear situation, a simple and physically sound assumption is to introduce an additional rate constant that describes the removal of $CO_{ad}$ from active sites by nucleation, that is, without reacting with a neighboring

$OH_{ad}$. For the initial formation of free active sites, a nucleation process is assumed to remove $CO_{ad}$ from active sites at rate

$$\nu_N = k_N (1 - \theta_\xi) \tag{10.12}$$

where $k_N$ is the rate constant of nucleation.

In the limit of high $CO_{ad}$ mobility, one would expect the results of the kMC simulations to coincide with those of the MF approach. This statement is in general not valid on a structured surface (see Ref. 99 for a detailed discussion). The MF approach neglects structural correlations between the positions of active sites on structured surfaces. The kMC simulations, on the other hand, explicitly account for correlations due to the clustering of active sites. Hence, good agreement between MF and kMC approaches can be expected for sufficiently high rates of nucleation $z_{aa}k_{ox} \leq k_N$, where $k_{ox}$ is the rate of the recombination step [Eq. (10.10)]; this condition indicates suppressed correlations between active sites. Both methods give identical transients only for $z_{aa} = 0$, that is, for uncorrelated active sites.

In Reference 26, the kMC and MF approaches were compared with experimental chronoamperometric current transients for particles with sizes 1.8, 3.3, and $> 5$ nm and electrode potentials, $E$ (vs. RHE), ranging from 0.68 to 0.85 V. The fitting procedure started with the analysis of experimental transients with the MF model. This allowed the use of straightforward least-squares routines, in contrast to the intricate curve-fitting procedure with the kMC method. The results of fitting in the MF approach were satisfactory for the large, electrodeposited particles ($> 5$ nm) over the whole potential range. No further improvement of the fits was achieved when the surface mobility of $CO_{ad}$ was allowed to vary using the kMC method. This result complies with experimental evidence for fast $CO_{ad}$ diffusion on large particles. For the intermediate particles (3.3 nm), the MF model provided good fits at $E < 0.8$ V, whereas the fits at $E \geq 0.8$ V could be substantially improved when the diffusion rate of $CO_{ad}$ was allowed to vary in the kMC model. For small particles (1.8 nm), the kMC model with finite surface diffusion had to be employed at all potentials $E$ to achieve good agreement with experimental data. Overall, consistent sets of parameters could be determined for all three particle sizes and at all potentials using the appropriate approach. Figure 10.6a exemplifies results of the fitting for particles with intermediate size (3.3 nm). As shown in Figure 10.6b, results of the active-site model are in better agreement with experimental data than are homogeneous mean-field [101,102,107,108] and nucleation and growth models [109–111]. These homogeneous surface models cannot reproduce the strong asymmetry in transients and the current tailing at longer times for small particle sizes and high step potentials.

From the fits of experimental data with the active-site model we extracted structural parameters as well as rate constants and charge transfer coefficients for the various electron transfer steps. The equilibrium potential of $OH_{ad}$ formation by water splitting was found at $E^{eq}_{OH} = 0.68$ V for pure Pt particles. In a similar model-based analysis of chronoamperometric transients of $CO_{ad}$ electrooxidation at bimetallic PtRu, we found $E^{eq}_{OH} = 0.37$ V. This difference explains the promotional role of

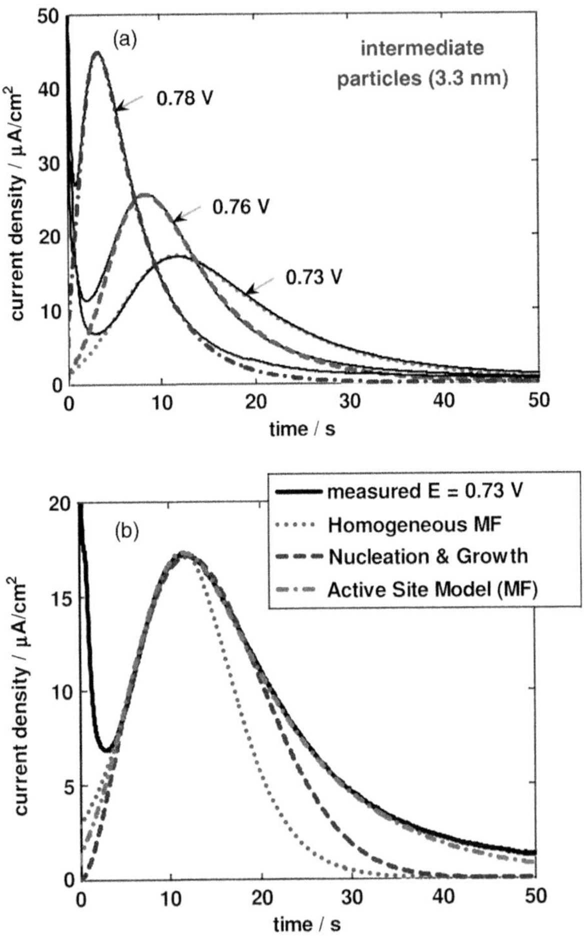


**FIGURE 10.6** (a) Fits of chronoamperometric transients of $CO_{ad}$ monolayer electrooxidation for Pt nanoparticles with intermediate size (3.3 nm) at various applied electrode potentials (vs. SHE) by active-site model. Main current peaks and pronounced tailing at long times in experimental data (thin black lines) are reproduced well in the model (thick colored lines). (b) Comparison of typical current transient with active-site model, which is in good agreement with experimental data, and with homogeneous surface models, namely, homogeneous mean field and nucleation and growth approaches.

Ru in activating $H_2O$, allowing the oxidation of $CO_{ad}$ to $CO_2$ at lower potentials at PtRu particles.

A central result of the modeling study was the strongly restricted surface mobility of $CO_{ad}$ on the smallest nanoparticles (1.8 nm), which corresponds to a surface diffusion coefficient $D_{CO} \approx 10^{-16}\ cm^2/s$. The overall kinetics are thus limited by diffusion of $CO_{ad}$ at these particles. For the large particles ($> 5$ nm), diffusion coefficients are at least two orders of magnitude higher, so that the overall kinetics are controlled by the recombination step. For the intermediate particles (3.3 nm), the overall activity is limited by on-site kinetics at small $E$ and by diffusion at large $E$; the transition occurs at $E \approx 0.8$ V. These findings for the particle size effect on $CO_{ad}$ mobility are in line with NMR experiments [97,98].

Another key finding from this study was the low fraction of active sites on small nanoparticles (1.7 and 3.3 nm), for which we found $\xi_{Pt} \simeq 0.1$. This value increased only for larger particles ($> 5$ nm) with a more corrugated surface structure. The low-active-site fraction is irreconcilable with the fractions of edge and corner atoms, which are often alluded to as "prospective active sites." For the 1.8-nm particles, edge and corner sites account for $\sim$50% of the surface sites, whereas the corners alone make up for $\sim$ 12%. The corresponding fractions for the 3.3 nm particles are 28% and 2%, respectively. Therefore, it seems unlikely that edge and/or corner sites act as active sites.

These studies confirmed that electrochemical reactivity at nanoparticles depends strongly on the heterogeneous atomic and electronic structure at the surface. The nature, abundance, and distribution of active sites should be accounted for in reactivity models. The main challenge is that, so far, there is no consistent geometric interpretation of active sites. Refined models of nanoparticle reactivity require electronic structure calculations for systems that mimic nanoparticle surfaces with heterogeneous geometric and electronic structure. Such calculations should elucidate reaction paths, surface mobility of adsorbates, and rates and transfer coefficients of charge transfer steps at nanoparticles. It is clear that the role of specific surface sites not only varies with particle size and shape but also depends on the considered reaction, the substrate material, the electrolyte, and the adsorption state.

### 10.2.4 Ab Initio Study of Oxygen Adsorption Energy at Nanoparticles

The surface heterogeneity of nanoparticles calls for detailed spatial maps of the energetics and kinetics of elementary surface processes. Han et al. [112] studied O and OH adsorption energies on spherical cubooctahedral Pt nanoparticles with sizes of 1 and 2 nm using DFT. For both adsorbents, they found that adsorption energies change dramatically from the Pt(111) surface to the nanoparticles in a similar manner. The highest adsorption energies were found at undercoordinated edge atoms. However, this study did not provide a separate discussion of Pt–Pt bond distortion and particle relaxation effects. We performed a similar study, in which we calculated the detailed map of oxygen adsorption energies at hemispherical cuboctahedral nanoparticles with (111) basal planes. These particles provide the highest variability of coordination structures of surface sites compared to other particle shapes in Figure 10.1. They are a suitable starting point for exploring correlations between surface morphology, surface electronic structure, and electrocatalytic activity. Figure 10.7 shows the three distinct surface facets, namely, one (111) top facet, three equivalent (111) side facets and three equivalent (100) side facets and their adsorption sites with different coordination structures. The (111) basal plane was not considered.

We selected nanoparticles with $N = 37$ (size $\sim$ 1 nm) and $N = 92$ (size $\sim$ 1.5 nm). In order to probe the heterogeneity of the surface electronic structure, we calculated

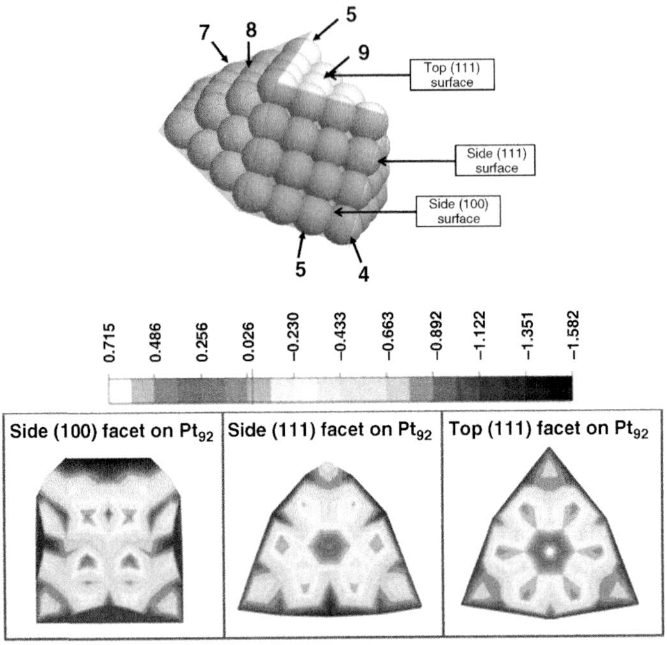


**FIGURE 10.7** Hemispherical cuboctahedral $Pt_{92}$ nanoparticle with distinct facets and distinct local coordination structures of surface atoms, indicated by their nearest-neighbor atom numbers (top). Contour plots of oxygen adsorption energies for the distinct surface facets of hemispherical cuboctahedral $Pt_{92}$ nanoparticle (bottom). (Reprinted with permission from *J. Phys. Chem. C*, 2009, 113 (42), 17989–17996. Copyright 2009 American Chemical Society.) (See insert for color representation.)

the potential energy associated with the dissociative chemisorption of oxygen. Pt–O interactions are crucial for the ORR and water splitting. Two covalent bonds can be formed between the two unpaired valence electrons in an oxygen atom and metal electrons. Variations of the nearest-neighbor (NN) structure of distinct adsorption sites on the metal (which depend on the number of neighboring Pt atoms, bond distances, and spin states) creates the well-known selectivity of O adsorption among fcc, hcp, bridge, and top sites on Pt(111) and among on-top, bridge, and fourfold hollow sites on Pt(100) [113–115]. The topological heterogeneity of nanoparticle surfaces is expected to strongly amplify this site selectivity.

On adsorption at an extended surface, the charge distributions in the metal as well as in the adsorbate reorganize, resulting in a lowering of the total energy. These relaxation phenomena are more pronounced on nanoparticles. In order to distinguish between the effects of electronic structure reorganization and nanoparticle relaxation on the adsorption energy, we performed calculations of oxygen adsorption energies in two steps. In the first step, we locked the positions of Pt atoms at the optimum configuration of the relaxed bare particle. We positioned the oxygen atom above a well-defined adsorption site on the particle and kept its lateral position fixed. In the DFT-based optimization calculation for a given adsorption site, the oxygen atom was allowed to relax only in the direction perpendicular to the surface. This procedure was repeated for all relevant adsorption sites on each of the distinct surface facets: on-top, bridge, and fourfold hollow sites on the Pt(100) facet and top, bridge, hpc, and fcc sites on the Pt(111) facets. From these results, we generated contour plots of the potential energy of oxygen adsorption for the unrelaxed particles. In the second step, full optimizations were performed with relaxation of all atoms in the system. The oxygen atom was again positioned initially at all high-symmetry sites on the surface. Thereby we quantified the contribution of particle relaxation to the total potential energy of adsorption.

The potential energy of adsorption after either of these steps is calculated with respect to the energy of the free $O_2$ molecule,

$$E_{\text{ads}} = E_{\text{Pt-O}} - E_{\text{Pt}} - \tfrac{1}{2} E_{O_2} \tag{10.13}$$

where $E_{\text{Pt-O}}$ is the total energy of the Pt particle with the adsorbed oxygen; $E_{\text{Pt}}$, the total energy of the pristine Pt particle ($E_{\text{Pt}} = -479.5$ eV for $Pt_{92}$ and $E_{\text{Pt}} = -181.61$ eV for $Pt_{37}$); and $E_{O_2}$, the total energy of the oxygen molecule in the gas phase. Because of the complicated electronic structure of $O_2$, the binding energy of $O_2$ cannot be determined precisely from DFT [116]. The total energy of $O_2$ was found to be $-9.975$ eV by using the experimental reaction Gibbs energy of the reaction $H_2O \rightarrow H_2 + \frac{1}{2}O_2$, $\Delta G = 2.46$ eV [117], and the Gibbs energies of $H_2O$ and $H_2$ in the gas phase, which can be readily obtained from DFT: $G_{H_2O} = 14.280$ eV, $G_{H_2} = 6.83$ eV.

Contour plots of $E_{\text{ads}}$ for fixed particle configurations are shown in Figure 10.7 (bottom panels). The complex topological distribution of adsorption energies reflects the underlying periodic arrangement of surface atoms as well as the pronounced nanoscale confinement effects. The highest absolute values of adsorption energies occur at the edges. Overall, we obtain stronger adsorption on $Pt_{37}$ compared to $Pt_{92}$, as expected.

In further detailed studies, we explored the effects of local site-averaged bond distances and coordination numbers of surface atoms on $E_{\text{ads}}$. From studies at extended surfaces, it is well known that undercoordination of surface atoms induces $d$-band narrowing, which causes an increase in the adsorption energy. A lattice contraction, on the other hand, induces a broadening of the $d$ band, which causes a decrease in the adsorption energy. Both of these effects compete on nanoparticle surfaces [118]. We did not observe any systematic trend for the net effect. This result attests to the strong spatial fluctuations of the local electronic structure on the particle surface.

On extended Pt(111), $\bar{z}_l$ and $\bar{d}_l$ are constant for one type of high-symmetry site (top, bridge, hpc, or fcc sites), corresponding to a single $E_{\text{ads}}$ value. In order to compare our results for adsorption energies on unrelaxed hemispherical cuboctahedral particles with the values for an unrelaxed, extended Pt(111) surface, we performed DFT-GGA-based calculations of oxygen atomic adsorption on a $3\times3$ unit cell of Pt(111) with four layers with oxygen coverage of 0.1 ML (i.e., 10% of a monolayer). The surface Brillouin zone was sampled by a Monkhorst–Pack $\vec{k}$-point set of $3\times3\times1$ [119].

Figure 10.8 shows the normalized site density distribution of geometrically equivalent adsorption sites versus the potential energies of oxygen adsorption for $Pt_{92}(111)$. The dashed lines correspond to adsorption at the same topological sites on

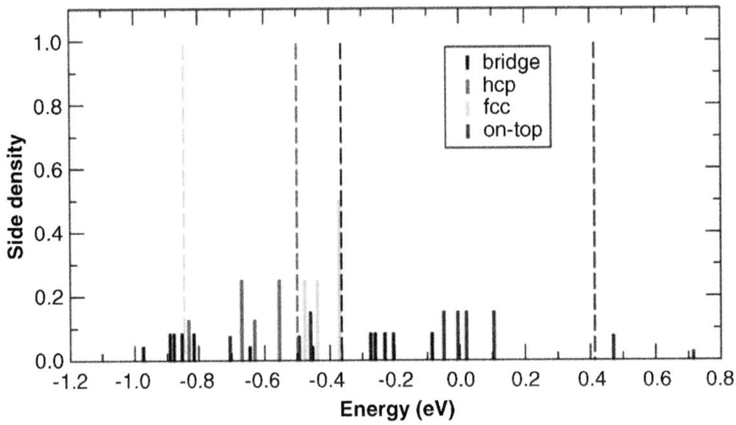


**FIGURE 10.8** Normalized site density distribution of geometrically equivalent adsorption sites versus the potential energies of oxygen adsorption for Pt92(111). The dashed lines correspond to adsorption at the same topological sites on an unrelaxed extended Pt(111) surface. (Reprinted with permission from *J. Phys. Chem. C*, 2009, 113(42), 17989–17996. Copyright 2009 American Chemical Society.)

an unrelaxed extended Pt(111) surface. Nanoscale confinement causes a pronounced broadening of the adsorption site density distributions. The preferred adsorption sites change from fcc on extended Pt(111) to bridge and hcp on $Pt_{92}(111)$. The shift in adsorption energies between nanoparticle and extended surface is particularly striking for the fcc site, which becomes a less favorable site for adsorption on the nanoparticle. However, for a number of bridge, hcp, and on-top sites, nanoscale effects lead to significantly stronger adsorption, namely, enhanced $|E_{ads}|$, in comparison to the corresponding site on the extended surface. The maximal nanoscale enhancements of $|E_{ads}|$ are 0.62, 0.33, and 1.11 eV for bridge, hcp, and on-top sites, respectively, which render bridge sites the most favorable for adsorption. We expect the site density distributions to become narrower with increasing particle size, approaching the limiting values for the analogous sites on an extended surface.

To evaluate whether the trends in chemisorption energies on Pt nanoparticles are consistent with those of the *d*-band model, we projected out the *d*-band density of states on the different adsorption sites and determined their corresponding *d*-band centers. No correlation was observed between the adsorption energies and site-specific *d*-band centers. Even though the metal nanoparticles possess a continuous electronic band structure, suggesting metal-like electronic properties, their catalytic surface properties are not controlled by band structure effects but by the local electronic structure of the adsorption sites. The catalytic properties of supported metal nanoparticles are not predictable from the properties of the corresponding bulk material.

Finally, we allowed the full relaxation of Pt particles and extended surfaces in the presence of the adsorbed oxygen atom. Thereby, we determined the maximum oxygen adsorption energy for every facet and the contribution due to particle relaxation. For $Pt_{37}$, the most favorable adsorption site, shown in Figure 10.9, is at the top corner of the Pt(100) facet; for $Pt_{92}$ it is at the bottom corner of the side facet Pt(111). The net effect

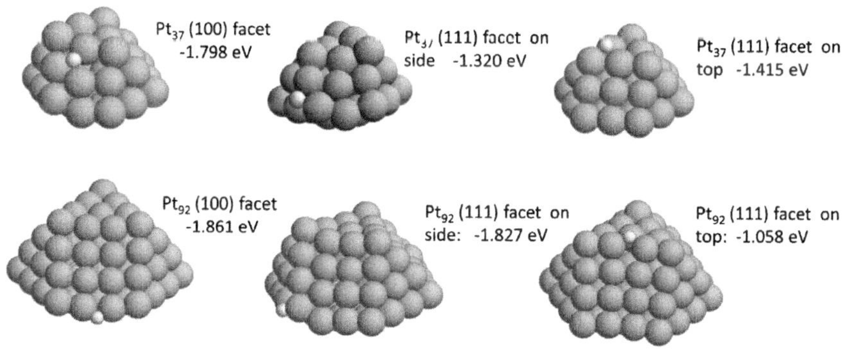


**FIGURE 10.9** Fully relaxed geometries of $Pt_{37}$ and $Pt_{92}$ nanoparticles with one adsorbed oxygen atom, showing the most favorable adsorption sites on different facets. (Reprinted with permission from *J. Phys. Chem. C*, 2009, 113(42), 17989–17996. Copyright 2009 American Chemical Society.)

of particle relaxation is smaller for $Pt_{37}$ than for $Pt_{92}$ because of the larger number of atoms that are involved in the geometry relaxation of $Pt_{92}$. Due to the particle relaxation, $|E_{ads}|$ on Pt(100) facet, side Pt(111) facet, and top Pt(111) facet of the $Pt_{37}$ particle increased by 0.62, 0.52, and 0.14 eV, respectively. For the $Pt_{92}$ particle, these increases are 0.28, 0.74, and 0.41 eV, respectively.

### 10.2.5 Reaction Pathway of ORR at $Pt_4$ Nanocluster

The ORR on Pt surfaces is a multistep process that has been widely studied in experiment and theory [120–123]. The two mechanisms proposed for ORR are the dissociative and the associative mechanisms. In the dissociative mechanism, the oxygen molecule first adsorbs on the metal and then dissociates by breakage of the O–O bond. The dissociation is followed by the transfer of two electrons and two protons to form two adsorbed $OH^-$. In the associative mechanism, $O_2$ first adsorbs on the metal surface, followed by transfer of an electron from the metal to the adsorbed oxygen molecule, with the O–O bond remaining intact. The adsorbed oxygen molecule is reduced further to $H_2O_2$ or $H_2O$ by subsequent transfers of $1e^-$ or $3e^-$, respectively [123].

Extensive theoretical studies based on DFT have been performed for both mechanisms [34,36,38,121,124–126]. For ORR on an extended Pt(111) surface with a full monolayer of water, DFT calculations have shown that the formation of OH* (where the asterisk "*" indicates an adsorbed species) is the rate-determining step of the dissociative mechanism at equilibrium potential [127]. For the associative mechanism, the formation of OOH* has highest barrier energy due to the electron transfer process, $O_2 + e^- \rightarrow O_2^-$, which precedes the OOH* formation [128,129].

We conducted a study of the associative mechanism of ORR at a tetrahedral $Pt_4$ cluster, in order to understand differences in the reaction path at nanoclusters or nanoparticles in comparison to extended surfaces. A preliminary calculation showed that the oxygen dissociation energy at $Pt_4$ is very high, $E_{O_2}^{diss} > 1.2$ eV and therefore the dissociative mechanism is unlikely to occur. Moreover, we restricted our study to the $4e^-$ transfer mechanism, which means that we neglected $H_2O_2$ formation as an intermediate process. The associative mechanism of ORR on the extended surface proposed by Rossmeisl et al. [116] consists of four elementary steps,

$$O_2^* + H^+ + e^- \rightarrow OOH^* \tag{10.14}$$

$$OOH^* + H^+ + e^- \rightarrow H_2O + O^* \tag{10.15}$$

$$O* + H^+ + e^- \rightarrow OH* \tag{10.16}$$

$$OH* + H^+ + e^- \rightarrow H_2O* \tag{10.17}$$

The proposed mechanism (10.14)–(10.17) suggests that all steps proceed at a single active site. The mechanism of ORR on nanoparticles is expected to deviate from the mechanism on the extended surface due to nanoconfinement effects involving the shifting and coarsening of the electronic structure as well as breakage of the spatial symmetry on the particle surface.

In our study we have included four extra water molecules surrounding the $Pt_4$ cluster. These water molecules mediate elementary reaction steps between neighboring active Pt sites on the cluster by providing pathways for proton transfer along the surface. Moreover, the water molecules stabilize polar adsorbates on the $Pt_4$ cluster and provide stable $H_3O^+$ in the first hydration shell. We explored the following reaction path for the overall reaction $O_2^{aq} + 4\,H^+ + 4\,e^- \leftrightarrow 2\,H_2O^{aq}$,

$$H_2O* \leftrightarrow H_2Oaq \tag{10.18}$$

$$O_2^{aq} \leftrightarrow O_2^* \tag{10.19}$$

$$O_2^* + H^+ + e^- \leftrightarrow OOH^* \tag{10.20}$$

$$OOH* + OH* \leftrightarrow H_2O^{aq} + 2O* \tag{10.21}$$

$$O* + H^+ + e^- \leftrightarrow OH* \tag{10.22}$$

$$O* + H^+ + e^- \leftrightarrow OH* \tag{10.23}$$

$$OH* + H^+ + e^- \leftrightarrow (H_2O)* \tag{10.24}$$

Superscripts "aq" denote species in solution. The main difference between the preceding mechanism and the mechanism proposed for an extended surface is that the reaction step (10.21) does not involve electron and proton transfer. However, this step involves proton transfer between active sites mediated by proton translocation along a chain of hydrogen-bonded water molecules, as illustrated in Figure 10.10a.

In the initial configuration of the $Pt_4$ cluster, two Pt sites are occupied by adsorbed oxygen atoms. The remaining two sites contain an adsorbed OH* and an adsorbed water molecule; these will be the active sites that participate in surface reactions. The initial configuration of the cluster is specified as $2O{-}Pt_4{-}OH{-}H_2O$ (with the two passive sites on the left side and the state of the active sites indicated on the right side).

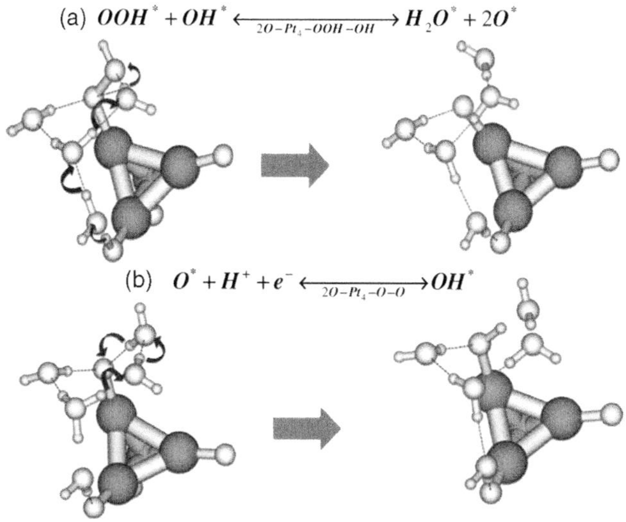


**FIGURE 10.10** The initial and final states of reaction step (10.21) in (a) and (10.22) in (b) in the reaction sequence of oxygen reduction at the Pt4 cluster with four surrounding water molecules. (Reproduced from *Phys. Chem. Chem. Phys.* 2010, 12, 614–620, by permission of the PCCP Owner Societies.)

The preadsorbed OH* and $H_2O$* species are required for performing the steps (10.18) and (10.21). The two preadsorbed oxygen species prevent water molecules in solution from adsorbing on the cluster during the optimization. Basically, one can select different preadsorbed species for these two sites, which may lead to slightly different activation and reaction energies.

We calculated the Gibbs free energy for each elementary step from $\Delta G = \Delta E + \Delta G_U$, where $\Delta E$ is the internal energy and $\Delta G_U$ is the energy shift due to an applied electrode potential. We neglected the effects of entropy changes and of zero-point energies. We determined the change in internal energy for each reaction by performing optimized geometry calculations for the initial and final states. For each reaction step (10.18–10.24), we assumed that the reorganization energy of water molecules to prepare the initial configuration is negligible. Therefore water molecules in each initial configuration of elementary reactions (10.20)–(10.24) are prearranged to form the optimum initial configuration. Figure 10.10 shows the initial and final states of reactions (10.21) and (10.22). The initial and final states of (10.23) and (10.24) are similar to those of (10.22) and therefore are not shown here. In order to find the minimum-energy reaction path we used the "nudged elastic band" (NEB) method [130].

Equilibrium corresponds to zero reaction Gibbs energy, $\Delta G_r^{total} = 0$, of the overall process, $O_2^{aq} + 4\,H^+ + 4\,e^- \leftrightarrow 2\,H_2O^{aq}$, with a corresponding overpotential $\eta = U - U_{eq} = 0$ V. We assume that application of an overpotential $\eta = U - U^{eq} \neq 0$ shifts the reaction Gibbs energy of each reaction step that involves electron transfer between an excess proton and the $Pt_4$ cluster by an equal amount $\Delta G_U = e\eta$, where $e$ is the elementary charge.

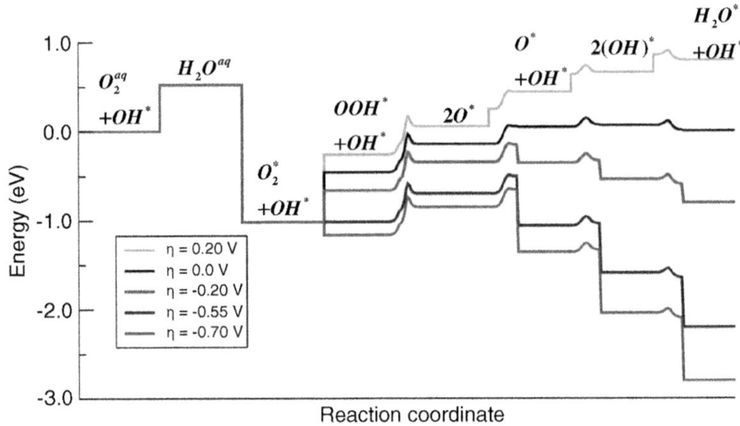


**FIGURE 10.11** Gibbs energy along reaction path of oxygen reduction at the $Pt_4$ for various overpotentials, indicated in the legend. Plateaus correspond to the initial states of the reaction sequence (10.18)–(10.24). (Reproduced from *Phys. Chem. Chem. Phys.* 2010, 12, 614–620, by permission of the PCCP Owner Societies.)

The constructed the Gibbs energy diagram along the reaction path is shown in Figure 10.11 for various applied overpotentials. The diagram connects the initial state $O_2^{aq}$ with the final state, in which a preadsorbed water molecule at an active Pt site and one newly formed water molecule have been released from the cluster and another newly formed water molecule sits at the active surface site of the $Pt_4$ cluster. The plateaus in Figure 10.11 correspond to intermediate states of the reaction sequence (10.18) to (10.24). The bumps visible at transitions between intermediate states correspond to the activation barriers determined by NEB. In the ORR direction (left to right), the initial water desorption step is endothermic, $\Delta G_r = \Delta G_a = 0.56$ eV, and it forms the largest barrier for the ORR sequence. This process is a necessary step to provide a free site for oxygen adsorption. Another high, potential-independent barrier is encountered by the transformation of OOH* in step (10.21), $\Delta G_a = 0.43$ eV.

At equilibrium, $\eta = 0$ V (black solid line in Fig. 10.11), the processes in which OOH*, O*, and OH* are formed are endothermic, whereas $H_2O$* formation in the final step [reaction (10.24)] is slightly exothermic. Figure 10.11 shows that the activation energy for OOH* formation at equilibrium is significantly larger than the reaction energy for the formation of any other intermediate species. The required energy to overcome this barrier is larger than 0.5 eV. For overpotentials $\eta > -0.55$ V, $O_2$* is the most stable intermediate species. As the overpotential becomes more negative, the activation energy for OH* and OOH* formation decreases. At $\eta = -0.55$ V, the activation energy for OOH* formation disappears. At large cathodic overpotentials, $\eta \leq -0.55$ V, the formation of OOH* becomes exothermic, rendering OOH* the most stable adsorbed intermediate species.

In the water decomposition process (from right to left), at overpotential $\eta = 0.2$, all intermediate steps are exothermic, except for the removal of the reaction product, $O_2^*$, from the cluster, which is strongly endothermic. We compared our result with the result for Pt(111) at low coverage. For Pt(111) at equilibrium the formation processes of all species are exothermic except OOH* while for $Pt_4$ we found that only the formation of O* and $H_2O$* are exothermic. Moreover, due to the strong interaction between the $Pt_4$ cluster and $O_2$* and $H_2O$*, the overall reaction rate will be limited by desorption of $O_2$* and $H_2O$* in anodic and cathodic direction of the net process, respectively.

## 10.3 SUBSTRATE EFFECTS

### 10.3.1 Requirements on the Substrate

The bulk of research in electrocatalysis for PEFCs has focused on understanding and improving the properties of the catalyst material. The carbon-based support material in conventional catalyst layers is seen primarily as an electronic conductor with no catalytic effect. The main concern is the lack of stability of carbon-based materials.

The requirements on the support material in fuel cells are, however, very critical, particularly in the cathode side. The functional requirements for materials that constitute CLs are structural support of the catalyst, electrical conductivity, and mass transport (i.e., porosity). The support material needs to provide a high surface area, with enough accessibility for reactants. It should be highly durable under oxidizing and acidic conditions that prevail in operating PEFCs and form stable bonds with Pt nanoparticles to stabilize the catalyst against dissolution and sintering/agglomeration. Designing all these functions into a single layer is a major challenge in current fuel cell research.

Conventional support layers using meso- and nanometric structures of carbon offer affordability and ease of fabrication. Self-organization of ionomer and carbon/Pt in the colloidal catalyst layer ink leads to the formation of phase-segregated, agglomerated morphologies [5]. Carbon agglomeration provides bimodal pore size distributions with primary pores between primary C/Pt particles inside agglomerates and secondary pores between agglomerates. The relative portions of primary and secondary pores and their wetting properties control the water distribution in the layer and thus the wetted fraction of Pt particles, which is an important factor in the exchange current density, [see Eq. (10.1)][11]. Moreover, the porous structure and the water distribution steer the interplay between vaporization exchange and gaseous reactant transport. The structures provided by carbon-based supports are beneficial in view of the water management requirements in PEFCs.

Carbon-based catalyst layers are thermodynamically unstable under the conditions that prevail in operating PEFCs. Electrochemical corrosion of high-surface-area carbon blacks according to the mechanism $C_{(s)} + 2\,H_2O \rightleftharpoons CO_2 + 4\,H^+ + 4\,e^-$ reduces the performance and the lifetime of the fuel cell. Corrosion of the cathode electrocatalyst (carbon and platinum) has been detected by FEG-SEM, TEM, and electrochemical techniques [131]. The performance decline is seen in the loss of power density that the fuel cell could provide [132]. Substitution of carbon materials by thermodynamically stable metal oxides ($TiO_2$, $NbO_2$, WO, $SnO_2$-doped $In_2O_3$) or ceramics (e.g., WC) has been widely explored as a means to overcome durability limitations of conventional catalyst support layers. $TiO_2$ and $NbO_2$ are attractive materials because of their excellent chemical stability in acidic solutions [133]. The study by Sasaki et al. revealed, moreover, a markedly enhanced ORR mass activity of niobium oxide supported Pt electrocatalysts over commercial Pt/C electrocatalysts.

The systematic understanding of effects of the support material on electrochemical properties, performance, and stability of catalyst layers must account for a hierarchy of effects. Conductivity is the main criterion in the preselection of suitable catalyst support materials. An estimate of the minimal electronic conductivity required can be obtained using a simple model of the interplay between the distributed rate of current generation and transport of electrons with conductivity $\sigma_{el}$ in the support phase. The underlying model is identical to models that evaluate the interplay of proton transport limitations and reaction in CCL, discussed, for instance, in Reference 14. Imposing a requirement that the voltage drop due to the finite conductivity of the conducting phase be below a certain maximal value, $\Delta\eta_s < \Delta\eta_{max}$, it follows that

$$\sigma_{el} > L\frac{j_0}{2b}\exp\left(-\frac{\Delta\eta_{max}}{b}\right),$$

where $L$ is the thickness, $b = RT/(\alpha_c F)$ is the Tafel parameter, and $j_0$ is the operating current density. For instance, at $L = 10\,\mu m$, $j_0 = 1\,A/cm^2$, $\alpha_C = 1$, and $T = 333$ K, specifying an upper limit $\Delta\eta_{max} = 1$ mV on the voltage drop in the electronically conducting phase, demands a minimal conductivity of $\sigma_{el} > 0.02\,S/cm$. In an ultrathin catalyst layer with thickness $L = 100$ nm this lower bound is even lower: $\sigma_{el} > 2 \cdot 10^{-4}\,S/cm$. Overall, conductivity requirements on the substrate are relatively modest; they can be met by most of the currently considered materials. Other relevant properties for the selection of support materials are the corrosion resistance and the pore size distribution, which, together with the thickness, determines the available interfacial area and transport properties of CLs.

To further qualify support materials, modeling studies using DFT and force-field-based molecular dynamics simulations should focus on systems consisting of a single nanoparticle or nanocluster on the support material, evaluated in vacuo. The properties and processes of interest include the molecular mechanism of particle–substrate bonding, the Gibbs energy of bonding, the amount of charge transferred between particle and substrate on bond formation, and the electron density distribution for the particle–substrate system. Taking it one step further, it will be vital to evaluate these properties in the presence of water and to characterize their responses to variations in the metal potential. Accounting for the effects of water and metal potential is essential in order to study the influence of the support material on ORR activity and the dissolution kinetics of the supported Pt nanoparticles. DFT-based calculations, similar to those presented in Section 10.2.5, can be employed for exploring reaction sequences, reaction and activation energies for elementary steps, and the stability of adsorbed intermediates. With these calculations, which are in progress in our group, the effect of the support on the Gibbs energy diagram along the reaction path could be evaluated. The reaction sequence can be implemented into a kinetic Monte Carlo (kMC) model to study ORR activity.

The next step in the structural hierarchy leads to mesoscopic arrays of nanoparticles or nanoislands on the substrate. For a single nanoparticle of Pd supported on Au(111), the electrocatalytic activity for the hydrogen evolution reaction (HER) was found to be strongly enhanced in comparison to bulk Pd [134]. Experimentally, the enhancement of electrocatalytic activity of ORR, HER, and other reactions at thin overlayers of Pd on Au(111) is a well-known effect [135]. In theory, electronic structure effects studied by DFT-based calculations can consistently explain the observed trends in activity for overlayer systems [62]. A pseudomorphically grown Pd monolayer on Au(111) exhibits a lattice expansion of close to 5%, which causes an upward shift of the *d*-band center of Pd. This effect leads to increased chemisorption energies of hydrogen during HER or of the reaction intermediates of other reactions. The activity enhancement becomes weaker with increasing number of monolayers, due to the surface properties approaching those of bulk Pd [136].

The most remarkable increase in activity of HER is observed at submonolayer coverages of Pd on the Au(111) substrate. Tafel plots of [136] reveal that the activity normalized to the Pd surface area increases by a factor $> 100$ when the Pd coverage on the Au substrate decreases from 100% to $< 1\%$ of a monolayer. This effect cannot be explained alone by the electronic structure effect discussed above. Indeed, the electronic structure effect is expected to be relatively independent of Pd coverage at submonolayer coverage. Stated differently, the current density of HER increases with the free gold surface. This suggests an alternative explanation for the reactivity enhancement, which involves an active role of the substrate surface due to the spillover effect [137].

The kinetic model of HER at a single Pd nanoparticle supported on Au(111), [61] employed the spillover effect to explain the high turnover rates at the smallest Pd nanoparticles. It was argued that a scenario in which surface processes proceed exclusively at the catalyst particles is irreconcilable with the observed hydrogen evolution current; the hydrogen desorption rate from the nanoparticle is too small in comparison to the potential-controlled rate of $H_{ad}$ formation by proton discharge, and the available surface area on the nanoparticle would be largely insufficient. $H_{ad}$ thus spills over from the Pd particle and diffuses on the Au surface from where it will be subsequently released. A bare Au surface does not promote the hydrogen evolution reaction, since the chemisorption of hydrogen on Au is a strongly endothermic process. The Pd nanoparticles are the sources of $H_{ad}$. As a result of the relatively weak binding of $H_{ad}$ to Au, the surface mobility of chemisorbed hydrogen will be high, with an activation energy in the range $E_d \approx 0.1-0.2$ eV. The expansion of the $H_{ad}$ diffusion front on the substrate is controlled by the ratio of surface diffusion rate to desorption rate. Thereby, the support provides a large storage capacity for adsorbed hydrogen.

It was demonstrated that this mechanism complies with the high turnover rates at the smallest catalyst particles and characteristic timescales observed in experimental current transients. The fitting of experimental data with the model was used to determine the rates of charge transfer, surface diffusion, and desorption. In the following text, we describe how the spillover approach can be adapted to Pd/Au overlayer systems at submonolayer coverage.

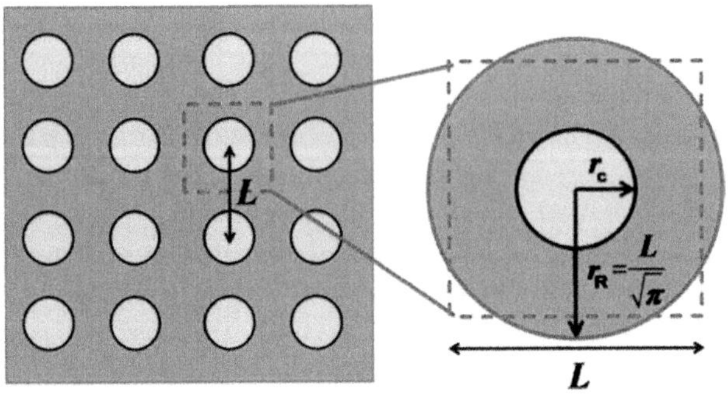


**FIGURE 10.12** Regular array of catalyst nanosislands (light color) on the substrate (dark color) as a model for the reactivity of Pd/Au overlayer structures at submonolayer coverage. The figure on the right indicates the simplified representation of the array by an effective circular disk area with the catalyst nanoisland in the center, adopted for the mean-field approximation.

### 10.3.2 Spillover Effect in Nanoparticle Arrays

For submonolayer coverages of Pd on the Au support, we represent the catalyst–substrate surface in the model as a regular array of monodisperse round nanoislands with radius $r_c$ and separation distance (or lattice constant) $L$. The model geometry is depicted in Figure 10.12. The parameters $r_c$ and $L$ can be adjusted to reproduce the catalyst coverage and the average island size determined in experiment.

The situation depicted in Figure 10.12 represents a many-body problem with interactions between nanoislands that are mediated by the spillover effect. We assume that all islands have identical circular shape with radius $r_c$. We employ a mean-field (MF) approximation to rationalize the impact of many-body effects on the current generated by the array. The basic idea of this MF approach is to represent all interactions and correlations between islands by a single island with effective interactions with the surrounding medium. Because of the periodicity of the array, the effective interaction environment for every nanoisland is similar to a Wigner–Seitz primitive cell. Applying the simplest variant of a cellular method, the area occupied by the cell is approximated by a "muffin-tin disk", as shown in Figure 10.12b. The separation distance $L$ between two neighboring Pd nanoislands and the radius of the effective disk area on the substrate $r_R$ are related by $L = \sqrt{\pi} r_R$. In this model, the Pd coverage of the substrate is defined by $\Gamma_c = r_c^2/r_R^2$.

The kinetic model for studying the reactivity of the hydrogen evolution reaction (HER) at the nanoparticle array is similar to that considered for the single supported catalyst nanoparticle [61]. The measured current density depends on elementary steps of adsorption and electrochemical discharge of protons, spillover and surface diffusion of $H_{ad}$, recombination of $H_{ad}$ with another $H_{ad}$ (Tafel mechanism) or with a proton from solution (Heyrovsky mechanism), and desorption of $H_2$. The change in local hydrogen coverage on the substrate, $\theta(r, t)$, expressed as the number of adsorbed hydrogen atoms divided by number of adsorption sites on the surface, is balanced by diffusion and desorption,

$$\frac{\partial\theta(r,t)}{\partial t} = \frac{D_s}{r}\frac{\partial}{\partial r}\left[r\frac{\partial\theta(r,t)}{\partial r}\right] - k_s^d\theta^\nu(r,t) \tag{10.25}$$

where $\nu = 1$ for first-order desorption (Heyrovsky mechanism) and $\nu = 2$ for second-order desorption (Tafel mechanism). $\theta(r,t)$ is a function of radial coordinate $r$ and time $t$. $D_s$ is the surface diffusion coefficient of adsorbed hydrogen and $k_s^d$ is the desorption rate constant of hydrogen from the substrate. On the catalyst nanoisland, that is, for $r \leq r_c$, we assumed $\theta(r,t) = \theta(r_c,t)$. At the boundary of catalyst particle and substrate we employed a flux boundary condition

$$-\left.\frac{\partial\theta(r,t)}{\partial r}\right|_{r=r_c} = \frac{r_c}{2D_s}k_c^a c_b[1-\theta(r_c,t)] - \frac{r_c}{2D_s}k_c^d\theta^\nu(r_c,t) \tag{10.26}$$

where $c_b$ is the proton concentration in the bulk solution and $k_c^a$ and $k_c^d$ denote the hydrogen adsorption and desorption rate constant on the catalyst surface, respectively. At the boundary of the muffin-tin disk, symmetry requires that the hydrogen flux be zero,

$$\left.\frac{\partial\theta(r,t)}{\partial r}\right|_{r=r_R} = 0 \tag{10.27}$$

We assumed $\nu = 1$ and obtained the analytical solution for $\theta(r,t)$ using Laplace transformation

$$\theta(r,t) = \theta_s(r) - \sum_{n=1}^{\infty} e^{-(k_s^d-\alpha_n^2)t} f(J_n(r), Y_n(r)) \tag{10.28}$$

where $\theta_s(r)$ is the steady-state hydrogen coverage. The complex function $f(J_n(r), Y_n(r))$ is composed of Bessel functions of the first and second kinds.

Hydrogen coverages as a function of dimensionless radial coordinate are displayed in Figure 10.13a at different times. Hydrogen coverage on the substrate reaches to a stationary state $\theta_s(r)$ within a relaxation time $\tau_s$, which depends on $D_s$, $k_c^a$, $k_s^d$, $r_C$, and $r_R$. Among these parameters, $k_s^d$ exerts the dominant effect on $\tau_s$.

The total current is the sum of contributions from the catalyst surface,

$$i_c = e_0\gamma_c\pi r_c^2\{k_c^a c_b(1-\theta_s(r_c)) + k_c^d\theta_s(r_C)\} \tag{10.29}$$

and the substrate surface (for $\nu = 1$, Heyrovsky mechanism),

$$i_s = e_0\gamma_s k_s^d 2\pi\int_{r_C}^{r_R}\theta_s(r)r\,dr \tag{10.30}$$

where $\gamma_c$ and $\gamma_s$ are the surface atom densities on the catalyst and the substrate, respectively. The total Faradaic current density normalized to the catalyst surface

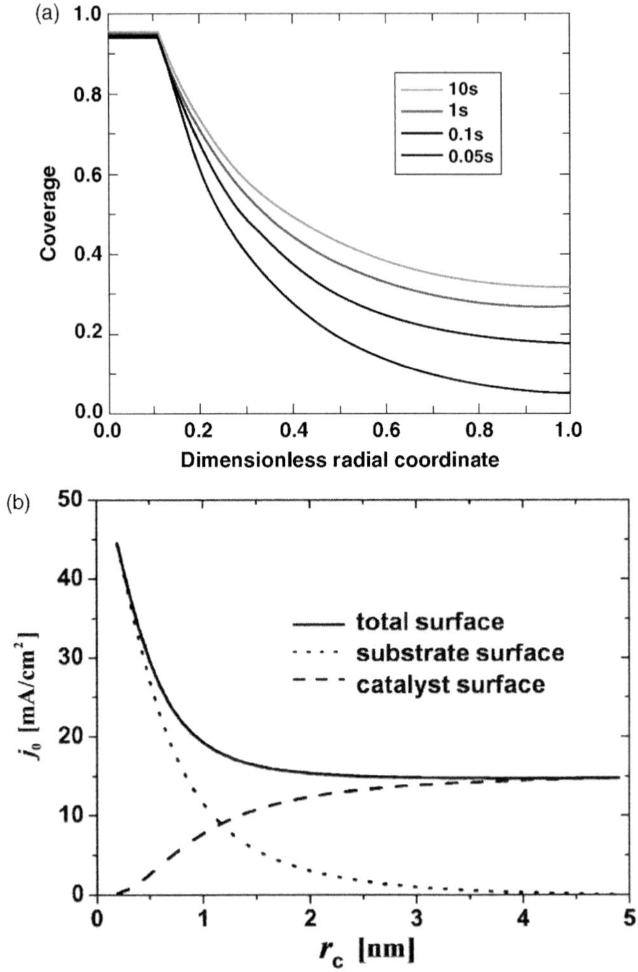


**FIGURE 10.13** (a) Hydrogen coverage on the Pd/Au surface, represented by the model in Figure 10.12b (right) as a function of radial position at various times. (b) Effect of the radius $r_c$ of catalyst nanoislands on the Faradaic current density $j_0$ normalized to the catalyst surface area at $r_R = 5$ nm. Current contributions from substrate and from catalyst nanoisland are shown separately. The main effect is the dependence of the contribution generated from the substrate on $r_c$.

area is $j_0 = (i_c + i_s)/(\pi r_C^{2c})$. Figure 10.13b shows that $j_o$ decreases dramatically with increasing $r_c$ (at fixed $r_R = 5$ nm). Assumptions of our model imply that the current density $j_c = i_c/(\pi r_c^2)$ generated from the catalyst surface is independent of $r_c$. Therefore, the strong effect of $r_c$ on $j_0$ is due to the change of the current generated by the substrate, $i_s$.

For catalyst nanoislands with fixed radius, the catalyst coverage on the substrate depends on the interparticle distance $L$. The limit $L = \sqrt{\pi} r_c$ corresponds to a full monolayer of the catalyst. The asymptotic limit $L \rightarrow \infty$ corresponds to the case of a single-catalyst nanoisland. With increasing catalyst coverage, that is, decreasing $L$, interactions between neighboring nanoislands due to overlapping diffusion fronts become more pronounced. Figure 10.14a shows the effect of $r_R = L/\sqrt{\pi}$ on $j_0$. At $r_R < 40$ nm, $j_0$ increases steeply with increasing $r_R$. In this range, $H_{ad}$ on the substrate strongly enhances the apparent surface reactivity. The maximum in $j_c$ at $r_R^{max}$ in Figure 10.14a is a result of two competing effects: (1) for $r_R < r_R^{max}$ increasing $r_R$ increases the spillover flux of adsorbed $H_{ad}$, which enhances the net rate of adsorption at the catalyst particle; and (2) for $r_R > r_R^{max}$, the prevailing effect on increasing $r_R$ is a dilution of the $H_{ad}$ coverage on the substrate due to the decreasing overlap of

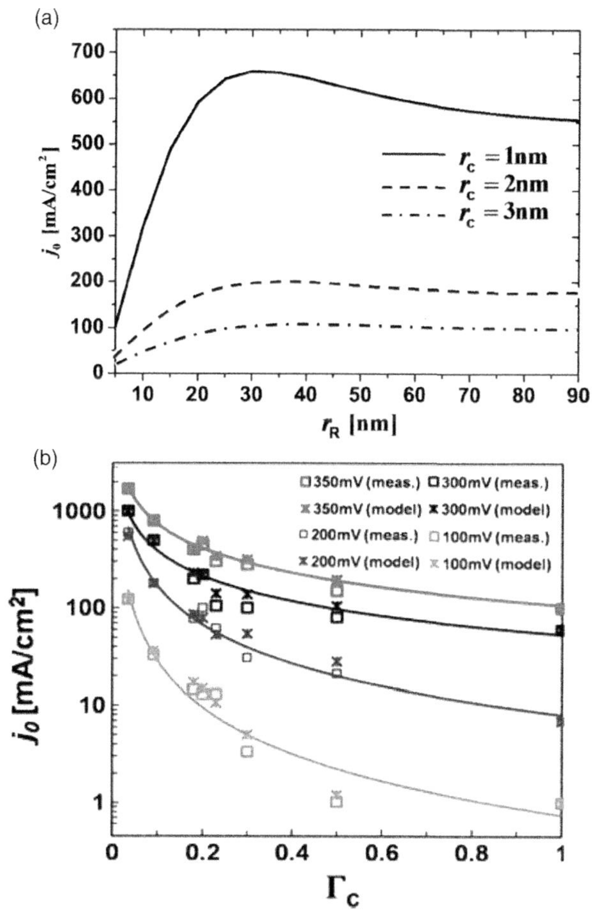


**FIGURE 10.14** (a) Effect of radius of substrate area per particle $r_R = \frac{1}{2} l_p$ on hydrogen oxidation current for various particle radii; (b) fit of the model of the nanoparticle array to experimental data for the hydrogen oxidation current per Pd surface area versus Pd surface coverage at various electrode potentials.

diffusion fields from neighboring catalyst particles, which diminishes the net rate of desorption from the substrate.

We have applied our model of the nanoisland array to experimental data of Pandelov and Stimming [136] for Au(111)-supported Pd islands with different submonolayer coverages of Pd at various electrode potentials. Figure 10.14b demonstrates that the data can be reproduced closely within the model.

At the same absolute values (but opposite signs) of overpotential, the current density enhancement reported as being due to the decreased Pt coverage is much higher for HER than for the hydrogen oxidation reaction (HOR). The discrepancy can be explained by inferring that the geometry of the particle arrangement on the substrate and particle correlations mediated by the spillover effect significantly affect the HER reactivity of supported catalyst nanoislands, whereas the effect of nanoisland arrangement on HOR reactivity is negligible. In the mechanism of HER, hydrogen desorption is the rate-determining step. Spillover and diffusion onto the substrate enhance the reactivity by facilitating hydrogen desorption. On the contrary, in the case of the HOR, the spillover effect is negligible since the rate-determining step occurs only on the catalyst nanoisland.

In general, intrinsic particle size and spillover effects both contribute to the reactivity of supported nanoparticle or nanoisland systems. Fitting of experimental data with the model allows us to evaluate the relative contributions of both effects to the overall enhancement of the current on varying sizes and packing densities of particles. The solution of the kinetic model for HER at the nanoisland array shows that hydrogen coverage reaches the stationary state after a short relaxation time, which depends strongly on the desorption rate of $H_{ad}$ from the substrate. At low catalyst coverage, relaxation times on the substrate are much longer than the relaxation time of hydrogen coverage on the catalyst nanoislands. This provides a means for distinguishing the substrate and particle contributions to overall reactivity in transient measurements. According to our calculation, desorption rate, and hydrogen diffusion coefficient on the substrate are the key kinetic parameters that influence the current density. The hydrogen diffusion coefficient is the primary parameter that determines the optimum separation distance between Pd clusters. By applying our model to experimental data, we can determine kinetic parameters and charge transfer coefficients of processes involved in HER.

## 10.4 NANOPORE CONFINEMENT AND SURFACE ROUGHNESS EFFECTS IN CATALYST LAYERS

### 10.4.1 Nanoscale Phenomena in Pores

Nanoscale phenomena in substrate pores lie between the well-defined realms of macrohomogeneous and ab initio approaches and thus are not studied well by either of them. They include (1) nanopore confinement effects, (2) surface heterogeneity of internal pore surfaces, and (3) geometric surface roughness effects. *Nanopore confinement* refers to double-layer overlap within nanopores with radii smaller than the Debye length, which can affect the proton, potential, and current density distributions. Atomic-scale variations in electronic structure arising from crystallographic disorder of the catalyst material, the distinct phases of catalyst and substrate, and their interaction effects give rise to surface heterogeneity of internal pore surfaces. Experimental studies of the potential of zero total charge (PZTC) show that the Pt metal surface charge, and hence its electrochemical behavior, varies greatly with step density [138] and to a lesser extent, with particle size [51]. Frequency dispersion of the capacitance of Pt(111) has also been attributed to atomic-scale inhomogeneities [139]. In supported catalysts, geometric surface roughness arises from the Pt nanoparticle size distribution and the arrangement of nanoparticles on the substrate. The effect of geometric roughness on capacitance, beyond an enhancement corresponding to the real/apparent surface area ratio, is suggested by theory [140–142]. The experimental exploration of geometric roughness effects, however, is complicated by the coexistence of geometric and crystallographic inhomogeneities [143].

The systematic study of nanoscale phenomena is the theme of a long-term research program. Ultimately, the goal is to understand how nanoporosity and surface roughness affect the effectiveness of Pt utilization within the electrode, beyond simply increasing the electrochemically active surface area. The nanopores of interest are contained in the ionomer-free agglomerates of carbon/Pt particles in conventional CLs [5,27] and in ionomer-free, ultrathin catalyst layers [144], as illustrated in Figure 10.15. Pore sizes are in the range of 2–50 nm, and pore walls consist of an electronically conducting material, such as carbon in conventional catalyst layers or Pt in sputter-deposited ultrathin layers. Assuming that pore walls are predominantly hydrophilic, the pores should be filled with liquid water. It will also be assumed that pores are ionomer-free, given more recent results from coarse-gained molecular dynamics simulations [5]. The roughness and heterogeneity of the pore walls depend on the catalyst-layer type and fabrication method. Figure 10.16 shows schematics of pores with different types of pore walls, one of an electrocatalytically inert, electronically conducting substrate covered with an array of Pt nanoparticles (found in conventional CLs), and one consisting of a continuous Pt layer whose substrate walls could be of an insulating material (in UTCLs).

The subsequent treatment focuses on cylindrical water-filled pores in ultrathin catalyst layers because of the more recent attention to the design and operation of such catalyst layers and their demonstrated ability to significantly enhance the power density per mass of Pt [145,146]. Presented concepts apply equally well to agglomerate pores in conventional layers. The basic question that we seek to address is: What is the response in the electrochemical current and effectiveness of Pt utilization to a change in the applied metal potential? The response function will be determined by the channel geometry (radius, length), materials properties (exchange current density, electronic structure represented by potential of zero charge, Fermi level, or workfunction), surface structure (including adsorbate effects), and surface roughness. A major challenge in this endeavor is that surface roughness and heterogeneity are not mere geometric effects but also influence the electronic structure at the surface.

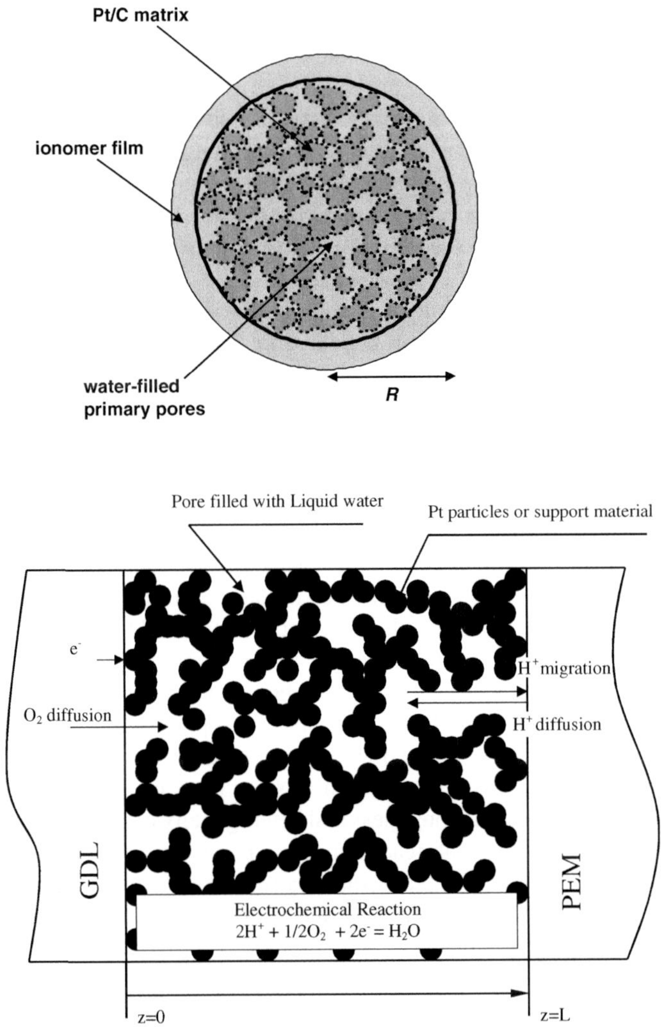


**FIGURE 10.15** Schematic illustrations of water-filled spherical agglomerate in conventional catalyst layer (top) and of ultrathin catalyst layers with water-filled nanochannels.

A major difference between the ionomer-free model and macrohomogeneous models of the conventional CL is that, in the latter, the proton concentration is assumed to be fixed by the type and loading of ionomer. This means that in conventional CLs, when nonuniformities in proton distribution at the agglomerate level are negligible, the proton concentration can be considered a materials property that is independent of the current drawn or the electrode potential applied. In the ionomer-free nanochannels of UTCL, the proton concentration varies along the channel axis as well as in the direction normal to the internal channel surface; the distribution of protons is determined by their electrostatic interaction with the charged channel walls.

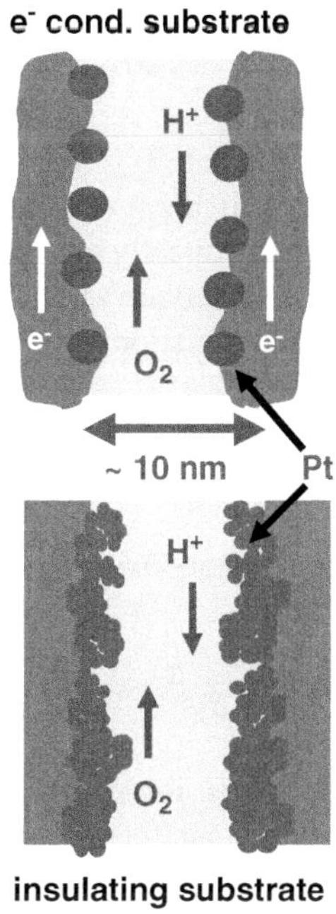


**FIGURE 10.16** Schematic depiction of pore wall of an electronically conducting substrate decorated with an array of Pt nanoparticles, and one covered with a continuous Pt layer.

In the following paragraphs, we first discuss the electrostatic effects in UTCLs for a simple 1D model. Thereafter, we will present a preliminary account of a 2D model of a cylindrical channel with smooth walls.

### 10.4.2 One-Dimensional Model of Ultrathin Ionomer-Free Catalyst Layers

Previously, transport and reaction in ionomer-free ultrathin catalyst layers with water-filled nanochannels were modeled using a 1D, macrohomogenous approach. This approach utilized an implicit assumption that the nanoscale phenomena arising from nanoporosity and rough, heterogeneous interfaces in the substrate material could be averaged out. We shall scrutinize this assumption below. The ultrathin two-phase catalyst layer was modeled as a homogenous slab [144]. Reactant, proton, and current density distributions were considered to vary only along the axis normal to the slab. While we discuss only the slab geometry here, the general approach and conclusions of the water-filled agglomerate model are very similar [27].

The potential and proton distributions within the ultrathin layer are given by Poisson–Nernst–Planck theory, which has been widely applied to ionic transport in biological membrane channels [147,148]. The flux equation accounts for ion transport via diffusion and migration, and is coupled to a current sink term,

$$-D_{H^+}\frac{d}{dz}\left(\frac{dc_{H^+}}{dz}+\frac{c_{H^+}F}{RT}\frac{d\eta}{dz}\right)=-\frac{Sj_F}{F} \tag{10.31}$$

where $z$ is the coordinate normal to the slab, $D_{H^+}$ is the effective diffusion coefficient of protons in water-filled pores, $c_{H^+}$ is the proton concentration, $\eta$ is the cathode overpotential $E^{eq}-E$, $S$ is the electrochemically active Pt area per unit volume, and $j_F$ is the Faradaic current density per real surface area. The relation of overpotential to concentration is given by the Poisson equation

$$\frac{d^2\eta}{dz^2}=-\frac{c_{H^+}F}{\varepsilon} \tag{10.32}$$

where $\varepsilon$ is the dielectric constant of water. Oxygen diffusion is given by Fick's law,

$$-D_{O_2}\frac{d^2c_{O_2}}{dz^2}=-\frac{Sj_F}{4F} \tag{10.33}$$

where $D_{O_2}$ is the effective oxygen diffusion coefficient in water-filled pores. The Faradaic current density (per unit real surface area of the catalyst) is given by the Butler–Volmer equation,

$$j_F=j^0\left(\frac{c_{O_2}}{c_{O_2}^{ref}}\right)^{\gamma_{O_2}}\left(\frac{c_{H^+}}{c_{H^+}^{ref}}\right)^{\gamma_{H^+}}\exp\left(\frac{\alpha_c F\eta}{RT}\right) \tag{10.34}$$

where the $\gamma_i$ terms are reaction orders and $\alpha_c$ is the cathode transfer coefficient. In References 27 and 144 it was assumed that $\gamma_{O_2}=\gamma_{H^+}=1$.

The effectiveness of Pt utilization can be defined as the ratio of the integral current density to the ideal current density corresponding to uniform reactant distribution,

$$\Gamma=\frac{\int_0^L j_F\,dz}{Lj_{ideal}} \tag{10.35}$$

where $z=0$ and $z=L$ correspond to the GDL and PEM interfaces, respectively. Nonuniform potential and reactant distributions lead to $\Gamma\neq 1$.

The effectiveness factor gives a modified Tafel law for the voltage–current response of the catalyst layer (assuming $\gamma_{O_2}=\gamma_{H^+}=1$):

$$j_0=\Gamma LSj^0\frac{c_{O_2}^0\,c_{H^+}^L}{c_{O_2}^{ref}\,c_{H^+}^{ref}}\exp\left(\frac{\alpha_c F}{RT}\eta_L\right) \tag{10.36}$$

Analytical solutions and an explicit expression for the effectiveness factor can be obtained at the limit of low current density, $j_0<10^{-1}\ \mathrm{A/cm^2}$, where uniform oxygen concentrations can be assumed, and the sink term in the Nernst–Planck equation, Eq. (10.31), can be ignored. The potential and proton distributions are then simply obtained as the solution of an electrostatic problem, and the PNP equations simplify to the Poisson–Boltzmann equation,

$$\begin{aligned} \frac{d^2\eta}{dz^2} &= -\frac{c_{\mathrm{H}^+}^{\mathrm{ref}}\, e}{\varepsilon} \exp\left(-\frac{F}{RT}(\eta-\eta_L)\right) \\ c_{\mathrm{H}^+} &= c_{\mathrm{H}^+}^{\mathrm{ref}} \exp\left(-\frac{F}{RT}(\eta-\eta_L)\right) \end{aligned} \tag{10.37}$$

The effectiveness factor in this case is

$$\Gamma = \frac{1}{L}\int_0^L dz \cos^{2(1-\alpha_c)}\left[\frac{L}{\sqrt{2}\lambda_{\mathrm{D}}}\exp\left(\frac{\eta_L-\eta_0}{2}\right)\right]\cos^{-2(1-\alpha_c)}\left[\frac{1}{\sqrt{2}\lambda_{\mathrm{D}}}\exp\left(\frac{\eta_0-\eta_L}{2}\right)z\right] \tag{10.38}$$

where $\lambda_{\mathrm{D}}$ is the Debye length $\lambda_{\mathrm{D}} = [(\varepsilon RT)/(F^2 c_{\mathrm{H}^+}^{\mathrm{ref}})]^{1/2}$. Equation (10.38) shows the importance of the cathode transfer coefficient $\alpha_{\mathrm{c}}$; using $\alpha_{\mathrm{c}} = 1$ (and $\gamma_{\mathrm{H}^+} = 1$) gives $\Gamma = 1$ at any $L$, corresponding to a uniform reaction rate distribution across the entire catalyst layer. For a good correspondence between the 1D model and experimental polarization curves, $\alpha_{\mathrm{c}}$ should, indeed, be close to 1 [144].

The general case requires that Equations (10.32)–(10.35) be solved numerically. Figure 10.17a shows the normalized proton and oxygen concentrations $c_{\mathrm{H}^+}/c_{\mathrm{H}^+}^{\mathrm{L}}$ and $c_{\mathrm{O}_2}/c_{\mathrm{O}_2}^0$ along the normalized length $z/L$ of the ultrathin layer, at layer thicknesses in the range from 0.02 to 0.4 μm and a current density of 1.0 A/cm$^2$. Without proton-conducting ionomer, the proton concentration drops off rapidly from the PEM interface within a sublayer of thickness that roughly corresponds to the Debye length; the proton concentrations are relatively more uniform at smaller thicknesses. On the other hand, oxygen diffusion is not a limiting process; at a thickness of 0.4 μm, the $c_{\mathrm{O}_2}$ decreases to only 80% of the value at the gas diffusion layer (GDL) interface.

The steep drop in proton concentration is partially compensated for by the rise in overpotential from the polymer electrode membarne (PEM) to GDL, as shown in Figure 10.17b, the normalized overpotential profile $\eta/\eta_L$. The opposite trends in proton concentration and overpotential as a function of the position $z$ counterbalance each other in the corresponding Faradaic current density,

$$j_{\mathrm{F}}(z) \propto c_{\mathrm{H}^+}(z)\exp\left(\frac{\alpha_c F\eta(z)}{RT}\right).$$

This partially mitigates the steep drop in reaction rate that would otherwise occur with a steep decrease in proton concentration. The net effect of the variations in proton concentration and potential on the reaction rate distribution is illustrated in the

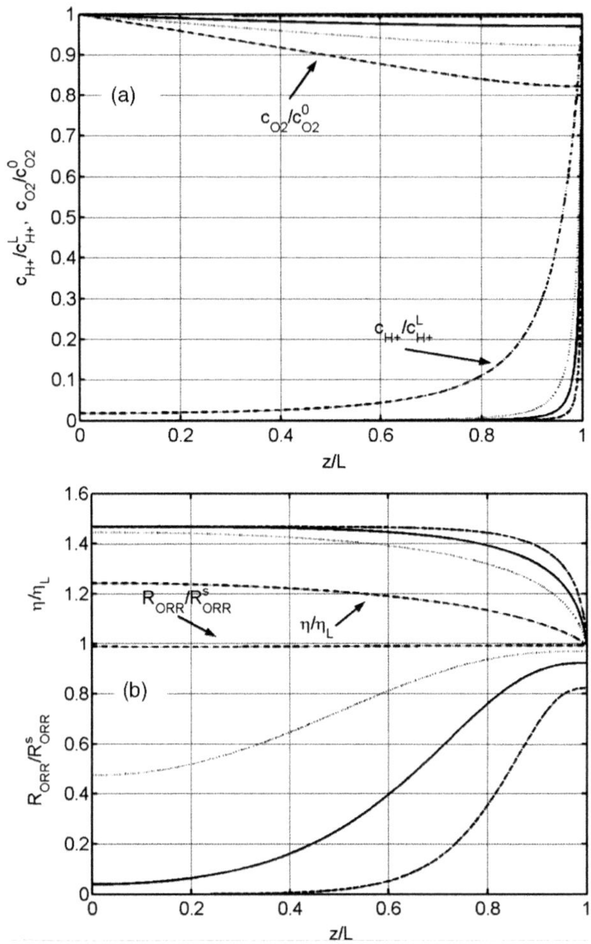


**FIGURE 10.17** (a) Normalized concentration profiles of oxygen; (b) normalized overpotential and reaction rate profiles at 1 A/cm$^2$ (dashed line, $L = 0.4$; solid line, $L = 0.2$; dotted line, $L = 0.1$; dashed–doted line, $L = 0.02$ μm).

reaction rate profiles $R_{\mathrm{ORR}}/R^{\mathrm{s}}_{\mathrm{ORR}}$ in Figure 10.17b, where $R^{\mathrm{s}}_{\mathrm{ORR}}$ is the ideal reaction rate corresponding to vanishing transport limitations; in calculating $R_{\mathrm{ORR}}/R^{\mathrm{s}}_{\mathrm{ORR}}$, $\alpha_c = 1$ was assumed. In the thinnest layer considered ($L = 0.02$ μm), the reaction rate across the layer is uniform. On the other hand, in the thickest layer ($L = 0.4$ μm), which shows the least uniform proton distribution, there is a dramatic drop in reaction rate across the layer, with essentially zero current produced over more than half the layer.

As a result of low oxygen diffusion in water-filled pores of ultrathin CLs, the thickness $L$ emerges as a key parameter determining the effectiveness factor $\Gamma$. At a given current density $j_0$, the smaller is $L$, the more uniform the reaction rate

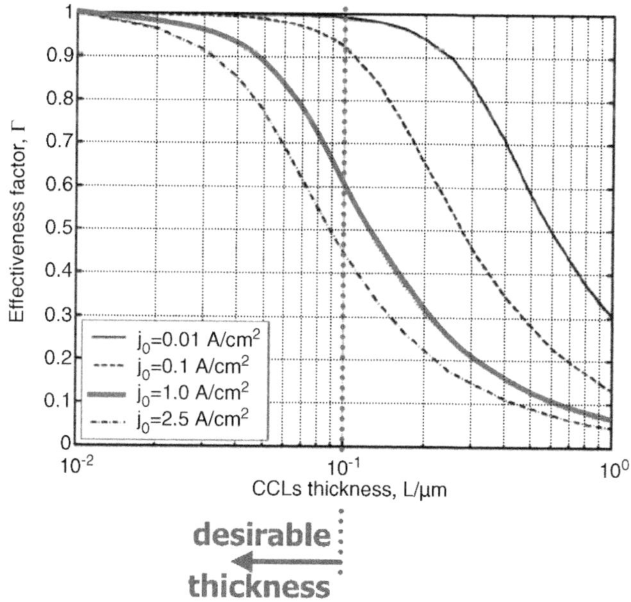


**FIGURE 10.18** The effect of thickness $L$ and current density $j_0$ on the effectiveness factor.

distribution. Figure 10.18 shows $\Gamma$ as a function of $L$ for various $j_0$ at $\alpha_c = 1$. For operation at a given $j_0$, a desired minimal value of $\Gamma$ will be exceeded if $L$ is below a certain limiting value. At higher $j_0$ this upper bound on $L$ is lower. For example, at $j_0 = 1\ \text{A/cm}^2$, an effectiveness factor $\Gamma > 0.6$ could be achieved only if the thickness is $L \leq 100$ nm.

The 1D macrohomogeneous model provided several interesting insights. First, proton and potential distributions in the UTCL are mainly electrostatic phenomena, due to the low exchange current density of ORR. The cathode transfer coefficient $\alpha_c$ determines the degree to which the rise in potential from the PEM mitigates the effect of the steep drop in proton concentration on the local reaction rate. Furthermore, the layer thickness determines the degree of mass transport limitations, and hence should not exceed a certain value; for a typical set of parameters, this value is about 100 nm. Finally, the porosity should be high to ensure a high catalyst surface area.

### 10.4.3 Two-Dimensional Model of Water-Filled Nanopores

The 1D macrohomogeneous model identified cathode transfer coefficient $\alpha_c$, thickness $L$, and porosity as the main factors that determine the effectiveness of Pt utilization and the performance of UTCL. However, its ability to reproduce experimental polarization curves hinges on the assumption of $\alpha_c = \gamma_{H^+} = 1$ over the entire relevant electrode potential range, which is inconsistent with ORR kinetic data from the last extensive pH dependent study [149]. More importantly, the 1D model approach completely discounts the presence of the metal phase and ignores the

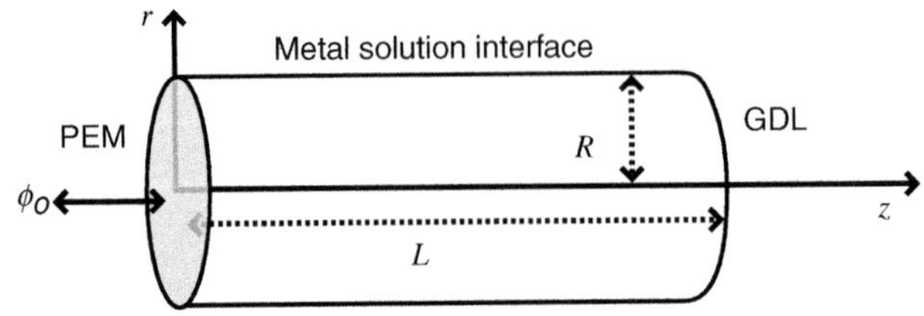


**FIGURE 10.19** Schematic representation of model cylindrical pore.

nanoscale phenomena arising from the morphology of the water channels. The metal interfacial charge density and double-layer overlap within nanosized pores can, however, both affect the reaction site potential and proton concentrations, and hence influence the overall effectiveness of Pt utilization. The focus of our current work is to develop a model of these nanoscale effects within water-filled nanopores.

A UTCL nanopore is modeled as a straight cylindrical water-filled pore spanning the thickness of the layer, as illustrated in Figure 10.19. Protons are supplied from the proton-conducting polymer electrolyte membrane (PEM), which fixes the proton concentration and electrolyte potential at one pore boundary. Oxygen is supplied from the opposite side, with oxygen concentration fixed at the other pore opening. The curved internal pore surface could be charged through the application of an electrical potential to the surrounding metal phase. As in the 1D model, the Poisson–Nernst–Planck (PNP) equations are used to obtain the proton and potential distributions, Fick's law is employed to account for oxygen diffusion, and the Butler–Volmer equation is used for the rate of charge transfer processes, corresponding to the following system of equations,

$$\vec{N}_{\mathrm{H}^+} = -D_{\mathrm{H}^+}\left[\nabla c_{\mathrm{H}^+} + \frac{c_{\mathrm{H}^+}F}{RT}\nabla\phi\right] \tag{10.39}$$

$$\nabla^2\phi = -\frac{c_{\mathrm{H}^+}e}{\varepsilon} \tag{10.40}$$

$$\vec{N}_{\mathrm{O}_2} = -D_{\mathrm{O}_2}\nabla c_{\mathrm{O}_2} \tag{10.41}$$

$$j_{\mathrm{F}} = j^0\left(\frac{c_{\mathrm{O}_2}}{c_{\mathrm{O}_2}^{\mathrm{ref}}}\right)^{\gamma_{\mathrm{O}_2}}\left(\frac{c_{\mathrm{H}^+}}{c_{\mathrm{H}^+}^{\mathrm{ref}}}\right)^{\gamma_{H^+}}\exp\left(-\frac{\alpha_c F}{RT}\left[\phi_{\mathrm{M}}-\phi^{\ddagger}-E_{\mathrm{eq}}^{\mathrm{ref}}\right]\right) \tag{10.42}$$

Here, $\vec{N}_i$ gives the flux of the $i$th species; $\phi_{\mathrm{M}}$, the metal potential; and $\phi^{\ddagger}$, the outer Helmholtz plane potential. Since current is generated only at the pore boundaries, the proton and oxygen fluxes within the pore are conserved:

$$\nabla\cdot\vec{N}_{\mathrm{H}^+} = \nabla\cdot\vec{N}_{\mathrm{O}_2} = 0 \tag{10.43}$$

At the metal/solution interface, the reactant fluxes are coupled to the current densities, and the potential is the outer Helmholtz plane potential corresponding to the assumed distance of closest approach of hydrated protons,

$$\vec{N}_{H^+}\Big|_{r=R}\cdot\hat{n}=\frac{j_F}{F},\qquad \vec{N}_{O_2}\Big|_{r=R}\cdot\hat{n}=\frac{j_F}{4F},\qquad \phi|_{OHP}=\phi^{\ddagger}. \tag{10.44}$$

At the PEM boundary, $z=0$, the concentration and potential is set to that of PEM, and the flux of oxygen over this boundary is zero, that is, $c_{H^+}=c^0_{H^+}$, $\phi=\phi_0$, $\vec{N}_{O_2}\cdot\hat{z}=0$. At the GDL boundary, $z=L$, the concentration of oxygen is assumed to be constant, and diffusion and migration of protons are zero, that is, $c_{O_2}=c^0_{O_2}$, $\nabla\phi\cdot\hat{z}=0$, $\nabla c_{H^+}\cdot\hat{z}=0$.

While Equations (10.39)–(10.42) are solvable numerically, it is possible to reduce them to a limiting case with a tangible analytic solution with a few simplifying assumptions. Given the sluggish oxygen reduction reaction, a simple estimate with the exchange current density on Pt, $j^0=1.2\cdot10^{-8}\,\mathrm{A/cm^2}$ [150], the diffusion coefficient of oxygen, $D_{O_2}=9.19\cdot10^{-5}\,\mathrm{cm^2/s}$ at a typical fuel cell operation temperature of 350 K [151], a typical operating overpotential $<0.45$ V, and a realistic nanopore length $L_p<200$ nm, we see that oxygen diffusion should not be limiting and so its concentration can be assumed to be uniform throughout the pore. Since the proton diffusion coefficient is a factor $\sim$2.5 larger than the oxygen diffusion coefficient, the distributions of proton concentration and electrostatic potential in the pore deviate only negligibly from equilibrium. Equation (10.39) thus simplifies to the Poisson–Boltzmann equation. The 2D Poisson–Boltzmann equation, in turn, simplifies to a 1D problem in the radial direction after a characteristic proton penetration depth along the axial direction. This Debye length of the proton distribution along the axial direction depends on the pore radius and the charge density at the curved metal|solution interface; in the relevant ranges of pore radii and surface charges it is small compared to the pore length $L_p$. In other words, beyond this decay region, the PEM boundary is essentially shielded from the rest of the pore.

The proton and potential distributions in the 1D radial problem is

$$\frac{1}{\gamma}\frac{d}{d\gamma}\left(\gamma\frac{d\Phi}{d\gamma}\right)=-K\exp(-[\Phi-\Phi_0]) \tag{10.45}$$

$$c_{H^+}=c^0_{H^+}\exp(-[\Phi-\Phi_0]) \tag{10.46}$$

where $\gamma=r/R$ is the dimensionless pore radius, $\Phi=e\phi/(k_BT)$ is the dimensionless potential, $K=R^2/\lambda_D^2=(R^2n_0e^2)/(\varepsilon k_BT)$, where $\lambda_D=[(\varepsilon RT)/(F^2c^0_{H^+})]^{1/2}$ is the Debye length and $c^0_{H^+}$ and $\Phi_0$ (set to zero) are the PEM interface concentration and potential, respectively. The boundary conditions are set by the excess charge density at the interface $\gamma=1$ and the symmetry the channel center $\gamma=0$,

$$\frac{d\Phi}{d\gamma}\bigg|_{\gamma=1}=\frac{\sigma eR}{\varepsilon k_bT}=\sum,\qquad \frac{d\Phi}{d\gamma}\bigg|_{\gamma=0}=0 \tag{10.47}$$

Note that the excess surface charge density must be negative in order to drag protons into the pore; this is evident from the 1D Poisson–Boltzmann equation [Eq. (10.45)] and the symmetry condition within the pore center. As this surface charge density approaches zero, protons are expelled from the pore. The solution is

$$\Phi - \Phi_0 = 2\ln(1-b\gamma^2) + \Phi_c, \tag{10.48}$$

$$b = \frac{\Sigma}{\Sigma-4}, \qquad \Phi_c = \ln\left(\frac{KR^2(\Sigma-4)}{8\Sigma}\right), \tag{10.49}$$

where $\Phi_c$ is the dimensionless potential through the pore center. The outer Helmholtz plane potential is

$$\Phi - \Phi_0 = \ln\left(\frac{2KR^2}{\Sigma(\Sigma-4)}\right). \tag{10.50}$$

With the outer Helmholtz plane potential, the effectiveness factor of Pt utilization, in the limit of fast oxygen and proton transport, can be written as a function of pore radius and net surface charge,

$$\Gamma = \exp\left([\alpha_c - \gamma_{H^+}][\Phi^{\ddagger} - \Phi_0]\right) = \left(\frac{2e\varepsilon k_B T c^0_{H^+} R}{\sigma(\sigma e R - 4k_B T)}\right)^{\alpha_c - \gamma_{H^+}} \tag{10.51}$$

The key parameters determining the effectiveness factor are the pore radius $R$, the excess surface charge density $\sigma$, and the difference between the transfer coefficient and proton reaction order, $\alpha_c - \gamma_{H^+}$. A pH-dependent ORR kinetic study [149] shows that, while $\alpha_c$ and $\gamma_{H^+}$ both change at the transition from Temkin to Langmuirian adsorption of oxygen species on platinum, their difference is $-\frac{1}{2}$ over the range of relevant pH and potentials. Figure 10.20 shows $\Gamma$ as a function of $\sigma$ in the limit of $R \rightarrow \infty$ down to $R = 1$ nm. For $R <\sim 10$ nm, the effectiveness increases substantially with decreasing $R$ at a given $\sigma$ due to double-layer overlap. Effectiveness factors higher than one correspond to the case where the surface charge is sufficiently negative that the proton concentration at the pore surface exceeds that at the PEM boundary.

Since $R$ and $\sigma$ have a key impact on the effectiveness factors, and neither of these are considered in the 1D model, an obvious question is: under what conditions can the 2D model be reduced to the 1D one? The 1D model ignores the effect of metal potential and corresponding interfacial charge density on the effectiveness of Pt utilization. This corresponds to an oversimplified special case of the 2D model with (1) large $R$, where the pore potential through the pore centre reduces to that from the 1D model; and (2) large absolute value of the negative surface charge density, where the pore boundary potential drops rapidly to the pore center potential, due to a very short lateral Debye length. These stringent restrictions put to question the validity of the 1D model altogether. Results from the 1D model remain strictly valid only for the specific values $\alpha_c = \gamma_{H^+} = 1$ and negligible proton flux, $\vec{N}_{H^+} \simeq 0$ in Equation (10.39); under these

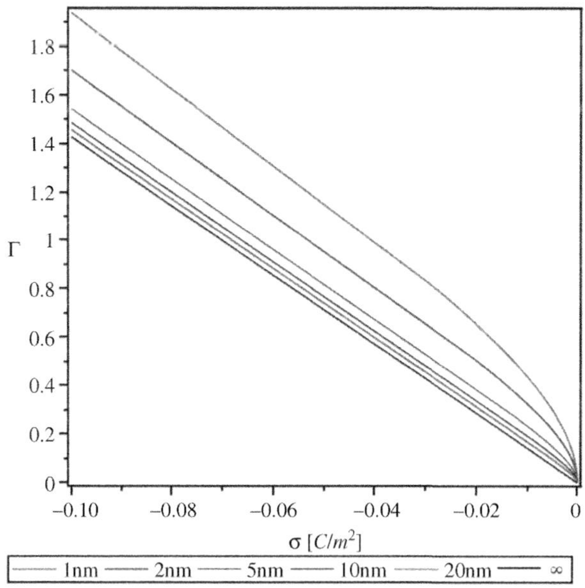


**FIGURE 10.20** Effectiveness factor of a cylindrical nanopore as a function of $\sigma$ for pore radii in the range $R = 1-10$ nm, in the limit of fast reactant transport ($\varepsilon = 80 \cdot \varepsilon_0$, $T = 348$ K, $\alpha_c - \gamma_H = -\frac{1}{2}$).

very specific conditions the opposite trends in distributions of proton concentration and electrostatic potential of the electrolyte phase (relative to the metal phase) exactly compensate each other. The values $\alpha_c = \gamma_{H^+} = 1$ are, however, inconsistent with ORR kinetic data [149].

With an expression for $\Gamma$ as a function of $\sigma$ and $R$, the key challenge at present in developing the nanopore model is to find $\sigma$ and hence the current density $j$ as a function of electrode potential $E$. Precisely, this surface charge density should include both the excess metal-free charge density as well as the charge density arising from any partially discharged adsorbates. In principle, this could be estimated from capacitance data from impedance spectra if partial adsorbate charge transfer can be ignored, but there are several complications to this approach that need to be resolved. The measurement of the double-layer capacity at relevant electrode potentials is obscured by adsorption processes and, thus, requires model assumptions to interpret impedance data [152,153]. The porous catalyst layer, with neither a normal supporting electrolyte nor well-defined crystal planes at its surface, should not necessarily have capacitances similar to those of model systems studied in the literature. The Pt potential of zero charge has been shown to vary with both particle size and surface step density [51,138] and on single-crystal surfaces, different techniques of measurements have yielded a wide range of results [154–156]. A further challenge would be to incorporate surface modulations at the metal–solution boundary and build on earlier work on the theory of rough electrodes [140–142]. Work is underway to sort out these pieces of the puzzle.

## 10.5 CONCLUSIONS

Physical and chemical phenomena at the nanometer scale are of central importance to the operation of catalyst layers in polymer electrolyte fuel cells. Ultimately, we have to understand these phenomena in order to design nanostructured catalyst layers with optimized electrochemical activity and selectivity for desired fuel cell reactions as well as a high utilization of the catalyst material. The research themes described in this chapter focus on systems containing Pt nanoparticles, based on the premise that Pt-based catalyst layers offer tremendous reserves for the improvement of mass-specific activities.

Physical theory and molecular modeling can address in part the scientific challenges in these systems. The size, shape, and composition of catalyst nanoparticles determine the local geometry of surface sites and the surface electronic structure. These relations can be studied by DFT calculations. Even for facet-type nanoparticles with well-defined shapes, the electronic surface structure turns out to be highly heterogeneous, as apparent from contour plots of oxygen adsorption energies in Section 10.2.4. This suggests that the catalyst surface will exhibit a high site selectivity for the complete sequences of fuel cell reactions: adsorption, charge transfer, desorption, and surface diffusion.

Site selectivity has been studied by DFT for elementary adsorption processes. Independently, site selectivity has been earmarked as an important factor in nanoparticle reactivity in a kinetic model of $CO_{ad}$ monolayer oxidation at Pt nanoparticles. The simple active-site model, discussed in Section 10.2.3, reproduces experimental trends very well; its implementations in a kinetic Monte Carlo program or in a mean-field model have shed light on the surface structure and kinetic parameters of surface processes. However, neither DFT calculations of oxygen adsorption nor the kinetic active-site model of surface reactivity allow, for a simple geometric interpretation of the nature and abundance of active sites. The full sequence of the oxygen reduction reaction has been studied by DFT calculations at a $Pt_4$ cluster with a dual-active-site structure, using the nudged elastic band (NEB) method and thermochemical modeling of elementary reaction steps. This study provided insights into the role of active sites for the ORR mechanism. It establishes a counterpoint to studies of the same reaction at extended surfaces.

Currently, it is not known whether there exist simple systematic trends between nanoparticle structure and reactivity. Because of their irregular surface structures, reactivity mechanisms on nanoparticles are fundamentally different from those at extended surfaces. At nanoparticles, the geometric effects of atom arrangement and local electronic structure dominate over electronic band structure effects. Ultimately, elucidating these complex nanoparticle structure–reactivity relations will require systematic DFT studies of elementary surface processes that allow for site selectivity, multiple active sites, and surface diffusion of adsorbed intermediates. Kinetic models of surface reactivity can then be parametrized using the results of such studies.

In arrays of supported catalyst nanoparticles, the fractional catalyst coverage on the substrate, or the nearest-neighbor separation distance of nanoparticles, constitutes an additional parameter in structure–reactivity relations. For overlayer systems or nanoisland arrays at submonolayer coverages, the substrate exerts an electronic structure effect that can be rationalized by the *d*-band model. In the case of sphere-like nanoparticles bonded to the substrate, future studies will look at the particle–substrate bonding mechanism (directly or via linker groups), the bond strength, charge transfer on bonding, and the charge distribution in the bond region. Correlation effects between nanoparticles over mesoscopic separations are caused by the spillover effect. For reactions that involve adsorbate spillover and surface diffusion between catalyst and substrate surface, the model presented in Section 10.3.2 or its extensions could be used as a tool for determining the ideal particle arrangement on the substrate.

The model of a water-filled, ionomer-free nanopore in Section 10.4.2 addressed the effects of substrate nanoporosity on electrochemical performance. It links (1) fundamental parameters of microscopic reaction mechanisms, such as electronic transfer coefficients and reaction orders; and (2) materials properties, such as exchange current density and potential of zero charge, with the rates of current conversion in pores and the overall effectiveness of Pt utilization. A key to further development of this approach will be to understand and incorporate relations between metal potential and surface charge density at pore walls. Surface heterogeneity and roughness effects at pore walls, which we plan to include in future refinements, affect these relations.

It is a widely accepted view that major improvements in the performance-to-cost ratio of polymer electrolyte fuel cells could be accomplished with the design and fabrication of advanced catalyst layers, which optimize the utilization and the mass-specific activity of the catalyst. Tremendous efforts worldwide focus, therefore, on the selection of materials with favorable properties in view of charge and reactant transport, electrocatalytic processes, and stability. The electrocatalyst is the material of primary interest in this realm. The oxygen reduction reaction is a multistep process with competing requirements for different elementary reaction steps. On one hand, the oxidation of the catalyst surface is an indispensable intermediate step; on the other hand, extensive oxidation of the catalyst surface blocks the access of further reactants to active surface sites. It is a grand challenge to understand whether any material could be found to overcome this catalyst conundrum better than platinum. So far, Pt-based catalysts, pure or mixed with other metals, are the best.

A viable route for achieving improvements in catalytic activity would be the manipulation of the catalyst surface structure at atomistic scale to achieve desired site selectivity for intermediate steps of the total reaction sequence. Coined specifically for the oxygen reduction reaction, this means that improved activity should be sought among complex catalysts with heterogeneous surface structure, which provide the "right" amounts and distributions of sites that get oxidized easily and sites with low affinity for oxide formation, exhibiting weak bonds with oxygen. This is suggested by reactivity studies at nanoparticles, which exhibit intrinsic site selectivity due to their heterogeneous surface morphology. Systematic studies of electronic structure effects and reactivity mechanisms at heterogeneous catalyst surfaces with functionalized surface sites for particular reaction steps are cumbersome to conduct because of the vast number of possible surface conformations. Moreover, catalyst surfaces with desired functional distinction of surface sites would be difficult to fabricate in a reproducible manner. Nevertheless, this route could be rather fruitful. Exploring it in theory will require a concerted effort, including ab initio and force-field-based molecular dynamics studies, Monte Carlo simulations, and kinetic modeling to establish predictive reactivity models.

In nanoparticle-based catalyst layers, the properties of the substrate material are as important as those of the catalyst itself. Interactions between catalyst and supporting substrate determine stability and electronic contact. An optimized nanoporosity of the substrate should provide a large catalyst surface area and a high effectiveness factor of catalyst utilization.

As demonstrated in this chapter, physical theory and molecular modeling provide a full hierarchy of tools to guide the selection of catalyst and support materials and to define the parameter space for the structural design of catalyst layers.

## REFERENCES


1. R. E. Smalley, *MRS Bull.* **30**, 412 (2005).
2. E. W. Vielstich, A. Lamm, and H. Gasteiger, eds., *Handbook of Fuel Cells: Fundamentals, Technology, Applications*, Wiley-VCH, Weinheim, 2003.
3. M. Eikerling, A. A. Kornyshev, and A. R Kucernak, *Physics Today* **59**, 38, 2006.
4. E. J. Carlson, *DOE Hydrogen Program Annual Merit Review*, May 16–19, 2006.
5. K. Malek, M. Eikerling, Q. Wang, T. Navessin, and Z. Liu, *J. Phys. Chem. C* **111**, 13627 (2007).
6. A. Wieckowski, E. R. Savinova, and C. G. Vayenas, *Catalysis and Electrocatalysis at Nanoparticle Surfaces*, Marcel Dekker, New York, 2003.
7. V. R. Stamenkovic, B. S. Mun, K. J. J. Mayrhofer, P. N. Ross, and N. M. Markovic, *J. Am. Chem. Soc.* **128**, 8813 (2006).
8. V. R. Stamenkovic, B. S. Mun, M. Arenz, K. J. J. Mayrhofer, C. A. Lucas, G. Wang, P. N. Ross, and N. M. Markovic, *Nature Mater.* **6**, 241 (2007).
9. V. R. Stamenkovic, B. Fowler, B. S. Mun, G. Wang, P. N. Ross, C. A. Lucas, and N. M. Markovic, *Science* **315**, 493 (2007).
10. Z. Xia, Q. Wang, M. Eikerling, and Z. Liu, *Can. J. Chem.* **86**, 657 (2008).
11. M. Eikerling, *J. Electrochem. Soc.* **153**, E58 (2006).
12. B. Andreaus and M. Eikerling, Catalyst layer operation in PEM fuel cells: From structural pictures to tractable models, in *Device and Materials Modeling in PEM Fuel Cells*, K. Promislow and S. Paddison, eds., Springer, 2009; repr. in Top. *Appl. Phys.* **113**, 41.
13. M. Eikerling, K. Malek, and Q. Wang, Catalyst Layer Modeling: Structure, Properties, and Performance, in *PEM Fuel Cell Electrocatalysts and Catalyst Layers*, J.J. Zhang, ed., Springer, 2008, pp. 381–446.
14. M. Eikerling, A. A. Kornyshev, and A. A. Kulikovsky, Physical modeling of cell components, cells and stacks, in *Encyclopedia of Electrochemistry*, Vol. 5, D. D. Macdonald and P. Schmuki, eds., Wiley-VCH, Weinheim, 2007, Section 8.2, pp. 447–543.
15. M. Eikerling and K. Malek, Electrochemical materials for PEM fuel cells: Insights from physical theory and simulation, in *Modern Aspects of Electrochemistry*, vol. 43, M. Schlesinger, ed., Springer, 2009.

16. J. Liu and M. Eikerling, *Electrochim. Acta* **53**, 4435 (2008).
17. M. Uchida, Y. Aoyama, E. Eda, and A. Ohta, *J. Electrochem. Soc.* **142**, 463 (1995).
18. J. Xie, D. L. Wood, III, K. L. More, P. Atanassov, and R. L. Borup, *J. Electrochem. Soc.* **152**, A1011 (2005).
19. Z. Xie, T. Navessin, Z. Shi, R. Chow, Q. Wang, D. Song, B. Andreaus, M. Eikerling, and Z. Liu, *J. Electrochem. Soc.* **152**, A1171 (2005).
20. K. Malek, T. Mashio, and M. Eiberling, Submitted to *J. Chem. Phys.*
21. G. M. Whitesides and G. W. Crabtree, *Science* **315**, 796 (2007).
22. J. X. Wang, N. M. Markovic, and R. R. Adzic, *J. Phys. Chem. B* **108**, 4127 (2004).
23. U. A. Paulus, A. Wokaun, G. G. Scherer, T. J. Schmidt, V. Stamenkovic, V. Radmilovich, N. M. Markovic, and P. N. Ross, *J. Phys. Chem. B* **106**, 41081 (2002).
24. M. R. Tarasevich, A. Sadkowski, and E. Yeager, Kinetics and mechanisms of electrode processes, in *Comprehensive Treatise of Electrochemistry*, B. E. Conway, J. O'M. Bockris, E. Yeager, S. U. M. Khan, and R. E. White, eds., Plenum Press, New York, 1983, Vol. **7**, pp. 301–398.
25. K. Kinoshita, *Electrochemical Oxygen Technology*, Wiley, & New York, 1992.
26. B. Andreaus, F. Maillard, J. Kocylo, E. Savinova, and M. Eikerling, *J. Phys. Chem. B* **110**, 21028 (2006).
27. Q. Wang, M. Eikerling, D. Song, and Z. Liu, *J. Electroanal. Chem.* **573**, 61 (2004).
28. M. Eikerling, A. S. Ioselevich, and A. A. Kornyshev, *Fuel Cells* **4**, 131 (2004).
29. T. E. Springer, M. S. Wilson, and S. Gottesfeld, *J. Electrochem. Soc.* **140**, 3513 (1993).
30. M. Eikerling and A. A. Kornyshev, *J. Electroanal. Chem.* **453**, 89, (1998).
31. S. Torquato, *Random Heterogeneous Materials: Microstructure and Macroscopic Properties*, Springer-Verlag, New York, 2002.
32. Q. Wang, M. Eikerling, D. Song, S. Liu, T. Navessin, Z. Xie, and S. Holdcroft, *J. Electrochem. Soc.* **151**, A950 (2004).
33. R. A. Sidik and A. B. Anderson, *J. Electroanl. Chem.* **528**, 69 (2002).
34. J. K. Nørskov, J. Rossmeisl, A. Logadottir, L. Lindqvist, J. R. Kitchin, T. Bligaard, and H. Jonsson, *J. Phys. Chem. B* **108**, 17886 (2004).
35. V. Stamenkovic, B. S. Mun, K. J. J. Mayrhofer, P. N. Ross, N. M. Markovic, J. Rossmeisl, J. Greeley, and J. K. Nørskov. *Angew. Chem Int. Ed.* **45**, 2897 (2006).
36. G. S. Karlberg, J. Rossmeisl, and J. K. Nørskov, *Phys. Chem. Chem. Phys.* **9**, 5158 (2007).
37. A. U. Nilekar and M. Mavrikakis, *Surf. Sci.* **602**, L89 (2008).
38. M. J. Janik, C. D. Taylor, and M. Neurock, *J. Electrochem. Soc.* **156**, B126 (2009).
39. M. J. Yacamán, J. A. Ascencio, H. B. Liu, and J. Gardea-Torresdey, *J. Vac. Sci. Technol. B* **19**, 1091 (2001).
40. E. Roduner, *Nanoscopic Materials: Size Dependent Phenomena*, RSC Publishing, Cambridge, UK, 2006, p. 285.
41. W. P. Halperin, *Rev. Mod. Phys.* **58**, 533 (1986).
42. W. Vielstich, A. Lamm, and H. A. Gasteiger, *Handbook of Fuel Cells*, Wiley, Chichester, UK, 2003.
43. M. Eikerling, A. A. Kornyshev, and A. R. Kucernak, *Phys. World* **20**, 32 (2007).
44. F. Maillard, M. Eikerling, O. V. Cherstiouk, S. Schreier, E. Savinova, and U. Stimming, *Faraday Discuss.* **125**, 357 (2004).

45. M. Bayati, J. M. Abad, C. A. Bridges, M. J. Rosseinsky, and D. J. Schiffrin, *J. Electroanal. Chem.* **623**, 19 (2008).
46. S. S. Mulla, N. Chen, L. Cumaranatunge, G. E. Blau, D. Y. Zemlyanov, W. N. Delgass, and F. H. Ribeiro, *J. Catal.* **241**, 389 (2006).
47. S. Mukerjee. *J. Appl. Electrochem.* **20**, 537 (1990).
48. S. Mukerjee and J. McBreen, *J. Electroanal. Chem.* **448**, 163 (1998).
49. G. A. Somorjai, *Introduction to Surface Chemistry and Catalysis*, Wiley, New York, 1994.
50. J. S. Bradley, in *Clusters and Colloids*, ed. G. Schmid, VCH: Weinheim, (1994).
51. K. J. J. Mayrhofer, B. B. Blizanac, M. Arenz, V. R. Stamenkovic, P. N. Ross, and N. M. Markovic, *J. Phys. Chem. B* **109**, 14433 (2005).
52. K. Kinoshita, *J. Electrochem. Soc.* **137**, 845 (1990).
53. T. S. Ahmadi, T. C. Green, A. Henglein, and M. A.El-Sayed, *Science* **272**, 1924 (1996).
54. I. Lisiecki. *J. Phys. Chem. B* **109**, 12231 (2005).
55. Y. Xiong, B. J. Wiley and Y. Xia, *Angew. Chem. Int. Ed.* **46**, 7157 (2007).
56. I. Lee, R. Morales, M. A. Albiter, and F. Zaera, *Proc. Natl. Acad. Sci. USA* **105**, 15241 (2008).
57. C. Susut, T. D. Nguyen, G. B. Chapman, and Y. Y. Tong, *Electrochim. Acta* **53**, 6135 (2008).
58. E. A. Meulenkamp, *J. Phys. Chem. B* **102**, 7764 (1998).
59. M. L. Sattler and P. N. Ross, *Ultramicroscopy* **20**, 21 (1986).
60. P. N. Ross, in *Handbook of Fuel Cells*, W. Vielstich, A. Lamm, and H. Gasteiger, eds., Wiley, Chichester, UK, 2003, Vol. **2**, pp. 465–480.
61. M. Eikerling, J. Meier, and U. Stimming, *Z. Phys. Chem.* **217**, 395 (2003).
62. B. Hammer and J. K. Nørskov, *Adv. Catal.* **45**, 71 (2000).
63. B. Hammer and J. K. Nørskov, *Surf. Sci.* **343**, 211 (1995).
64. C. A. Menning and J. G. Chen, *J. Chem. Phys.* **128**, 164703 (2008).
65. M. Neurock, *J. Catal.* **216**, 73 (2003).
66. K. Klabunde, *Free Atoms, Clusters and Nanoscale Particles*, Academic Press, London, 1994.
67. G. Rupprechter, *Catalysis on Well-Defined Surfaces: From Single Crystals to Regular Nanoparticles*, Kluwer Academic/Plenum Publishers, New York, 2007.
68. J. K. Jiang, G. Oberdörster, A. Elder, R. Gelein, P. Mercer, and P. Biswas. *Nanotoxicology* **2**, 33 (2008).
69. W. F. Lin, J. M. Jin, P. A. Christensen, F. Zhu, and Z. G. Shao, *Chem. Eng. Commun.* **195**, 147 (2008).
70. J. Meier, K. A. Friedrich, and U. Stimming, *Faraday Discuss.* **121**, 365 (2002).
71. N. Lopez, T. V. W. Janssens, B. S. Clausen, Y. Xu, M. Mavrikakis, T. Bligaard, and J. K. Nørskov, *J. Catal.* **223**, 232 (2004).
72. P. Claus and H. Hofmeister, *J. Phys. Chem. B* **103**, 2766 (1999).
73. A. I. Frenkel. C. W. Hills, and R. G. Nuzzo, *J. Phys. Chem. B* **105**, 12689 (2001).
74. C. Burda, X. Chen, R. Narayanan, and M. A. El-Sayed, *Chem. Rev.* **105**, 1025 (2005).
75. R. M. Rioux, H. Song, M. Grass, S. Habas, K. Niesz, J. D. Hoefelmeyer, P. Yang, and G. A. Somorjai, *Top. Catal.* **39**, 167 (2006).

76. R. Narayanan, C. Tabor, and M. A. El-Sayed, *Top. Catal.* **48**, 60 (2008).
77. R. Narayanan and M. A. El-Sayed, *Top. Catal.* **47**, 15 (2008).
78. P. Wynblatt and N. A. Gjostein, *Acta Metallogr.* **24**, 1165 (1976).
79. Z. Kuntová, Z. Chvoj, V. Šíma, and M. C. Tringides, *Phys. Rev. B.* **71**, 12415 (2005).
80. X. Lin, N. J. Ramer, A. M. Rappe, K. Hass, W. F. Schneider, and B. L. Trout, *J. Phys. Chem. B* **105**, 7739 (2001).
81. D. R. Lide, *Handbook of Chemistry and Physics*, 71st ed., CRC Press, Boca Raton, FL, 1990–1991.
82. R. G. Linford, in *Solid State Surface Science*, M. Green, ed., Marcel Dekker, New York, 1973.
83. S. Ijima and H. Ichihashi, *Phys. Rev. Lett.* **56**, 616 (1986).
84. R. A. van Santen and M. Neurock, *Molecular Heterogenous Catalyst*, Wiley-VCH, Weinheim, 2006.
85. M. A. Rigsby, W. P. Zhou, A. Lewera, H. T. Duong, P. S. Bagus, W. Jasegermann, R. Hunger, and A. Wieckowski, *J. Phys. Chem. C* **112**, 15595 (2008).
86. M. Mavrikakis, B. Hammer, and J. K. Nørskov, *Phys. Rev. Lett.* **81**, 2819 (1998).
87. H. J. Wasserman and J. S. Vermaak, *Surf. Sci.* **32**, 168 (1972).
88. Q. Jiang, L. H. Liang, and D. S. Zhao, *J. Phys. Chem. B* **105**, 6275 (2001).
89. A. Roudgar and A. Gross, *Surf. Sci.* **559**, L180 (2004).
90. G. A. Somorjai, *Introduction to Surface Chemistry and Catalysis*, Wiley, New York, 1994.
91. J. S. Bradley, in *Clusters and Colloids*, G. Schmid, ed., VCH, Weinheim, 1994.
92. M. Arenz, K. J. J. Mayrhofer, V. Stamenkovic, B. B. Blizanac, T. Tomoyuki, P. N. Ross, and N. M. Marković, *J. Am. Chem. Soc.* **127**, 6819 (2005).
93. O. V. Cherstiouk, P. A. Simonov, and E. R. Savinova, *Electrochim. Acta* **48**, 3851 (2003).
94. F. Maillard, S. Schreier, M. Heinzlik, E. R. Savinova, S. Weinkauf, and U. Stimming, *Phys. Chem. Chem. Phys.* **7**, 385 (2005).
95. F. Maillard, E. R. Savinova, and U. Stimming, *J. Electroanal. Chem.* **599**, 221 (2007).
96. S. Gilman, *J. Phys. Chem.* **68**, 70 (1964).
97. L. R. Becerra, C. A. Klug, C. P. Slichter, and J. H. Sinfelt, *J. Phys. Chem.* **97**, 12014 (1997).
98. T. Kobayashi, P. K. Babu, L. Gancs, J. H. Chung, E. Oldfield, and A. Wieckowski, *J. Am. Chem. Soc. Commun.* **127**, 14164 (2005).
99. B. Andreaus and M. Eikerling, *J. Electroanal. Chem.* **607**, 121 (2007).
100. C. Lu, C. Rice, R. I. Masel, P. K. Babu, P. Waszczuk, H. S. Kim, E. Oldfield, and A. Wieckowski, *J. Phys. Chem. B.* **106**, 9581 (2002).
101. A. V. Petukhov, *Chem. Phys. Lett.* **277**, 539 (1997).
102. M. T. M. Koper, A. P. J. Jansen, R. A. van Santen, J. J. Lukkien, and P. A. J. Hilbers, *J. Chem. Phys.* **109**, 6051 (1998).
103. C. Saravanan, N. M. Markovic, M. Head-Gordon, and P. N. Ross, *J. Chem. Phys.* **114**, 6404 (2001).
104. C. Korzeniewski and D. Kardash, *J. Phys. Chem. B.* **105**, 8663 (2001).
105. D. T. Gillespie, *J. Comput. Phys.* **22**, 403 (1976).
106. A. P. J. Jansen, *Comput. Phys. Commun.* **86**, 1 (1995).

107. N. P. Lebedeva, M. T. M. Koper, J. M. Feliu, and R. A. van Santen, *J. Phys. Chem. B* **106**, 12938 (2002).
108. M. T. M. Koper, N. P. Lebedeva, and C. G. M. Hermse, *Faraday Discuss.* **121**, 301 (2002).
109. B. Love and J. Lipkowski, in *Electrochemical Surface Science. Molecular Phenomena at Electrode Surfaces*; M. P. Soriaga, Ed., ACS Symp. Series vol. 378, Washington, DC, 1988, p. 484.
110. C. McCallum and D. Pletcher, *J. Electroanal. Chem.* **70**, 277 (1976).
111. J. Jiang and A. Kucernak, *J. Electroanal. Chem.* **543**, 187 (2003).
112. B. C. Han, C. R. Miranda and G. Ceder, *Phys. Rev. B* **77**, 075410 (2008).
113. T. Jacob, B. V. Merinov, and W. A. Goddard, III, *Chem. Phys. Lett.* **385**, 374 (2004).
114. P. J. Feibelman, *Phys. Rev. B* **56**, 10532 (1997).
115. A. Eichler, F. Mittendorfer, and J. Hafner, *Phys. Rev. B* **62**, 4744 (2000).
116. J. Rossmeisl, A. Logadottir, and J. K. Nørskov, *Chem. Phys.* **319**, 178 (2005).
117. M. R. Tarasevich, A. Sadkowski, and E. Yeager, in *Comprehensive Treatise of Electrochemistry*, B. Conway, J. O. Bockris, E. Yeager, and S. U. M. Khan, eds., Plenum Press, New York, 1983, Chapter 9, Vol. 7.
118. A. Gross, *J. Comput. Theor. Nanosci.* **5**, 894 (2008).
119. H. J. Monkhorst and J. D. Pack, *Phys. Rev. B* **13**, 5188 (1976).
120. Z. Gu and P. B. Balbuena, *J. Phys. Chem. Lett. A.* **110**, 9783 (2006).
121. A. Panchenko, M. T. M. Koper, T. E. Shubina, and S. J. Mitchell, *J. Electrochem. Soc.* **152**, A2016 (2004).
122. J. X. Wang, F. A. Uribe, T. E. Springer, J. Zhang, and R. R. Adzic, *Faraday Discuss.* **140**, 347 (2008).
123. N. M. Markovic, T. J. Schmidt, V. Stamenkovic, and P. N. Ross, *Fuel Cells* **1**, 105 (2001).
124. Y. Wang and P. B. Balbuena, *J. Phys. Chem. B* **109**, 14896 (2008).
125. H. A. Hansen, J. Rossmeisl, and J. K. Nørskov, *Phys. Chem. Chem. Phys.* **10**, 3722 (2008).
126. J. Rossmeisl and J. K. Nørskov, *J. Phys. Chem. B* **110**, 21833 (2006).
127. T. Jacob, *Fuel Cell* **6**, 159, (2006).
128. A. Damjanovic, M. A. Genshaw, and J. O. M. Bockris, *J. Chem. Phys.* **45**, 4057, (1966).
129. C. Hartnig and M. T. M. Koper, *J. Electroanal. Chem.* **532**, 165 (2002).
130. Henkelman and H. Jónsson, *J. Chem. Phys.* **113**, 9978 (2000).
131. E. Guilminot, A. Corcella, F. Charlot, F. Maillard, and M. Chatenet, *J. Electrochem. Soc.* **154**, B96 (2007).
132. P. J. Ferreira, G. J. La O', Y. Shao-Horn, D. Morgan, R. Makharia, S. Kocha, and H. A. Gasteiger, *J. Electrochem. Soc.* **152**, A2256 (2005).
133. K. Sasaki, L. Zhang, and R. R. Adzic, *Phys. Chem. Chem. Phys.* **10**, 159 (2008).
134. J. Meier, J. Schiotz, P. Liu, J. K. Nørskov, and U. Stimming, *Chem. Phys. Lett.* **390**, 440 (2004).
135. M. Baldauf and D. M. Kolb, *Electrochim. Acta* **38**, 2145 (1993).
136. S. Pandelov and U. Stimming, *Electrochim. Acta* **52**, 5548 (2007).
137. V. P. Zhdanov and B. Kasemo, *Surf. Sci. Rep.* **39**, 25 (2001).
138. R. Gomez, V. Climent, J. M. Feliu, and M. J. Weaver, *J. Phys. Chem.* **104**, 597 (2000).
139. Z. Kerner and T. Pajkossy, *Electrochim. Acta* **46**, 207 (2000).

140. L. I. Daikhin, A. A. Kornyshev, and M. Urbakh. *Phys. Rev. E* **53**, 6192 (1996).

141. L. I. Daikhin, A. A. Kornyshev, and M. Urbakh. *Electrochim. Acta* **42**, 2853 (1997).

142. L. I. Daikhin, A. A. Kornyshev, and M. Urbakh, *J. Chem. Phys.* **108**, 1715 (1998).

143. E. Lust, A. Jänes, V. Sammelselg, and P. Miidla, *Electrochim. Acta* **46**, 185 (2000).

144. Q. Wang, M. Eikerling, D. Song, and S. Liu, *J. Electrochem. Soc.* **154**, F95 (2007).

145. M. Saha, A. F. Gull, R. J. Allen, and S. Mukerjee, *Electrochim. Acta* **51**, 4680 (2006).

146. D. Gruber and J. Muller, *J. Power Sources* **171**, 294 (2007).

147. J. Keener, *Mathematical Physiology*, Springer-Verlag, New York, 1998.

148. R. D. Coalson and M. G. Kurnikova, Poisson-Nernst-Planck theory of ion permeation through biological channels, in *Biological Membrane Ion Channels: Dynamics, Structure, and Applications*, S. H. Chung, O. S. Anderson, and V. Krishnamurthy, eds. Springer, New York, 2007.

149. D. B Sepa, M. V. Vojnovic, and A. Damjanovic, *Electrochimica Acta* **26**, 781 (1981).

150. A. Parthasarathy, S. Srinivasan, and A. J. Appleby, *J. Electrochem. Soc.* **139**, 2530 (1992).

151. C. Marr and X. Li, *J. Power Sources* **77**, 17 (1999).

152. T. Pajkossy and D. M. Kolb, *Electrochem. Commun.* **5**, 283 (2003).

153. W. G. Pell, A. Zolfaghari, and B. E. Conway, *J. Electroanal. Chem.* **532**, 13 (2002).

154. K. A. Friedrich, W. Daum, F. Dederichs, and W. Akemann, *Z. Phys. Chem.* **217**, 527 (2003).

155. V. Climent, N. Garcia-Araez, E. Herrero, and J. Feliu, *Russ. J. Electrochem.* **42**, 1145 (2006).

156. U. W. Hamm, D. Kramer, R. S. Zhai, and D. M. Kolb, *J. Electroanal. Chem.* **85**, 414, (1996).

CHAPTER 11

# Fuel Cells with Neat Proton-Conducting Salt Electrolytes

DOMINIC GERVASIO

Department of Chemical and Environmental Engineering, University of Arizona, Tucson, Arizona

The first report of the syntheses of neat proton-containing (protic) salt electrolytes for use as proton conductors in fuel cells was in the year 2003. "Neat" in this context means that only the protic salt—with no water or other molecular solvent—is needed for proton conduction. Having a water-free proton conductor leads to many favorable consequences for fuel cell electrodes and systems. This chapter is meant to be an evocative—not comprehensive—account of how to make such a protic salt. The design, synthesis, and characterization of protic ionic liquids (PILs) are discussed first in order to develop an understanding of proton conduction in protic salts. Then concepts learned and developed from studies of the PILs are extended to solid protic ionic membranes (PIMs). The PIMs are targeted for making more efficient and compact proton electrolyte membrane (PEM) fuel cells. Examples of PILS and PIMs were selected from recent literature and compared to conventional aqueous acidic liquid and solid polymer electrolytes. It is hoped the reader will be encouraged to develop new salt electrolytes, because there are increasingly better prospects for overcoming the barriers that have impeded the implementation of fuel cells as practical power sources, as better and better protic salt electrolytes are found.

## 11.1 INTRODUCTION

Electrical power is ideal for most applications. The clean efficient generation of electrical power is now more important than ever. Today a fuel cell power source system can generate electrical power with > 40% efficiency and with no noxious

*Fuel Cell Science: Theory, Fundamentals, and Biocatalysis*,
Edited by Andrzej Wieckowski and Jens K. Nørskov

emissions such as NOx (nitrogen oxides), ozone, and carbon monoxide. This is done by converting petroleum to a hydrogen-rich fuel to feed the anode and drawing air to feed the cathode. The fuel cell system makes electrical power much more efficiently than does burning fuel in an internal-combustion engine that drives a magnetic induction generator. So, using fuel cell power sources would virtually eliminate noxious emissions while extending global energy reserves and reducing carbon dioxide emissions in direct proportion to the efficiency gain. Although 40% chemical-to-electrical conversion efficiency is good, the efficiency of the fuel cell is presently only 65%, although it can theoretically approach 95%. The loss of efficiency is due to the shortcomings of the materials used in the key components of the fuel cell. Presently these components are costly as well. The key components in a fuel cell are the electrolyte and electrode, which work together to induce charge transfer reactions generating electricity. Flowing fuel (hydrogen) is directly oxidized, and flowing oxidant (oxygen from air) is directly reduced to spontaneously generate electrical power. For a long time, there have been many attempts to use fuel cells as power sources. However, making a fuel cell power source practical has been largely elusive because of the shortcomings of conventional *aqueous* electrolyte and catalyst combinations. These limit not only efficiency of the fuel cell but also system performance. Using aqueous electrolyte requires humidifiers and operation below 100°C to preserve the high conductivity that is required for efficient operation of the fuel cell. As a result, there is a loss in system efficiency due to parasitic power needed for the humidifier, and the power density of the system suffers from the bulk of the humidifier and the need for excessively large radiators to reject waste heat. One largely overlooked approach for overcoming these limitations is to use a nonaqueous salt electrolyte. The main advantages of using a nonaqueous salt electrolyte in a fuel cell are that this type of electrolyte can have high ionic conductivity with low water activity over a broad range of temperatures. This eliminates the humidifier, reduces the size of the radiator, and improves the efficiency of the air cathode that polarizes in the presence of high-activity ("bulk") water at high potentials (0.9–1.2 V vs. NHE).

Although water ($H_2O$) is vital to the health of most living things, water in an electrolyte can interfere with electrochemical processing by adsorbing on or reacting with the electrode, reactants, or products. For example, it is not possible to electroplate a metal salt to metal (M) in an aqueous solution when the heat of formation of the corresponding metal oxide (MO) is highly exothermic. The mechanism of interference by water can be illustrated by the "local cell" process, which is summarized in reactions (11.1), (11.2), and (11.3) and Figure 11.1.

$$M + H_2O \rightarrow MO + 2e^- + 2H^+ \quad (11.1)$$

$$2e^- + 2H^+ \rightarrow H_2 \quad (11.2)$$

$$M + H_2O \rightarrow MO + H_2 \quad (11.3)$$

Metal and bulk water spontaneously react to form metal oxide and protons ($H^+$) on one part of the metal surface [as shown in reaction (11.1)], while at the same time, $H^+$ (from bulk water) is reduced on another part of the metal surface forming hydrogen gas

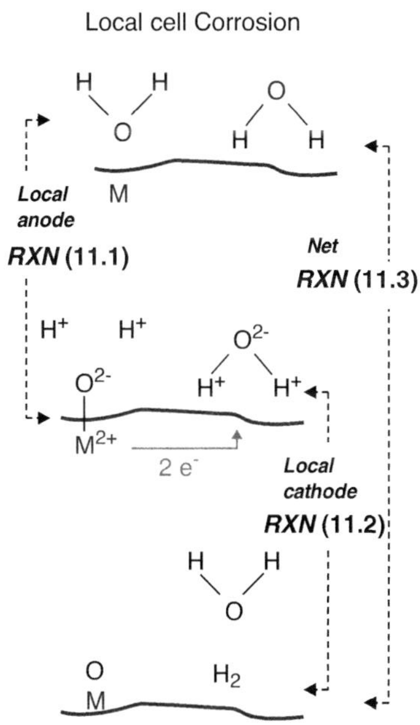


**FIGURE 11.1** Local cell corrosion on a metal surface under a layer of water.

($H_2$) and balancing charge [as shown in reaction (11.2)]. The net result is the loss of metal and water and the formation of metal oxide and hydrogen gas [as shown in reaction (11.3)].

The more exothermic that metal oxide formation is, the greater is the interference by water. For example, when some metals (e.g., aluminum, silicon) are electroplated from aqueous solution, as soon as metal forms on a cathode, it will react with water and revert to its oxide ($Al_2O_3$, $SiO_2$, etc.) Furthermore, the metal oxide (MO) that forms can be very persistent and electrically resistive, thereby passivating the electrode and halting cell activity. Platinum oxide forms on a fuel cell cathode in aqueous electrolytes. This oxide does not revert to catalytically active metallic platinum until high cathode overpotentials $\eta$ (overpotentials) $= \sim 0.4$ V or oxide reduction potential $= \sim 0.8$ V vs RHE. As a result, the cell voltage drop and chemical-to-electrical conversion efficiency in the fuel cell is only 70% of the theoretically allowed values. Using salt electrolytes obviates the need for water and can lead to more efficient fuel cells and fuel-cell system operation.

## 11.2 SALT ELECTROCHEMISTRY

### 11.2.1 High-Temperature Electrolytes

The first practical application of salt electrolytes was the electrosynthesis of aluminum. In 1886, aluminum metal was as costly as the precious metal, silver, even though aluminum oxide is one of the most abundant ores on the earth. This fact motivated Charles Martin Hall to make aluminum inexpensively [1]. Hall recognized that an alternative process for smelting aluminum metal was needed because aluminum is roasted by a local cell reaction with water [reaction (11.1)] or in a similar local cell involving oxygen (in the reaction $2e^- + \frac{1}{2}O_2 + H_2O \rightarrow 2OH^-$, leading to the net reaction $M + \frac{1}{2}O_2 \rightarrow MO$; both of these reactions are continuations of reactions (2) and (3), respectively) at a relatively higher rate than when it is smelted by carbon in a traditional high-temperature furnace. Roasting is driven by the high heat of formation of aluminum oxide (−4000 cal/g) compared to a furnace-smeltable metal like silver ($\Delta H^f$ of silver oxide = −32 cal/g). Hall's alternative process was to melt aluminum oxide in halide salt electrolyte at around 900°C and quickly electroplate aluminum metal on a cathode at the bottom of a cell while a carbon anode on top formed carbon dioxide over the cell to prevent entry of water and oxygen from air. The net reaction is $2\,Al_2O_3 + C \rightarrow 4\,Al + CO_2$. By 1914, the cost of aluminum was 200 less than silver.

### 11.2.2 Low-Temperature Electrolytes

Fifty years later, Hurley first exhibited aluminum deposition at low temperatures using a low-melting salt [2]. This low-temperature aluminum electroplating was also developed by Osteryoung [3].

## 11.3 GENERAL CHARACTERISTICS OF SALT ELECTROLYTES

### 11.3.1 Properties of Ionic Liquids

Ionic liquids (ILs) are molten salts with melting points below 100°C. They are composed of over 99.99% ions, have low vapor pressures, are highly conductive with excellent solvating properties, and are thermally stable, for both high- and low-temperature uses. These general physical properties of ionic liquids have been well reviewed by Anthony et al. [4]. As was discussed in their review [4], the ionic liquids can be low melting and have low viscosity by choosing a cation and an anion that do not pack into a stable lattice. One special distinction is given to "a salt that is liquid at or below room temperature"; this is called a *room-temperature ionic liquid* (RTIL). Room-temperature ionic liquids (RTILs) consist of bulky and asymmetric positive ions (cations), such as 1-alkyl-3-methylimidazolium, 1-alkylpyridinium, *N*-methyl-*N*-alkylpyrrolidinium, and other alkyl ammonium ions. The negative ions (anions) can include halides ($Cl^-$, $F^-$, $Br^-$, $I^-$), which generally give higher-melting salts, as well as bulky inorganic anions such as tetrafluoroborate, hexafluorophosphate, and large organic anions, including: bistriflylateimide, triflate, tosylate, formate, alkylsulfate, alkylphosphate, and glycolate. Since ionic liquids are in fact neat salts that are liquid, and not diluted with water or any other solvents, their constituent ions are free to translate, which is consistent with their high conductivity. Presently lithium ion batteries use electrolytes consisting of lithium salts dissolved in flammable and volatile organic compounds (VOCs), such as propylene carbonate and ethylene oxides. These other organic ether solvents are noxious and hazardous. On the other hand, ionic liquids have been made from lithium salts with low viscosity, high stability, and low volatility. These ILs can be used as lithium electrolytes, because some have ionic conductivities that are as high as or even higher than ions dissociated in aqueous and organic electrolytes. Accordingly, these ionic liquids are being considered as advanced electrolytes for lithium ion batteries and electrolytic capacitors [5].

Less well studied is the use of ionic liquids in fuel cells. This is surprising, since the first ionic liquid made by Paul Walden over 95 years ago, ethyl ammonium nitrate (EAN) is well suited for a hydrogen/air fuel cell use [6]. EAN is well suited for this fuel cell, because it contains a proton within the salt; that is, EAN is a protic ionic liquid (PIL). Why is a PIL a good choice for the fuel cell? In a hydrogen/air fuel cell, the ionic product of the hydrogen anode is proton ($H^+$) and the product of the oxygen cathode is oxygen dianion ($O^{2-}$). To sustain steady operation, the fuel cell needs to conduct one of these ionic products to complete the electrical circuit. Since hydrogen is lighter, its ionic product, proton, would be the logical choice, since lighter species are generally more mobile. When proton ($H^+$) is conducted to the cathode, it combines with the oxygen cathode product, oxygen dianion ($O^{2-}$), to make water, which is the net chemical product of the fuel cell. In general, an ion conductor contains the ion being conducted. Since EAN contains a proton, this is why EAN is a suitable fuel cell electrolyte.

Well known advantageous qualities of PIL electrolytes, like any ionic liquid, are that PILs are non-volatile and do not need solvent water for high conductivity. These qualities allow PILs to be used from room temperature to their decomposition temperature, which is often greater than 200°C. But a less well known and intriguing advantage is that a PIL can have low water activity when used as an electrolyte at high temperature. Low water activity helps avoid reaction with water and metal oxide formation [shown in reaction (11.1)] on cathode catalyst (like Pt) surfaces at high electrode potentials. The advantage of avoiding Pt oxide formation is Pt is in the catalytically active metallic form at higher potentials, so the cathode reduces oxygen at more positive potentials. This is conducive to higher power and more efficient fuel cell operation. There is much yet to be learned about the overlooked *electrochemistry in low-water-activity salt electrolytes*, particularly at the cathode in electrochemical power sources and electrosynthetic reactors.

#### 11.3.1.1 From Aqueous to Salt Electrolytes

*Proton Conduction* There has been recent interest in making a solid proton-conducting electrolyte membrane (PEM) for a fuel cell that operates at high temperature for portable applications. Proton-conducting acidic electrolytes are preferred because carbon dioxide, from impure hydrogen and air, which can adversely affect the fuel cell, is rejected by acidic electrolytes. For many years, only aqueous acid electrolytes have been used in fuel cells. However, with aqueous electrolytes, fuel cell *system* power density is limited and its efficiency drops from 95% to no more than 40%. How to eliminate the need for water can be understood by considering proton conduction in liquid phosphoric acid. Electrons are conducted through solid metals via a conduction band of electronic states, according to the quantum mechanics. Proton ($H^+$) conduction in an analogous band of protonic states would be highly desirable, but the proton is so massive that this conductor has never been found. Instead, protons move through electrolytes via translations, rotations, and vibrations as dictated by statistical mechanics. Conventional proton-conducting electrolytes usually involve water, which plays two roles: (1) ionization of a dissolved acid and (2) a medium for ion transport. There are two kinds of ion transport: (1) a slower translational diffusion of protons and anions and (2) a faster "proton hopping" (i.e., a coordinated rotational and vibrational transport). Liquid concentrated phosphoric acid is one of the few oxy acid electrolytes known to conduct only proton with no cotransport of other species, like water. The proton hopping mechanism for phosphoric acid is illustrated in Figure 11.2.

Phosphoric acid serves as an example of how to make a synthetic electrolyte that conducts proton with no water. Orthophosphoric acid ($H_3PO_4$) autoionizes into different proton states, such as hydronium phosphate ($H_4PO_4{}^+$) and dihydrogen phosphate ($H_2PO_4{}^-$). The energy of these proton states can be quantified by their dissociation constant ($K$). The difference in the logarithm of the dissociation constant ($-\log_{10}K_a$) for two dissociation reactions (see Table 11.1) is directly proportional to the free-energy difference ($\Delta G^o = -2.3\,RT \log K_a$) for those two proton occupancies.

The difference in the dissociation constants ($\Delta pK$ values) for these states is about 5. The difference in energy for these proton states is on the order of thermal energy. Proton transfers between states are thermally accessible, so one of these states can act as a proton donor (acid) and another as a proton acceptor (base). Figure 11.2 shows proton transport can occur by rotational and vibrational transfer of proton between two available species (proton states), phosphoric acid ($H_3PO_4$), and phosphonium ion ($H_4PO_4{}^+$). The proton is passed by rotational and vibration motions. Phosphoric acid and its various ionic forms form tetrahedrally, and any two species have both the suitable energy separation (proper $\Delta pK$) for proton transfer and symmetry for the required vibrations and rotations to perform the proton transfer. This leads to a general rule. Proton conduction can occur when two conditions are met: (1) there is a suitable energy separation ($\Delta pK$) between proton acceptor and donor species and (2) these species have high symmetry so that effective rotational and vibrational motions are not impeded. When a proton-containing salt is made to satisfy these two conditions, this rule predicts that the salt will be useful as proton-conducting electrolytes.

#### *11.3.1.2 Protic Ionic Liquids (PILs)*

The simplest example of a proton-conducting salt electrolyte is the liquid salt called a *protic ionic liquid* (PIL). A PIL forms when proton transfers in an equimolar mixture of

**TABLE 11.1 Acid Dissociation Constants ($K_a$) for Different Proton States of Phosphoric Acid**

| Dissociation Reaction | $K_a$ | $-\log_{10}K_a$ |
|---|---|---|
| $H_4PO_4{}^+ + H_2O_{(l)} \rightarrow H_3O^+{}_{(aq)} + H_3PO_{4(aq)}$ | $K_{a0} = \sim 10^2$ | $pK_{a0}$ = -2 |
| $H_3PO_4 + H_2O_{(l)} \rightarrow H_3O^+{}_{(aq)} + H_2PO_4{}^-{}_{(aq)}$ | $K_{a1} = 7.5 \times 10^{-3}$ | $pK_{a1} = 3$ |
| $H_2PO_4{}^- + H_2O_{(l)} \rightarrow H_3O^+{}_{(aq)} + HPO_4{}^{2-}{}_{(aq)}$ | $K_{a2} = 6.2 \times 10^{-8}$ | $pK_{a2} = 7$ |
| $HPO_4{}^{2-} + H_2O_{(l)} \rightarrow H_3O^+{}_{(aq)} + PO_4{}^{3-}{}_{(aq)}$ | $K_{a3} = 2.14 \times 10^{-13}$ | $pK_{a3} = 12.7$ |

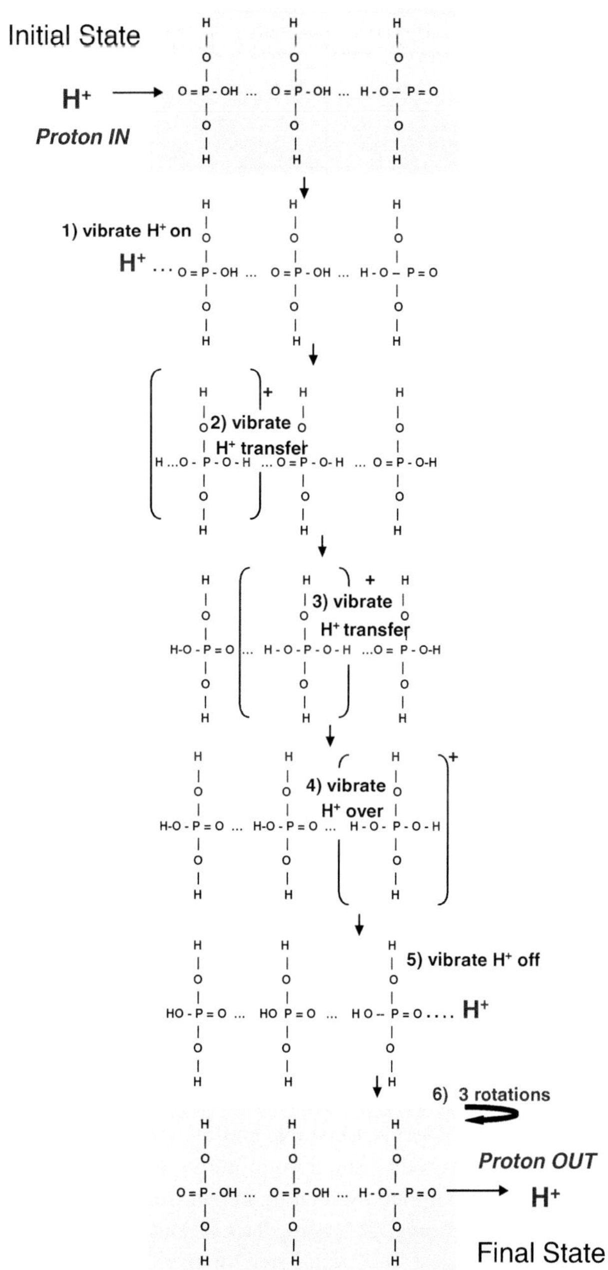


**FIGURE 11.2** Proton hopping in phosphoric acid.

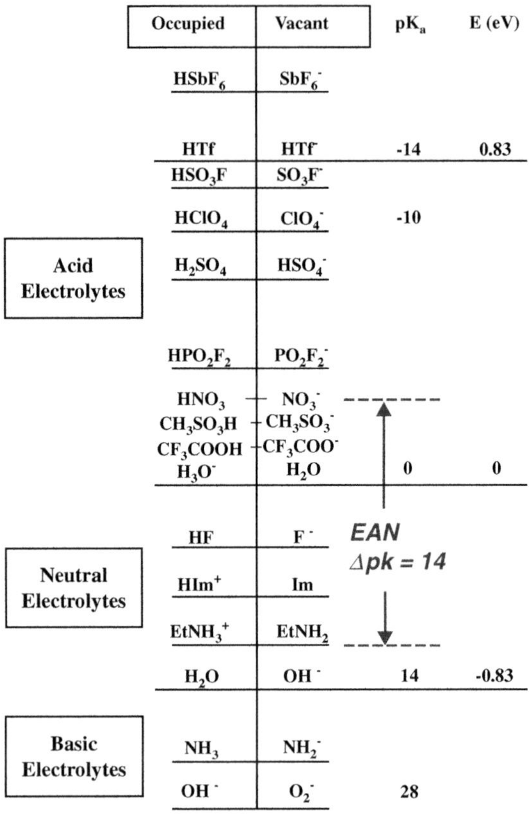


**FIGURE 11.3** Gurney proton energy diagram.

an acid and a base. The energetics of the proton transfer process was organized by R. Gurney [7].

The Gurney proton energy-level diagram describes the energy change as a proton moves from one species to another. For any pair of levels, the stable entities are right upper and left lower, for example, see the levels for ethyl ammonium nitrate with $\Delta pK = 14$.

Figure 11.3 shows a Gurney plot for acids and bases that can be used to make PIL electrolytes [8]. For any pair of levels, the stable entities are right upper and left lower; for example, see the levels for ethyl ammonium nitrate with $\Delta pK = 14$. The proton energy levels are typically compiled by relating known values of proton dissociation constants in water, $K_a$, to free energy, $\Delta G^o$, using the relationship $\Delta G^o = -2.3\,RT \log K_a$. This is an approximation, because the $K_a$ values are for dissociation of acids and bases in water, and the hydration energy is not likely to be the same for the acid and the base.

The mechanism of ion conduction can be classified by the *Walden plot*, named after Paul Walden, who first reported this kind of plot [9]. The heavy line represents the ideal Walden line. Figure 11.4 shows such a plot for a number of ionic compounds, including the protic ionic liquid, EAN [5a,10]. The heavy solid diagonal line is the ideal line from

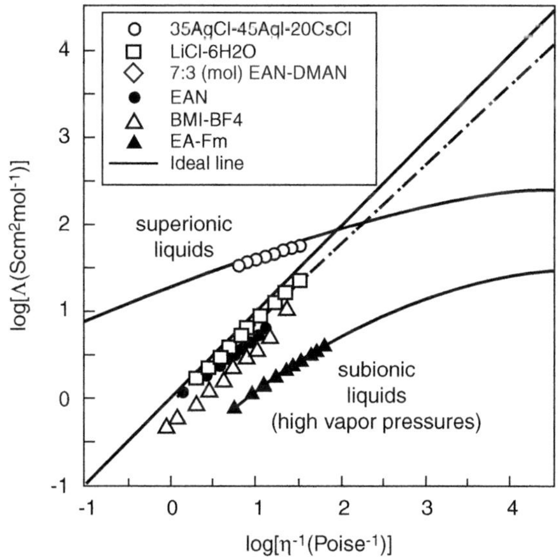


**FIGURE 11.4** Walden plot for classifying the mechanism of ion conduction.

data for 1 M aqueous KCl solution at ambient temperature, and the temperature dependence of conductivity is set by the fluidity (inverse of viscosity). In charge-concentrated systems like those with very high vapor pressure, there is strong interaction between the ions, and interionic friction typically causes loss of mobility, which is more important at higher temperatures. This friction gives rise to the below-ideal slope. This behavior is found for all ILs, protic or aprotic, including EAN. When there is a special mechanism for conductance, the Walden plot falls above the ideal line. An example is for superionics, like 35AgCl–45AgI–20CsCl, in which ions hop instead of translate through the condensed phase (e.g., Grotthus "hopping" mechanism). A superionic PIL is of particular interest for choosing the ionic moieties to attach to a polymer for forming a solid proton-conducting polymer electrolyte membrane (PEM).

*Fuel Cells* Watanabe et al. first reported using a PIL as a fuel cell electrolyte [11]. A number of PILs were made by selecting one of a number of organic bases and combining in equimolar ratios the selected base always with one particular acid, bis(trifluoro methane sulfonyl) amide (TFMSA). The mixture of imidazole and TFMSA is shown in Figure 11.5. Figure 11.5 also shows how these salts were used as the electrolyte for electrooxidation of $H_2$ and electroreduction of $O_2$ on Pt electrodes with no humidification of the electrolyte. Watanabe and coworkers showed that these protic ionic liquid electrolytes could stably conduct proton with no humidification of the electrolyte at elevated temperatures. The ionic liquids of Watanabe displayed good stability, but the resulting cathodes and fuel cells gave only marginal performance in terms of cell voltage and power density.

In the short-term measurements, a fuel cell that uses liquid EAN gives fuel cell performance much better than that using 85% $H_3PO_4$ as the electrolyte (see Fig. 11.6).

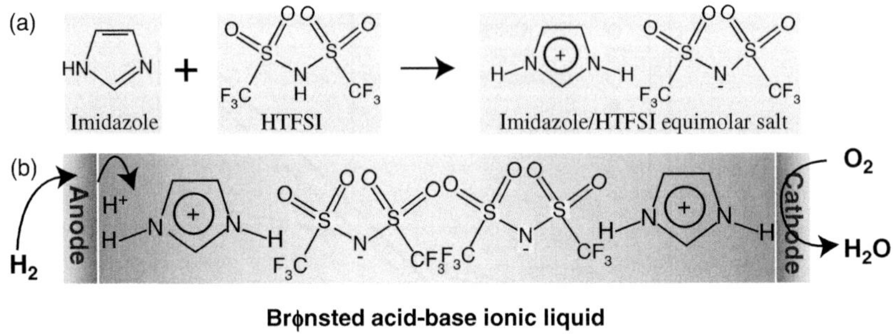


**FIGURE 11.5** A PIL made from 1 mol of imidazole and 1 mol of TFMSA (top). $H_2$ electrooxidation and $O_2$ electroreduction on Pt with no humidification (bottom).

Ironic, considering that EAN was one of the first protic ionic liquids made by Walden almost 100 years ago. The fuel cell data shown in Figure 11.6 were obtained using a cell that consisted of a U-tube-containing electrolyte and two low-surface-area Pt wires wrapped around gas bubblers (see Fig. 11.7). On one side hydrogen gas was bubbled over the Pt for the anode and on the other side, oxygen for the cathode. The fuel cell with EAN electrolyte shows higher open-circuit voltage and cell voltage under load than with 85% phosphoric acid as the electrolyte. In fact, there is hardly any polarization of the fuel cell with EAN electrolyte until the mass-transfer-limited current. The cell used in Figure 11.7 is very useful for comparing the relative fuel cell performance of liquid electrolytes on a metal catalyst, like Pt. However, the absolute current generated in this kind of cell is very low. To generate higher current and power densities, fuel cells typically use Pt-catalyzed porous gas-fed electrodes as is shown in Figure 11.8.

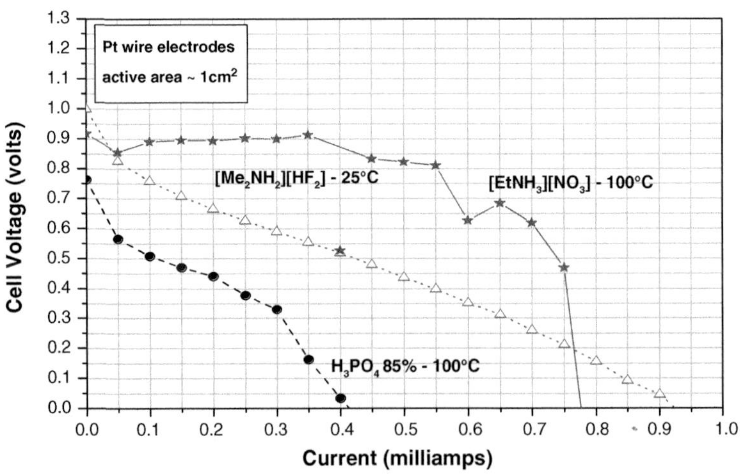


**FIGURE 11.6** Polarization curves for fuel cell with Pt wires fed hydrogen and oxygen in EAN (red star), $Me_2NH_2$:$HF_2$ (green triangle), and 85% $H_3PO_4$ (black circle).

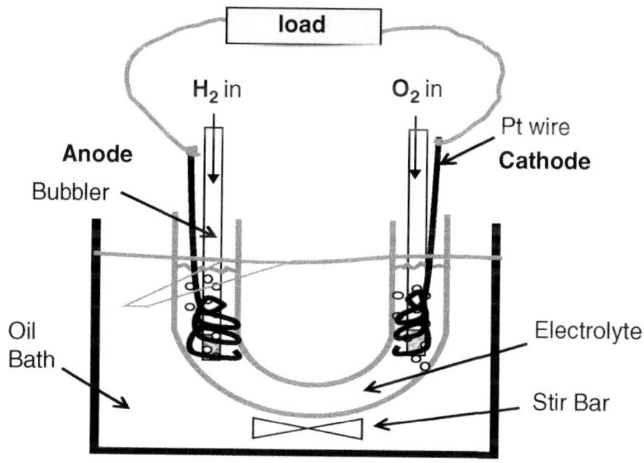


**FIGURE 11.7** Setup for evaluating liquid fuel cell electrolytes.

Figure 11.9 shows how the performance for the fuel cell with a Pt-catalyzed porous gas-fed hydrogen anode and oxygen cathode (shown in Figure 8) changes when it is filled with one of three different liquid electrolytes. The open-circuit voltage and the cell voltages at low current densities (up to a 2 mA/cm$^2$) are both higher with EAN (solid triangle) than with 85% phosphoric acid (open triangle) as the electrolyte. At current densities higher than 3 mA/cm$^2$, the fuel cell with phosphoric acid shows performance better than that with EAN as the electrolyte. The relative dropoff in performance with EAN at higher current densities occurs because EAN wets Teflon. EAN floods the pores of the electrodes, which impedes the transport of hydrogen and oxygen to the Pt in Teflon-bound gas-fed electrodes. Electrode wetting and flooding were greatly attenuated when 1% ethyl ammonium phosphate (EAP) was added to EAN (open circle in Fig. 11.9). With the 99% EAN–1%EAP electrolyte, the fuel cell open-circuit voltage approaches the thermodynamic limit of 1.2 V and the cell voltage is 20% higher than the fuel cell with phosphoric acid as the electrolyte from the open circuit up to a current density of 10 mA/cm$^2$. This is a remarkable result. However, at

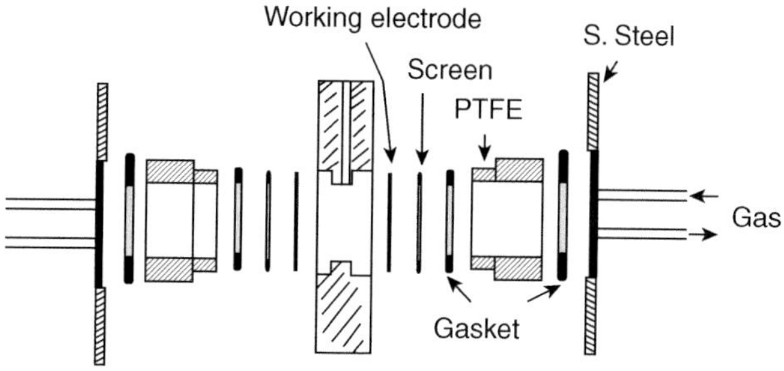


**FIGURE 11.8** Schematic diagram of the setup for evaluating liquid fuel cell electrolytes with Pt-catalyzed porous gas-fed electrodes.

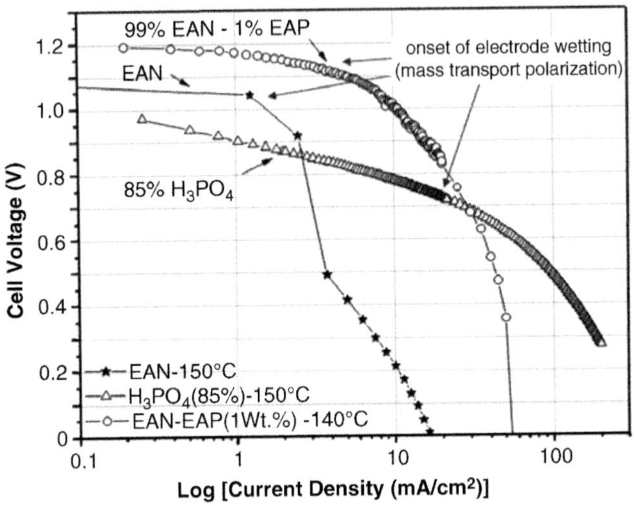


**FIGURE 11.9** Polarization curves for $H_2/O_2$ fuel cell with EAN, 99% EAN/1%EAP, and 85% $H_3PO_4$ electrolytes on Pt-catalyzed porous gas-fed ETEK V3 electrodes; Pt loading = 0.5 mg Pt/cm$^2$.

current densities greater than 10 mA/cm$^2$, wetting of the electrode by the 99% EAN–1%EAP electrolyte impedes the mass transfer of hydrogen and oxygen to the Pt in the porous gas-fed electrode enough so that the voltage drops.

Furthermore EAN has limited oxidative stability on Pt at high potentials. This is clear from Figure 11.10, which shows the voltammetry of Pt in EAN equilibrated with

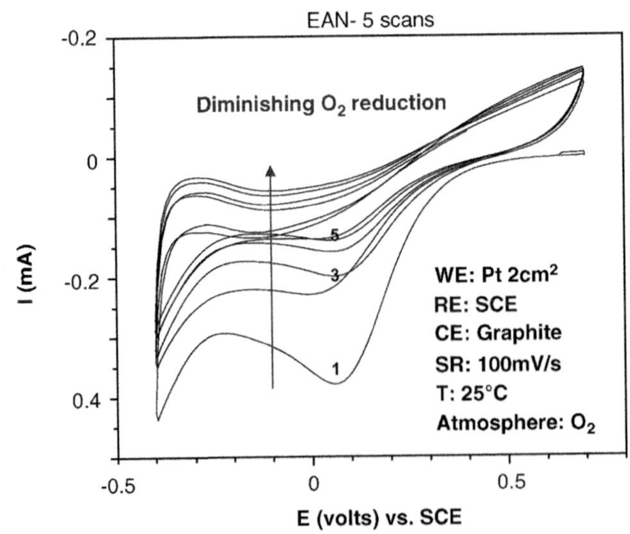


**FIGURE 11.10** Voltammetry of Pt in EAN equilibrated with oxygen (Pt area = 2 cm$^2$).

**TABLE 11.2 Requirements of a Fuel Cell Electrolyte**

| | |
|---|---|
| 1. | Conductor for protons, conductivity for $H^+ > 10^{-2}$ S/cm |
| 2. | Insulator for electrons, conductivity for $e^- < 10^{-9}$ S/cm |
| 3. | Operation at temperatures from −50°C to 230°C |
| 4. | Stable to alkalinity (generated at cathode) |
| 5. | Stable to acidity (generated at anode) |
| 6. | No strong adsorption on electrocatalysts |
| 7. | No wetting or flooding of gas-fed electrode (mass transport limitation) |
| 8. | Stable under reducing potentials (in presence of $H_2$) on Pt |
| 9. | Stable under oxidizing potentials (in presence of $O_2$) on Pt catalysts |

oxygen gas. The first scan shows high currents for oxygen reduction, but the current rapidly diminishes with every subsequent scan. So the 20% improvement at low current density lasts only a short time, because the EAN decomposes. In fact, it is difficult to distinguish whether the 20% fuel cell improvement with EAN versus phosphoric acid is due solely to electrolyte properties of EAN or if EAN is a reactant or is generating radicals to initiate a radical-catalyzed hydrogen anode or cathode. Ethyl ammonium nitrate (EAN) has some outstanding liquid electrolyte attributes. It is a virtually nonvolatile proton conductor at high temperature with no humidification. A hydrogen/oxygen fuel cell with EAN electrolyte gives strikingly better short-term fuel cell performance than with one of the best liquid fuel cell electrolytes, concentrated phosphoric acid. Still, EAN is not an acceptable fuel cell electrolyte. It wets Teflon and has limited oxidative stability.

In general, an electrolyte must fulfill many requirements for use in an acid fuel cell. These criteria are listed in Table 11.2.

Next-generation PILs have been made to determine whether the deficiencies of EAN could be overcome, while retaining the outstanding attributes of EAN. Since oxidative stability is an important issue, the logical approach is to make the protic ionic liquid (PIL) from an acid and base, each of which having good stability on Pt at high electrode potentials. The most likely candidates are inorganic materials and fluorocarbons.

New liquid proton transfer ionic liquid (PIL) electrolytes were formed by mixing more stable inorganic acid and base moieties [12]. Figure 11.11 shows the temperature-dependent conductivity with no humidification up to 120°C. Phosphoric acid (98%) and EAN are also shown for comparison. The conductivity of the new liquid electrolytes can be quite high.

Figure 11.12 shows the short-term polarization curves of fuel cells with porous Pt-catalyzed gas-fed electrodes fed hydrogen to the anode and oxygen to the cathode (setup shown in Fig. 11.8) when each of the new PILs was used as the electrolyte. Phosphoric acid and EAN are also shown for comparison.

The best fuel cell performance with a PIL electrolyte is when EAN is the electrolyte. A fuel cell with one new electrolyte, 60 ammonium nitrate : 40 ammonium triflate ($NH_4NO_3$–$NH_4Tf$ 60 : 40 mol%), gives performance comparable to that of EAN at low current densities. But the fuel cell with the new $NH_4NO_3$–$NH_4Tf$ PIL electrolyte

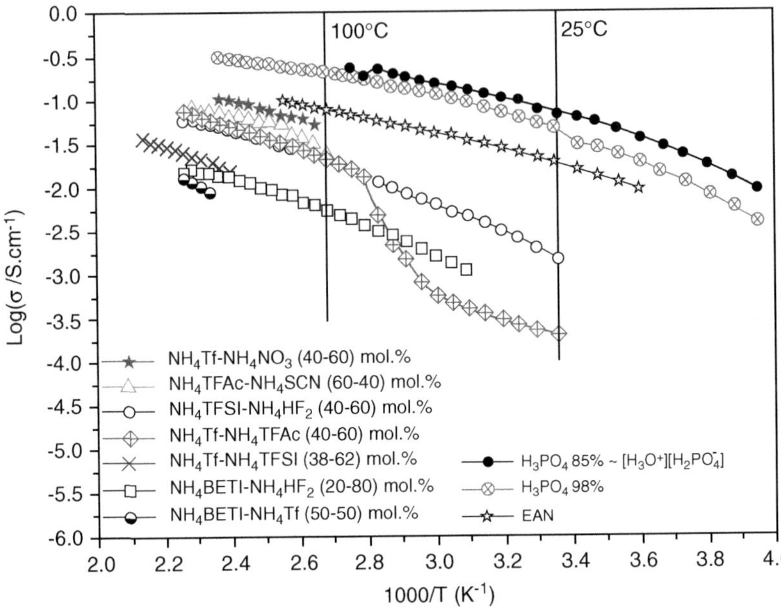


**FIGURE 11.11** Arrhenius plot of the conductivity of new stable ammonium salt mixtures.

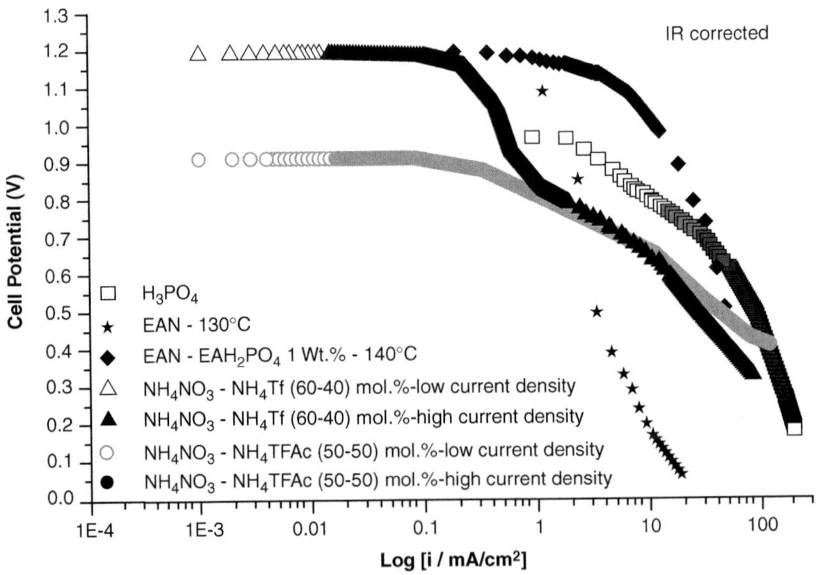


**FIGURE 11.12** Polarization curves for a series of inorganic binary ammonium salts, an organic ammonium salt, and phosphoric acid. The binary ammonium salt mixtures are trifluoromethane sulfonate (triflate, Tf) + trifluoroacetate (TFAc), trifluoroacetate + nitrate, triflate + nitrate.

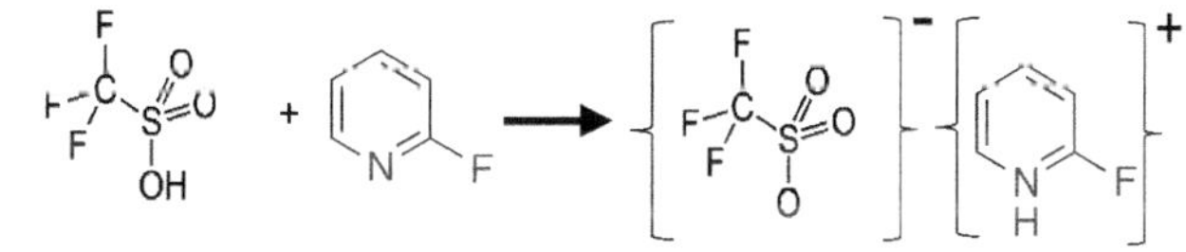


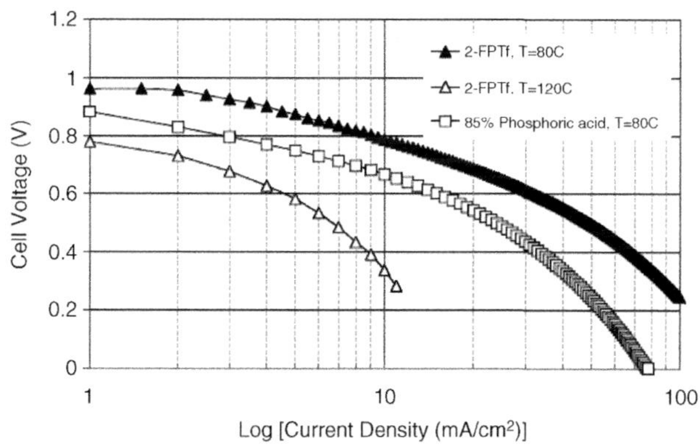


**FIGURE 11.13** *I*/*V* curves for $H_2$ and $O_2$ fed to Pt-catalyzed porous electrodes in 2-FPTf electrolyte at 80°C and 120°C and 85% phosphoric acid electrolyte at 80°C; $s$(2-FPTf) $= 4 \times 10^{-3}$ S/cm, $A = 0.5\ \mathrm{cm}^2$, $t^{\mathrm{electrolyte}} = 0.3$ cm.

suddenly loses cell voltage when the current density is only 0.1 mA/cm$^2$. Since the electrolyte is stable, the most likely reason for the drop in voltage is adsorption of the electrolyte on Pt (requirement 6 in Table 11.2).

A new fluorinated PIL, 2-fluoropyridinium triflate (2-FPTf), is the first PIL electrolyte in a fuel cell that outperforms one with concentrated phosphoric acid electrolyte over all current densities [13]. The 2-FPTf was made by mixing equimolar amounts of 2-fluoropyridine and triflic acid.

Figure 11.13 shows a Pt-catalyzed fuel cell fed dry hydrogen and oxygen gases that was stable at 80°C with neat 2-FPTf as the electrolyte. The fuel cell with 2-FPTf is more than 15% more efficient (>0.1 V cell voltage gain) compared to an identical fuel cell except with aqueous 85% phosphoric acid as the electrolyte. Although 2-FPTf was stable under fuel cell operating conditions at 80°C for an extended time (days), at 120°C the polarization curve for 2-FPTf degraded significantly. The stable performance with 2-FPTf is attributed to two factors: (1) fluorine is substituted for H in reactive CH bonds and (2) it does *not* wet Teflon. The enhanced fuel cell performance is attributed to lower polarization for oxygen reduction, because the salt has little water, resulting in Pt being metallic and catalytically active even at potentials as high as 1.2 V versus RHE.

The lower oxygen reduction polarization is corroborated by the voltammetry of Pt in 2-fluorpydinium triflate at 80°C (Fig. 11.14). The onset of oxygen reduction is near

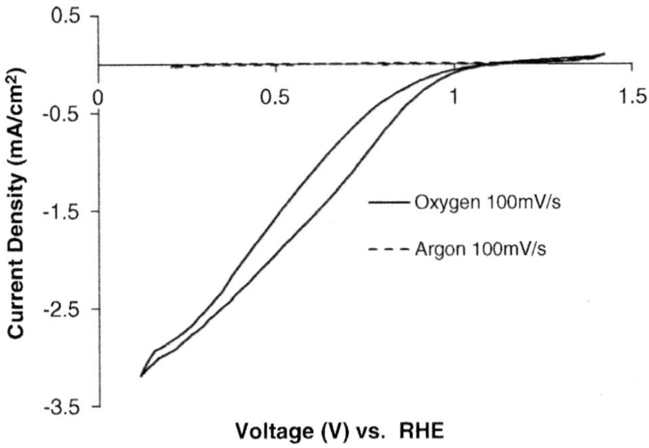


**FIGURE 11.14** Cyclic voltammogram of Pt in 2-FPTf at 80°C under dry oxygen and argon; $P = 1$ atm, scan rate = 100 mV/s; electrode area = 0.1 $cm^2$.

the reversible potential of 1.2 V versus RHE. Unlike that of EAN (Fig. 11.10), the voltammetry of Pt in neat 2-FPTf salt electrolyte equilibrated with 1 atm of argon, and oxygen gas at 80°C was stable for many hours at 80°C. The results with 2-fluoropyridinium triflate electrolyte shows that high fuel cell performance in PIL electrolytes is not dependent on the presence of nitrate in the electrolyte, so a free-radical mechanism (outer-sphere electron transfer) is probably not the origin of high oxygen reduction activity and good fuel cell performance in EAN. Low water activity in the electrolyte, which hinders Pt oxide formation, is the most likely reason for the high performance (potential) of the oxygen cathode and the reduced polarization (high voltage) of the fuel cell.

### 11.3.2 Salt Membranes

There are issues that make using liquid fuel cells unacceptable, especially for portable applications. Liquids can seep out of cells, causing electrical short circuits between cells stacked in series. This short-circuiting of cells leads to voltage and power density drops. Liquids soften structural materials, thus accelerating failure of a fuel cell power source.

All of these issues can be corrected by using solid electrolytes. Thinner solid electrolyte layers keep gases better separated than do thicker liquid electrolyte layers, so a fuel cell stack can be made smaller to ensure a higher power density. Solid electrolytes do not readily soften or corrode cell structures. Solid electrolytes do not leak, preventing power-robbing short circuits. Clearly, solid electrolyte membranes are desired. The new issue is how to conduct proton with no water. Having a solid proton conductor that does not need water would avoid the parasitic power losses that come with humidifiers and allow higher-temperature operation, thus avoiding the need for bulky radiators of waste heat. A water-free solid proton-conducting electrolyte

**TABLE 11.3 Extra Requirements with a Fuel Cell Membrane Electrolyte**

| | |
|---|---|
| 1. | Low permeability to fuel ($H_2$ gas) and oxidant ($O_2$ gas) |
| 2. | Not soluble in water (product of fuel cell) |
| 3. | No electroosmosis (e.g., transfer of water with proton) |
| 4. | Dimensionally stable with change of hydration state |
| 5. | Dimensionally stable with temperature (from −50°C to 230°C) |
| 6. | Good cohesion and adhesion to electrode |
| 7. | Pinhole-free (no electrode short-circuiting or gas blowthrough) |

would also avoid Pt oxide formation and passivation of the Pt catalyst in the cathode that rob cell voltage, power, and efficiency.

In Section 11.3.1.2, there were examples of electrolytes that conducted proton at high temperature with no humidification. Whether similar solid salt electrolytes can be made is the subject of this section.

All the requirements of liquid electrolytes given in Table 11.2 hold for solid electrolytes. There are extra requirements for solid electrolyte membranes; these are listed in Table 11.3.

Perhaps the greatest challenge to meet is requirement 3, for water-free proton conduction, since this is an intrinsic material property. For this reason, the remainder of this chapter stresses solid proton-conducting membranes operating at temperatures higher than the boiling point of water with no humidification of feed gases. To some degree the other requirements can be met with ingenious processing. Still, when making a material for an electrolyte membrane, all seven requirements should be kept in mind.

A solid salt electrolyte membrane would avoid the problems with liquid electrolytes and aqueous electrolyte. A solid salt electrolyte cannot conduct proton or ionic charge by a vehicular mechanism. Some kind of "hopping" mechanism must be operative. Examples of such membranes were virtually nonexistent at modest temperatures until quite recently. There are a few examples of solid protic salt membranes that are similar to the protic ionic liquids discussed in Section 11.3.1.2. It remains a challenge to learn from these examples in order to make a proton-conducting membrane that has all the qualities needed for use in a fuel cell, especially a membrane that has suitable mechanical properties, such as flexibility and toughness, while operating over a broad range of temperatures without the need for humidification.

*The Problem* The most popular fuel cell membrane today is made from Nafion, a perfluorinated organic polymer with pendant sulfonic acid groups covalently attached to the polymer. Inspection of Figure 11.15 reveals that water is needed for proton conduction in this membrane.

Water ionizes the sulfonic acid, and water provides a path for proton conduction between sulfonic acid groups [14]. The need for water in the membrane for proton conduction gives rise to many negative consequences. With too little water, there is no

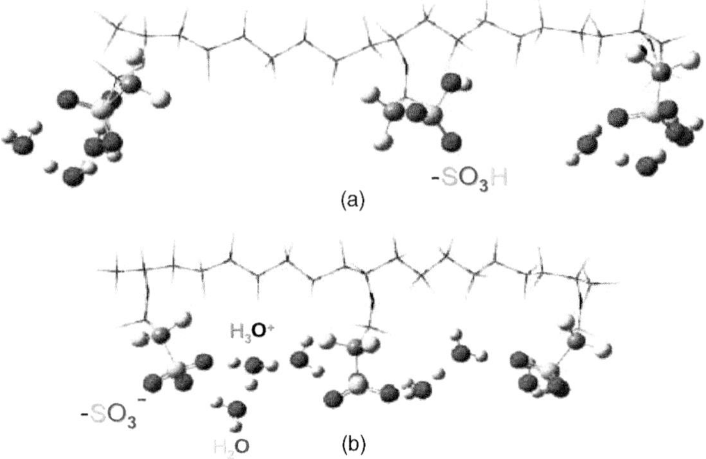


**FIGURE 11.15** Perfluorocarbon sulfonic acid polymer in (a) the dry nonconducting low-water form and (b) proton-conducting high-water (hydrated) form.

proton conduction, because the pendant sulfonic acid groups are neither all ionized nor bridged by water molecules, as shown in Figure 11.15a. With three (or more) waters per acid group, the membrane conducts proton, because water bridges completely ionized acid groups, as shown in Figure 11.15b. Three waters per acid means that *bulklike water* is in the membrane, so the fuel cell operation temperature must be 80°C or lower at atmospheric pressure, and humidification of feed gases is required to retain solvent water; a large radiator is needed to reject waste heat from 80°C to room temperature, and the performance of the platinum-catalyzed air cathode is poor. Pt catalysis suffers, because at high potentials, platinum oxides form in the presence of the bulklike water, which inhibits air–cathode activity until the metallic Pt is restored when the potential drops to 0.8 V or lower, so the fuel cell efficiency can be no higher than ~65% or lower under load.

*Polymeric Salt Membrane* The need for proton-conducting electrolytes for fuel cells operating at temperatures of 120°C has prompted the investigation of "water-free" polymeric materials. Nayanan and all have reported on noteworthy water-free proton-conducting membranes prepared from high-molecular-weight polymeric organic amine salts, made from bisulfates and dihydrogenphosphates adducts with poly-2-vinylpyridine (P2VP) and poly-4-vinylpyridine (P4VP), and polyvinylimidazoline (PVI), whose structures are shown in Figure 11.16.

Membranes consisting of the polymers shown in Figure 11.16 were investigated over the temperature range of 25–180°C [15]. These membranes conducted proton at temperatures above the boiling point of water without external hydration. The polyvinylpyridine phosphates were the most conductive of the polymers shown, and their conductivity as a function of temperature is shown in Figure 11.17.

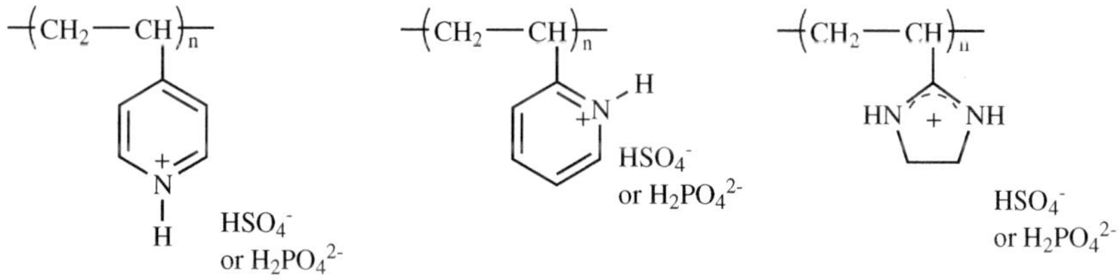


**FIGURE 11.16** Structures of bisulfates and dihydrogenphosphates of poly-4-vinylpyridine [(a), P4VP], poly-2-vinylpyridine [(b) P2VP], and polyvinylimidazoline [(c), PVI].

All polymer membranes were more conductive at higher temperatures, where polymer mobility is no longer a substantial barrier to proton transport. The activation energy for ionic conduction was found to decrease with increasing temperature, and this is attributed the increased polymer mobility and ionization of the proton. This is confirmed by the high degree of motional narrowing that is observed in proton NMR experiments. Narayanan and coworkers reported that the addition of nanoparticulate silica particles increases the conductivity substantially at temperatures where the polymer backbone is not mobile, suggesting that the acid–base properties and the chainlike arrangement of nanoparticulate particles can be used to lower the barriers for proton conduction when polymer motion is hindered.

*Ceramic Salt Membranes* Nagao, et al. have reported that solid ceramic membranes based on tin phosphate doped with various trivalent ions, such as $In^{3+}$, exhibit exceptionally high proton conductivity ($> 0.1\ \Omega^{-1}\,cm^{-1}$) from room temperature to 200°C [16]. Most interesting is the report of a composite of 90% of

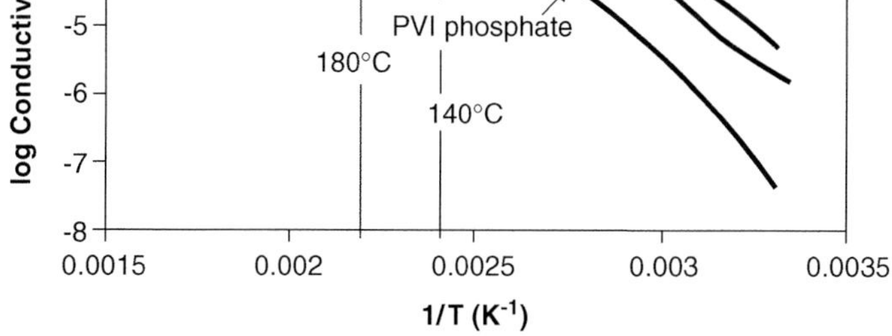


**FIGURE 11.17** Ionic conductivity of dihydrogen phosphates of P2VP, P4VP, and PVI as a function of temperature.

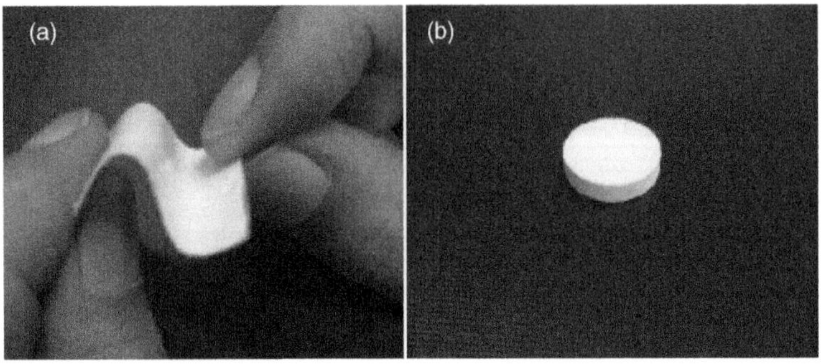


**FIGURE 11.18** Photographs of membrane of (a) 10 wt% organics and 90% ITP composite and (b) a pure ITP pellet.

$Sn_{0.9}In_{0.1}P_2O_7$ (ITP) blended with 10% of an equimolar mixture of two organic compounds. This composite is formed into a flexible membrane with high proton conductivity over a wide range of temperatures with no humidification [17]. The two organic compounds are 1,8-bis(triethoxysilyl)octane (TES-Oct) and 3-(trihydroxysilyl)-1-propane sulfonic acid [(THS)Pro-$SO_3H$].

Figure 11.18 shows this flexible inorganic/organic composite membrane in curves (a) and the brittle membrane consisting of 100% ceramic ITP in curves (b). The inorganic/organic composite membrane shown in Figure 11.18 has a conductivity of about 0.04 S/cm at 150–200°C with no humidification. More organic component gives more flexibilty to the composite. More ITP gives higher conductivity of the membrane. The membrane with 90 wt% ITP and 10 wt% organic component represents the optimal ratio of inorganic ITP and organic organic component for making a membrane with both good flexibility and the highest conductivity.

The conductivity of the organic component with an equimolar blend of the TES-Oct and (THS)Pro-$SO_3H$ components is humidity-dependent [18]. It is intriguing to consider what the performance of an inorganic ITP/organic composite membrane would be like if the organic component in the composite were a humidity-independent organic polymer membrane, like the polyvinylpyridine phosphates of Narayanan and all [15].

Figure 11.19 shows a fuel cell made of a flexible composite membrane consisting of 90 wt% of $Sn_{0.9}In_{0.1}P_2O_7$ and 10 wt% of an equimolar mixture of TES-Oct and (THS) Pro-$SO_3H$ operating at 150°C using unhumidified $H_2$ at the anode and unhumidified air at the cathode both with and without an intermediate layer between the membrane and electrodes. The peak power density at 0.4 cell voltage is 149 mW/cm$^2$ when the electrode is directly pressed on the membrane and is 197 mW/cm$^2$ when there is an intermediate layer between the membrane and electrodes. This is an impressive result, as the fuel cell operates at high temperature and ambient pressure with no humidification. There is still a long way to go toward development of a practical fuel cell, but this is a good start and suggests that more research on solid salt electrolyte membranes is warranted.

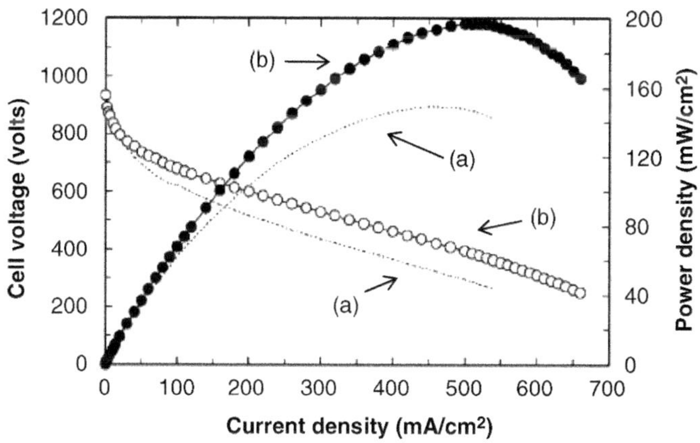


**FIGURE 11.19** Cell voltage and power density versus current density of fuel cells with the composite membrane of 90 wt% of $Sn_{0.9}In_{0.1}P_2O_7$ and 10 wt% of an equimolar mixture of TES-Oct and (THS)Pro-$SO_3H$ at 150°C. Unhumidified $H_2$ was supplied to the anode and air to the cathode at a flow rate of 30 mL/min. The electrolyte thickness was 60 μm. The light lines (a) indicate no intermediate layer between electrodes and membrane; the lines with circle markers (b) indicate an intermediate layer of 20 parts ITP: 1 part organic between electrodes and membrane. Pt loading = 0.6 mg/cm$^2$.

## 11.4 CONCLUSIONS

Protic salt electrolytes are formed by proton transfer from an acid to a base whose (1) $\Delta pK$ is greater than 4 but less than 14 and (2) symmetry is sufficiently high that rotational and vibrational motions are not impeded. Salts that are made to satisfy these two conditions can be used as proton-conducting electrolytes over a broad range of temperatures with no need for solvent water.

Better performance was obtained for a hydrogen/oxygen fuel cell with a liquid salt electrolyte, 2-fluoropyridinium triflate (2-FPTf), as the electrolyte instead of phosphoric acid. A solid organic polymer, poly(vinylpyridinium phosphate) (PVPP), has an immobilized salt yet has a proton conductivity of $5 \times 10^{-3}$ S/cm at 140°C with no "free" salt or solvent water in the membrane. Solid ceramic indium tin phosphate (ITP) and ITP blended with organic additive protons in fuel cells operate over an even wider range of temperatures (100–200°C). The organic/inorganic membrane is flexible and at a cell voltage of 0.4 V demonstrate a peak power density of 149 mW/cm$^2$ when the electrode is directly pressed on the membrane and 197 mW/cm$^2$ when there is an intermediate layer between the membrane and the electrodes.

These results with water-free solid salt electrolytes are encouraging in that they well meet requirement 3 of Table 11.3, which is that there is no need for water transfer during flow of proton current. There are still many requirements that need to be addressed before a protic salt membrane appears in a commercial fuel cell including the requirements pertaining to mechanical strength, optimization of the electrode/membrane interface, and adhesion. These "water-free" electrolytes could lead to better air cathode and fuel cell performance by impeding Pt oxide formation and retaining metallic Pt at high cathode potentials provided the membrane adsorption on catalyst is weak.

Since these electrolytes are less corrosive than aqueous electrolytes, using them "opens the door" to using fewer noble metal catalysts than needed for Pt. This is a particularly interesting topic that has hardly been explored.

The solid salt electrolyte need not be soluble in water, so investigating cold starting or operation in the presence of liquid water would seem wise. However, normal operation at temperatures above 120°C is not only possible but more desirable, as this simplifies the fuel cell system and improves its efficiency. Above 120°C, fluid handling is simplified, since only free-flowing gases (hydrogen, air, and water vapor) are involved. Thermal management is simplified because waste heat is readily rejected from the cell at 120°C to room-temperature air. The parasitic losses are reduced, since the power-demanding humidifier, used in room-temperature fuel cells, is no longer needed.

The use of salt electrolytes in fuel cells is new, and there are prospects for better electrolytes and fuel cells as better salt electrolytes are found.

## REFERENCES


1. C. M. Hall, Process of Reducing Aluminium by Electrolysis, US Patent 400,766, filed July 9, 1886, patented April 2, 1889; Paul T. Héroult, of Paris, FR 175,711; April 23, 1886.
2. (a) F. H. Hurley, Electrodeposition of Aluminum, US Patent 4,446,331 (1948); (b) F. H. Hurley and T. P. Wier, The electrodeposition of aluminium from nonaqueous solutions at room temperature, *J. Electrochem. Soc.* **98**, 207–212 (1951); (c) F. H. Hurley and T. P. Wier, Electrodeposition of metals from fused quaternary ammonium salts, *J. Electrochem. Soc.* **98**, 203–206 (1951); (d) T. P. Wier, Jr. and F. H. Hurley, US Patent 4,446,349 (1948).
3. R. A. Osteryoung, and B. J. Welch, Electrochemical studies in low temperature molten salt systems containing aluminum chloride, *J. Electrochem. Soc.* **118**, (1981), 455.
4. J. L. Anthony, J. F. Brennecke, J. D. Holbrey, et al., Physicochemical properties of ionic liquids, in *Ionic Liquids in Synthesis*, P. Wasserscheid and T. Welton, eds., Wiley-VCH Verlag GmbH, Weinheim, 2002, Chapter 3, pp. 41–126.
5. (a) W. Xu and C. A. Angell, Solvent-free electrolytes with aqueous solution-like conductivities, *Science* **302**, 422–425, (2003); (b) J.-H. Shin, W. A. Henderson, and S. Passerini, *Electrochem. Commun.* **5**(12),1016–1020 (2003).
6. P. Walden, *Bull. Acad. Imper. Sci. St. Petersbourg*, **68**, 405–422 (1914).
7. R. W. Gurney, *Ionic Processes in Solutions*, Dover Publications, New York, 1953.
8. J.-P. Beleieres and C. A. Angell, Protic ionic liquids: Preparation, characterization, and the proton transfer energy level diagram, *J. Phys. Chem. B* **111**(18), 4926–4937 (2007).
9. P. Walden, *Bull. Acad. Imper. Sci. St. Petersbourg* **68**, 1800 (1914).
10. C. A. Angell, Fast ion motion in glassy and amorphous materials, *Solid State Ion.*, **9/10**, 3 (1983).
11. M. A. B. H. Susan, A. Noda, S. Mitsushima, and M. Watanabe, Brønsted acid–base ionic liquids and their use as new materials for anhydrous proton conductors, *Chem. Commun.* **2003**, 938–939 (2003).

12. J.-P. Belieres, D. Gervasio, and C. A. Angell, Binary Inorganic salt mixtures as high conductivity electrolytes for >100°C fuel cells", *Chem. Commun.* 4799–4801 (2006).

13. J. Thomson, P. Dunn, L. Holmes, J.-P. Belieres, C. A. Angell, and D. Gervasio, A fluorinated ionic liquid as a high-performance fuel cell electrolyte, *Electrochem. Soc. Trans.* **13**(28), 21 (2008).

14. S. J. Paddison and J. A. Elliott, Selective hydration of the "short-side-chain" perfluorosulfonic acid membrane. An ONIOM study, *Solid State Ion.* **178**(7–10), 561–567 (2007).

15. S. R. Narayanan, S.-P. Yen, L. Liu, and S. G. Greenbaum, Anhydrous proton-conducting polymeric electrolytes for fuel cells, *J. Phys. Chem. B* **110**, 3942–3948 (2006).

16. (a) M. Nagao, A. Takeuchi, P. Heo, T. Hibino, M. Sano, and A. Tomita, A proton conducting $In^{3+}$ doped $SnP_2O_7$ electrolyte for intermediate-temperature fuel cells, *Electrochem. Solid State Lett.* **9**, A105 (2006); (b) P. Heo, H. Shibata, M. Nagao, T. Hibino, and M. Sano, Performance of an intermediate-temperature fuel cell using a proton-conducting $Sn_{0.9}In_{0.1}P_2O_7$ electrolyte, *J. Electrochem. Soc.* **153**, A897 (2006); (c) M. Nagao, T. Kamiya, P. Heo, A. Tomita, T. Hibino, and M. Sano, Proton conduction in $In^{3+}$ doped $SnP_2O_7$ at intermediate temperatures, *J. Electrochem. Soc.* **153**, A1604 (2006); (d) M. Nagao, P. Heo, T. Kamiya, T. Hibino, M. Sano, and A. Tomita, Proton conduction in $In^{3+}$ doped $SnP_2O_7$ with various P/(Sn + In) ratios, *Electrochem. Soc. Trans.* **2**, 43 (2007); (e) A. Tomita, Y. Namekata, M. Nagao, and T. Hibino, Room-temperature hydrogen sensors based on an $In^{3+}$ doped $SnP_2O_7$ proton conductor, *J. Electrochem. Soc.* **154**, J172 (2007); (f) A. Tomita, N. Kajiyama, T. Kamiya, M. Nagao, and T. Hibino, Intermediate-temperature proton conduction in $Al^{3+}$ doped $SnP_2O_7$, *J. Electrochem. Soc.* **154**(12), B1265–B1269 (2007); (g) A. Tomita, Y. Namekata, M. Nagao, and T. Hibino, Room-temperature hydrogen sensors based on an $In^{3+}$ doped $SnP_2O_7$ proton conductor, *J. Electrochem. Soc.* **154**(5), J172–J176 (2007).

17. P. Heo, M. Nagao, T. Kamiya, M. Sano, A. Tomita, and T. Hibino, $Sn_{0.9}In_{0.1}P_2O_7$ -based organic/inorganic composite membranes application to intermediate temperature fuel cells, *J. Electrochem. Soc.* **154**, B63–B67 (2007).

18. O. Nishikawa, T. Sugimoto, S. Nomura, K. Doyama, K. Miyatake, H. Uchida, and M. Watanabe, *Electrochim. Acta* **50**, 667 (2004).

CHAPTER 12

# Vibrational Spectroscopy for the Characterization of PEM Fuel Cell Membrane Materials

CAROL KORZENIEWSKI

Department of Chemistry and Biochemistry, Texas Tech University, Lubbock, Texas

Applications of vibrational spectroscopy techniques in the study of proton exchange membrane (PEM) materials are discussed. The focus is on approaches for probing ionomer structure, membrane hydration state, and the environment within membrane pores and channels. The use of steady-state mid-infrared and Raman methods and ultrafast dynamic vibrational lifetime measurements are highlighted.

## 12.1 INTRODUCTION

Proton exchange membrane (PEM) fuel cells continue to be of interest for their potential as high-efficiency, zero-emission electrical power sources capable of operation at relatively low ($<\sim 120°C$) temperatures [1–4]. Improving the properties of the membrane that separates the anode and cathode compartments has been among the challenges for advancing PEM fuel cells. Membranes must possess high proton conductivity and mechanical stability, and limit the crossover of reactant fuels [1–3]. In striving to attain these properties, historically investigators have tended to focus on perfluorosulfonic acid ionomers (Scheme 12.1). Investigations of Nafion [5], in particular, have laid foundations on which the effort to develop new materials has been built [1,2,4].

To support advances in PEMs, several physical characterization methods have been employed, enabling structure–function relationships to be probed across length scales

*Fuel Cell Science: Theory, Fundamentals, and Biocatalysis*,
Edited by Andrzej Wieckowski and Jens K. Nørskov

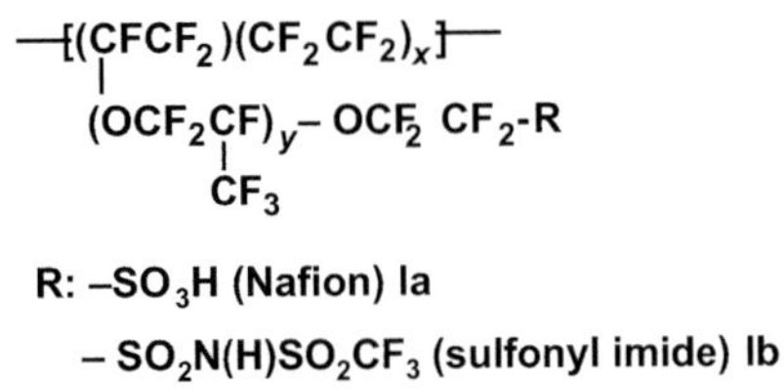


**SCHEME 12.1** Structures for the protonated forms of Nafion (a) and bis[(perfluoroalkyl)-sulfonyl] imide (PFSI) (b) ionomers.

(atomic to macroscopic). Mauritz and Moore described approaches for the characterization of ionomer materials in a review of Nafion [5]. Vibrational spectroscopy techniques were mentioned in terms of their ability to enable nondestructive probing of membranes and to provide insights into the environment surrounding chemical functional groups and the properties of imbibed water. The present chapter considers advantages, limitations, and prospects for greater use of vibrational spectroscopy in the development of new fuel cell membranes. Steady-state mid-infrared and Raman methods and ultrafast dynamic vibrational lifetime measurements are discussed.

## 12.2 OVERVIEW OF PEM MATERIALS

As several reviews bring to light, the first PEMs were based on hydrocarbon polymers that suffered degradation during fuel cell operation [1–3]. The development of fluorinated ion exchange polymers was a major step forward for PEM fuel cell technology [1–3]. Scheme 12.1a shows the general structure for a class of perfluorosulfonic acid ionomers, of which Nafion ($y = 1$ and $x \sim 5$–14) is among the most common forms. With the aim of increasing membrane durability and performance, a wide range of other materials have been investigated [1,2,6–8]. As depicted in Scheme 12.1b, one type of variation builds on the perfluorosulfonic acid structure, but substitutes for the sulfonate group other acidic functionality [7]. For the majority of alternative membranes, variation in the polymer backbone has been explored [2,6]. The introduction of aromatic, ether, ketone, sulfone, and imide, among other groups, has been investigated [2], along with copolymers, blends, and organic–inorganic composites [4,6,8].

A step in the characterization of fuel cell membrane materials is the determination of equivalent weight (EW), a figure of merit that provides information about ionomer acid site content [2,5]. For perfluorosulfonic acid ionomers, EW is calculated as the grams of dry proton exchanged polymer per mole of sulfonic acid groups [5]. The latter quantity is often determined by titration. Nafion membranes with EW 1100 g/mol, which is consistent with $x = 6$–7 in Scheme 12.1a, have a long history of use in PEM fuel cell studies [3,5]. Discussions in this chapter focus on Nafion because of the central role it has played in the development of PEM fuel cells [1–3] and its use as a model in studies of membrane physicochemical properties [5,9–11]. The general role of vibrational spectroscopy in the characterization of new PEM materials, however, should not be overlooked [6,12–14].

## 12.3 EXPERIMENTAL APPROACHES

Most often, vibrational spectra of fuel cell membrane materials have been obtained through the use of mid-infrared spectroscopy combined with either attenuated total reflection (ATR) or transmission sampling [5,9]. In early work, Falk pointed out the advantages of transmission relative to ATR measurements when quantitative information in relation to spectral band properties (i.e., intensities, positions, shapes) is desired [9]. In particular, bands in ATR spectra suffer distortion due to the wavelength-dependent depth of radiation penetration into samples [15–17]. However, transmission methods can be limited, because absorbance bands for the fundamental vibrational modes of ionomer materials become saturated when the membrane thickness is greater than $\sim$1 μm [9,19,20]. ATR methods enable ionomer spectra to be recorded from membrane samples that have thicknesses >1 μm, including those that have more practical use in fuel cells where the thickness ranges from tens to hundreds of micrometers [2,5]. Furthermore, versatile ATR cells have been developed that allow rapid exchange of vapors with the membrane [10,16,18]. The sections below discuss the development of strategies for infrared transmission and ATR sampling of PEM materials. Also mentioned are more specialized approaches that include in situ techniques for probing membrane–electrode assemblies [21–23], Raman measurements [17,24], and ultrafast infrared vibrational lifetime measurements [25].

### 12.3.1 Infrared Transmission Sampling

In early studies of Nafion membranes, Falk and coworkers applied transmission sampling to investigate water uptake [9,26,27]. Although the membrane was too thick ($\geq$50 μm) to enable observation of vibrational bands arising from the fundamental modes of Nafion ionomer, it was possible to detect water present inside membrane pores and channels without limitations from absorbance saturation. In addition to spectral features of condensed, bulklike water, bands for water interacting with polymer functional groups appeared. In subsequent work, descriptions were reported for infrared transmission cells that permitted control of the humidity [11,19,24,25,28] and pressure [24] of atmospheric gases surrounding the membrane under study and enabled measurements to be performed on either freestanding or thin cast membrane samples [11,19,24,25,28]. More recently, advances were made in methods for equilibration and infrared probing of membranes in atmospheres of controlled humidity, including allowance for $D_2O$ additions to produce isotopically labeled water vapor [11,25].

A drawing of a transmission cell configured for use with freestanding membrane is shown in Figure 12.1. Components stack inside an aluminum ring attached to the backplate of a commercial infrared transmission cell [20,29,30]. The windows are held parallel in the cell and separated a fixed distance ($\sim$1 cm) by a pair of spacers. In addition to positioning the windows, the spacers also sandwich the membrane and hold it stationary in the infrared beam path. The spacers depicted include a port for entry of a 23-gauge syringe needle and a shallow well beneath the port to hold a droplet of humectant solution [19,28]. To ensure that the area of polymer sampled completely

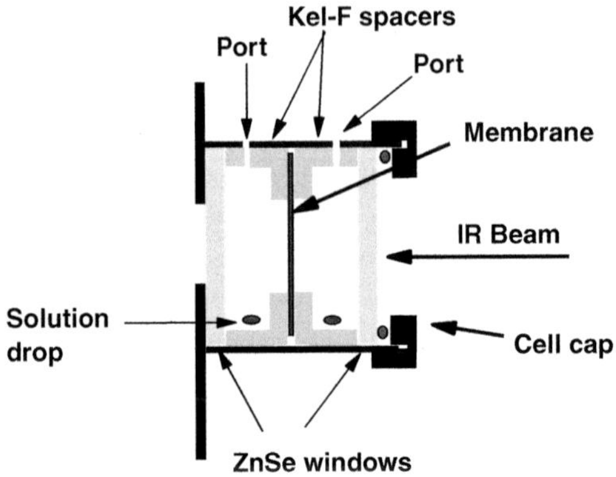


**FIGURE 12.1** Diagram of a cell employed for transmission infrared sampling of freestanding membrane materials.

fills the infrared beam, a circular aperture constructed from aluminum is placed adjacent to the backplate, where it will be in close proximity to the infrared beam focus when the cell is placed in the spectrometer sample compartment. Once assembled, a threaded aluminum cap mounts to the ring and secures the components of the cell.

The cell described in Figure 12.1 was designed around 32-mm-diameter × 3-mm-thick windows, although thinner, and smaller diameter windows can be utilized, provided the dimensions of all other cell components are scaled to match. A variety of widow materials can be employed. $CaF_2$ and ZnSe are often used, because they provide a wide spectral range and are insoluble in water and common organic solvents. In addition to use with freestanding membrane, the cell in Figure 12.1 can also be adapted for measurements on thin (<1-μm) films prepared by casting directly onto one of the windows [19,20].

### 12.3.2 Infrared Attenuated Total Reflectance (ATR) Sampling

Conventional ATR sampling techniques are frequently applied to obtain spectra of PEM materials [10,17,18,31]. Solutions of ionomer can be cast directly onto the ATR crystal to form an adherent film [10]. Alternatively, freestanding membrane can be pressed against the ATR crystal; however, care must be taken to ensure that the surfaces maintain contact and to avoid inducing stress in the membrane [16,17,18]. Also, if spectral measurements are to be performed with the membrane under controlled atmosphere, in securing the sample to the ATR crystal a method must be used that does not block the sample–fluid interface and slow the rate of equilibration. Advances have been made in recent years in developing ATR cells for the study of fuel cell membrane [16,18].

Blanchard and Nuzzo demonstrated an ATR flow cell for investigating cast ionomer films [10]. A uniform film was deposited on one side of the ATR crystal by spin coating. After the cell was assembled, infrared spectra were recorded to monitor film structure

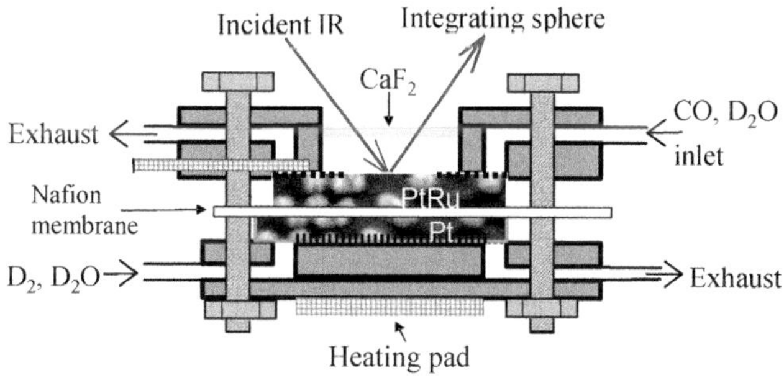


**FIGURE 12.2** Experimental arrangement for performing in situ infrared spectal measurements on operating PEM fuel cells. Source radiation is directed into the cell through a $CaF_2$ window that mounts over an opening in the gas diffusion layer. Diffuse reflectance optics are used to collect radiation emitted from the membrane–electrode assembly. (From Ref. 34, with permission.)

as gas of controlled humidity passed over the sample. Films were made thin ($\sim$1 μm) to allow rapid exchange of ions prior to experiments and equilibration of gases during spectral measurements. More recently, Hallinan and Elabd [16] and Grosmaire and coworkers [18] described ATR flow cells for use with freestanding membrane. The cell developed by the latter group was constructed around a stainless-steel chamber that enabled spectra to be recorded under controlled temperature and pressure [18].

### 12.3.3 Infrared Diffuse Reflectance Sampling

Smotkin and coworkers pioneered an approach for obtaining infrared spectra from membrane electrode assemblies (MEAs) in operating fuel cells [23,32,33]. A thermostatted single fuel cell was modified to accommodate an infrared transparent window (Fig. 12.2) [32,34]. The window mounts over an opening in the gas diffusion layer, allowing an incident infrared beam access to the MEA. Diffuse reflectance optics are used to collect radiation emitted from the MEA. The researchers have used the cell mainly to investigate reactions over Pt and Pt alloy anode catalysts during direct methanol fuel cell operation [23,32–35]. While studies have focused on the observation of spectral bands arising from methanol oxidation byproducts, the system also provides a means to probe the membrane–catalyst–gas three-phase boundary [34].

### 12.3.4 In Situ Infrared Spectroelectrochemical Sampling

Traditional reflectance infrared spectroelectrochemistry and surface-enhanced infrared spectroscopy (SEIRAS) techniques have been applied to probe properties of Nafion cast as a thin layer on Pt electrodes [21,22]. Infrared reflectance spectroscopy measurements of thin organic layers are potentially limited by optical dispersion [36–38]. The resulting band distortions can complicate spectral interpretation

[36,37]. However, it has been shown that careful analysis of spectra from reflectance infrared spectroelectrochemistry measurements provides insights into ionomer–catalyst interactions and effects of applied potential on membrane ion and solvent content [21].

The SEIRAS method is potentially powerful for in situ studies of ionomer films on metal electrodes [22]. As has been discussed in several reports on the SEIRAS technique [39–43], the ATR-based optical sampling arrangement lifts restrictions on mass transport imposed by the mandatory thin-layer cell configuration required for in situ reflectance sampling. SEIRAS can provide a means to probe changes in ionomer structure accompanying the flux of solvent and ions into the film as electrode potential is varied. SEIRA spectra of Nafion cast onto Pt have been reported and shown to have many features in common with standard transmission and ATR spectra of Nafion membrane materials [22]. A possible drawback of SEIRAS, however, is that the electromagnetic enhancement is confined to within a few nanometers ($\sim$8 nm) of the metal [44,45], which may result in nonuniform probing of thin films.

### 12.3.5 Vibrational Lifetime Measurements

Fayer and coworkers have been applying vibrational lifetime measurements to probe local environments inside ionomer membrane [11,25]. The ultrafast pump–probe measurements determine the vibrational population relaxation with subpicosecond time resolution (see Section 12.4.3). Their experiments examine the O–D stretching fundamental of HOD present at a few percent in $H_2O$. In studies of Nafion fuel cell membrane, analysis of the vibrational excited-state relaxation responses for HOD as a function of membrane water content showed evidence of water molecules in multiple types of environments inside Nafion, and the characteristics of the water ensembles changed with membrane hydration state. Modeling of the population decays provided greater insight into the distribution of water environments inside Nafion pores and channels than was obtained by steady-state infrared spectroscopy.

### 12.3.6 Raman Measurements

Applications of Raman spectroscopy in the study of fuel cell membrane materials have been limited. High-signal-to-noise ratio (SNR) Raman spectra have been reported for Nafion 117 samples [17,24]. In these cases, excitation at 1064 nm was used. An advantage of the 1064-nm source is that it greatly reduces interference from background fluorescence [46]. However, Raman instruments capable of operation at 1064 nm, which are typically Fourier transform Raman spectrometers, are not widely available. Instruments that achieve the multiplex advantage through the use of array detectors sensitive to visible wavelengths are becoming more popular. As advances are made in near-infrared detector technology, access to Raman instruments equipped with near-infrared sources will likely improve and stimulate more widespread use of Raman spectroscopy in investigations of fuel cell membranes.

## 12.4 VIBRATIONAL SPECTROSCOPY OF NAFION AND RELATED IONOMER MEMBRANE MATERIALS

### 12.4.1 Nafion Mid-Infrared Spectra

A notable characteristic of perfluorosulfonate membrane mid-infrared spectra is the bands for the fundamental polymer modes appear at low energies, below ~1450 $cm^{-1}$. The region above 1450 $cm^{-1}$ is mostly free of polymer bands and enables observation of features from permeating vapors [10,19,20,26,27]. An example is shown in Figure 12.3. The spectrum labeled "dry" in Figure 12.3 was recorded after a freshly cast Nafion film was dried under conditions that remove most of the water. The bands below 1450 $cm^{-1}$ arise from polymer vibrational modes. Above 1450 $cm^{-1}$, the baseline is fairly flat and contains only a few weak features that can be assigned to vibrations of trace water in the membrane and combination bands of the polymer. After the spectrum labeled "dry" was recorded, a water droplet was added to the cell to produce a water-vapor-saturated atmosphere adjacent to the membrane. The top spectrum in Figure 12.3 was obtained 20 min after exposing the Nafion film to the humid atmosphere. Bands in the region between 3000 and 3800 $cm^{-1}$ and near 1647 $cm^{-1}$ arise from vibrations of water and are similar to those of bulk liquid water [20,24].

Spectral bands of Nafion and imbibed water are sensitive to the hydration level and nature of the cation exchanged into the membrane [9,10,17,19,20,24,26–30,47–50]. Traditional assignments for the main Nafion polymer bands are summarized in Table 12.1. Although Nafion vibrational band assignments were reported almost

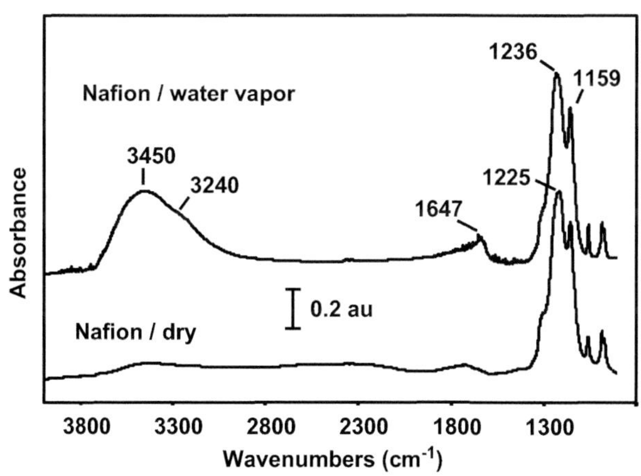


**FIGURE 12.3** Transmission infrared spectra of a thin (~1 μm) Nafion film cast onto a ZnSe window. Spectra were recorded of the film at room temperature after drying at 110 °C and ambient pressure for approximately16 h to remove water (lower spectrum) and subsequent exposure (20 min) to water vapor saturated air (upper spectrum). (From Ref. 20, with permission.)

**TABLE 12.1 Traditional Nafion Infrared Band Assignments**

| Band Region ($cm^{-1}$) | Assignment[a] |
|---|---|
| 3200 | O–H stretching ($-SO_3H$ group) |
| 1435 | S=O antisymmetric stretching ($-SO_3H$ group) |
| 1320 | $-SO_3^-$ antisymmetric stretching + S=O symmetric stretching ($-SO_3^-$ group) |
| 1305 | $-CF_3$ antisymmetric stretching |
| 1231 | $-CF_2$ antisymmetric stretching |
| 1157 | $-CF_2$ symmetric stretching |
| 1057 | $-SO_3^-$ symmetric stretching |
| 982 | C–O–C symmetric stretching |
| 972 | C–O–C symmetric stretching (terminus)[b] |

[a] Based on assignments for Nafion (in Refs. 9,10,17,19,24,28,48, and 51).
[b] Sidechain ether group closest to the sidechain endgroup [48].

30 years ago [9], new insights have continued to emerge [10,47,48,51]. More recently, Warren and McQuillan made refinements through study of a model Nafion sidechain compound [51]. The work combined infrared spectral measurements and density functional theory (DFT) vibrational mode analysis. An important finding was that modes dominated by C–F stretching also contain coupled motion of atoms in the C–O–C and $-SO_3^-$ functional groups. The results help explain some of the complex features that appear in Nafion infrared spectra and the unexpected changes that have been observed in response to membrane hydration and dehydration. The sections below consider mid-infrared spectral features of Nafion and hydrating water molecules in greater detail.

## 12.4.2 Nafion Ionomer Vibrational Bands

Figure 12.4 shows transmission infrared spectra of Nafion between 1400 and 950 $cm^{-1}$. The major fundamental modes for perfluorosulfonate species are present in the region, although a few additional bands associated with low-energy vibrations of C–F groups appear in spectra that extend to energies below 950 $cm^{-1}$ [17,28,49]. The spectra in Figure 12.4 were obtained from a Nafion thin film exchanged by $Na^+$ ions. Related features are present in spectra of proton-exchanged Nafion [19]. A notable exception is in modes of the $-SO_3^-$ group, which shift into the $-SO_3H$ form at low hydration levels in proton-exchanged Nafion. The presence of $-SO_3H$ groups causes bands near 1410 $cm^{-1}$ and 910 $cm^{-1}$ associated with S–OH stretching to develop [19,24,29,49] (see Fig. 12.5 and discussion below). The series of spectra in Figure 12.4 were recorded during hydration of the Nafion film. The initial spectrum was recorded just prior to exposing the film to humid air. Arrows next to the major bands indicate the direction of intensity change with water uptake.

The strongest features in Figure 12.4 appear at 1235 and 1159 $cm^{-1}$ and are attributed primarily to C–F stretching modes of $-CF_2$ groups [9,10,17,21,28]. In the 1350–1200-$cm^{-1}$ region, the side features are believed to originate from the

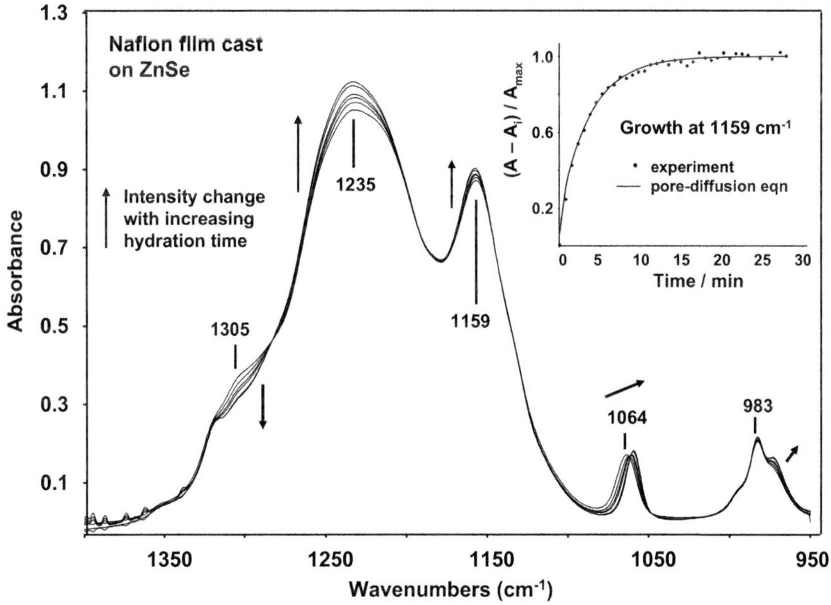


**FIGURE 12.4** Experimental infrared spectra of a thin (~1-μm) Nafion film cast onto a ZnSe optical window recorded during hydration in 100% relative humidity air starting from a dry state. Spectra were recorded at times of 0, 0.75, 1.5, 2.25, 6.75, and 28 min after exposure to humid air. The arrows indicate the direction of band intensity change with increasing hydration. The inset is a plot of the growth in the relative intensity in the band at 1159 $cm^{-1}$, where $A_i$ is the absorbance of the band in the initially dry membrane and $A_{max}$ is the absorbance at maximum intensity. The points in the plot are from the experiment and the solid line is the fit to Equation (12.1), assuming a linear (Beer's law) relationship between $N$ and the absorbance changes at 1159 $cm^{-1}$. (From Ref. 30, with permission.)

asymmetric S–O stretching modes of the $-SO_3^-$ groups [9,10,20,21,28]. With water uptake, the 1159-$cm^{-1}$ band grows, but the position remains fixed, while the 1235-$cm^{-1}$ band undergoes more complex changes [10,20]. The sensitivity to water uptake of the bands associated with hydrophobic fluorocarbon groups in Nafion has been discussed in terms of possible contributions from –CF moieties in the sidechains affected by the entry of water molecules into membrane regions rich in $-SO_3^-$ groups [10,20]. Near 1235 and 1300 $cm^{-1}$, hydration of the $-SO_3^-$ groups is likely a factor influencing the changes [9,10,20,21,28]. The coupling of atomic motions of hydrophilic functional groups into C–F vibrations, as discussed by Warren and McQuillan [51], may also play a role.

The band near 1064 $cm^{-1}$ arises from the symmetric S–O stretching mode of $-SO_3^-$ groups [9,19,47,48]. The spectral region was the focus of early infrared measurements that probed hydration in perfluorosulfonate ionomers [47]. Nafion membranes exchanged by alkali metal cations were studied. In the hydrated state, the band was present at ~1058 $cm^{-1}$ and shifted toward higher energy and became broader as the membrane water content decreased [47]. The trend is evident in Figure 12.4 for the

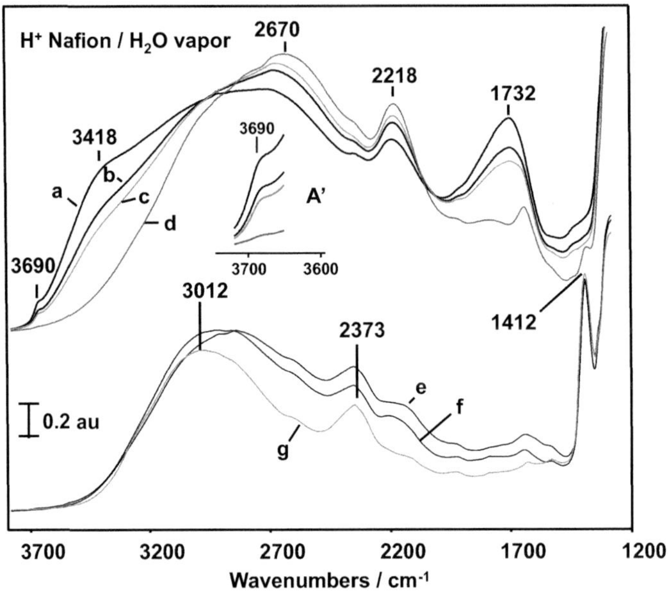


**FIGURE 12.5** Transmission infrared spectra of proton-exchanged Nafion 112 membrane (~50 μm thickness) in air at room temperature during dehydration. The high-frequency region is shown. After proton exchange and washing steps [29], the membrane was initially exposed to 95°C and ambient pressure for 10 min to remove surface water and subsequently treated according to the following sequential steps: held in dry air at ambient temperature for (curve a) 10 min; (curve b) 3.5 h; (curve c) 5.5 h; (curve d) placed under a few Torr of vacuum at 65°C for 1 h followed by dry air at ambient temperature for 5.5 h; (curve e) placed under a few Torr of vacuum at 95°C for 1 h followed by dry air at ambient temperature for 15 min; (curve f) placed under a few Torr of vacuum at 95°C for 25 h followed by dry air at ambient temperature for 15 min; (curve g) placed under a few Torr of vacuum at 95°C for 72 h followed by dry air at ambient temperature for 15 min. The inset (labeled A′) is an enlargement of the region near 3690 $cm^{-1}$ for spectra in curves (a)–(d) (progressing from top to bottom). (From Ref. 29, with permission.)

$Na^+$-exchanged material. The response was discussed in terms of electrostatic interactions between cations and $-SO_3^-$ groups and effects of changes in the water solvent shell surrounding the ions [47].

Another significant feature is present near 983 $cm^{-1}$ (Fig. 12.4). The pair of peaks has been assigned to stretching modes of the sidechain C–O–C groups [9,10,28,48]. The 983-$cm^{-1}$ band is insensitive to water uptake [48], but the lower-energy feature at ~972 $cm^{-1}$ is strikingly sensitive [48]. In experiments that employed model perfluorosulfonate ionomers, Cable and coworkers showed the ~972-$cm^{-1}$ band to be associated with vibrations of the ether group closest to the $-SO_3^-$ endgroup at the sidechain terminus in Nafion [48]. The electrostatic and hydrophilic interactions that affect the $-SO_3^-$ groups during membrane hydration are believed to contribute to the spectral changes at ~972 $cm^{-1}$ [48].

In in situ infrared studies of a Nafion thin film adsorbed to Pt, the vibrational modes of the Nafion sidechain $-SO_3$ and C–O–C groups were scrutinized to gain insights into membrane–catalyst interactions [21]. Shifts in the prominent band near 1058 cm$^{-1}$ for the symmetric S–O stretching mode of the $-SO_3^-$ group and in the band near 970 cm$^{-1}$ for stretching of the C–O–C group near the sidechain terminus were detected in response to changes in applied potential. Evidence was found for preferential binding of $-SO_3^-$ groups to Pt.

### 12.4.3 Infrared Spectra of Water in Nafion

The upper spectrum in Figure 12.3 includes bands for water condensed within pores and channels of a thin Nafion film. Just as for bulk water, features at 3450 and 3240 cm$^{-1}$ arise from O–H stretching vibrations, and the 1647-cm$^{-1}$ band is due to the H–O–H bending vibration [50]. Quantitative examination of spectra recorded during the experiment indicated that the film had reached equilibrium with the water vapor in the cell and was in a highly hydrated state with about 15 water molecules per $-SO_3^-$ group [20]. When transmission infrared measurements are performed on membranes that have thicknesses much greater than that of the sample used in the experiments reported in Figure 12.3, probing water in high-hydration-state samples becomes more difficult, because the absorbance bands for water, particularly through the O–H stretching region, become saturated. To overcome this limitation, strategies have involved restricting membrane hydration levels through control of the humidity in the atmosphere surrounding the membrane [24,29,49,50], or use of isotopically labeled water to produce HOD, which, when present at a few percent in $H_2O$, can be selectively and quantitatively detected in fuel cell membranes in high hydration states [25–27].

For Nafion exchanged by metal cations, the water vibrational bands are similar to those of bulk water [24,29,49,50]. In Nafion exchanged by protons, the water features are much more complicated and reflect the presence of solvated proton species, $(H_2O)_nH^+$, where $n$ can vary from 1 to large values [24,29,49]. An example is shown in Figure 12.5 for a proton-exchanged Nafion 112 sample undergoing dehydration. Similar spectra measured in related experiments also have been reported [19,24,49]. The spectrum marked "a" (i.e., curve a) in Figure 12.5 is dominated by a broad peak that extends over the range between 2300 and 3750 cm$^{-1}$. A prominent shoulder, which lies close to the position expected for bulk water, appears near 3418 cm$^{-1}$. Intensity in this shoulder region diminishes (spectra labeled "a"–"c") as water is removed from the membrane during dehydration. The loss is offset by growth in a feature near 2670 cm$^{-1}$. Buzzoni and coworkers have discussed the complicated nature of vibrational modes for aqueous acid solutions and attributed the development of features toward the low-energy side of the water O–H stretching bands in spectra of hydrated Nafion to a continuum of $(H_2O)_nH^+$ clusters [24]. The 1732-cm$^{-1}$ band in Figure 12.5 has been discussed as arising from the bending modes of $(H_2O)_nH^+$ species [19,24,49].

On further drying of the Nafion 112 sample by placing the membrane under a few Torr of vacuum (spectrum "d"), peak intensity shifts to 2670 cm$^{-1}$ and the 3418-cm$^{-1}$

band is no longer evident. Between acquisition of spectra "d" and "e" in Figure 12.5, the membrane was heated to 95°C while under vacuum. Heating expelled additional water and led to several more changes. In particular, the broad O–H stretching feature has shifted toward 3012 cm$^{-1}$, the 1732-cm$^{-1}$ band is no longer present, and a prominent new feature close to 1412 cm$^{-1}$ is apparent. Bands at 3012 and 1412 cm$^{-1}$ are known to be characteristic of $-SO_3H$ groups in Nafion [19,24,49]. Spectrum "e" in Figure 12.5 indicates that placing the proton-exchanged membrane under a few Torr of vacuum at 95°C for 1 h can result in water loss sufficient to shift the ionized sulfonate groups into the $-SO_3H$ form. Spectra labeled "f" and "g" in Figure 12.5 show that performing the treatment over longer periods leads to further water loss as evidenced by the appearance of a more sharply defined band at 3012 cm$^{-1}$ and flattening of the region around 1732 cm$^{-1}$.

Bands from interfacial water have been observed in infrared spectra of Nafion for membrane in low to moderate hydration states [11,24–27,29,50]. In studies of Nafion 112, Falk ascribed sharp peaks in the region 3650–3720 cm$^{-1}$ on the high-energy side of the broad O–H stretching band to water interacting with functional groups in Nafion [26,27]. The sharp features are similar to those for the "free" OH stretching mode of water observed in vibrational sum frequency (VSF) spectroscopy studies of water interfaces [52–55]. The inset in Figure 12.5 enlarges the region near 3690 cm$^{-1}$ to highlight the band for water in proton-exchanged Nafion. Only a single interfacial water band has typically been observed for Nafion membrane exchanged by protons [24,49], while two or more bands are often present in spectra of Nafion exchanged by $Na^+$, or other metal cations [26,27,49,50]. The contrasting behavior of $H_3O^+$ relative to metal cations at salt–solution/air interfaces observed by VSF spectroscopy [55] has been suggested to explain the different interfacial water features observed for Nafion in the proton versus metal-cation-exchanged forms [29].

In combination with steady-state infrared experiments, additional details about the environment for water in Nafion membrane materials have been uncovered by dynamic vibrational lifetime measurements [11,25]. Figure 12.6 shows vibrational lifetime decays recorded as a function of Nafion hydration state. The water vapor hydrating the membrane contained a few percent $D_2O$, and the measurements probed at the peak of the main O–D stretching band for HOD. The vibrational lifetime is longer for water in Nafion than in bulk solution. In addition, the lifetime increases as the membrane water content becomes lower. As discussed in detail [25], the results indicate that water in Nafion is more highly constrained than in bulk solution and the degree of confinement increases with decreasing membrane water content. In modeling the decays, a biexponential function was required to fit the full temporal range of the measurements. The decays did not fit well to a simpler function that reproduced the hydration-state-dependent changes in the steady-state infrared spectral band for the O–D stretching mode of HOD [25]. The vibrational lifetime decays provided new insights into the range of environments occupied by water in ionomer membrane and served as a complement to traditional steady-state infrared spectral measurements.

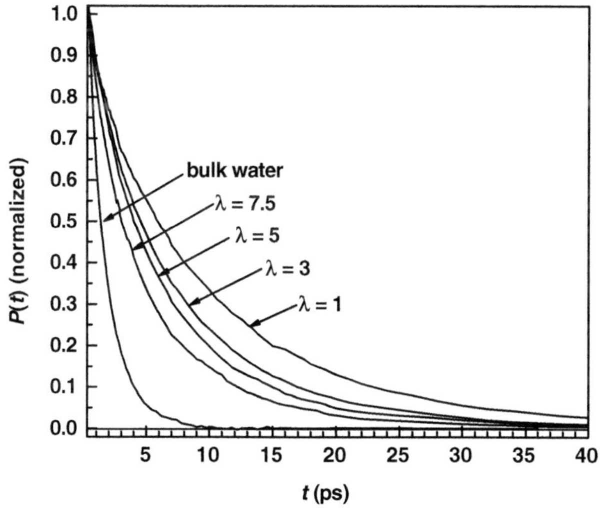


**FIGURE 12.6** Vibrational lifetime decays measured for water in Nafion 117 membranes as a function of the membrane water content, where λ is the ratio of water molecules to sulfonate groups. (From Ref. 25, with permission.)

### 12.4.4 Least-Squares Modeling

Least-squares modeling is a data analysis technique that involves decomposing a set of experimental spectra into a few basis spectra, linear combinations of which reconstruct the original dataset [30,56,57]. The basis spectra provide information about the components of the system under study. In studies of Nafion membrane hydration, least-squares modeling revealed unexpected differences in the responses of the hydrophilic and more hydrophobic ionomer functional groups during water uptake [30].

As an initial step in the analysis, a data matrix (**D**) is constructed by writing the absorbance values of each experimental spectrum as a column vector followed by assembling the vectors into **D**. Then, standard matrix algebra techniques are used to manipulate the equation **D** = **AC**. The columns of **A** contain the basis spectra absorbance values. Each row of **C** represents a state of the experiment, and the elements ($c(i,j)$) define the linear combinations of elements in **A** that reconstruct **D**. Details of the theory and application of least-squares modeling are discussed in References 30, 56, and 57.

In work on Nafion, least-squares modeling was applied to the data shown in Figure 12.4. An important advance in the study was learning that the elements in **C** could be predicted from the mass transport equation for diffusion in a cylindrical pore in the long time limit [57–60]. This diffusion equation is known to describe the transport of water and ions within Nafion membrane [58–60]. The equation has the form

$$\frac{N}{N_{\max}} = \left[1 - \sum_{k=0}^{\infty} \frac{8}{(2k+1)^2\pi^2} \exp\left(\frac{-(2k+1)^2\pi^2 t}{\tau}\right)\right] \tag{12.1}$$

where $N$ is the number of species in the film undergoing transport at time $t$, $N_{max}$ is the maximum $N$ attained after long measurement periods, and $\tau$ is a time constant given by the ratio of the pore length $L$ (usually taken as the membrane thickness), and the diffusion coefficient for analyte in the pore $D$ ($\tau = L^2/D$) [57–60]. The inset in Figure 12.2 shows the intensity of the 1159-cm$^{-1}$ band changing according to Equation (12.1) during water uptake. The agreement suggested that Equation (12.1) may be an appropriate model for predicting the **C** matrix and led to tests of that equation in analysis of the full data range in Figure 12.4 [30].

The spectral changes in Figure 12.4 were treated in terms of the transformation of the membrane from an initially "dry state" to a "hydrated state." The $\mathbf{c}(i,j)$ values that describe the growth of the hydrated state were given by $N/N_{max}$ in Equation (12.1) calculated at time $t(j)$ with $\tau = \tau_h$, where $\tau_h$ is the time constant representing the appearance of the hydrated state. Similarly, the $\mathbf{c}(i,j)$ values that describe the loss of the dry polymer state were calculated as $(1 - N/N_{max})$ for time $t(j)$ and $\tau = \tau_d$, where $\tau_d$ is the time constant for disappearance of the dry state. For accuracy at short times, the sum in Equation (12.1) was carried out to $k = 50$.

To begin the transformation of **D**, a starting **C** matrix was calculated from an estimate of $\tau_h$ and $\tau_d$. **D** was then decomposed to give a matrix of estimated basis spectra ($\hat{\mathbf{A}}$) as follows

$$\mathbf{D}\,\mathbf{C}^{\mathbf{T}}[\mathbf{C}\mathbf{C}^{\mathbf{T}}]^{-1} = \hat{\mathbf{A}} \tag{12.2}$$

where $\mathbf{C}^{\mathbf{T}}$ is the transpose of **C**. The parameters $\tau_h$ and $\tau_d$ in the model were optimized by using a simplex algorithm [61] to minimize $|\mathbf{D} - \hat{\mathbf{A}}\mathbf{C}|^2$.

Following identification of the best $\hat{\mathbf{A}}$ and **C** matricies, the $\hat{\mathbf{A}}\mathbf{C}$ product was calculated, and the resulting vectors were compared to the experimental spectra. An example is shown in Figure 12.7 for the regions encompassing the sidechain C–O–C and $-SO_3^-$ groups. For the ether group modes, the agreement between the experimental and calculated spectra suggests that the structural changes occurring about the C–O–C groups are in step with diffusion of water into the membrane pores and channels during hydration. A similar trend was observed for modes of the $-CF_2$ groups [30]. However, Figure 12.7 shows that the agreement is not as close for the symmetric S–O stretching mode of the $-SO_3^-$ group near 1064 cm$^{-1}$. The more complex relationship between experiment and calculation indicates that the simple two-state modeling function does not completely account for changes that take place in the vicinity of the $-SO_3^-$ group during water uptake by the membrane. The results have been discussed in terms of the complex structure for water and cations surrounding the electronically charged $-SO_3^-$ group [30]. The study demonstrated that least-squares modeling is a potentially powerful technique for analyzing infrared spectra of PEM materials and mapping structural changes taking place under the influence of a perturbation to specific ionomer functional groups.

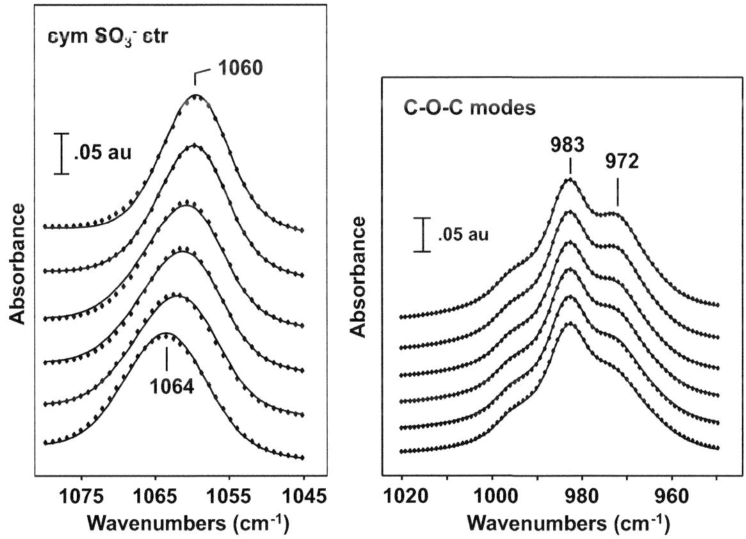


**FIGURE 12.7** Symbols indicate data from the experimental spectra in Figure 12.4 with focus on the spectral regions encompassing modes of the sidechain C–O–C and $-SO_3^-$ groups. A point is plotted every 1 $cm^{-1}$. Solid lines represent spectra calculated from the $\hat{\mathbf{A}}\mathbf{C}$ product [see Eq. (12.2) and related text] following optimization. (From Ref. 30, with permission.)

## 12.5 OUTLOOK

In addition to routine functional group identification, infrared spectroscopy is likely to find greater use in the coming years for probing physical and chemical interactions within ion-conducting regions of PEM materials. Spectrometers for performing steady-state measurements across the mid-infrared spectral range are widely available and simple to operate. However, advances are needed in defining the limits for the quantitative determination of water and ionomer functional group content of membranes [50]. In addition, studies that provide more detailed knowledge of vibrational mode assignments, such as those that combine computational methods and spectral measurements on model compounds [51], should continue, as they enable greater understanding to be derived from infrared spectra in relation to structure and function in PEM materials. Further use and development of in situ infrared methods for probing membrane–catalyst interactions in operating fuel cells [23,32,33] and on electrodes [21,22] would also be valuable.

In considering ultrafast, time-resolved infrared measurements, studies have been limited to probing the vibrational lifetime of water molecules in Nafion membrane. Although not routine, some ability to extend the technique to investigate water in different types of PEM materials would provide insights into effects of ionizable groups and polymer backbone variations on the environments occupied by water molecules. For example, in more recent studies of bis[(perfluoroalkyl)sulfonyl] imide (PFSI) (Scheme 12.1b) membranes, steady-state infrared spectra indicated that the sidechain endgroup undergoes weaker perturbations than the $-SO_3^-$ group in Nafion when progressing from dry to fully hydrated states [14]. Greater electronic charge delocalization over the sulfonyl imide moiety relative to the $-SO_3^-$ group and shielding of water by the terminal $-CF_3$ group in the PFSI materials were suggested as important factors contributing to the differences. It would be insightful to probe the environments for water in PFSI membranes, in addition to a variety of other PEM materials, through vibrational lifetime measurements in a manner analogous to studies that have been carried out on Nafion [11,25]. Vibrational lifetimes interrogate nanoscopic environments in ionomer, and understanding relationships between lifetimes for populations of water molecules in PEMs and fuel-cell-relevant properties would be valuable for guiding the synthesis of new membrane materials.

Although Raman spectroscopy has not been used extensively for characterization of ionomer membrane, there is potential to employ the technique more frequently and adapt it for in situ measurements on PEM fuel cells. Compared to in situ diffuse reflectance infrared sampling [23,32,33], Raman measurements can be less invasive. To develop the full potential of Raman spectroscopy for the study of fuel cell membrane materials, there is a need for greater access to instruments with sources that effectively reduce background fluorescence.

Finally, it would be valuable to explore further the potential of data analysis techniques, such as least-squares modeling, for identifying different responses among functional groups in ionomer materials undergoing changes in hydration state, or experiencing other types of stress. In particular, approaches are needed that enable connections to be made between vibrational spectral characteristics and properties relevant to membrane performance during fuel cell operation (e.g., durability, proton conductivity).

## 12.6 CONCLUSIONS

Vibrational spectroscopy techniques are useful for probing the environment within PEMs and have potential to be utilized more extensively, particularly in the development of new ionomer materials. Steady-state mid-infrared measurements have been performed in the majority of applications. Among mid-infrared sampling methods, transmission is preferred when quantitative information is sought, because the spectral bands suffer the least distortion from optical effects. Transmission sampling has been successfully applied to detect changes in the water content and sidechain functional group vibrations of fuel cell PEMs during hydration and dehydration. However, transmission measurements easily become limited by absorbance saturation through the spectral region encompassing the strong ionomer backbone modes (C–F stretching modes in Nafion materials). When ionomer vibrational modes of practical PEMs are of primary interest, ATR sampling can be very useful. In interpreting ATR spectra, however, care must be taken to account for the wavelength-dependent depth of radiation penetration into samples and the need to maintain contact between the membrane and ATR crystal. In the case of diffuse reflectance and in situ spectroelectrochemical sampling, these methods yield complex spectra of

PEM materials, but provide a means to probe samples under the influence of an applied voltage.

The demonstration of mid-infrared dynamic measurements of vibrational lifetimes for water in Nafion fuel cell membrane has opened new directions for vibrational spectroscopy in the study of PEMs. The ultrafast, time-resolved technique revealed water molecules present in a wider range of environments than had been detected by steady-state infrared measurements. Future applications that investigate water in Nafion and other types of ionomer membrane will provide a more complete picture of membrane–water interactions and their effects on hydration and proton conductivity.

The use of Raman spectroscopy in the study of fuel cell membrane has been hampered by interference from background fluorescence. Advances will be aided by improved access to Raman spectrometers equipped to operate in the near-infrared spectral region (i.e., 1064 nm). Raman sampling is versatile and has potential application in in situ measurements on fuel cells. Thus, studies in the area are worthwhile to pursue.

The possibility for scrutinizing spectral datasets for correlations in the response of different functional groups over the course of an experiment is common to all vibrational spectroscopy techniques. Data modeling is beginning to support vibrational spectral analysis in work on PEM materials. The more in-depth information that can be attained relative to simple functional group identification will increase the utility of vibrational spectroscopy techniques in the development of new fuel cell membrane materials.

## ACKNOWLEDGMENTS

The US National Science Foundation and Department of Energy are gratefully acknowledged for support of the author's research on applications of vibrational spectroscopy in electrochemistry.

## REFERENCES

1. D. Dunwoody and J. Leddy, *Interface* **14**, 37 (2005).
2. M. A. Hickner, H. Ghassemi, Y. S. Kim, B. R. Einsla, and J. E. McGrath, *Chem. Rev.* **104**, 4587 (2004).
3. S. Gottesfeld, and T. A. Zawodzinski, Polymer electrolyte fuel cells, in *Advances in Electrochemical Science and Engineering*, R. C. Alkire, H. Gerischer, D. M. Kolb, and C. W. Tobias, eds., Wiley-VCH, New York, 1997, Vol. 5 p. 195.
4. J. Jugar-Grodzinski, *Polym. Adv. Technol.* **18**, 785 (2007).
5. K. A. Mauritz, and R. B. Moore, *Chem. Rev.*, **104**, 4535 (2004).
6. J. Roziere, and D. J. Jones, *Annu. Rev. Mater. Res.* **33**, 503 (2003).
7. M. W. Perpall, D. W. Smith, Jr., D. D. Desmarteau, and S. E. Creager, *J. Macromol. Sci., C Polym. Rev.* **46**, 297 (2006).

8. J. Zhang, Z. Xie, J. Zhang, Y. Tang, C. Songa, T. Navessin, Z. Shi, D. Songa, H. Wang, D. P. Wilkinson, Z.-S. Liu, and S. Holdcroft, *J. Power Sources* **160**, 872 (2006).
9. M. Falk, Infrared spectra of perfluorosulfonated polymer and of water in perfluorosulfonated polymer, in *Perfluorinated Ionomer Membranes*, A. Eisenberg and H. L. Yeager, eds., American Chemical Society, Washington, DC, 1982, Vol. 180, pp. 139.
10. R. M. Blanchard and R. G. Nuzzo, *J. Polym. Sci. B Polym. Phys.* **38**, 1512 (2000).
11. D. E. Moilanen, D. B. Spry, and M. D. Fayer, *Langmuir* **24**, 3690 (2008).
12. B. Lafitte and P. Jannasch, *J. Polym. Sci. A Polym. Chem.* **45**, 269 (2007).
13. Y. S. Kim, F. Wang, M. Hickner, T. A. Zawodzinski, and J. E. McGrath, *J. Membr. Sci.* **212**, 263 (2003).
14. C. K. Byun, I. Sharif, D. D. DesMarteau, S. E. Creager, and C. Korzeniewski, *J. Phys. Chem. B* **113**, 6299 (2009).
15. (a) N. Harrick, *Internal Reflection Spectroscopy*, Interscience, New York, 1967. (b) P. R. Griffiths and J. A. de Haseth, *Fourier Transform Infrared Spectrometry*, 2nd ed., Wiley, Hoboken, NJ, 2007.
16. D. T. Hallinan, Jr. and Y. A. Elabd, *J. Phys. Chem. B* **111**, 13221 (2007).
17. A. Gruger, A. Regis, T. Schmatko, and P. Colomban, *Vib. Spectrosc.* **26**, 215 (2001).
18. L. Grosmaire, S. Castagnoni, P. Huguet, P. Sistat, M. Boucher, P. Bouchard, P. Bebin, and S. Deabate, *Phys. Chem. Chem. Phys.* **10**, 1577 (2008).
19. M. Ludvigsson, J. Lindgren, and J. Tegenfeldt, *Electrochim. Acta*, **45**, 2267 (2000).
20. C. Korzeniewski, D. Snow, and R. Basnayake, *Appl. Spectrosc.* **60**, 599 (2006).
21. D. Malevich, V. Zamlynny, S.-G. Sun, and J. Lipkowski, *Z. Phys. Chem.* **217**, 513 (2003).
22. Y. Ayato, K. Kunimatsu, M. Osawa, and T. Okada, *J. Electrochem. Soc.* **153**, A203 (2006).
23. S. Sanicharane, A. Bo, B. Sompalli, B. Gurau, and E. S. Smotkin, *J. Electrochem. Soc.* **149**, A554 (2002).
24. R. Buzzoni, S. Bordiga, G. Ricchiardi, G. Spoto, and A. Zecchina, *J. Phys. Chem.* **99**, 11937 (1995).
25. D. E. Moilanen, I. R. Piletic, and M. D. Fayer, *J. Phys. Chem. A*, **110**, 9084 (2006).
26. M. Falk, *Can. J. Chem.* **58**, 1495 (1980).
27. S. Quezado, J. C. T. Kwak, and M. Falk, *Can. J. Chem.* **62**, 958 (1984).
28. M. Laporta, M. Pegoraro, and L. Zanderighi, *Phys. Chem. Chem. Phys.* **1**, 4619 (1999).
29. R. Basnayake, G. R. Peterson, D. J. Casadonte, Jr., and C. Korzeniewski, *J. Phys. Chem. B* **110**, 23938 (2006).
30. C. Korzeniewski, E. Adams, and D. Liu, *Appl. Spectrosc.* **62**, 634 (2008).
31. Z. X. Liang, W. M. Chen, J. G. Liu, S. L. Wang, Z. H. Zhou, W. Z. Li, G. Q. Sun, and Q. Xin, *J. Membr. Sci.* **233**, 39 (2004).
32. Q. Fan, C. Pu, and E. S. Smotkin, *J. Electrochem. Soc.* **143**, 3053 (1996).
33. A. Bo, S. Sanicharane, B. Sompalli, Q. Fan, B. Gurau, R. Liu, and E. S. Smotkin, *J. Phys. Chem. B*, **104**, 7377 (2000).
34. E. S. Smotkin, FTIR and X-ray absorption spectroscopy of operating fuel cells, in *In-Situ Spectroscopic Studies of Adsorption at the Electrode and Electrocatalysis*, S. -G. Sun, P. A. Christensen, and A. Wieckowski, eds., Elsevier, Oxford, UK, 2007, p. 247.
35. K. L. Ley, R. Liu, C. Pu, Z. Fan, N. Leyarovska, C. Segre, and E. S. Smotkin, *J. Electrochem. Soc.* **144**, 1543 (1997).

36. M. D. Porter *Anal. Chem.* **60** 1143A (1988).
37. D. D. Popenoe, S. M. Stole, and M. D. Porter, *Appl. Spectrosc.* **46**, 79 (1992).
38. S. M. Stole, D. D. Popenoe, and M. D. Porter, Infrared spectroelectrochemistry: A probe of the molecular architecture of the electrochemical interface, in *Electrochemical Interfaces. Modern Techniques for In-Situ Interface Characterization*, H. D. Abruna, ed., VCH New York 1991, p. 339.
39. M. Osawa, In-situ surface-enhanced infrared spectroscopy of the electrode/solution interface, in *Advances in Electrochemical Science and Engineering*, Vol. X, In Situ Spectroscopic and Diffraction Methods, R. Alkire, D. M. Kolb, J. Lipkowski, and P. N. Ross, eds., Wiley-VCH, Weinheim, 2006, Vol. 9, p. 269.
40. M. Osawa and K. Yoshii, *Appl. Spectrosc.* **51**, 512 (1997).
41. Y. X. Chen, A. Miki, S. Ye, H. Sakai, and M. Osawa, *J. Am. Chem. Soc.*, **125**, 3680 (2003).
42. K. Ataka, Y. Hara, and M. Osawa, *J. Electroanal. Chem.* **473**, 34 (1999).
43. C. Korzeniewski, Recent advances in in situ infrared spectroscopy and applications in single crystal electrochemistry and electrocatalysis, in *Advances in Electrochemical Science and Engineering*, Vol. X, *In situ Spectroscopic and Diffraction Methods*, R. Alkire, D. M. Kolb, J. Lipkowski, and P. N. Ross, eds., Wiley-VCH, Weinheim, 2006, p. 233.
44. M. Osawa, Surface-enhanced infrared absorption spectroscopy, in *Handbook of Vibrational Spectroscopy*, J. M. Chalmers, and P. R. Griffiths, eds. Wiley, Chichester, UK, 2002, Vol. 1 p. 785.
45. K. Ataka, F. Giess, W. Knoll, R. Naumann, S. Haber-Pohlmeier, B. Richter, and J. Heberle, *J. Am. Chem. Soc.*, **126**, 16199 (2004).
46. R. L. McCreery, *Raman Spectroscopy for Chemical Analysis*, Wiley, New York. 2000.
47. S. R. Lowry and K. A. Mauritz, *J. Am. Chem. Soc.*, **102**, 4665 (1980).
48. K. M. Cable, K. A. Mauritz, and R. B. Moore, *J. Polym. Sci. B Polym. Phys.*, **33**, 1065 (1995).
49. R. Iwamoto, K. Oguro, M. Sato, and Y. Iseki, *J. Phys. Chem. B*, **106**, 6973 (2002).
50. Y. Wang, Y. Kawano, S. R. Aubuchon, and R.A. Palmer, *Macromolecules*, **36**, 1138 (2003).
51. D. S. Warren and A. J. McQuillan, *J. Phys. Chem. B*, **112**, 10535 (2008).
52. M. G. Brown, E. A. Raymond, H. C. Allen, L. F. Scatena, and G. L. Richmond, *J. Phys. Chem. A* **104**, 10220 (2000).
53. L. F. Scatena, M. G. Brown, and G. L. Richmond, *Science* **292**, 908 (2001).
54. M. J. Shultz, S. Baldelli, C. Schnitzer, and D. Simonelli, *J. Phys. Chem. B* **106**, 5313 (2002).
55. M. Mucha, T. Frigato, L. M. Levering, H. C. Allen, D. J. Tobias, L. X. Dang, and P. Jungwirth, *J. Phys. Chem. B* **109**, 7617 (2005).
56. D. Rivera, P. E. Peterson, R. H. Uibel, and J. M. Harris, *Anal. Chem.* **72**, 1543 (2000).
57. D. Rivera, and J. M. Harris, *Anal. Chem.* **73**, 411 (2001).
58. A. Goswami, A. Acharya, and A. K. Pandey, *J. Phys. Chem. B* **105**, 9196 (2001).
59. G. Suresh, Y. M. Scindia, A. K. Pandey, and A. Goswami, *J. Membr. Sci.* **250**, 39 (2005).
60. G. Suresh, Y. M. Scindia, A. K. Pandey, and A. Goswami, *J. Phys. Chem. B* **108**, 4104 (2004).
61. W. H. Press, B. P. Flannery, S. A. Teukolsky, and W. T. Vetterling, *Numerical Recipes*, Cambridge Univ. Press, Cambridge, UK, 1989.

CHAPTER 13

# Ab Initio Electrochemical Properties of Electrode Surfaces

ISMAILA DABO and YANLI LI
Université Paris-Est, CERMICS, Projet Micmac ENPC-INRIA, France

NICÉPHORE BONNET and NICOLA MARZARI
Department of Materials Science and Engineering, Massachusetts Institute of Technology, Cambridge, Massachusetts

The electrical response of electrochemical convertors (e.g., fuel cells, batteries, electrochemical capacitors) is to a large extent governed by polarization and charge separation phenomena that take place at the electrode–electrolyte double-layer interface in the presence of an applied voltage. Therefore, predicting the electrical properties of electrochemical systems from first principles entails understanding the influence of the macroscopic voltage on the microscopic electronic and ionic charge distributions at the surface of the electrode. Despite considerable progress in large-scale quantum chemistry calculations, determining the potential dependence of the interfacial charge represents a challenging problem, due to the prohibitively large length scales that characterize electrostatic screening in the electrolyte. To overcome these limitations, we present an atom-continuum method that enables the description of solvated electrode surfaces immersed in semiinfinite ionic solvent media in both the high- and low-ionic-concentration regimes. We apply the method to describe the electronic structure and to predict the surface energy of electrode–electrolyte interfaces as a function of the applied voltage.

## 13.1 INTRODUCTION

The interaction between the molecules and ions of an electrolyte with a solvated electrode causes important electronic and atomic structure modifications at the

*Fuel Cell Science: Theory, Fundamentals, and Biocatalysis*,
Edited by Andrzej Wieckowski and Jens K. Nørskov

electrode surface. Although it is difficult in general to precisely identify and deconvolute the different solvation contributions because of the complexity of physical interactions in the liquid state, it is common and useful in solvation theories to establish a distinction between specific and nonspecific interactions [1]. *Specific interactions* are those solvation interactions that result from the formation of chemical bonds (e.g., covalent, ionic, hydrogen, and van der Waals bonds), whereas *nonspecific interactions* do not entail chemical bonding between the electrode and the electrolyte (e.g., electrostatic polarization, dispersion–repulsion [2,3], pressure, and cavitation contributions).

Since electrochemical catalysis involves the breaking of chemical bonds at electrode–electrolyte interfaces, the computational study of electrochemical catalytic processes requires a precise description of specific interactions. These interactions, which originate from nonclassical electronic effects, are most accurately captured at the level of quantum theory using first-principles approaches, such as density functional theory methods [4]. Clearly, however, nonspecific solvation interactions are also fundamental in understanding the electrochemistry of solvated electrodes and, because these interactions are typically much less local than specific bonding contributions, their description greatly increases the computational burden of first-principles calculations. As a matter of comparison, the polarized solvent and counterion distribution at the electrode–electrolyte interface, which is directly related to the applied macroscopic voltage, can extend as far as thousands of nanometers for dilute electrolyte solutions, while the size of typical density functional simulations do not exceed tens of nanometers [5].

In order to surmount these important limitations, one can exploit the fact that nonspecific interactions are very often accurately treated using classical or semiclassical theories, as they depend much less on the electronic characteristics of the electrochemical system than do specific interactions. This motivates the use of implicit continuum descriptions of the solvent, which represent interesting alternatives to account for nonspecific contributions from the liquid phase [1,6,7]. Such continuum solvation models have a long successful tradition in the field of modern ab initio molecular chemistry but have been, until quite recently, rarely employed for the ab initio study of electrochemical systems.

In this chapter, we focus on the prediction of the electrical response and differential capacitance of electrochemical interfaces. The interface capacitance is related to the ability of the electrolyte medium to screen electrostatic interactions, which results from the solvent polarizability and the presence of ions in solution. Therefore, one needs to understand and, to some extent, be able to predict the influence of the solvent and counterions in the vicinity of the metal surface. To this end, we present an atom-continuum approach for the description of electrochemical interfaces embedded in effective semiinfinite ionic solvent media.

## 13.2 THEORY OF ELECTROCHEMICAL INTERFACES

### 13.2.1 Electrical Potential of Metal Electrodes

While the extraction of an electron from a charged metal surface isolated in vacuum (i.e., in the absence of counterelectrode) requires an infinite expense of energy, the removal energy of an electron from a charged electrode immersed in an ionic solution is always finite. This difference is due to the accumulation of counterions at the metal–solution interface, which forms a local effective counterelectrode and screens the surface electrostatic potential that would otherwise diverge linearly at large distances from the surface.

The finite electron removal energy is referred to as the *electrode potential*. Explicitly, the potential of an electrode $E$ is defined as [8]

$$\mathcal{E} = \mathcal{W} - \epsilon_C = \mathcal{V}_S - \mathcal{V}_M - \epsilon_C \tag{13.1}$$

where $\mathcal{W} = \mathcal{V}_S - \mathcal{V}_M$ denotes the electrostatic work done in moving an electron from the metal to the solvent, and $\epsilon_C$ stands for the chemical energy of an electron in the metal. The chemical energy $\epsilon_C$ equals the sum of all the nonelectrostatic contributions to the energy of the highest-energy metal electrons [8]. In the context of density functional theory (DFT), these contributions can be directly identified as the kinetic and exchange correlation energies of the metal electronic states at the Fermi level. As a consequence, the chemical energy corresponds to the difference between the Fermi energy $\epsilon_F$ and the electrostatic potential of the electron in the metal:

$$\epsilon_C = \epsilon_F - \mathcal{V}_M \tag{13.2}$$

Substituting Equation (13.2) into Equation (13.1), we can express the electrode potential as the difference between the electrostatic energy of an electron in the bulk of the solvent and the Fermi energy of the metal:

$$\mathcal{E} = \mathcal{V}_S - \epsilon_F \tag{13.3}$$

While this expression defines the absolute potential of an electrode, experimental measurements of the electrode potential are generally referenced to the potential of a noble metal electrode under electrochemical equilibrium with a solvent in standard conditions of temperature, partial pressure, and ionic activity. The referenced electrode potential $\mathcal{E}'$ is related to the absolute electrode potential $\mathcal{E}$ through

$$\mathcal{E}' = \mathcal{E} - \mathcal{E}^\circ \tag{13.4}$$

where $\mathcal{E}^\circ$ is the absolute potential of the reference electrode. A commonly used reference electrode is the *standard hydrogen electrode* (SHE), which typically consists of a platinum surface in contact with an aqueous solution. The accurate determination of the absolute SHE potential is still the subject of active research. Current estimations of the absolute SHE potential $\mathcal{E}^\circ(H_2/H^+)$ lie in the energy range [8]

$$4.44\ \text{eV} < \mathcal{E}^\circ(H_2H^+) < 4.78\ \text{eV} \tag{13.5}$$

We emphasize, once more, that evaluating the electrode potential from first-principles simulations requires determining the value of the effective electrostatic energy of an electron in the solvent $\mathcal{V}_S$. Because of the dynamical evolution of the solvent

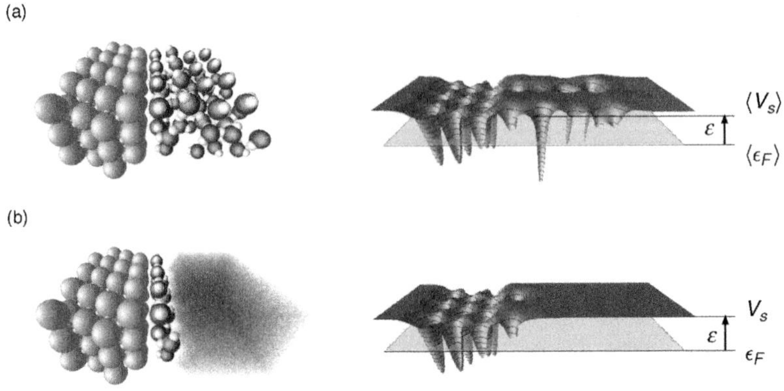


**FIGURE 13.1** Determination of the absolute electrode potential $\mathcal{E}$ using (a) an explicit solvent model, and (b) an implicit solvent model. Within the explicit model, the potential $\mathcal{E}$ is calculated from the dynamical averages of the solvent electrostatic energy $\langle \mathcal{V}_S \rangle$ and of the metal Fermi energy $\langle \epsilon_F \rangle$. (See insert for color representation.)

molecules, the definition of $\mathcal{V}_S$ in the context of explicit solvent models is not straightforward [9]. In contrast, implicit solvent models allow a direct determination of $\mathcal{V}_S$ as the asymptotic value of the electrostatic potential in the bulk of the ionic continuum (Fig. 13.1).

Note that Equation (13.3) is similar to the definition of the workfunction $\phi$ for neutral surfaces in vacuum

$$\phi = \mathcal{V}_0 - \epsilon_F \tag{13.6}$$

where $\mathcal{V}_0$ is the vacuum electrostatic energy measured at a large distance from the surface. The electrode workfunction is related to the potential of a neutral electrode—the potential of zero charge $\mathcal{E}_{PZC}$—through

$$\mathcal{E}_{PZC} = \phi - \chi_S \tag{13.7}$$

where $\chi_S$ is commonly referred to as the *dipole potential* (or *surface potential*) of the solvent and corresponds to the supplementary electrostatic work for taking a metal electron across the polarized solvent layer. Experimental correlations between the potential of zero charge and the workfunction for various single-crystal electrodes indicate that the solvent dipole potential $\chi_S$ is moderately affected by surface orientation but depends on some general electronic specificities of the electrode material (in particular, on whether the electrode is made of a *s, p,* or *d* metal) [5,10].

### 13.2.2 Capacitance of Metal Electrodes

The capacitance of an electrode–electrolyte interface relates the local state of charge of the electrode surface to the applied external potential. The specific differential capacitance of a double-layer interface $C$ is defined as

$$C = \frac{\partial \sigma}{\partial \mathcal{E}} \tag{13.8}$$

where $\sigma$ is the surface charge and $\mathcal{E}$ is the electrode potential.

Experimental measurements of electrode capacitances indicate that the electrical response of electrode–electrolyte interfaces is strongly nonlinear. In particular, at low ionic concentrations, the differential capacitance typically shows a sharp minimum at the potential of zero charge. This steep decrease can be explained, to some extent, by considering the electrical response of the ionic solvent. Three standard theories for describing the ionic contribution to the electrode capacitance—namely, the Helmholtz–Perrin, the Gouy–Chapman, and the Gouy–Stern double-layer theories—are presented in the next section.

### 13.2.3 Standard Double-Layer Models

In the conventional double-layer models described below, the metal electrode is represented by a polarizable continuum of infinite dielectric constant $\epsilon_M = +\infty$. Consequently, the electrostatic potential is equal to $\mathcal{V}_M$ everywhere inside the metal. The chemical energy contribution $\epsilon_C$ is set to zero. The ionic solution is modeled by a countercharge distribution immersed in a uniform solvent medium of dielectric constant $\epsilon_S$.

The simplest model is the Helmholtz–Perrin model (Fig. 13.2a), which consists of representing the ionic countercharges by a thin charged layer, the outer Helmholtz plane (OHP) located at a distance $L_H$ from the surface. The intermediate solvent region

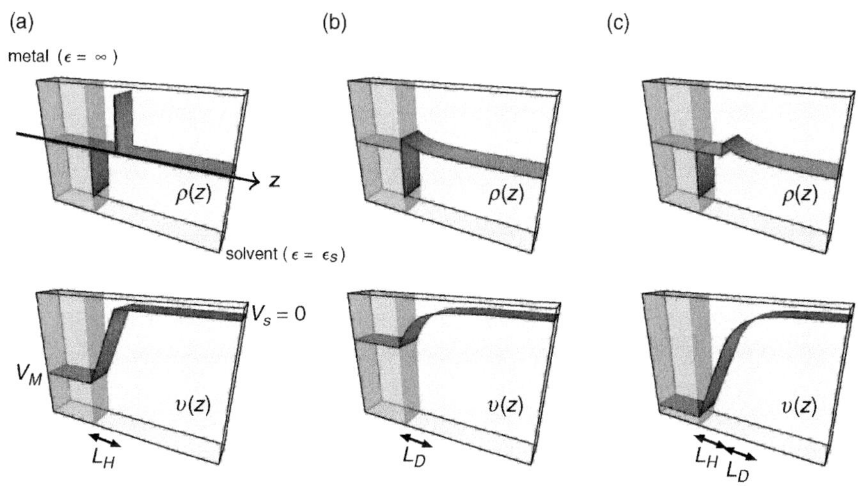


**FIGURE 13.2** Charge density and electrostatic potential profiles for the Helmholtz–Perrin (a), Gouy–Chapman (b), and Stern (c) double-layer models.

that separates the metal layer from the outer Helmholtz plane is referred to as the *Helmholtz layer*. Hence, the differential capacitance of the Helmholtz–Perrin model equals

$$C_{\mathrm{H}} = \frac{\epsilon_{\mathrm{S}}}{4\pi L_{\mathrm{H}}} \tag{13.9}$$

This rough estimate can be fruitful in inferring orders of magnitude, such as the value of the electric field at the interface and its impact on adsorbate energies [11,12]. However, the main limitation of the Helmholtz–Perrin model is that it does not capture the influence of the counterion concentration and always predicts a simple linear dependence of the surface charge as a function of the applied voltage regardless of experimental conditions.

Introducing a refined description of the counterion distribution, the Gouy–Chapman model (Fig. 13.2b) consists of representing the countercharge profile as a diffuse Boltzmann distribution of ions in thermal equilibrium with its electrostatic surrounding. Explicitly, the diffuse ionic distribution $\rho_{\mathrm{d}}$ can be expressed in terms of the electrostatic potential $v$ as

$$\rho_{\mathrm{d}} = z_{\mathrm{d}} c_{\mathrm{d}} \left(e^{-z_{\mathrm{d}} v/k_{\mathrm{B}} T} - e^{z_{\mathrm{d}} v/k_{\mathrm{B}} T}\right) \tag{13.10}$$

where $c_{\mathrm{d}}$ is the concentration of the ionic solution, $z_{\mathrm{d}}$ is the ionic valence (the anion and cation valences are supposed to be equal here), $k_{\mathrm{B}}$ is the Boltzmann constant, and $T$ is the solvent temperature. As a consequence, the potential in the ionic solution satisfies the electrostatic equation

$$\frac{d^2 v}{dz^2}(z) = \frac{8\pi}{\epsilon_{\mathrm{S}}} z_{\mathrm{d}} c_{\mathrm{d}} \sinh \frac{z_{\mathrm{d}} v(z)}{k_{\mathrm{B}} T} \tag{13.11}$$

Integrating Equation (13.11) with vanishing electric field boundary conditions, we obtain

$$\frac{dv}{dz}(z) = -\left(\frac{32\pi c_{\mathrm{d}} k_{\mathrm{B}} T}{\epsilon_{\mathrm{S}}}\right)^{1/2} \sinh \frac{z_{\mathrm{d}} v(z)}{2 k_{\mathrm{B}} T} \tag{13.12}$$

Then, using Gauss' law to determine the magnitude of the electric field at the electrode–electrolyte interface ($z = 0$) as a function of the surface charge, the differential capacitance of the diffuse-layer model can be expressed as

$$C_{\mathrm{D}} = C_{\mathrm{D,PZC}} \cosh \frac{z_{\mathrm{d}} \mathcal{E}}{2 k_{\mathrm{B}} T} \tag{13.13}$$

where $C_{\mathrm{D,PZC}} = [(\epsilon_{\mathrm{S}} c_{\mathrm{d}} z_{\mathrm{d}}^2)/(2\pi k_{\mathrm{B}} T)]^{1/2}$ is the value of the diffuse-layer capacitance at zero charge. Alternatively, the capacitance at the point of zero charge can be written as

$C_{D,PZC} = \epsilon_S/4\pi L_D$, where the Debye length $L_D$ is the characteristic screening length in the double layer, which results from the polarizability of the solvent and the presence of counterions:

$$L_D = \left( \frac{\epsilon_S k_B T}{8\pi c_d z_d^2} \right)^{1/2} \tag{13.14}$$

The Debye screening length can be considerably large (typically 10–10$^3$ Å for ionic concentrations in the range $10^{-4}$–$10^{-1}$ M), which makes the explicit first-principles representation of the screening counterion distribution difficult and expensive.

It should be noted that the Gouy–Chapman model predicts a nonlinear dependence of the surface charge as a function of the electrode potential. The double-layer capacitance reaches a characteristic minimum at zero charge, thus reproducing experimental trends. Nevertheless, the Gouy–Chapman differential capacitance is largely overestimated when the electrode potential deviates from the potential of zero charge. The Stern double-layer model—also referred to as the *modified Gouy–Chapman theory* (Fig. 13.2c)—improves on the Gouy–Chapman description by inserting a Helmholtz layer of thickness $L_H$ between the metal and the diffuse layer, thereby excluding ions from the direct proximity of the metal surface. Hence, the Stern double layer can be identified as consisting of a Helmholtz solvent layer placed in series with a Gouy–Chapman diffuse layer. From this analogy, the Stern capacitance $C_S$ can be expressed as

$$\frac{1}{C_S} = \frac{1}{C_H} + \frac{1}{C_D} \tag{13.15}$$

As a result, the dependence of the Stern capacitance as a function of the electrode potential can be written as

$$C_S = \frac{\epsilon_S}{4\pi} \left( L_H + \frac{L_D}{\cosh(z_d \mathcal{E}/2k_B T)} \right)^{-1} \tag{13.16}$$

Note that the capacitance $C_S$ reaches a minimum at the potential of zero charge, similarly to the Gouy–Chapman model. For large deviations from the potential of zero charge, the capacitance of the Stern double layer equals that of the Helmholtz layer $C_H$, reflecting the rapid decay of the electric field in the ionic layer at high electrode voltages.

The original Gouy–Chapman model of the diffuse layer requires corrections for regimes of high fields or high ionic concentrations because it ignores the finite size of the component ions. Fawcett and colleagues extended an analytical derivation of the potential drop across the diffuse layer on the basis of definite integrals leading to a Taylor series in the Gouy–Chapman estimate of the same quantity [13,14]. Their model gives an excellent description of Monte Carlo data, and presents the advantage of providing an analytical correction to the Gouy–Chapman model. Moreover, steric effects from the ions have been invoked to explain the possible nonmonotonic behavior of the diffuse layer capacitance beyond a threshold potential [15].

It is also important to observe that contrary to the idealized picture proposed by Stern, the capacitance contribution from the metal and from the Helmholtz layer is not negligible. In order to take into account the electrode potential dependence of $C_S$, a microscopic description of the metal surface covered by solvent overlayers in the presence of adsorbates and solvated ions is necessary. This brings motivation to determine electrode capacitances from first principles by capturing how the interface electronic structure is affected by the solvent.

## 13.3 SOLVATION MODEL

### 13.3.1 Electrostatics in Dielectric Media

As explained previously, in the absence of specific chemical interactions, or if those are negligible, the discrete solvent molecules can be appropriately modeled by a dielectric medium. The electrostatic potential $v$ induced by the electrode charge distribution $\rho$ includes an additional screening contribution that arises from the polarization of the solvent

$$\nabla^2 v = -4\pi(\rho - \rho_p) \tag{13.17}$$

where $\rho_P = -\nabla \cdot \mathbf{P}$ is the charge distribution corresponding to the polarization density $\mathbf{P}$ [16]. In the linear response regime, the induced polarization is related to the electric field $\mathbf{E} = -\nabla v$ through

$$\mathbf{P} = \chi \mathbf{E} \tag{13.18}$$

where $\chi$ denotes the susceptibility of the medium. Substituting Equation (13.18) into Equation (13.17), we obtain the generalized Poisson equation

$$\nabla \cdot \epsilon \nabla v = -4\pi\rho \tag{13.19}$$

where $\epsilon = 1 + 4\pi\chi$ is the dielectric constant of the polarizable continuum.

The microscopic interpretation of the permittivity $\epsilon$ is not straightforward. In fact, on the basis of results from ab initio molecular dynamics, Sharma, Resta and Car showed that the static dielectric constant of water is due equally to the intrinsic water molecule polarizability and the local reorientation of molecular dipoles in the presence of a field [17]. However, at the mesoscopic level, only the value of $\epsilon$ is important. In the linear regime, the permittivity $\epsilon$ is field-independent but space-dependent, since the absence of solvent inside the solute region translates into a permittivity equal to 1. Various models differ according to the way they treat the solute–solvent interface [7,18]. In the polarizable continuum model (PCM), the interface is materialized by interlocking spheres centered on the solute atoms [19,20]. The electrostatic reaction field from the solvent is then represented by point charges placed at the solvation interface. Spurious discontinuities may be avoided by introducing Gaussian charge distributions instead [21]. This kind of distribution is well suited for Gaussian-type orbitals methods, but it may be inappropriate for other schemes, like those using plane waves. The model developed by Fattebert and Gygi intends to accommodate those more general methodologies [16]. The dielectric constant is taken to be locally dependent on the solute charge density $\rho$. It is equal to $\epsilon_S$ in regions of low charge density, while it approaches unity at higher density. As proposed by Fattebert and Gygi, the density dependence of the dielectric constant can be chosen to be a smooth analytical function of the form

$$\epsilon(\rho) = 1 + \frac{\epsilon_S - 1}{2}\left(1 + \frac{1-(\rho/\rho_0)^{2\beta}}{1+(\rho/\rho_0)^{2\beta}}\right) \tag{13.20}$$

where $\rho_0$ is the charge density cutoff defining the transition between the solvent and the metal and $\beta$ is the exponent characterizing the smoothness of the transition. The analytical behavior of the dielectric constant is illustrated in Figure 13.3. An appropriate value for $\beta$ is chosen by fitting solvation energies predicted for various polar or unpolar molecules to their experimental values [22]. It should be noted that a pure continuum description may be insufficient if the solvent molecules adsorb specifically on the surface or if dielectric saturation effects arise because of the strong electric field in the Helmholtz layer [23]. In this case, a combined approach is to include a representative number of explicit solvent molecules in the region of chemical interactions and then to embed the entire system in a polarizable continuum.

In the context of density functional calculations, the charge density dependence of the dielectric constant contributes an additional term to the effective electronic

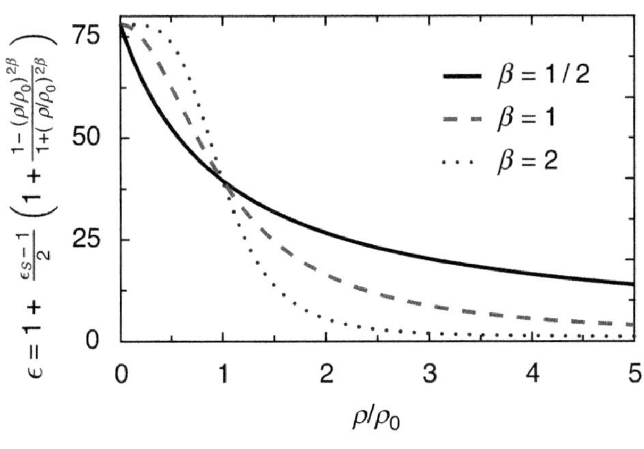


**FIGURE 13.3** Dielectric constant $\epsilon$ as a function of the reduced density $\rho/\rho_0$ varying the value of the smoothness exponent $\beta$.

Hamiltonian. This additional contribution $v_S$ can be expressed as [16]

$$v_S(\mathrm{r}) = -\frac{1}{8\pi}|\nabla v(\mathrm{r})|^2 \frac{\partial \epsilon}{\partial \rho}(\mathrm{r}) \tag{13.21}$$

This additional contribution to the Hamiltonian brings about a sharp singularity across the solvation shell because the derivative $\partial\epsilon/\partial\rho$ is large in that region. As a result of this singularity, electrons tend to escape from the metal surface into the bulk of the solvent, which impairs the numerical convergence of the self-consistent energy minimization. To eliminate this numerical instability, we apply a real-space preconditioning to the Hamiltonian that consists of adding a potential barrier in the solvent region to prevent electrons from leaking out. Another possible alternative is to redefine the dielectric function as a function of atomic coordinates to eliminate the dependence of the dielectric constant with respect to the self-consistent charge density [24,25].

### 13.3.2 Corrective Solvation Potential

The Poisson equation in the presence of a surrounding polarizable medium can be solved using multigrid techniques [26,27]. In Figure 13.4, we compare the calculated potential to that of a periodically repeated platinum slab in vacuum, obtained using conventional fast Fourier transform (FFT) techniques.

Inspired by the analysis presented in Reference 28, it is useful to introduce a corrective electrostatic potential $v^{\mathrm{corr}}$, which is defined as the difference between the potential of the solvated systems and its periodic counterpart:

$$v^{\mathrm{corr}} = v - v' \tag{13.22}$$

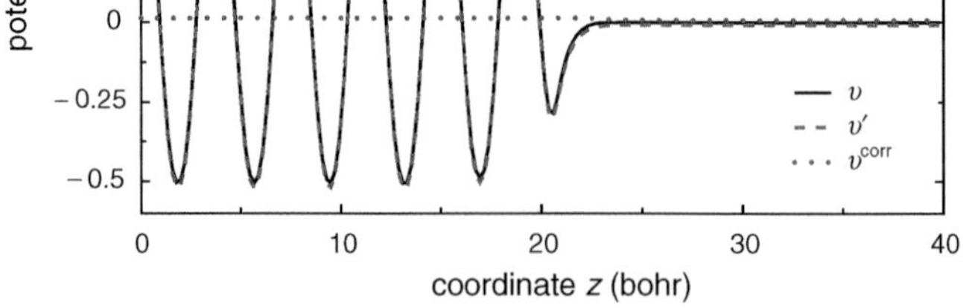


**FIGURE 13.4** Electrostatic potential in a solvent $v$, periodic potential $v'$, and electrostatic correction $v^{\mathrm{corr}}$ averaged in the $xy$ plane parallel to the interface for a solvated platinum slab. The parameters of the solvation model are $\beta = 1.25$ and $\rho_0 = 0.0005$ bohr$^{-3}$.

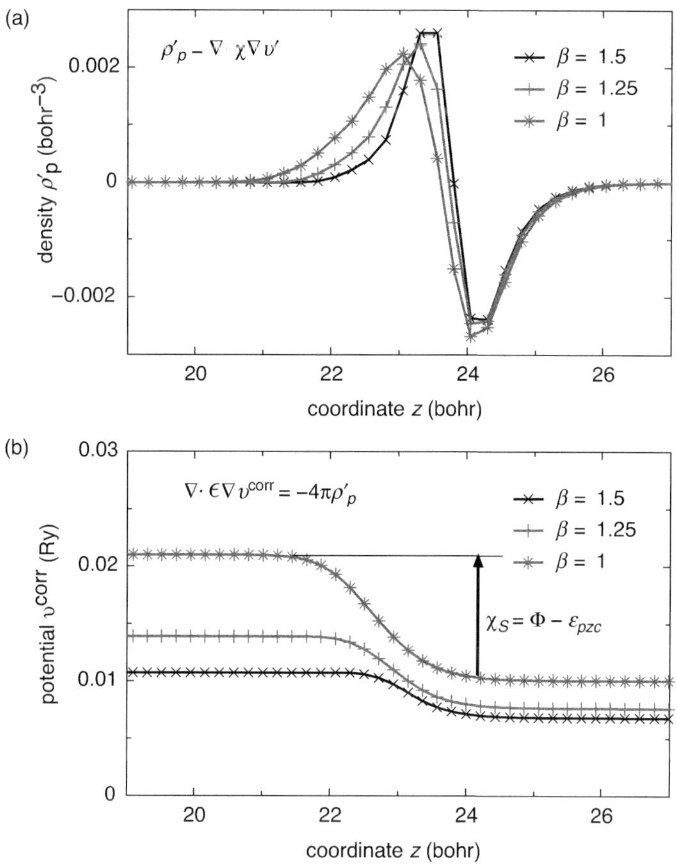


**FIGURE 13.5** Corrective polarization charge density $\rho'_p$ and corrective solvation potential $v^{corr}$ for a solvated platinum surface varying the value of the exponent $\beta$ defining the smoothness of the dielectric transition. The value of the charge density cutoff $\rho_0$ is set to 0.0005 bohr$^{-3}$.

Figure 13.4 shows that the potential $v^{corr}$ is constant inside the slab and in the bulk of the solvent. Thus, the electrostatic contribution from the solvent environment does not affect the electronic and atomic structures deep inside the electrode surface, as expected. It is also seen that the two constant-$v$ regions are connected by a smooth transition. This transition reflects the contribution from the solvent surface dipole $\chi_S$ (defined in Section 13.2), which directly depends on the value of the exponent parameter $\beta$. Indeed, as shown in Figure 13.5, decreasing the solvation parameter $\beta$ tends to raise the dipole potential $\chi_S = \phi - \mathcal{E}_{PZC}$. Therefore, the value of $\beta$ is directly connected to a relevant measurable electrochemical observable.

In order to characterize further the analytical behavior of the corrective potential $v^{corr}$, let us first write the Poisson equation for the periodic potential [28]

$$\nabla^2 v' = -4\pi(\rho - \langle\rho\rangle) \tag{13.23}$$

where $\langle\rho\rangle$ denotes the average of the total charge density inside the calculation cell, which corresponds to the contribution from the compensating jellium background. Substituting Equation (13.23) into Equation (13.19) and using the definition of the corrective potential, we obtain the following differential equation for the corrective solvation potential

$$\begin{cases} \nabla\cdot\epsilon\nabla v^{\text{corr}} = -4\pi(\langle\rho\rangle + \rho'_{\text{p}}) \\ \rho'_{\text{p}} = \nabla\cdot\chi\nabla v' \end{cases} \tag{13.24}$$

where $\rho'_{\text{p}}$ is the polarization charge density induced by the periodic potential. Typical profiles of the polarization density are shown in Figure 13.5 for different values of the solvation exponent parameter $\beta$. These profiles clearly show that $\rho'_{\text{p}}$ vanishes inside both the solid phase and the liquid phase, and contributes a dipole layer at the metal–solution interface. The additional density contribution $\langle\rho\rangle$ is there to cancel that from the uniform jellium background $\rho_{\text{jell}} = -\langle\rho\rangle$. Because the total corrective density $\rho^{\text{corr}} = \langle\rho\rangle + \rho'_{\text{P}}$ is constant inside the solid and liquid phases, the corrective potential exhibits a very smooth behavior, except at the metal–solution interface, due to the dipole-layer contribution that arises from polarization of the solvent.

It now remains to incorporate one last ingredient: the electrostatic contribution from the solvated counterions. This can be done by simply adding the potential-dependent diffuse ionic distribution $\rho_{\text{d}}$ to the source term of the generalized electrostatic problem:

$$\nabla\cdot\epsilon\nabla v = -4\pi(\rho + \rho_{\text{d}}) \tag{13.25}$$

A simple parametrization of countercharge distribution consists in defining $\rho_{\text{d}}$ as

$$\rho_{\text{d}} = \begin{cases} z_{\text{d}}c_{\text{d}}\left(e^{-z_{\text{d}}v/k_{\text{B}}T} - e^{z_{\text{d}}v/k_{\text{B}}T}\right) & \text{if} \quad \rho < \rho_1 \\ 0 & \text{if} \quad \rho \geq \rho_1 \end{cases} \tag{13.26}$$

where the parameter $\rho_1$ sets the the boundary of the countercharge region and can be selected to reproduce typical Helmholtz layer thicknesses, which are experimentally found to be equal to 3–5 Å (i.e., approximately the thickness of a water bilayer) regardless of the nature of the electrode compound [5]. Note that more representative ionic parametrization may also be employed to account for finite-size steric interactions between counterions [29] and the discrete dipolar nature of the polarizable medium [30].

As a consequence of including $\rho_{\text{d}}$, the electrostatic equation for the total corrective potential now reads

$$\nabla\cdot\epsilon\nabla v^{\text{corr}} = -4\pi(\langle\rho\rangle + \rho'_{\text{p}} + \rho_{\text{d}}) \tag{13.27}$$

Note that although the counterion density $\rho_{\text{d}}$ introduces some long-range spatial dependence inside the solvent region of the source term $\rho^{\text{corr}} = \langle\rho\rangle + \rho'_{\text{p}} + \rho_{\text{d}}$, the corrective potential $v^{\text{corr}}$ is still much smoother than $v$. This fact allows for the inexpensive computation of the corrective solvation potential $v^{\text{corr}}$ on coarse-grained meshes using multigrid techniques in the spirit of the density countercharge method [28].

It should be mentioned that, because of the nonlinearity that arises from the self-consistent dependence of the counterion diffuse density $\rho_d$ as a function of the potential $v$ (and, thus, as a function of the potential $v^{corr}$), we choose to employ a nonlinear multigrid solver for the efficient solution of Equation (13.27). In addition, solving the *modified Poisson-Boltzmann* (MPB) electrostatic problem in a finite simulation cell with, for example, periodic or homogeneous Dirichlet boundary conditions results in significant errors in the electrode voltage. To correctly extrapolate the slowly vanishing electrostatic potential, we choose to impose fictitious electrochemical boundary conditions (EBCs). Those can be obtained from Equation (13.12), which directly relates the potential $v$ to its derivative inside the bulk of the solvent. Thus, assuming that the distance between the boundaries of the cell and the electrode surface is sufficiently large for the monodimensional approximation to be valid, we can impose the following electrochemical boundary conditions

$$\nabla v \cdot \mathbf{n}_z = -\sqrt{\frac{32\pi c_d k_B T}{\epsilon_S}} \sinh \frac{z_d v}{2 k_B T} \tag{13.28}$$

where $\mathbf{n}_z$ denotes the external surface vector. This computational procedure can be efficiently incorporated in most electronic structure schemes with a moderate increase in computational cost; in our slab calculations, the additional computational cost is found to be lower than 10%.

A comparison of the convergence of the electrode potential for an 11-layer platinum (100) slab of surface charge 0.1 C/m$^2$ immersed in a monovalent ionic solution of concentration 0.0001 M is reported in Figure 13.6. We employ ultrasoft pseudopotentials [31] with plane-wave energy cutoffs of 40 and 400 Ry. The Brillouin zone is

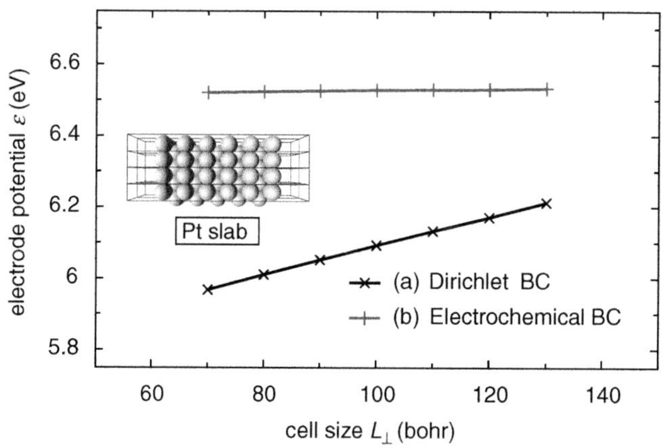


**FIGURE 13.6** Electrode potential as a function of transverse cell size for a platinum surface immersed in a dilute ionic solvent imposing (a) homogeneous Dirichlet boundary conditions ($v(\pm L/2) = 0$), and (b) electrochemical boundary conditions [Eq. (13.12)].

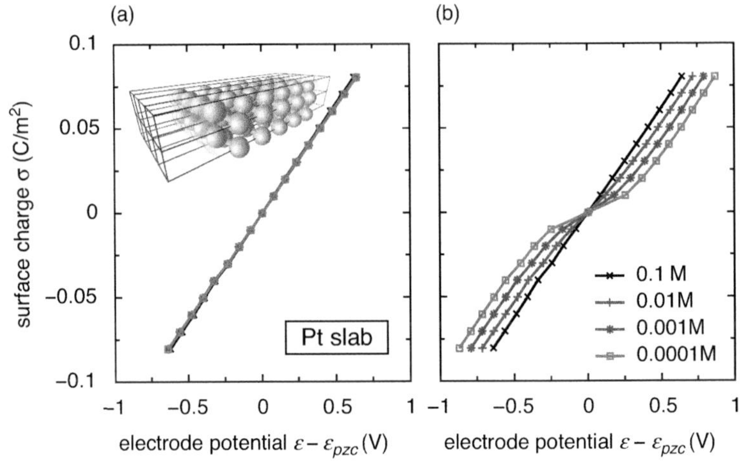


**FIGURE 13.7** Surface charge as a function of the electrode potential for a platinum surface immersed in an ionic solvent imposing (a) homogeneous Dirichlet boundary conditions and (b) electrochemical boundary conditions.

sampled using a $5 \times 5 \times 1$ mesh with cold-smearing occupations (the smearing temperature is 0.05 Ry). The parameters of the electrochemical model are selected to be $\epsilon_S = 78$, $k_B T = 0.0025$ Ry, $\rho_0 = 0.0005$ bohr$^{-3}$, $\rho_1 = 0.0001$ bohr$^{-3}$, and $\beta = 1.25$. We observe that while the error in $\mathcal{E}$ is on the order of 0.5 eV using homogeneous Dirichlet boundary conditions, the potential rapidly converges within a few millielectronvolts using electrochemical boundary conditions. As an additional computational validation, we study the charge–voltage characteristics of a platinum interface as a function of the ionic concentration (in Fig. 13.7). The supercell size in the transverse direction is 80 bohr. Using homogeneous Dirichlet boundary conditions, we observe that the electrical response of the platinum electrode remains linear for all concentrations in the range 0.0001–0.1 M (Fig. 13.7a). This behavior contrasts with the significant decrease in the charge–voltage slope occurring at the potential of zero charge for electrodes subject to electrochemical boundary conditions, which reflects a marked drop in electrode differential capacitance $\mathcal{C}$ from 0.11 F/m$^2$ at 0.1 M to less than 0.04 F/m$^2$ at 0.0001 M, in qualitative accordance with experimental observations.

### 13.3.3 Surface Electronic Structure and Surface Energy of Electrode–Electrolyte Interfaces

We now apply the atom-continuum approach to understand the modifications of the electronic structure at an electrode–electrolyte interface in the presence of an applied voltage (Fig. 13.8). All quantities are averaged in the $xy$ = surface plane. We denote by $\langle \Delta\rho_e \rangle_{xy}$ the difference in electronic density when the metal is brought from neutrality to a surface charge of $-10\,\mu$C/cm$^2$. The $xy$-averaged potential $\langle \Delta v_{\text{coul}} \rangle_{xy}$ is the resulting difference in electrostatic potential. Calculations were done on a five-layer infinite aluminum slab. The perpendicular $z$ axis lies along the (111) direction. The

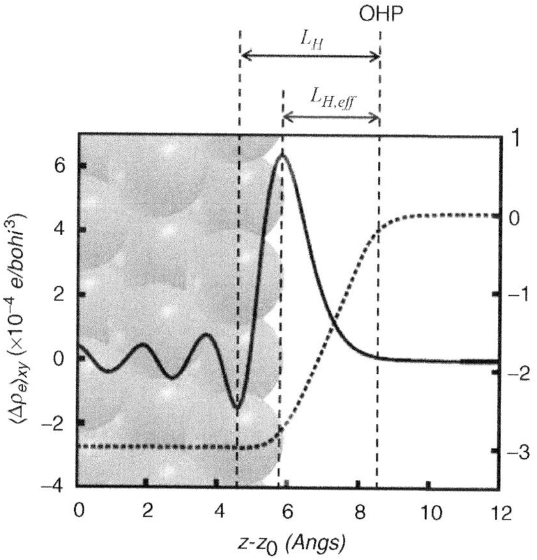


**FIGURE 13.8** Electronic structure at electrode–electrolyte interface.

center of the slab is positioned at $z_0$. Here, the diffuse-layer capacitance is infinite (i.e., the ionic concentration is very high). That is why the potential levels off directly after the outer Helmholtz plane (OHP). Even at the quantum level, the electrostatic behavior of the metal is correctly described by the classical Helmholtz model. $\langle \Delta v_{\text{coul}} \rangle_{xy}$ is flat inside the metal, then increases linearly, and is flat again beyond the OHP. However, quantum effects from the metal are important quantitatively. We observe that the effective Helmholtz length $L_{\text{H,eff}}$ is reduced because of an electronic spillover extending up to 1 Å from the metal into the solvent region.

Changing the charge affects the potential along with the surface energy. Results displayed in Figure 13.9 are for a (100) aluminum surface immersed in a 0.1 M ionic

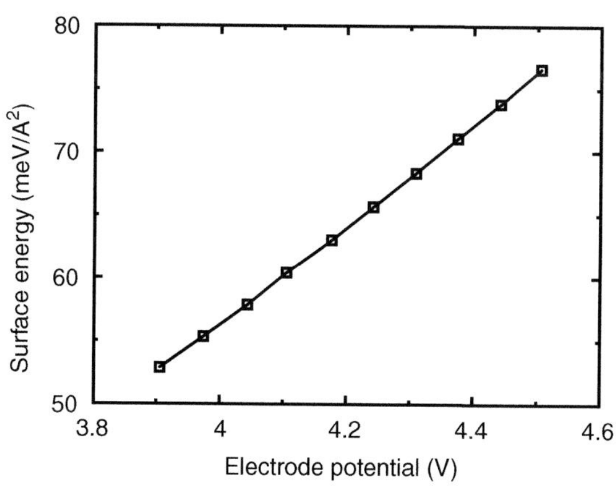


**FIGURE 13.9** Surface energy as a function of the electrode potential.

solution. The surface energy is calculated from its value in vacuum [32], to which is added the solvent effect treated using the solvation continuum model. The Helmholtz length is 4 Å. The electrode potential is swept across a 0.6 V range by bringing the surface charge from 0 to 9 μC/cm$^2$, from which the surface capacitance is calculated to be equal to 15 μF/cm$^2$. Meanwhile, the surface energy increases by 24 meV/Å$^2$. The absolute potential at zero charge, which is also the electron chemical potential, is equal to 3.9 V. Therefore, a perturbative analysis predicts a surface energy change equal to the product $3.9\,\text{V} \times 9\,\mu\text{C/cm}^2 = 22\,\text{meV/Å}^2$, which is very close to the value of 24 meV/Å$^2$ calculated using the fully self-consistent solvation scheme.

## 13.4 CONCLUSION

In this chapter, we have presented an atom-continuum model for studying electrified interfaces under electrochemical conditions, particularly focusing on length-scale limitations in the description of the ionic electrolyte. We have shown that the electrostatic contribution from the ionic solvent can be calculated independently at low computational cost using nonlinear multigrid techniques and taking advantage of the periodic solution of the electrostatic problem. This approach allows us to account for the influence of an applied electrode voltage in both the high- and low-ionic-molarity regimes, and to describe the electrical response of electrochemical interfaces for the interpretation of electrochemical trends.

In our electrochemical calculations, we have employed a canonical (i.e., constant-charge) framework, which consists of controlling the electrical state of the system via the total number of electrons and determining the applied voltage from the calculated Fermi level. A possible alternative is to work in the grand-canonical regime by directly fixing the value of the applied voltage in the self-consistent minimization of the total energy. Although the constant-charge and constant-potential approaches are exactly equivalent for macroscopic systems, it is important to note that they can significantly differ in describing local surface phenomena. Indeed, while one may safely employ the canonical framework for predicting phenomena that are known to modify the electrical state of the whole electrode (e.g., voltage-induced collective desorption, surface reconstruction), one must work at constant potential for the study of a local electrochemical reaction that does not affect the entire system. Consequently, the development of a constant-voltage computational scheme would be very valuable for future applications of the electrochemical model, including the study of electrochemical reactions at electrode–electrolyte interfaces and the interpretation of local electrochemical spectroscopy experiments.

## ACKNOWLEDGEMENTS

This work was supported in part by the MRSEC Program of the National Science Foundation under award number DMR-0819762 and by the *Agence Nationale de la Recherche* under grant number ANR-06-CIS-014 (SIRE).

## REFERENCES


1. B. Mennucci and R. Cammi, *Continuum Solvation Models in Chemical Physics*, Wiley, 2007.
2. F. Floris and J. Tomasi, *J. Comput. Chem.* **10**, 616 (1989).
3. F. Floris, J. Tomasi, and J. L. Pascual Ahuir, *J. Comput. Chem.* **12**, 784 (1991).
4. M. C. Payne, M. P. Teter, D. C. Allan, T. A. Arias, and J. D. Joannopoulos, *Rev. Mod. Phys.* **64**, 1045 (1992).
5. N. Sato, *Electrochemistry at Metal and Semiconductor Electrodes*, Elsevier, 1998.
6. J. Tomasi and M. Persico, *Chem. Rev.* **94**, 2027 (1994).
7. A. V. Marenich, C. J. Cramer, and D. G. Truhlar, *J. Chem. Theory Comput.* **4**, 877–887 (2008).
8. J. O'M. Bockris, A. K. N. Reddy, and M. Gamboa-Adelco, *Modern Electrochemistry*, 2nd ed. Vol. Electrodics 2A, Kluwer Academic/Plenum Publishers, 1998.
9. C. D. Taylor, S. A. Wasileski, J.-S. Filhol, and M. Neurock, *Phys. Rev. B* **73**, 165402 (2006).
10. J. O'M. Bockris and S. U. M. Khan, *Surface Electrochemistry*, Plenum Press, 1993.
11. J. K. Nørskov et al., *J. Phys. Chem. B* **108**, 17886 (2004).
12. G. S. Karlberg, J. Rossmeisl, and J. K. Nørskov, *Phys. Chem. Chem. Phys.* **9**, 5158 (2007).
13. W. R. Fawcett and T. G. Smagala, *Electrochim. Acta* **53**, 5136 (2008).
14. W. R. Fawcett, *Electrochim. Acta* **54**, 4997 (2009).
15. M. S. Kilic, M. Z. Bazant, and A. Ajdari, *Phys. Rev. E* **75**, 021503 (2007).
16. J.-L. Fattebert and F. Gygi, *Int. J. Quantum Chem.* **93**, 139 (2003).
17. M. Sharma, R. Resta, and R. Car, *Phys. Rev. B* **98**, 247401 (2007).
18. B. Mennucci, E. Cancès, and J. Tomasi, *J. Phys. Chem. B* **101**, 10506–10517 (1997).
19. S. Miertus, E. Scrocco, and J. Tomasi, *J. Chem. Phys.* **55**, 117 (1981).
20. V. Barone, M. Cossi, and J. Tomasi, *J. Chem. Phys.* **107**, 3210 (1997).
21. D. M. York and M. Karplus, *J. Phys. Chem. A* **103**, 11060 (1999).
22. D. A. Scherlis, J.-L. Fattebert, F. Gygi, M. Cococcioni, and N. Marzari, *J. Chem. Phys.* **124**, 74103 (2006).
23. T. Pajkossy and D. M. Kolb, *Electrochim. Acta* **46**, 3063 (2001).
24. R. Jinnouchi and A. B. Anderson, *Phys. Rev. B* **77**, 245417 (2008).
25. V. Sánchez, M. Sued, and D. A. Scherlis, *J. Chem. Phys.* **131**, 174108 (2009).
26. M. Holst and F. Saied, *J. Comput. Chem.* **14**, 105–113 (1993).
27. M. Holst and F. Saied, *J. Comput. Chem.* **16**, 337–364 (1995).
28. I. Dabo, B. Kozinsky, N. E. Singh-Miller, and N. Marzari, *Phys. Rev. B* **77**, 115139 (2008).
29. I. Borukhov, D. Andelman, and H. Orland, *Phys. Rev. Lett.* **79**, 435 (1997).
30. A. Abrashkin, D. Andelman, and H. Orland, *Phys. Rev. Lett.* **99**, 077801 (2007).
31. D. Vanderbilt, *Phys. Rev. B* **41**, 7892–7895 (1990).
32. N. E. Singh-Miller and N. Marzari, *Phys. Rev. B* **80**, 235407 (2009).

CHAPTER 14

# Electronic Structure and Reactivity of Transition Metal Complexes

HEATHER J. KULIK and NICOLA MARZARI

Department of Materials Science and Engineering, Massachusetts Institute of Technology, Cambridge, Massachusetts

Transition metals are ever-present as reactive centers in biological and inorganic catalytic cycles. However, the open-shell character that gives $3d$ transition metals unique reactive properties also makes transition metal complexes challenging to study using first-principles approaches, especially when using local or semilocal approximations to density functional theory (DFT). We describe here an approach based on Hubbard $U$ corrections—widely used in the solid-state community to describe strongly correlated systems—and show how it helps achieve predictive accuracy in DFT calculations of transition metal complexes. The success of this approach comes from counteracting the tendency of $3d$ electrons to delocalize, driven by the imperfect cancellation of electrostatic self-interactions in common exchange correlation approximations. Since the Hubbard term $U$ is calculated through a linear response formulation—recently extended by us to allow for self-consistency—it represents a fully *ab initio*, nonempirical approach. We analyze the performance of the DFT+$U$ formulations on a few paradigmatic test cases, with special attention to the structure, electronic structure, and potential energy surface of $Fe_2$ dimers and the addition–elimination reaction of hydrogen or methane on $FeO^+$. Thanks to negligible computational overheads, we also show how the approach can be effortlessly applied to large-scale simulations, such as the case presented here of functionalized cobalt porphyrins on a metal support.

## 14.1 INTRODUCTION

Transition metals are ever-present as reactive centers in molecules active in biological and inorganic catalytic cycles. In biological systems, mid-row $3d$ transition metals

*Fuel Cell Science: Theory, Fundamentals, and Biocatalysis*,
Edited by Andrzej Wieckowski and Jens K. Nørskov

facilitate reactions as diverse as methane-to-methanol conversion at an antiferromagnetically coupled dimetal center [1,2], unactivated alkane halogenation by a high-energy, high-spin ferryl-oxo center in the halogenase SyrB2 [3], and oxygen binding at iron porphyrins in hemoglobin [4]. In inorganic catalysts, many transition metal complexes have been designed to carry out synthetic pathways that mimic active biological catalytic cycles. However, the open-shell character that gives 3*d* transition metals unique reactive properties also makes transition metal complexes quite difficult to study using density functional theory (DFT) approaches, which are particularly attractive as their efficiency gives access to systems of several hundreds of atoms in size. Common implementations of DFT, however, including the local density approximation (LDA) and the generalized-gradient approximation (GGA), utilize mean-field approximations that are ill-suited to studying 3*d* transition metals with highly localized, open-shell orbitals [5]. Following the success of LDA+*U* (and GGA+*U*) in accurately reproducing the ground state of transition metal oxide solids [5,6], we showed for the first time how this approach can also be applied to single- or few-site molecular systems [7]. The LDA+*U* approach is particularly suitable for application to single- or few-site molecules because the *U* term plays a specific role in correcting exchange self-interaction errors and tuning the hybridization in the local bonding interactions.

The LDA+*U* or GGA+*U* method in its original formulation is an approximation that corrects LDA or GGA in order to better describe strong electronic correlations. It is based on a Hubbard model approach for treating strongly correlated systems [8,9]. As a result, electronic structure approaches have incorporated Hubbard *U* terms in order to better treat correlated solids for quite some time [9]. A simplified, rotationally invariant LDA+*U* functional was introduced by Cococcioni and de Gironcoli [5]. In addition, in their approach, the *U* term may be calculated as a property of the system from linear response theory. The general form of the LDA+*U* functional (where LDA can really be extended to any exchange correlation functional) is as follows:

$$E_{\mathrm{LDA}+U}[\mathbf{n}(r)] = E_{\mathrm{LDA}}[\mathbf{n}(r)] + E_{\mathrm{Hub}}[n^{I\sigma}_{mm'}] - E_{\mathrm{dc}}[n^{I\sigma}] \tag{14.1}$$

In this equation $\mathbf{n}(r)$ is the electron density, while $n^{I\sigma}_{mm'}$ is a generalized occupation matrix for atom *I*, spin σ, which is considered the source of the strongly correlated manifold, usually given by the localized *d* or *f* orbitals of the free atom. The $E_{\mathrm{LDA}}[\mathbf{n}(r)]$ term represents the contribution of the energy from the LDA or GGA functional, and, in this work, we actually employ GGA in our examples [10]. Since the GGA exchange correlation functional already includes some mean-field treatment of on-site electronic interactions, when we introduce a correction to model properly the on-site correlation (in this case $E_{\mathrm{Hub}}[n^{I\sigma}_{mm'}]$), it is also necessary to subtract out the mean-field correlation in LDA using a double-counting term, $E_{\mathrm{dc}}[n^{I\sigma}]$. To recover the proper electron–electron interactions, $E_{\mathrm{Hub}}[n^{I\sigma}_{mm'}]$ is broken down into a Coulomb (the parameter *U*) and exchange interaction (the parameter *J*), which have been modeled formally borrowing from atomic Hartree–Fock Slater integrals [5,9].

By neglecting the exchange term, or essentially modeling an effective interaction $U_{\mathrm{eff}}$ that is equal to $U - J$, it is possible to greatly simplify previous expressions for the

LDA+$U$ energy functional. The double-counting term $E_{\rm dc}[n^{I\sigma}]$ is approximated in terms of the $U$ and $J$ parameters as

$$E_{\rm dc}\left[n^I\right] = \sum_I \frac{U}{2} n^I(n^I-1) - \sum_I \frac{J}{2}\left[n^{I\uparrow}(n^{I\uparrow}-1) + n^{I\downarrow}(n^{I\downarrow}-1)\right] \qquad (14.2)$$

leading to the final form for the Hubbard $+U$ functional [5]:

$$\begin{aligned} E_U\left[n^{I\sigma}_{mm'}\right] &= E_{\rm Hub}\left[n^{I\sigma}_{mm}\right] - E_{\rm dc}\left[n^{I\sigma}\right] \\ &= \frac{U}{2}\sum_I \sum_{m,\sigma} n^{I\sigma}_{mm} - \sum_{m'} n^{I\sigma}_{mm'} n^{I\sigma}_{m'm} \\ &= \frac{U}{2}\sum_{I,\sigma} \mathrm{Tr}\left[\mathbf{n}^{I\sigma}(1-\mathbf{n}^{I\sigma})\right] \end{aligned} \qquad (14.3)$$

If the occupation matrix is diagonalized, this expression simplifies even further to an $E_U[n^{I\sigma}_{mm'}]$ of the form $(U/2)\sum_{I,\sigma}\sum_i \lambda_i^{I\sigma}(1-\lambda_i^{I\sigma})$, where $\lambda_i^{I\sigma}$ is simply the occupation of an orbital of type $i$, spin $\sigma$, on site $I$. In general, projector operators $P^I_{mm'}$ for the localized orbital manifold on site $I$ may be used to build the occupation matrix. The projection operators may be defined by any number of bases, but we use the atomic $3d$ orbitals, which are encoded in the pseudopotential of the atom. It is particularly key to ensure that the atomic orbitals in the pseudopotential are normalized and the best possible approximation for the appropriate charge state and electron configuration of the molecule. An alternative approach is to utilize Wannier functions as the projection manifold, but these should be kept frozen when comparing different structures [11].

In order to explain why a Hubbard approach can be useful in describing the electronic structure of molecules, it is worthwhile to recall [5] that the LDA energy of an atom in contact with a reservoir of electrons will essentially behave parabolically as a function of electron occupation, with the minimum often situated at a fractional occupation number. The true energy of this model system is instead a statistically weighted, piecewise-linear function [12]. Therefore, the $+U$ term that is added back to an LDA functional may be regarded as the correction that offsets the unphysical curvature of the original LDA energy (see Fig. 1 in Ref. 5). This curvature can be calculated from first-principles theory as the second derivative of the energy with respect to occupations. It should be noted, though, that in addition to the unphysical curvature in energy of the fully screened, interacting system, there exists some nonlinear variation in the energy of the Kohn–Sham, noninteracting system due to rehybridization after a shift in occupation number. The effective curvature that one wants to extract and cancel out should be the difference between the screened, interacting $U$ and the bare, noninteracting one.

This viewpoint allows us to forego supercell constrained-density functional calculations [5,9], and to instead apply a linear response approach in order to straightforwardly calculate $U$. This technique is implemented in the open-source electronic structure package, QUANTUM-ESPRESSO [13]. Using Janak's theorem, we may equate the eigenvalue of state $i$ with the derivative of the energy with respect to the occupation of the state, to recast the second-derivative problem as a first-derivative, linear response property. Thus, the partial derivative of the energy of the state with occupation $q_J$ with respect to $q_I$ is $-\alpha_I$, and an analogous expression is true for the bare Kohn–Sham system. The second derivative of the energy with respect to occupations, therefore, is simply the derivative of the potential $\alpha_I$ with respect to occupations $q_I$. Such a change of variables is key because it is difficult to constrain occupations of the localized orbitals of a system. In order to further simplify the practical details of the calculation, a Legendre transform to the energy can be used to have as independent variables a potential $\alpha_I$ of the localized orbitals. The new energy expression can then be written as a function of $\alpha_I$:

$$E[\alpha_I] = \min_{\mathbf{n}(r)} E[\mathbf{n}(r)] + \sum_I \alpha_I n_I \tag{14.4}$$

An analogous equation may be written for the bare system. A single potential shift takes the form $\Delta V = \sum_I \alpha_I P^I$, where $P^I$ are the projector operators, and thus affects each individual localized orbital. The linear response functions of this system with respect to a local perturbation $\Delta V$ are then

$$\begin{aligned} \chi_{IJ} &= \frac{\partial^2 E}{\partial \alpha_I \, \partial \alpha_J} = \frac{\partial n_I}{\partial \alpha_J} \\ \chi^0_{IJ} &= \frac{\partial^2 E^0}{\partial \alpha^0_I \, \partial \alpha^0_J} = \frac{\partial n_I}{\partial \alpha^0_J} \end{aligned} \tag{14.5}$$

where $I$ and $J$ are neighboring sites containing localized orbitals or differing manifolds of localized orbitals on the same atomic site. Response functions are obtained by measuring the change in occupations from several small potential shifts to an already converged single point density, and the associated computational cost is negligible (see Fig. 14.1). The $U$ associated with a single site $I$ is then simply $U = \chi_0^{-1} - \chi^{-1}$.

In this approach, special care should be taken in the inversion of the response functions. As response functions approach asymptotically small numbers, their inverses will diverge. The difference of the response functions, therefore, may appear to fall within a reasonable range ($U = 0 - 10$ eV), while the value is simply an artifact of the calculation. It is key to examine the magnitude of the response functions before inversion to ensure that the linear response values are not of the same order as the numerical noise of the calculation.

Normally, we consider only 3$d$ states in transition metals to be relevant for determining $U$, but the value of $U$ is strongly dependent on the mode of bonding. If, as is true in the case of transition metal hydrides, 4$s$–3$d$ hybridization is strong or 4$s$ orbitals dominate molecular bonding, then the response of this manifold is also relevant. In this case the response functions $\chi_0$ and $\chi$ should be expressed as matrices. These matrices are then inverted and their diagonal terms are subtracted to provide

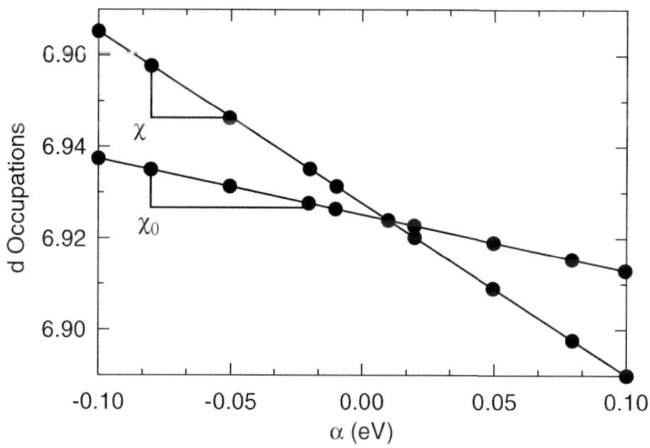


**FIGURE 14.1** Examples of bare $\chi_0$ and converged $\chi$ linear response functions for the quartet transition-state ($^4$TS-1) structure in the addition–elimination reaction of $H_2$ on $FeO^+$.

Hubbard parameters, $U_{3d}$ and $U_{4s}$, appropriate to the respective manifolds. In solids, interaction and hybridization of $3d$–$4s$ is seldom relevant, but in molecules strong mixing, particularly of $3d_{z^2}$ and $4s$ orbitals, often occurs.

While the linear response formulation has been shown to be quite useful, some careful considerations are in order to increase the reliability and accuracy of the results derived. In the solid state, LDA+$U$ often opens a gap in Mott insulators, which are predicted by LDA to be metals [5]. Analogously for molecules, LDA may predict the wrong ground-state spin and symmetry when compared to LDA+$U$. In either case, the LDA wavefunctions will differ greatly from those obtained using LDA+$U$, and we expect that the linear response of the LDA ground state will not work well to characterize the true LDA+$U$ system. We have proposed [7] an approach that overcomes this difficulty, and in which the Hubbard $U$ may be found in a fully self-consistent (referred to as $U_{\text{scf}}$) manner. To clarify our approach, we first identify in the GGA+$U$ functional the electronic terms that have quadratic dependence on the occupations, and express these as a function of a projection matrix that has been diagonalized:

$$
\begin{aligned}
E_{\text{quad}} &= \frac{U_{\text{scf}}}{2}\sum_I\left[\sum_i \lambda_i^I\left(\sum_j \lambda_j^I - 1\right)\right] \\
&+ \frac{U_{\text{in}}}{2}\sum_I\sum_i \lambda_i^I(1-\lambda_i^I)
\end{aligned}
\tag{14.6}
$$

The first term represents the contribution already contained in the standard GGA functional, modeled here as a double-counting term, while the second term is the customary+$U$ correction. Therefore, $U_{\text{scf}}$ represents the effective on-site electron–electron interaction already present in the GGA energy functional for the GGA+$U$ ground state when $U$ is chosen to be $U_{\text{in}}$. Consistency is enforced by choosing $U_{\text{in}}$ to be equal to $U_{\text{scf}}$.

The $U$ obtained from linear response [5] (labeled here $U_{out}$) is also obtained by differentiating Equation (14.6) with respect to $\lambda_T^I$:

$$U_{out} = \frac{d^2 E_{quad}}{d(\lambda_T^I)^2} = U_{scf} - \frac{U_{in}}{m} \tag{14.7}$$

where $m = 1/\sum_i (a_i^I)^2$ can be interpreted as an effective degeneracy of the orbitals whose population is changing during the perturbation {to linear order, $\delta\lambda_i^I = a_i^I \delta\lambda_T^I$ with $\sum_i a_i^I = 1$ and $d^2/d(\lambda_T^I)^2 = \sum_{ij} a_i^I a_j^I [d^2/(d\lambda_i^I d\lambda_j^I)]$}. Even if in principle $U_{scf}$ depends on $U_{in}$, we find it to be constant over a broad interval, as apparent from Figure 14.2: $U_{out}$ is linear in $U_{in}$ for the relevant range of $U_{in} \sim U_{scf}$. Thus, from few linear response calculations for different $U_{in}$ ground states, we are able to extract the $U_{scf}$ that should be used. This approach is particularly useful for transition metal dimers, as illustrated in Figure 14.2. Otherwise, the effective difference between $U_0$ and $U_{scf}$ is often small and probably minimized by the constraints we employ. In fact, for molecular systems, we constrain the total spin of the system and enforce fixed occupations without a smearing or electronic temperature in most cases. We therefore calculate the value of $U$ for each spin and symmetry independently and average over the low-lying states. Generally, $U$ is highly dependent on both spin and symmetry as well as nuclear identity and configuration and should always be calculated for the state considered. On the other hand, when comparing energetics, a single, averaged value of $U$ must be applied to all systems, as will be discussed later for the low-lying states of the iron dimer and the addition–elimination reactions of methane and molecular hydrogen on $FeO^+$. We note that the apparent tendency of the LDA+$U$ functional to disproportionately favor integer occupations competes with the LDA or GGA tendency to favor fractional occupations. Under most conditions, the bonds may become slightly elongated with the addition of a+$U$ term, but energetics are

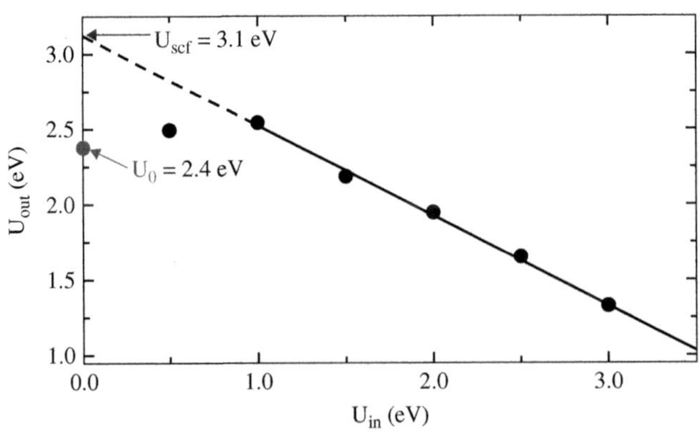


**FIGURE 14.2** Linear response $U_{out}$ calculated from the GGA+$U_{in}$ solution for ${}^7\Delta_u$ $Fe_2$, together with the extrapolated $U_{scf}$. $U_0$ is $U_{out}$ calculated for $U_{in} = 0$.

significantly improved. The overelongation is partially due to the fact that the projection manifold we choose is that of the isolated atomic orbitals.

## 14.2 RESULTS

Transition metal dimers and small clusters present a formidable challenge for both experimental and theoretical techniques. These systems are of fundamental importance in our understanding as both a model for dimetal enzyme active sites [1] and as the building blocks of nanoparticle catalysts that accelerate growth of nanotubes and nanowires [14]. We'll first consider the iron dimer as a test case for our approach. For nearly 30 years the ground state of neutral $Fe_2$ has been identified as $^7\Delta_u(3d^{14}: \sigma_g^2\pi_u^4\delta_g^3\pi_g^{*,2}\delta_u^2\sigma_u^*, 4s^2 : \sigma_g^2)$ (where subscripts $g$ and $u$ denote *gerade* and *ungerade*, respectively) in numerous theoretical studies [15–24] and some experimental studies [25,26]. However, the configuration and the structural and vibrational properties of this electronic state are not consistent with available experimental findings. Early density functional studies lacked accuracy in exchange correlation functional approximations, which may have overstabilized certain electronic states [17,19]. A tradeoff between basis set size and active space size in early configuration interaction studies limited the accuracy of the relative energetics for the hundreds of states of various spins and symmetries considered [22,23]. Despite the need to reconsider the discrepancy between experimental and theoretical findings since the late 1970s, even recent comprehensive studies of transition metal diatomic molecules [24,27] continue to assume the $^7\Delta_u$ assignment to be the correct one for the ground state of iron dimer. There have, however, been some more recent high-level quantum chemistry calculations—multireference configuration interaction (MRCI) [28] as well as large basis set, single-reference coupled cluster [CCSD(T)] calculations [29]—which provide evidence for a more consistent alternative ground state $^9\Sigma_g^-$ for the iron dimer. All of this prompted us to examine how GGA+$U$, a method already successful at improving calculations on correlated-electron transition metal solids, would describe the low-lying states of the iron dimer. A GGA+$U$ approach is particularly appealing because it can also be extended to much larger systems because it scales favorably in comparison to much more limited post-Hartree–Fock approaches.

While experimental results are far from complete, they have provided significant clues that help to limit the scope of potential electronic states for the low-lying states of the iron dimer. Several spin multiplicities are good candidates for the dimer; these include anything from an antiferromagnetically coupled singlet to a high-spin nonet. The large variety of states is due to the close spacing of the two lowest states of the neutral iron atom. The ground state of the Fe atom, $^5$D (valence electrons: $3d^64s^2$), is experimentally only 0.87 eV lower in energy than the excited $^5$F state with $3d^74s^1$ configuration. Additionally, the lowest triplet Fe state, $^3$F $3d^74s^1$, lies only 1.50 eV above the ground state. While states derived by unifying some combination of these three lowest symmetries are likely the best candidates for the lowest-lying states of the iron dimer, this still leaves a large number of possible electron configurations to consider varying from $3d^{14}4s^2$ to $3d^{12}4s^4$ and with any number of spin multiplicities. Stern–Gerlach experiments have shown the dimer to have a relatively large magnetic moment, thus ruling out an antiferromagnetic state [30], and interpretations of Mössbauer hyperfine interactions confirm the likelihood of a high-spin state [31,32]. Experimental Mössbauer results also indicated a $4s$ occupation per atom of 1.47 in the dimer by comparing isomer shift results with those for the iron atom, which has a known ground-state configuration of $3d^6 4s^2$ [33], but it is worth noting that $4s^3$ configurations have been largely overlooked in nearly all theoretical studies. Both the bond length and frequency have been measured experimentally in rare-gas (both Ne and Ar) matrices and found to be 2.02 Å and 300 cm$^{-1}$, respectively [34]. This low fundamental frequency suggests a fairly weak bond order, also in contradiction with the proposed $^7\Delta_u$ state, which has been suggested to have a formal bond order of 3 [35].

The piece of experimental evidence that provides the most information about the relative energies of electronic states is the photoelectron spectrum of $Fe_2^-$. This spectrum is remarkably simple, with just two peaks corresponding to transitions to $Fe_2$ [26]. This simplicity is also in direct opposition to the myriad of low-lying iron dimer electronic states that have been studied theoretically [22,23]. Experimental photoelectron spectra typically access only the bound electronic states that differ from the anion by a single electron detachment, and, in this case, the spectrum of $Fe_2^-$ displays two prominent peaks, one at 1.0 eV and the other 0.53 eV above the first peak. Structurally, the neutral states have identical properties within the error of the experiment ($r_e = 2.02 \pm 0.02$ and $\omega_e = 300 \pm 20\ \text{cm}^{-1}$) and are more bound than the anion, which has a bond length of $2.10 \pm 0.04$ Å and a vibrational frequency of $250 \pm 20\ \text{cm}^{-1}$, indicating an increase in bond order from the anion to the neutral dimer. By studying both the low-lying states of the anion and the corresponding allowed neutral states of the neutral dimer, a more recent MRCI study has identified an excitation scheme in which the lowest-lying anion state is $^8\Sigma_u^-$ and the two neutral peaks correspond to $^9\Sigma_g^-$ and $^7\Sigma_g^-$, respectively. This MRCI study also finds the previously proposed neutral iron dimer ground state, $^7\Delta_u$, to be 0.7 eV above the true ground state of the dimer. While both the alternative states $^9\Sigma_g^-$ and $^7\Sigma_g^-$ have been considered previously, less accurate methods found these states to be significantly excited with respect to $^7\Delta_u$ [23,24,33].

In order to build a GGA and GGA+$U$ description of the states present in the experimental photoelectron spectrum, we identify the lowest-lying states of the anion as $^8\Delta_g(3d^{14} : \sigma_g^2\pi_u^4\delta_g^3\pi_g^{*,2}\delta_u^2\sigma_u^*, 4s^3 : \sigma_g^2\sigma_u^*)$ and $^8\Sigma_u^-(3d^{13} : \sigma_g^2\pi_u^4\delta_g^2\pi_g^{*,2}\delta_u^2\sigma_u^* 4s^4 : {}_{\sigma}^{g2}\sigma_u^{*,2})$. While $^8\Delta_g$ is the lowest anion state as determined with GGA by 0.52 eV, a self-consistent GGA+$U$ approach favors the $^8\Sigma_u^-$ state by more than 0.38 eV. The difference in state splittings between the two methods is nearly an electronvolt, and the GGA+$U$ results are in much better agreement with the previous MRCI study. For all calculations, we obtain the Hubbard $U$ from the GGA ground state ($U_0$) as well as with a self-consistent approach ($U_{scf}$) and numerically average over all states and structures of interest, giving an overall $U_0$ of 2 eV and a $U_{scf}$ of 3 eV. Strictly speaking, the iron dimer has two independent iron sites within the formalism of GGA+$U$, and thus a $U$ for each iron atom should be determined by obtaining a $2 \times 2$ matrix for the linear response of each iron site. In practice, the $U$ we calculate here (correctly) treats the iron atoms as identical sites, and $U$ is obtained from the total response of the *3d* electrons of both iron sites. The net difference in the contribution to $E_U$ between the case where the iron sites are treated separately ($U_{\mathrm{sep}}$) versus the case where they are treated identically ($U_{\mathrm{ident}}$) is negligible; that is, the $U$ calculated differs, where $U_{\mathrm{sep}} = 2U_{\mathrm{ident}}$, because the $U$ is related to the inverse of the derivative in occupations and in the latter case the derivative is twice as large as the former. However, this difference cancels with the expression for $E_U$ because there is only a single site with half the occupations in $U_{\mathrm{ident}}$, while in the $U_{\mathrm{sep}}$ case one sums over the two sites. The only significant difference in these approaches is that the practical implementation of a separate $U$ formalism tends to enhance convergence to broken-symmetry, antiferromagnetic states. As such states are, for the most part, not highly relevant at low energies for $Fe_2$, enforcing the identical nature of the iron states is more suitable here. Naturally, in other multisite molecules, if antiferromagnetic character or multiple valence is present, a separate $U$ scheme should always be employed.

Beginning with the GGA anion ground state ${}^8\Delta_g$, the two lowest neutral dimer states that differ by a single electron loss are ${}^7\Delta_u$ and ${}^9\Delta_g$, respectively. The ${}^7\Delta_u$ state $(3d^{14} : \sigma_g^2\pi_u^4\delta_g^3\pi_g^{*,2}\delta_u^2\sigma_u^*, 4s^2 : \sigma_g^2)$ is produced through ionization of the spinup $4s$ $\sigma_u$ molecular orbital. The ${}^9\Delta_g$ state $(3d^{13} : \sigma_g^2\pi_u^4\delta_g^2\pi_g^{*,2}\delta_u^2\sigma_u^*, 4s^3 : \sigma_g^2\sigma_u^*)$ is produced by the loss of the spindown $3d$ $\sigma_g$ orbital and thus is expected to possess structural properties different from those of ${}^7\Delta_u$, which is inconsistent with experiments. The electronic configuration of these states is illustrated in Figure 14.3, and the calculated potential energy curves (see Fig. 14.4) of these states confirm the bond order analysis. By using a Makov–Payne estimate [36] for the true, open-boundary conditions, total energy of our charged $Fe_2^-$ system, we are able to estimate the energy required to ionize the ${}^8\Delta_g$ anion to the neutral state ${}^7\Delta_u$ to be 1.5 eV with the GGA approach. The GGA

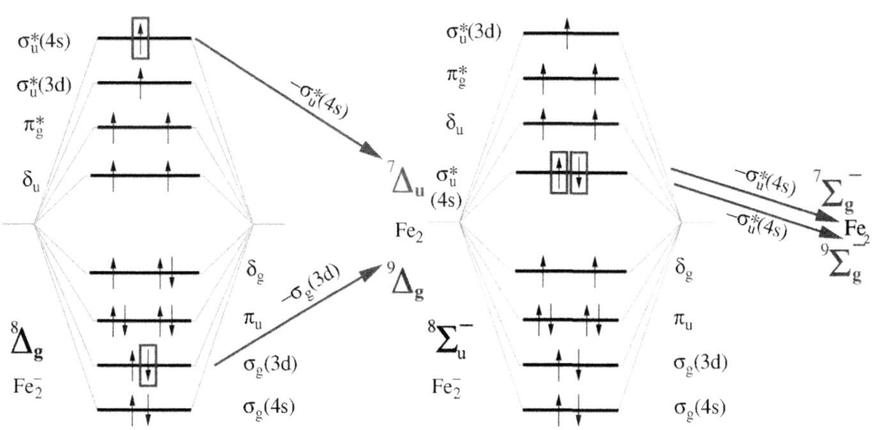


**FIGURE 14.3** Two proposed excitation schemes for the two lowest neutral states derived from known lowest-lying anion states, ${}^8\Delta_g$ (left, GGA) and ${}^8\Sigma_u^-$ (right, GGA+$U$) by a single-electron ionization (indicated by arrows and associated orbitals).

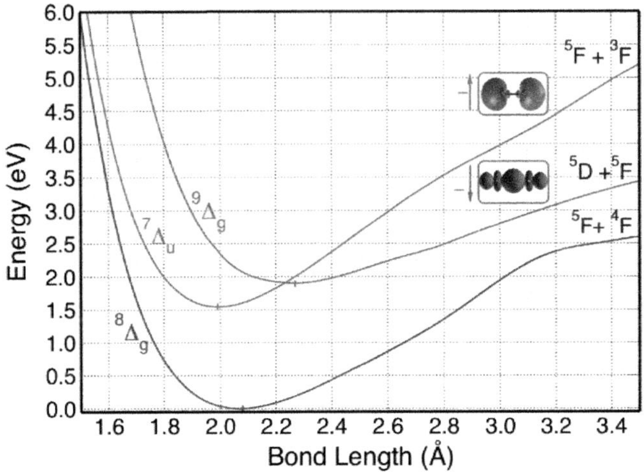


**FIGURE 14.4** Potential energy curves for the proposed GGA ground-state anion $^8\Delta_g$ and the two related low-lying neutral states $^9\Delta_g$ and $^7\Delta_u$.

ionization energy fails to agree with experiments, as does the splitting of the two lowest neutral states (0.40 eV). The hybrid functional, B3LYP, which is also a commonly employed exchange correlation functional, performs poorly in a manner similar to that of GGA. Even though B3LYP correctly identifies the neutral ground state, $^9\Sigma_g^-$, splittings between that state and the other neutral states as well as anion state splittings are too severely underestimated to provide any qualitative agreement with experiment (see Table 14.1). For GGA, the bond length of $^7\Delta_u$ is decreased ($r_e = 1.99$ Å) by 0.21 Å with respect to $^8\Delta_g$ ($r_e = 2.20$ Å), while the $^9\Delta_g$ state exhibits an increased bond length ($r_e = 2.26$ Å) with respect to the anion. Not surprisingly, the trend in fundamental frequencies for these states also fails to fit the experimental picture, with the $^7\Delta_u$ frequency of 413 cm$^{-1}$ being markedly increased from the $\omega_e$ of 360 cm$^{-1}$ for the anion, while $^9\Delta_g$ has a reduced frequency of 285 cm$^{-1}$.

The GGA+$U$ approach provides a much more consistent prediction of the excitation scheme with respect to both experiment and the highly accurate

**TABLE 14.1 Spin Splittings (in eV) for $Fe_2^-$ and $Fe_2$ at Several Levels of Theory**

| State | B3LYP | GGA | $+U_0$ (2 eV) | $+U_{scf}$ (3 eV) | CCSD(T) | MRCI[a] |
|---|---|---|---|---|---|---|
| $^8\Sigma_u^-$ | 0.00 | 0.00 | 0.00 | 0.00 | 0.00 | 0.00 |
| $^8\Delta_g$ | 0.14 | −0.52 | 0.04 | 0.38 | 0.40 | 0.45 |
| $^9\Sigma_g^-$ | 0.00 | 0.00 | 0.00 | 0.00 | 0.00 | 0.00 |
| $^7\Sigma_g^-$ | 0.34 | 0.65 | 0.66 | 0.60 | 0.55 | 0.62 |
| $^7\Delta_u$ | 0.18 | −0.12 | 0.48 | 0.72 | 0.86 | 0.69 |
| $^9\Delta_g$ | 0.36 | 0.28 | 0.36 | 0.41 | 0.38 | 0.45 |

[a] Data from Reference 28.

CCSD(T) [37] and MRCI results. Beginning with the GGA+$U$ ground-state anion $^8\Sigma_u^-(3d^{13}:\sigma_g^2\pi_u^4\delta_g^2\pi_g^{*,2}\delta_u^2\sigma_u^*4s^4:\sigma_g^2\sigma_u^{*,2})$, the two lowest states that differ from ionization of a single electron are $^9\Sigma_g^-$ and $^7\Sigma_g^-$, in accordance with previous findings in the MRCI study [28]. The two states have nearly identical valence electron configurations (both are $3d^{13}4s^3$) but differ simply by the spin of the $4s\ \sigma_u^*$ orbital ionized, spin down in the case of the nonet and spin up in the case of the septet as shown in Figure 14.3. Therefore, we may expect that both bond length and frequency of the two neutral states should be nearly identical to each other. The loss of the antibonding orbital suggests a decrease in bond length and increase in frequency with respect to the anion, consistent with experiment. To obtain the most accurate relative energetics of the three states, we use an average $U_{scf}$ of 3 eV that, we note, provides improved energetics with respect to $U_0$ from GGA, which is only 2 eV. The potential energy curves of this system, shown in Figure 14.5, are consistent with experiment. The ionization energy of the $^8\Sigma_u^-$ state is approximately 1.05 eV after inclusion of the Makov–Payne correction. The splitting of the two lowest neutral states is roughly 0.6 eV, in good agreement with the experimental splitting of 0.53 eV. The trend in structural properties is also consistent with experiment. The two neutral states exhibit nearly identical structural properties (see Table 14.2). The bond length $r_e$ is observed to decrease by 0.08 Å from the anion to the neutral states (from 2.20 to 2.12 Å), while the fundamental frequency is observed to increase by roughly 35 cm$^{-1}$. While the absolute values of the bond length and frequency are not identical to those obtained by experiment, we note that these discrepancies are likely due to (1) experimental uncertainty and (2) the anharmonic effects that cannot be fully eliminated in the experimental measurement of $\omega_0$ but are excluded in our theoretical measurement of the harmonic frequency.

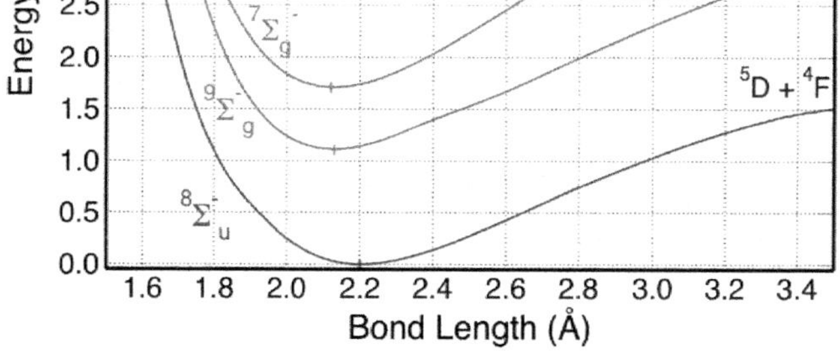


**FIGURE 14.5** Potential energy curves for the proposed GGA+$U$ ground-state anion $^8\Sigma_u^-$ and the two related low-lying neutral states, $^9\Sigma_g^-$ and $^7\Sigma_g^-$.

**TABLE 14.2 Bond Lengths $r_e$ (Å) and Harmonic Frequencies $\omega_e$ ($cm^{-1}$), for $Fe_2^-$ and $Fe_2$, compared to Experiment[a]**

| State | GGA | GGA+$U_{scf}$ | CCSD(T) | MRCI[b] | Experiment[c] |
|---|---|---|---|---|---|
| $^8\Sigma_u^-$ | 2.20, 305 | 2.20, 301 | 2.24, 276 | 2.23, 272 | 2.1, 250 |
| $^8\Delta_g$ | 2.07, 360 | 2.08, 355 | 2.12, 321 | 2.4, — | — |
| $^9\Sigma_g^-$ | 2.11, 339 | 2.13, 335 | 2.17, 296 | 2.18, 299 | 2.0, 300 |
| $^7\Sigma_g^-$ | 2.10, 335 | 2.12, 331 | 2.16, 304 | 2.17, 310 | 2.0, 300 |
| $^7\Delta_u$ | 1.99, 413 | 2.00, 419 | 2.00, 404 | 2.25, 195 | — |
| $^9\Delta_g$ | 2.26, 285 | 2.26, 280 | 2.28, 220 | 2.35, — | — |

[a] Fundamental frequencies $\omega_0$.
[b] Data from Reference 28.
[c] Data from Reference 26.

One additional piece of experimental evidence worth considering is the bond strength or dissociation energy of the anion and neutral iron dimers. While the bond order appears to nominally increase from the anion to the neutral states, corresponding to a shorter bond length, the experimental bond strength decreases roughly from $1.53 \pm 0.17$ eV for the anion to 0.78 eV $\pm 0.17$ eV for the lowest neutral state. On first glance at Figure 14.5, it appears that the bond strength, roughly 1.5 eV for the anion, is in agreement while the neutral state demonstrates a much higher bond strength of roughly 1.9 eV. However, we note that the GGA+$U$ anion state dissociates into a neutral $^5$D Fe atom and a negatively charged $^4$F Fe ion, both in their ground-state. The proposed dissociation for the lowest neutral state, $^9\Sigma_g^-$ dissociates into a ground-state neutral $^5$D Fe and an excited-state $^5$F Fe. Once we incorporate the strong likelihood of an avoided crossing with a nonet state that has a dissociation limit of two $^5$D neutral Fe atoms, we recover a bond strength of roughly 0.8 eV, in excellent agreement with experiment. Overall, our GGA+$U$ results are in remarkable agreement with both experiment and high-level quantum chemistry in the form of both single-reference CCSD(T) and multireference MRCI (see Tables 14.1 and 14.2). These calculations on the iron dimer demonstrate the potential of GGA+$U$ to systematically improve results obtained by commonly employed exchange correlation functionals and to help systematically predict and explain experimental spin-state orderings.

The gas-phase reactions of bare diatomic $FeO^+$ cations with hydrogen, methane, nitrogen gas, and other species have been widely studied as fundamental processes that can provide clues into the behavior of larger metal-oxo catalytic systems [38]. These paradigmatic reactions have been studied by numerous mass spectrometric techniques in detail [38–41], but initial experimental results were surprisingly discordant with classical transition-state theories [39]. The reaction of $FeO^+$ with $H_2$ has been under particularly extensive study for some time now both theoretically [42–48] and experimentally [38,39,41,49] as the most fundamental example of the two-state reactivity paradigm. Additionally, the reaction of $FeO^+$ with $CH_4$ has been studied in detail, primarily with DFT [50–56] and, to a lesser extent, experiment [38,57–59]. Methane oxidation by reaction with $FeO^+$ is relevant because of methane's greater utility as a fuel on conversion to methanol [60,61], as well as the relevance to enzymatic systems such as methane monooxygenase [2,62].

Reactions of $FeO^+$ with $CH_4$ and $H_2$ are both known to be exothermic by about 10 and 37 kcal/mol, respectively. However, these reactions are very inefficient: only about 10% of all collisions lead to products in the case of methane, and 1% of all collisions lead to products for the hydrogen case [39]. This inefficiency is particularly surprising because the same-spin nature of the ground-state reactant and product species, $^6\Sigma^+$ $FeO^+$ and $^6D$ $Fe^+$, make these reactions spin-allowed [43]. There is an observed decrease in rate constant with increasing energy or temperature for both reactions, which is also in direct contradiction with the expected Arrhenius-like behavior for a spin-allowed reaction [44]. The resulting reaction inefficiency was proposed to be due to the crossing of multiple relevant spin surfaces for the reactions [44,45], a feature now expected to be relevant in larger systems that include the enzyme horseradish peroxidase [63] and model non-heme inorganic iron alkane hydroxylation catalysts [64]. It has thus been proposed that the spin crossover from the steep sextet surface to a shallow but excited quartet spin state surface[1] may be the true source of inefficiency. It has also been found that radical side products can form by dissociation of the reacting complex to $FeOH^+$ and radical H or $CH_3$, which are weakly endothermic or exothermic by roughly 3 kcal/mol, respectively. The relative exothermicity of the dissociated radical side reaction is key in interpreting and predicting branching ratios between the two types of reactions.

While theoretical considerations pointed to crossings between the sextet and quartet potential energy surfaces as the source for this reaction inefficiency [44,45], it is nevertheless notable that subsequent studies [41,44–47], which employed density functional techniques, have failed to agree with experimental results or highly accurate quantum chemistry calculations. While most implementations of DFT predict a qualitatively correct steeper sextet surface in accordance with an experimentally observed low-efficiency, exothermic reaction, the quartet surface is not predicted to be nearly shallow enough and quantitative barrier heights are incorrect. Common exchange correlation functionals incorrectly predict a quartet $Fe^+$ product to be stabilized with respect to the sextet [44], and by using some rough estimates to correct this erroneous splitting, researchers have suggested that a correct qualitative picture of the spin crossover between quartet and sextet surfaces could be recovered at both the entrance and exit channels.

We review here the reactions of $FeO^+$ with $H_2$ and $CH_4$ as a challenging test for a Hubbard $U$ density functional approach, which we have introduced to improve semi-local or hybrid density functional descriptions of transition metal complexes [7]. We specifically consider the generalized-gradient approximation (GGA) in density functional theory (DFT) for its efficiency and widespread use in the community [10]; nevertheless, the Hubbard correction introduced here could be applied in conjunction with other exchange correlation approximations. First and foremost, we find that the GGA+$U$ approach systematically improves the description of the electronic

[1]We refer to the two species by their spin multiplicities, defined as $2S+1$, where $S$ is the net spin of the molecule, $\frac{5}{2}$ for the sextet and $\frac{3}{2}$ for the quartet.

structure of $FeO^+$ in both spin states, but it also improves the electronic structure description of intermediates that perturb the underlying diatomic molecule's electronic structure to varying degrees throughout the reaction coordinate. In order to gauge the success of our approach on both of these reactions, we compare DFT+$U$ (GGA+$U$) results with both standard GGA results and highly accurate quantum chemistry approaches. Post-Hartree–Fock (post-HF) techniques are able to recover nearly all of the correlation energy and may be used as an accurate benchmark or reference for the DFT calculations, but these approaches suffer from poor scaling, which limits them to very small sizes such as the reactions under consideration here (i.e., no more than 10 atoms). The reaction of $FeO^+$ with molecular hydrogen is compared against CCSD(T) results [37], while the reaction with methane is compared against MRCI results [65].

The value of $U$ calculated for each intermediate and transition state for the reactions with both hydrogen and methane provides a means for identifying the broad similarities between the two cases. In order to obtain a global reaction coordinate, we must use energies calculated with a single value of Hubbard $U$, which we choose as an average, $U_{scf,av}$, over all states. Since total energies at differing values of $U$ are not comparable, it is important to pay attention to large deviations in $U_{scf}$ at any stationary point from the global average. The overall $U_{scf,av}$ that we obtain for the $H_2$ reaction is 4.93 eV, and the value for the $CH_4$ reaction is very similar at 5.09 eV. The $U_{scf}$ is on average larger than the value calculated from GGA ($U_0$) for most geometries. We find the smallest discrepancy between the self-consistent and non-self-consistent values of $U$ for reaction intermediates, while the differences for transition states are as large as 1.5 eV, as is in the case of $^6$TS-$2_{CH_4}$. For most single points in these two reactions, the individual $U_{scf}$ values are within 1 eVof the $U_{scf,av}$. In typical cases (e.g., intermediate splittings and sextet barriers), employing GGA+$U$ with a global, average value of $U_{scf,av}$ that is in a range 1 eV too small or too large for the structure considered yields energy differences in the range of 0.01–0.1 eV; this uncertainty can, in many cases, be decreased by examining splittings using a $U_{scf}$ averaged "locally" on the two intermediates and the transition state of interest.

The addition–elimination reactions of hydrogen and methane with $FeO^+$ share common intermediate and transition state structures. Namely, at the entrance channel, molecular hydrogen or methane weakly binds to the iron center (Int-1), and abstraction of a hydrogen from molecular hydrogen or methane to be shared over the iron–oxo bond (TS-1) is the first reaction step. Following the bond-breaking event, a second intermediate (Int-2) is stabilized in which one hydrogen is bound solely to the oxygen while the remainder of the reactant complex is still bound to the iron atom. The second reaction step involves the transfer of the reactants from the iron center to the oxygen over the iron–oxo bond (TS-2), which ultimately facilitates the formation of a radical–reactant complex bond with oxygen as well. On completion of the transfer of the reactants to oxygen, the newly formed water or methanol molecule is weakly bound to the cationic iron center (Int-3) and may subsequently dissociate to products. While the ground state of the intermediates is generally a sextet, shallower quartet surfaces make the quartet reaction barriers and transition states highly relevant.

The similarities of the values of $U$ for the hydrogen and methane reactions highlight the fact that coordination and electronic configuration contribute most strongly to the calculated $U$ value. Overall, we observe a larger $U_{\text{scf}}$ (in the range of 5.5–6.5 eV) at the entrance channel, where the iron center is weakly three- or two-fold coordinated in the Int-1 structure. At the first reaction barrier, $U_{\text{scf}}$ decreases monotonically for all TS-1 and Int-2 structures, and this change is concomitant with a reduction in $FeO^+$ bond order and coordination. At the second reaction barrier, the quartet and sextet surfaces exhibit different linear response properties. This behavior of $U_{\text{scf}}$ is strongly affected by differences in the chemical coupling of the quartet and sextet $Fe^+$ centers with the organic ions in the two spin surfaces. The $3d^6 4s^1$-like sextet's two minority spin electrons are equally divided between the $3d$ and $4s$ manifolds, and this results in an increased prominence of the $U_{\text{scf},4s}$ for $^6\text{TS-2}_{H_2}$ and an overall lower $U_{\text{scf},3d}$ of ~3–4 eV for both reactions. While the extension to include $4s$ states, and thus a $U_{4s}$, is straightforward, it plays a role exclusively in the class of weakly bound molecules where the $4s$–$3d$ hybridization dominates and $3d$- and $4s$-derived molecular states are close in energy. The sextet transition state, especially in the methane case, is quite weakly bound, but the quartet transtion state, which has a $3d^7$ character, is more tightly bound and the $U_{\text{scf}}$ of $^4$TS-2 is close to $U_{\text{scf,av}}$. All Int-3 structures at the exit channel closely resemble a weakly bound $Fe^+$ ion with the product water or methanol. Consequently, the $Fe^+$ charge density is increasingly atom-like, and the values of the Hubbard $U$ are significantly lower at this point in the reaction surface for the quartet at around 2 eV (the sextet displays a $U_{\text{scf}}$ considerably closer to the average). Since these two reactions involve unsaturated metal centers for which the number and character of metal–ligand bonds change across various spin states, they provide a far more challenging probe of the robustness of our method compared to more typical problems that involve saturated metals or fewer changes in metal–ligand bonding.

In addition to being system-dependent, the value of the self-consistent, linear response $U$ is dependent on the definition of the projection manifold. It is evident that the value of $U$ should vary if a different qualitative form of projection is used such as atomic orbitals versus maximally localized Wannier functions. However, we show here that differences in pseudopotentials can affect the value of Hubbard $U$, and, to a lesser extent, the calculated properties of a transition metal molecule. In practice, we shall show that the value of $U$ obtained for the same structure but with differing pseudopotentials may easily differ by as much as 2–3 eV, particularly if the pseudopotentials were generated in different oxidation states. Although this value might seem high, it reflects the changes that atomic orbitals undergo as a function of the total atomic charge. As a key example relevant to results for the addition–elimination reactions, we provide here comparison of $FeO^+$ sextet and quartet-state properties calculated using pseudopotentials generated in various oxidation states. The pseudopotentials generated in the $Fe^{2+}$ $(d^{5.5}s^{0.5})$, $Fe^{0.5+}$ $(d^{6.5}s^{1.0})$, and Fe($d^6s^2$, used throughout) states had values of $U_{\text{scf}}$ of 7.0, 4.9, and 5.5 eV, respectively. As the electronic configuration of $FeO^+$ states is either $d^6s^1$ or $d^7s^0$, it is reasonable that the second pseudopotential produces the lowest value of $U$.

**TABLE 14.3 Comparison for Several Pseudopotentials of Structural Properties Including Equilibrium Bond Length [$r_e$(Å)], Harmonic Frequency [$\omega_e$(cm$^{-1}$)], and State Splittings (in eV) of the $^6\Sigma^+$ and $^4\Phi$ States of FeO$^{+a}$**

| GGA ($U=0$ eV) | $Fe^{2+}$ ($d^{5.5}s^{0.5}$) | $Fe^{0.5+}$ ($d^{6.5}s^1$) | Fe ($d^6s^2$) |
|---|---|---|---|
| $^6\Sigma^+$ $r_e$(Å), $\omega_e$ (cm$^{-1}$) | 1.59, 902 | 1.63, 906 | 1.62, 901 |
| $^4\Phi$ $r_e$(Å), $\omega_e$ (cm$^{-1}$) | 1.52, 1033 | 1.57, 1040 | 1.56, 1038 |
| $\Delta E_{6\rightarrow4}$ (eV) | 0.76 | 0.82 | 0.84 |
| **GGA + $U_{scf}$ (eV)** | **7.0** | **4.9** | **5.5** |
| $^6\Sigma^+$ $r_e$(Å), $\omega_e$ (cm$^{-1}$) | 1.65, 745 | 1.66, 751 | 1.66, 749 |
| $^4\Phi$ $r_e$(Å), $\omega_e$ (cm$^{-1}$) | 1.76, 606 | 1.75, 613 | 1.75, 612 |
| $\Delta E_{6\rightarrow4}$ (eV) | 0.50 | 0.52 | 0.54 |

[a]The GGA + $U$ results are obtained at the respective $U_{scf}$ of each pseudopotential.

In comparing LDA+$U$ or GGA+$U$ calculations of different pseudopotentials, total energies are not comparable.[2] However, the properties derived from each spin and symmetry's ground-state density, including frequencies, bond lengths, and state splittings, may be compared. The GGA results for each of the three cases show significant differences for some structural properties and energetic splittings (see Table 14.3). While the harmonic frequencies for $^4\Phi$ and $^6\Sigma^+$ of 1040 and 900 cm$^{-1}$ are identical for all pseudopotentials within the accuracy of the fitting procedure, more significant differences are observed in the bond lengths. The most ionic pseudopotential ($d^{5.5}s^{0.5}$) yields the shortest GGA value with $^4\Phi$ and $^6\Sigma^+$ at 1.52 and 1.59 Å, respectively, while the other two pseudopotentials yield bond lengths within 0.01 Å of each other. The GGA spin splittings are quite close at 0.82 and 0.84 eV for the $Fe^{0.5+}$ and Fe pseudopotentials, respectively, while the more ionic $Fe^{2+}$ is 0.76 eV. On augmentation of a Hubbard term using the $U_{scf}$ appropriate to each respective pseudopotential, the bond lengths of all three pseudopotentials are brought into agreement with each other within 0.01 Å, the harmonic frequencies agree within 6–7 cm$^{-1}$, and the spin splittings for all three pseudopotentials agree within 0.04 eV. Overall, a much smaller spread in values is observed for GGA+$U$ than for GGA (see Table 14.3). Even a pseudopotential whose reference state is very different from the one considered seems to remedy itself so long as the proper value of $U$, $U_{scf}$, is used. These observations demonstrate, nevertheless, that while careful pseudopotential selection is not mandatory; the best pseudopotential choice is one that clearly reflects the oxidation states and charges of the physical system of interest.

Looking at the details of the addition–elimination reaction of molecular hydrogen on $FeO^+$, we see that the GGA+$U$ reaction coordinate provides a substantial, systematic improvement over GGA when compared against experiment or highly accurate CCSD(T) reference calculations. The GGA potential energy surface (see Figs. 14.6 and 14.7) is in direct contradiction to available experimental evidence and theory because it (1) underestimates the exothermicity by over 1 eV, (2) the quartet

[2]This also means that great caution should be taken in comparing band gap values for different values of $U$, since the same arbitrary shift in energy affects eigenvalues.

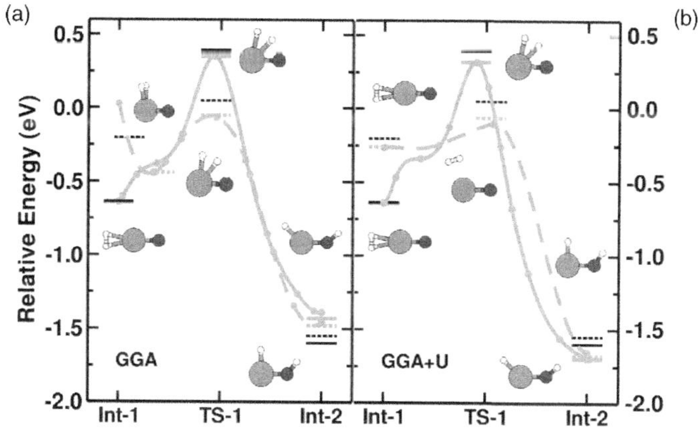


**FIGURE 14.6** The TS-1 NEB minimum energy paths (blue) for GGA (a) and GGA+U (b) are compared against CCSD(T) (black) reference energies for both quartet (dashed line) and sextet (solid line) surfaces of the reaction of $FeO^+$ with $H_2$. Here, $U$ is the globally averaged $U_{scf,av}$ of 5 eV.

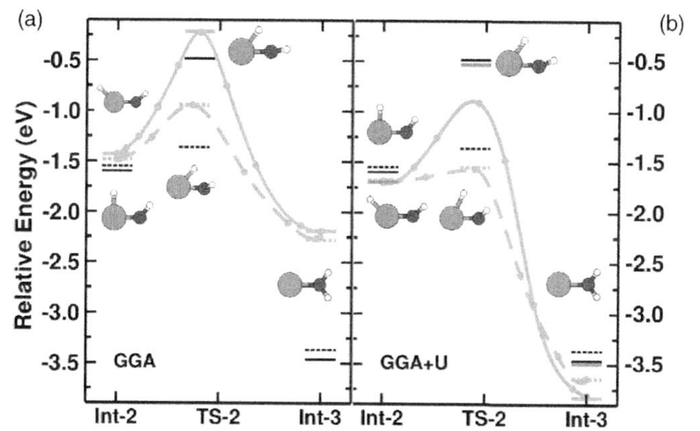


**FIGURE 14.7** The TS-2 NEB minimum-energy paths (blue) for GGA (a) and GGA+U (b) (both in blue) are compared against CCSD(T) (black) reference energies for both quartet (dashed line) and sextet (solid line) surfaces of the reaction of $FeO^+$ with $H_2$. The GGA+$U$ results were determined at the globally averaged $U_{scf,av}$ of 5 eV, excluding the points calculated at locally averaged $U_{3d} = 4$ eV and $U_{4s} = 4$ eV (shown in green).

surface is too steep to allow for a "shallow, barrierless" reaction [39,45], (3) spin splittings at intermediates show large errors, and (4) there is no spin crossover at the exit channel.[3] The majority of these errors in the GGA reaction coordinate are localized to the second reaction barrier (see Fig. 14.7), with the exception of the errors

[3]The potential energy surfaces calculated with different methods have been aligned at the $^6$Int-1 states. The curves associated with GGA and GGA + $U$ reaction coordinates are interpolations of the points obtained for the minimum-energy path as determined by NEB calculations.

in the quartet Int-1 structures and barriers in the first barrier surface (see Fig. 14.6). While the first GGA sextet TS barrier largely agrees with CCSD(T) results, the second GGA barrier exhibits much greater errors for the exothermicity and state ordering at Int-3. In contrast, the GGA+$U$ (where $U = 5$ eV) reaction coordinate (see Figs. 14.6 and 14.7) correctly estimates and predicts the reaction exothermicity, the shallow but excited quartet surface, barrier heights, and spin crossover at the exit channel (see Fig. 14.7). Systematic improvements on GGA + $U_{\text{scf,av}}$ results in which $U$ deviates locally from $U_{\text{scf,av}}$ (e.g., at the hydrogen $^6$TS-2 barrier) were achieved by including explicitly the 4$s$ manifold, highly relevant for isolated transition metal complexes, or by locally averaging the value of $U_{\text{scf}}$. Additionally, the portions of the reaction coordinate that are properly estimated by GGA in the first TS barrier (see Fig. 14.6), namely, the sextet barrier steepness and structures, are preserved in GGA+$U$, while the errors of the GGA approach in the second TS barrier (see Fig. 14.7) are greatly improved upon.

In order to further measure the quantitative accuracy of the hydrogen addition–elimination reaction, we estimated equilibrium CCSD(T) geometries for each intermediate and compared against the GGA and GGA+U relaxed structures.[4] We also averaged the error of the GGA+$U$ and GGA splittings for each intermediate with respect to the CCSD(T) splittings for that intermediate. The average error for the five spin splittings along the reaction coordinate is 0.04 eV in GGA+$U$, to be compared with GGA, which exhibits an average error of 0.20 eV, or B3LYP that performs even more poorly with an average error of 0.30 eV in the splittings. Not only does GGA+$U$ improve errors by fivefold over GGA, but using the CCSD(T) equilibrium geometries, we found from interpolating a potential energy scan, we also observe that GGA+$U$ on average produces improved geometries. Mean errors for GGA+$U$ geometries are reduced from 4.3 pm (for GGA) to 2.2 pm, and are only slightly worse than B3LYP geometries (average error of 1.3 pm). Although B3LYP shows a slight improvement in geometries, we consider the sevenfold improvement in spin splittings of GGA+$U$ to be much more key. The difference in errors between B3LYP and GGA+$U$ geometries could be attributed largely to systematic ~1-pm errors in bonds between H and the other atoms, Fe and O, which are present in GGA and GGA+$U$ but not in B3LYP.

The GGA description of $FeO^+$ on methane demonstrates even greater discordance with theoretical considerations and known experimental results. In fact, using GGA, both the reactions to create methanol and to form radical side products turn out to be endothermic by 0.2 and 0.4 eV, respectively (see Fig. 14.8). Additionally, the sextet TS-2 barrier resides slightly above the reactant energies. The remaining GGA discrepancies are comparable to those for the reaction with hydrogen and include spin splittings at intermediates, barrier estimates, and the absence of a spin crossover at the exit channel. When we instead consider the GGA+$U$ results and compare them to highly accurate MRCI calculations, we observe quantitative agreement between the two methods (see Fig. 14.9). The GGA+$U$ quartet surface is shallower but not barrierless, in agreement with MRCI results.

[4]We note that we obtained CCSD(T) barrier height estimates by calculating single-point energies of GGA and GGA+$U$ TS structures at the CCSD(T) level.

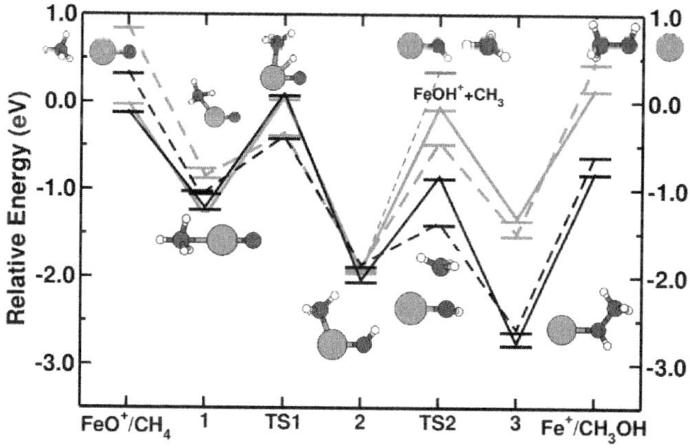


**FIGURE 14.8** Comparison of GGA (blue) and MRCI (black) reaction coordinates for quartet (dashed) and sextet (solid) surfaces of the reaction of $FeO^+$ with $CH_4$.

We see in our calculations an improved estimate of exothermicity of both main and side reactions, the spin crossover at the exit channel is now preserved, and the description of the quartet Int-1 geometry is also improved. A comparison of the accuracy of GGA+$U$ geometries with MRCI is beyond the scope of this study because of the higher dimensionality of the PES over that for the four-atom reaction.

We can further examine the accuracy of GGA+$U$ in treating the energetics of these reactions by extracting the exothermicity for each reaction step and multiplicity, as summarized in Table 14.4. While GGA produces errors as large as 1.4 eV for the methane reaction and 1.1 eV for the hydrogen reaction, GGA+$U$ is in very good agreement with quantum chemistry. The largest errors, such as the 0.30–0.32 eV

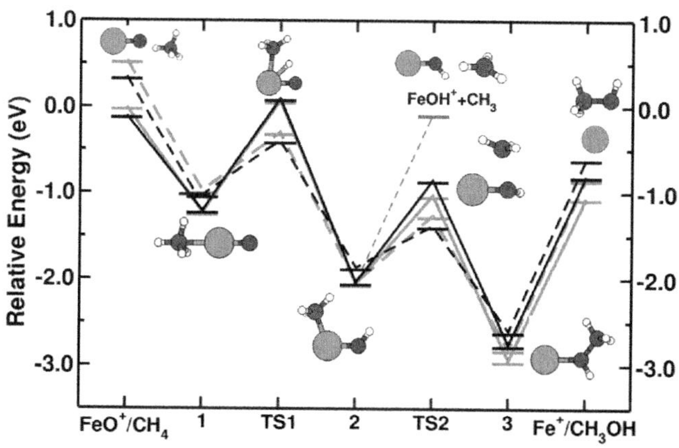


**FIGURE 14.9** Comparison of GGA+$U$ ($U$=5 eV) (blue) and MRCI (black) reaction coordinates for quartet (dashed) and sextet (solid) surfaces of the reaction of $FeO^+$ with $CH_4$.

**TABLE 14.4 Net Energy Required for Reaction Steps (in eV) for GGA, GGA+$U$ ($U$=5 eV), and Post-HF Methods, CCSD(T) for the $H_2$ Reaction, and MRCI for the $CH_4$ Reaction**

| | TS−1$^4$ | | TS−1$^6$ | | TS−2$^4$ | | TS−2$^6$ | |
|---|---|---|---|---|---|---|---|---|
| Method | $CH_4$ | $H_2$ | $CH_4$ | $H_2$ | $CH_4$ | $H_2$ | $CH_4$ | $H_2$ |
| GGA | −1.14 | −1.04 | −0.54 | −0.59 | 0.46 | −0.92 | 0.61 | −0.79 |
| GGA+$U$ | −1.07 | −1.42 | −0.71 | −1.05 | −0.84 | −2.02 | −0.87 | −2.19 |
| Post-HF | −0.99 | −1.35 | −0.75 | −0.96 | −0.87 | −1.82 | −0.80 | −1.87 |

underestimate at the sextet TS-2$_{H_2}$ barrier, can be systematically improved by extensions that include either a $U_{4s}$ or a locally averaged $U_{scf}$.

## 14.3 CONCLUSIONS

Using the GGA approximation, we find that activation barriers, stationary-point spin splittings, and reaction energies exhibit errors as large a 1.4 eV for both the methane and the hydrogen systems discussed. By augmenting the GGA exchange correlation functional with a Hubbard correction, these errors are reduced by an order of magnitude, to 0.1 eV on average with respect to the best available quantum chemical methods [such as CCSD(T) and MRCI] as well as experimental values. We stress that the Hubbard term is not used as a fitting parameter, but it is a true first-principles linear response property of the transition metal complex that can augment any exchange correlation density functional. The practical limitations to this approach stem from the complexity of the systems studied; evolution of the coordination environment along a global reaction coordinate may result in local deviations from a globally averaged Hubbard $U$ or bring to the forefront the role of 4$s$ contributions.

Overall, the DFT+$U$ results (here, GGA+$U$) have been shown to provide for systematic improvement over GGA for all systems considered thus far by treating, for the first time within a DFT framework, the energetics of differing spin surfaces and electronic states with the same level of accuracy. The inexpensive linear response $U$ calculation also acts as a probe for the relative importance of the DFT+$U$ approach for a given transition metal complex; that is, if the Hubbard $U$ calculated is small or nearly zero, a standard description is sufficient. We believe that this approach paves the way for large-scale simulations of systems that contain transition metal complexes with both accuracy and efficiency.

As an example, in Figure 14.10 we show the results of GGA+$U$ calculations on self-assembled monolayers of tetrabromophenyl porphyrins with cobalt metal centers (TBrPP-Co), which form on the (111) surface of copper [66,67]. The unit cell, which consists of the porphyrin and two layers of a Cu slab, has a total of 177 atoms (77 from the molecule, 100 from the copper slab) and 1357 electrons (257 from the molecule, 1100 from the copper). Spin-polarized calculations using either the GGA or the GGA+$U$ approach are of comparable cost and were carried out in about 24 h on six

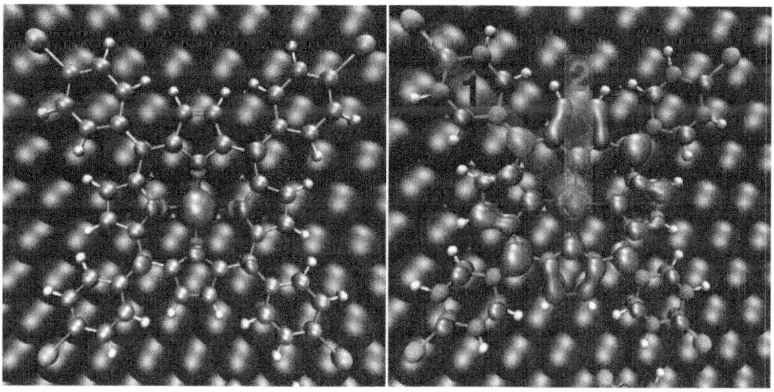


**FIGURE 14.10** Spin density profiles for GGA (a) and GGA+$U$ (b) of TBrPP-Co molecule on Cu(111) surface. The GGA+$U$ density increases toward the outer ring of the molecule, where the density increase is higher along direction 1 than direction 2, while GGA density is primarily at the center of the molecule and equivalently small for all directions radiating outward.

Intel Core2 2.4 GHz CPUs. These GGA+$U$ calculations show enhanced predictive accuracy over GGA and successfully identify a spin density induced throughout the molecule fully consistent with experimental results, which is not observed in GGA.

## ACKNOWLEDGMENTS

This work was supported in part by an NSF graduate research fellowship and grants from the DOE (DE-AC04-94AL850000, by Sandia's LDRD program). Computational facilities were provided through PNNL Grant EMSL-UP-9597.

## REFERENCES

1. K. Yoshizawa and T. Yumura, *J. Inorg. Biochem.* **96**(1), 257–257 (2003).
2. K. Yoshizawa, *J. Inorg. Biochem.* **78**(1), 23–34 (2000).
3. L. C. Blasiak, F. H. Vaillancourt, C. T. Walsh, and C. L. Drennan, *Nature* **440**(7082), 368–371 (2006).
4. S. P. D. Visser, F. Ogliaro, Z. Gross, and S. Shaik, *Chem. Eur. J.* **7**(22), 4954 (2001).
5. M. Cococcioni and S. De Gironcoli, *Phys. Rev. B* **71**(3), 035105 (2005).
6. V. I. Anisimov, I. V. Solovyev, M. A. Korotin, M. T. Czyzyk, and G. A. Sawatzky, *Phys. Rev. B* **48**(23), 16929 (1993).
7. H. J. Kulik, M. Cococcioni, D. A. Scherlis, and N. Marzari, *Phys. Rev. Lett.* **97**(10), 103001 (2006).
8. V. I. Anisimov, J. Zaanen, and O. K. Andersen, *Phys. Rev. B* **44**(3) 943–954 (1991).
9. A. I. Liechtenstein, V. I. Anisimov, and J. Zaanen, *Phys. Rev. B* **52**(8), R5467–R5470 (1995).

10. J. P. Perdew, K. Burke, and M. Ernzerhof, *Phys. Rev. Lett.* **77**(18), 3865–3868 (1996).
11. S. Fabris, S. de Gironcoli, S. Baroni, G. Vicario, and G. Balducci, *Phys. Rev. B* **71**(4), (2005).
12. J. P. Perdew and M. Levy, *Phys. Rev. Lett.* **51**(20), 1884–1887 (1983)
13. P. Giannozzi, et al. *J. Phys. Condens. Matt.* [arXiv: 0906.2569v2 (cond-mat.mtrl-sci)] (2009).
14. A. M. Cassell, J. A. Raymakers, J. Kong, and H. Dai, *J. Phys. Chem. B* **103**(31), 6484 (1999).
15. P. Ballone and R. Jones, *Chem. Phys. Lett.* **233**(5–6), 632 (1995).
16. M. Castro and D. R. Salahub, *Phys. Rev. B* **49**(17), 11842 (1994).
17. J. Chen, C. Wang, K. A. Jackson, and M. R. Pederson, *Phys. Rev. B* **44**(12), 6558 (1991).
18. S. Dhar and N. Kestner, *Phys. Rev. A* **38**(3), 1111 (1988).
19. J. Harris and R. Jones, *J. Chem. Phys.* **70**(2), 830 (1979).
20. H. Nagarathna, P. A. Montano, and V. M. Naik, *J. Am. Chem. Soc.* **105**(10), 2938 (1983).
21. T. Noro, C. Ballard, M. H. Palmer, and H. Tatewaki, *J. Chem. Phys.* **100**(1), 452 (1994).
22. I. Shim and K. A. Gingerich, *J. Chem. Phys.* **77**(5), 2490 (1982).
23. M. Tomonari and H. Tatewaki, *J. Chem. Phys.* **88**(3), 1828 (1988).
24. S. Yanagisawa, T. Tsuneda, and K. Hirao, *J. Chem. Phys.* **112**(2), 545 (2000).
25. E. Rohlfing, D. Cox, A. Kaldor, and K. Johnson, *J. Chem. Phys.* **81**(9), 3846 (1984).
26. D. G. Leopold and W. Lineberger, *J. Chem. Phys.* **85**(1), 51 (1986).
27. C. J. Barden, J. C. Rienstra-Kiracofe, and H. F. Schaefer III, *J. Chem. Phys.* **113**(2), 690 (2000).
28. O. Hubner and J. Sauer, *Chem. Phys. Lett.* **358**(5–6), 442 (2002).
29. A. Irigoras, M. Michelini, E. Sicilia, N. Russo, J. Mercero, and J. Ugalde, *Chem. Phys. Lett.* **376**(3–4), 310 (2003).
30. D. Cox, D. Trevor, R. Whetten, E. Rohlfing, and A. Kaldor, *Phys. Rev. B* **32**(11), 7290 (1985).
31. P. Montano, P. Barrett, and Z. Shanfield, *J. Chem. Phys.* **64**(7), 2896 (1976).
32. D. Guenzburger, and E. M.B. Saitovitch, *Phys. Rev. B* **24**(5), 2368 (1981).
33. T. McNab, H. Micklitz, and P. Barrett, *Phys. Rev. B* **4**(11), 3787 (1971).
34. H. Purdum, P. Montano, G. Shenoy, and T. Morrison, *Phys. Rev. B* **25**(7), 4412. (1982).
35. D. Leopold, J. Almlof, W. Lineberger, and P. Taylor, *J. Chem. Phys.* **88**(6), 3780 (1988).
36. G. Makov and M. C. Payne, *Phys. Rev. B* **51**(7), 4014 (1995).
37. M. J. Frisch, et al., Gaussian03, rev. 02 (2004); see `http://www.gaussian.com`.
38. D. Schroder, H. Schwarz, D. E. Clemmer, Y. M. Chen, P. B. Armentrout, V. I. Baranov, and D. K. Bohme, *Int. J. Mass Spectrom. Ion Process.* **161**(1–3), 175–191 (1997).
39. D. Schroder, A. Fiedler, M. F. Ryan, and H. Schwarz, *J. Phys. Chem.* **98**(1), 68–70 (1994).
40. D. E. Clemmer, Y. M. Chen, F. A. Khan, and P. B. Armentrout, *J. Phys. Chem.* **98**(26), 6522–6529 (1994).
41. J. M. Mercero, J. M. Matxain, X. Lopez, D. M. York, A. Largo, L. A. Eriksson, and J. M. Ugalde, *Int. J. Mass Spectrom.* **240**(1), 37–99 (2005).
42. K. Andersson, P. A. Malmqvist, B. O. Roos, A. J. Sadlej, and K. Wolinski, *J. Phys. Chem.* **94** (14), 5483–5488 (1990).

43. A. Fiedler, D. Schroder, S. Shaik, and H. Schwarz, *J. Am. Chem. Soc.* **116**(23), 10734–10741 (1994).
44. D. Danovich and S. Shaik, *J. Am. Chem. Soc.* **119**(7), 1773–1786 (1997).
45. M. Filatov and S. Shaik, *J. Phys. Chem. A* **102**(21), 3835–3846 (1998).
46. S. Chiodo, O. Kondakova, M. D. Michelini, N. Russo, E. Sicilia, A. Irigoras, and J. M. Ugalde, *J. Phys. Chem. A* **108**(6), 1069–1081 (2004).
47. A. Irigoras, J. E. Fowler, and J. M. Ugalde, *J. Am. Chem. Soc.* **121**(37), 8549–8558 (1999).
48. M. Rosi and C. W. Bauschlicher, *J. Chem. Phys.* **90**(12), 7264–7272 (1989).
49. F. Aguirre, J. Husband, C. J. Thompson, K. L. Stringer, and R. B. Metz, *J. Chem. Phys.* **119**(19),10194–10201 (2003).
50. K. Yoshizawa, Y. Shiota, and T. Yamabe, *Chem. Eur. J.* **3**(7), 1160–1169 (1997).
51. K. Yoshizawa, Y. Shiota, and T. Yamabe, *Organometallics* **17**(13), 2825–2831 (1998).
52. M. Bronstrup, D. Schroder, and H. Schwarz, *Chem. Eur. J.* **5**(4), 1176–1185 (1999).
53. K. Yoshizawa, Y. Shiota, and T. Yamabe, *J. Chem. Phys.* **111**(2), 538–545 (1999).
54. K. Yoshizawa, A. Suzuki, and T. Yamabe, *J. Am. Chem. Soc.* **121**(22), 5266–5273 (1999).
55. Y. Shiota, and K. Yoshizawa, *J. Am. Chem. Soc.* **122**(49), 12317–12326 (2000).
56. Y. Shiota and K. Yoshizawa, *J. Chem. Phys.* **118**(13), 5872–5879 (2003).
57. F. Aguirre, J. Husband, C. J. Thompson, K. L. Stringer, and R. B. Metz, *J. Chem. Phys.* **116** (10), 4071–4078 (2002).
58. D. Schroder and H. Schwarz, *Angew. Chem. Int. Ed. Eng.* **29**(12), 1433–1434 (1990).
59. H. Schwarz, *Int. J. Mass Spectrom.* **237**(1), 75–105 (2004).
60. D. K. Bohme and H. Schwarz, *Angew. Chem. Int. Ed.* **44**(16), 2336–2354 (2005).
61. H. Schwarz and D. Schroder, *Pure Appl. Chem.* **72**(12), 2319–2332 (2000).
62. A. C. Rosenzweig, C. A. Frederick, S. J. Lippard, and P. Nordlund, *Nature* **366**(6455), 537–543 (1993).
63. E. Derat, S. Shaik, C. Rovira, P. Vidossich, and M. Alfonso-Prito, *J. Am. Chem. Soc.* **129**(20),6346 (2007).
64. H. Hirao, D. Kumar, L. Que, and S. Shaik, *J. Am. Chem. Soc.* **128**(26), 8590–8606 (2006).
65. H. J. Werner, et al., MOLPRO 2006 (2006); see `http://www.molpro.net`.
66. V. Iancu, A. Deshpande, and S.-W. Hla, *Nano Lett.* **6**, 820–823 (2006).
67. U. G. E. Perera, H. J. Kulik, V. Iancu, L. G. G. V. Dias da Silva, S. E. Ulloa, N. Marzari, and S. W. Hla, *Phys. Rev. Lett.* (in press).

CHAPTER 15

# Quantitative Description of Electron Transfer Reactions

PATRICK H.-L. SIT

Center for Molecular Modeling, Department of Chemistry, University of Pennsylvania, Philadelphia and Institute for Computational Molecular Science, College of Science and Technology, Temple University, Philadelphia, Pennsylvania

AGOSTINO MIGLIORE

Center for Molecular Modeling, Department of Chemistry, University of Pennsylvania, Philadelphia, Pennsylvania

MICHAEL L. KLEIN

Institute for Computational Molecular Science, College of Science and Technology, Temple University, Philadelphia, Pennsylvania

NICOLA MARZARI

Department of Materials Science and Engineering, Massachusetts Institute of Technology, Cambridge, Massachusetts

In this chapter we present a fully ab initio approach for the evaluation of the thermodynamic and electronic properties of electron transfer reactions. Diabatic free-energy surfaces are obtained from extensive first-principles molecular dynamics simulations of the active centers and their surrounding aqueous media. The reaction coordinate, chosen as the excitation energy corresponding to the transfer of an electron at any given ionic configuration, is accurately assessed within ground-state density functional theory via a novel penalty functional that appropriately acts on the electronic degrees of freedom alone. The diabatic free-energy surfaces are then constructed from umbrella sampling on large ensembles of configurations; the self-exchange reaction between ferrous and ferric ions in water is taken here as a paradigmatic test case.

*Fuel Cell Science: Theory, Fundamentals, and Biocatalysis*,
Edited by Andrzej Wieckowski and Jens K. Nørskov

We then present a more recently developed approach to calculate the transfer integral between the solvated ions, based on fractional occupation numbers (FONs). This approach is crucial in obtaining the correct ground-state wavefunction. The computed transfer integral and reorganization energy provide a theoretical estimate of the electron self-exchange rate. Once Hubbard $U$ corrections are included—to deal with the strongly localized $d$ electrons of the transition metal ions—we obtain a theoretical rate constant in excellent agreement with the experimental estimates obtained from kinetic measurements.

## 15.1 INTRODUCTION

Electron transfer is one of the most ubiquitous processes in organic and inorganic redox reactions. It covers processes and applications as diverse as solar energy conversion in the early steps of photosynthesis, oxidation–reduction reactions between a metallic electrode and solvated ions, and the $I$–$V$ (current–voltage) characteristics of molecular electronics devices [1]. According to the seminal work by Marcus [2], fluctuations of the environment (solvent) are crucial in mediating the transfer of an electron from the reactants to the products. Reaction rates, in the general scenario of Marcus theory [3,4], have a thermodynamic contribution (the classical Franck–Condon factor, broadly related to the free-energy cost of a nuclear fluctuation that causes the donor and acceptor levels to degenerate in energy), and an electronic structure, tunneling contribution (the Landau–Zener term, related to the overlap of the initial and final states).

The theory of electron transfer in polar solvents was first developed by R. A. Marcus in the 1950s [2–6]. Since its introduction, the Marcus theory has become the framework for the study of electron transfer reactions. Marcus, in recognition of his contributions to the field, was awarded the Nobel Prize in 1992. While Marcus theory provides a macroscopic understanding of polarization effects on electron transfer reactions, many pioneering microscopic studies [7–11] have been performed since the late 1970s. In particular, the semiclassical trajectory approach [8] has shown success in free-energy studies of electron transfer reactions in polar solvents, and in providing microscopic-level understanding of these reactions [7–10]. However, these studies were based on classical force field parameters. Full ab initio approaches were introduced relatively recently and have been shown to provide excellent quantitative descriptions of the systems studied [12–14].

We discuss the electron transfer mechanism briefly here. Consider an electron that transfers from a donor species $D$ to an acceptor species$A$ in solution. First, $D$ and $A$ must diffuse through the solution and come into proximity to form a complex $DA$, in which the donor and acceptor are separated by a distance at which the electronic coupling matrix element is significant. In this scenario (Fig. 15.1), when the complex is formed, the system has two states, corresponding to the two relevant electronic states (reactant and product states) $DA$ and $D^+A^-$. The potential energy of the reactant state (or the product state) and the surrounding medium is a function of all of these nuclear coordinates, and defines a multidimensional potential energy surface.

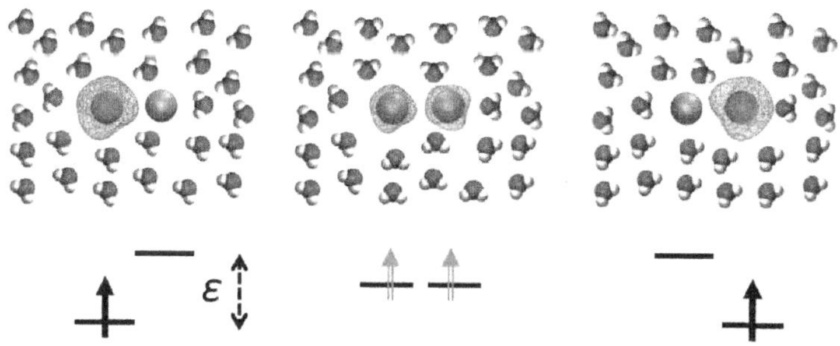


**FIGURE 15.1** Schematic diagram of the electron-transfer process. The solvent fluctuations mediate the transfer process, and the transfer occurs when the reactant and the product states are degenerate.

In order for the electron transfer to occur, two principles have to be satisfied; (1) the Franck–Condon principle and (2) the conservation of energy. According to the Franck–Condon principle, the electron, which is a light particle, transfers from the donor site to the acceptor site instantaneously so that the ions do not have time to change either their positions or momenta. The conservation of energy requires that the product and reactant be degenerate during the transfer process. Since the ions are stationary during the transfer process, the conservation of energy requires, in a picture of independent electrons, degeneracy for the transferring electron to be at the donor or the acceptor site. This degeneracy is usually achieved owing to thermal fluctuations of the solvent and of the donor and the acceptor molecules. This temporary reorganization creates a situation where the electron resonates between the sites. If there were a finite coupling between the two electronic states, transfer could occur, and relaxations of the solvent molecules would then localize the electron on the new site.

As is the case with all chemical reactions, it is important to define a reaction coordinate to characterize the role of ionic motions on electron transfer reactions. However, the solvent fluctuation effects on the reactant and product energies is a collective quantity depending on all the solvent ionic coordinates. In order to capture these collective contributions, the reaction coordinate ($\epsilon$) of a particular ionic configuration is defined as the potential energy difference (energy gap) between the product and the reactant state at that configuration, as suggested by Warshel et al. [8].

Figure 15.2 shows a schematic diagram of the free-energy surfaces of an electron transfer reaction as a function of this reaction coordinate. As mentioned above, the electron transfer occurs when the reactant and product are degenerate, that is, when $\epsilon = 0$. Once the system reaches the intersection of the two curves in Figure 15.2, the probability of going from the reactant to the product free-energy surface depends on the extent of coupling of the electronic orbitals for the two reactants (the transfer integral), which, in turn, depends on the separation distance $r$ of the reactants.

The electron transfer rate constant can be written as

$$\begin{aligned} k_{\mathrm{ET}} = {} & \text{Rate of system reaching intersection of free-energy surfaces} \\ & \times \text{probability of electron tunneling when at intersection} \end{aligned} \tag{15.1}$$

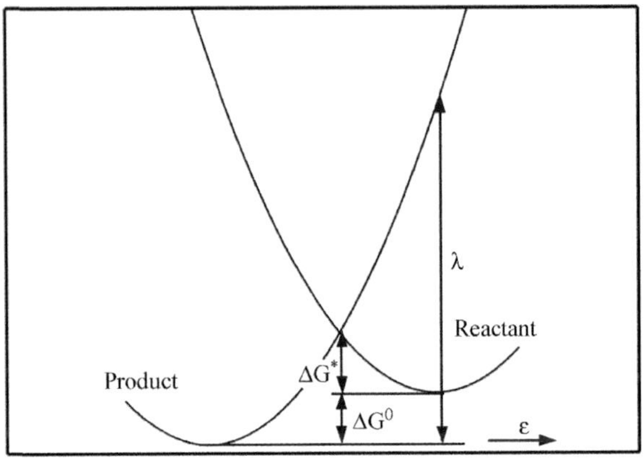


**FIGURE 15.2** Diabatic free-energy surfaces for a typical electron transfer reaction.

The rate of the system reaching the intersection depends on temperature and the solvent relaxation time, while the transfer integral is directly related to the electronic coupling matrix element between the reactant and the product states. Depending on the value of the transfer integral, the electron transfer reaction can be classified into adiabatic and diabatic processes.

In adiabatic reactions, the coupling is large so that the time for tunneling to occur is small compared to the time needed for the system to reach the intersection of the free-energy surfaces. Therefore, electron tunneling occurs whenever the interaction is reached, and the probability of electron tunneling is 1. In this regime, the rate constant can be written as

$$\begin{aligned} k_{\mathrm{ET}} &= \text{rate of system reaching intersection of curves} \\ &= A\, e^{-\Delta G^*/k_{\mathrm{B}}T} \end{aligned} \tag{15.2}$$

that is, the Arrhenius probability of reaching the intersection from the bottom of the reactants' free-energy surface, where $G^*$ is the free-energy barrier. An adiabatic process usually occurs when the donor and the acceptor are in close contact, as in the case of some inner-sphere electron transfer processes.

In the diabatic limit, the transfer integral is small and the system has to diffuse up to the intersection several times before the electron tunnels. Such a weak coupling case can be treated quantum mechanically using Fermi's golden rule. In the diabatic case, $k_{\mathrm{ET}}$ can be written as

$$k_{\mathrm{ET}} = \frac{2\pi}{\hbar} {V_{\mathrm{IF}}}^2 \rho_{\mathrm{FC}} \tag{15.3}$$

where $V_{\mathrm{IF}}$ is the transfer integral between the initial and final states; $\rho_{\mathrm{FC}}$ is the density of states weighted by the Franck–Condon factor and thermally averaged. This term becomes $[\exp(-(\Delta G^*/k_{\mathrm{B}}T)]/[(4\pi\lambda k_{\mathrm{B}}T)^{1/2}]$ in the classical case and leads to the original Marcus rate equation $k_{\mathrm{ET}} = (k_{\mathrm{el}} k_{\mathrm{B}} T/h)\exp(-(\Delta G^*/k_{\mathrm{B}}T))$ with $V_{\mathrm{el}} \ll 1$.

If we assume the free-energy surfaces to be parabolic, we can express $\Delta G^*$ as

$$\Delta G^* = \frac{(\Delta G^\circ + \lambda)^2}{4\lambda} \tag{15.4}$$

where $\lambda$, the reorganization energy, is the free-energy cost to change from the equilibrium configurations of the product into the equilibrium configurations of the reactant without the electron transfer having taken place.

There have been numerous pioneering classical molecular dynamic studies [7,8,11] of electron transfer reactions. However, quantitative agreement has not been achieved with classical force fields. This discrepancy is due to the inaccuracy of classical potentials in dealing with processes where there is a significant change in the electronic structure. It is thus of great interest to test and employ a fully ab initio molecular dynamics approach to study electron transfer reactions, noting that previous theoretical estimates of the transfer integral are semiempirical [15] or used several approximations in the quantum chemical description of the system [11,16].

In order to provide a better quantitative picture of electron transfer reactions, we first present a novel approach to allow the study of electron transfer reactions from ab initio molecular dynamics. This approach involves extensive umbrella sampling of the solvent phase space and the introduction of a novel energy functional to calculate accurately both the ground state and first excited state of electron transfer reactions [14]. We then present a more recently developed method for the theoretical estimate of transfer integrals [44]. A Hubbard $U$ correction [17] is also introduced to better correct for the electronic self-interaction of $d$ electrons. With aqueous ferrous–ferric self-exchange as a paradigmatic example, we show how accurate diabatic free-energy surfaces, transfer integrals, and reaction rates can be obtained, leading to excellent agreement with the experimental results.

## 15.2 ELECTRON TRANSFER FREE-ENERGY SURFACES FROM MOLECULAR DYNAMICS SIMULATIONS

In this section, we discuss the calculation of the free-energy surfaces of the ferrous–ferric self-exchange reaction. We first describe previous studies that used classical molecular dynamics, followed by our own approach based on ab initio molecular dynamics, umbrella sampling, and a penalty functional to force the system in the ground and excited diabatic states. With the ferrous–ferric self-exchange reaction as an example, we studied the two cases of the two ions either separated infinitely or at a finite distance apart. We discuss the challenges in performing umbrella sampling ab initio molecular dynamics when the two ions are a finite distance apart and introduce a penalty functional that appropriately constrains the electrons to the diabatic states.

### 15.2.1 Electron Transfer Free-Energy Surfaces from Classical Molecular Dynamics

According to the Marcus theory of electron transfer, if the environment responds linearly to the electric field [5,7], the free-energy surfaces are parabolic. In order to both qualitatively and quantitatively understand electron transfer reactions, there have been a number of classical molecular dynamic studies [7–11] supporting this assumption in the case of reactions between aqueous metal ions. For example, Figure 15.3 shows the free-energy surfaces of ferrous–ferric self-exchange reaction taken from Reference 11, using classical molecular dynamics with umbrella sampling.

Despite the qualitative consistency, quantitative agreement has not been achieved with classical force fields. The reorganization energy λ (i.e., the free-energy cost to move from the equilibrium configurations of the product to those of the reactant without electron transfer) for the aqueous $Fe^{2+}$–$Fe^{3+}$ self-exchange reaction was found to be 83.2 kcal/mol or 3.6 eV for ions 5.5 Å apart [11], while experimentally [20] (at the slightly shorter separation of 5.32 Å) it is found to be 2.1 eV. Although there have been studies of electron transfer reactions including electronic polarization in classical force field potentials [9,10], fully ab initio studies are required to describe these reactions realistically and quantitatively. More recently, an elegant grand-canonical density functional approach has been introduced to address this class of problems [12,18,19]. This approach is, however, targeted at half-reactions for a donor or an acceptor in contact with an electron reservoir, and the free-energy surfaces and free-energy barriers cannot be calculated using this approach. The need to qualitatively and quantitatively describe electron transfer reactions has prompted us to explore the possibility of performing ab initio molecular dynamics studies of full electron transfer reactions.

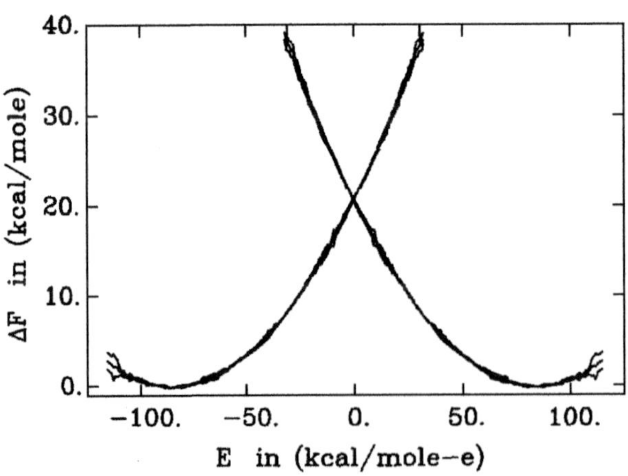


**FIGURE 15.3** Diabatic free-energy surfaces for $Fe^{2+}$–$Fe^{3+}$ electron transfer reaction calculated using classical molecular dynamics simulations when the ions are 5.5 Å apart. (Figure from Ref. 11.)

### 15.2.2 Static Density Functional Calculation of Reorganization Energy

As depicted in Equation (15.4), the reorganization energy is an important parameter in the reaction calculation. In Marcus' original paper of electron transfer [2], a dielectric continuum treatment was introduced to calculate the reorganization energy contributions from the solvent (outer-sphere contributions $\lambda_{os}$), assuming a linear response of the environment to the electric field. This original dielectric continuum treatment is in widespread use and can be sufficiently accurate, provided some reliable estimate of the effective radii of the reactants can be made. However, this method does not take into account the reorganization of the reactant molecules (inner-sphere contribution $\lambda_{is}$) due to a change of oxidation states during electron transfer, and this contribution can sometimes be significant. Theoretical approaches to estimate $\lambda_{is}$ are surprisingly few, and the best approach is still a matter of debate. Klimkans and Larsson [21,22] suggested that $\lambda_{is}$ could be calculated in a convenient way from the total energies of the isolated reactant molecules in their ground and excited electronic states. In this section, we adopt this method to calculate $\lambda_{is}$, and together with the dielectric continuum treatment to obtain $\lambda_{os}$, we are able to compute the total reorganization energy for the ferrous–ferric self-exchange reaction.

In the case of iron ions solvated in water, the inner sphere can be taken to consist of the ion and the first solvation shell water molecules. As shown in Figure 15.4, the reorganization energy contributions are divided into an inner-sphere (shaded) regions and an outer-sphere (remaining) region. $\lambda_{is}$ is calculated using isolated clusters of aqua iron coordinated with six water molecules, as it has been suggested experimentally [23,24] that both ferrous and ferric ions are sixfold-coordinated in the first solvation shell by water molecules when solvated in water.

Figure 15.5a shows schematically this procedure to calculate $\lambda_{is}$ from ab initio. First, the ionic geometry of hexaaqua ferrous cluster is optimized in vacuum. One electron is then removed from the system and the total energy of ferric ion in the optimized geometry of the ferrous ion cluster is calculated. A similar procedure is

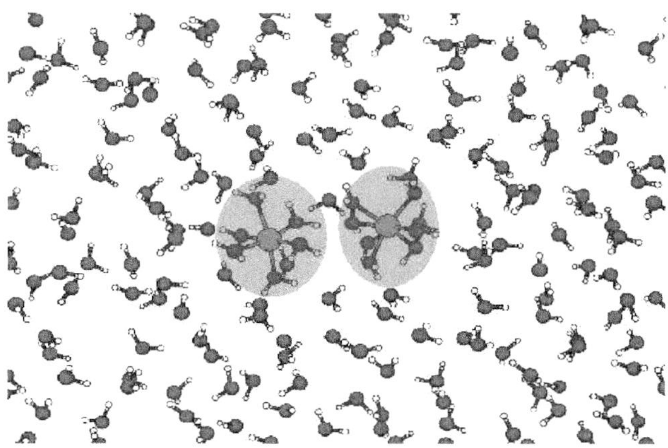

**FIGURE 15.4** Schematic diagram showing the division of the reorganization energy contributions into inner and outer spheres.

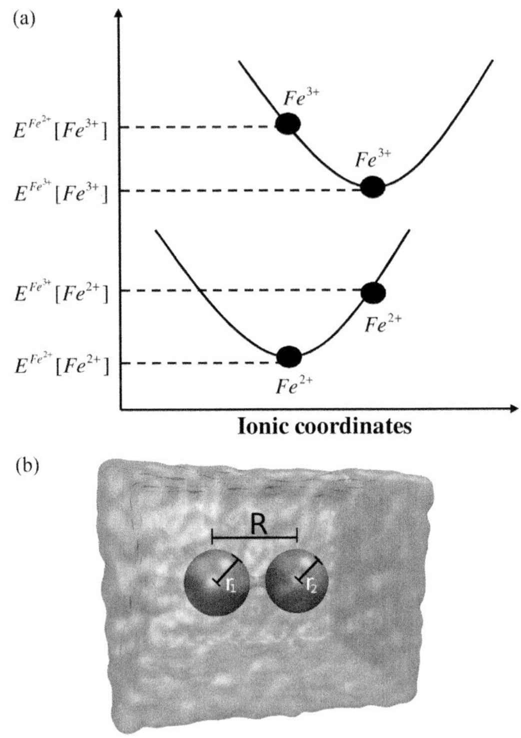


**FIGURE 15.5** Procedures for calculating (a) inner-sphere contributions to reorganization energy using four separate ab initio calculations; and (b) outer-sphere contributions to reorganization energy using the dielectric continuum theory. $r_1$ and $r_2$ are the radii of the sphere occupied by the ferrous and ferric ion clusters, respectively, taken as approximately the distance between the ion and the hydrogen atoms in the first solvation shell. $R$ is the center-to-center distance between the two clusters.

performed with the hexaaqua ferric cluster, and one extra electron is added to the system after ionic geometry optimization. $\lambda_{is}$ can then be obtained from

$$\lambda_{is} = E^{Fe^{3+}}[Fe^{2+}] + E^{Fe^{2+}}[Fe^{3+}] - E^{Fe^{2+}}[Fe^{2+}] - E^{Fe^{3+}}[Fe^{3+}] \qquad (15.5)$$

where $E^{Fe^{2+}}[Fe^{2+}]$ ad $E^{Fe^{2+}}[Fe^{3+}]$ are the total energies of the ferrous or ferric ion in the optimized geometry of the ferrous ion cluster. Similar notation is used for $E^{Fe^{3+}}[Fe^{2+}]$ and $E^{Fe^{3+}}[Fe^{3+}]$. In plane-wave codes employing periodic boundary conditions, comparing energies between systems with different charges always introduces errors due to the Coulomb interactions between periodic images. However, in the above procedure, these Coulomb interactions cancel out when calculating the difference in energies.

We perform these calculations with the $PW_{scf}$ package [25], using ultrasoft pseudopotentials [26] for iron, oxygen, and hydrogen atoms at Perdew-Burke-Ernzerhof generalized gradient approximation (PBE-GGA) level [27]. The ultrasoft pseudopotentials for O and H are from the standard $PW_{scf}$ distribution (`H.pbe-rrkjus.UPF` and `O.pbe-rrkjus.UPF`). We generated a 16-electron iron ultrasoft pseudopotential (`Fe.pbe-sp-vanmit.UPF`) using the Vanderbilt code [28]. Plane-wave cutoffs for wavefunction and charge density are 25 and 200 Ry (rydbergs), respectively. The ion cluster is in a unit cell of side length 15.9 Å. Using the described procedure, the inner-sphere contribution ($\lambda_{is}$) equals 0.50 eV.

Contributions to the reorganization energy from the outer solvent molecules can be estimated from the dielectric continuum model [2,5]. In this model, the ions and the first solvation shell water molecules are replaced by conducting spheres with volumes approximately equal those occupied by the hexaaqua iron clusters as shown in Figure 15.5b. $\lambda_{os}$ can be calculated as

$$\lambda_{os} = \frac{e^2}{4\pi\varepsilon_0}\left(\frac{1}{2r_1} + \frac{1}{2r_2} - \frac{1}{R}\right)\left(\frac{1}{\varepsilon_\infty} - \frac{1}{\varepsilon_{static}}\right) \tag{15.6}$$

where $r_1$ and $r_2$ are the radii of the sphere occupied by the ferrous and ferric ion clusters, respectively, taken as approximately the distance between the ion and the hydrogen atoms in the first solvation shell; $R$ is the center-to-center distance between the two clusters; and $\epsilon_\infty$ and $\epsilon_{static}$ are the optical and static dielectric constants of water. Taking $r_1 = 2.66$ Å, $r_2 = 2.66$ Å, R = 5.5 Å, $\epsilon_\infty = 1.8$ and $\epsilon_{static} = 78$, respectively, the outer-sphere contribution, $\lambda_{os}$, is 1.54 eV.

The total reorganization energy ($\lambda = \lambda_{is} + \lambda_{os}$) equals 2.04 eV, which is in very good agreement with the experimental value of 2.1 eV. However, this static calculation does not take into account entropic effects, and does not provide the free-energy surfaces for the reaction, and therefore the free energy barrier. It also depends on the choice of $r_1$, $r_2$, and $R$. Despite these deficiencies, it is a clear indication that ab initio calculations can provide good quantitative agreement with the experiments.

### 15.2.3 Diabatic Free-Energy Surfaces from Ab Initio Molecular Dynamics

In this section, we introduce a novel ab initio molecular dynamics technique that involves some algorithmic advances that allow us to quantitatively describe the thermodynamic properties of electron transfer reactions, with ferrous–ferric self-exchange as a paradigmatic example. In all the simulations discussed in this section, we use Car–Parrinello molecular dynamics [29–31] and spin-polarized DFT in the PBE-GGA approximation. The ferrous and ferric ions are kept in the high-spin states, as confirmed by the literature [32,33] and our own calculations. The ultrasoft pseudopotentials for Fe, O, and H are the same pseudopotentials used in Section 15.2.2. The wavefunctions and charge density cutoffs are 25 and 200 Ry, respectively. The deuterium mass is used in place of hydrogen mass to allow for a larger time step of integration. The Car–Parrinello fictitious mass ($\mu$) and timestep are 450 and 5 au, respectively. A Nose–Hoover thermostat is used on both electrons and ions. Previous ab initio molecular dynamics studies [34,35] show that, at PBE level, water is in a glassy state at room temperature. Therefore, the ionic temperature is set at

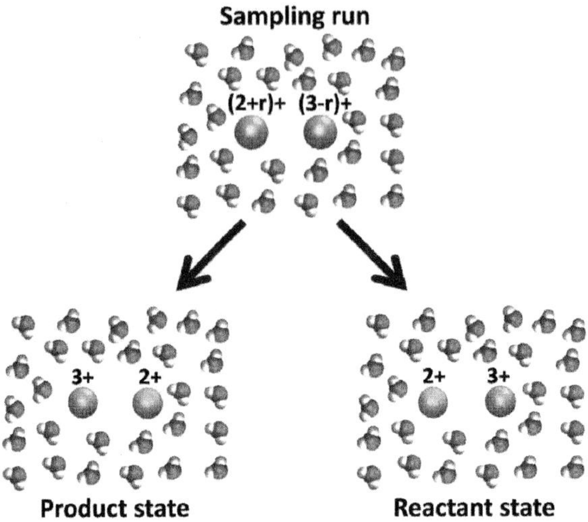


**FIGURE 15.6** Procedure used to calculate the diabatic free energy surfaces for electron transfer reactions from ab initio molecular dynamics. The reaction coordinate at each timestep is calculated from the energy difference between the product and reactant states in the ionic configuration provided by the sampling run. The phase space is explored efficiently via the umbrella sampling parameter *r*.

400 K, at which the structural and dynamic properties agree well with the experimental data at room temperature [35].

Figure 15.6 shows schematically the sampling procedure used, following the lines of Reference 11 for classical simulations. We first generate an ionic trajectory with the ions in the $(2+r)$ and $(3-r)$ states of charge, respectively. The desired charge state of the ions can be obtained by introducing a fractional number of electrons in the systems in the case of two ions infinitely apart or by means of a penalty functional when two ions are at a finite distance apart. $r$ is an "umbrella-sampling" parameter used to force the system to explore different regions of the phase space that turns out to perform this task very efficiently. We then perform two separate runs with the electronic state constrained in the reactant or in the product electronic configuration, and with the ions constrained to follow the afore-generated ionic trajectories. The reaction coordinate $\epsilon$ at every timestep is thus given by the difference between the energies of the product and the reactant state. The probability distribution $P(\epsilon)$ is then calculated as

$$P(\epsilon) = \frac{\sum_{\tau} \delta_{\epsilon'(\tau),\epsilon} \exp\{-\beta[E_r(\tau) - E_s(\tau)]\}}{\sum_{\tau} \exp\{-\beta[E_r(\tau) - E_s(\tau)]\}} \quad (15.7)$$

where $\epsilon'(\tau) = E_p(\tau) - E_r(\tau)$ is the reaction coordinate at time $\tau$; $E_r(\tau), E_p(\tau)$ and $E_s(\tau)$ are the energies of the system in the reactant, product and sampling oxidation states, respectively, at time $\tau$; and $\delta_{\epsilon'(\tau),\epsilon}$ is the Krönecker delta. The exponential term in the expression above restores the correct thermodynamical sampling according to the energy surface $E_r$ compared to our non-Boltzmann sampling of trajectory $E_s$. The free energy $F(\epsilon)$ is derived from the probability distribution $P(\epsilon)$ as $F(\epsilon) = -k_B T \ln(P(\epsilon))$; each sampling run will lead to a different section of $F(\epsilon)$, to be assembled in one continuous curve.

However, because of the lack of self-interaction correction in common exchange correlation functionals, the transferring electron (3*d* minority spin in the case of ferrous–ferric self-exchange reaction) will unphysically split between the two ions. Moreover, to compute the energy gap at each timestep, we need to accurately calculate the total energy when the minority spin electron localizes at either the reactant or product site for any given ionic configuration. In order to address these central problems, we consider two cases: (1) when two ions are infinitely apart and (2) when the two ions are at a finite distance from each other.

### 15.2.4 Special Case When Two Ions Are at an Infinite Distance Apart

We first consider the simple case when we can trivially control the ions' oxidation states, that is, when the two ions are infinitely apart in this case. The two ions can be studied in separate simulation cells, and the oxidation states are controlled by simply changing the total number of electrons. For this special case, we performed runs using $Fe^{(2+r)+}$ and $Fe^{(3-r)+}$ (with $r=0.0$, 0.25, 0.50, 0.75, and 1.0), each solvated with 31 water molecules in the unit cell. We then carried out one $Fe^{2+}$ run and one $Fe^{3+}$ run in the trajectory generated with $Fe^{(2+r)+}$, and one $Fe^{2+}$ run and one $Fe^{3+}$ run in the trajectory generated with $Fe^{(3-r)+}$. Because of the symmetry of the reactant and the product, the sampling ensembles of $r=0$, 1 were taken from the same trajectories; similarly for $r=0.25$, 0.75 (as mentioned in Section 15.2.3, when calculating energies of charged systems in periodic boundary conditions, the Coulomb interactions of charges with their periodic images should be removed [36]. In practice, these errors cancel out on calculation of the energy gap, which is the difference in energy between systems with the same charge.

The initial configuration for each simulation is prepared by replacing seven water molecules from the last configuration of a well-equilibrated 32-water-molecule simulation at 400 K with an hexaaqua iron ion. A rescaling of the unit cell volume to 95% of the original size is needed to account for the difference in volumes occupied by seven water molecules and an hexaaqua iron ion. A 5-ps equilibration run with the 31 water molecules and a ferrous ion is then carried out. The last ionic configuration of this equilibration run is taken as the initial configuration for each simulation with different oxidation states for the ion. Each simulation consists of a production run of about 5 ps after 1 ps of thermalization. To assess whether the systems prepared this way are well equilibrated, we show in Figure 15.7 the Fe–O radial distribution functions for simulations performed with different ionic oxidation states. The black and the gray curves show the radial distribution functions calculated from, respectively, the first and the second halves of the 5 ps production runs. As seen in the figure, the solvation structures do not change significantly after 1 ps of thermalization. This observation is in good agreement with Bader et al. [37] suggesting that a full relaxation occurs within 1 ps after the change in oxidation state of the aqueous iron ion.

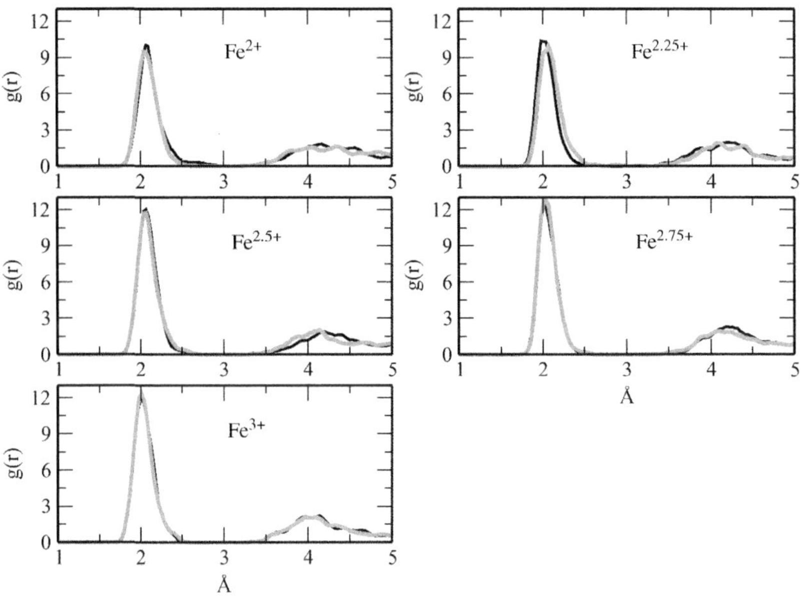


**FIGURE 15.7** The Fe–O radial distribution functions for $Fe^{(2+r)+}$ solvated in water with $r = 0,0.25,0.5,0.75,1$. Each simulation lasts about 5 ps after dropping the first 1 ps. The black and the gray curves show the radial distribution functions calculated from the first and second halves of different simulations, respectively.

Figure 15.8 shows the resulting diabatic surfaces; the final result is obtained by integrating [38]

$$F(\epsilon) = \frac{\sum_r F_r(\epsilon) g_r(\epsilon)}{\sum_r g_r(\epsilon)} \tag{15.8}$$

where $F_r(\epsilon)$ is the slope of the free-energy curve in the different sections, each characterized by an umbrella sampling parameter $r$, and the weighting factor $g_r(\epsilon)$ is $\langle\delta(\epsilon - \epsilon(\tau))\rangle_r$. For this special case of infinitely apart ions, the trajectories generated with $Fe^{(2+r)+}$ and $Fe^{(3-r)+}$ are independent; since each of them provides $n$ data points, there will be $n^2$ energy gaps, providing excellent statistics and a very smooth free-energy curve that fits accurately a parabola, with a coefficient of determination ($R^2$) of 0.9996, and providing a reorganization energy of 1.77 eV. Note that at the tail ends of each sampling region the statistical accuracy becomes lower—this might explain the slight deviations from a parabola seen in Figure 15.8.

### 15.2.5 Control of Oxidation States through a Penalty Functional

Common exchange correlation functionals used in DFT contain self-interaction errors that can affect the accuracy of the theoretical predictions. In particular, such errors can lead to a dramatic qualitative failure in describing ions in different oxidation states,

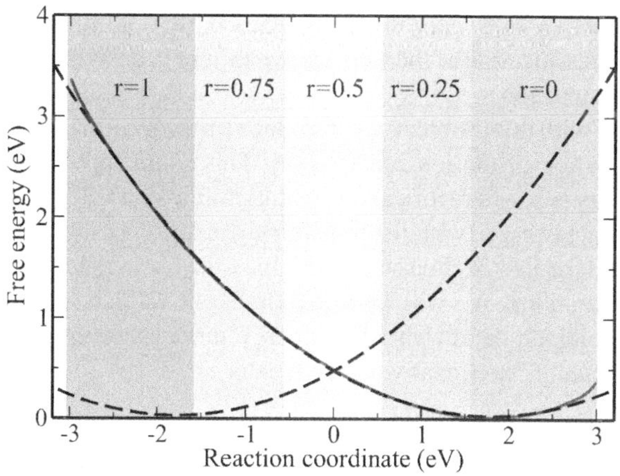


**FIGURE 15.8** Diabatic free-energy surfaces for ferrous–ferric electron transfer obtained in the special case of two ions infinitely apart from result in Reference [14]. The solid curve has been obtained from ab initio molecular dynamics. The dashed curves are mirror images, and correspond to a parabolic fit of the data. Different shades indicate portions of the diabatic surface sampled with $r = 0, 0.25, 0.5, 0.75, 1$.

when more than one ion is present. This failure can be easily seen in the case when two iron ions in the $+2$ and $+3$ oxidation states are studied in the same simulation cell. Out of the transition coordinates, we expect a localization of the HOMO on one iron center, with an appreciable tail on the other ion only close enough to the transition state. Figure 15.9 shows the HOMO charge density of the system with two hexaaqua ions of different local structures side by side, using a PBE exchange correlation functional. As a result of self-interaction, the HOMO electron abnormally delocalizes over the two iron centers to decrease the electrostatic interaction with itself. This behavior takes place irrespective of the chemical environment of the two iron centers; we observe it

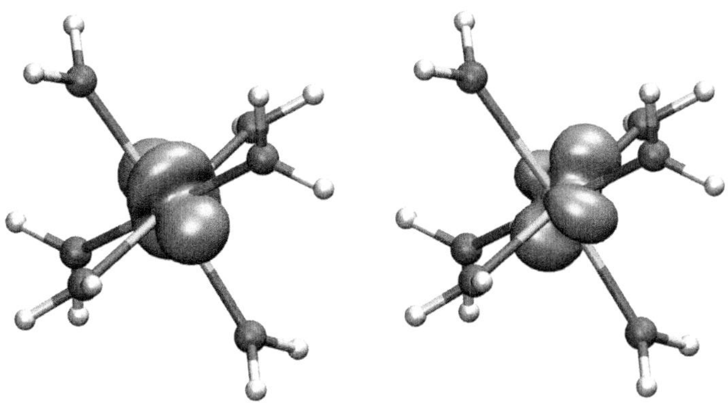

**FIGURE 15.9** The HOMO orbital charge density distribution in the case of hexaaqua ferrous and hexaaqua ferric ions in the same supercell. The minority-spin electronic charge density (HOMO) splits between two ions as a result of the electronic self-interaction.

for two isolated ions, two hexaaqua iron complexes, or two ions fully solvated; in many ways it is just a generalization of the well-known failure in describing the dissociation of the $H_2{}^+$ molecular ion.

In order to deal with this problem, we introduce a penalty cost to ground states with noninteger occupation of the ion centers [14]. This same approach is also used to calculate the energy gap, where for a given configuration we need to determine both the correct ground-state energy (with the transferring electron in the reactant electronic configuration) and the first excited state (with the transferring electron in the product electronic configuration). We will first quantitatively validate this functional, and then apply it to compute the diabatic free-energy surfaces using ab initio molecular dynamics. The penalty functional we introduce is

$$E[\{\psi_i\}] \rightarrow E[\{\psi_i\}] + \sum_I \frac{P^I}{\sigma_I\sqrt{2\pi}} \int_{-\infty}^{f_0^I - f_{\{\psi_i\}}^I} \exp\left(-\frac{x^2}{2\sigma_I^2}\right) dx \qquad (15.9)$$

where $f_{\{\psi_i\}}^I$ is the largest eigenvalue of the minority-spin occupation matrix $(n_{mm'}^{I,\sigma})$ on ion $I$, calculated in this work by projecting the minority-spin Kohn–Sham orbitals on the $3d$ atomic orbitals of iron

$$n_{mm'}^{I,\sigma} = \sum_i \langle \psi_i^\sigma | \phi_{m'}^I \rangle \langle \phi_m^I | \psi_i^\sigma \rangle \qquad (15.10)$$

($\psi_i^\sigma$ are the Kohn-Sham orbitals and $\phi_m$ are the $3d$ atomic orbitals of iron), while $f_0^I$ is the target value of $f_{\{\psi_i\}}^I$ (the procedure for determining its actual value is discussed later). In the case of ultrasoft pseudopotentials, this occupation matrix is generalized to

$$n_{mm'}^{I,\sigma} = \sum_i \langle \psi_i^\sigma | S(\{R_I\}) | \phi_{m'}^I \rangle \langle \phi_m^I | S(\{R_I\}) | \psi_i^\sigma \rangle \qquad (15.11)$$

where the overlap operator $S(\{R_I\})$ is a function of the augmentation charges of ultrasoft pseudopotentials. There is no unique way to define the occupation matrix, $n_{mm'}^{I,\sigma}$ and the choice should depend on the systems studied. In this work, we chose to project the Kohn–Sham orbitals on localized atomic $d$ orbitals; this definition works well in systems where the transferring electron does not participate in any chemical bonding. Molecular orbitals, Wannier functions, or Lowdin charges [13] can, instead, be used if the donor or acceptor ion is chemically bonded to the ligands.

Note that the definition of $f_{\{\psi_i\}}^I$ is independent of any unitary transformations between the $\psi_i^\sigma$ terms or any other definition of the $3d$ atomic orbitals also obtained by an unitary transformation between the $\phi_m^I$ terms. The penalty is in the form of a step function with the smearing relating to the value of $\sigma$. When $P^I > 0$, a penalty will be incurred on the total energy if $f_{\{\psi_i\}}^I$ is smaller than $f_0^I$. Therefore, the ion is forced to have $f_{\{\psi_i\}}^I > f_0^I$. Similarly, if $P^I < 0$, then $f_{\{\psi_i\}}^I < f_0^I$. By changing the sign of $P^I$ and the target value $f_0^I$ on each reaction site, the transferring electron, that is, the minority-spin $d$ electron in this case, can be forced to localize at the desired site (Fig. 15.10). A fraction of an electron, instead of the whole one, can also be localized at one site or the other; this fractional charge localization will be useful for our umbrella sampling procedure.

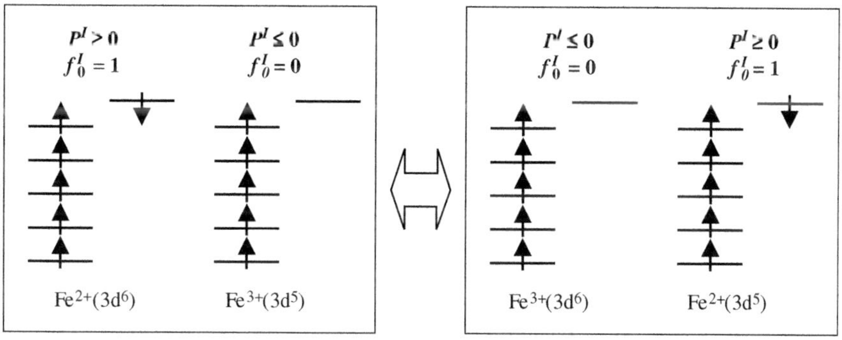


**FIGURE 15.10** The transferring electron can be localized at the desired location with the appropriate choice of $P^I$ and $f_0^I$.

In performing Car–Parrinello molecular dynamics, the wavefunctions and ions evolve according to the new equations of motion:

$$-m_I \frac{d^2R^I}{dt^2} = \frac{\partial E[\{\psi_i\}]}{\partial R^I} + \sum_I \frac{P^I}{\sigma_I\sqrt{R\pi}} \exp\left[-\frac{(f_0^I - f_{\{\psi_i\}}^I)^2}{2\sigma_I^2}\right]\left(-\frac{\partial f_{\{\psi_i\}}^I}{\partial R^I}\right) \tag{15.12}$$

$$-\mu_I \frac{d^2\psi_i}{dt^2} = \frac{\delta E[\{\psi_i\}]}{\delta\psi_i^*} + \sum_I \frac{P^I}{\sigma_I\sqrt{2\pi}} \exp\left[-\frac{(f_0^I - f_{\{\psi_i\}}^I)^2}{2\sigma_I^2}\right]\left(-\frac{\delta f_{\{\psi_i\}}^I}{\delta\psi_I^*}\right) \tag{15.13}$$

$f_{\{\psi_i\}}^I$ can be written as,

$$f_{\{\psi_i\}}^I = \sum_{m,m'} U_{5,m}^+ n_{m,m'}^{I,\sigma} U_{m',5} \tag{15.14}$$

where the unitary matrix $U_{hl}$ diagonalizes $n_{m,m'}^{I,\sigma}$ and the eigenvalues are arranged in ascending order (we note in passing that an additional fictitious dynamics of the $U_{hl}$ matrices should be added to the Car–Parrinello Lagrangian and equations of motion; in practice we used the simplified expressions above with no observable ill effects). The right-hand side (RHS) of Equation (15.12) becomes

$$\sum_i \psi_i | \frac{\partial \hat{H}}{\partial R^I} - \sum_I \frac{P^I}{\sigma_I\sqrt{2\pi}} \exp\left[-\frac{(f_0^I - f_{\{\psi_i\}}^I)^2}{2\sigma_I^2}\right] \times \left(\sum_{m,m'} U_{5,m}^+ \frac{d|S(\{R_I\})|\phi_{m'}^I\rangle\langle\phi_m^I|S(\{R_I\})|}{dR^I} U_{m',5}\right)|\psi_i \tag{15.15}$$

where $\hat{H}$ is the Hamiltonian corresponding to the original DFT functional. In this expression, both the overlap operator $S(\{R_I\})$ and the atomic orbitals $\phi_m^I$ depend analytically on the positions of the ions, $R^I$.

In practice, when the ions are solvated in water, the electronic structure as well as the occupation numbers are expected to be altered by the solvation process. To determine the optimal parameters used in the penalty functional, the reference states at which the electronic structure should resemble the case of full solvation were studied. In this case, we separately calculated the minority-spin occupation matrix for either a ferrous or a ferric hexaaqua ion embedded in a dielectric continuum ($\epsilon = 78$) [39]. Then, we determine the parameters in the penalty functional applied on the ions so that the occupation matrices of the ferrous or ferric clusters are accurately reproduced once the two are studied in the same unit cell. We obtained $P^I = 0.54$ eV, $f_0^I = 0.95$, and $\sigma_I = 0.01$ on the ferrous ion and $P^I = -0.54$ eV, $f_0^I = 0.28$, and $\sigma_I = 0.01$ on the ferric ion. We note that the target occupations for the minority spin are not chosen exactly one or zero, since the orbital hybridization between the iron $3d$ orbitals and the lone pairs of the water molecules contributes to the projection onto atomic orbitals. The contributions to the total energy from the penalty functional are close to zero on the ferrous ion and close to $-0.54$ eV on the ferric ion. When we calculate the energy gap, these contributions are not included in the total energy; in any case, these effects are negligible since these contributions cancel out when calculating energy differences. Different constraints or penalties have been implemented in density functional calculations [13,14,40]; we found our choice particularly robust, but several variations on the theme can be envisioned.

## 15.3 VALIDATION OF THE PENALTY FUNCTIONAL

A first, qualitative validation of this penalty functional is performed examining the charge density difference obtained by subtracting the charge density of an isolated ferrous hexaaqua ion, and of a ferric hexaaqua ion from that of a ferrous hexaaqua ion and a ferric hexaaqua ion in the same unit cell (the hexaaqua ions are 10 Å apart, and a dielectric continuum surrounds the two clusters to remove long-range electrostatic interactions between them). The charge density difference is shown in Figure 15.11.

As anticipated before, in the absence of penalty, the transferring electron splits between the two ions; once the penalty is applied, the charge density around the ions reorganizes itself so that it produces a charge density that closely matches the superposition of that obtained from the two independent calculations.

We can make the validation quantitative by calculating the energy gap for the system described, using two penalty functional calculations that require the HOMO electron to localize first on one, then on the other ion. This energy gap can also be calculated exactly with PBE-GGA using the four-point approach [21,22], provided again that all long-range electrostatic interactions are screened out. The four calculations involve $Fe^{2+}$ in two $Fe(H_2O)_6$ geometries ($A$ and $B$), and $Fe^{3+}$ in the same geometries; the energy gap is $[E_A(Fe^{2+}) + E_B(Fe^{3+}) - E_B(Fe^{2+}) - E_A(Fe^{3+})]$. The calculation follows the same lines as the computation of $\lambda_{is}$ in Section 15.2.2. We choose one configuration in which the hexaaqua ions are fully relaxed, and three carved out from random steps in the molecular dynamics simulations. Table 15.1 shows the energy gaps calculated with the penalty functional and with the four-point

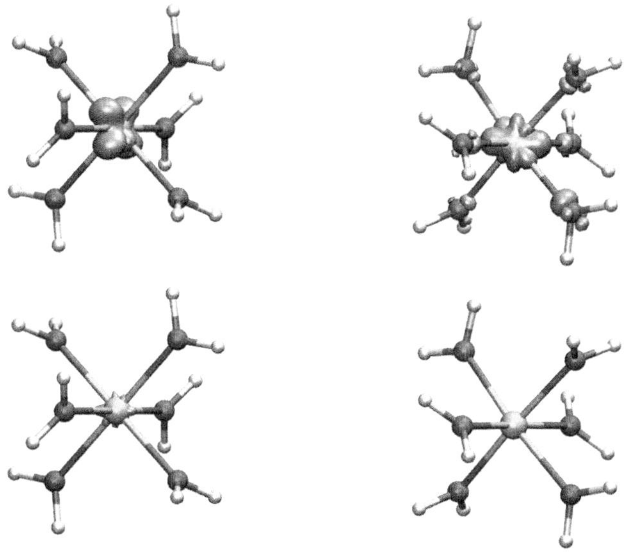

**FIGURE 15.11** Charge density difference compared to the correct DFT ground-state charge density (calculated from ferrous and ferric ion clusters in separate cells), without (above) and with (below) penalty functional. The unphysical splitting of the HOMO electron is clearly visible in the first case. The contours are plotted at 70% of the maximum density difference of the calculation with the penalty functional. The maximum value obtained with the penalty functional is 8.8% of that without the penalty.

method. The energy gaps obtained with the penalty are in excellent agreement with the four-point values. It is worth mentioning that the energy gap is an excited-state property of the system, and thus, in principle, outside the scope of DFT, which is a ground-state theory. However, since charge densities of HOMO and LUMO do not overlap, we can argue that all that is required is a description of the charge density that is locally correct (the excited state has an electron locally in equilibrium around an iron, oblivious of the other iron ion where it could sit more favorably). A more formal justification can be offered on the basis of "nearsightedness" in electronic matter [41,42].

**TABLE 15.1 Energy Gaps Calculated with Penalty Functional and "Four-Point" Approach for Different Random Cluster Geometries**

| | Penalty Functional (eV) | "Four-Point" Approach (eV) |
|---|---|---|
| I | 0.632 | 0.622 |
| II | 0.569 | 0.542 |
| III | 0.769 | 0.769 |
| IV | 1.027 | 1.012 |

*Source:* Reference 14.

### 15.3.1 Diabatic Free-Energy Surfaces When Two Ions Are at a Finite Distance Apart

With these tools, we determined the diabatic free-energy surfaces for two iron ions separated by 5.5 Å and solvated in 62 water molecules, in periodic boundary conditions. A distance of 5.5 Å was suggested [4,43] as optimal for electron transfer. The initial configuration for the simulations were prepared by carving out two spheres, each containing seven water molecules, from the last configuration of a well-equilibrated 64-water-molecule simulation at 400 K and replacing them by two hexaaqua iron ions, 5.5 Å apart. An equilibration run of 7 ps was then performed without the penalty functional. The sampling runs with different $r$ were carried out from the last configuration of this equilibration run. Each sampling run has about 4.5 ps production time after 1 ps of thermalization. We show our results in Figure 15.12, together with a parabolic fit to the data. Different color shades indicate portions of the diabatic surface sampled $r = 0, 0.5, 1$. The reorganization energy ($\lambda$) that we obtain is 1.93 eV, in good agreement with the experimental value of 2.1 eV [20]. As discussed later, this value is further improved by the introduction of Hubbard $U$ corrections. The energy barrier $\Delta G$ is 0.49 eV, about a quarter of $\lambda$, as expected if the free-energy surfaces are parabolic. As a final note, we point out that the energy gap calculation takes the difference between periodic boundary calculations with the same $+5$ charge—thus charge–charge interactions cancel out. Higher-order effects are negligible for the ground state since water screens periodic images with a large static dielectric constant, but will be somewhat relevant for the excited state, whose periodic images are incompletely screened by the ions. We estimate this error to be around 0.18 eV for our supercell size if we assume that the excited state periodic images are screened only by the electrons.

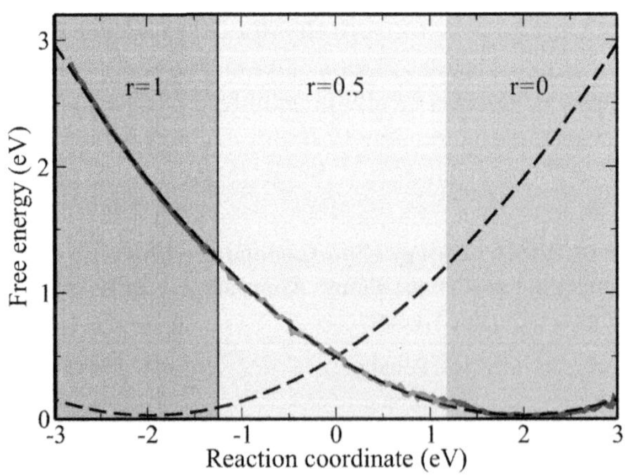


**FIGURE 15.12** Diabatic free-energy surface for ferrous–ferric electron transfer when the two ions are 5.5 Å apart from values in Reference [14]. Different color shades indicate portions of the diabatic surface sampled $r = 0, 0.5, 1$. The right dashed-line curve represents the parabolic fit of the data and the left dashed-line curve, its mirror image.

## 15.4 AB INITIO CALCULATIONS OF TRANSFER INTEGRAL

In this section, we introduce a newly developed method in the two-state model to calculate transfer integrals that allows a full-electron description of the electron transfer system. This method does not resort to any empirical parameter and does not require knowledge of the exact transition state coordinate. Therefore, it appears particularly suitable to be combined with the theoretical calculation of the reorganization energy described previously, in order to obtain a fully ab initio estimate of the electron transfer rate constant to be compared with the experiment.

### 15.4.1 Theoretical Approach

The transfer integral of the ferrous–ferric redox system can be calculated using the two-state approximation from [44]

$$V_{IF} = \left| \frac{\Delta E_{IF} ab}{a^2 - b^2} \right| \tag{15.16}$$

Here, $\Delta E_{IF}$ is the energy difference between the ET diabatic states $I$ and $F$, and $a$ and $b$ are the amplitudes of $I$ and $F$, respectively, in the expression of the adiabatic ground state, which in a two-state model is expressed as $|\Psi\rangle = \mathrm{a}|\Psi\rangle_I + \mathrm{b}|\Psi\rangle_F$. In fact, according to the two-state model, only the two electronic states $I$ and $F$ need to be considered, while the coupling with other electronic state is disregarded. Let us stress that Equation (15.16) provides directly the value of the effective electronic coupling, which is the quantity entering the expression for the ET rate constant in Equation (15.3). In the two-state model, it is related to the electronic coupling $H_{IF} = \langle\Psi_I|H|\Psi_F\rangle$, and the quantities $H_{II} = \langle\Psi_I|H|\Psi_I\rangle, H_{FF} = \langle\Psi_F|H|\Psi_F\rangle$, and $S_{IF} = \langle\Psi_I|\Psi_F\rangle$ via the relation [45]

$$V_{IF} \equiv \frac{1}{1-S_{IF}^2} H_{IF} - S_{IF} \frac{H_{II} + H_{FF}}{2} \tag{15.17}$$

Equation (15.16) has been implemented by some of us [44,46,47] in a spin-polarized DFT approach, although it can be used in other computational schemes. It allows us to compute transfer integrals starting from a few electronic quantities directly derived from ground-state self-consistent field (SCF) calculations on the overall redox system and its isolated donor ($D$) and acceptor ($A$) groups. In fact, in many physical systems, the $D$ and $A$ sites are not connected by a covalent bridge, or the covalent bridge is not relevant to the ET process. This is the case, for example, for the sugar–phosphate backbone in DNA, which affects the ET rate through the conformational changes of the nucleobase stacking, thus requiring a structural average over $V_{IF}$, but does not intervene in the ET process through the nucle-obases. In the ferrous–ferric system, $D$ and $A$ are not connected through a covalent bridge; in fact, they can be identified each with a solvated ion. Then, the initial state vector is defined as $|\Psi_I\rangle = |D\rangle|A\rangle$ and the final state vector as $|\Psi_F\rangle = |D^+\rangle|A^-\rangle$, where $|D\rangle(|D^+\rangle)$

denotes the reduced (oxidized) ground state of the isolated $D$ group and $|A^-\rangle\rangle(|A\rangle)$, the reduced (oxidized) ground state of the isolated $A$ group. The relevant wavefunctions are thus constructed as single Slater determinants of the lowest-lying Kohn–Sham spin orbitals. The feasibility of using DFT wavefunctions is argued in Reference 44 (and other sources cited therein), where it is also shown, in the limit of small $S_{IF}$, that Equation (15.16) can be used even when the two-state approximation does not exactly hold.

The quantity $\Delta E_{IF}$ is calculated as

$$\Delta E_{IF} = (E_D + E_A) - (E_{D^+} + E_{A^-}) + W_{D-A} - W_{D^+-A^-} \tag{15.18}$$

where $E_D(E_{D^+})$ is the ground-state energy of the isolated donor group in its reduced (oxidized) state of charge, $E_A\,(E_{A^-})$ is the same as for the acceptor group in its oxidized (reduced) state, and $W_{D-A}$ and $W_{D^+-A^-}$ are the energies of (essentially electrostatic) interactions between the $D$ and $A$ groups in the $I$ and $F$ states, respectively. Note that if the diabatic states are computed with techniques constraining electronic states like the penalty functional [14] described above or constrained DFT [13], the diabatic state energies $E_I = H_{II}$ and $E_F = H_{FF}$ and their difference $\Delta E_{IF}$ can be directly obtained from the SCF calculations yielding the $I$ and $F$ states.

The SCF calculations required by Equation (15.16) (as well as in the calculation of other physical quantities) in transition metal systems are quite problematic. A crucial source of error is the unphysical self-interaction of the relevant electronic charge with itself, arising from the fact that the approximations to the exchange correlation (XC) energy are independent of the electrostatic repulsion term [48]. For open-shell atoms, a further complication stems from the multiplet problem, because many of the currently available XC functionals do not provide the proper spin and spatial symmetry behavior to correctly describe multiplet systems [49–51]. Such a problem is exacerbated by the simultaneous presence of two transition metal ions, since the molecular orbitals centered on the two ions come to be virtually degenerate. One solution consists in the use of a linear combination of determinants to describe the wavefunction of the system [48]. Finally, a computational issue comes from the oscillations in the iterative self-consistent calculations that intervene whenever some orbitals close to the Fermi level are alternatively occupied and unoccupied from one iteration to the next. Often, these can be controlled by allowing fractional occupations [52,53].

The DFT calculations on the overall ferrous–ferric system are affected by a combination of the abovementioned issues. As a consequence, ground-state SCF calculations can be difficult to converge. Even when convergence is achieved, it generally shows the failure of common XC functionals in the description of the electronic ground state, with the valence electron charge unphysically shared between the two iron centers [14]. Indeed, poor convergence is a common feature for self-consistent field calculations on open-shell transition metal systems, both within the unrestricted or open-shell restricted Hartree–Fock (HF) and the unrestricted DFT approaches. This point is clearly addressed the article by Neese [54].

It should be pointed out that, at the transition state coordinate, the two redox sites are degenerate and the valence electron charge has to be shared by the two ionic centers. In fact, at that coordinate, the coefficients $a$ and $b$ in the expression of the ground-state wavefunction become equal and the energy of the ground state is lowered, relative to degenerate energies of the two diabatic states, by an amount equal to the transfer integral (and the first excited state is raised by the same amount). On the other hand, given the small value of the electronic coupling in the system under consideration, for a small displacement from the transition state coordinate the valence charge becomes almost fully localized around one ion, with a small tail on the other ion. This tail is essential in determining the value of the transfer integral. But when it is very small, its evaluation by DFT is largely affected by the unphysical electron self-interactions, even if these do not produce significant relative changes in the charge density of the donor group. The transfer integral is correspondingly affected. Using Equation (15.16), this means that one of the $a$ and $b$ parameters, which is significantly smaller than the other one, is strongly overestimated by DFT calculations because of self-interaction. Moreover, as discussed previously, the self-interaction problem leads to ground-state wavefunctions almost equally spread over the two iron centers [14]. As shown in Reference 55, a correct localization of the ground-state wavefunction can be achieved through a suitable DFT approach with fractional occupation numbers (FON-DFT) and self-consistent Hubbard $U$ corrections to the electronic self-interaction. In the FON approach, states close to the Fermi level are partially occupied according to a given smearing parameter. However, only states below the Fermi level are chosen to construct the ground-state wavefunction used to calculate the overlap parameters $a$ and $b$ in Equation (15.16). It has been shown [55] that the FON approach introduces only a few percent error to the valuation of transfer integrals using Equation (15.16). Although the ground-state wavefunction is accurately obtained, this FON approach, together with the Hubbard $U$ correction, rarely provides the correct ground-state electron density. In fact, the calculated electron density, characterized by fractional occupations, can still exhibit an unphysical splitting of the valence electron; nevertheless, it is not used in Equation (15.16).

### 15.4.2 Hubbard *U* Corrections

The self-interaction correction in transition metal complexes rests mainly on the appropriate treatment of the electronic exchange and correlation effects on the valence charge distribution. Such effects can be expected to cancel, to a large extent, in the calculation of the energy difference $\Delta E_{IF}$ between the diabatic states. However, since the expected transfer integral is much smaller than $\Delta E_{IF}$, the delocalization of the transferring electron, and thus the overlaps $a$ and $b$ of Equation (15.16), can still be considerably affected.

In order to improve the description of the valence electron density, an orbital-dependent correction functional $E_U$, adapted from the Hubbard model [56] (dealing with strongly correlated electron systems), is added to the standard DFT functional $E_{DFT}$. According to the LDA/GGA + $U$ approach, first introduced by Anisimov and coworkers [57,58] a few localized orbitals, for example, the $d$ orbitals for transition metals, are selected and the corresponding correlation is treated in a special way. The magnitude of $E_U$, and hence the amount of the corrective electron correlation, is controlled by the Hubbard $U$ parameter, which measures the screened on-site Coulomb interaction. The total energy functional is given by

$$E_{\mathrm{DFT}+U}[n] = E_U[\{n_{mm'}^{I\,s}\}] + E_{\mathrm{DFT}}[n], \tag{15.19}$$

where $I$ identifies the atomic site that experiences the Hubbard-like interaction (i.e., the ferric and ferrous species in the system under consideration), $s$ denotes the electron spin, $m$ is the magnetic quantum number, and $n^{Is}$ is the atomic orbital occupation matrix, which describes the degrees of freedom, associated with the strongly correlated electrons, on which the Hubbard $U$ acts. In our work, we use a rotationally invariant LDA/GGA + $U$ scheme [17,59], where the $U$ parameter is obtained from ab initio calculations, by using a linear response approach internally consistent with the chosen definition for the occupation matrix [59]. The expression for $E_U$ is

$$E_U[\{n_{mm'}^{I\,s}\}] = \frac{U}{2}\sum_{I,s} \mathrm{Tr}[n^{I\,s}(1-n^{I\,s})] \tag{15.20}$$

which is explained in detail by Cococcioni and de Gironcoli [59]. Here we only mention that the ab initio value of the Hubbard $U$ parameter appropriate for a given system is derived from the bare and screened linear responses of the system to variational changes in the occupation numbers; its application and relevance for the case of transition-metal complexes has been described in detail in References 50 and 51.

### 15.4.3 Reorganization Energy with Hubbard U Corrections

In the case of ferrous and ferric ions solvated in water, the Hubbard $U$ functional, depending on the localized Fe $d$ orbitals, is expected to be minimally affected by the water molecules outside the first solvation shell. For the same reason, this correction will have little effect on the physical properties beyond the first solvation shell. Therefore, the Hubbard $U$ term should change the inner-sphere contribution ($\lambda_{\mathrm{is}}$) only when the reorganization energy is separated into inner- and outer-sphere contributions under the same scheme discussed in Section 15.2.2. In view of this, the Hubbard $U$ values for the simpler systems of isolated hexaaqua ion clusters are calculated, obtaining $U = 5.6$ eV for both hexaaqua ferrous and ferric ions. This value of $U$ is used to calculate the inner-sphere contributions to the reorganization energy $\lambda_{\mathrm{is}}$, using the four-point method described previously. With $U = 5.6$ eV, we obtain $\lambda_{\mathrm{is}} = 0.75$ eV, compared to $\lambda_{\mathrm{is}} = 0.50$ eV calculated previously without Hubbard $U$. The difference of 0.25 eV can be directly translated in the total reorganization energy since the contributions due to the extra water molecules are expected to be unaffected by the $U$ correction. For the case when two ions are 5.5 Å apart, the total reorganization energy obtained from ab initio molecular dynamics is 1.93 eV. After the Hubbard $U$ correction, $\lambda$ increases to 2.18 eV, in even closer agreement with the experimental value of 2.1 eV.

### 15.4.4 Transfer Integrals

All the transfer integral calculations [55] are performed with the $PW_{scf}$ program in the QUANTUM-ESPRESSO distribution [25] using plane-wave spin-polarized FON-DFT with wavefunction and charge density cutoffs of 25 and 200 Ry, respectively, and ultrasoft pseudopotentials to represent the atomic cores. The overlaps between the individual spin orbitals, necessary to derive the overlap integrals $a$ and $b$, are computed through the DTI program [60], and the interaction energies in Equation (15.18) through full electrostatics calculations using a Poisson solver and the quantum electrostatic potential.

The redox system considered is $Fe(H_2O)_6{}^{2+}$-$Fe(H_2O)_6{}^{3+}$, which consists of hexaaqua ferrous and ferric ions. In fact, the effects of the outside water on $V_{IF}$ can be considered relatively minor [16,61], also on general considerations about the tunneling nature of the electronic coupling between redox partners [62]. According to this perspective, inclusion of the external water would consistently change the values of both $\Delta E_{IF}$ and $a,b$ in (15.16), while essentially preserving the value of $V_{IF}$. The two groups are the localized $D$ and $A$ species, whose structural identity is maintained throughout the (outer-sphere) reaction [61]. The $D$ and $A$ groups are assumed to be in contact around the transition state, at the ET optimal interionic distance of 5.5 Å [4,11,14,16,43]. The transfer integrals are computed for selected nuclear configurations along a Car–Parrinello MD (CPMD) run from Reference 14; the timestep is 5 au and the selected snapshots are 3000 steps apart ($\approx$0.36 ps). In the (FON-)DFT + $U$ calculations we use the same $U$ value for obtaining the electronic structures of the whole system and of the isolated $D$ and $A$ groups in their oxidation states, as required by (15.16). The dependence of the electronic charge distribution, and thus of the electronic exchange and correlation, on the set of nuclear coordinates is taken into account by separately evaluating the $U$ interaction parameter for each selected configuration. The resulting range of $U$ is 5.5–6.2 eV.

The effective electronic couplings computed at the PW91-GGA [63] and PBE-GGA [27] levels are reported in Figure 15.13a [55]. The supercell size is 15 × 15 × 20 Å [47], where the longest cell length is in the ion–ion direction. The ground-state calculations are performed by constraining the total magnetization, so that the total numbers of the majority-spin and minority-spin electrons are fixed, and the system is in the correct high-spin state [61,64], as in Reference 14.

As shown in Figure 15.13a, PW91 and PBE yield very similar results, which can be expected even if the two approximations are not equivalent [65]. Each estimate $V_{IF}$ of the effective electronic coupling can be associated with a maximal error $CV_{IF}/(|a|-|b|)$ [4], which gives an upper bound for the uncertainty due to the departure from the two-state model. Hence, a maximal error can also be associated with the mean transfer integral by using independent-error propagation [66]. The estimates corresponding to the configurations after 33,000 and 51,000 simulation steps have been rejected because of a strong failure of the two-state approximation (i.e., $a^2 + b^2 > 0.75$, from which $c > 0.5$) resulting from the localization of the valence electron charge on different $d$-like molecular orbitals in the ground state and in the diabatic states of the system. The final number of snapshots is statistically meaningful, as displayed by the cumulative

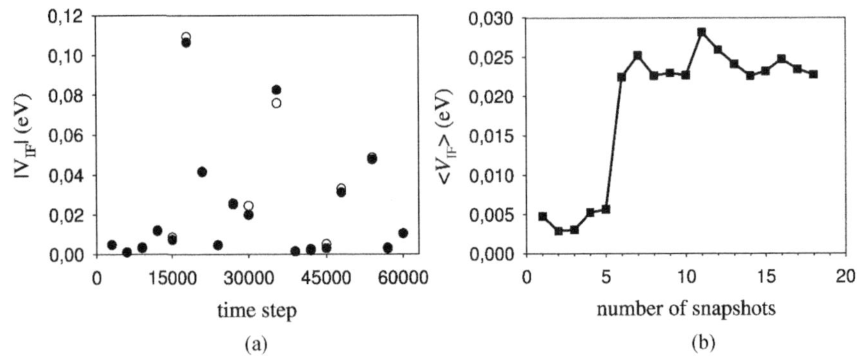


**FIGURE 15.13** (a) Transfer integral versus CPMD step number, for the PW91 (empty circles) and PBE (full circles) calculations; (b) dependence of the PBE mean transfer integral on the number of configurations used to compute the average. (The data are taken from Ref. 55.)

average in Figure 15.13b [55]. Within the accuracy determined by the maximal error, the PW91 and PBE exchange correlation functionals yield the same mean value and root-mean-square (rms) value for the electronic coupling, that is, $\langle V_{IF}\rangle_{\text{PW91}} = V_{IF}\rangle_{\text{PBE}} = (23 \pm 3) \times 10^{-3}\,\text{eV}$ and $\sqrt{\langle V_{IF}\rangle^2_{\text{PW91}}} \equiv (V_{IF})_{\text{rms,PW91}} = (V_{IF})_{\text{PBE,rms}} = (37 \pm 7) \times 10^{-3}\,\text{eV}$, respectively.

The difference between the mean and the rms values, statistically quantified by the standard deviation $\sigma_V = \sqrt{\langle V^2_{IF}\rangle - \langle V_{IF}\rangle^2} = 0.029\,\text{eV}$, is a consequence of the spread of the results and denotes a breakdown of the Condon approximation [67] due to the dynamical nature of the water medium, since water displacements important for ET are allowed within the solvation cages. In fact, the $V_{IF}$ values corresponding to different nuclear coordinates can be viewed as "independent estimates" of the quantity $\langle V_{IF}\rangle$ or $(V_{IF})_{\text{rms}}$. Therefore, $\langle V_{IF}\rangle$ or $(V_{IF})_{\text{rms}}$ can be endowed with a statistical error of $\sigma_{\langle V\rangle} = \sigma_{\langle V\rangle}/\sqrt{18} = 0.007\,\text{eV}$, as an estimate of the mean value obtained, in principle, after an infinite number of evaluations. Nevertheless, the best value of $(V_{IF})_{\text{rms}}$ is significantly larger than the experimental estimate of $7.3 \times 10^{-3}\,\text{eV}$ derived from the ET Marcus equation (see Table 15.2). Such an overestimation arises from the incorrect asymptotic decay of the XC potential, associated with the spurious electron self-interaction [13,68], which determines, for each give nuclear configuration, an excessive spread of the valence charge toward the higher-energy redox site.

The correction of the charge distribution around the ions by means of the Hubbard $U$ term, and FON whenever required, leads to the electronic couplings $V^U_{IF}$ shown in Figure 15.14. Only the PBE functional has been used in conjunction with the Hubbard $U$ correction (due to the observed good match between the results by PW91 and PBE).

For four of the selected MD configurations, the GGA + $U$ scheme yielded the effective electronic coupling also without FON. On the other hand, the strong interdependence between energies and occupation numbers of the orbitals implicated in the Hubbard $U$ correction [see Eq. (15.20)] makes the convergence of self-consistent calculations in some MD configurations difficult. Ultimately, we obtained the transfer

**TABLE 15.2 Comparison between Transfer Integral Best Estimates from Experimental Rates and from Different Theoretical Approaches**

| Method | (Model) System | $V_{IF}$ ($10^{-3}$ eV) | Reference |
|---|---|---|---|
| Experimental, rms value | | | |
| From Equation (15.2) and $\lambda^{(exp)}$ | $(Fe_{aq}–Fe_{aq})^{5+}$ | 7.1 | [20,69] |
| From Equations (15.21)–(15.22) and $\lambda^{(exp)}$ | $(Fe_{aq}–Fe_{aq})^{5+}$ | 7.3 | [20,69] |
| Theoretical | | | |
| FON-DFT + $U$, rms value | $[Fe(H_2O)_6–Fe(H_2O)_6]^{5+}$ | $11.0 \pm 2.6$ | [55] |
| FON-DFT, rms value | $[Fe(H_2O)_6–Fe(H_2O)_6]^{5+}$ | $34 \pm 8$ | [55] |
| Spin-restricted open-shell HF | $[Fe(H_2O)_6–Fe(H_2O)_6]^{5+}$ | 12.2 | [16] |
| Fitting to Schrödinger equation solution | $Fe^{2+} \rightarrow Fe^{3+} + e$ | 15.3 | [11] |

integrals for the nine snapshots reported in Figure 15.14. In general, $V_{IF}^{U}$ is found to be smaller than the corresponding $V_{IF}$. Moreover, as shown in Reference 55, although the decrease in the transfer integral varies with the nuclear configuration, the values for $V_{IF}$ and $V_{IF}^{U}$ are highly correlated. Therefore, the relative values of the transfer integral obtained from the "bare" FON-DFT computational scheme remain meaningful, on average, for the system under study. This suggests the possibility of using the present approach even without self-interaction corrections for determining various useful ET properties of the ferrous–ferric system, such as the decay constant of the transfer integral and the decay factor for the ET rate, as a function of the distance between the two irons. Indeed, it is worth noting that the direct implementation of Equation 15.16

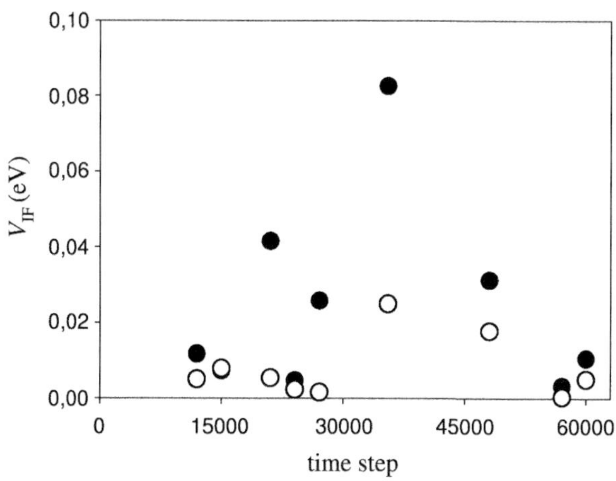


**FIGURE 15.14** Transfer integral versus CPMD step number from Reference 55. The full circles represent the transfer integrals by the FON-DFT approach, using PBE. The empty circles are obtained after the introduction of the $U$ correction term.

in Reference 44 yielded a decay constant of the ET rate for Azurin dimers in good agreement with the empirical average packing density model [70].

The (FON-)DFT + $U$ approach gives $\langle V_{IF}^{U} \rangle = (7.8 \pm 2.6) \times 10^{-3}$ eV and $(V_{IF}^{U})_{\text{rms}} = (11.0 \pm 2.6) \times 10^{-3}$, to be compared to the estimates of $\langle V_{IF} \rangle = (24 \pm 8) \times 10^{-3}$ eV and $(V_{IF})_{\text{rms}} = (34 \pm 8) \times 10^{-3}$, respectively, obtained without the Hubbard $U$ correction on the reduced set of nine nuclear configurations (and essentially equal to those based on 18 snapshots). The transfer integral is decreased considerably by the improved description of the valence electron through the Hubbard $U$ interaction term. In fact, the discrepancy between our best value for $(V_{IF})_{\text{rms}}$ and the current best estimate from the experiment amounts to 1.4 standard deviations, which is statistically not significant for the number of sampled configurations.[1]

As shown in Table 15.2, our ab initio $(V_{IF}^{U})_{\text{rms}}$ value lies between the experimental value and two previous benchmark theoretical estimates in the literature [11]: the single-point value of $V_{IF} = 98\ \text{cm}^{-1} = 12.2$ meV, which refers to a model system including only the water ligands between the two ions at a common FeO distance [16], and the estimate of $V_{IF} = 124\ \text{cm}^{-1} = 15.3$ meV, which uses a simple model consisting of one electron in the pseudopotential field of two ferric ions [11]. Note that the discrepancy with the latter value is statistically significant and that the rms transfer integral obtained in Reference [55] more closely approaches the experimental estimate. Furthermore, it allows a statistical analysis of the remaining gap, also addressing the issue of the Condon approximation through the theoretical analysis on water configurations along the MD trajectories [55].[2]

In conclusion, introduction of the self-consistent Hubbard $U$ correction significantly decreases the standard deviation associated with the rms transfer integral, as displayed by the reduced spread of the relevant data points (empty circles) in Figure 15.14 and by the flat cumulative average transfer of Figure 15.15. However, the statistical error remains high, as observed with high-level quantum chemical methods [47,71]. On the other hand, the removal of systematic errors (e.g., self-interaction errors, whose quantitative effect on the transfer integral generally depends on the given nuclear coordinates) allows us to identify the spread of the data as coming exclusively from the non-Condon features of the system. Finally, let us note that $(V_{IF}^{U})_{\text{rms}}$ has a larger relative statistical error than $(V_{IF})_{\text{rms}}$ against a smaller absolute statistical error, which is simply a consequence of the physical lower limit in the spread of the couplings in conjunction with the overestimate of the rms transfer integral in the absence of a Hubbard $U$ correction.

[1]The two estimates differ by about 1.4 standard deviations. The one-tail probability of obtaining a discrepancy that is at least 1.4 standard deviations is 8%. In other words, by assuming that our average value complies with a normal distribution centered on the expected (i.e., experimental) transfer integral, the probability that our single valuation of the rms electronic coupling gives a result at least as large as $11 \times 11^{-3}$ is 8%. Therefore, according to the usual 5% criterion, the discrepancy between the two values is not significant.

[2]In Ref. 16 the Condon approximation is parametrically tested on the apex-to-apex conformation, by translating the reactants along the ion-to-ion direction.

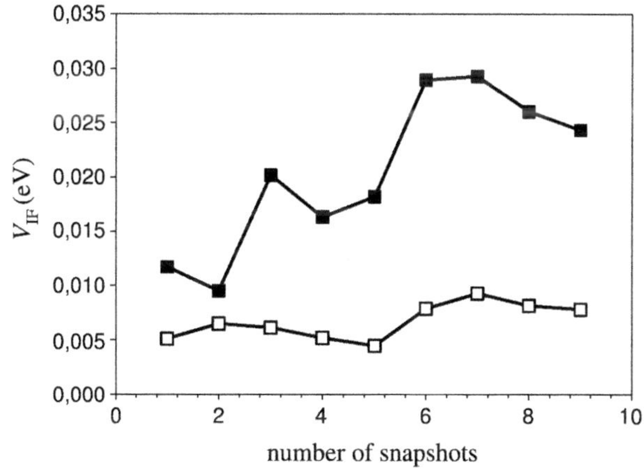


**FIGURE 15.15** Dependence of the PBE mean transfer integral on the number of configurations used to compute the average. Empty and full squares denote the results with and without Hubbard $U$ correction, respectively. (Data are taken from Ref. 55.)

### 15.4.5 Electron Transfer Rate Constant

According to Marcus' ET theory, the experimental estimate of the rms transfer integral for the electron self-exchange reaction in the ferrous–ferric system is obtained from Equations (15.3) and (15.4) once the experimental value of the reorganization energy is inserted. The best values of ET rate constant and reorganization energy from experimental data are [20,69] $k_{\mathrm{ET}}^{(\mathrm{exp})} = 7.9 \times 10^2 \mathrm{s}^{-1}$ and $\lambda^{(\mathrm{exp})} = 2.1$ eV, respectively (see Table 15.3). The resulting transfer integral is $(V_{IF})_{\mathrm{rms}}^{(\mathrm{exp})} = 7.1 \times 10^{-3}$ eV. A slightly larger value $[(V'_{IF})_{\mathrm{rms}}^{(\mathrm{exp})} = 7.3 \times 10^{-3}$ eV$]$ can be derived from the more refined expression for the self-exchange rate constant ($\Delta G°$) [72]:

$$k_{\mathrm{ET}} = \nu_n \frac{1-\exp(-\nu_{\mathrm{el}}/2\nu_n)}{1-\frac{1}{2}\exp(-\nu_{\mathrm{el}}/2\nu n}\exp\left(-\frac{\lambda}{4k_{\mathrm{B}}T}\right) \tag{15.21}$$

where

$$\nu_{\mathrm{el}} = \frac{\langle V_{IF}^2\rangle}{h}\sqrt{\frac{\pi}{\lambda k_{\mathrm{B}}T}} \tag{15.22}$$

**TABLE 15.3 Comparison between Transfer Integral Best Estimates from Experimental Rates and from Different Theoretical Approaches**

| Method | PBE | PBE + $U$ | Experiment [20,69] |
|---|---|---|---|
| $\lambda$ (eV) [14,73] | 1.93 | 2.18 | 2.11 |
| $(V_{IF})_{\mathrm{rms}}$ (meV) [55] | 37.0 | 11.0 | 7.1 |
| $k_{\mathrm{ET}}$ ($\mathrm{s}^{-1}$) | $1.2 \times 10^5$ | $8.4 \times 10^2$ | $7.9 \times 10^2$ |

is an electronic frequency for the electron transition within the activated complex and $\nu_n$ is an effective frequency for the nuclear motion along the reaction coordinate. By using the DFT estimates $\lambda$ and $(V_{IF})_{rms}$ without Hubbard corrections, taken from References 14 and 55, respectively, the resulting theoretical value of the ET rate constant is more than two orders of magnitude larger than the experimental one. This overestimation is shared almost equally between the underestimation of the reorganization energy and the overestimation of the rms transfer integral. The changes in the electron and nuclear factors resulting from the Hubbard $U$ correction lead to an estimate of the ET rate constant very close to the experiment (values in bold in Table 15.3). Note that the experimental estimate of the transfer integral is derived from $k_{ET}$ and $\lambda$. Therefore, the quantity $k_{ET}$ offers a more direct comparison between theory and experiments. In conclusion, according to our results, the Hubbard $U$ correction to DFT proposed in References 50 and 59 yields a reliable description of the electronic states around the ferrous and ferric ions, thus leading to a correspondingly reliable estimate of the experimentally relevant ET rate constant.

## 15.5 CONCLUSION

We developed a novel method to calculate free-energy surfaces in electron transfer reactions from ab initio molecular dynamics, using extensive umbrella sampling and by controlling the oxidation states of ions. Because of the self-interaction problem in common DFT functionals and the need to perform excited-state calculations to sample the reaction coordinate, the free-energy surfaces of a full electron transfer reaction from ab initio molecular dynamics have not been calculated before. We first considered the case when two ions are infinitely apart, in which the oxidation states of the ions can be controlled trivially. With the ferrous–ferric self-exchange reaction as a paradigmatic example, we found that the free-energy curves are parabolic, validating the linear solvation model first proposed by Marcus [2]. To study electron transfer reactions between the ions at a finite distance apart, we introduced a penalty functional, which solves both the self-interaction error and the excited-state challenge. We calculated fully ab initio the free-energy surface for the reaction when the ions are 5.5 Å apart. The reorganization energy obtained is 1.93 eV, which is in good agreement with the experimental value of 2.1 eV, a great improvement in quantitative description of electron transfer reactions. A further improved value of 2.18 eV is obtained by including Hubbard $U$ correction to accurately describe the strongly localized $d$ electrons. Such a correction turns out to be very essential in calculating transfer integrals, which determine the electronic factor in the self-exchange rate constant. In fact, it yields a decrease in the root mean-square transfer integral from 37.0 to 11.0 meV, which corresponds to a correction of the excessive spread of the transferring electron due to self-interaction. These calculations allowed us to estimate the electron transfer reaction rate constant fully ab initio; the theoretical value of $8.4 \times 102\ s^{-1}$ is in excellent agreement with the experimental estimate of $7.9 \times 102\ s^{-1}$. It is hoped that this approach will allow an increasing number of ab initio studies of such reactions, since it can be successfully applied to a wide class of oxidation–reduction reactions in solution (as often happens in electrochemistry or biochemistry) or in the solid state (intervalence charge transfer).

## ACKNOWLEDGMENT

We gratefully acknowledge support from the Croucher Foundation (PH-LS), MURI Grant N. DAAD 19-03-1-0169 (NM), the MRSEC Program of the National Science Foundation under Grant DMR 02-13282 (PH-LS), and National Institutes of Health Grant GM 067689 (PH-LS, AM, and MLK).

## REFERENCES

1. A. Nitzan and M. A. Ratner, *Science* **300**, 1384 (2003).
2. R. A. Marcus, *J. Chem. Phys.* **24**, 966 (1956).
3. R. A. Marcus and N. Sutin, *Biochem. Biophys. Acta* **881**, 265 (1985).
4. M. D. Newton and N. Sutin, *Annu. Rev. Phys. Chem.* **35**, 437 (1984).
5. R. A. Marcus, *J. Chem. Phys.* **24**, 979 (1956).
6. R. A. Marcus, *Rev. Mod. Phys.* **65**, 599 (1993).
7. A. Warshel and W. W. Parson, *Annu. Rev. Phys. Chem.* **42**, 279 (1991).
8. A. Warshel, *J. Phys. Chem.* **86**, 2218 (1982).
9. A. Warshel and J. Hwang, *J. Chem. Phys.* **84**, 4938 (1986).
10. G. King and A. Warshel, *J. Chem. Phys.* **93**, 8682 (1990).
11. R. A. Kuharski, J. S. Bader, D. Chandler, M. Sprik, M. L. Klein, and R. W. Impey, *J. Chem. Phys.* **89**, 3248 (1988).
12. I. Tavernelli, R. Vuilleumier, and M. Sprik, *Phys. Rev. Lett.* **88**, 213002 (2002).
13. Q. Wu and T. Van Voorhis, *Phys. Rev. A* **72**, 024502 (2005).
14. P. H.-L. Sit, M. Cococcioni, and N. Marzari, *Phys. Rev. Lett.* **97**, 028303 (2006).
15. Y. Bu, Y. Wang, F. Xu, and C. J. Deng, *Mol. Struct. (THEOCHEM)* **453**, 43 (1998).
16. J. Logan and M. D. Newton, *J. Chem. Phys.* **78**, 4086 (1983).
17. M. Cococcioni, A. Dal Corso and S. de Gironcoli, *Phys. Rev. B* **67**, 094106 (2003).
18. J. Blumberger, L. Bernasconi, I. Tavernelli, R. Vuilleumier, and M. Sprik, *J. Am. Chem. Soc* **126**, 3928 (2004).
19. Y. Tateyama, J. Blumberger, M. Sprik, and I. Tavernelli, *J. Chem. Phys.* **122**, 234505 (2005).
20. K. M. Rosso and J. R. Rustad, *J. Phys. Chem. A* **104**, 6718 (2000).
21. A. Klimkans and S. Larsson, *Chem. Phys.* **189**, 25 (1994).
22. S. F. Nelsen, S. C. Blackstock, and Y. Kim, *J. Am. Chem. Soc.* **109**, 677 (1987).
23. G. J. Herman and G. W. Neilson, *J. Phys. Condens. Matt.* **4**, 649 (1992).
24. G. J. Herman and G. W. Neilson, *J. Phys. Condens. Matt.* **4**, 627 (1992).
25. P. Giannozzi et al., *J. Phys. Condens. Matt.* [`arXiv: 0906.2569v2 (condmat.mtrl-sci)`] (2009).

26. D. Vanderbilt, *Phys. Rev. B* **41**, 7892 (1990).
27. J. P. Perdew, K. Burke, and M. Ernzerhof, *Phys. Rev. Lett.* **77**, 3865 (1996).
28. See http://www.physics.rutgers.edu/dhv/uspp/index.html.
29. R. Car and M. Parrinello, *Phys. Rev. Lett.* **55**, 2471 (1985).
30. K. Laasonen, A. Pasquarello, R. Car, C. Lee, and D. Vanderbilt, *Phys. Rev. B* **47**, 10142 (1993).
31. D. Marx and J. Hutter, *Ab initio* molecular dynamics: Theory and implementation, in *Modern Methods and Algorithms in Quantum Chemistry*, Forschungzentrum Juelich, NIC Series, Vol. 1, 2000 (http://www.fz-juelich.de/nic-series/Volume1/marx.pdf).
32. A. Fouqueau, S. Mer, M. E. Casida, L. M. L. Daku, A. Hauser, T. Mineva, and F. Neese, *J. Chem. Phys.* **120**, 9473 (2004).
33. D. Harris, G. H. Loew, and A. Komornicki, *J. Phys. Chem. A* **101**, 3959 (1997).
34. E. Schwegler, J. Grossman, F. Gygi, and G. Galli, *J. Chem. Phys.* **121**, 5400 (2004).
35. P. H-L. Sit and N. Marzari, *J. Chem. Phys.* **122**, 204510 (2005).
36. G. Makov and M. C. Payne, *Phys. Rev. B* **51**, 4014 (1995).
37. J. S. Bader and D. Chandler, *Chem. Phys. Lett.* **157**, 501 (1989).
38. J. S. Bader and D. Chandler, *J. Phys. Chem.* **96**, 6423 (1992).
39. D. A. Scherlis, J.-L. Fattebert, F. Gygi, M. Cococcioni, and N. Marzari, *J. Chem. Phys.* **124**, 074103 (2006).
40. M. d'Avezac, M. Calandra, and F. Mauri, *Phys. Rev. B* **71**, 205210 (2005).
41. W. Kohn, *Phys. Rev. Lett.* **76**, 3168 (1996).
42. E. Prodan and W. Kohn, *Proc. Natl. Acad. Sci. USA* **102**, 11635 (2005).
43. B. L. Tembe, H. L. Friedman, and M. D. Newton, *J. Chem. Phys.* **76**, 1490 (1982).
44. A. Migliore, S. Corni, R. Di Felice, and E. Molinari, *J. Chem. Phys.* **124**, 064501 (2006).
45. A. Farazdel, M. Dupuis, E. Clementi, and A. Aviram, *J. Am. Chem. Soc.* **112**, 4206 (1990).
46. A. Migliore, S. Corni, R. Di Felice, and E. Molinari, *J. Phys. Chem. B* **110**, 23796 (2006).
47. A. Migliore, S. Corni, R. Di Felice, and E. Molinari, *J. Phys. Chem. B* **111**, 3774 (2007).
48. W. Koch and M. C. Holthausen, *A Chemist's Guide to Density Functional Theory*, 2nd ed., Wiley-VCH Verlag GmbH, New York, 2000.
49. C. Jamorski, A. Martinez, M. Castro, and D. R. Salahub, *Phys. Rev. B* **55**, 10905 (1997).
50. H. J. Kulik, M. Cococcioni, D. A. Scherlis, and N. Marzari, *Phys. Rev. Lett.* **97**, 103001 (2006).
51. H. J. Kulik and N. Marzari, *J. Chem. Phys.* **129**, 134314 (2008).
52. M. Springborg, R. C. Albers, and K. Schmidt, *Phys. Rev. B* **57**, 1427 (1998).
53. N. Marzari, D. Vanderbilt, and M. C. Payne, *Phys. Rev. Lett.* **79**, 1337 (1997).
54. F. Neese, *Chem. Phys. Lett.* **325**, 93 (2000).
55. A. Migliore, P. H-L. Sit, M. L. Klein, *J. Chem. Theory Comput.* **5**, 307 (2009).
56. J. Hubbard, *Proc. R. Soc. London A* **276**, 238 (1963).
57. V. I. Anisimov, J. Zaanen, and O. K. Andersen, *Phys. Rev. B* **44**, 943 (1991).
58. A. I. Liechtenstein, V. I. Anisimov, J. Zaanen, *Phys. Rev. B* **52**, R5467 (1995).
59. M. Cococcioni and S. de Gironcoli, *Phys. Rev. B* **71**, 035105 (2005).

60. (a) A. Ferretti, A. Ruini, G. Bussi, E. Molinari, and M. J. Caldas, *Phys. Rev. B* **69**, 205205 (2004); (b) A. Ferretti, DTI program, (2005). (http://www.s3.infm.it/dti).

61. M. D. Newton, *J. Phys. Chem.* **92**, 3049 (1988).

62. A. A. Stuchebrukhov, *Theor. Chem. Acc.* **110**, 291 (2003).

63. J. P. Perdew, J. A. Chevary, S. H. Vosko, K. A. Jackson, M. R. Pederson, D. J. Singh, and C. Fiolhais, *Phys. Rev. B* **46**, 6671 (1992).

64. P. P. Knops-Gerrits, P. A. Jacobs, A. Fukuoka, M. Ichikawa, F. Faglioni, and W. A. GoddardIII, *J. Mol. Catal. A Chem.* **166**, 3 (2001).

65. A. E. Mattsson, R. Armiento, P. A. Schultz, and T. R. Mattsson, *Phys. Rev. B* **73**, 195123 (2006).

66. J. R. Taylor, *An Introduction to Error Analysis: The Study of Uncertainties in Physical Measurements*, 2nd ed., University Science Books, 1996.

67. A discussion of the failure of the Condon approximation and its relation to the use of Equation (15.16) can be found in Ref. 55.

68. E. Ruiz, D. R. Salahub, and A. Vela, *J. Phys. Chem.* **100**, 12265 (1996).

69. N. Sutin, Theory of electron transfer reactions, in *Electron-Transfer and Electrochemical Reactions. Photochemical Reactions and Other Energized Reactions*, J. J. Zuckermann, ed., VCH, New York, 1986, Vol. 15, p. 16.

70. C. C. Page, C. C. Moser, X. Chen, and P. L. Dutton, *Nature* **402**, 47 (1999).

71. S. S. Skourtis, I. A. Balabin, T. Kawatsu, and D. N. Beratan, *Proc. Natl. Acad. Sci. USA* **102**, 3552 (2005).

72. N. Sutin, *Acc. Chem. Res.* **15**, 275 (1982).

73. P. H-L. Sit, M. Cococcioni, and N. Marzari, *J. Electroanal. Chem.* **607**, 107112 (2007).

CHAPTER 16

# Understanding Electrocatalysts for Low-Temperature Fuel Cells

PETER FERRIN and MANOS MAVRIKAKIS

Department of Chemical and Biological Engineering, University of Wisconsin, Madison, Wisconsin

JAN ROSSMEISL and JENS K. NØRSKOV

Department of Physics, Center for Atomic-Scale Materials Design, Technical University of Denmark, Lyngby, Denmark

Electronic structure calculations modeling the complex electrocatalytic environment can be very challenging. In this chapter we present an approximate theoretical framework whereby simple corrections, for instance, for potential bias and pH, are utilized to more standard vapor-phase/metal interface density functional theory calculations to make them relevant for electrocatalysis. The success of this framework in elucidating the fundamental reaction mechanism relevant to methanol electrooxidation and to oxygen reduction reaction is then discussed, along with emerging opportunities for novel alloy catalyst design with improved electrocatalytic properties for these and other electrocatalytic reactions.

## 16.1 INTRODUCTION

In this introductory section we discuss the theoretical framework for utilizing density functional theory for electrocatalysis.

### 16.1.1 Surface Science under Water

Simulations of electrocatalytic reactions at the atomic level involve a number of challenges. Some of these are shared with other branches of surface science, especially heterogeneous catalysis. The major challenge for atomistic simulations for all

*Fuel Cell Science: Theory, Fundamentals, and Biocatalysis*,
Edited by Andrzej Wieckowski and Jens K. Nørskov

branches of surface science is to obtain an accurate description of the interaction between the reactant and the surface, while keeping the computational requirements within reasonable limits. Density functional theory within the generalized-gradient approximation (DFT-GGA) offers a good compromise between system size that can be treated with a realistic amount of resources and accuracy in derived electronic and thermodynamic properties. Simulations using this method have shown predictive power with respect to trends in catalytic activity and selectivity. Heterogeneous catalysts designed from first-principles DFT calculations are beginning to appear for several applications [1].

Atomistic simulations of electrocatalysis face special challenges because of the unique components involved in the model. These are related to: (1) the electrolyte, which is the ion-conducting media between the electrodes; (2) the electrodes over which there is a potential bias; and (3) the charged interface between the solid electrode and the liquid electrolyte, exactly the place where the electrocatalytic reactions are taking place.

In this section we present a framework for including some of the physics of the electrochemical environment; some features are not considered in this approach. In Section 16.2 we discuss how to apply this framework to two different electrocatalytic reactions: methanol electrooxidation (the direct methanol fuel cell anode reaction) and the oxygen reduction reaction (ORR).

One of the first challenges in electrocatalysis is that the reactions take place in the presence of a solvent, very often water. This introduces the problem of finding the structure of adsorbates in the presence of water at the surface. By itself, this is a huge challenge since the structure of water depends on the electrode surface, adsorbates, temperature, and pH. Adding to the challenge is that standard GGA-DFT has problems in simulating hydrogen bonds, which have some van der Waals character. However, strides have been made in modeling the solvent–reactant interactions on the surface. In the following sections we restrict our discussion to (111) surfaces at low temperature. Under ultra-high-vacuum (UHV) conditions, it is well known that water forms a hexagonal ice layer on many metals [2,3] (see Fig. 16.1). This ice layer can be used as a model for the solvent, since (1) it is a physically realized structure of water and (2) on the basis of both experiments and simulations, that some icelike order seems to be present under electrochemical conditions [4,5].

At high potentials, water will dissociate, first forming a half-dissociated water layer, where every second water molecule has donated a proton to the solvent and an electron to the electrode, meaning that there will be OH and water at the electrode surface. This structure has also been observed in UHV environments; if a surface has been precovered with oxygen before it is exposed to water, water will donate one of its hydrogen atoms to oxygen, forming two OH* groups. Depending on the coverage of oxygen, different structures will form [6–9]. At low potentials, hydronium ions will transfer a proton to the surface, forming H*. This hydrogen cannot participate in the hydrogen bonded network as OH* can, and hence the interaction between water and H* at the surface is weak.

Thus, by using the hexagonal ice layer as a model, one can approximately simulate the electrocatalytic surface in aqueous media and therefore overcome the challenge of modeling the entire electrolyte phase.

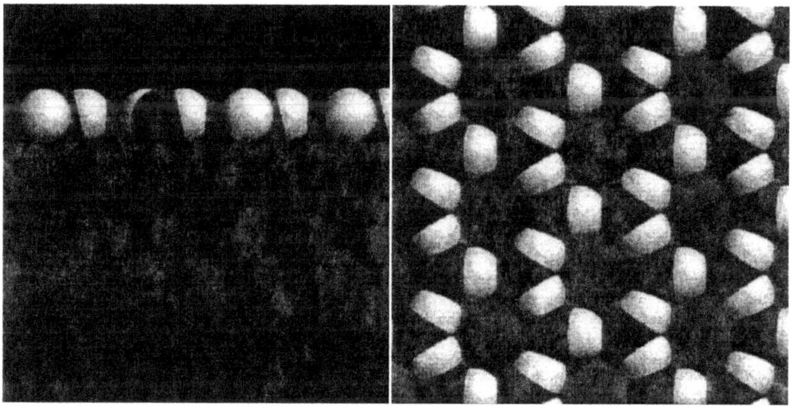

(a) (b)

**FIGURE 16.1** The hexagonal half-dissociated water layer on Pt (111): sideview (a) and topview (b). All hydrogen bonds in the plane of the surface are saturated. White spheres indicate hydrogen; red, oxygen, and pink, Pt. (See insert for color representation.)

### 16.1.2 Including Potential and pH to Zeroth Order

Simulating reactions involving hydronium ions introduces the next challenge, since the surfaces described above are formed via reactions of the kind

$$H_2O \rightarrow OH^* + H^+(aq) + e^-$$

We have to assign a chemical potential to ions in solution and electrons at a given electrochemical potential in order to calculate at which potential the reaction will occur. The chemical potential of $H^+$ (aq) will depend on the pH of the solvent, and the chemical potential of the electron in the electrode will depend on the bias.

Assuming for the moment that the binding energy of OH* is independent of potential, we can calculate the change in the free energy of the reaction above as a function of the chemical potential of $H^+$ and $e^-$. To link this to an electrochemical scale of potential, we apply the standard hydrogen electrode:

$$\tfrac{1}{2}H_2(g) \leftrightarrow H^+(aq) + e^-$$

Under the conditions used to define the standard hydrogen electrode (SHE) potential, $T = 300$ K, $p_{H_2} = 1$ bar, and pH = 0, the reaction free energy of this reaction, $\Delta G$, is zero. Hence, the free energies of $\frac{1}{2}H_2(g)$ and, $H^+(aq) + e^-$, are equal. This defines potential $U = 0$ V. At finite potentials, the chemical potential of the electrons, $e^-$, is changed by a factor of $-eU$ with respect to $H_2$ in the gas phase (where e is the charge on the electron). Taking this into account, we can obtain the reaction energy of, for example

$$H^+(aq) + e^- \rightarrow H^*$$

which is

$$\Delta G(U) = \Delta G_0 + eU$$

where $\Delta G_0$ is the free energy of reaction for

$$\frac{1}{2}\, H_2 \leftrightarrow H^*$$

at standard conditions. This free energy can be directly calculated using DFT and standard molecular tables via

$$\Delta G_0 = \Delta E + \Delta(\text{ZPE}) - T\,\Delta S$$

Here $\Delta E$ and $\Delta$(ZPE) are the differential adsorption energy for H* and the difference in zero-point energy between $H_2$ and H*. $T$ is temperature, and $\Delta S$ is the difference in entropy. At a pH different from zero, the entropy for the $H^+$ ion will change. We can correct for this effect by adding $\Delta G(\text{pH}) = -k_B T \ln[H^+]$, where $[H^+]$ is in moles per liter. Since this reaction equilibrium defines the SHE potential, we now have a direct link between quite standard density functional theory (DFT) calculations for the metal–gas interface and the electrode potential. In a similar way, we can now calculate potential-dependent reaction free energies for other reactions, such as $O^* + H^+ + e^- \rightarrow OH^*$ or $OH^* + H^+ + e^- \rightarrow H_2O$.

### 16.1.3 The Charged Solid–Liquid Interface

The approach described above is valid if the binding of OH* is independent of potential and pH. However, in reality, both pH and potential will affect the field in the surface region. Since there is a bias between the two electrodes, this bias difference is also present over the electrolyte. Since the electrode is an electron conductor, no field can penetrate in the electrode, and since the electrolyte is an ion conductor, the bias difference will be screened over a very short distance from the surface, called the *outer Helmholtz plane* (OHP). For a perfect ion conductor the whole field can be screened by a single, charged water layer. In this region the field is huge: of the order 0.1 V per angstrom (see Fig. 16.2). A simple way to approximate the inclusion of the OHP is to apply an external electric field. This will, however, give only the relative dependence on the field [10,11]. It is not trivial to assign a given field to a specific potential since potential and field in general do not have a common zero point related to potential of zero electronic charge of the electrode surface. Fortunately, in general, we find that the binding energies vary only slightly with respect to the field, which in many cases makes it a good assumption to neglect the field effects.

### 16.1.4 Charge Transfer Reactions

We have now established the link between gas-phase/metal binding energies and the respective reaction energies in an electrochemical environment. However, the gas-phase approach cannot be applied when calculating reaction barriers and especially a

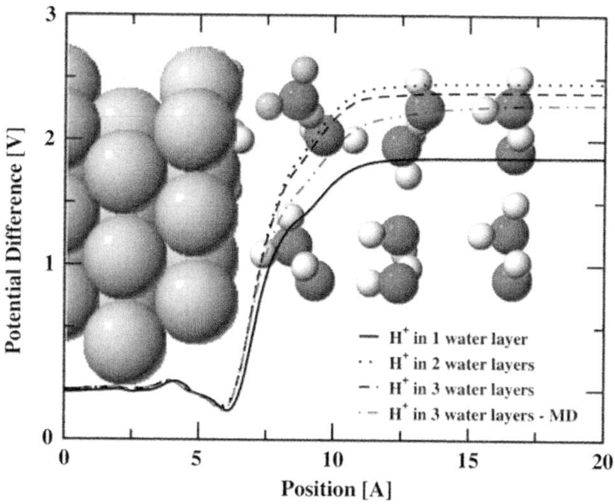


**FIGURE 16.2** The charged solid–liquid interface: charged Pt (111) slab with three water layers outside and one solvated hydronium ion per unit cell. The electrode potential, due to the charged interface averaged parallel to the surface, is shown for systems with one, two, and three relaxed water layers along with results from a DFT molecular dynamics simulation of a proton solvated in three water layers. (Reproduced from Ref. 17.) (See insert for color representation.)

barrier of a charge transfer reaction (e.g., a proton in the electrolyte reacting with an electron in the metal forming hydrogen adsorbed at the surface). In this case, the potential, pH, field, surface charge, and proton concentration cannot be viewed as independent parameters and will all affect the reaction barrier. Since 2001, a number of approaches for including the effect of potential on reaction barriers has been suggested [12–15]. All of these methods use the surface charge to vary the potential and field. For periodic simulations the countercharge has to be distributed in the supercell to conserve overall charge neutrality. Common to all of these methods is that the chemical potential of the protons via pH and proton concentration is varied independently of the potential. A method where the concentration of protons in the outer Helmholtz layer is the key parameter or descriptor [16] has been suggested. The advantage of this approach is that the relation between the electrode potential and the chemical potential of proton is included in the natural way. The challenge, however, is that over the course of a charge transfer reaction step, there is a change in the chemical potential of the proton/electrode pair. To assign a reaction barrier to a specific potential, the challenge is to keep the potential constant during a charge transfer reaction. However, by performing the same simulation in supercells with different surface areas, it is possible to extrapolate to the limit of constant potential [17].

### 16.1.5 Surface Pourbaix Diagrams

The so-called pressure gap in heterogeneous catalysis, between UHV conditions and real catalyst operation conditions, spans several order of magnitude in terms of pressure, which changes the reaction energies by $k_B T \ln(p/p_0)$. In electrocatalysis, however, the potential directly influences the reaction energies in more complex ways. This means that coverage effects are extremely important in electrocatalysis. In other words, the catalysis on a clean metal surface or an oxide surface of the same metal might be very different. With the assumptions above it is possible to calculate the potentials and pH at which H*, OH*, O*, or the oxide will form at the surface, and thereby construct surface Pourbaix diagrams showing the most stable surface at given conditions [18].

It is important to note that many effects of the electrochemical environment described above are important for understanding details of the reaction mechanism, both adsorption thermochemistry and barriers [17]. However, it seems that for these systems many trends and relative features can be obtained within a somewhat simpler framework [19]. Thus, in the following two case studies (methanol electroxidation and oxygen reduction) we concentrate on the effect of the surface and its electronic structure to capture trends in reactivity across catalytic materials, and thereby give some guidelines for improving catalyst performance through a first-order materials design.

## 16.2 DIRECT METHANOL FUEL CELL ANODE CATALYSTS

### 16.2.1 The Need for New DMFC Anode Catalysts

Direct methanol fuel cells (DMFCs) have garnered considerable attention as mobile power generators for high-performance applications. They have many of the same advantages as $H_2$-PEM fuel cells, such as high theoretical efficiency, high power density, low operating temperature, and few moving parts, without some of the challenges of such fuel cells, notably that of hydrogen storage [20–22]. Unfortunately, DMFCs generally have lower efficiencies and are less powerful than equivalent-sized $H_2$ fuel cells. Improving the overall efficiency of a DMFC is important to realizing their potential as power sources.

The anode reaction of the DMFC is one of the limiting factors to the efficiency of the overall cell [21–23]. The overall reaction taking place at the DMFC anode is methanol electrooxidation, $CH_3OH + H_2O \rightarrow CO_2 + 6H^+ + 6e^-$, where methanol and water react in an electrochemical environment to form six proton–electron pairs, with carbon dioxide as a byproduct. Even with state-of-the-art Pt–Ru-based catalysts [24–29], the applied potential necessary for the anode reaction to take place is higher than that of the corresponding hydrogen fuel cell anode, which lowers the efficiency of the overall fuel cell. This overpotential is necessary even though the equilibrium potential of the overall methanol electrooxidation reaction is very similar to that of the anode reaction in a hydrogen fuel cell [22]. Related to the overpotential problem is the challenge of CO* poisoning. CO* is a natural intermediate to methanol decomposition, but it is a strong poison for the anode catalyst, significantly reducing the catalytic activity and fuel cell efficiency, thus reducing the overall power that can be produced by the fuel cell. In addition to the problems of efficiency, the use of a precious metal as the catalyst increases the cost of the fuel cell significantly. Mitigating these serious problems through materials design is one of the main challenges in DMFC research.

### 16.2.2 Mechanism for Methanol Electrooxidation

In order to optimize the DMFC anode catalyst, we need to understand the anode reaction mechanism. There has been some debate in the literature as to which mechanism is active for a given catalyst under operating conditions. In general, two major classes of mechanisms have been proposed for this electrooxidation reaction: the indirect mechanism and the direct mechanism [30–36]. In the indirect mechanism, methanol is first partially oxidized to CO* and the resulting surface-bound CO* is further oxidized to $CO_2$. In the direct mechanism, methanol does not go through a CO* intermediate; rather, it is directly oxidized to $CO_2$. Specifically, more recent studies [31,35] have shown that the overall reaction takes place on some surfaces at potentials too low to oxidize CO*, which suggests that under certain conditions the direct pathway is operational for methanol electrooxidation. For both mechanisms, there are multiple reaction intermediates and pathways, a subset of which are illustrated in Figure 16.3.

Density functional theory has been very successful in elucidating reaction pathways for many catalytic vapor-phase reactions [37–42]. In these cases, the thermochemistry of all possible reaction pathways can be calculated on the catalytic surface of interest. The relative thermochemistry of each pathway can give some insights as to which pathway is likely to dominate under different reaction conditions. Using the reference method described in the previous section, an analysis similar to that for the aqueous-phase electrooxidation of methanol can be performed [43]. First, the stability of all possible reaction intermediates for methanol electrooxidation on the catalyst surface of interest is calculated. Then, as described above, the stability of all intermediates is referenced to the standard hydrogen electrode (SHE). This allows the calculation of the free-energy change for all reaction steps at the SHE and therefore as a function of potential bias. Using this information, it is possible to construct the possible pathways for electroxidation at a given potential. Figure 16.4 gives the free energy of all reaction intermediates described in Figure 16.3 at standard conditions

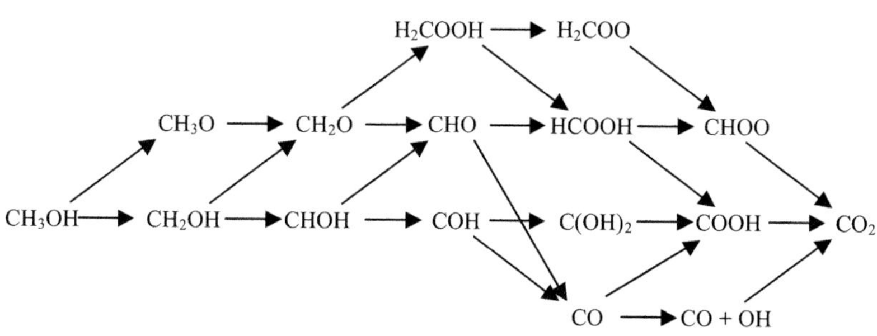


**FIGURE 16.3** Schematic representation of the reaction paths and possible intermediates for methanol electrooxidation. Arrows to the right implicitly include the generation of a proton/electron pair (not shown) from either the carbonaceous species or the surrounding $H_2O$ (the latter implies the addition of a hydroxyl group). Liquid-phase reactions can take place for $H_2CO$ and HCOOH, which are not included. (Reproduced from Ref. 43.)

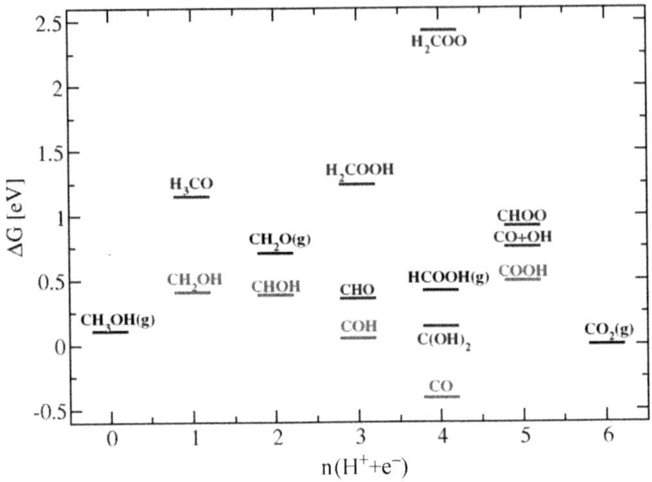


**FIGURE 16.4** Free energies for the different methanol electrooxidation intermediates on Pt(111), with $H_2$(g), $CO_2$(g), and $H_2O$(g) as a reference. The *x* axis indicates how many proton/electron pairs have been created from the original reactants [e.g., $CH_3OH + H_2O \rightarrow HCOOH$ (g) $+ 4H^+ + 4e^-$].

(0 V vs. SHE, 298 K, 0 pH) on Pt(111), as Pt is the most active monometallic catalyst for methanol electrooxidation.

With this database of free energies on Pt, one can also find the potential at which *all* reaction steps along some pathway from methanol to $CO_2$ are downhill in free energy. Since application of the potential bias U lowers the chemical potential of electrons by $eU$ (where $e$ is the charge on an electron), the free energy of any given mechanistic step in Figure 16.3 is described by

$$\Delta G(U) = \Delta G_{\mathrm{SHE}} - eU$$

where $\Delta G_{\mathrm{SHE}}$ is the free-energy change for that step at 0 V versus the SHE reference. In order for the overall reaction to take place, the $\Delta G$ values for all reaction steps along at least one pathway must be negative. Since $\Delta G$ is a function of potential, there exists an onset potential ($U_{\mathrm{onset}}$), defined as the lowest potential at which every step in the overall reaction for at least one pathway is downhill in free energy. This can be mathematically described as

$$U_{\mathrm{onset}} = \min\left(\max\left(\frac{\Delta G^i_{\mathrm{SHE}}}{e}\right)\right)_j$$

where $i$ refers to elementary steps along a given pathway and $j$ are all possible pathways from methanol to $CO_2$. Additionally, this analysis provides the mechanistic pathway by which the reaction will occur at the onset potential, as well as the potential-determining step (PDS). The latter is analogous to the rate-limiting step in a kinetic analysis of, for instance, a vapor-phase catalytic reaction; this is the step along the minimum-potential reaction pathway that has a $\Delta G_{SHE}$ equivalent to the onset potential times the electron charge. In order to maximize fuel cell efficiency, one would like to run a fuel cell at the lowest potential possible; thus, having a lower onset potential is desirable. While a low-onset potential is not sufficient to show a good anode catalyst, running at a reasonably low potential is a necessary condition for an efficient fuel cell.

The onset potential for methanol electrooxidation on Pt(111) at standard conditions calculated in this manner is 0.37 V, where the PDS is the formation of carboxyl from $C(OH)_2$* (initial methanol activation to $CH_2OH$* also requires a potential within 0.1 V of this onset potential). The pathway followed at this potential is a direct one, which is in agreement with recent experimental and theoretical studies [21,31]. The overall pathway forms a COH* intermediate by sequential elimination of carbonic hydrogen atoms in methanol, followed by the addition of an hydroxyl group from a nearby water molecule to form $C(OH)_2$*, and then further dehydrogenation to carboxyl (COOH) and then to carbon dioxide ($CO_2$). The predicted initial carbonic hydrogen abstraction steps in this pathway are in agreement with the conclusions of multiple experimental electrochemical studies [22]. The relative stability of CO* as compared with other intermediates is also in agreement with several experimental and theoretical studies [31,39,44].

This analysis can be extended beyond Pt(111) to other metal surfaces, allowing one to identify trends for the onset potential and the nature of the PDS across the periodic table. In Figure 16.5, the onset potential for the close-packed surfaces of 12 transition metals is shown, together with the potential-determining step for each surface. The range of onset potentials across the 12 transition metals shown is quite large, between 0.35 and 1.28 V. While many of these surfaces have onset potential similar to that of Pt, none has a significantly lower onset potential for methanol electrooxidation. Additionally, the PDS varies across metals. Initial methanol activation to methoxy ($CH_3O$*) is the PDS for Ag, Cu, and Rh, whereas methanol activation to $CH_2OH$* is the PDS for Au, Pd, and Ir. Metals binding oxygen relatively strongly (Re, Ru, Os, Co, Ni) require a large potential to remove methoxy from the surface (forming formaldehyde). The PDS for Pt is the formation of a carboxyl intermediate, although, as noted above, initial methanol activation requires a potential that is very close to the overall onset potential.

While the onset potential for many of these metals is relatively low, there is an additional problem on all of the non–group 11 metals discussed above. While the thermochemical pathway to $CO_2$ from methanol is open at low potentials, the strong adsorption of CO* on these surfaces may prevent a high degree of activity. To illustrate this, we use the following argument. Assume that there is a high coverage of CO* on the surface, as CO* is the most stable intermediate and is also more stable than the end product, $CO_2$(g), over a large potential range. Further, assume steady state among surface intermediates. The rate of the direct mechanism is proportional to the coverage of the intermediate in the step producing HCOOH or $C(OH)_2$, for example, CHO*. Since we have assumed steady state of all intermediates with CO*, the coverage of

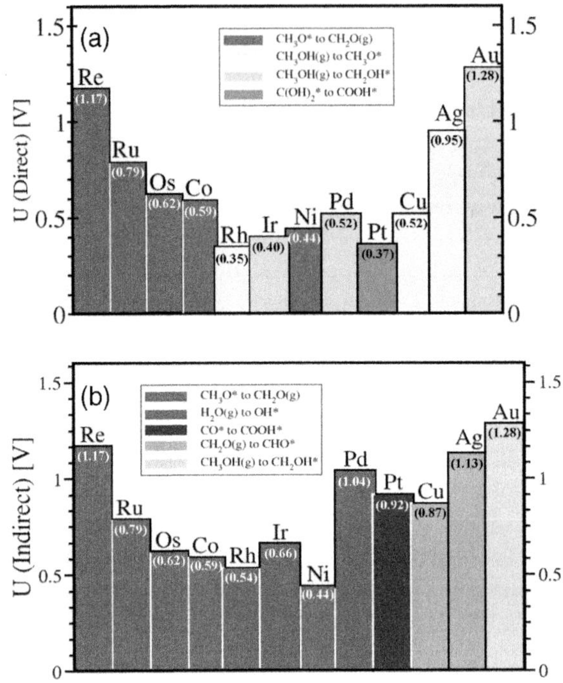


**FIGURE 16.5** (a) The lowest potential needed to ensure that all steps along the *direct* mechanism path are downhill in free energy on 12 close-packed metal surfaces. For Os, the final step ($CO^* + OH^* \rightarrow CO_2$) has a $\Delta G$ within 0.1 eV of that of the potential-determining step. For Pt, initial methanol activation ($CH_3OH \rightarrow CH_2OH$) has a $\Delta G$ within 0.1 eV of that of the potential-determining step. (b) The lowest potential needed to ensure that all steps along the *indirect* mechanism path are downhill in free energy on 12 close-packed metal surfaces. For Os, the final step ($CO^* + OH^* \rightarrow CO_2$) has a $\Delta G$ within 0.1 eV of that of the potential-determining step. The color code is related to the nature of the potential-determining step, as described in the legend in the figure. (See insert for color representation.)

these reactive intermediates is proportional to the difference in free energy between $CO^*$ and the intermediate in question, for instance, $\theta_{CHO} \propto \exp(-(G_{CHO^*} - G_{CO^*})/kT)\theta_{CO}$ for $G_{CO^*} < G_{CHO^*}$. Also, the rate of, for example, $CHO^* + H_2O \rightarrow HCOOH + H^+ + e^-$ is proportional to $\theta_{CHO} \times \exp(-\Delta G/kT)$, where $\Delta G = G_{HCOOH} - G_{CHO^*}$. Assuming a high coverage of $CO^*$ ($T_{CO} \sim 1$), this, in turn, means that the rate is effectively proportional to $\exp(-(G_{HCOOH} - G_{CO^*})/kT)$, the exponential of the difference between the free energies of $CO^*$ and HCOOH(g). An analogous argument can be made for pathways through other non-CO intermediates, such as $C(OH)_2^*$. Thus, the rate of the overall reaction, even going through the direct mechanism, will depend on the stability of $CO^*$, which binds very strongly to non–group 11 metals under these conditions. Therefore, and despite the reasonably low onset potential, the activity at these potentials is too low for practical fuel cell applications.

Owing to the large effect of CO* poisoning, as a practical matter, for materials with strong CO* binding, one can use the indirect mechanism (where CO* is oxidized to $CO_2$) as a guide to a realistic overpotential for fuel cell applications. At potentials relevant to the indirect mechanism, the problem of CO* poisoning is negated by using it as a reactive intermediate, allowing the surface to reactively clean itself continuously. By using only the reaction pathways going through the indirect mechanism (shown in green in Fig. 16.3), one can calculate the onset potential of the indirect mechanism for each of the 12 transition metal surfaces; this is shown in Figure 16.5b.

Examining the differences in the onset potentials for the direct and indirect mechanisms on the same metal surface is instructive. In all cases, the indirect mechanism has an onset potential equivalent to or greater than that of the direct mechanism. The difference is greatest for Pd and Pt, which have a difficult time activating water to form carboxyl or hydroxyl groups on the surface. The PDS for the indirect mechanism on the close-packed surface of Rh and Ir is also the formation of hydroxyl from water, although the onset potential for this mechanism is lower than that of Pt or Pd. Ag and Cu have very high onset potentials for the indirect mechanism, due to the difficulty in activating the formaldehyde intermediate to CHO*. On Au, the direct and indirect mechanisms are equally difficult and have very high onset potentials, due to the difficult initial activation of methanol. On very reactive metals (Ni, Co, Os, Ru, and Re), the PDS for both the direct and indirect mechanisms is the same; thus their onset potentials are the same. In fact, the onset potential for both mechanisms on Ni(111) is quite low (0.44 V). Unfortunately, Ni, as well as several other metals studied (e.g., Os, Re), have the drawback of either oxidation or dissolution in electrochemical environments, thereby disqualifying them as good DMFC anode catalysts. For metals that are stable in typical electrochemical environments, the onset potential for the indirect mechanism is generally significantly higher than the direct mechanism, and always higher than desirable for an efficient DMFC operation. Reducing the onset potential for the indirect mechanism, therefore, is of interest in reducing the operating potential of the fuel cell anode and increasing overall cell efficiency.

Despite the inability to directly use the very active metals as DMFC anodes for practical reasons, their study is helpful in establishing reactivity trends. Interestingly, there is a volcano-type dependence to the onset potential across the periodic table. For the direct mechanism, more reactive metals (Re, Os) have a very high onset potential, as do the most noble metals (Ag, Au). Metals with more intermediate reactivity have the lowest onset potential. If one examines trends in the PDS for the direct mechanism across the periodic table, one sees that more noble catalysts, (e.g., Au) have difficulty binding intermediates, and thus the potential-determining steps involve the formation of surface intermediates. On the other hand, more reactive metals (e.g., Re and Ru) have potential-determining steps related to the removal of species from the surface. For the onset potential of the indirect mechanism, the pattern is not as obvious, but the general trend of intermediate-binding catalysts having the lowest onset potentials is still evident. As with the direct mechanism, the PDS for the indirect mechanism changes from removal of surface species to the left of the periodic table to the formation of surface species to the right of the periodic table.

In addition to clarifying the trends in methanol electrooxidation pathways across the periodic table, this thermochemical analysis can also explain the experimentally observed structure sensitivity of this reaction. Several groups [21,31,35,36,45] have found that at intermediate potentials, oxidation on the (100) and (110) facets of Pt occurs through the indirect mechanism, while on the (111) facet oxidation occurs through the direct mechanism. If one uses an analysis on the more open facets similar to the one that has already been presented on Pt(111), one can rationalize these differences. There is a significant difference in the onset potential of the indirect mechanism on Pt(111) and Pt(100)–$U_{onset} = 0.92$ eV for Pt(111); $U_{onset} = 0.56$ eV for Pt(100) [46]. At moderately low potentials, therefore, the direct mechanism is the only one accessible on Pt(111), whereas on Pt(100), the indirect mechanism is also possible. Clearly, this method can be applied to understanding trends in structure sensitivity of MeOH electroxidation on metals other than Pt.

### 16.2.3 Reactivity Descriptors for Bimetallic DMFC Anodes

As demonstrated above, changing the metal changes the onset potential for methanol electrooxidation on monometallic catalysts. The obvious next step is to use this method to search the space of bimetallic and ternary alloy catalysts in an effort to minimize onset potential. However, because of the large number of possible alloy combinations, the computational requirements for a simply systematic study would be intense. To reduce the computational time and effort in finding promising systems, it behooves us to find correlations between the adsorption free energies for the various reactive intermediates involved in this reaction. In the present case, there are correlations between the free energy of all intermediate species studied and the free energy of one of two species: CO* and OH*. Free energies of species binding through a carbon atom scale linearly with the free energy of CO* on the surface, and all free energies for species binding through an oxygen atom scale linearly with the free energy of OH* [43]. The origins of these remarkable correlations can be explained using arguments similar to bond order conservation arguments [47–49].

These linearly scaling correlations, then, can be used to reduce the complexity of the onset potential from being a function of multiple free energies of adsorption to only two *reactivity descriptors*: $G_{CO*}$ and $G_{OH*}$. Using the above-described correlations, therefore, the onset potential for methanol electrooxidation (either the direct or the indirect mechanism) can be mapped across the entire phase space of these two variables. Mapping out the phase space of these two variables is done for the indirect mechanism on the close-packed surfaces in Figure 16.6. This figure illustrates the Sabatier principle [50] for the onset potential of methanol electrooxidation. The ideal methanol electrooxidation catalyst, in terms of onset potential, is a compromise between binding intensity of two species: CO* and OH*. Excessively strong CO* and OH* binding makes it too difficult to remove these intermediates from the surface. Weak OH* binding makes it very difficult to activate water (to either OH* or COOH*); however, excessively strong OH* binding makes it difficult to remove OH* and methoxy from the surface. Extremely weak CO* binding makes it difficult to activate formaldehyde or methanol intermediates in the overall methanol electrooxidation.

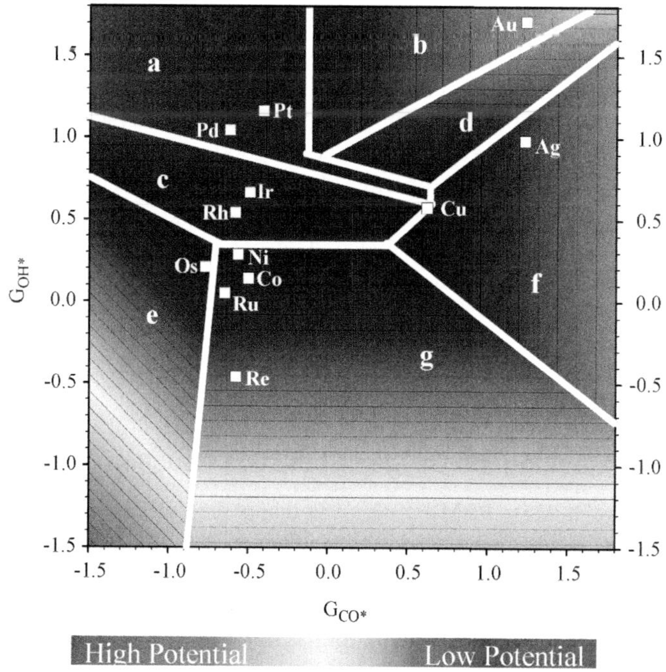


**FIGURE 16.6** The onset potential and potential-determining steps (PDSs) for the *indirect* mechanism of methanol electrooxidation on close-packed hcp and fcc surfaces plotted using $G_{CO^*}$ and $G_{OH^*}$ as descriptors. These two adsorption free energies describe the overall reaction landscape and can be used to estimate the potential through the linear relations, as described in Reference 43. The PDSs (without showing proton/electron transfer explicitly) for each region are as follows: (a) $CO^* \rightarrow COOH^*$; (b) $H_3COH \rightarrow H_2COH^*$; (c) $H_2O \rightarrow OH^*$; (d) $H_3COH \rightarrow H_3CO^*$; (e) $CO^* + OH^* \rightarrow CO_2$; (f) $H_2CO \rightarrow CHO^*$; (g) $H_3CO^* \rightarrow H_2CO$. Isopotential lines are included for reference. Each line represents a difference of 0.1 V. Because of scatter in the correlations used, the potential-determining region for the individual metal does not always match the actual PDS for that metal, which is given in Figure 16.5b. (See insert for color representation.)

The best catalysts are a compromise between these extremes, with moderate binding of both OH* and CO*. As shown in Figure 16.6, Ni(111) actually falls very close to the optimal predicted region for methanol electrooxidation by the indirect mechanism. Unfortunately, as noted previously, the dissolution and oxidation of Ni make it impossible to use as a stable methanol electrooxidation catalyst. Thus, alloy catalysts that strike the desirable reactivity descriptors compromise should be sought.

This first-principles-based framework to understanding what makes a low-onset potential a catalyst has promise in identifying bimetallic catalysts for use as DMFC anodes. By understanding the limitations of each catalyst (according to their positions in the phase space provided in Fig. 16.6), one can identify combinations of metals that may have synergistic effects. The well-known Pt–Ru alloy is a nice illustration of the power of this framework to screen for promising alloy catalysts. By inspection of the relative positions of Pt and Ru in Figure 16.6, one can see that Pt has difficulty activating $H_2O$ to form carboxyl, while Ru binds O-containing species too strongly. An alloy of the two metals may be expected to have intermediate properties, and therefore should have a lower onset potential than either Pt or Ru alone, allowing for a more efficient DMFC anode. A catalyst containing more similar metals, such as Pt–Pd, may not be as good a methanol electroxidation catalyst, because of the similar properties of the two constituent metals, which may limit the synergistic effect. While Pt–Ru is already a well-known combination in the DMFC literature, this framework can also identify and screen novel combinations for DMFC anodes that may be cheaper than and as effective as current state-of-the-art DMFC anode catalysts.

## 16.3 OXYGEN REDUCTION CATALYSTS

### 16.3.1 Understanding the Oxygen Reduction Reaction

The cathode side of both $H_2$-PEM and direct methanol fuel cells use a Pt-based catalyst to reduce the $O_2$ reactant. One of the largest challenges in proton exchange membrane fuel cell catalysis concerns the high overpotentials required to drive the oxygen reduction reaction (ORR) [22,51]. Sluggish ORR kinetics account for the majority of the voltage drop in $H_2$-PEMFCs, limiting state-of-the-art systems to operating voltages of only $\sim 0.7$ V, far from the equilibrium potential of $\sim 1.2$ V. For direct methanol and other direct alcohol fuel cells, the necessary cathode overpotential adds on top to the anode overpotential described in the previous section, leading to even lower overall cell efficiencies. A second, more practical problem of current ORR catalysts is the high cost associated with the Pt metal.

The high potential and acidic environment of the cathode make dissolution of the catalyst a huge problem; among the pure metals only Pt and Au are relatively stable at potentials close to the equilibrium potential. Thus, one way forward is to improve the ORR activity of Pt by modifying the surface Pt atoms. This will, of course, not solve the problem related to the cost of Pt completely, but it is important to note that a more active catalyst means that the loading of Pt can be significantly reduced without compromising the power density of the fuel cell.

### 16.3.2 Measuring ORR Activity

Despite its relative simplicity, the exact ORR reaction mechanism is difficult to pinpoint for a given catalytic surface. The following are possible elementary reaction steps for ORR:

$$O_2 + * \rightarrow O_2* \tag{16.1}$$

$$O_2* + * \rightarrow 2O* \tag{16.2}$$

$$O_2* + (H^+ + e^-) \rightarrow O_2H* \tag{16.3}$$

$$O_2H* + (H^+ + e^-) \rightarrow H_2O_2* \tag{16.4}$$

$$OOH^* + * \rightarrow O^* + OH^* \tag{16.5}$$

$$H_2O_2^* + * \rightarrow 2OH^* \tag{16.6}$$

$$O^* + (H^+ + e^-) \rightarrow OH^* \tag{16.7}$$

$$OH^* + (H^+ + e^-) \rightarrow H_2O \tag{16.8}$$

Three distinct mechanisms can be isolated from these eight steps. These three mechanisms are differentiated by the degree of hydrogenation of the $O_2$ species prior to cleavage of the oxygen–oxygen bond. The first mechanism, commonly referred to as the *dissociative mechanism*, is $O_2$ adsorption [step (16.1)], followed by O–O bond scission prior to any hydrogenation [step (16.2)], followed by sequential hydrogenation of the resulting products [steps (16.7) and (16.8)]. The second mechanism involves hydrogenation of the adsorbed $O_2$ molecule to hydroperoxy (OOH*) [step (16.3)] before O–O bond scission [step (16.5)], and subsequent hydrogenation [steps (16.7) and (16.8)]. The last mechanism involves hydrogenation of adsorbed oxygen to $H_2O_2$* [steps (16.3) and (16.4)] before O–O bond scission and subsequent hydrogenation [steps (16.6) and (16.8)]. The latter two mechanisms are generally termed *associative mechanisms* in the literature. All oxygen–oxygen bond scissions [steps (16.2)(16.5) and (16.6)] are not proton transfer steps, whereas the hydrogenation steps [(16.3) (16.4) (16.7) and (16.8)] are.

Using this information and the methods described previously in this chapter, one can calculate the free-energy changes of all mechanistic steps at relevant ORR potentials, assuming standard temperature and pressure. At 1.23 V (the equilibrium potential for oxygen reduction), all proton transfer steps are quite endothermic and are the most difficult steps in oxygen reduction on Pt catalysts, regardless of the mechanism chosen (see Fig. 16.7). Thus, for the ORR reaction to occur, $H_2$ and direct alcohol fuel cell cathodes must run at a lower potential than the equilibrium potential. At lower bias (e.g., 0.8 V), the proton transfer steps become more competitive with O–O bond scission on Pt, because the decrease in the potential bias lowers the electron chemical potential. On the other hand, O–O bond cleavage is unaffected by potential within the framework of these calculations, as there is no proton–electron transfer associated with those steps.

In addition to identifying the mechanism for oxygen reduction at relevant potentials, one can extend this analysis to get a more quantitative picture of the mechanism on the catalytic surfaces of interest and calculate the trends in the overall activity across surfaces using a simple Sabatier model [52]. If one assumes optimum coverages and ensembles, the Sabatier principle states that the maximal activity toward a reaction mechanism is

$$A = kT\min_i\left(\log\left(\frac{k_i}{k_0}\right)\right)$$

where $A$ is a normalized measure of activity, $k_i (= \nu_i e^{-E_a^i/kT})$ is the rate constant for the forward direction of the elementary step $i$, and $k_0$ normalizes the rate of nonactivated

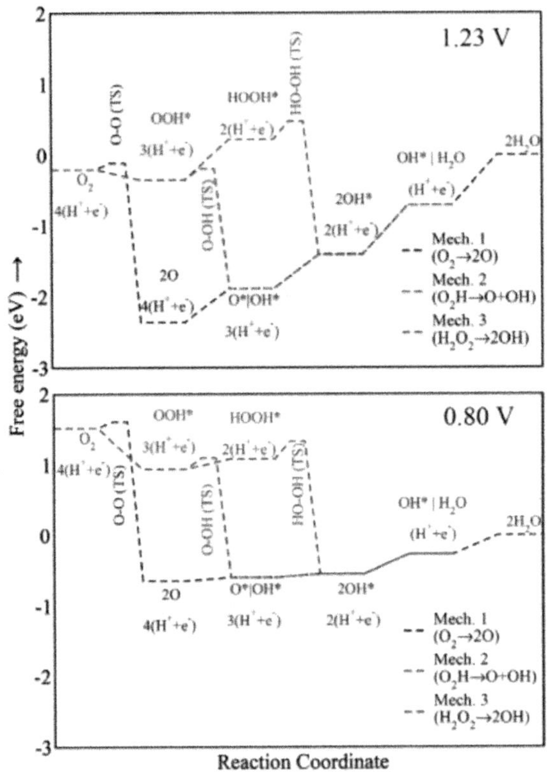


**FIGURE 16.7** Free-energy diagram for ORR on Pt(111) via three different mechanisms at two different potentials. (Taken from Ref. 16.)

proton–electron transfer steps to zero, with an appropriate normalization for differences in prefactors between $O_2$ bond scission and hydrogenation [52]. Thus, one can compare activities of different surfaces to each other in a simple way.

### 16.3.3 Understanding Oxygen Reduction Reaction Catalysts

Now that a measure of normalized activity has been introduced, the next step is to identify a descriptor for the overall reaction that can capture trends in catalytic activity. As all intermediates in oxygen reduction interact with the catalyst surface via oxygen, the adsorption energy of oxygen is an obvious candidate for a descriptor. Using only the dissociative mechanism and without including charge transfer kinetic barriers, it was found that among the pure metals Pt has the highest activity (see Fig. 16.8). From this simple picture, however, we see that a catalyst having an oxygen binding somewhat weaker than that on Pt would possess better activity [52]. This is in agreement with the observation that pure Pt ORR catalysts suffer from poisoning by hydroxyl (OH) species, a necessary intermediate for the ORR, which inhibits that reaction [22,53].

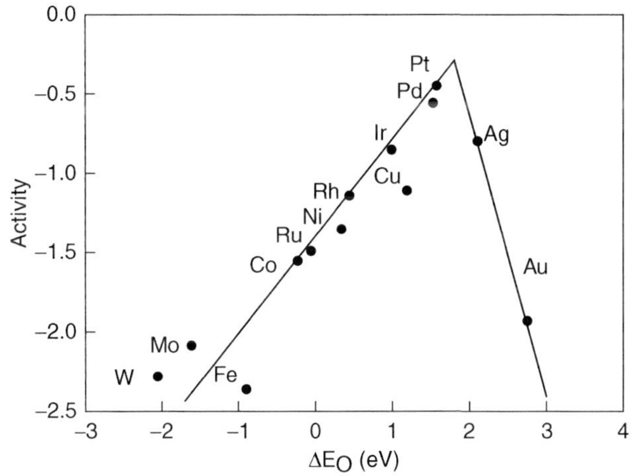


**FIGURE 16.8** Calculated trends in electrocatalytic oxygen reduction reaction on close-packed surfaces of pure metals, as a function of oxygen binding. (Taken from Ref. 52.)

In order to avoid problems of instability under electrochemical conditions discussed previously, we want to keep the Pt surface and only tune its catalytic activity for ORR via a decrease in oxygen binding. On the basis of insights from electronic structure calculations, this can be achieved either by introducing another metal in the subsurface layer or by modifying the lattice constant; in most cases involving two or more metals in the catalysts, both effects are simultaneously present. To date, a number of different approaches have been taken to improve the activity of Pt; three in particular have been shown to be effective: (1) alloying Pt with other transition metals, such as Co, Ni, or Fe [54–56]; (2) modifying Pt's lattice constant via Pt (or Pt alloy) overlayer structures [57–64]; and (3) overlaying other metals, such as Au, onto Pt supports [65].

As argued above, the binding energy of oxygen is, in general, a good descriptor for ORR activity of the close-packed metal surfaces [52], defects [66], Pt alloys [56,57], and for catalysts where a monolayer of Pt atoms is deposited on a substrate of a second late transition metal [57–62]. Plotting the logarithm of the reported specific kinetic current density relative to Pt versus the difference in binding energy of oxygen to the surface minus that on Pt gives rise to a similar volcano as for the pure metals (see Fig. 16.9) [68]. Materials with a slightly weaker binding of oxygen than Pt show an increased ORR activity compared to pure Pt; however, when the oxygen binding is much weaker than that of Pt, the ORR activity decreases again. Thus an optimal catalyst should bind the intermediates not so weakly that it becomes difficult to activate the reactants and not so strongly that the products poison the catalyst. In other words, on the weak binding side of the volcano, the catalyst is limited by the reaction steps that form bonds to the surface and on the strong binding side, the catalyst is limited by the reaction steps where bonds to the surface are broken.

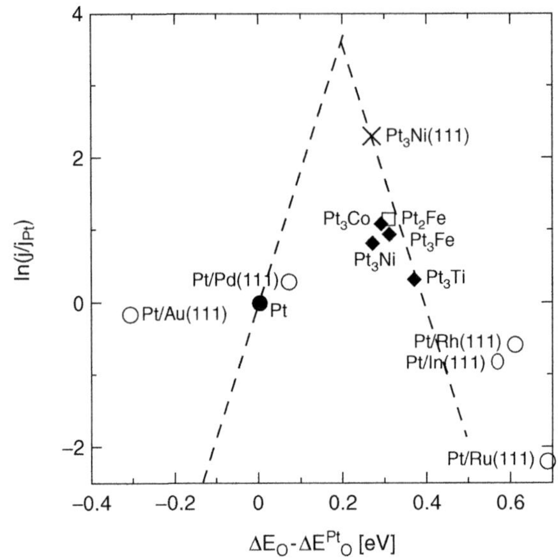


**FIGURE 16.9** Measured kinetic current density relative to that of Pt versus calculated oxygen binding energies for different Pt containing catalytic materials: Pt skins on transition metal (111) surfaces (○), Pt skin on $Pt_3Ni(111)$ surface (x) and polycrystalline surfaces of Pt alloys (◆). The dashed line represents the theoretically predicted Sabatier volcano.

Under electrochemical conditions the weak binding leg of the volcano is defined by reaction step (16.3) (the formation of OOH*) and the strong binding leg is defined by reaction step (16.8) (reduction of OH* to $H_2O$). The resulting theoretical volcano is represented by the dashed lines in Figure 16.9. We note that the prediction of the volcano was first published in 2004 [52], whereas all the experimental data, also shown in Figure 16.9, are from later studies. Furthermore, according to a kinetic model compared with detailed experimental data, it is possible to determine that the top of the volcano should be somewhere between Pt and $Pt_3Ni$, as these two catalysts are on opposite sides of the volcano curve [67]. We note here that the best catalyst of those shown in Figure 16.9 denotes a catalyst with a Pt skin monolayer overlaid on top of the $Pt_3Ni(111)$ alloy surface.

Finally, it is instructive to notice the nice qualitative agreement between predicted relative activity and that found in experiments. More importantly, it also illustrates the tradeoff between hydrogenation-type (breaking of bonds to the surface) steps and "bond forming to the surface" steps with oxygen binding energies within the electrochemical system. This confirms the Sabatier analysis that a moderate O binding corresponds to the most active catalyst under typical ORR conditions. Our analysis also illustrates the usefulness of even relatively simple vapor-phase DFT calculations to elucidate reaction trends in more complicated environments, as long as the major reactivity factors are captured in the simpler models. One can apply the approach outlined here to other novel electrocatalysts as well, to probe their activity for the ORR reaction and for other electrocatalytic reactions.

## 16.4 CONCLUSION

The electrocatalytic environment provides unique challenges to atomistic modeling due to the complexity of the reaction environment and the effects of environmental factors on various aspects the catalytic cycle. Over the past few years (as of 2010), significant strides have been made in accounting for these factors and thereby making such simulations more accurate. Specifically, accounting for solvent, potential, charge transfer, and coverage effects allows for a very detailed picture of surface chemistry at electrocatalytic surfaces. Using these models allows for a better understanding of the mechanism of electrocatalytic reactions under experimental conditions.

While such complex models have been constructed and provide good accuracy for electrocatalytic phenomena, one should not discount the power of even relatively simple DFT calculations. Such calculations may be used in at least a qualitative way to better understand the trends in electrocatalytic activity across metal surfaces. In many cases, such as those of methanol electrooxidation and oxygen reduction illustrated here, these simpler models can identify a few simple parameters (such as the binding of key intermediates) that can be used to predict important chemical properties, such as onset potential and reactivity. In turn, such "reactivity descriptors" can guide experimental synthesis and testing of new electrocatalysts with improved catalytic properties.

## REFERENCES

1. J. K. Nørskov, T. Bligaard, J. Rossmeisl, and C. H. Christensen, *Nature Chem.* **1**, 37 (2009).
2. H. Ogasawara, B. Brena, D. Nordlund, M. Nyberg, A. Pelmenschikov, L. G. M. Pettersson, and A. Nilsson, *Phys. Rev. Lett.* **89**, 276102 (2002).
3. M. A. Henderson, *Surf. Sci. Rep.* **46**, 5–308 (2002).
4. S. Schnur and A. Groβ, *New J. Phys.* **11**, 125003 (2009).
5. H. Noguchi, T. Okada, and K. Uosaki, *Faraday Discuss.* **140**, 125–137 (2008).
6. G. S. Karlberg and G. Wahnstrom, *J. Chem. Phys.* **122**, 194705 (2005).
7. C. Clay, S. Haq, and A. Hodgson, *Phys. Rev. Lett.* **92**, 046102 (2004).
8. K. Bedurftig, S. Volkening, Y. Wang, J. Wintterlin, K. Jacobi, and G. Ertl, *J. Chem. Phys.* **111**, 11147–11154 (1999).
9. P. A. Thiel and T. E. Madey, *Surf. Sci. Rep.* **7**, 211–385 (1987).
10. J. Rossmeisl, J. K. Nørskov, C. D. Taylor, M. J. Janik, and M. Neurock, *J. Phys. Chem. B* **110**, 21833–21839 (2006).
11. G. S. Karlberg, J. Rossmeisl, and J. K. Nørskov, *Phys. Chem. Chem. Phys.* **9**, 5158–5161 (2007).

12. M. Otani and O. Sugino, *Phys. Rev. B* **73**, 115407 (2006).
13. J. S. Filhol and M. Neurock, *Angew. Chem. Int. Ed.* **45**, 402–406 (2006).
14. A. Y. Lozovoi, A. Alavi, J. Kohanoff, and R. M. Lynden-Bell, *J. Chem. Phys.* **115**, 1661–1669 (2001).
15. R. Jinnouchi and A. B. Anderson, *Phys. Rev. B* **77**, 245417 (2008).
16. E. Skulason, G. S. Karlberg, J. Rossmeisl, T. Bligaard, J. Greeley, H. Jonsson, and J. K. Nørskov, *Phys. Chem. Chem. Phys.* **9**, 3241–3250 (2007).
17. J. Rossmeisl, E. Skúlason, M. E. Björketun, V. Tripkovic, and J. K. Nørskov, *Chem. Phys. Lett.* **466**, 68–71 (2008).
18. H. A. Hansen, J. Rossmeisl, and J. K. Nørskov, *Phys. Chem. Chem. Phys.* **10**, 3722–3730 (2008).
19. G. S. Karlberg, T. F. Jaramillo, E. Skulason, J. Rossmeisl, T. Bligaard, and J. K. Nørskov, *Phys. Rev. Lett.* **99**, 126101 (2007).
20. E. Reddington, A. Sapienza, B. Gurau, R. Viswanathan, S. Sarangapani, E. S. Smotkin, and T. E. Mallouk, *Science* **280**, 1735–1737 (1998).
21. A. Hamnett, *Catal. Today* **38**, 445–457 (1997).
22. J. Lipkowski and P. N. Ross, eds., *Electrocatalysis*, VCH, New York, 1998.
23. H. S. Liu, C. J. Song, L. Zhang, J. J. Zhang, H. J. Wang, and D. P. Wilkinson, *J. Power Sources* **155**, 95–110 (2006).
24. S. R. Brankovic, J. X. Wang, and R. R. Adzic, *Electrochem. Solid State Lett.* **4**, A217–A220 (2001).
25. P. Liu, A. Logadottir, and J. K. Nørskov, *Electrochim. Acta* **48**, 3731–3742 (2003).
26. C. Lu, C. Rice, R. I. Masel, P. K. Babu, P. Waszczuk, H. S. Kim, E. Oldfield, and A. Wieckowski, *J. Phys. Chem. B* **106**, 9581–9589 (2002).
27. P. Strasser, *J. Combin. Chem.* **10**, 216–224 (2008).
28. S. Surampudi, S. R. Narayanan, E. Vamos, H. Frank, G. Halpert, A. LaConti, J. Kosek, G. K. S. Prakash, and G. A. Olah, *J. Power Sources* **47**, 377–385 (1994).
29. P. Waszczuk, J. Solla-Gullon, H. S. Kim, Y. Y. Tong, V. Montiel, A. Aldaz, and A. Wieckowski, *J. Catal.* **203**, 1–6 (2001).
30. V. S. Bagotzky, Y. B. Vassiliev, and O. A. Khazova, *J. Electroanal. Chem.* **81**, 229–238 (1977).
31. D. Cao, G. Q. Lu, A. Wieckowski, S. A. Wasileski, and M. Neurock, *J. Phys. Chem. B* **109**, 11622–11633 (2005).
32. J. Kua and W. A. Goddard, *J. Am. Chem. Soc.* **121**, 10928–10941 (1999).
33. N. M. Markovic, H. A. Gasteiger, P. N. Ross, X. D. Jiang, I. Villegas, and M. J. Weaver, *Electrochim. Acta* **40**, 91–98 (1995).
34. Y. X. Chen, A. Miki, S. Ye, H. Sakai, and M. Osawa, *J. Am. Chem. Soc.* **125**, 3680–3681 (2003).
35. T. H. M. Housmans, A. H. Wonders, and M. T. M. Koper, *J. Phys. Chem. B* **110**, 10021–10031 (2006).
36. S. Sriramulu, T. D. Jarvi, and E. M. Stuve, *Electrochim. Acta* **44**, 1127–1134 (1998).
37. A. A. Gokhale, J. A. Dumesic, and M. Mavrikakis, *J. Am. Chem. Soc.* **130**, 1402–1414 (2008).
38. L. C. Grabow, A. A. Gokhale, S. T. Evans, J. A. Dumesic, and M. Mavrikakis, *J. Phys. Chem. C* **112**, 4608–4617 (2008).

39. J. Greeley, and M. Mavrikakis, *J. Am. Chem. Soc.* **126**, 3910–3919 (2004).
40. S. Kandoi, A. A. Gokhale, L. C. Grabow, J. A. Dumesic, and M. Mavrikakis, *Catal. Lett.* **93**, 93–100 (2004).
41. A. Logadottir and J. K. Nørskov, *J. Catal.* **220**, 273–279 (2003).
42. A. Logadottir, T. H. Rod, J. K. Nørskov, B. Hammer, S. Dahl, and C. J. H. Jacobsen, *J. Catal.* **197**, 229–231 (2001).
43. P. Ferrin, A. U. Nilekar, J. Greeley, M. Mavrikakis, and J. Rossmeisl, *Surf. Sci.* **602**, 3424–3431 (2008).
44. M. S. McGovern, P. Waszczuk, and A. Wieckowski, *Electrochim. Acta* **51**, 1194–1198 (2006).
45. T. Iwasita, *Electrochim. Acta* **47**, 3663–3674 (2002).
46. P. Ferrin and M. Mavrikakis, *J. Am. Chem. Soc.* **131**, 14381–14389 (2009).
47. F. Abild-Pedersen, J. Greeley, F. Studt, J. Rossmeisl, T. R. Munter, P. G. Moses, E. Skulason, T. Bligaard, and J. K. Nørskov, *Phys. Rev. Lett.* **99**, 016105 (2007).
48. A. T. Bell, and E. Shustorovich, *J. Catal.* **121**, 1–6 (1990).
49. E. Shustorovich, *Surf. Sci.* **176**, L863–L872 (1986).
50. P. Sabatier, *Ber. Dtsch. Chem. Gesell.* **44**, 1984 (1911).
51. H. A. Gasteiger, S. S. Kocha, B. Sompalli, and F. T. Wagner, *Appl. Catal. B Environ.* **56**, 9–35 (2005).
52. J. K. Nørskov, J. Rossmeisl, A. Logadottir, L. Lindqvist, J. R. Kitchin, T. Bligaard, and H. Jonsson, *J. Phys. Chem. B* **108**, 17886–17892 (2004).
53. J. X. Wang, N. M. Markovic, and R. R. Adzic, *J. Phys. Chem. B* **108**, 4127–4133 (2004).
54. V. R. Stamenkovic, B. S. Mun, M. Arenz, K. J. J. Mayrhofer, C. A. Lucas, G. F. Wang, P. N. Ross, and N. M. Markovic, *Nature Mater.* **6**, 241–247 (2007).
55. V. R. Stamenkovic, B. Fowler, B. S. Mun, G. F. Wang, P. N. Ross, C. A. Lucas, and N. M. Markovic, *Science* **315**, 493–497 (2007).
56. V. Stamenkovic, B. S. Mun, K. J. J. Mayrhofer, P. N. Ross, N. M. Markovic, J. Rossmeisl, J. Greeley, and J. K. Nørskov, *Angew. Chem. Int. Ed.* **45**, 2897–2901 (2006).
57. J. L. Zhang, M. B. Vukmirovic, Y. Xu, M. Mavrikakis, and R. R. Adzic, *Angew. Chem. Int. Ed.* **44**, 2132–2135 (2005).
58. J. L. Zhang, M. B. Vukmirovic, K. Sasaki, A. U. Nilekar, M. Mavrikakis, and R. R. Adzic, *J. Am. Chem. Soc.* **127**, 12480–12481 (2005).
59. M. B. Vukmirovic, J. Zhang, K. Sasaki, A. U. Nilekar, F. Uribe, M. Mavrikakis, and R. R. Adzic, *Electrochim. Acta* **52**, 2257–2263 (2007).
60. A. U. Nilekar, Y. Xu, J. L. Zhang, M. B. Vukmirovic, K. Sasaki, R. R. Adzic, and M. Mavrikakis, *Top. Catal.* **46**, 276–284 (2007).
61. A. U. Nilekar and M. Mavrikakis, *Surf. Sci.* **602**, L89–L94. (2008).
62. R. R. Adzic, J. Zhang, K. Sasaki, M. B. Vukmirovic, M. Shao, J. X. Wang, A. U. Nilekar, M. Mavrikakis, J. A. Valerio, and F. Uribe, *Top. Catal.* **46**, 249–262 (2007).
63. S. Koh and P. Strasser, *J. Am. Chem. Soc.* **129**, 12624 (2007).
64. S. Koh, J. Leisch, M. F. Toney, and P. Strasser, *J. Phys. Chem. C* **111**, 3744–3752 (2007).
65. J. Zhang, K. Sasaki, E. Sutter, and R. R. Adzic, *Science* **315**, 220–222 (2007).
66. J. Greeley, J. Rossmeisl, A. Hellman, and J. K. Nørskov, *Z. Phys. Chem. Int. J. Res. Phys. Chem. Chem. Phys.* **221**, 1209–1220 (2007).

67. J. Rossmeisl, G. S. Karlberg, T. Jaramillo, and J. K. Nørskov, *Faraday Discuss.* **140**, 337–346 (2008).

68. J. Rossmeisl and J. K. Nørskov, *Surf. Sci.* **602**, 2337–2338 (2008).

69. J. Rossmeisl, A. Logadottir, and J. K. Nørskov, *Chem. Phys.* **319**, 178–184 (2005).

70. J. K. Nørskov, T. Bligaard, A. Logadottir, S. Bahn, L. B. Hansen, M. Bollinger, H. Bengaard, B. Hammer, Z. Sljivancanin, M. Mavrikakis, Y. Xu, S. Dahl, and C. J. H. Jacobsen, *J. Catal.* **209**, 275–278 (2002).

CHAPTER 17

# Operando XAS Techniques: Past, Present, and Future

CHRISTINA ROTH
Institute for Materials Science, Technische Universität Darmstadt, Germany

DAVID E. RAMAKER
Chemistry Department, George Washington University, Washington, DC

Sir William Robert Grove's groundbreaking discovery, which demonstrated that hydrogen and oxygen could be combined to produce water in an electrochemical cell, thereby generating "green" or environmentally friendly energy, has been around since 1839. Nevertheless, for low-temperature polymer electrolyte membrane fuel cells (PEMFCs), the transition step from promising concept to areawide market launch has not yet been made. Two severe problems have prevented this: the need for both expensive and rare noble metal catalysts (e.g., mostly Pt and Pt-based alloys) and their limited long-term stability in the harsh conditions of the fuel cell environment.

However, it is also these issues that prompt interest in the use of spectroscopic techniques, which may provide detailed insight into the catalyst's behavior during operation. A very powerful approach in this respect is X-ray absorption spectroscopy (XAS), which requires neither UHV nor long-range structural order. Combining the conventional EXAFS analysis with the novel $\Delta\mu$ XANES technique offers the unique and powerful advantage of providing both structural information along with in situ adsorbate coverage and binding-site information with one experimental technique.

In this chapter, the use of *operando* XAS techniques in fuel cell research is reviewed with the more recent experimental and analytical developments, and the different operating fuel cell modes, and XAS capabilities are highlighted. In order to understand the differences between in situ and operando measurements, as well as to select the best setup for a specific research aim, the evolution of sample and experimental

*Fuel Cell Science: Theory, Fundamentals, and Biocatalysis*,
Edited by Andrzej Wieckowski and Jens K. Nørskov

designs is summarized and the differences, advantages, and limitations of these designs are discussed. More recent XAS results from studies on the standard Pt–Ru anode catalyst, which is commonly applied when fuels with CO impurities are used, and at the Pt cathode in operating fuel cells are highlighted. Finally, possible future XAS developments are discussed as they relate to operando fuel cells.

## 17.1 INTRODUCTION

Since Sir William Robert Grove's discovery in 1839, fuel cells have attracted attention, although the concurrently developed combustion engine "seized the day" and significantly slowed progress in fuel cell development. In the 1960s, applications in air and space within the American Gemini and Apollo missions spurred more interest. Then, in the 1990s the interest in fuel cells soared when the looming fossil fuel limits became more evident. Consequently, research in fuel cells and related topics began already in 1965, but took off only in the 1990s, and since then has resulted in an output of more than 200 fuel-cell-related papers per month. Figure 17.1 shows the number of publications rising from 171 in 1990 to 4179 in 2008. Likewise, the number of publications linked to the use of X-ray absorption spectroscopy (XAS) increased significantly, starting from a mere 190 per year in 1995 to almost twice as much (372 per year) in 2008.

The shear number of fuel-cell-related papers and XAS papers is overwhelming (Fig. 17.1), and combining the search string to "EXAFS AND fuel cells" still results in a number of search hits that is too large to be considered in this chapter. Consequently, this review limits its examination of XAS studies primarily to operating fuel cells. Since such fuel cell studies can vary widely, we will first summarize the evolution of electrode designs and experimental techniques over the years, focus on some milestones in the development and use of in situ and operando XAS techniques, and then cite some exemplary operando studies.

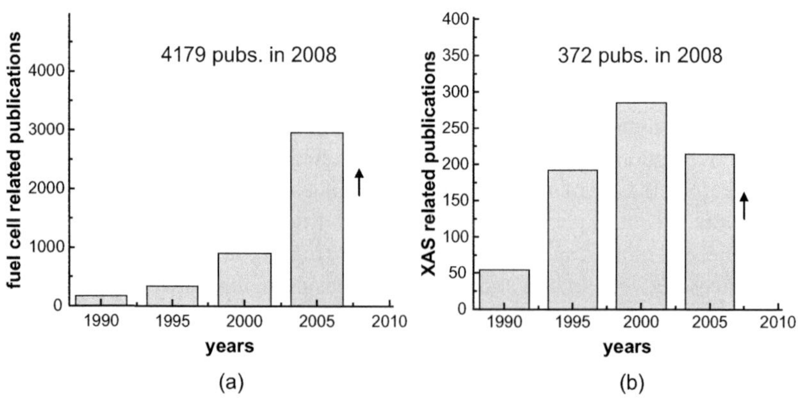


**FIGURE 17.1** (a) Number of publications found when the keyword "fuel cell*" is used in an otherwise open search on "Web of Science"; (b) number of publications linked to X-ray absorption spectroscopy.

## 17.2 SELECTION OF EXPERIMENTAL SETUP

For those readers not very familiar with general electrochemical methodology and tools, the wealth of different approaches might be confusing. However, the choice of the most appropriate instrumental setup is often dictated by the specific research objective. The methodologies basically differ in the complexity of the electrode structure, the electrode–electrolyte interface, and the potential control of the electrodes; for instance, the CO oxidation reaction was studied on various single-crystal faces as model surfaces [1–3], and also at carbon-supported Pt alloy catalysts [4–6] and sometimes in fuel cells using hydrogen feeds with CO impurities, the latter closer to reality [7,8].

Since the review in this chapter is limited in part to application of XAS to operando studies, it seems appropriate to discuss the differences between what we refer to as in situ versus true operando applications, and the additional questions, problems, and complexities that can arise with the latter. We further hope this may help to explain why so many different approaches have been adopted for the development of in situ and/or operando fuel cell experiments, in which a compromise between precise experimental conditions versus the less precise and more complex but realistic operando situation has to be made.

### 17.2.1 Complexity of Electrode Structure and Electrode–Electrolyte Interface

The techniques that have been developed to study reaction mechanisms and catalyst performance in an electrochemical environment differ in part from the perspective of electrode structure. Figure 17.2 shows the evolving complexity of different electrode

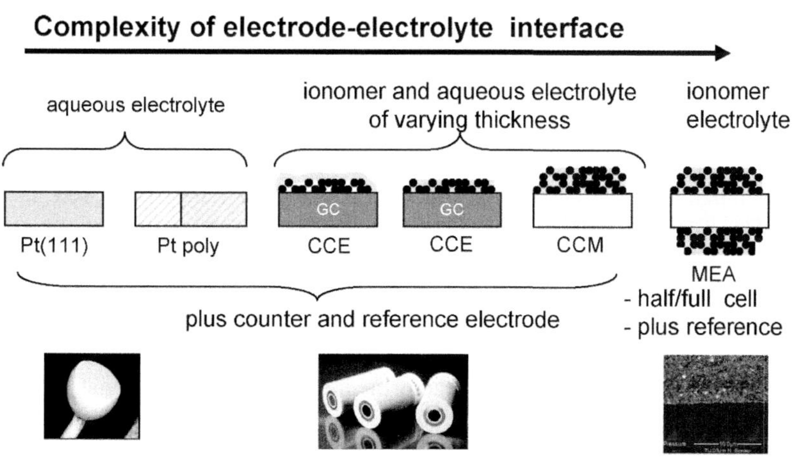


**FIGURE 17.2** Schematic illustration of various electrode structures with the complexity increasing from left to right. The bottom pictures illustrate a Pt disk electrode, RDE electrodes, and a TEM image of a full MEA.

designs. Early research began with single-crystal studies in aqueous electrochemical cells. Since then, CO oxidation on different Pt crystal surfaces [e.g., Pt(111)] is among the most widely studied topics in electrocatalysis [1–3,9]. In addition, stepped surfaces and single-crystal surfaces decorated by a second element have been investigated [10,11] with the intention to also study the effects of edge and kink sites and cocatalysis. For instance, numerous papers deal with Ru-decorated Pt facets as suitable model systems for the investigation of the bifunctional mechanism [12] and the ligand effect in fuel cell catalysis [13]. The advantage of these studies is in the great amount of control that can be exerted on the variables; for example, the measurements can be conducted in a three-electrode electrochemical setup under well-defined conditions at precisely controlled potentials on basically mirror-polished smooth surfaces in direct contact with a liquid electrolyte. Thus, the results obtained are very reproducible. A disadvantage of this approach, however, is that catalyst support effects, the influence of the electrode structure and electrode–electrolyte interface, and in particular the electrode's porosity and ionomer distribution cannot be accessed or otherwise mimicked easily; all these aspects are important to surface diffusion, the concentration of adsorbates on the surface, and hence the performance of a fuel cell.

In order to account for catalyst–support interactions in electrocatalysis, investigators began in the 1990s to look at carbon-supported catalysts in real fuel cell applications. The catalyst powders were either coated on an electrochemically inactive electrode [e.g., highly oriented pyrolitic graphite (HOPG) or glassy carbon (GC)] or on a well-characterized polished electrode surface (Au) and tested by cyclic voltammetry (CV) and rotating-disk experiments (RDEs). In 1998, Schmidt et al. [14] reported a preparation technique, which has since been widely adopted for real catalyst studies, using an ink prepared from a catalyst powder and an appropriate amount of Nafion ionomer. In a slightly modified approach, the catalyst powder is dusted onto a mirror-polished glassy carbon electrode and covered with a thin film of Nafion. In both cases, it is important that the ionomer coating be thin enough to not interfere with the hydrogen adsorption kinetics [15,16]. To avoid the uncertainties linked to the unknown coating thickness of the ionomer, Arenz [17] replaced the ionomer in the ink by ultrapure water. These samples are called *catalyst-coated electrodes* (CCEs) in the following text.

In an effort to mimic the complex structure of real electrodes, as they are applied in low-temperature fuel cell systems, porous gas diffusion electrodes were manufactured by coating much thicker (in the order of micrometers) ink films on substrates or even membranes, either just on one side [referred to here as *catalyst-coated membranes* (CCMs)] or on both sides as in so-called membrane electrode assemblies (MEAs). For the preparation of such gas diffusion electrodes, airbrushing, printing, or hot-pressing techniques are used to obtain porous catalyst networks of a specific thickness on a suitable substrate, or directly on the proton exchange membrane. Such porous electrode structures might lead to mass transport limitations and other issues caused by the (counter)diffusion of liquid and gaseous reactants, and therefore need to be studied.

The importance of the electrolyte must also be emphasized. This has been analyzed in part in a paper by Uchida et al. [18] Single-crystal studies usually utilize a liquid electrolyte, such as aqueous perchloric or sulfuric acid, while the electrolyte in fuel cell measurements is usually a polymer, as, for example, the perfluorinated polysulfonic acid, Nafion. In both CCE and CCM studies in an electrochemical cell, however, the situation is even more complex, since the porous gas diffusion electrode is formed by a catalyst–ionomer network, and in addition the aqueous electrolyte will soak into the electrode pores. Consequently, both a liquid electrolyte and a solid polymer electrolyte phase will coexist during the measurement, a situation that will have severe consequences for the formation of the so-called three-phase boundary regions where the reaction takes place. Uchida et al. [18] compared the pore structure and electrolyte distribution of an electrode in liquid electrolyte to an electrode "coated" by the polymer electrolyte. In this work, the authors discussed primary and secondary pores, where the primary pores were filled with liquid electrolyte (thus not occurring in MEAs) and the secondary pores would provide the gas channels. Consequently, the gaseous fuels would have to diffuse through either the liquid electrolyte, the polymer electrolyte, or a mixture of both, where the solid polymer electrolyte might cover the active sites and severely hamper the reaction from proceeding. This is expected to significantly affect the access of gases to the catalytic active sites, and thus has to be considered in the choice of experimental setup, in particular, if precise kinetic studies are planned.

In fuel cell systems, the supply of reactants often occurs in gaseous form [PEMFC, indirect methanol fuel cell (IMFC)], whereas the products formed could be either gaseous or liquid depending on the operating conditions. In direct alcohol (methanol or ethanol) fuel cells (DAFC; DMFC, DEFC) the respective alcohol is directly fed into the fuel cell in liquid form. Thus, counterdiffusion of gaseous/liquid products and gaseous reactants becomes an important aspect, which can be studied in a full-cell measurement only using an MEA.

### 17.2.2 Electrode Potential Control

Not only can the electrode structure vary widely; the experimental design can vary from a simple three-electrode electrochemical cell, to half-cell experiments to full-cell measurements. Figure 17.3 depicts these different experimental designs and schematically shows the potential change with current of the anode and cathode for the reactions indicated. It also acknowledges that there is often a connection between the electrode control and electrode structure. Single crystals, as well as CCE and CCM electrodes, will be applied mostly in three-electrode electrochemical cells in aqueous electrolyte, while MEAs are usually used in either a half-cell or a full-cell arrangement. In a half-cell measurement, the cathode will be flooded with hydrogen to work as a hydrogen reference electrode, against which the anode potential can be measured with good precision. In contrast, most real fuel cells are operated in a two-electrode configuration, where only the cell potential can be measured so that accurate anode and cathode potentials cannot be obtained, although certain techniques are available to estimate these potentials [19]. Furthermore, there have been a few reports in the literature in which a third reference electrode was implemented in the fuel cell setup [20].

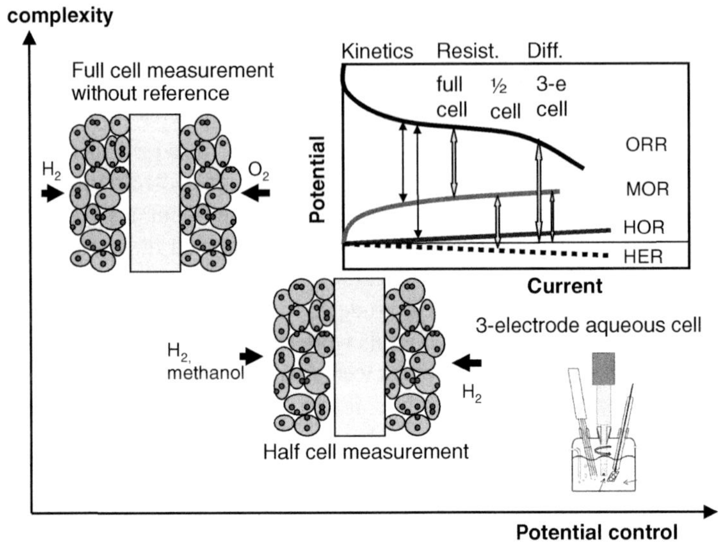


**FIGURE 17.3** Illustration of full-cell, half-cell, and three-electrode setups and the usual correlation between the complexity of the elecrode–electrolyte interface and experimental design. The insert schematically shows the current occurring at the anode [for either methanol oxidation (MOR) or hydrogen oxidation (HOR)] and at the cathode [oxygen reduction (ORR) or hydrogen evolution (HER)]. Regions are also indicated where the kinetics, cell resistance, and diffusion limit effects dominate. The shaded arrows relate to the potentials measured for the respective full, half, and three-electrode cells as indicated. The line arrows relate to the discussion given with regard to Figure 17.13, when the difference gives the MOR anode potential.

As illustrated in Figure 17.3, half-cell measurements may be preferred over full-cell studies, as they allow for a more controlled experiment, since precise anode potentials against the $H_2$-purged cathode will be obtained. This, of course, helps one to study fundamental reaction mechanisms in a well-defined way. However, unlike the case in full-cell experiments, no water will be produced, which plays an important role in catalyst degradation phenomena [21,22] and Ru corrosion from a PtRu anode catalyst in a direct methanol fuel cell [23]. These effects can thus be considered only in a limited way, when opting for a CCE, CCM, or MEA half-cell experiment.

## 17.3 XAS APPLICATIONS TO FUEL CELLS: THE PAST

### 17.3.1 In Situ versus Operando Experiments

When discussing XAS applications, we must stress the most important difference between in situ and operando studies. The term *in situ* here is limited to the electrochemical environment, perhaps a three-electrode aqueous electrochemical cell.

In contrast to the above-defined in situ studies, *operando* is a term coined exclusively for those investigations, which are carried out in a real system under operating conditions. In contrast to the structural characterization of fuel cell catalysts, the identification and characterization of reaction products and intermediates (usually adsorbates on the catalysts) as well as of fundamental reaction mechanisms is relevant to real systems only in an operando electrochemical environment. Further, since the XAS data normally take 20–40 min per spectrum, when either the end product or reactant is an adsorbate (such as $H_2O$ activation to OH/Pt, $O_2$ reduction to O/Pt, or O/Pt reactant to $O_2$ gas), the results reflect the conditions when current has stopped flowing and the coverage is at the equilibrium result for that potential. Even when this is not the case, such as in ORR (i.e., $O_2$ reduction to $H_2O$), the number of reactive Pt sites for this reaction is determined by the equilibrium of the reactions that poison the catalyst's surface. Thus the major difference between in situ and operando results will be the diffusion rates and bulk concentrations of the reactants and the coverage of adsorbed poisons and/or reactants. In the operando fuel cell, the adsorbate coverages are determined by gas diffusion rates at the *three-phase* boundary [e.g., at the $O_2$(gas)–Pt interface], as opposed to that in an in situ cell where they are determined by gas diffusion rates in the liquid electrolyte (e.g., $O_2$ in $H_2O$). Below we compare results in each to emphasize this difference.

Since the late 1980s the use of operando spectroscopic techniques that provide detailed insight into the catalyst's structure and/or adsorbate coverage and binding sites, has gathered momentum. It is generally believed that to solve the electrocatalysts costs and durability issues, a more detailed understanding of the underlying fundamental mechanisms is required. The goal is the controlled and systematic design of optimized catalyst systems, with the necessary properties to function in a real operational environment at lower costs and for a longer period of time. This goal might be reached only by operando investigations, where the catalysts can be studied in an experiment as close to reality as possible.

In principle, there exists a multitude of methods, which can be applied for in situ investigations in order to (1) pursue structural changes in the catalyst and/or (2) follow coverage with different adsorbates revealing fundamental mechanisms for catalytic activity. However, only a few techniques render themselves suitable for operando studies, as additional requirements often will apply. The main prerequisites for a characterization technique to become an operando method are that it might be carried out under any kind of atmosphere and that the timescale of the measurements is faster than the studied effects. Very often, however, the choice of suitable methods is, indeed, significantly limited by the fact that many techniques require UHV conditions or long-range structural order in the catalysts, so that they do not provide realistic operando fuel cell measurements. In the fuel-cell-relevant literature, in situ X-ray diffraction [24,25], small-angle X-ray scattering [26], Mössbauer, and X-ray absorption spectroscopy (XAS) measurements have been reported. With respect to structural changes, in situ X-ray diffraction (XRD) could be a method of choice [24,25], whereas techniques such as Fourier transform infrared (FTIR) [27] and optical sum frequency generation (SFG) directly reveal CO and other molecular adsorption sites [28].

To follow both structure and adsorbates concurrently, only X-ray absorption spectroscopy (XAS) has proved to be an appropriate technique. X-ray absorption spectroscopy (XAS) is best described as a very versatile tool, which requires neither UHV conditions nor long-range order in the material [29]. These enviable features make it a suitable method for operando studies of the kind of nanostructured materials, which are used in heterogeneous catalysis and electrocatalysis [30,31]. By analysis of the extended X-ray absorption fine structure (EXAFS) in the absorption spectrum, information on the catalyst's geometric structure can be obtained and expressed by the number of nearest neighbors (NNs) of the absorber atom and their distance ($R$) from it. Focusing on the systematic analysis of the X-ray absorption near-edge structure (XANES) region, the novel $\Delta\mu$ XANES approach [32,33] furthermore provides information about adsorbates on the catalyst surface that can be applied under operando conditions, thus shedding light on the kinetics of the oxygen reduction reaction (ORR) or methanol oxidation reaction (MOR) process, as well as changes due to diffusion limitations and other factors.

A select choice of in situ XAS investigations are reported below giving lively proof that a variety of problems in fuel cell research can be addressed by the XAS approach. Thus, for instance, structural changes in the catalysts under different atmospheres and at various temperatures are examined, and in an electrochemical environment, the impact of H, OH, and CO adsorption at different electrochemical potentials on the catalyst structure is followed. Finally, more recent operando results are presented.

### 17.3.2 Selected Fuel-Cell-Related Ex Situ and In Situ XAS Investigations

A typical example of early work in different gas atmospheres (we refer to these as *ex situ* since they are in fact not in an electrochemical environment) has been reported by Nashner and coworkers [34], who applied XAS to study carbon-supported Pt–Ru nanoparticles synthesized from the molecular cluster precursor $PtRu_5C(CO)_{16}$. Their work revealed a pronounced sensitivity of the bimetallic system toward exposure to $H_2$ and $O_2$ resulting in significant structural fluctuations. On the basis of ex situ EXAFS measurements in combination with transmission electron microscopy (TEM), a detailed structural model of the nanoparticles was proposed, and these results were compared with data under hydrogen and oxygen atmosphere. Under $H_2$ atmosphere, Pt appeared to show a pronounced preference for segregation to the particle surface, as suggested by an increase in the Pt–Pt coordination seen the EXAFS data. After chemisorption of oxygen at room temperature, a significant reduction in this amplitude was observed along with the formation of metal–oxygen bonds. The exact origin of these changes was unclear in those early XAS experiments. On the basis of today's knowledge, it might be assumed that the changes observed by Nashner et al. were due to morphology changes during gas adsorption, as reported and further discussed elsewhere [35,36]. Further investigations at various temperatures [37] showed that the Pt–Ru nanoparticles were initially present in a highly disordered structure, in which platinum is found preferentially at the core of the condensing particle. The nanoparticles adopted the inverted structure described above only when subjected to higher temperatures.

In another early ex situ XAS study, catalyst–support effects were systematically investigated by Hills et al. [38] using carbon black, fullerene soot, or desulfurized carbon black, respectively, as different support materials. EXAFS data recorded under hydrogen and oxygen atmosphere indicated that the nanoparticle behavior does not change significantly with the carbon support used. The authors claimed that the influence of the support materials can only be seen by the presence or absence of Ru–C interactions. In the case of the fullerene soot, the only visible catalyst–support interaction was the proposed formation of a Ru carbide phase at the nanoparticle–support boundary.

In 1999 Lampitt et al. [39] published results for the combination of both gas-phase measurements and experiments in an electrochemical environment with a view to studying the structure of carbon-supported Pt electrocatalysts for the methanol oxidation reaction. Pt $L_3$-edge EXAFS data indicated that the selected synthesis procedure involving oxidation of the $Na_6Pt(SO_3)_4$ precursor formed highly dispersed $PtO_x$ species on the carbon support. After reduction in hydrogen in a reactor or potentiostatically in an electrochemical cell with aqueous electrolyte, the $PtO_x$ species was reduced, forming small platinum particles, as could be seen from the presence of Pt–Pt neighbors in the first coordination shell. A Pt–C contribution from the catalyst support was also evident in the first coordination shell.

In situ investigations in an electrochemical environment in fact dominate the fuel-cell-relevant literature, with many fewer ex situ studies in the gas phase at various temperatures and atmospheres. In 1995, McBreen and coworkers [40] investigated changes in the structure of Pt–Ru electrocatalysts under applied potential in an aqueous electrochemical cell using 1 M $HClO_4$ as supporting electrolyte. Spectra of the Pt $L_3$ edge confirmed an alloyed structure with the addition of ruthenium increasing the Pt *d*-band vacancies (i.e., electrons moved from the Pt toward the Ru atoms), as visible in the white-line feature, and decreasing the Pt–Pt bond distances. Data obtained from the Ru *K* edge were less conclusive, largely because of a significant excess of unalloyed ruthenium in the catalyst, as is not uncommon in Pt–Ru systems as reported in other studies [41,42]. XANES results at various potentials suggested that alloying with Ru alters (most likely reduces) the adsorption of hydrogen on platinum. Furthermore, an indication of $RuO_x$ formation was observed on analysis of the EXAFS region. With respect to their results, the authors suggested that the bifunctional mechanism needs to be modified to take into account the effect of the increase in Pt *d*-band vacancies on the adsorption of H and CO residues.

Different carbon-supported Pt–M electrocatalysts (M = Cr, Mn, Fe, Co, Ni) for the hydrogen oxidation reaction were also studied in situ by Mukerjee et al. in 1996 [43]. In comparison with a pure Pt catalyst, alloyed Pt–M showed changes in the *d*-band vacancies along with a pronounced lattice contraction. After repeated cycling between the double-layer region ($\sim$ 0.5 V vs. RHE) and 0 V versus RHE, significant changes in the XAS spectra of the pure Pt catalyst were observed, whereas no changes were identified for the alloyed systems. XANES data of the Pt catalyst, however, indicated the generation of empty electronic states in the vicinity of the Fermi level due to the adsorption of hydrogen. Alloying of the Pt with first-row transition metals was found to suppress both the electronic and structural effects at 0 V.

In early work by Aberdam et al. [44] and more recent studies by Maniguet and coworkers [45], the sensitivity of the near-edge region for probing adsorbate structures was discovered and subsequently applied to study the electrooxidation of CO on Pt(Ru) nanoparticles. Aberdam et al. observed only small changes of the XANES region with electrode potential in CO, whereas potential cycling under inert-gas atmosphere initiated severe changes in the shape of the spectra. The authors assigned these changes to differences in the electronic and geometric environment predominantly of the Ru atoms. In contrast, the work of Maniguet et al. [45], focused more closely on the potential dependence of the EXAFS region measured for a carbon-supported Pt electrocatalyst exposed to CO. The data were analyzed using the difference file method in order to separate the dominant contributions of the Pt neighbors from those of the adsorbed species. By detailed examination of the remaining non-Pt neighbors, the presence of adsorbed CO with a Pt–C distance of 1.85 Å at potentials below 0.5 V versus RHE was indicated. Increasing the potential above 0.5 V resulted first in the removal of the adsorbed CO and—at more positive potentials—in the formation of an oxide layer, as indicated by the presence of a Pt–O coordination shell at 2.00 Å. However, contributions from the C neighbors of the carbon support prevented the accurate determination of the effect of the applied potential on the CO bond length, as well as the independent discrimination between the various possible adsorption sites, such as linear or bridge-bonded CO.

### 17.3.3 The First Operando XAS Experiments

O'Grady and coworkers [46] were the first in 2001 to investigate catalysts in real operating fuel cells with XAS. They compared ex situ XANES data of two different PtRu catalysts with both the respective in situ spectra at a potential of 0.1 V versus $Pd/H_2/H^+$ in an aqueous electrochemical cell and after operation in a direct methanol fuel cell (DMFC). The as-prepared ex situ catalysts were found to be predominantly oxidized. However, after being placed in an electrochemical cell and held at methanol oxidation potentials, as well as after DMFC operation, XANES data indicated the reduction of these metal oxides to their metallic form. Similar results were obtained by Maniguet et al. [45], who showed that for a well-mixed PtRu catalyst both Pt and Ru were metallic at potentials where hydrogen oxidation occurs. Both authors suggested that characterization of as-prepared ex situ samples has comparatively little relevance to the methanol oxidation activity of PtRu materials, and strategies for the synthesis and design of new tailor-made fuel cell catalysts.

The first operando studies on working fuel cells were reported in 2002, when Viswanathan et al. [47] published XAS data on the anode of a working reformate-air fuel cell. This was followed shortly after by Roth et al. [48,49] using XAS on both operating $H_2/O_2$ and direct methanol fuel cells. Although using different cell designs, both studies reported that under the reducing conditions of normal fuel cell operation, the standard Pt–Ru anode catalyst existed in a predominantly metallic state. Since then, many more in situ studies on fuel-cell-relevant systems have been published, most prominently work by the group of Mukerjee at Northeastern University [50,51] and the Russell group from Southampton [52,53] while operando investigations were still scarce (mostly work by the Smotkin group [54–56] and our own work from the Ramaker and Roth groups). The focus of these investigations has more recently shifted from conventional Pt and PtM electrode materials to newer electrode materials, and from a more structure-related view to the analysis of adsorbates on the electrode surface. Operando adsorbate coverages can be crucial to unraveling the fundamental mechanisms operational in real fuel cell operation. In the future, we expect the emphasis to be on the further development of the experimental techniques and instrumentation in order to obtain additional insight into time- and site-dependent processes on operating fuel cells.

## 17.4 XAS METHODS

### 17.4.1 Instrumentation

As can be concluded from the discussion above, X-ray absorption spectroscopy (XAS) is a very powerful approach providing detailed insight into the working fuel cell, without the need for ultrahigh vacuum (UHV) or long-range structural order. However, since the measurements require a radiation source that is both intense and tunable in energy, they can be conducted only at large-scale synchrotron radiation facilities. This limits their availability and largely prevents their application as a routine analysis tool.

One less obvious advantage of XAS is the method's flexibility. A typical setup for a transmission experiment consists of three ionization chambers in order to monitor the beam intensity before the sample, after the sample, and after a reference foil, which is used for energy calibration (Fig. 17.4, center part). In a fluorescence experiment, the intensity of the X rays before the sample is measured with the first ionization chamber. However, in contrast to the transmission experiment, the sample is placed at a 45° angle to the path of the incident X rays, so that the maximum solid angle of the fluorescence may be collected at the solid state detector. Because of the very flexible

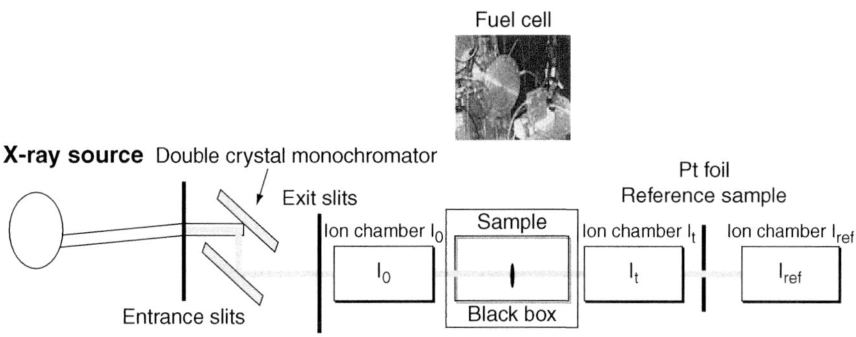


**FIGURE 17.4** Illustration of the placement of the three ion chambers relative to sample and Pt foil to provide transmission XAS signal for Pt in the fuel cell and for a Pt foil to provide energy calibration.

beamline setup, ample space is available between the first and the second ionization chambers thus allowing for bulky equipment such as a fuel cell test bench (Fig. 17.4). This allows the researcher to freely choose many independent parameters, enabling anode and cathode catalysts to be investigated in a low-temperature operating PEMFC using hydrogen reformate ($H_2$ re-formed from a carbon base fuel) and liquid or vaporized methanol or ethanol as anode feeds under a wide range of temperature, gas flow, and potential conditions—conditions comparable to the setup applied for fuel cell tests in the laboratory.

Some limitations still apply, however, and some compromises have to be made in cell design, metal loading, liquid fuels, and counterelectrodes. When making measurements on CCMs in an aqueous electrolyte for example, thin-layer geometries are favored to minimize the contribution of the absorbance by the liquid electrolyte. In their review paper, Russell and Rose [52] give a nice overview of the different cell designs that have been used in the past. These include gas treatment cells for transmission experiments, thin-layer electrochemical cells, and different operando cell designs for real fuel cell measurements. The design of the operando cell is crucial for obtaining quality spectra. For example, bubble formation must be avoided, as it can cause spectral noise and abrupt changes in absorbance. This is true for bubbles in the liquid electrolyte as well as bubbles emerging during fuel cell operation in direct methanol mode. In the former case, bubbles can be prevented by keeping the whole assembly under compression. However, in the DMFC when the methanol oxidation is catalyzed, $CO_2$ bubbles will evolve and push the liquid fuel aside and thus lead to changes in the absorbance. This drawback is still not completely solved.

Metal loadings higher than normal are often utilized to improve the signal-to-noise ratio (SNR) in the XAS spectra. Under these high loadings, the catalyst utilization in the membrane electrode assemblies begins to become an issue. This is a particularly important consideration given that EXAFS is an averaging technique; the whole catalyst layer should be in the same electrochemical or chemical state to facilitate the interpretation of the results. This is not as critical in the $\Delta\mu$ technique because nonactive catalyst (i.e., catalyst not exposed to electrolyte) will not change with potential, and therefore its contribution to $\mu$ will simply cancel when taking the difference. Currently, thin-film electrodes with Pt loadings as low as 0.1 mg Pt/cm$^2$ are used in fuel cells in an effort to maximize utilization. One way to approach these loadings for operando spectroscopic experiments, while still maintaining a sufficient SNR in the XAS spectra, is to operate the cell in fluorescence mode. In 2005, Wiltshire and coworkers [53] developed a miniature PEM fuel cell enabling XAS data to be collected from catalyst layers with low metal loadings and high catalyst utilization.

There is still another obstacle to be solved when studying an MEA in a real fuel cell in transmission geometry. In most cases, both the anode and the cathode catalyst contain platinum. In a Pt edge spectrum, contributions from the anode and the cathode will be measured simultaneously and cannot be distinguished. Different approaches have been reported in the literature, starting from removal of a small portion of the counterelectrode in the beam path, application of a fluorescence setup,

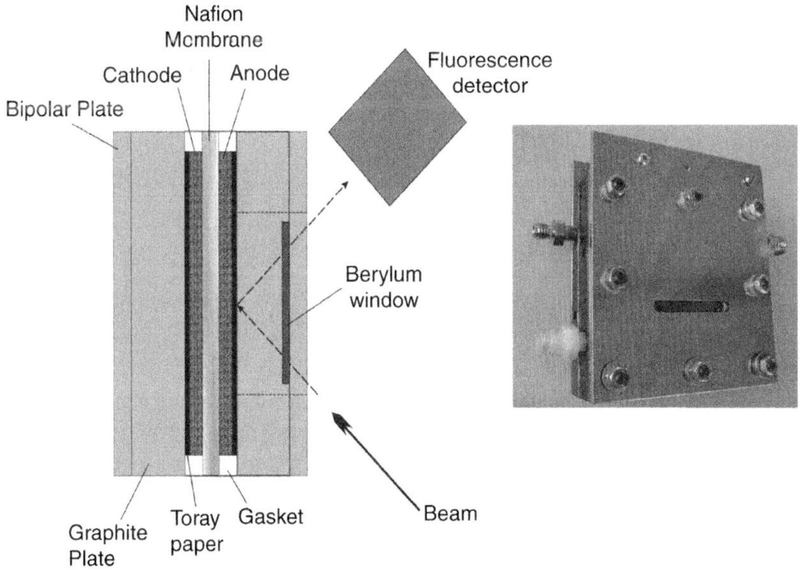


**FIGURE 17.5** Illustration of cell design for fluorescence-mode XAS and photograph of assembled cell. (From Ref. 49.)

where the spectrum should be dominated by the face electrode, and replacement of the Pt/C at the cathode by a Pd/C catalyst leaving only Pt/C at the anode. However, some of these modifications may alter the results. A review of different cell designs can be found in Reference 52 as well as detailed background information in a review by Principi et al. [57].

In the work summarized below, we focus on operando experiments in a fuel cell environment as close to real working conditions as possible. The cell design, illustrated in Figure 17.5, has been modified to allow for experiments in transmission and fluorescence mode with the optimization process described above. Although it is true that similar information can sometimes be obtained more conveniently in an electrochemical cell or half-cell, significant differences in the environment of the electrode surface can exist in these systems as summarized above. These include the extent of hydration of the electrode surface, the interface with Nafion, introduction of the $H_2$ gas via a gas membrane versus through the electrolyte, the acidity of the electrolyte, and the dynamic flow of current. We particularly emphasize the latter current effect on the adsorbate coverage below.

### 17.4.2 Data Analysis: $\Delta\mu$ XANES, FEFF Calculations, and EXAFS

New data analysis techniques allow us not only to extract structural information such as catalyst particle size, composition, morphology, and oxidation state from the extended region after the absorption edge (EXAFS) but also offer the unique and powerful advantage of providing in situ adsorbate coverage and binding-site information by a novel subtractive–normalization technique in the near-edge region, the so-called Δμ XANES approach [58,59]. Combining the information from both evaluation techniques, in situ information on oxidation states, adsorbates, and structure can be obtained in just one experiment, nicely complementing such standard ex situ methods as X-ray diffraction (XRD; structure) and X-ray photoelectron spectroscopy (XPS; oxidation states) and in situ–capable Fourier-transformed infrared spectroscopy (FTIR; adsorbed species).

#### *17.4.2.1 Δμ XANES Technique*

Full details of the data processing and analysis of the spectra are beyond the scope of this chapter; therefore, only a summary is given here. The interested reader is referred to Reference [60] or the original works for full details.

The Pt $L_3$, or other metal core-level, XANES is sensitive to the adsorption of various adsorbates, as is shown schematically in the analysis procedure in Figure 17.6b. The total absorption signal from XAS after preedge background removal is normally referred to as μ. The preedge background is removed, followed by normalization over the 50–150 eV range (relative to the edge) typically for XANES analysis. The normalized data are then energy-calibrated using foil data of the same material studied (Pt foil, Ru foil, etc. for Pt or Ru edges, etc.). This energy calibration is the most crucial for the success of the Δμ XANES technique, as full cancellation of the atomic contribution (i.e., that contribution coming from the atom itself and not containing any

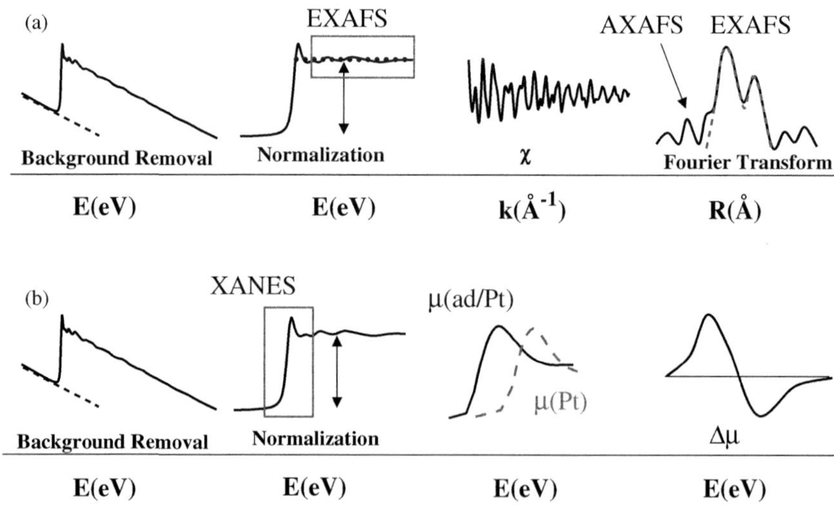


**FIGURE 17.6** (a) Schematic showing the removal of background and normalization of the raw XAS data, extraction of the EXAFS signal (χ), and finally least-squares fit of model functions to the Fourier transform of χ to obtain the EXAFS parameters. Also indicated here is the AXAFS feature at low *R*. (b) Schematic showing the removal of background and normalization, and then the difference between μ taken with and without adsorbate in the XANES region to give the Δμ signal. (From Ref. 63.)

bonding information) in the XANES must be achieved when taking the difference to obtain $\Delta\mu$. This atomic contribution dominates the spectrum so that the isolated $\Delta\mu$ is typically only about 3–5% of the total $\mu$ signal. However, the small $\Delta\mu$ is seldom problematic here, since the noise level after smoothing is generally much smaller (around 0.2%). The oscillations utilized in EXAFS are even smaller than those in the XANES, particularly at large $k$ ranges, and yet EXAFS is a well-established technique, which was first developed in the early 1980s or before, when less brilliant light sources were available. As the high brilliance of modern synchrotrons becomes more readably accessible, the noise levels and reproducibility of the data are expected to improve, making the $\Delta\mu$ technique even more powerful.

Figure 17.6b shows two different XANES spectra. Here $\mu$(Ad/Pt) is the Pt $L_3$ edge spectrum in the presence of the adsorbate and $\mu$(Pt) the $L_3$ edge spectrum at some different potential or current, where the adsorbate coverage is either much smaller, or at least very different from the initial spectrum so that the effect of the adsorbate can be isolated. The $\Delta\mu$ technique can therefore be considered a subtractive technique with the difference $\Delta\mu$, given as $\Delta\mu = \mu(\mathrm{Ad/M}) - \mu(\mathrm{M_{clean}})$ (see Fig. 17.6b). Essentially, careful normalization [61,62] and subtraction of the XANES signals at different applied potentials from the "clean" potential (i.e., the double layer) will result in a spectrum that has completely eliminated the underlying chemically un-reactive bulk signal. Hence the resulting $\Delta\mu$ line shapes reflect the chemistry, site symmetry, and magnitude of adsorbed species on the surface. It should be further noted that contributions in $\mu$ coming from surface sites not available for adsorption (e.g., those in contact with the support or with neighboring particles, those covered by surface islands or just not in the region of the required electrochemical three-phase boundary) will cancel out when taking the difference, so the $\Delta\mu$ technique can even give an indication of the accessible surface Pt sites as shown previously [63,64].

The resulting experimental difference spectra obtained by the $\Delta\mu$ technique do not offer much information by themselves. Indeed, they require proper interpretation by comparing theoretical $\Delta\mu$ spectra to the experimentally derived curves, such as illustrated in Figure 17.7. Good agreement of the experimental $\Delta\mu$ signature with the theoretical $\Delta\mu$ signature confirms the validity of the entire technique, and sometimes allows the binding site to be determined, since a strong dependence on binding site is found in specific cases. To generate the theoretical $\Delta\mu$ spectrum, first a theoretical XANES spectrum of a 3D model of the "clean" (adsorbate-free) cluster is generated using the structural information gleaned from a full analysis of the experimental EXAFS spectrum of the material of interest. The closer the model structure is to the experimental one, the more likely that the theoretical "signature" will match the experimental data. XANES spectra obtained by covering the aforementioned "clean" cluster model with differing adsorbed species in different adsorption geometries, such as atop or bridge-bonded, are then subtracted from the clean theoretical XANES spectrum to obtain the theoretical $\Delta\mu$. Comparison between experiment and theory then allows for interpretation of the experimental $\Delta\mu$ to provide adsorbate information such as the coverage and binding-site geometry.

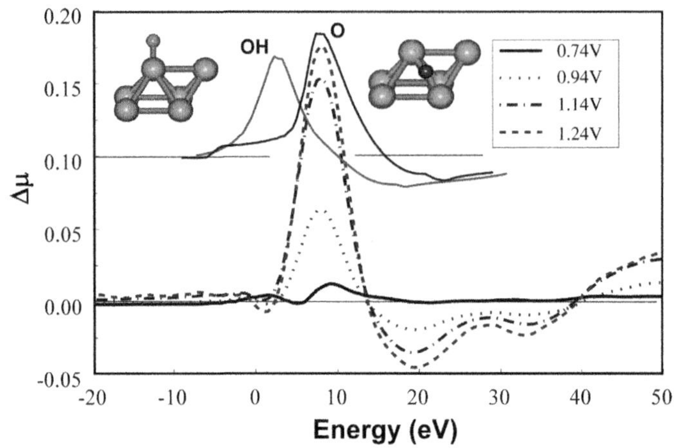


**FIGURE 17.7** Pt $L_3$ edge $\Delta\mu$ for a Pt electrode in $HClO_4$ at the indicated potentials using 0.54 V as reference. Theoretical signatures obtained from FEFF8.0 in the manner described in the text are shown for OH (atop) and O (threefold) binding sites. Note the increasing magnitude of $\Delta\mu$ with potential reflecting the increasing coverage with potential and the changeover from a mix of OH and O at 0.74 V to mostly O at higher potential. (From Ref. 58.)

Theoretical calculations using the FEFF8 [65] code are performed to calculate $\mu$ for the model clusters and prove the significance of the experimental $\Delta\mu$ results by comparison with the resultant theoretical $\Delta\mu$. FEFF8 performs real-space full multiscattering calculations utilizing a muffin-tin potential calculated with the Hedin–Lundquist exchange correlation approximation, and implements self-consistent field (SCF) potentials for determination of the Fermi level and the charge transfer. Thus, the difference $\Delta\mu$ between a spectrum with and without an adsorbate can be calculated by performing the same $\Delta\mu$ difference as obtained experimentally, that is, $\Delta\mu = \mu(Ad/Pt_6) - \mu(Pt_6)$, where, for illustrative purposes here, $Pt_6$ is a cluster of six Pt atoms as shown in Figure 17.7.

The $Pt_6$ cluster has been used most often [66], as this highly asymmetric cluster provides for all of the possible common binding sites (fcc, hcp, bridged, and atop), does not introduce any "surface resonances" that might arise from a more symmetric cluster, and yet is sufficiently large to account for the change in Pt–Pt scattering and new Pt–Ad scattering introduced by the adsorber atom. Agreement between the theoretical and experimental signatures, although seldom quantitative, is sufficient to determine the binding site. Since the late 1990s the $\Delta\mu$ technique has been applied to XANES data obtained by several different research groups on several different synchrotrons (NSLS, HASYlab, Grenoble, etc.) from well over 50 different Pt and $Pt_nM$ (M = Ru, Mo, Sn, Fe, Co, Ni, etc.) catalysts having a wide range of particle sizes (0.5–5 nm), morphologies (geometric shapes and M atom distributions), and supports (Vulcan carbon, metal oxides, etc.). As a result, generally reliable and reproducible $\Delta\mu$ signatures, sufficient for easy distinction between H, O, OH, and CO in different binding sites, have been identified. Some of these "characteristic" $\Delta\mu$ signatures are summarized below.

### *17.4.2.2 Characteristic Δμ Signatures*

In this section the Δμ technique is illustrated utilizing one of the most common systems in electrochemistry, a polycrystalline Pt electrode in acidic electrolyte as investigated in an electrochemical cell. Of course, this situation is also very much similar to a porous gas diffusion electrode, which is generally applied in a low-temperature fuel cell, and will also be discussed below. In these experiments, several adsorbates could come down onto the Pt surface, depending on potential and other working conditions, and compete for adsorption sites including H, O, OH, and CO. A strength of the Δμ XANES technique is that the competition between these different adsorbates for free Pt sites can now be followed during operation.

*O and OH Adsorption in an Electrochemical Cell* As an example of some Δμ signatures, the theoretical results in Figure 17.7 show the importance of changes in the Pt $L_3$ XANES region due to the influence of chemisorbed O with increasing coverage. When O species are adsorbed in an atop position versus the three-fold (fcc) surface sites, the Pt–Pt scattering changes in a different way as discussed above, so that the total becomes sensitive to the specific adsorption site. As the theoretical calculations show, it is possible to distinguish between atop and $n$-fold O by their so-called fingerprints. Oxygen in the atop position has been assumed to correspond to OH (since the scattering from H is negligible), while oxygen in an $n$-fold position is treated as O. This is consistent with chemical intuition and density functional theory (DFT) calculations [6,7], which show that OH prefers to be singly coordinated, whereas O prefers a doubly or triply coordinated position. Unfortunately, for bridged, fcc, and hcp sites, overlapping fingerprint curves are obtained (not shown), making it impossible to distinguish among these sites. However, the sensitivity of the Δμ XANES technique is sufficient to distinguish adsorbed atop OH species from $n$-fold O, which is critical in fuel cell work. Figure 17.7 also shows how the amplitude of the Δμ signature increases with increasing O coverage, and for the most part we assume that the Δμ magnitude is representative of relative adsorbate coverage, although we have discussed deviations from this previously [64,68].

It is also possible to identify the neighbor atoms to the OH adsorption sites. It has been observed in previous work that the signature curves are significantly affected when atop OH is adsorbed on a Pt atom near another "foreign" metal atom such as Ni, Co, or Ru. The fingerprint is found to be shifted downward in energy compared to OH adsorbed at a Pt site surrounded only by Pt atoms (see Fig. 17.8). Thus, the Δμ XANES can provide the following information: (1) it indicates the coverage and kind of adsorbates found on the surface, (2) it indicates the binding site (e.g., atop OH or $n$-fold O), and (3) in the case of atop OH, it even indicates whether the Pt site has another atom (e.g., Ru) as a neighbor instead of only Pt neighbors. The latter feature renders the Δμ XANES a suitable technique for unraveling complex reaction mechanisms and synergetic actions, as they are often active in bimetallic catalysts.

*CO, H, and OH Signatures on an Active Fuel Cell Catalyst* The utility of the Δμ method depends on the availability of a good reference spectrum, most favorably a

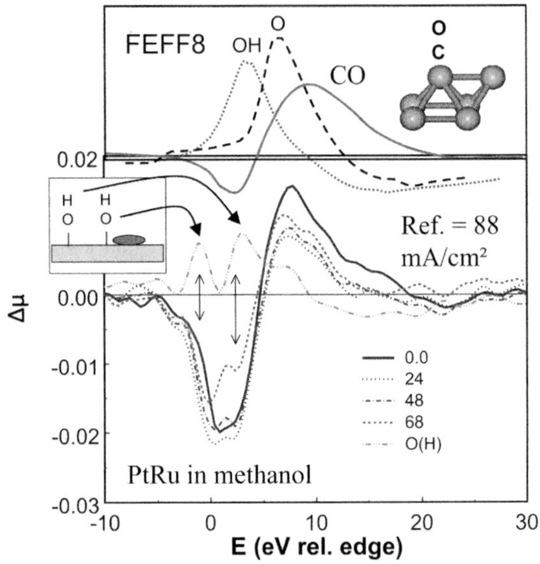


**FIGURE 17.8** $\Delta\mu = \mu(i) - \mu(88\ \text{mA/cm}^2)$ obtained from Pt $L_3$ XAS data at a PtRu anode in methanol at the indicated currents in mA/cm$^2$. Theoretical $\Delta\mu = \mu(\text{ad/Pt}_6) - \mu(\text{Pt}_6)$ results obtained from FEFF8 calculations for the indicated adsorbates on a $Pt_6$ cluster are also shown. The features arising from atop CO, fcc O/Pt, and atop OH/Pt near a Ru island and away from it are illustrated. (From Ref. 63.)

spectrum obtained from the blank, adsorbate-free metallic surface. Unfortunately, for operando measurements in a working fuel cell the situation is not so straightforward, as it is difficult to find an appropriate reference spectrum where no species, or at least a controlled amount, are adsorbed on the surface. The specific measures taken for methanol and reformate operation in a working fuel cell, that is, when there is CO, H, and O[H] coadsorption, are described below. More specifically, we illustrate the $\Delta\mu$ XANES technique to follow the coverage of specific adsorbates on carbon-supported Pt and different Pt/Ru bimetallic systems in pure $H_2$, in simulated reformate (150 ppm CO in $H_2$), and in direct methanol operation as a function of current in an operating PEM fuel cell.

Figure 17.8 shows Pt $L_3$ $\Delta\mu$ data for a PtRu catalyst in 0.3 M methanol using the data obtained at 88 mA/cm$^2$ as the reference, and also shows theoretical signatures for CO, OH, and O in order to identify the changes seen in the experimental results with potential. Clearly, CO is leaving as O species build up on the surface. Note the three-peak structure between 0 and 6 eV. From the discussion above concerning OH and O, the peaks correspond respectively to atop OH/Pt near a Ru island, atop OH/Pt away from the islands (as illustrated), and bridged or fcc/hcp bonded O/Pt as illustrated. These three O(H) peaks can be separated from the CO signature by simply scaling out the CO contribution as determined around 10 eV. Then the magnitudes of the isolated O(H) signature can be estimated from the individual peak heights and their buildup followed with potential.

In simulated reformate ($H_2$ with 100 ppm CO), determination of the simultaneous relative abundances of $H_{opd}$, $H_{upd}$, and CO, is even more complicated than in methanol operation. In the case of an operating fuel cell using reformate, the anodic surface is always at low potential, comparable to the situation in the known H adsorption region. Therefore, we assume that the surface has to be covered to some extent with $H_{upd}$ at all currents. When using pure $H_2$ as fuel, the only adsorbate seen with the $\Delta\mu$ technique under these conditions is H in an atop position on an already H $n$-fold-covered surface (i.e. $H_{opd}$ covered with $H_{upd}$). This is because at low potential the $H_{upd}$ contribution cancels out in the $\Delta\mu$, since the $H_{upd}$ coverage is relatively constant. Further, the highest operation current is chosen as the reference current, as this is the current at which the potential is highest, and therefore the $H_{opd}$ coverage should be lowest. In the simulated reformate case, however, CO is found to displace $n$-fold H, resulting in an experimental signature for $\Delta\mu$ depending on the extent of this exchange. Consequently, the experimental $\Delta\mu$ observed in reformate operation needs to be fit to all three species, H $n$-fold, CO, and H ontop, with linear regression in the range 0–35 eV. Details on this regression technique have been presented elsewhere [68]. Results from this work are shown in Figures 17.9 and 17.10.

### 17.4.2.3 Methodology for EXAFS and AXAFS Techniques

Since the EXAFS analysis technique is a much better known and more established method, we shall present far fewer details on this approach. Many standard reference texts [69,70], papers [71], and Websites [72] exist fully describing this technique and well-known software packages, such as the IFEFFIT package and others [73], are available for data analysis. Figure 17.6a shows the basic essentials, whereby again the XAS data must have a background removed, then normalized, and energy calibrated much as described for the XANES. FEFF8 calculations are again performed on model clusters, and the many parameters such as the mean free path, scattering amplitudes,

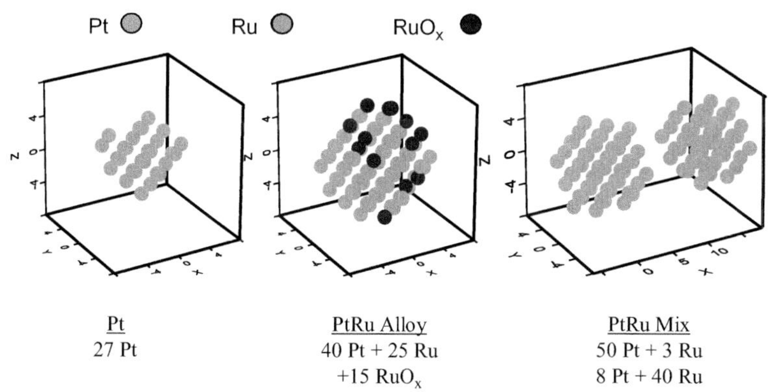


**FIGURE 17.9** Structural models estimated from EXAFS analysis for the indicated electrocatalysts prepared as described in the text. The PtRu mix catalyst was prepared by mechanically mixing Pt/C and Ru/C powder, with some atoms of the "other" element thereby being deposited on each cluster surface. (From Ref. 68.) (See insert for color representation.)

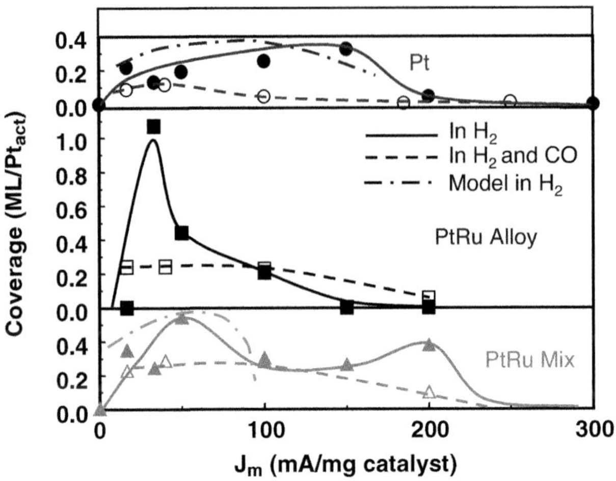


**FIGURE 17.10** Estimated experimental Hopd coverage (at $t =$ steady state) on Pt on the three catalysts illustrated in Figure 17.9, as a function of fuel cell current. The Hopd experimental coverage estimates were obtained from the $\Delta\mu$ XANES analysis described previously [68]. Model coverages were obtained from the kinetic model in the text and scaled for optimum agreement with experiment. (From Ref. 68.)

and phases are provided internally. In this work, we limit ourselves to two paths: Pt–Pt and Pt–M scattering for PtM bimetallic catalysts. Usually four variables are obtained for each path, namely, the Pt–M coordination number $N_{\rm PtM}$, the bond length $R_{\rm PtM}$, the Debye–Waller factor $\sigma$, and the $E_{\rm o}$ parameter, which establishes the inner potential. Assuming that both the Pt and M atom XAS data were measured, one can then obtain four coordination numbers, two at each edge, namely, the Pt–Pt, Pt–M, M–Pt, and M–M coordination numbers. These are the critical parameters that have been used to propose an average nanoparticle morphology [64,68] and given below for one case.

The total absorption cross section in XAS is normally written as $\mu = \mu_{\rm emb}(1 + \chi_{\rm EX})$, where $\chi_{\rm EX}$ is the EXAFS function. The absorbing embedded atom (i.e., the atom bonded to its environment) accounts for the major part $\mu_{\rm emb}$ of the absorption coefficient. The structure in $\chi_{\rm EX}$ enables the local geometry about the absorbing atom to be determined [69]. Holland et al. [74] and more recently Rehr et al. [75] noted that $\mu_{\rm emb}$ is not a totally smooth function, but can also have structure $\chi_{\rm AX}$ due to scattering from the periphery of the absorbing atom itself. Thus the atomic cross section of the embedded atom can be expressed as $\mu_{\rm emb} = \mu_{\rm free}(1 + \chi_{\rm AX})$ with $\mu_{\rm free}$ as the absorption cross section for the free atom in contrast to that for the embedded atom $\mu_{\rm emb}$. The $\chi_{\rm AX}$ function, called "atomic" XAFS [AXAFS], offers the possibility of obtaining electronic structure information utilizing a procedure comparable to EXAFS. The two scattering phenomena (AXAFS and EXAFS) are schematically illustrated in Figure 17.6a. The AXAFS represents the change in scattering by the absorbing atom itself relative to the free atom (see discussion below). This change is due to bonding of the absorbing atom with its environment; consequently, any changes in the environment, which changes the electronic charge or structure of the absorbing atom, should be reflected in the AXAFS. Changes in the Pt AXAFS magnitude due to adsorption of O or H [76,77] to increasing potential in an electrochemical cell [78], and to changes in the Pt particle support have been demonstrated previously [79]. Readers are directed to this original work for more details on isolation of the AXAFS from the data, as well as more details as to its source.

## 17.5 OPERANDO XAS STUDIES IN FUEL CELL RESEARCH: THE PRESENT

As noted above, the major difference between in situ and operando results is the coverage of adsorbed reactants and/or posions, which are determined by the diffusion rates and nearby gas concentrations. Below we give three examples of operando results using XAS. The first examines the anode electrocatalyst in $H_2$ and $H_2 + CO$ (reformate) operation, where the reactant intermediate is believed to be $H_{opd}$. The second case examines the fuel cell cathode where the intermediate is OH in the ORR. Finally, we show AXAFS results, which track polarization of the anode in methanol and $H_2$ as a function of the current, illustrating the importance of current flow in the operando experiment.

### 17.5.1 $H_{opd}$ On the Anode

The typical voltammogram of a polycrystalline Pt electrode in acidic solution shows characteristic features at potentials close to 0 V versus RHE, which arise from the hydrogen evolution reaction (HER; negative-going cycle) and the hydrogen oxidation reaction (HOR; positive-going cycle), together involving the overpotential deposited hydrogen ($H_{opd}$), where the hydrogen molecule gets adsorbed on the platinum surface and dissociates into two hydrogen atoms. In the underpotential region ($H_{upd}$) between ~0 and 0.3 V versus RHE, the electron transfer is catalysed, turning the adsorbed hydrogen atom into a proton and an electron.

So far, the exact nature of the adsorbed $H_{opd}$ involved in the HOR/HER reactions on the Pt surface has been subject to controversial discussions. Some studies corroborated the idea that the $H_{upd}$ and $H_{opd}$ are in the same Pt binding site, but with $H_{opd}$ having a weaker binding energy [80,81]. However, spectroscopic evidence for an atop H binding site appearing at higher H coverage has been reported by several other investigators [82–84]. Thus, strong evidence exists from optical spectroscopy that the $H_{opd}$ is ontop H (meaning atop H in the presence of full coverage of fcc H) in a binding site different from the $H_{upd}$. In a study [85] using the $\Delta\mu$ XANES, H was found in an ontop binding site on Pt and attributed to the $H_{opd}$ intermediate directly involved in the hydrogen evolution and oxidation.

In fuel cell operation, the different H adsorbates should also be present on the Pt and Pt/Ru surfaces, although it is expected that the reaction kinetics (Volmer–Tafel and Volmer–Heyrovsky kinetics) are affected by variables such as surface morphology (smooth well-defined surfaces vs. real supported powder catalysts) and experiment design (liquid electrolyte vs. polymer electrolyte). Consequently, a study in the exact same sample environment as in the real fuel cell setup is expected to provide some additional insight, which seems indispensable. When pure hydrogen is replaced by either reformate or methanol as anode feed in the fuel cell, additional adsorbates will be found on the surface. Depending on the potential, hydrogen will have to compete with both CO and O[H] for adsorption sites. CO attaches strongly to platinum surfaces, which is well known as the CO poisoning effect on electrocatalytically active surfaces. Since CO adsorbs in an atop site, both hydrogen and O [H] are expected to compete with the stronger adsorbate CO for adsorption sites or to interact with each other. On Ru sites, competition between CO and O[H] for adsorption sites should already start at lower electrode potentials, since Ru is reported to catalyze the water activation at potentials significantly more negative than Pt. Moreover, like Pt, ruthenium sites should also be active for H and CO oxidation, although less so than Pt.

The $\Delta\mu$ XANES technique has been used to study the competition between H and CO for adsorption sites in two different PtRu catalysts and a pure Pt sample for comparison. Figure 17.9 shows the estimated average morphology of these three catalysts, which were applied to probe the hydrogen oxidation reaction (HOR) in 100 ppm CO in a working fuel cell. As the plain Pt reference, a standard carbon-supported Pt catalyst from E-TEK (20 wt% on Vulcan XC-72) was chosen and compared with both a traditional carbon-supported PtRu bimetallic electrocatalyst also from E-TEK (20 wt% Pt : Ru (1 : 1) on Vulcan XC-72; *PtRu alloy*) as well as a so-called carbon-supported PtRu mixture catalyst (*PtRu mix*). The latter was produced by mechanically mixing carbon-supported Pt and carbon-supported Ru (both from E-TEK) in an agate mortar, resulting in Pt nanoparticles with a small Ru surface component and similarly composed Ru particles with some Pt on the surface. The corresponding estimated morphologies presented in Figure 17.9 were obtained from the calculated Pt–Pt, Pt–Ru, Ru–Pt, and Ru–Ru coordination numbers using the EXAFS cluster modeling technique described elsewhere [68,86].

The specific average nanoparticle morphology has been found to be crucial to the catalyst's activity. For example, CO poisoning has been shown to strongly depend on the Ru distribution and island size. Scott et al. [86] have shown that PtRu catalysts with large Ru islands are mostly oxidized to $RuO_x$ and therefore are inert to CO adsorption but exert a large ligand effect on the nearby Pt. This means CO oxidation occurs primarily from OH adsorbed near the Ru islands, which we refer to as the ligand mechanism. Small Ru islands, on the other hand, experience a ligand effect reverse from that of Pt, are therefore much less reactive and primarily Ru metal in character. They therefore can adsorb CO and OH at low potentials and CO oxidiation therefore primarily involves OH on the Ru islands, which is generally referred to as the *bifunctional CO oxidation mechanism* [68,86]. Thus it is critical to know the nanoparticle morphology to understand behavior in a fuel cell; unfortunately, this is often lacking. Figure 17.9 shows that the PtRu mix has very small Ru islands, whereas the PtRu alloy catalysts have much larger $RuO_x$ islands. This will be critical to understanding the differences in $H_{opd}$ coverages below.

In order to fully understand the $H_{opd}$ coverage versus current, the kinetics must be carefully examined. We follow the kinetic analysis of Conway et al. [87]. The HER/HOR on Pt can be considered to proceed via the following steps with the rate constants indicated:

$$\begin{array}{llll} & \text{HER} \rightarrow \leftarrow \text{HOR} & & \\ \text{Volmer}: & H^+ + e^- \leftrightarrow H_{opd}/\text{Pt} & \text{rates } k_v/k_{-v} & \\ \text{Heyrovsky}: & H^+ + H_{opd}/\text{Pt} + e^- \leftrightarrow H_2 & \text{rates } k_h/k_{-h} & \\ \text{Tafel}: & 2H_{opd}/\text{PT} \leftrightarrow H_2 & \text{rates } k_t/k_{-t}, & \end{array} \quad (17.1)$$

The HOR reaction occurring at the anode in a fuel cell on Pt can be considered to follow either of two steps, consisting of the Tafel–Volmer (TV) and/or Heyrovsky–Volmer (HV) sequence. Conway et al. [87] and others [88] have solved the kinetic equations and given the coverage of $H_{opd}$ as a function of current, but these complex expressions do not need to be given here. It is often assumed [87,88] that the Heyrovsky rate is negligible, and the Volmer rate is small relative to the Tafel rate. Under these conditions the $H_{opd}$ coverage decays exponentially as one goes from 0 V to higher potentials [88,89], but as we will see below, this is not necessarily valid for different samples and situations. Furthermore, the equilibrium $H_{opd}$ coverage in electrochemical systems is found to decrease exponentially with potential from 0 to 0.2 V [32].

The results in Figure 17.10 indicate a very different coverage dependence, namely, an initial increase of coverage at low current, followed by a decrease at higher current. How can this be accounted for? It turns out that only for Pt(111), when the $k_t/k_v$ ratio is, indeed, very small (0.08) does the coverage decrease exponentially [85]. As this ratio increases, the fractional coverage in the HER region decreases and more importantly the ratio $\theta_{HOR}/\theta_{HER}$ increases. In addition, the Heyrovsky rates become important on small particles, and this increases the $H_{opd}$ coverage in the HOR region. The presence of Ru further increases the Heyrovsky rate and apparently even alters the $H_2$ diffusion limit near the islands. With reasonable rate coefficients, the $H_{opd}$ coverages shown in Figure 17.10 can, indeed, be explained as indicated by the model results shown [85].

In Figure 17.10, all samples show a decrease in $H_{opd}$ coverage above 200 mA/mg catalysts in either pure $H_2$ or $H_2$ + CO, as expected, because the Volmer reaction rate increases with potential (current) to eventually overcome the diffusion rate and Tafel adsorption rates. At lower currents, the Pt has a $H_{opd}$ coverage one-half or one-third that of the two PtRu samples; this is true even though the fractional CO coverage is near one in all cases. This arises as indicated above because the ligand effect from the $RuO_x$ islands increases the rate of the Heyrovsky reaction, which encourages higher $H_{opd}$ coverage. Note that although the PtRu alloy has a much higher $H_{opd}$ coverage in pure $H_2$ than does the PtRu mix (consistent with the greater ligand effect on the Heyrovsky reaction in the PtRu alloy from the bigger $RuO_x$ islands), CO poisoning decreases this $H_{opd}$ coverage more (consistent with the smaller reactivity of CO on the big $RuO_x$ islands). Finally, the PtRu mix catalysts shows two maxima (at 50 and 300 mA/mg cat.), the first reflecting more PtRu character, the second indicating more pure Pt character, consistent with the nonhomogeneity of the Ru island distribution in this mechanically mixed catalysts. Taken together, this explains the better behavior of the PtRu mix catalyst in reformate; the effects of CO in the reformate are less dramatic, and the HV sequence (not particularly desirable) is kept to a minimum.

The results above show that the rates of the individual reactions depend strongly on the particle morphology (size and presence of other metals such as $RuO_x$ islands and their size), and perhaps many other variables that can change the rate constants and hence relative rates of the TV and HV reactions and the $H_{opd}$ coverage. Only operando results, which include an estimate of reactant coverage, such as those in Figure 17.10, can allow a full sorting out of these many variables. Much more work needs to be done, but these results illustrate the importance of operando results, which is the critical point we want to make here.

### 17.5.2 OH on the Cathode

Another reaction highlighting the necessity of operando studies is the oxygen reduction reaction (ORR). As noted above, the primary adsorbed reactant intermediate in ORR is OH. The kinetic equations derived from the double-trap kinetic model of Wang et al. [89,90] for the 4e ORR in acidic media, include four elementary reactions:

$$\tfrac{1}{2}O_2 \rightarrow O_{ad} \qquad [\text{dissociative adsorption (DA)}] \tag{17.2}$$

$$\tfrac{1}{2}O_2 + H^+ + e^- \rightarrow OH_{ad} \qquad [\text{reductive adsorption (RA)}] \tag{17.3}$$

$$O_{ad} + H^+ + e^- \rightarrow OH_{ad} \qquad [\text{reductive transition (RT)}] \tag{17.4}$$

$$OH_{ad} + H^+ + e^- \rightarrow H_2O \qquad [\text{reductive desorption (RD)}] \tag{17.5}$$

These reactions can be summarized even more basically as follows:

$$\tfrac{1}{2}O_2 \rightarrow OH_{ad}(\text{via RA or via DA} + \text{RT}) \rightarrow H_2O\ (\text{via RD}) \tag{17.6}$$

and as shown by Wang et al., the RA mechanism is dominant below 0.7 V RHE. Thus the adsorbed OH intermediate is the most important.

To illustrate the differences between operando results and the in situ equilibrium results, Figure 17.11 shows $\Delta\mu$ magnitudes reflecting O(H) coverage as illustrated in Figure 17.7, but now in different environments. Experiments were performed in an electrochemical cell with either nitrogen or oxygen purging and the results compared to those from an operando fuel cell. The in situ $N_2$-purged results reflect the coverage of atop OH and fcc O due only to activation of water when moving the potential in the anodic direction. Below 0.7 V, OH exists primarily on corners and edges of the Pt clusters, and above 0.7 V fcc O is observed. Purging with $O_2$ shows the additional effect of $O_{ad}$ coming from reaction (17.2) above, causing the presence of $O_{ad}$ now beginning at 0.6 V and a slightly reduced coverage of OH. The PEM fuel cell operando results, now obtained at the cathode, show a much larger coverage of OH extending down to well below 0.3 V. To demonstrate that these results are consistent with the ORR kinetics explained above, we show theoretical model results from

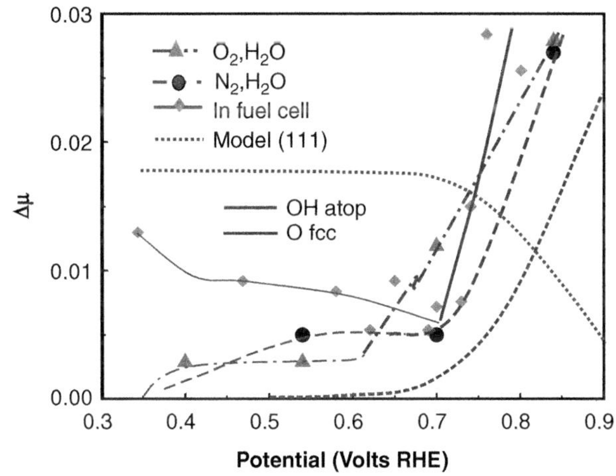


**FIGURE 17.11** Comparison of $\Delta\mu$ magnitudes [indicating O(H) coverage] with potential for Pt catalysts in water purged with $N_2$ or $O_2$ in the anodic direction and that in a working fuel cell at the cathode, along with coverage given by the Wang et al. model [89] for Pt(111) in a fuel cell. Coverage of atop OH indicated by the thin lines (mostly below 0.6 V) and coverage of fcc O by the thick lines as indicated by the $\Delta\mu$ signatures. The model results have been arbitrarily scaled to fit on the $\Delta\mu$ scale.

Wang et al. [88,89] using the double-trap model in reactions (17.2)–(17.5) for Pt(111). Unlike the experimental results, the model coverages are separated into OH and O. From the model, a transition between atop OH and O is proposed at about 0.8 V. In the experiment, however, transition points below 0.7 V are observed. The higher transition point between the dominant OH versus O coverage at 0.8 V is believed to reflect the lower reactivity of a Pt(111) surface compared to the 1–3-nm-supported Pt particles, which have been used in the experiments. Otherwise, the agreement in these results is remarkable, even though the model results given in ML have been arbitrarily weighted to fit on the $\Delta\mu$ scale in Figure 17.11. The increase in $\Delta\mu$ intensity below 0.4 V in the experimental PEMFC results is believed to arise from the adsorption of $H_2O_2$ (i.e., the breakdown in the 4e ORR producing some 2e ORR) due to the crowding of the surface from the adsorption of $H_2$ and other anions at these low potentials.

Figure 17.12 shows the typical Tafel plot of $V$ versus log $i$, and the two slopes, 125 V/dec below 0.8 V and 60 V/dec above. Wang et al. [90] have attributed this slope change to the change from DA + RT to RA dominance in the production of OH. Figure 17.12 shows that right at the slope change, the dominant coverage obtained from the $\Delta\mu$ XANES results switches from primarily OH to O on the surface consistent with this model. Figure 17.12 also shows that the sharp deviation from the 60 V/dec slope around 550 mV occurs because additional oxygen-containing species, in particular peroxide ($H_2O_2$), begin to come on the surface at this point. As mentioned above, the onset of $H_2$ adsorption crowds the surface and encourages the 2e ORR, which causes the current to fall with potential at an even faster rate. Thus, the operando fuel cell measurements support the validity of the Wang et al. model, and further allow us to directly study by a

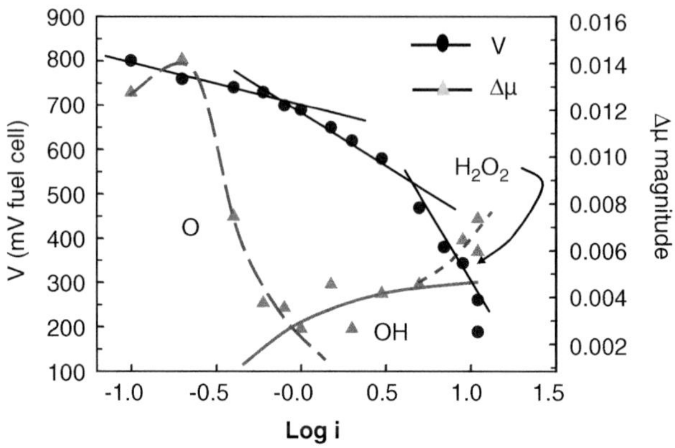


**FIGURE 17.12** Tafel plot ($V$ vs. log $i$) for a Pt fuel cell in $H_2$ along with the $\Delta\mu$ magnitudes (right axis) reflecting fcc O, atop OH, and $H_2O_2$ coverage on the Pt cathode. Note the change in slopes right where the dominant adsorbate changes character. (From Ref. 95.)

spectroscopic method the coverages existing in the diffusion-limited region and/or when other adsorbates such as $H_2$, $H_2O_2$, and other anions are on the surface. Such studies are critical to understanding the complex electrode–electrolyte interface, which determines the surface diffusion, the concentration of other adsorbates on the electrode surface, and hence the performance of a fuel cell.

### 17.5.3 Charge on the Pt Particles

One disadvantage of the conventional two-electrode full fuel cell measurements is a missing reference electrode as summarized in Figure 17.3 above. In that case, only the anode–cathode potential difference, $\Delta V = V_a - V_c$, is measured as a function of current. However, if one has measured the currents for the same fuel cell in both $H_2$ and methanol operation, one can estimate the anode potential in methanol by taking the difference, $\delta V = \Delta V(\text{methanol}) - \Delta V(\text{hydrogen}) = V_a(\text{methanol}) - V_a(\text{hydrogen})$, where it is assumed that the potential of the cathode or oxygen electrode cancels out since it is constructed similarly for both cases (this is illustrated in Figure 17.3). Further, since it is known that the hydrogen anode only varies between 0 and 0.03 V with current [91], the $\delta V$above is a reasonable approximation to the anode potential in methanol. This $V$–$I$ plot is given in Figure 17.13 (solid line).

We have shown previously that the magnitude of the AXAFS tracks with the charge on an electrode [78]. If, indeed, the increase in the AXAFS occurs as a result of polarization of the electrode, then the AXAFS in methanol should track with the anode potential. By plotting the AXAFS and the estimated anode potential in methanol on the same graph, this exact scenario can be seen in Figure 17.13. The remarkable correlation of the AXAFS amplitude with the estimated anode potential does, indeed, verify this interpretation for the AXAFS variation, namely, due to the polarization of the anode. Note the much smaller increase in AXAFS for $H_2$, consistent with the much smaller change in anode potential with current ($V < 30$ mV).

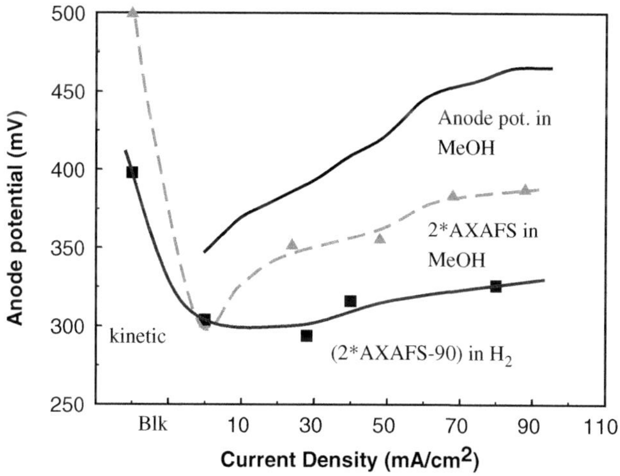


**FIGURE 17.13** Plot of the anode potential versus current density obtained as described in the text along with the AXAFS amplitude for anodes in $H_2$ and methanol. The AXAFS magnitudes have been scaled and shifted as indicated to place them on the potential scale. The values for "Blk" correspond to the AXAFS magnitudes obtained prior to exposure to the fuel, and illustrate the dramatic increase in the AXAFS when O is still present on the anodes. (From Ref. 63.)

The AXAFS magnitude (Fig. 17.13) is very large when taken before the electrode is introduced to the fuel (labeled blank in Fig. 17.13). As discussed above, before exposure to the fuel, the Pt electrode is highly oxidized. The O atoms strongly polarize the Pt via the field effect. The strong effect of O on the Pt atoms near it, and consequent enhancement of the AXAFS amplitude, has been discussed extensively before [76,77,79]. The remarkable agreement between the measured potential and variation in AXAFS suggests that the polarization effect is dominant under operando conditions, so that the change in coverage of $H_{opd}$ (or even possibly CO in the case of methanol) with current is not altering the AXAFS nearly as much as oxidation does. The results in Figure 17.13 are limited to the "kinetics" region at the anode (these regions are identified in Fig. 17.3). Studies of what happens at the cathode, or at higher currents at either electrode in the diffusion limited region, might be interesting and reveal conditions at the complex cathode–electrolyte interface as discussed in Section 17.2.1, but we are not aware of such studies to date.

## 17.6 THE FUTURE OF OPERANDO XAS

X-ray absorption spectroscopy is an attractive in situ and operando technique, as shown by both past and present work in the field. More recently, the more conventional EXAFS data analysis technique has been complemented by the newer $\Delta\mu$ XANES technique. With the latter it is now possible to determine adsorbate, specific adsorption sites, and adsorbate coverage, which helps a great deal in unraveling the kinetic mechanisms operating in working fuel cell systems. Atop CO, opd and upd H, bridge-bonded or fcc O, and even OH, all possibly present on a Pt surface, can now be monitored individually. However, the strength of XAS lies in particular in its ability to provide information on both nanoparticle morphology and adsorbate coverages simultaneously with just one technique. Thus, it enables us to directly probe the structure–activity relationship even in bimetallic catalysts, such as a state-of-the-art PtRu system.

The combined $\Delta\mu$ XANES–EXAFS studies have more recently been reported on other more complex electrocatalysts, including RuS [92], RhSe [93], Au/$SnO_x$ [94], Pt/$NbO_x$ [95], and Pt/TPPTP/C (TPPTP = triphenyl phosphine triphosphate) [96] catalysts. The $\Delta\mu$ XANES has also been used to study the effects of anion adsorption on Pt, including bisulfate [97] and Cl [98], differentiate the various oxidation states of S during the electrochemical oxidation of S on Pt [99], and identify acetaldehyde-type intermediates during ethanol oxidation on Pt [100]. In all of these studies, unprecedented new details on the coverage and binding sites of adsorbates in situ or operando have been revealed. Further, similar XAS studies in gas-phase catalytic reactors have been reported, including H on supported Pt clusters in zeolites and various flat supports [101] as well as ethylene on Pt and its different molecular geometries (atop π-bonded, bridged di-σ bonded, and a dehydrogenated vertical ethylidyne species) as the temperature was increased, revealing the ability of the $\Delta\mu$ technique to differentiate between different bonding geometries [102] on Pt, and even ethylene-type intermediates in the hydroepoxidation of propene over Au [103]. With the complexity of the systems and reactions that can be studied with the $\Delta\mu$ technique increasing more recently, biofuel cell investigations and biocatalysis experiments could come into focus soon, with some experiments already underway [104]. We expect a much wider application of the combined $\Delta\mu$ XANES–EXAFS techniques in the future.

The ultimate goal is to study shortlived intermediates on the surface of a catalyst to obtain even more information on the kinetics occurring in a complex system, such as, for example, a fuel cell. This will require XAS data to be taken at a much faster rate than the normal data summarized above (about 30–50 min). Time-resolved data can be obtained by either quick EXAFS (QEXAFS) or by energy dispersive XAS (EDXAS), both needing a dedicated beamline setup with specific additional components. Quick EXAFS (taken in seconds up to minutes depending on scan rate and applied monochromator setup) has been reported [105], but even the high speeds possible on the super-XAS beamline at the Swiss Light Source, as reported by Frahm et al. [106] are still too slow for most shortlived intermediates. Tada et al. [107] have reported operando time-resolved results on a Pt/C cathode in a fuel cell using a "time-gated" QEXAFS (TG-QEXAFS) procedure giving a time resolution of 1 s. Figure 17.14 shows a schematic of how the QEXAFS and potentiostat system were synchronized with a trigger signal sent from the potentiostat to the QEXAFS system allowing the data to be taken in 1-s intervals and effectively "splicing" together 15 different spectra. The procedure described assumes that the fuel cell does not change with each 1-s cycle, but the results suggest that this is not a problem. Using this procedure, Tada et al. were able to watch the Pt–Pt and Pt–O coordination numbers change over a 120 s time period

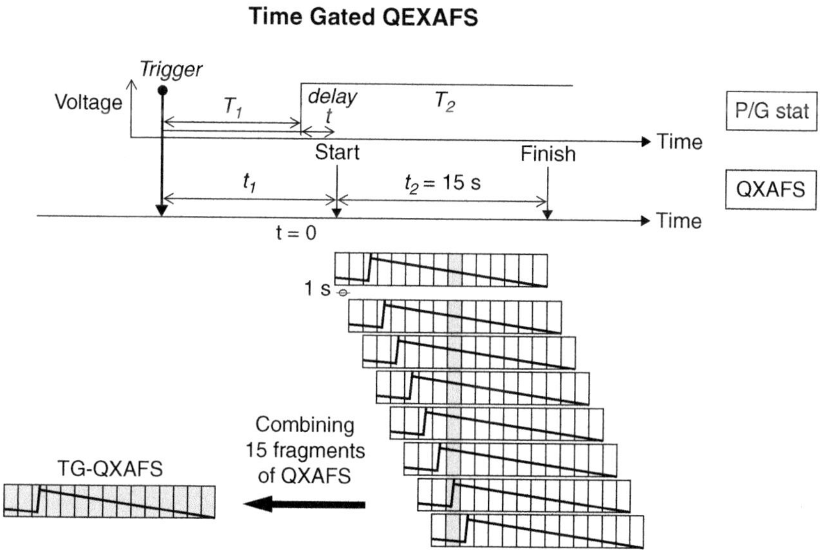


**FIGURE 17.14** Illustration of the time-gated QEXAFS procedure utilized by Tada et al. [107].

at a 1–10 s time resolution, and they used their insights to propose a model for the changing nanoparticle structure when cycling between 0.4 and 1 V.

Still shorter time intervals are allowed using the energy dispersive EXAFS (ED-EXAFS) technique with time resolution on the order 0.1–10 s. However, to date the signal-to-noise ratio has been too small for good EXAFS or $\Delta\mu$ analysis, in particular when considering more inhomogeneous catalyst samples and less well-defined in situ studies [108–110]. However, this is expected to improve in the future. We also expect spatially resolved data to be utilized in the future, since degradation processes in fuel cells, and reactor beds in heterogeneous catalysis, would especially benefit from site-dependent information. In particular, exposed regions, for example, at the fuel inlet in a direct methanol fuel cell, could be identified, and novel, more stable catalysts could be probed. With sufficient resolution and using a micrometer-sized beam, spatially resolved $\Delta\mu$ XANES results seem possible, to follow, for example, variations in the electrode–electrolyte water content and its effect on the catalyst's long-term performance. Corrosion effects in the catalysts linked to water content might lead to changes in the predominant mechanism, which could then be followed. We are convinced that the future will hold many more interesting methodical developments, since increasingly advanced catalysts and system designs will force the methods to keep pace.

## ACKNOWLEDGMENT

The authors would like to thank the staff of beamlines BM29 and ID24 at ESRF, Grenoble; beamline X1 at HASYLAB, Hamburg; beamline XAS at ANKA, Karlsruhe; and beamlines X-11A,B at NSLS, Brookhaven for their kind support.

## REFERENCES


1. V. Stamenkovic, B. N. Grgur, P. N. Ross, and N. M. Markovic, *J. Electrochem. Soc.* **152**, A277 (2005).
2. G. Q. Lu, J. O. White, and A. Wieckowski, *Surf. Sci.* **564**, 131 (2004).
3. (a) K. A. Friedrich, K.-P. Geyzers, U. Linke, U. Stimming, and J. Stumper, *J. Electroanal. Chem.* **402**, 123 (1996); (b) W. Chrzanowski and A. Wieckowski, *Langmuir* **14**, 1967 (1998).
4. K. Kunimatsu, T. Sato, H. Uchida, and M. Watanabe, *Langmuir* **24**, 3590 (2008).
5. M. Arenz, V. Stamenkovic, B. B. Blizanac, K. J. J. Mayrhofer, N. M. Markovic, and P. N. Ross, *J. Catal.* **232**, 402 (2005).
6. (a) G. Garcia, J. A. Silva-Chong, O. Guillen-Villafuerte, J. L. Rodriguez, E. R. Gonzalez, and E. Pastor, *Catal. Today* **116**, 415 (2006); (b) J. Kaiser, L. Colmenares, Z. Jusys, R. Mortel, H. Bonnemann, G. Kohl, H. Modrow, J. Hormes, and R. J. Behm, *Fuel Cells* **6**, 190 (2006).
7. M. Mazurek, N. Benker, C. Roth, and H. Fuess, *Fuel Cells* **6**, 208 (2006).
8. L. P. L. Carrette, K. A. Friedrich, M. Huber, and U. Stimming, *Phys. Chem. Chem. Phys.* **3**, 320 (2001).
9. E. Herrero, K. Franaszczuk, and A. Wieckowski, *J. Phys. Chem.* **98**, 5074 (1994).
10. V. Del Colle, A. Berna, G. Tremiliosi, E. Herrero, and J. M. Feliu, *Phys. Chem. Chem. Phys.* **10**, 3802 (2008).
11. G. Garcia and M. T. M. Koper, *Phys. Chem. Chem. Phys.* **10**, 3802 (2008).
12. M. Watanabe and S. Motoo, *J. Electroanal. Chem.* **60**, 275 (1975).
13. (a) P. Liu and J. K. Nørskov, *Phys. Chem. Chem. Phys.* **3**, 3814 (2001); (b) Y. Y. Tong, H. S. Kim, P. K. Babu, P. Waszczuk, A. Wieckowski, and E. Oldfield, *J. Am. Chem. Soc.* **124**, 468 (2002).
14. T. J. Schmidt, M. Noeske, H. A. Gasteiger, R. J. Behm, P. Britz, and H. Bönnemann, *J. Electrochem. Soc.* **145**(3), 925 (1998).
15. S. L. Gojkovic and T. R. Vidakovic, *Electrochim. Acta* **47**, 633 (2001).
16. J. H. Jiang and A. Kucernak, *J. Electroanal. Chem.* **567**, 123 (2004).
17. M. Arenz, Technische Univ. München, personal communication, 2008.
18. M. Uchida, Y. Aoyama, N. Eda, and A. Ohta, *J. Electrochem. Soc.* **142**, 4143 (1995).
19. C. Roth, N. Benker, Th. Buhrmester, M. Mazurek, M. Loster, H. Fuess, D. C. Koningsberger, and D. E. Ramaker, *J. Am. Chem. Soc.* **127**(42), 14607 (2005).
20. D. Gerteisen, *J. Appl. Electrochem.* **37**, 1447 (2007).
21. X. Yu and S. Ye, *J. Power Sources* **172**, 145 (2007).
22. S. C. Ball, S. L. Hudson, D. Thompsett, and B. Theobald, *J. Power Sources* **171**, 18 (2007).
23. A. V. Virkar and Y. Zhou, *J. Electrochem. Soc.* **154**(6), B540 (2007).
24. M. Loster, D. Balzar, K. A. Friedrich, and J. Garche, *J. Phys. Chem. C* **111**, 9583 (2007).
25. W. Vogel, *J. Phys. Chem. B* **112**, 13475 (2008).
26. H.-G. Haubold, P. Hiller, H. Jungbluth, and T. Vad, *Jpn. J. Appl. Phys.* **38**, 36 (1999).
27. S. R. Brankovic, N. S. Marinkovic, J. X. Wang, and R. R. Adžić, *J. Electroanal. Chem.* **532**, 57 (2002).

28. G. Q. Lu, J. O. White, and A. Wieckowski, *Surf. Sci.* **564**, 131 (2004).
29. B. M. Weckhuysen, in *In-situ Spectroscopy of Catalysts*, B. M. Weckhuysen, ed., American Scientific Publishers, 2004, pp. 1–10.
30. C. Roth, N. Benker, M. Mazurek, F. Scheiba, and H. Fuess, *Appl. Catal. A* **319**, 81 (2007).
31. A. E. Russell, S. C. Ball, S. Maniguet, and D. Thompsett, *J. Power Sources* **171**, 72 (2007).
32. M. Teliska, W. E. O'Grady, and D. E. Ramaker, *J. Phys. Chem. B* **109**, 8076 (2005).
33. D. C. Koningsberger, M. K. Oudenhuijzen, J. de Graaf, J. A. van Bokhoven, and D. E. Ramaker, *J. Catal.* **216**, 178 (2003).
34. M. S. Nashner, A. I. Frenkel, D. L. Adler, J. R. Shapley, and R. G. Nuzzo, *J. Am. Chem. Soc.* **119**, 7760 (1997).
35. P. L. Hansen, J. B. Wagner, S. Helveg, J. R. Rostrup-Nielsen, B. S. Clausen, and H. Topsoe, *Science* **295**, 2053 (2002).
36. M. Newton, *Chem. Soc. Rev.* **37**, 2644 (2008).
37. M. S. Nashner, A. I. Frenkel, D. Somerville, C. W. Hills, J. R. Shapley, and R. G. Nuzzo, *J. Am. Chem. Soc.* **120**, 8093 (1998).
38. C. W. Hills, M. S. Nashner, A. I. Frenkel, J. R. Shapley, and R. G. Nuzzo, *Langmuir* **15**, 690 (1999).
39. R. A. Lampitt, L. P. L. Carrette, M. P. Hogarth, and A. E. Russell, *J. Electroanal. Chem.* **460**, 80 (1999).
40. J. McBreen and S. Mukerjee, *J. Electrochem. Soc.* **142**(10), 3399 (1995).
41. A. S. Arico, P. Creti, H. Kim, R. Mantegna, N. Giordano, and V. Antonucci, *J. Electrochem. Soc.* **143**, 3950 (1996).
42. (a) C. Roth, N. Martz, and H. Fuess, *Phys. Chem. Chem. Phys.* **3**, 315 (2001); (b) A. S. Arico, P. Creti, H. Kim, R. Mantegna, N. Giordano, and V. Antonucci, *J. Electrochem. Soc.* **143**(12), 3950 (1996).
43. S. Mukerjee and J. McBreen, *J. Electrochem. Soc.* **143**(7), 2285 (1996).
44. D. Aberdam, R. Durand, R. Faure, F. Gloaguen, J. L. Hazemann, E. Herrero, A. Kabbabi, and O. Ulrich, *J. Electroanal. Chem.* **398**, 43 (1995).
45. S. Maniguet, R. J. Mathew, and A. E. Russell, *J. Phys. Chem. B* **104**, 1998 (2000).
46. W. E. O'Grady, P. L. Hagans, K. I. Pandya, and D. L. Maricle, *Langmuir* **17**, 3047 (2001).
47. R. Viswanathan, G. Hou, R. Liu, S. R. Bare, F. Modica, G. Mickelson, C. U. Segre, N. Leyarovska, and E. S. Smotkin, *J. Phys. Chem. B* **106**, 3458 (2002).
48. C. Roth, N. Martz, Th. Buhrmester, and H. Fuess, *Phys. Chem. Chem. Phys.* **4**, 3555 (2002).
49. C. Roth, N. Martz, M. Mazurek, F. Scheiba, and H. Fuess, *Adv. Eng. Mater.* **7**, 952 (2005).
50. J. M. Ziegelbauer, V. S. Murthi, C. O'Laoire, A. F. Guila, and S. Mukerjee, *Electrochim. Acta* **53**, 5587 (2008).
51. S. Mukerjee, R. C. Urian, S. J. Lee, E. A. Ticianelli, and J. McBreen, *J. Electrochem. Soc.* **151**, A1094 (2004).
52. A. E. Russell and A. Rose, *Chem. Rev.* **104**, 4613 (2004).
53. R. J. K. Wiltshire, C. R. King, A. Rose, P. P. Wells, M. P. Hogarth, D. Thompsett, and A. E. Russell, *Electrochim. Acta* **50**, 5208 (2005).

54. S. Stoupin, H. Rivera, Z. R. Li, C. U. Segre, C. Korzeniewski, D. J. Casadonte, H. Inoue, and E. S. Smotkin, *Phys. Chem. Chem. Phys.* **10**, 6430 (2008).
55. S. Stoupin, E. H. Chung, S. Chattopadhyay, C. U. Segre, and E. S. Smotkin, *J. Phys. Chem. B* **110**, 9932 (2006).
56. R. Viswanathan, R. Liu, and E. S. Smotkin, *Rev. Sci. Instrum.* **73**, 2124 (2002).
57. E. Principi, A. Di Cicco, A. Witkowska, and R. Marassi, *J. Synchr. Radiat.* **14**, 276 (2007).
58. M. Teliska, W. E. O'Grady, and D. E. Ramaker, *J. Phys. Chem. B* **109**, 8076 (2005).
59. D. C. Koningsberger, M. K. Oudenhuijzen, J. de Graaf, J. A. van Bokhoven, and D. E. Ramaker, *J. Catalysis* **216**, 178 (2003).
60. C. Roth and D. E. Ramaker, in *Modern Aspects of Electrochemistry* (in press).
61. D. C. Koningsberger, B. L. Mojet, G. E. van Dorssen, and D. E. Ramaker, *Top. Catal.* **10**, 143 (2000).
62. G. E. van Dorssen, D. C. Koningsberger, and D. E. Ramaker, *J. Phys. Condens. Matt.* **14**, 13529 (2002).
63. C. Roth, N. Benker, Th. Buhrmester, M. Mazurek, M. Loster, H. Fuess, D. C. Koningsberger, and D. E. Ramaker, *J. Am. Chem. Soc.* **127**, 14607 (2005).
64. F. J. Scott, S. Mukerjee, and D. E. Ramaker, *J. Phys. Chem. C* **114**, 442 (2010).
65. A. L. Ankudinov, B. Ravel, J. J. Rehr, and S. D. Conradson, *Phys. Rev.* **58**, 7565 (1998).
66. E. Janin, H. von Schenck, M. Gothelid, U. O. Karlsson, and M. Svensson, *Phys. Rev. B Condens. Matt. Mater. Phys.* **61**, 13144 (2000).
67. D. E. Ramaker, M. Teliska, Y. Zhang, A. Yu. Stakheev, and D. C. Koningsberger, *Phys. Chem. Chem. Phys.* **5**, 4492 (2003).
68. F. J. Scott, C. Roth, and D. E. Ramaker, *J. Phys. Chem. C* **111**, 11403 (2007).
69. D. C. Koningsberger and R. Prins, eds., *X-ray Absorption: Principles, Applications, Techniques of EXAFS, SEXAFS, and XANES*, in Chemical Analysis Series, Vol. 92, Wiley, New York, 1988.
70. E. A. Stern and S. M. Heald, Basic principles and applications of EXAFS, in *Handbook of Synchrotron Radiation* E. E. Koch, ed., North-Holland, Amsterdam, 1983, Chapter 10, pp. 995–1014.
71. D. C. Koningsberger, B. L. Mojet, G. E. van Dorssen, and D. E. Ramaker, *Top. Catal.* **10**, 143 (2000).
72. Tutorials and other EXAFS training on the Web: G. Bunker, `http://gbxafs.iit.edu/training/tutorials.html;` tutorial from Daresbury Lab, UK; `http://srs.dl.ac.uk/XRS/courses/,` advanced course by B. Ravel, `http://leonardo.phys.washington.edu/ravel/course.`
73. Software packages can be found at `http://www.esrf.fr/computing/scientific/exafs/, http://cars9.uchicago.edu/IXS-cgi/XAFS,Programs, and http://leonardo.phys.washington.edu/feff.`
74. B. W. Holland, J. B. Pendry, R. F. Pettifer, and J. J. Bordas, *Phys. Rev. C* **11**, 633 (1978).
75. J. J. Rehr, C. H. Booth, F. Bridges, and S. I. Zabinsky, *Phys. Rev. B* **49**, 12347 (1994).
76. D. E. Ramaker, B. L. Mojet, D. C. Koningsberger, and W. E. O'Grady, *J. Phys. C Condens. Matt.* **10**, 8753 (1998).
77. D. E. Ramaker, G. E. van Dorssen, B. L. Mojet, and D. C. Koningsberger, *Top. Catal.* **10**, 157 (2000).

78. W. E. O'Grady, D. E. Ramaker, and X. Qian, *J. Phys. Chem. B* **101**, 5624 (1997).
79. D. E. Ramaker, J. de Graaf, J. A. R. van Veen, and D. C. Koningsberger, *J. Catal.* **203**, 7 (2001).
80. J. Barber, S. Morin, and B. E. Conway, *J. Electroanal. Chem.* **446**, 125 (1998).
81. N. M. Markovic, B. N. Grgur, and P. N. Ross, *J. Phys. Chem. B* **101**, 5405 (1997).
82. A. Bewick and J. W. Russell, *J. Electroanal. Chem. Interface Electrochem.* **132**, 329 (1982).
83. H. Ogasawara and M. Ito, *Chem. Phys. Lett.* **221**, 213 (1994).
84. A. Peremans and A. Tadjeddine, *J. Chem. Phys.* **103**, 7197 (1995).
85. D. E. Ramaker, F. J. Scott, and C. Roth, manuscript in preparation.
86. F. J. Scott, S. Mukerjee, and D. E. Ramaker, *J. Electrochem. Soc.* **154**, A396–A406 (2007).
87. B. E. Conway, J. Barber, and S. Morin, *Electrochim. Acta* **44**, 1109 (1998).
88. J. X. Wang, T. E. Springer, and R. R. Adzic, *J. Electrochem. Soc.* **153**, A1732 (2006).
89. J. X. Wang, T. E. Springer, P. Liu, M. Shao, and R. R. Adzic, *J. Phys. Chem. C* **111**, 12425–12433 (2007).
90. J. X. Wang, J. Zhang, and R. R. Adzic, *J. Phys. Chem. A* **111**, 12702 (2007); J. X. Wang, F. A. Uribe, T. E. Springer, J. Zhang, and R. R. Adzic, *Faraday Discuss* 140 (2008).
91. S. Mukerjee, R. C. Urian, S. J. Lee, E. A. Ticianelli, and J. McBreen, *J. Electrochem. Soc.* **151**, A1094 (2004); S. Mukerjee and R. C. Urian, *Electrochim. Acta* **47**, 3219 (2002); R. C. Urian, PhD thesis, Northeastern Univ., 2003.
92. J. M. Ziegelbauer, D. Gatewood, A. F. Gullá, D. E. Ramaker, S. Mukerjee, M. J.-F. Guinel, and F. Ernst, *J. Phys. Chem. C* **113**, 6955 (2009).
93. D. E. Ramaker, B. Shyam, T. Arruda, S. Mukerjee, and A. Wieckowski, manuscript in preparation.
94. D. Gatewood, D. E. Ramaker, K. Sasaki, and K. E. Swider-Lyons, *J. Electrochem. Soc.* **155**, B834 (2008).
95. A. Korovina, K. Swider-Lyons, and D. E. Ramaker, manuscript in preparation.
96. D. S. Gatewood, T. L. Schull, O. Baturina, J. J. Pietron, Y. Garsany, K. E. Swider-Lyons, and D. E. Ramaker, *J. Phys. Chem. B* **112**, 4961 (2008).
97. M. Teliska, V. S. Murthi, S. Mukerjee, and D. E. Ramaker, *J. Phys. Chem. C* **111**, 9267 (2007).
98. T. Arruda, B. Shyam, J. Ziegelbauer, S. Mukerjee, and D. E. Ramaker, *J. Phys. Chem.* **112**, 18087 (2008).
99. D. E. Ramaker, D. Gatewood, A. Korovina, Y. Garsany, and K. E. Swider-Lyons, *J. Phys. Chem. C* in press (2010).
100. J. Melke, A. Schoekel, D. Dixon, C. Cremers, D. E. Ramaker, and C. Roth, *J. Phys. Chem. C* **114**, 5914 (2010).
101. Y. Ji, V. Koot, A. M. J. van der Eerden, B. M. Weckhuysen, D. C. Koningsberger, and D. E. Ramaker, *J. Catal.* **245**, 415 (2007).
102. E. Bus, D. E. Ramaker, and J. A. van Bokhoven, *J. Am. Chem. Soc.* **129**, 8094 (2007).
103. T. A. Nijhuis, E. Sacaliucb, A. M. Beale, A. M. J. van der Eerden, J. C. Schouten, and B. M. Weckhuysen, *J. Catal.* **258**, 256 (2008).
104. T. Aruda, B. Shyam, D. Ramaker, and S. Mukerjee, manuscript in preparation.

105. D. Lutzenkirchen-Hecht, R. Wagner, U. Haake, A. Watenphul, and R. Frahm, *J. Synchr. Radiat.* **16**, 264 (2009).

106. R. Frahm, personal communication, Grenoble, 2009.

107. M. Tada, S. Murata, T. Asaoka, K. Hiroshima, K. Okumura, H. Tanida, T. Uruga, H. Nakanishi, S. Matsumoto, Y. Inada, M. Nomura, and Y. Iwasawa, *Angew. Chem. Int. Ed.* **46**, 4310 (2007).

108. P. G. Allen, S. D. Conradson, M. S. Wilson, S. Gottesfeld, I. D. Raistrick, J. Valcrio, and M. Lovato, *J. Electroanal. Chem.* **384**, 99 (1995).

109. T. Ressler, O. Timpe, T. Neisius, J. Find, G. Mestl, M. Dieterle, and R. Schlögl, *J. Catal.* **191**, 75 (2000).

110. D. E. Ramaker, D. Gatewood, A. M. Beale, and B. M. Weckhuysen, *AIP Conf. Proc.* **882**, 619 (2007).

CHAPTER 18

# Operando X-Ray Absorption Spectroscopy of Polymer Electrolyte Fuel Cells

EUGENE S. SMOTKIN

Department of Chemistry and Chemical Biology, Northeastern University, Boston, Massachusetts

CARLO U. SEGRE

Physics Division, Illinois Institute of Technology, Chicago, Illinois

## 18.1 INTRODUCTION

Haldor Topsoe's teaching [1] that "the active state of a catalyst exists only during catalysis" is succinct rationale for *operando* characterization of catalytic devices. This review focuses on cell design for operando characterization of catalysts in polymer electrolyte membrane (PEM) electrode reactors. PEM reactors are used for organic synthesis [2–4], environmental remediation [5], and power generation [6]. Operando characterization of PEM catalytic layers requires steady-state flow of reactant streams to the anode and cathode in normal working environments. Figure 18.1 schematizes the end cell of a PEM electrode stack [7]. The membrane electrode assembly, a PEM (Nafion) sandwiched between anode and cathode catalytic layers, is the heart of the reactor. The catalytic layers are contacted to gas diffusion layers consisting of porous carbon paper (or cloth) optimized for transport of reactant streams between the flow field grooves and the full area of the catalytic layers. The gas diffusion layers are often rendered hydrophobic by Teflon, particularly at the cathode, where flooding occurs. The diffusion layers enable conduction of electrons between the catalytic layers and the flow field groove "lands" without blocking electrode area. Dispersions of catalyst particles in aqueous alcoholic solutions of solubilized Nafion (i.e., inks) can be

*Fuel Cell Science: Theory, Fundamentals, and Biocatalysis*,
Edited by Andrzej Wieckowski and Jens K. Nørskov

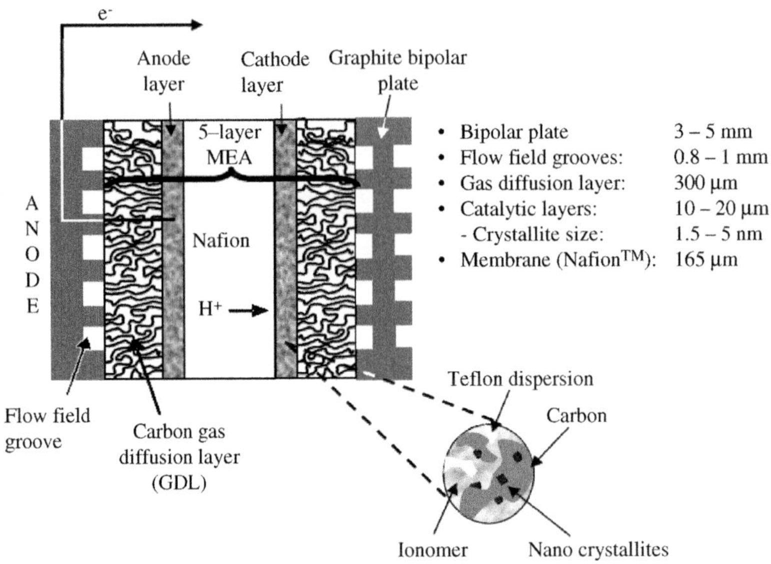


**FIGURE 18.1** Polymer electrolyte fuel cell schematic. (From Ref. 17.)

deposited directly onto the diffusion layers and then hot-pressed to the membrane (yielding a five-layer electrode assembly), or applied directly onto the membrane to yield a catalyst-coated membrane electrode assembly (i.e., a three-layer assembly). Teflon dispersion is often added to the cathode ink to mitigate flooding at the cathode. Membrane electrode assembly fabrication methods are thoroughly reviewed by Gottesfeld and Zawodzinski [8].

The need for operando characterization is better appreciated when considering the complex structure of the catalytic layer (Fig. 18.1) The catalyst layers include carbon supported (for $H_2$/air fuel cells) or metal black catalysts (for direct oxidation fuel cells) [9], Teflon dispersion (at the cathode), and proton-conducting ionomer (originating from the ink) to provide conductive paths for protons. A solid–gas electrolyte interface is not in the intended design; the catalytic particles are coated with an ultrathin overlayer of Nafion that conducts protons and enhances catalysis [10]. The superacidic Nafion environment has no mobile anions to contribute to anion poisoning. Aqueous acidic electrolytes (e.g., $H_2SO_4$ or $HClO_4$) not only contribute mobile anion adsorbates but also preclude operation of the cell at temperatures relevant to fuel cells (e.g., 40–90 °C). These are compelling arguments for operando analysis.

## 18.2 PARAMETERS AND APPLICATIONS OF OPERANDO SPECTROSCOPY

Proper cell design is the key challenge to operando spectroscopy. In the case of X-ray absorption spectroscopy, there are restrictions as to which components can be metallic. Stainless steel includes iron, nickel, and chromium, which fluoresce at energies similar

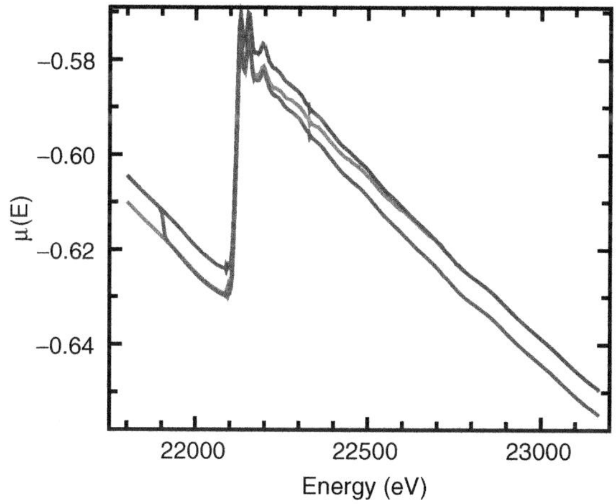


**FIGURE 18.2** Absorption transition events at a liquid feed DMFC anode. (From Ref. 11.)

to the edge energies of metal catalysts. The cell must accommodate both transmission and glancing angle measurements. Glancing angle fluorescence measurements minimize interference due to the other cell components. Operando spectroscopy requires careful water and vapor management. Mixed–phase flow causes fluctuations in the overall system absorption coefficient [11], reducing the amount of data available for signal averaging. Figure 18.2 shows absorption transition events often present in the raw data of direct methanol fuel cell (DMFC) X-ray absorption spectroscopy. These events are attributed to evolution of $CO_2$ bubbles in the methanol anode flow field (red line) at 21,870 eV that reduce the overall absorption coefficient, followed by the gradual expulsion of the bubble from 22,200 to about 22,750 eV (green line), resulting in an increase in absorption. Stoupin et al. discusses mitigation of such absorption transitions in DMFCs by application of slight backpressures at the anode outlet and operation at lower temperatures (i.e., lower conversion rates) [11].

Catalyst loadings should be limited to 4 mg/cm$^2$ for DMFCs and 0.1 mg/cm$^2$ for hydrogen air fuel cells. Higher catalyst loadings suffer from a $z$ dependence (normal to the catalyst-layer plane) in the current and potential profiles [12,13]. This is problematic for bulk averaging techniques such as X-ray absorption spectroscopy. Although challenging, these prerequisites are a necessity for realistic (i.e., operando) characterization of fuel cell catalytic layers. In addition to the specialized requirements for fuel cells, the standard requirements for proper electrochemical cell design apply, including equal resistance between all points of the working electrode surface and the auxiliary electrode surface, and a low-impedance reference electrode [14]. The electrolyte resistance between the anode and cathode catalytic layers of a membrane electrode assembly is governed by the thickness of the membrane (e.g., about 7 mil for dry Nafion 117). The fuel cell counter electrode can serve as both the auxiliary electrode and the reference electrode [15,16]. The hydrogen anode doubles as the reference electrode when one obtains cathode polarization curves. Within the kinetically controlled region of the fuel cell polarization curve, the hydrogen electrode polarization is negligible. Although use of the counter electrode as a hydrogen reference electrode has drawbacks at high currents, the alternative of developing a third electrode as a reference electrode is inferior to correcting for reference electrode polarization losses using current interrupt or impedance measurements. A hydrogen counter electrode is ideal as a reference electrode when obtaining DMFC anode polarization curves [11,16] Pure water can be used at the counterreference electrode when obtaining anode polarization curves. Hydrogen evolved at the counter-reference electrode, during anode polarization, immediately poises [17] the counter electrode; the hydrogen ion activity is set by the equivalent weight of the Nafion (e.g., 1100 g for Nafion 117). Nafion has aqueous phases with acidity roughly equivalent to half-molar sulfuric acid. The high acidity is attributed to the fluorine atoms of the $-CF_2$ group alpha to the exchange group [18]. The use of the fuel cell counter electrode as the reference electrode is exemplified in a fuel cell CO stripping analysis of the cathode of a DMFC MEA subjected to a lifetime study (Fig. 18.3). The cathode and anode were catalyzed with Johnson Matthey Pt black and PtRu black, respectively, both at 4 mg/cm$^2$. The cell was operated at 0.45 V for 700 h with 0.3 M methanol at 0.5 mL/min and air at ~200 mL/min. For the CO stripping voltammogram (20 mV/s), the cell was fed with humidified 5% CO in Ar for 30 min and then purged with humidified $N_2$ for 30 min. It is noteworthy that the CO stripping wave initiates at potentials negative of 400 mV versus the hydrogen electrode. This is a result of Ru crossover from the DMFC anode during 700 h of fuel cell operation. The unintended Ru cocatalyst at the cathode activates water at more negative potentials and thus reduces the overpotential required for CO stripping. This phenomenon was previously reported by Piela et al. [19].

Viswanathan [20,21] and Stoupin [11] were first to report operando X-ray absorption spectroscopy of hydrogen–air and liquid feed DMFCs, respectively. Figure 18.4 shows the fuel cell used in both studies. Palladium was used at the counter electrode to avoid interference with Pt edge energies of the Pt-based working electrodes of interest.

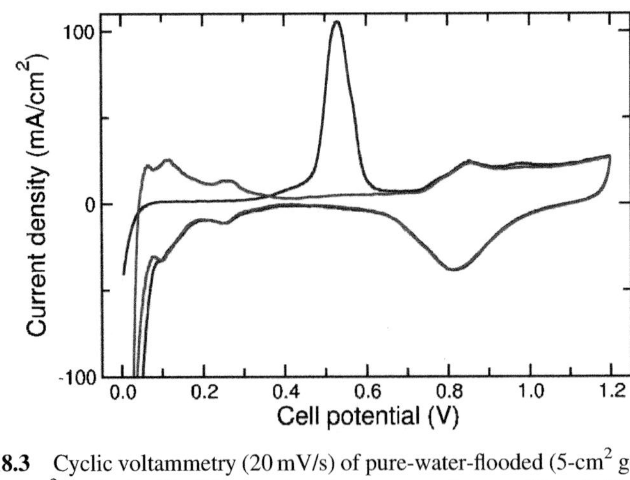


**FIGURE 18.3** Cyclic voltammetry (20 mV/s) of pure-water-flooded (5-cm$^2$ geometric) fuel cell Pt (4 mg/cm$^2$) cathode; $H_2$ (100 sccm) at the counterreference. Flow fields at 60°C.

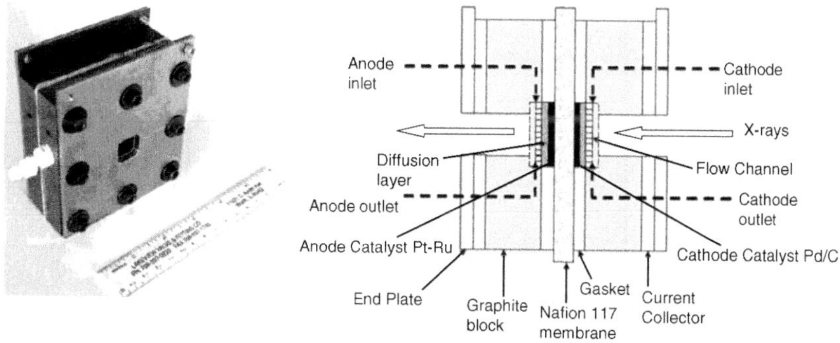


**FIGURE 18.4** Operando spectroscopy fuel cell and schematic. (From Ref. 21.)

Figure 18.5 shows X-ray absorption data [11] of Johnson Matthey PtRu black at the anode of a liquid feed DMFC at the Materials Research Collaborative Access Team (MRCAT) beamline at Argonne's Advanced Photon Source. In the Pt $L_{III}$-edge data on the left, the red line shows Pt XANES at 450 mV in an operating DMFC, while the green and blue lines are the as-received catalyst (mounted on scotch tape) and Pt foil XANES respectively, also measured in transmission.

The increased white-line intensity of the alloyed Pt (red) versus the pure Pt (blue) is due to alloy-induced *d*-band vacancies, as was observed in XANES from an arc-melted PtRu 80 : 20 alloy [20]. The large edge intensity of the green curve signifies extensive oxidation of the as-received catalyst. Stoupin et al. observed that the potential-dependent (250, 300, 350, 400, and 450 mV) EXAFS perfectly overlap, confirming that, within the DMFC operating potential window, the Pt signal is insensitive to potential, similarly to the Pt within supported PtRu catalyst in a $H_2$/air fuel cell. Figure 18.5b shows the excellent first shell fit of the 350 mV EXAFS data (blue) with

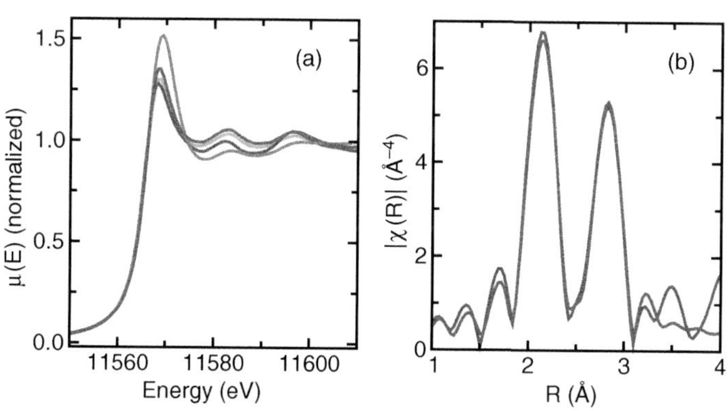


**FIGURE 18.5** (a) Pt $L_{III}$-edge XANES: in situ catalyst at 450 mV (red), in situ catalyst at 450 mV from another run (orange), Pt foil (blue), and as-received catalyst (green); (b) Pt EXAFS fit (red line) to in situ catalyst data at 450 mV (blue) as a totally metallic environment (only Pt and Ru nearest neighbors). Fit range: 1.5–3.1 Å. (From Ref. 11.)

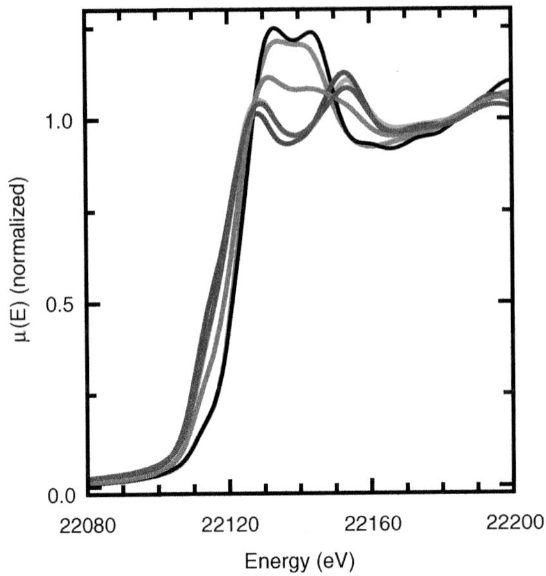


**FIGURE 18.6** Ru *K*-edge XANES: operating fuel cell anode at 450 mV (red; orange, another similar experiment), metallic Ru powder (blue), Ru oxide hydrate (pink), Ru oxide (black), and as-received JM PtRu (1 : 1) (green) standards mounted on scotch tape. (From Ref. 11.)

a model fit (red) simulating a totally metallic environment (fit range 1.5–3.1 Å) of mixed Pt and Ru atoms. Figure 18.6 shows the ex situ Ru *K*-edge XANES of metallic Ru powder (blue), in situ catalyst Ru at 350 mV (red), as-received PtRu (1 : 1) catalyst (green), Ru oxide (black), and Ru oxide hydrate (pink).

The coincidence of the catalyst Ru edge at 350 mV (red) with the metallic Ru powder (blue) confirms that within the DMFC operating potential window the Ru component is primarily metallic. The deviation of the red line from the blue line at 22,120 eV reflects the difference between the nearest-neighbor environments of pure Ru (hexagonal) and Ru in a PtRu catalyst (FCC). The as-received catalyst edge (green) is intermediate in energy between the metallic (blue and red) lines and the oxides of Ru (pink and black), confirming that the as-received catalyst is substantially oxidized. The near-edge data of the as-received catalyst are intermediate between the saddle point of the Ru oxide and the metallic catalyst line (red), again confirming a substantially oxidized as-received catalyst.

Table 18.1 summarizes the EXAFS structural parameters for a PtRu catalyst prepared sonochemically by Korzeniewski et al. in a DMFC operating with 0.1 molar methanol at the anode in comparison to the same catalyst in as-received conditioned [22,23]. In comparison to the XRD-determined Ru mole fraction, the smaller in situ EXAFS-determined Ru mole fractions suggest that the sizes of the nanoparticle crystalline domains under in situ and ex situ conditions are different. Indeed, the EXAFS determined mole fraction of the as-received (oxidized) catalyst ($0.14 \pm 0.07$) is somewhat closer to the XRD-determined mole fraction of the as-received catalyst.

**TABLE 18.1 EXAFS Structural Parameters of Sonochemical PtRu Catalyst**

| EXAFS Parameter | Notation | MEA | As-Received Catalyst |
|---|---|---|---|
| Ru around Pt fraction[b] | $y_{Ru}$ | 0.05(0.03) | 0.11(0.06) |
| Pt around Ru fraction | $y_{Pt}$ | 0.79(0.02) | 0.69(0.03) |
| Ru-Ru bond DWF[c] (Å$^2$) | $\sigma^2_{Ru-Ru}$ | 0.0021(0.0007) | 0.0042[d] |
| Ru–Pt bond DWF (Å$^2$) | $\sigma^2_{Ru-Pt}$ | 0.0053(0.0007) | 0.0038[d] |
| Pt–Pt bond DWF (Å$^2$) | $\sigma^2_{Pt-Pt}$ | 0.0073(0.0008) | 0.0064[d] |
| Ru–Ru bond length (Å) | $R_{Ru-Ru}$ | 2.632(0.005) | 2.63(0.02) |
| Ru–Pt bond length (Å) | $R_{Pt-Ru}$ | 2.695(0.004) | 2.69(0.01) |
| Pt–Pt bond length (Å) | $R_{Pt-Pt}$ | 2.726(0.005) | 2.71(0.02) |
| Total number of first-shell atoms | $N^{tot}$ | 9.2(0.4) | 5.0(0.5) |
| Ru–O bond DWF (Å$^2$) | $\sigma^2_{Ru-O}$ | 0.01[d] | 0.0057[d] |
| Ru–O bond length (Å) | $R_{Ru-O}$ | 1.88(0.03) | 1.97(0.01) |
| Ru–O coordination number | $N_{Ru-O}$ | 0.23(0.09) | 2.3(0.2) |
| Pt–O bond DWF (Å$^2$) | $\sigma^2_{Pt-O}$ | N/A | 0.0049[d] |
| Pt–O bond length (Å) | $R_{Pt-O}$ | N/A | 1.90(0.04) |
| Pt–O coordination number | $N_{Pt-O}$ | N/A | 1.3(0.2) |

[a] With simultaneous fit of the in situ Pt and Ru EXAFS at 0.1 M methanol at all potentials and fit with the same model of the as-received catalyst data. The uncertainties are given in parentheses.
[b] Fractional coordination numbers are defined as follows: $y_{Ru} = N^{Ru}_{Pt}/N^{tot}_{Pt}$; $y_{Pt} = N^{Pt}_{Ru}/N^{tot}_{Ru}$, where $N^{Ru}_{Pt}$ is the average number of Ru neighbors around a Pt atom and $N^{Pt}_{Ru}$ is the average number of Pt around Ru; $N^{tot}_{Ru} = N^{tot}_{Pt} = N^{tot}$ - (total number of fcc first-shell atoms). By definition, $N^{tot}_{Ru} = N^{Ru}_{Ru} + N^{Pt}_{Ru}$ and $N^{tot}_{Pt} = N^{Pt}_{Pt} + N^{Ru}_{Pt}$. These relationships enable derivation of coordination numbers for each pair of atoms using the above-fitted parameters: $N^{Ru}_{Ru} = N^{tot}(1-y_{Pt})$, $N^{Pt}_{Ru} = N^{tot}y_{Pt}$, $N^{Pt}_{Pt} = N^{tot}(1-y_{Ru})$, $N^{Ru}_{Pt} = N^{tot}y_{Ru}$.
[c] Debye–Waller factor.
[d] Fixed parameter in the fit.
*Source:* Reference 11.

A likely explanation is that modification of the catalyst morphology, such as restructuring of the metallic core might, take place in the reducing environment of the fuel cell anode. This restructuring is manifested by increases in the EXAFS first-shell total coordination number. For example, in Table 18.1, $N^{tot}$ increases from 5.0 in the as-received catalyst to 9.2 after the breakin period of the DMFC. The observed reduction of Ru content indicates that Pt atoms are primarily involved in the process; subsurface O is removed from the structure, and all of the Pt reduces to a metallic state and becomes incorporated into the core. In contrast, the Ru remains partially oxidized and unincorporated into the core. It is important to differentiate between reductions of the Ru mole fraction in the fcc lattice due to reconstruction versus a reduction of Ru due to corrosion of the catalyst. Our lifetime studies of DMFC catalysts over a period of >100 h of continuous operation show that Ru is lost from the anode and is deposited on the cathode side [24]. The change in the X-ray absorption of the operating DMFC discussed in this work occurs over a period of less than an hour after initiation of fuel delivery.

A number of operando X-ray absorption [11,20,21,25–28] studies have followed with many focused on the oxidation state of the metal components [11,20,25,26]. All found that at relevant fuel cell anode operating potentials, platinum is metallic. The oxidation of CO was studied both on Pt [29,30] and PtRu [27,28,31,32], confirming the electronic benefits of Ru as a cocatalyst for CO oxidation. Studies of the adsorption of oxygen reduction reaction intermediates on fuel cell catalysts show that adsorption is both potential-dependent and site-specific [27,33–35].

Operando FTIR is an ideal complement to X-ray absorption spectroscopy for electrode adsorbate studies. Surface adsorption can enhance dipole moments, disrupt the symmetry, and alter the selection rules and thus allow vibrational modes forbidden in the free molecule [36,37]. Because most fuel cell materials are nontransmissive in the IR region; operando cells must accommodate IR transparent windows to enable reflection techniques. For example, the outer layers of a five-layer membrane electrode assembly consist of carbon gas diffusion layers. A window the size of the IR-beam footprint must be cut out of the gas diffusion layer to expose the catalytic layer [7,38–41]. Fan introduced operando FTIR spectroscopy of fuel cells [42–45] using a diffuse IR fuel cell [45] inserted into a 1990s model Harrick Praying Mantis (Pleasantville, NY) diffuse IR accessory. The key design elements include a slot in the gas diffusion layer to expose the catalytic layer to the IR beam, the IR window, and graphite endplates with flow field grooves.

Figure 18.7 shows how the IR-XAS cell combines the features of the IR cell of Fan et al. [38] and the X-ray absorption spectroscopy fuel cell of Viswanathan [20]. Various operando methods can be applied to the same electrode surface without the need for extraction of the membrane electrode assembly and interchange of

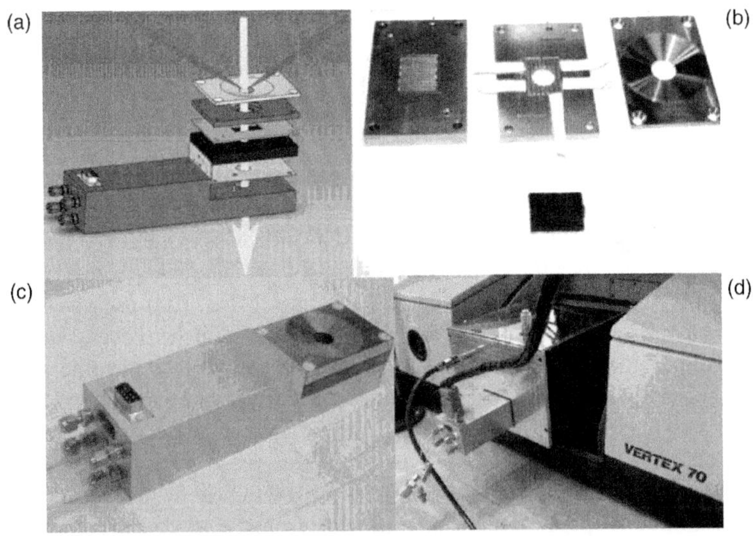


**FIGURE 18.7** IR-XAS fuel cell; (a) red path, reflectance IR, or fluorescence XAS; yellow path; transmission XAS; (b) bottom and top flow field with wicking material (white strips) adhered to upper flow field, MEA at bottom. (c) DE9 connector for electrodes, heater cartridge, and thermistoer; (d) cell installed in Pike diffuse reflectance accessory in a Bruker (Billerica, MA) Vertex 70 FTIR spectrometer.

spectroscopy cells. The complex humidity- and air-sensitive parameters required of the fuel cell make this feature particularly attractive. In Figure 18.7a the top flow field (i.e., second layer from top) houses a CaF window for IR studies. When the working electrode is functioning as an air-breathing cathode, the top plate and CaF window can be removed for transmission or fluorescence X-ray absorption spectroscopy of cathode catalysts. The advantage of fluorescence X-ray absorption is that beam intensity is not attenuated by the lower graphite flow field (fourth layer from top). The IR-XAS cell is designed to interface to commercially available diffuse reflection accessories for FTIR systems (e.g., Fig. 18.7d). Figure 18.7b shows the lower flow field, the top flow field where wicking material is incorporated to remove condensation at the cathode and the top plate used to secure the CaF window when the cell is used for IR studies. This wicking material enables orientation-independent operation. This is required for easy manipulation of the cell in the cramped environment of a synchrotron beamline experimental station.

To accommodate transmission spectroscopy, the slider has a wide bevel (not viewable) within a slot on the backside of the IR-XAS cell (Fig. 18.8a) to enable optimum angles for beam access from either side of the cell. Figure 18.8 shows the cell operating as an air-breathing fuel cell at the MRCAT beamline at the Argonne National Laboratory [46]. The beam from the source (center of Fig. 18.8a photo) passes through the cathode GDL on the front face of the cell (Fig. 18.8b photo). The Lytle detector is located at right side of the photo in Figure 18.8a. The slot on the backside of the IR-XAS cell accommodates the slide assembly of a Pike Technologies (Madison, WI) diffuse IR accessory. The cell slides into position under the integrating mirrors of the diffuse reflectance assembly (Fig. 18.7d).

The wide-beveled angle of 158° on the top plate maximizes the collection of the scattered signal. The upper flow field (just beneath the top plate) employs a pin-style flow field to optimize flow distribution around the $CaF_2$ window inset.

Figure 18.9 compares X-ray absorption spectra obtained with the Viswanathan cell (red lines) to those obtained with the IR-XAS cell (green lines). The hydrogen–air fuel cells were operated at 50°C. Five-layer membrane electrode assemblies were prepared by ink deposition onto the gas diffusion layers [47]. The cathode catalyst

(a) (b)

**FIGURE 18.8** IR-XAS cell at MRCAT beamline, Argonne National Laboratory.

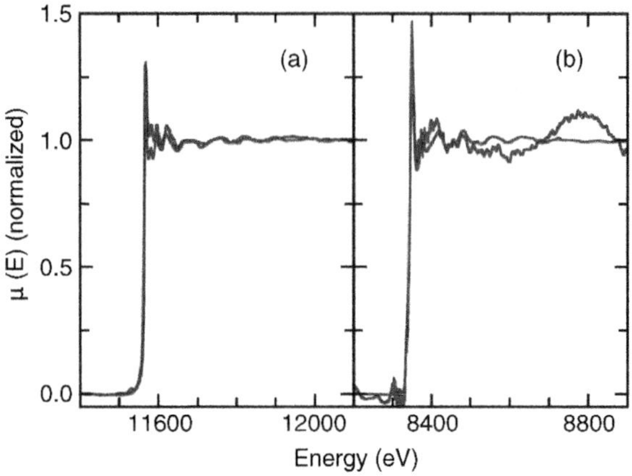


**FIGURE 18.9** X-ray absorption data acquired with the Viswanathan cell (red line) versus the IR-XAS cell (green line): (a) Pt $L_{III}$-edge $Pt_3Ni$ cathode; (b) Ni $K$-edge of the $Pt_3Ni$ cathode.

was acid-washed with 30% PtNi on carbon support (E-Tek, custom-made E1030523) at 0.9 mg/cm$^2$. The anode catalyst was Pd/C (Alfa Aesar, stock no. 38309, lot no. C02R014) at 1.0 mg/cm$^2$ loading. Palladium is used at the anode because the $L$-edge energy is much lower than that of Pt and its $K$-edge energy is twice as large as that of the Pt $L_{II-III}$ edge [20]. The Ni $K$-edge energy is substantially lower than that of Pt, enabling separate examination of Pt and Ni atoms. The spectra were acquired at the MRCAT beamline 10-ID-B at the Argonne National Laboratory Advanced Photon Source. The cell was aligned for fluorescence mode between the source and the Lytle detector. The spectra were collected by tuning the beamline undulator to the Pt $L_{III}$-edge energy range using the second harmonic of a Si(111) crystal monochromator. Higher harmonics were rejected using a Rh reflection mirror. The incident ion chamber was filled with $N_2$ gas. The spectra acquired with the new cell show a step improvement in signal-to-noise ratio (SNR) over the Viswanathan cell in transmission mode with similar catalyst loadings, as exemplified by the Ni and Pt edge data of the PtNi catalyst. The nickel edge data show a much cleaner EXAFS spectral region and the absence of observable preedge noise. At the Pt edge, the difference in SNR is not as distinct, but there is less fluctuation in the EXAFS region. It is noteworthy that the Pt and Ni loadings are below the mg/cm$^2$ range in standard membrane electrode assemblies.

The IR-XAS cell was used as an air-breathing fuel cell to compare Pt/C and PtNi (1 : 1)/C cathode catalysts at ~1 mg/cm$^2$. The objective was study of the potential dependent oxygen structure at the catalyst surface, and the role of Ni in cathode catalysis. Measurements were made at electrode potentials varying from 0 mV (short circuit) to 900 mV by taking a series of full spectra in ~2 min for at least one hour per potential to ensure that any short-time variations could be eliminated from the analysis. Figure 18.10 compares the Pt $L_3$ absorption edge from the Pt/C and PtNi/C at 0 mV and selected spectra that were subtractively normalized to spectra obtained with the electrode at 0 volts versus the reference [ Δ (XANES)] [48].

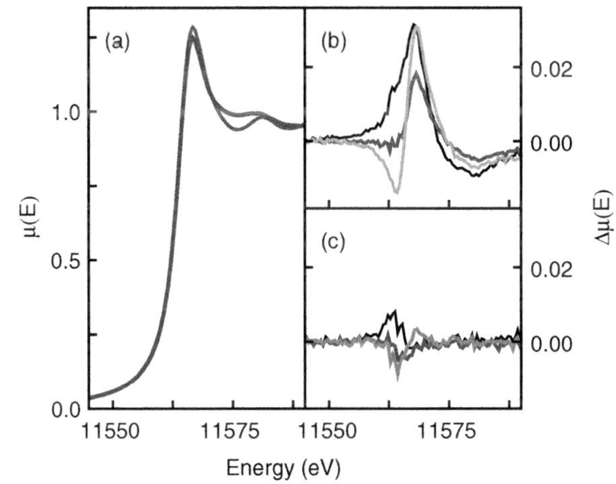


**FIGURE 18.10** Pt $L_3$ absorption edge (a) for Pt/C (black) and PtNi/C (light gray) at 0 mV and difference in normalized absorption for 530 mV (black), 780 mV (light gray) and 900 mV (dark gray) potentials in (b) Pt/C and (c) PtNi(1 : 1)/C catalyst. The Pt spectra at 0 mV were used as reference. (From Ref. 48.)

While the Δ(XANES) for the Pt/C catalyst shows a significant variation with potential, the presence of Ni in the PtNi/C catalyst nearly completely removes any potential variation of the Pt Δ(XANES), as well as lowering the white-line intensity of the Pt absorption edge. The effect of Ni on the performance of Pt in a cathode catalyst observed in the data of Figure 18.10 was quantitatively measured by fitting the EXAFS with a structural model including a metal cluster and Pt–O (and Ni–O) bonds. As with the previous results on the PtRu anode catalysts, the metallic nanoparticle core was found to be invariant with potential. Table 18.2 shows the result of the fits to both catalysts, including metal core fits and the metal–oxygen distances that are common

**TABLE 18.2 Fitting Results for Metal Cores and the Metal–Oxygen Distances of Pt/C and PtNi/C Nanoparticle Catalysts**

| | | PtNi/C | | | |
|---|---|---|---|---|---|
| Pt/C | | Pt | | Ni | |
| $N_{Pt}$ | $8.7 \pm 0.2$ | $N_{Pt}$ | $6.1 \pm 0.3$ | $N_{Ni}$ | $3.7 \pm 0.2$ |
| $\sigma_{Pt-Pt}{}^2 \times 10^3$ | $6.1 \pm 0.1$ | $\sigma^2_{Pt-Pt} \times 10^3$ | $6.1 \pm 0.7$ | $\sigma_{Ni-Ni}{}^2 \times 10^3$ | $9 \pm 1$ |
| $r_{Pt-Pt}$ (Å) | $2.749 \pm 0.001$ | $r_{Pt-Pt}$ (Å) | $2.692 \pm 0.003$ | $r_{Ni-Ni}$ (Å) | $2.572 \pm 0.006$ |
| | | $N_{Ni}$ | $3.4 \pm 0.1$ | $N_{Pt}$ | $8.9 \pm 0.5$ |
| | | $\sigma_{Pt-Ni}{}^2 \times 10^3$ | | $9 \pm 1$ | |
| | | $r_{Pt-Ni}$(Å) | | $2.635 \pm 0.004$ | |
| | | $N$ | $9.5 \pm 0.4$ | $N$ | $12.6 \pm 0.7$ |
| $r_{Pt-O}$ (Å) | $2.02 \pm 0.01$ | $r_{Pt-O}$ (Å) | $2.09 \pm 0.03$ | $r_{Ni-O}$ (Å) | $1.90 \pm 0.01$ |

*Source:* Reference 48.

for all potentials measured. The data taken at 0 mV for the PtNi catalyst provided an upper limit to the detectability of Pt–O bonds in this system.

The results in Table 18.2 indicate that Pt has the same number of metal near neighbors (NNs) in both the Pt/C and PtNi/C catalysts. The Ni, however, has approximately 12 NN metal atoms, indicating that the PtNi/C catalyst nanoparticles have Pt at the surface with Ni predominantly buried in the core, supporting the theory that Pt will segregate to the surface of a PtNi alloy, driven by negative segregation energies [49,50]. A second remarkable feature is the significant reduction in Pt–Pt bond distance in the PtNi/C catalyst (2.692 Å) as compared with the nominal Pt–Pt distances in Pt metal (2.775 Å). This is in agreement with previous reports and results in tensile force between Pt atoms, which acts to modify its electronic properties [51–54].

The results of Table 18.2 also show that the Pt–O distance is lengthened in PtNi/C compared to Pt/C, with Ni–O distance significantly shorter than in either of the others. The longer Pt–O distance in the PtNi/C catalyst is indicative of a Ni-induced weakened Pt–O bond [55]. The number of Ni–O and Pt–O bonds in the PtNi/C catalyst was found to be consistent with a preference for oxygen to adsorb on the Ni, leaving the active sites of Pt unencumbered to mediate the oxygen reduction reaction more effectively.

An interest in the interpretation of the derivativelike features (i.e., shifts of edge energies and white-line variations) of the Δ(XANES) line shapes motivates an in-depth analysis beyond the simple comparison of the amplitudes of the Δ(XANES) spectra of Figure 18.10b,c. Lewis et al. attempted to analyze such data using a theoretical Δ(XANES) method developed by Ramaker et al. [46]. Theoretical Δ(XANES) (Δμ signatures) data of $Pt_6$ Janin clusters, with and without adsorbates, were used for analysis of experimental Δ(XANES) (Δμ fingerprints) [27,32,33,56,57].

The six-atom Janin cluster accommodates four important oxygen absorption site configurations: atop, bridged, fcc, and hcp. The atop and bridged configurations are shown in Figure 18.11.

Ramaker calculates Δμ signatures using FEFF8, a full multiple scattering code that employs self-consistent-field calculations of local electronic structure [58,59]. To date, the Ramaker Δμ signatures for oxygen adsorption have been limited to the scattering of photoelectrons originating from only one adsorbing Pt atom (Fig. 18.11, blue atom). However, adsorbed oxygen on an isolated $Pt_6$ Janin cluster would modify the X-ray absorption properties of the adsorbate-free Pt atoms as well. Thus theoretical Δμ signatures that do not include X-ray absorption events from Pt atoms in close proximity to the adsorbing atom (limited absorber model) preclude consideration of charge transfer effects induced by adsorbate oxygen. However, inclusion of X-ray absorption events from all atoms of an isolated Janin cluster with adsorbate oxygen atom would overemphasize charge depletion of the NN X-ray-absorbing Pt atoms because they are artificially undercoordinated relative to bulk Pt atoms (i.e., CN = 12). Lewis et al. [46]. address this by embedding the Janin cluster into a larger reservoir of Pt atoms. Figure 18.11 summarizes the changes in the Δμ signatures, for the atop configuration and the bridged adsorption configuration, as the theory progressively increases the number of NN atoms that contribute X-ray absorption events. The atoms and lines are color-coded, and the *y* axes of the first and middle columns are spaced at

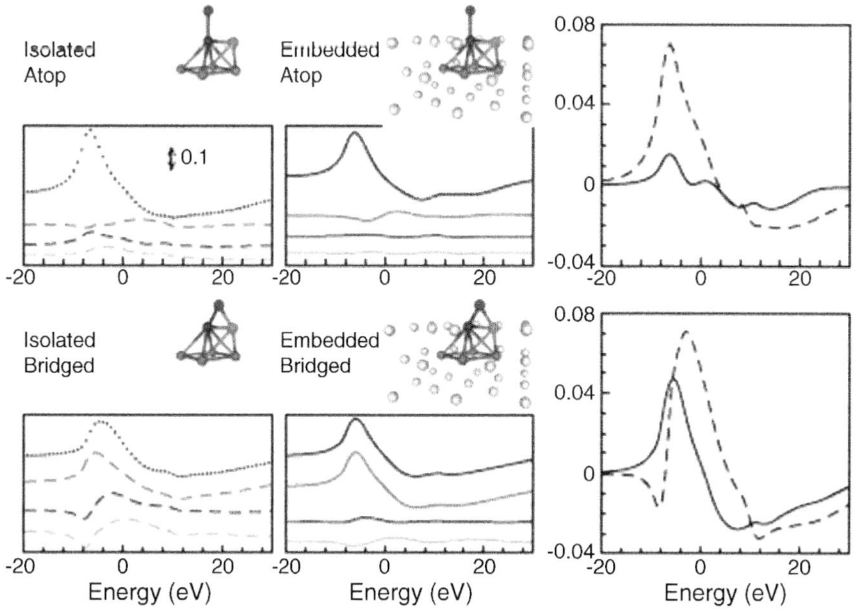


**FIGURE 18.11** (a) Atop isolated per atom signatures, (b) bridged isolated per atom signatures; (c) atop embedded per atom signatures; (d) bridged embedded per atom signatures; (e,f) configurational averages of the per atom signatures—dashed lines, isolated all-atom signatures; solid lines, embedded all-atom signatures: (e) atop configuration; (f) bridged configuration. (From Ref. 46.)

0.1 arbitrary absorption units. The dashed and dotted lines are the per atom signatures of the isolated Janin cluster. The solid lines represent the per atom signatures of the embedded Janin cluster. Figure 18.11a shows the isolated per atom $\Delta\mu$ signatures for the atop configuration. Panel (b) shows the per atom $\Delta\mu$ signatures for the bridged configuration. The blue atom $\Delta\mu$ signatures for the atop and bridged configurations (dotted blue lines) correspond to the Teliska limited-absorber signatures, which consider X-ray photoelectrons only from the blue atom. The primary effect of atop oxygen adsorption is an increase in the white line of the adsorbing atom (blue atom). Panels (c) and (d) show the embedded per atom $\Delta\mu$ signatures of the atop and bridged configurations. The atop adsorbing atom signature intensity decreases significantly [note different scaling of the $y$ axis on the rightmost column, i.e., in panels (e) and (f)] when the cluster is exposed to the Pt reservoir electron density, as expected. The embedded per atom $\Delta$(XANES) signatures demonstrate that only the surface atom 1 in the atop configuration remains charge depleted after charge compensation from the Pt reservoir [46]. All of the charge depletion resulting from charge transfer effects from subsurface atoms is compensated for by the Pt reservoir. However, in the case of the bridged bound oxygen, the subsurface atoms are not fully compensated by the Pt reservoir. The progressive improvement of the signature models, from the limited-absorber model, to an all-atom model of the isolated Janin cluster, and finally to an embedded all-atom model, elucidates the importance of charge transfer effects induced by the adsorbate. The overemphasis of charge depletion of the isolated Janin cluster all-atom model is attenuated by charge compensation resulting from embedding the Janin cluster into a 32-atom Pt reservoir. The reduced amplitudes of the $\Delta\mu$ signatures place the magnitudes of the negative dips in Figures 18.11e,f and 18.10b,c with noise levels. Given the 1 eV bandwidth of a Si(111) monochromator, and the inability to align edges to better and 0.1 eV, a detailed interpretation of the derivativelike features of experimental fingerprints is a bit risky. A safer strategy is full interpretation of the unnormalized XAS data, such as what was done to yield the analyses of Tables 18.1 and 18.2.

Figure 18.12b shows the operando potential-dependent $\Delta\mu$ fingerprints of a carbon-supported Johnson Matthey Pt cathode catalyst layer with the IR-XAS cell operated as a hydrogen–air fuel cell at 50°C, and the as-prepared membrane electrode assembly before installation into the fuel cell. The cathode gas diffusion layer was exposed to the beamline hutch atmosphere (i.e., $CaF_2$ window not installed) while humidified hydrogen was delivered to the Pd counterreference electrode. After a one-hour conditioning period [20,21], a reference XANES was obtained at 0 Vs. Potential-dependent X-ray absorption spectra were then sequentially acquired and subtractively normalized to the reference spectrum to yield $\Delta\mu$ fingerprints. At open-circuit voltage, the white-line intensity increases dramatically. The ex situ dry MEA fingerprint shows that the cathode is highly oxidized before exposure to the fuel cell environment, consistent with the results of Lewis et al. [60].

Figure 18.12a shows time-dependent fingerprints (at 530 mV) [46], each obtained as the average of three consecutive scans. The potential-dependent data are smoother than the time-dependent data because 30 scans were averaged at each potential-

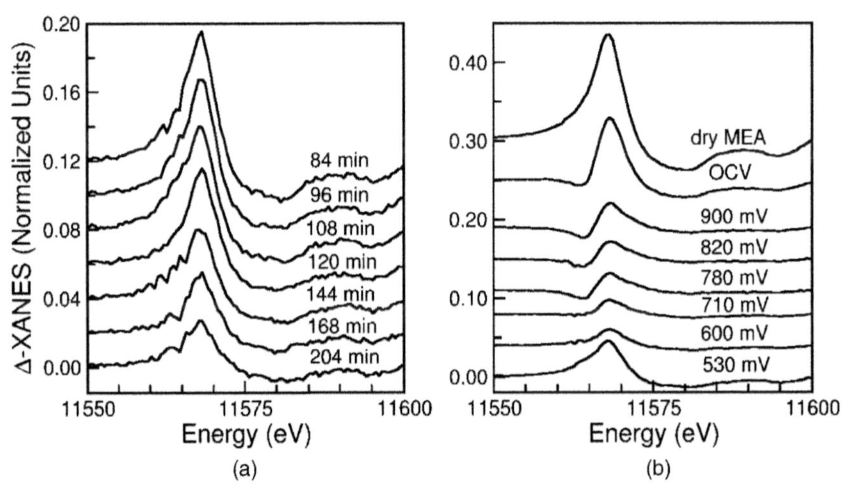


**FIGURE 18.12** XANES of an air-breathing cathode of the IR-XAS cell subtractively normalized to XANES at 0 V versus Pd hydrogen anode: (a) time-dependent data at 530 mV; (b) potential-dependent steady-state data.

dependent fingerprint. The time-dependent fingerprints decrease with time, as the Pt is reduced, with a time constant on the order of hours. The time constant was unexpectedly large and cannot be attributed only to reduction of chemisorbed oxygen. The reduction of subsurface oxygen species, which have been incorporated into the crystallite core structures, may be responsible for the long-time constant. This has ramifications concerning the use of $\Delta\mu$ signatures for analysis of $\Delta\mu$ fingerprints as discussed by Lewis et al. The $\Delta\mu$ fingerprints are affected by changes in the core structure of the catalyst, and not only by adsorbed species.

The potential-dependent data of Figure 18.12b exhibit negative dips around 11,564 eV at potentials of 780, 820, and 900 mV. Charge transfer would shift an XANES edge to higher energy if there were charge depletion at the atoms contributing the X-ray absorption events. This would result in derivativelike features such as observed in the $\Delta\mu$ fingerprints. Issues other than those introduced by Lewis, et al. complicate interpretation of derivativelike features of $\Delta\mu$ fingerprints. The-time dependent $\Delta\mu$ fingerprints obtained at 530 mV elucidate a time constant on the order of hours. The oxidation–reduction of platinum involves chemisorbed surface oxygen species as well as distinct oxide phases that extend into the core of the catalyst. Stoupin et al. have shown that in the case of PtRu, the oxide phase extends into the core of particle. The data of Figure 18.12 suggest that subtractively normalized XANES, with time and potential dependence, have enormous value for elucidation of trends. At this time, we have a poor understanding of the processes that contribute to the time constant(s) associated with the data of Figure 18.12b. Because the results were unexpected, we were unable to reorganize the design of experiments to more pointedly address these slower processes within the limited beamtime available. Further work is in progress.

## 18.3 CONCLUSION

The time constants probed by the work discussed above are intermediate, on the order of minutes to hours. Equally important to an understanding of operando fuel cell catalysts are time regimes shorter than a few seconds and those on the order of catalyst lifetimes. Tada et al. [61]. have reported a method for obtaining 1-s time-resolved full EXAFS spectra from a fuel cell being cycled between 0.4 and 1.0 V. They have observed separate time constants for charging and discharging, as well as the formation and dissociation of surface Pt–O bonds using a fuel cell design very similar to that of Viswanathan in transmission mode. Even shorter time constants will be possible using dispersive XAFS techniques, and lifetime studies using catalysts kept in operation for hundreds of hours will become possible as operando fuel cell become more available.

## REFERENCES

1. H. Topsoe, Developments in operando studies and in situ characterization of heterogeneous catalysts, *J. Catal.* **216**(1–2), 155–164 (2003).

2. L. Ploense, M. Salazar, B. Gurau, and E. S. Smotkin, Proton spillover promoted isomerization of n-butylenes on Pd-black cathodes/Nafion 117, *J. Am. Chem. Soc.* **119**(47),11550–11551 (1997).
3. L. Ploense, M. Salazar, B. Gurau, and E. S. Smotkin, Spectroscopic study of NEMCA promoted alkene isomerizations at PEM fuel cell Pd-Nafion cathodes, *Solid State Ion.* **136–137**, 713–720 (2000).
4. M. Salazar and E. S. Smotkin, Electrochemically promoted olefin isomerization reactions at polymer electrolyte fuel cell membrane electrode assemblies, *J. Appl. Electrochem.* **36**(11),1237–1240 (2006).
5. Z. Liu, R. G. Arnold, E. A. Betterton, and E. Smotkin, Reductive dehalogenation of gas-phase chlorinated solvents using a modified fuel cell, *Environ. Sci. & Technol.* **35**(21), 4320–4326 (2001).
6. E. Reddington, A. Sapienza, B. Gurau, R. Viswanathan, S. Sarangapani, E. S. Smotkin, and T. E. Mallouk, Combinatorial electrochemistry: A highly parallel, optical screening method for discovery of better electrocatalysts. *Science* **280**(5370), 1735–1737 (1998).
7. E. S. Smotkin, FTIR and X-ray absorption spectroscopy of operating fuel cells. in *In-Situ Spectroscopic Studies of Adsorption at the Electrode and Electrocatalysis*, S.-G. Sun, P. A. Christensen, and A. Wieckowski, eds., Elsevier, Oxford, UK, 2007, pp. 247–272.
8. S. Gottesfeld and T. A. Zawodzinski, Polymer electrolyte fuel cells, in *Abstracts of Papers, 234th ACS National Meeting,* Boston, Aug. 19–23, 2007, R. C. Alkire and H. Gerischer, et al., eds. Wiley-VCH; 1997, Vol. **5**, pp. 229–244.
9. L. Liu, C. Pu, B. Viswanathan, Q. Fan, R. Liu, and E. S. Smotkin, Carbon supported and unsupported Pt-Ru anodes for liquid feed direct methanol fuel cells, *Electrochim. Acta* **43**(24),3657–3663 (1998).
10. L. Liu, R. Viswanathan, R. Liu, and E. S. Smotkin, Methanol oxidation on Nafion spin-coated polycrystalline platinum and platinum alloys, *Electrochem. Solid-State Lett.* **1**(3),123–125 (1998).
11. S. Stoupin, E.-H. Chung, S. Chattopadhyay, C. U. Segre, and E. S. Smotkin, Pt and Ru X-ray absorption spectroscopy of PtRu anode catalysts in operating direct methanol fuel cells, *J. Phys. Chem. B* **110**(20), 9932–9938 (2006).
12. D. M. Bernardi and M. W. Verbrugge, A mathematical model of the solid-polymer-electrolyte fuel cell, *J. Electrochem. Soc.* **139**(9), 2477–2491 (1992).
13. D. M. Bernardi and M. W. Verbrugge, Mathematical model of a gas diffusion electrode bonded to a polymer electrolyte, *AIChE J.* **37**(8), 1151–1163 (1991).
14. P. T. Kissinger and W. R. Heineman, Solution resistance in potentiostatic circuits, in *Laboratory Techniques in Electroanalytical Chemistry*, 2 ed. , Marcel Dekker, New York, 1996, p. 199.
15. H. Rivera, J. S. Lawton, D. E. Budil, and E. S. Smotkin, Effect of sorbed methanol, current, and temperature on multicomponent transport in nation-based direct methanol fuel cells, *J. Phys. Chem. B* **112**(29), 8542–8548 (2008).
16. B. Gurau and E. Smotkin, Methanol crossover in direct methanol fuel cells: A link between power and energy density, *J. Power Sources* **112**, 339–352 (2002).
17. A. J. Bard and L. R. Faulkner, *Electrochemical Methods: Fundamentals and Applications*, 2nd ed., Wiley, New York, 2001, pp. 135, 253.
18. R. Buzzoni, S. Bordiga, G. Ricchiardi, G. Spoto, and A. Zecchina, Interaction of $H_2O$, $CH_3OH$, $(CH_3)(2)O$, $CH_3CN$, and pyridine with the superacid perfluorosulfonic acid

membrane Nafion—An IR and RAMAN study, *J. Phys. Chem.* **99**(31), 11937–11951 (1995).

19. P. Piela, C. Eickes, E. Brosha, F. Garzon, and P. Zelenay, Ruthenium crossover in direct methanol fuel cell with Pt-Ru black anode, *J. Electrochem. Soc.* **151**(12), A2053–A2059 (2004).
20. R. Viswanathan, G. Hou, R. Liu, S. R. Bare, F. Modica, G. Mickelson, C. U. Segre, N. Leyarovska, and E. S. Smotkin, In-situ XANES of carbon-supported Pt-Ru anode electrocatalyst for reformate-air polymer electrolyte fuel cells, *J. Phys. Chem. B* **106**(13), 3458–3465 (2002).
21. R. Viswanathan, R. Liu, and E. S. Smotkin, In situ x-ray absorption fuel cell, *Rev. Sci. Instrum.* **73**(5), 2124–2127 (2002).
22. R. Basnayake, Z. Li, S. Katar, W. Zhou, H. Rivera, E. S. Smotkin, D. J. Casadonte, Jr., and C. Korzeniewski, PtRu nanoparticle electrocatalyst with bulk alloy properties prepared through a sonochemical method, *Langmuir* **22**(25), 10446–10450 (2006).
23. S. Stoupin, H. Rivera, Z. Li, C. U. Segre, C. Korzeniewski, D. J. Casadonte, Jr., H. Inoue, and E. S. Smotkin, Structural analysis of sonochemically prepared PtRu versus Johnson Matthey PtRu in operating direct methanol fuel cells, *Phys. Chem. Chem. Phys.* **10**(42), 6430–6437 (2008).
24. Z.-B. Wang, H. Rivera, X.-P. Wang, H.-X. Zhang, P.-X. Feng, E. A. Lewis, and E. S. Smotkin, Catalyst failure analysis of a direct methanol fuel cell membrane electrode assembly, *J. Power Sources* **177**(2), 386–392 (2008).
25. R. J. K. Wiltshire, C. R. King, A. Rose, P. P. Wells, M. P. Hogarth, D. Thompsett, and A. E. Russell, A PEM fuel cell for in situ XAS studies, *Electrochim. Acta* **50**(25–26), 5208–5217 (2005).
26. C. Roth, N. Martz, T. Buhrmester, J. Scherer, and H. Fuess, In-situ XAFS fuel cell measurements of a carbon-supported Pt-Ru anode electrocatalyst in hydrogen and direct methanol operation, *Phys. Chem. Chem. Phys.* **4**(15), 3555–3557 (2002).
27. C. Roth, N. Benker, T. Buhrmester, M. Mazurek, M. Loster, H. Fuess, D. C. Koningsberger, and D. E. Ramaker, Determination of O[H] and CO coverage and adsorption sites on PtRu electrodes in an operating PEM fuel cell, *J. Am. Chem. Soc.* **127**(42), 14607–14615 (2005).
28. F. J. Scott, C. Roth, and D. E. Ramaker, Kinetics of CO poisoning in simulated reformate and effect of Ru island morphology on PtRu fuel cell catalysts as determined by operando X-ray absorption near edge spectroscopy, *J. Phys. Chem. C* **111**(30), 11403–11413 (2007).
29. S. Maniguet, R. J. Mathew, and A. E. Russell, EXAFS of carbon monoxide oxidation on supported Pt fuel cell electrocatalysts, *J. Phys. Chem. B* **104**(9), 1998–2004 (2000).
30. J. Singh, M. Tromp, O. V. Safonova, P. Glatzel, and J. A. v. Bokhoven, In situ XAS with high-energy resolution: The changing structure of platinum during the oxidation of carbon monoxide, *Catal. Today* (in press).
31. A. E. Russell, S. Maniguet, R. J. Mathew, J. Yao, M. A. Roberts, and D. Thompsett, In situ X-ray absorption spectroscopy and X-ray diffraction of fuel cell electrocatalysts, *J. Power Sources* **96**(1), 226–232 (2001).
32. F. J. Scott, S. Mukerjee, and D. E. Ramaker, CO coverage/oxidation correlated with PtRu electrocatalyst particle morphology in 0.3 M methanol by in situ XAS, *J. Electrochem. Soc.* **154**(5), A396–A406 (2007).

33. M. Teliska, W. E. O'Grady, and D. E. Ramaker, Determination of H adsorption sites on Pt/C electrodes in $HClO_4$ from Pt $L_{23}$ X-ray absorption spectroscopy. *J. Phys. Chem. B* **108**(7), 2333–2344 (2004).
34. M. Teliska, V. S. Murthi, S. Mukerjee, and D. E. Ramaker, Site-specific vs specific adsorption of anions on Pt and Pt-based alloys, *J. Phys. Chem. C* **111**(26), 9267–9274 (2007).
35. M. Teliska, D. E. Ramaker, V. Srinivasamurthi, and S. Mukerjee, In situ determination of O(H) adsorption sites on Pt based alloy electrodes using X-ray absorption spectroscopy, *Proc. Electrochem. Soc.* (issue on fundamental understanding of electrode processes), **2003–30**, 212–216 (2005).
36. C. Korzeniewski and S. Pons, In situ infrared spectroelectrochemistry, *J. Vac. Sci. Technol., B: Microelectron. Nanometer Struct* **3**(5), 1421–1424 (1985).
37. S. Pons, C. Korzeniewski, R. B. Shirts, and A. Bewicks, Field-induced infrared absorption in metal surface spectroscopy: The electrochemical Stark effect, *J. Phys. Chem.* **89**(11), 2297–2298 (1985).
38. Q. Fan, C. Pu, and E. S. Smotkin, In situ fourier transform infrared-diffuse reflection spectroscopy of direct methanol fuel cell anodes and cathodes, *J. Electrochem. Soc.* **143**(10)(1996).
39. K. L. Ley, R. Liu, C. Pu, Q. Fan, N. Leyarovska, C. Segre, and E. S. Smotkin, Methanol oxidation on single-phase Pt-Ru-Os ternary alloys, *J. Electrochem. Soc.* **144**(5), 1543–1548 (1997).
40. S. Sanicharane, A. Bo, B. Sompalli, B. Gurau, and E. S. Smotkin, In situ 50°C FTIR spectroscopy of Pt and PtRu direct methanol fuel cell membrane electrode assembly anodes, *J. Electrochem. Soc.* **149**(5), A554–A557 (2002).
41. A. Bo, S. Sanicharane, B. Sompalli, Q. Fan, B. Gurau, R. Liu, and E. S. Smotkin, In situ Stark effects with inverted bipolar peaks for adsorbed CO on Pt electrodes in 50°C direct methanol fuel cells, *J. Phy. Chem. B* **104**, 7337–7381 (2000).
42. I. Tkach, A. Panchenko, T. Kaz, V. Gogel, K. A. Friedrich, and E. Roduner, In situ study of methanol oxidation on Pt and Pt/Ru-mixed with Nafion anodes in a direct methanol fuel cell by means of FTIR spectroscopy. *Phys. Chem. Chem. Phys.* **6**(23), 5419–5426 (2004).
43. Q. Fan, C. Pu, K. L. Ley, and E. S. Smotkin, In situ FTIR-diffuse reflection spectroscopy of the anode surface in a direct methanol/oxygen fuel cell, *J. Electrochem. Soc.* **143**(2), L21–L23 (1996).
44. Q. Fan, C. Pu, and E. S. Smotkin, *In-Situ FTIR-Diffuse Reflection Study of Methanol Oxidation Mechanisms on Fuel Cell Anodes*, IEEE, Washington, DC, USA, 1996, pp. 1112–1116.
45. Q. B. Fan, C. Pu, and E. S. Smotkin, In situ Fourier transform infrared-diffuse reflection spectroscopy of direct methanol fuel cell anodes and cathodes, *J. Electrochem. Soc.* **143**(10),3053–3057 (1996).
46. E. A. Lewis, C. U. Segre, and E. S. Smotkin, Embedded cluster Δ-XANES modeling of absorption process on Pt, *Electrochim. Acta* **54**(28), 7181–7185 (2009).
47. B. Gurau, and E. S. Smotkin, Methanol crossover in direct methanol fuel cells: A link between power and energy density, *J. Power Sources* **112**(2), 339–352 (2002).
48. Q. Jia, E. A. Lewis, C. Grice, E. S. Smotkin, and C. U. Segre, In situ XAFS studies of the oxygen reduction reaction on carbon supported Pt and PtNi(1:1) catalysts, in *Proc XAFS14, J. Phys. Conf. Series, Camerino, Italy*, 2009.

49. I. A. Abrikosov, A. V. Ruban, H. L. Skriver, and B. Johansson, *Phys. Rev. B* **50**, 2039 (1994).

50. B. Hammer and J. K. Nørskov, *Adv. Catal.* **45**, 71 (2000).

51. S. Mukerjee, S. Srinivasan, and M. P. Soriaga, Role of structural and electronic properties of Pt and Pt alloys on electrocatalysis of oxygen reduction. An in situ XANES and EXAFS investigation, *J. Electrochem. Soc.* **142**(5), 1409–1422 (1995).

52. M. Mavrikakis, B. Hammer, and J. K. Nørskov, *Phys. Rev. Lett.* **81**, 2819 (1998).

53. J. K. Nørskov, T. Bligaard, A. Logadottir, S. Bahn, L. B. Hansen, M. Bollinger, H. Bengaard, B. Hammer, Z. Slivancanin, M. Mavrikakis, Y. Xu, S. Dahl, and C. J. H. Jacobsen, *J. Catal.* **209**, 275 (2002).

54. J. R. Kitchin, J. K. Nørskov, M. A. Barteau, and J. G. Chen, Modification of the surface electronic and chemical properties of Pt(111) by subsurface 3d transition metals, *J. Chem. Phys.* **120**(21), 10240–10246 (2004).

55. H. J. Kim, S. M. Choi, S. H. Nam, M. H. Seo, and W. B. Kim, *Catal. Today* (in press).

56. M. Teliska, W. E. O'Grady, and D. E. Ramaker, Determination of O and OH adsorption sites and coverage in situ on Pt Electrodes from Pt $L_{23}$ X-ray absorption spectroscopy, *J. Phys. Chem. B* **109**(16), 8076–8084 (2005).

57. M. K. Oudenhuijzen, J. A. van Bokhoven, J. T. Miller, D. E. Ramaker, and D. C. Koningsberger, Three-site model for hydrogen adsorption on supported platinum particles: influence of support ionicity and particle size on the hydrogen coverage, *J. Am. Chem. Soc.* **127**(5), 1530–1540 (2005).

58. A. L. Ankudinov, B. Ravel, J. J. Rehr, and S. D. Conradson, Real-space multiple-scattering calculation and interpretation of x-ray-absorption near-edge structure, *Phys. Rev. B* **58**(12), 7565 (1998).

59. A. L. Ankudinov, A. I. Nesvizhskii, and J. J. Rehr, Dynamic screening effects in x-ray absorption spectra, *Phys. Rev. B* **67**(11), 115120 (2003).

60. E. A. Lewis, Q. Jia, H. Rivera, C. Grice, S. Stoupin, M. Fuccillo, C. Segre, and E. S. Smotkin, Time and potential dependent in situ XANES and FTIR of operating fuel cells, paper presentated at *214th Meeting of the Electrochemical Society, Honolulu*, 2008.

61. M. Tada, S. Murata, T. Asasoka, K. Hiroshima, K. Okumura, H. Tanida, T. Uruga, H. Nakanishi, S. Matsumoto, Y. Inada, M. Nomura, and Y. Iwasawa, In situ time-resolved dynamic surface events on the Pt/C cathode in a fuel cell under operando conditions, *Angew. Chem. Int. Ed.* **46**, 4310–4315 (2007).

# CHAPTER 19

# New Concepts in the Chemistry and Engineering of Low-Temperature Fuel Cells

FIKILE R. BRUSHETT and PAUL J. A. KENIS

Department of Chemical and Biomolecular Engineering, University of Illinois at Urbana–Champaign, Urbana, Illinois

ANDRZEJ WIECKOWSKI

Department of Chemistry, University of Illinois at Urbana—Champaign, Urbana, Illinois

Exploration of alternative energy sources and of more efficient energy conversion methods is needed in light of dwindling traditional resources such as oil and natural gas and rising global energy demands. The continued development of fuel cell technologies plays a pivotal role in the complex structure of the evolving energy market, especially for portable electronics and automotive applications. In this chapter we report on recent advances in science and engineering of low-temperature fuel cells. First, we discuss new fuel cell configurations based on novel physicochemical concepts, including alkaline membrane, mixed-feed, and membraneless or laminar flow-based fuel cells. Second, we discuss progress in fuel cell catalyst development highlighting new ex situ and in situ analytical techniques. While over 200 references are cited, this chapter presents a representative, but not exhaustive, coverage of modern low-temperature fuel cell topics.

## 19.1 INTRODUCTION

### 19.1.1 Modern Energy Challenges

Global industrial development is approaching the point where new strategies must be implemented to manage energy supply and demand. Rising population and industrial

*Fuel Cell Science: Theory, Fundamentals, and Biocatalysis*,
Edited by Andrzej Wieckowski and Jens K. Nørskov

growth in the developing world, particularly in China and India, have led to increasing global energy consumption. Simultaneously, a concerted effort is needed to confront the fast-disappearing reserves of fossil fuels, namely, oil and natural gas. These fuels are often mined in regions of the world that may be considered unfriendly to North America, Europe, and Australia, which are still the main energy consumers. Furthermore, the inefficiency of and significant greenhouse gas emissions due to the direct combustion of fossil fuels (i.e. coal) makes this route increasingly undesirable. Therefore, a need exists for (1) discovery and large-scale production of alternative fuels and (2) high-efficiency utilization of these fuels via novel energy conversion systems such as fuel cells.

Fuel cells enable efficient conversion of chemical energy to electrical energy without the *Carnot cycle* limitations. Moreover, alternative fuels, specifically, hydrogen, formic acid, and bioalcohols (generated from green processes using renewable energy sources, such as biomass, wind, solar) can be utilized by fuel cells, depending on the anode catalyst choice. Thus, the continued development of fuel cell technologies is of vital importance to strategies aimed at addressing the world's current energy challenges.

### 19.1.2 Low-Temperature Fuel Cells

Low-temperature fuel cells (henceforth referred to as *fuel cells*) have been the focus of intensively industrial and academic research efforts for a wide range of energy conversion applications at temperatures no higher than 120°C [1–3]. Compared to modern batteries, fuel cells offer higher energy densities. Moreover, unlike batteries, fuel cells may be instantaneously recharged. Compared to internal combustion engines, fuel cells offer higher efficiencies and reduced emissions. Fuel cells are typically classified according to electrolyte type, although exceptions exist; in particular, direct liquid fuel cells (DLFCs) are often defined by the fuel used. Historically, fuel cells, such as phosphoric acid fuel cells (PAFCs) and alkaline fuel cells (AFCs), employed stationary aqueous electrolytes [1]. Modern fuel cells utilize submillimeter acidic polymer electrolyte membranes (PEMs) to selectively conduct ions and block the passage of unreacted fuels and reaction byproducts (Fig. 19.1). Nafion, a persulfonated polytetrafluoroethylene developed by DuPont, is the most widely used polymer electrolyte material, although other polyaromatic-sulfonate ionomers are also used [4–8]. PEM-based fuel cells can be broadly divided into two categories: hydrogen-fueled PEM-based fuel cells (PEMFCs) and (DLFCs) using small alcohols (i.e., methanol, ethanol), formic acid or ethers as fuels (Fig. 19.2) [1–4,9–16]. PEMFCs can be considered ecofriendly, zero-emission power sources, as the only generated byproduct is water. However, hydrogen fuel must be generated, ideally from water or biomass using the abovementioned renewable solar, nuclear, or wind energies. Among the most developed fuel cell technologies, PEMFCs have been successfully demonstrated in stationary, extraterrestrial, and, most recently, automotive applications [1–3]. For portable applications, safety concerns and practical issues associated with the onboard storage of hydrogen at high energy density have spurred the development of DLFCs, which benefit from the high energy density and

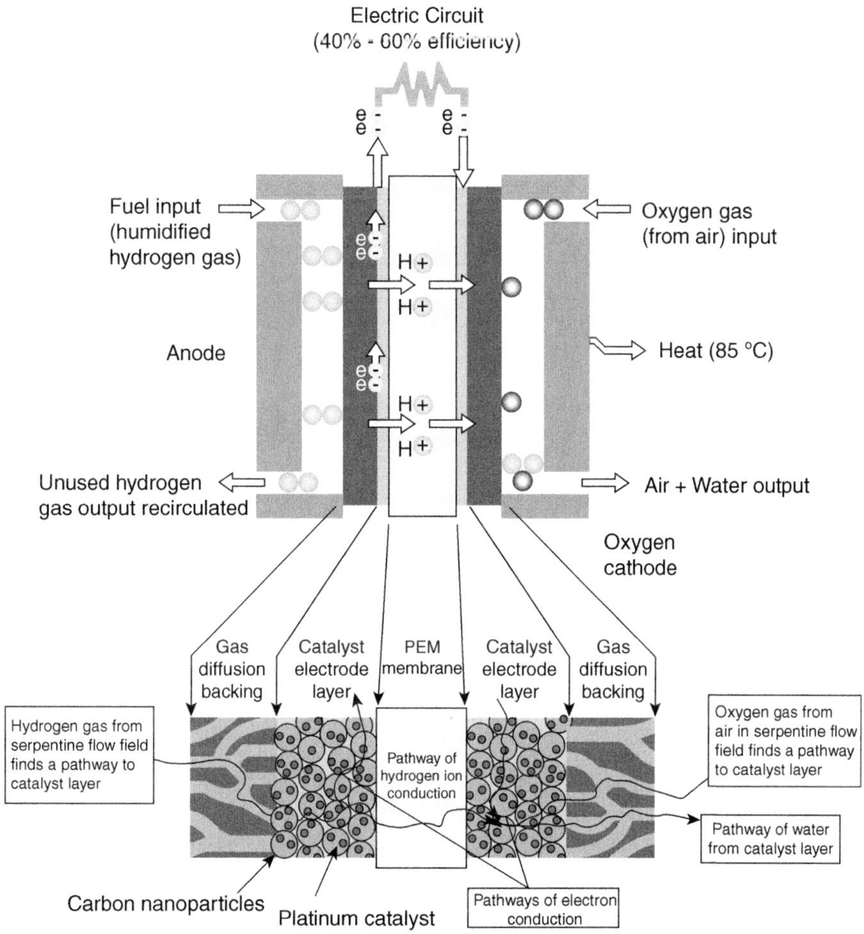


**FIGURE 19.1** Schematic of a zero-emission, hydrogen-fueled acidic polymer electrolyte membrane fuel cell (PEMFC). Hydrogen electrooxidizes on the anode, splitting into protons and electrons. These electrons travel around an external circuit powering a load whereas the protons migrate through a conductive PEM (i.e., Nafion) which separates the anode and cathode. On the cathode, oxygen combines with the protons and electrons and electroreduces to produce water. The water is then removed from the fuel cell by the now-depleted oxygen stream. Image courtesy of the Neutron Imaging Facility at the National Institute of Standards and Technology (NIST).

easier storage of organic fuels. Unlike PEMFCs, DLFCs produce the greenhouse gas carbon dioxide ($CO_2$), via organic fuel electrooxidation, but the emission rate per energy unit is significantly lower than the $CO_2$ production rate of conventional energy conversion systems because of the much higher efficiency of fuel-cell-based systems. Both direct methanol fuel cells (DMFCs) and direct formic acid fuel cells (DFAFCs) have experienced moderate success in niche applications, namely, military and telecommunications.

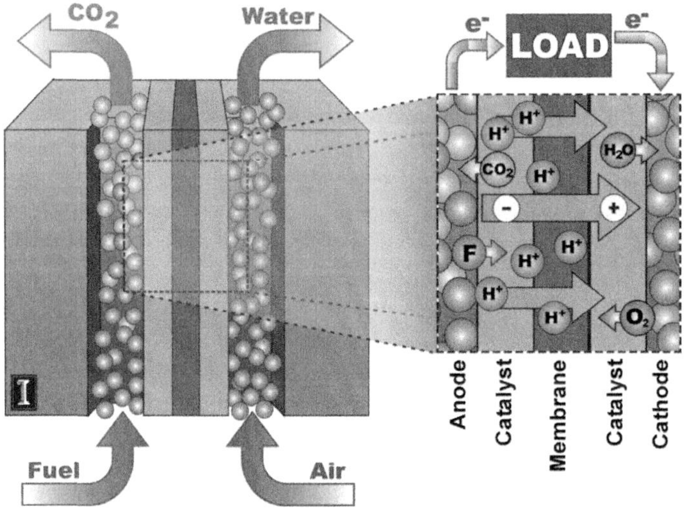


**FIGURE 19.2** Schematic of an acidic direct liquid fuel cell (DLFC) utilizing a small organic fuel (e.g., methanol, ethanol, formic acid). The fuel electrooxidizes to CO2 on the anode producing protons and electrons. These electrons travel around an external circuit powering a load whereas the protons migrate through a conductive PEM (i.e., Nafion) which separates the anode and cathode. On the cathode, oxygen combines with the protons and electrons combine and electroreduces to produce water. The water is then removed from the fuel cell by the now-depleted oxygen stream. Reprinted with the permission of Joshua Ackerman.

Despite some success, widespread market penetration of fuel cell technologies has yet to be realized mainly because of high system costs (particularly catalysts), poor durability, and—to a lesser extent—system performance limitations. First, the limited availability of expensive platinum (Pt) is a significant concern, as it is typically the main component of both anode and cathode fuel cell catalysts (see Section 19.3). As Pt is mass-produced in only two world regions, South Africa and Russia, its natural abundance is low. The continued participation of these countries in the free-market economy is not guaranteed. Furthermore, with future large-scale production of fuel cells, Pt prices would likely increase causing significant perturbations in the market. Thus, minimizing or eliminating the Pt content of fuel cell catalysts while maintaining acceptable power output and electrode durability remains a major challenge. In addition, despite continued advances in the synthesis and theory of conducting polymers, PEM materials remain prohibitively expensive.

Second, the insufficient long-term stability of the present PEM-based fuel cell technologies, coupled with the already-mentioned high costs, thus far have prevented widespread commercialization. For example, according to the US Department of Energy (DOE) guidelines, fuel cell systems for automotive applications must provide power for 5000 h and 150,000 mi to be comparable with modern internal combustion engines [3,17]. Presently, most PEM-based fuel cell systems operate efficiently for only a fraction of that required lifetime before power output decreases. Improved understanding of the complex electrochemical, transport, and degradation processes that govern the performance of individual components (catalyst layers, carbonaceous catalyst–support material, gas diffusion layers, membranes, etc.) *within an operating fuel cell* is key to designing more robust fuel cell configurations. Probing in situ degradation mechanisms is a significant challenge and requires the development of novel analytical techniques and platforms.

Third, the performance of PEM-based fuel cells is often limited by membrane-related issues, mainly water management, including anode dryout and cathode flooding, and fuel crossover. In acidic fuel cells, anode dryout occurs during operation at high current densities because of the osmotic drag of water molecules along with protons transported across the membrane, from the anode to the cathode. The osmotic drag, in combination with water formation, causes flooding of the cathode, which hampers oxygen transport to electrocatalytic sites. For AFCs these processes occur on the opposite electrodes, cathode dryout and, anode flooding (see Section 19.2.1). Also, operating fuel cells at elevated temperatures ($\geq$80°C) further complicates water management, as the PEM must stay fully hydrated to maintain sufficient conductivity and thus optimal performance. Several active (i.e., electroosmotic pumps) [18,19] and passive (i.e., component modification) [20] water management strategies have been proposed to overcome this limitation. Unfortunately, such strategies often require ancillary components that complicate fuel cell design and reduce overall system energy density. Fuel crossover occurs when unreacted fuel migrates (via electroosmosis and/or diffusion) through the membrane and reacts on the cathode causing mixed potentials, thereby reducing performance [21]. The crossover may be mitigated by diluting fuel streams, but this, in turn, reduces system energy density.

While significant scientific and technical challenges remain, fuel cell technologies benefit from enormous investments by the governments (and industries) of the United States, Japan, and Europe. Consequently, systematic increases in performance, durability, and concomitant cost reductions have been achieved for all fuel cell types since the early 1990s [10]. For this reason, one may look optimistically at the future development and widespread implementation of a range of fuel cell technologies in society.

In this chapter, our focus is twofold: (1) to highlight new fuel cell configurations based on novel physicochemical concepts (i.e., alkaline membrane, mixed feed, laminar flow) that overcome some present fuel cell design limitations and improve performance and (2) to discuss current progress in the theory and understanding of fuel cell reaction catalysis and the development of new analytical methods to investigate modern fuel cell catalysts.

## 19.2 TECHNOLOGY AND PROGRESS IN FUEL CELL DEVELOPMENT

At present, Nafion-based fuel cells are the "configuration of choice" for most low-temperature applications (i.e., portable electronics, automotive transport) [1–5,9–16,22,23]. However, challenges related to durability, cost, and overall system performance have prevented widespread market penetration to date [3]. Here, we review three more recently introduced configurations: alkaline membrane, mixed-feed, and membraneless or laminar flow-based fuel cells, which may offer new pathways forward.

### 19.2.1 Alkaline Membrane Fuel Cells

Operating fuel cells under alkaline conditions, as opposed to acidic conditions, is advantageous as enhanced fuel oxidation and oxygen reduction kinetics improves fuel cell efficiency and enables the use of cheaper non-precious metal catalysts (i.e., nickel (Ni)-based anodes, silver (Ag)-based cathodes), which dramatically reduces overall fuel cell system costs [24–28]. Furthermore, during operation in alkaline media, fuel crossover is reduced because the gradient of produced anions, such as hydroxyl ions, is opposite—from cathode to anode—compared to the proton gradient—from anode to cathode-in acidic fuel cells. Traditional AFCs used for vehicular studies in the 1950s and the National Aeronautics and Space Administration (NASA) Apollo space program in the 1960s were among the first successful applications of fuel cell technologies [1,2]. As mentioned previously, these AFC configurations utilized a stationary liquid alkaline electrolyte, usually highly concentrated potassium hydroxide (i.e. 30–45 wt% KOH) [1]. Major technical concerns that have hampered AFCs include (1) lack of long-term electrode durability in caustic environments, (2) carbonate formation when oxidizing organic fuels directly, and, to a lesser extent (3) water management at the electrodes (anode flooding and cathode dryout) [24,25,29,30]. Electrode degradation commonly occurs via radical chain destruction of hydrophobic binder polymers within the catalytic layer, which leads to "electrode weeping" where the electrolyte floods the entire electrode, and mechanical instability in the catalyst layer causing catalyst leaching [29]. Carbonate formation ($CO_3^{2-}/HCO_3^-$) occurs when the hydroxyl ions ($OH^-$) present in the electrolyte react with carbon dioxide ($CO_2$) from either organic fuel oxidation, hydrogen reformate obtained from biomass, or the quiescent environment. In the presence of mobile cations, the carbonates precipitate somewhere within the electrodes, eventually clogging the microporous architecture and blocking electrocatalytic sites, which reduces AFC performance and lifetime. Furthermore, carbonate formation reduces the hydroxyl ion concentration in a stationary electrolyte, thus gradually reducing electrode kinetics and electrolyte conductivity. AFC performance also can be hampered by water management at each electrode. At the anode, if all the hydrogen is consumed, then no gas remains to sweep the formed water away from the anode, causing flooding, which, in turn, hinders hydrogen transport to the catalytic sites. Thus, hydrogen must be in stoichiometric excess, or a carrier gas, such as nitrogen, must be used to limit anode flooding. Furthermore, at high current densities, significant amounts of water are consumed at the cathode, which cannot be replenished sufficiently quickly, leading to dryout. As a consequence of these adverse effects, traditional AFC technology was limited to applications where cost is not a concern, allowing for the use of the high-purity hydrogen and oxygen gas streams, with $CO_2$ concentrations below the parts per million (ppm) level.

To address some of the aforementioned issues, a number of AFCs with circulating electrolyte streams have been commercially developed [29–34]. The convecting electrolyte stream, again typically 30–45 wt% KOH, improves and lifetime compared to the stationary electrolyte [29]. The flowing aqueous electrolyte improves water management at each electrode by removing water to the anode and supplying water for the cathode. Furthermore, the electrolyte stream facilitates carbonate removal from the electrode surface, which reduces precipitation and enables higher $CO_2$ tolerance ($\sim$100 ppm) in AFCs [29]. A flowing electrolyte stream also enables the development of alkaline-based DLFCs, such as alkaline DMFCs, because any carbonates formed are immediately removed from the electrode surface by the dynamic stream [34]. While such flowing electrolyte configurations prolong AFC lifetimes, as compared to stationary electrolyte configurations, long-term electrode stability in the caustic environment remains a challenge [29]. In addition, the electrolyte solution must periodically be replenished or replaced to maintain conductivity and prevent carbonate precipitation due to saturation. Furthermore, circulating electrolytes require ancillary pumping and storage, which increases both parasitic losses and system complexity. Leakage of concentrated electrolyte can also be a safety concern [31].

Emerging alkaline anion exchange membrane (AAEM) fuel cell technologies may provide solutions to issues associated with operation in alkaline media while maintaining the electrocatalytic advantages of AFCs [25,35]. Several commercial alkaline anion exchange membrane technologies are available, including Morgane ADP membranes (Solvay SA, Belgium), AHA ammonium-type membranes, and NeoSepta ion exchange membranes (both from Tokuyama Corporation, Japan) [35]. In addition, a number of academic researchers have developed their own membrane technologies. Xing and Savadago investigated the performance of polybenzimidazole (PBI) membranes doped with various alkaline salt solutions in a PEMFC [36]. At the University of Surrey, Slade, Varcoe, and coworkers have pioneered relatively cheap fabrication methods for producing high-performance AAEMs via radiation-grafting techniques [25,37–46]. In this synthesis route, commercial polymer films are modified by grafting monomeric endgroups onto the preformed polymeric backbone, followed by animation and alkali exchange reactions to yield the desired AAEM. Compared to the more conventional membrane development methods, this technique benefits from the widespread availability of polymer films, which not only eliminates complex film formation steps but also provides a range of adjustable experimental parameters (e.g., film thickness, radiation conditions), enabling versatile design strategies. Lu et al. developed high-performance alkaline quaternary ammonium polysulfone (QAPS) membranes with high thermal stability ($T \sim 120°C$) and tunable conductivity [47]. Unlike the radiation-grafted AAEMs developed by Slade and Varcoe, QAPS membranes may be dissolved in certain solvents such as *N,N*-dimethylformamide enabling membrane deposition via casting, as well as ionomer addition to catalyst inks leading to impregnated catalyst layers with reduced interfacial resistances. For a detailed analysis of the state of the art in AAEM technology, including patent literature, the review by Varcoe and Slade is recommended [25].

In AAEMs, the mobile $OH^-$ ions are the active species and AAEMs are less susceptible to $CO_3^{2-}/HCO_3^-$ poisoning because no mobile cations exist *within the membrane*. Note that precipitation of carbonate can still occur if reactant streams or gas diffusion electrodes contain soluble metal ion contaminants [17]. As a result of these promising membrane developments, AAEM-based DLFCs may soon become viable alternative power sources for low-temperature applications [48–57]. However, in AAEM-based DLFCs, the formation of carbonate species remains a significant challenge. Wang et al. performed a comprehensive feasibility analysis of the thermodynamic disadvantages versus kinetic advantages of using AAEM-based direct methanol fuel cells (AAEM-DMFCs) by studying electrode reaction mechanisms in aqueous solution [57]. The authors demonstrated that, at room temperature, a substantial pH gradient (6.1 pH units) exists between the cathode (due to $OH^-$ production) and the anode (due to $CO_3^{2-}/HCO_3^-$ generation), leading to a large voltage loss (~360 mV). By operation of the AAEM-DMFC at elevated temperatures ($T \sim 80°C$), the thermodynamic drawbacks can be minimized because (1) the pH gradient and consequent voltage loss between electrodes is reduced and (2) the reaction kinetics at both electrodes is improved, which counters the voltage losses. Furthermore, the aforementioned reduction in methanol crossover due to an opposing ion gradient accounts for another voltage gain. Therefore, Wang et al. conclude that the kinetic voltage gains would roughly cancel out the thermodynamic voltage losses in an AAEM-DMFC operated at elevated temperatures [57]. Unfortunately, such operating conditions are challenging because of poor AAEM stability at higher temperatures and increased parasitic losses associated with heating the fuel cell [25]. To minimize the adverse effects of carbonate formation, several authors have introduced alkaline species into the aqueous anodic fuel stream to control the pH and to assist in the electrooxidation reaction [50,54–56]. Adams et al. reported a high-performance, carbon-dioxide-tolerant alkaline membrane that utilizes carbonates, rather the hydroxides, as the mobile ion [58]. Remarkably, this novel carbonate-based AAEM outperforms an equivalent hydroxide-based AAEM in both $H_2$/air and methanol/air fuel cells. The authors attribute this enhanced performance to improved oxygen reduction reaction kinetics in alkaline carbonates as compared to aqueous KOH solutions, although the exact reaction mechanism remains poorly understood [24,58]. Note that carbonate-based AAEM stability remains unknown as the continuous production of $OH^-$ at the cathode will reduce $CO_3^{2-}/HCO_3^-$ species (i.e., self-purging).

While the various AAEM technologies have been improved dramatically since the late 1990s, several research challenges remain. Improving the durability of AAEMs under fuel cell operating conditions, especially at elevated temperatures, remains a key challenge. In addition, AAEMs remain expensive. One possible strategy for reducing costs and improving environmental credentials would be the use of nonfluorinated polymeric films as the AAEM backbones [25]. Optimizing membrane electrode assembly (MEA) structure (e.g., minimizing interfacial resistances, nanostructuring electrocatalytic materials, developing porous high-surface-area catalyst support architectures), is also essential in developing efficient AAEM-based fuel cell systems. However, the performance of AAEM-based fuel cells, as with PEM-based fuel cells, remain hindered by identical membrane-related limitations, most notably water management at each electrode. Moreover, the inherent alkalinity of AAEMs limits ways to fine-tune the electrolyte composition with respect to optimizing individual electrode reaction kinetics. Consequently, the complete elimination of conducting polymer electrolyte membranes has been pursued in several new fuel cell architectures.

### 19.2.2 Mixed-Reactant Fuel Cells (MRFCs)

In conventional membrane-based fuel cells, both acidic and alkaline, the main functions of the polymeric membrane are to facilitate ion transport and to prevent reactant crossover. If fuel or oxidant crosses over to the cathode or anode, respectively, parasitic chemical reactions cause mixed potentials at each electrode, which reduces the overall fuel cell potential as well as its efficiency. Furthermore, unwanted side reactions can lead to catalyst poisoning and reduce fuel cell durability. This problem is particularly acute in DMFCs due to high methanol diffusivity in ionomeric membranes [59]. In practice, Pt cathodes experience a potential drop of a few hundred millivolts when methanol crossover occurs, due to competition between the methanol oxidation and the oxygen reduction reactions, as well as poisoning of the catalyst surface by carbon monoxide (CO, see Section 19.3). The poor selectivity and tolerance of common fuel cell electrocatalysts require strict segregation of the fuel and oxidant streams in most fuel cells, which, in turn, necessitates the use of bulky bipolar plates with strictly separated flow fields for the fuel and oxidants streams.

In contrast to conventional membrane-based fuel cells, mixed-reactant fuel cells (MRFCs) employ highly selective anode and cathode electrocatalysts to separate fuel oxidation and oxidant reduction reactions, respectively (Fig. 19.3) [60,61]. The earliest MRFC configurations were developed in the 1950s to generate electricity using hydrogen and oxygen produced by splitting water using nuclear fission processes. Utilizing an alkaline electrolyte, these original designs first reduced the oxygen from the gaseous mixture using a selective cathode (e.g., C, Ag, Au) and then oxidized the remaining hydrogen on the anode (e.g., Pt, Pd) [62]). In subsequent years, numerous gas- and liquid-phase MRFCs have been developed as comprehensively

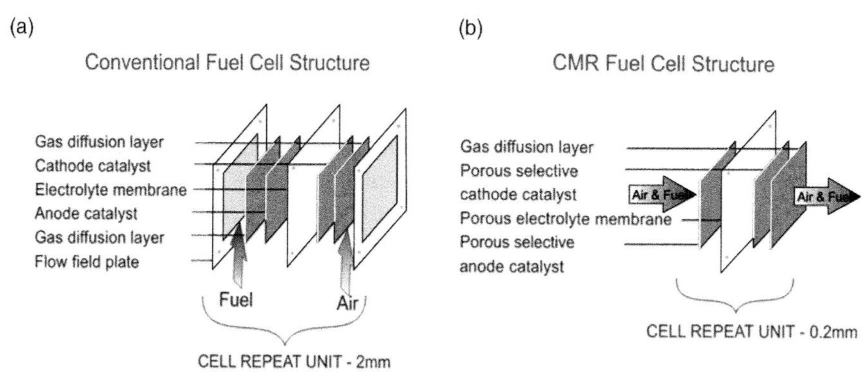


**FIGURE 19.3** Comparative schematics of (a) a conventional PEM-based fuel cell and of (b) a compact mixed reactant (CMR) fuel cell with a flow-through porous-MEA design.

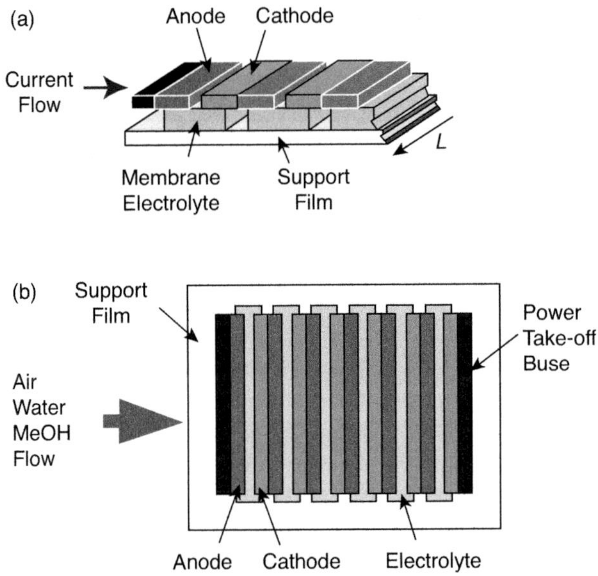


**FIGURE 19.4** Simplified schematic of a strip-cell mixed reactant direct methanol fuel cell from the (a) side-view and (b) top-down view. Reprinted from Ref. 63, 2001 by permission of Elsevier.

reviewed by Priestnall et al. [60]. Of the MRFC designs described in the literature, the two most common configurations are the strip cell and the porous membrane-electrode assembly (MEA)-based designs. Barton et al. described a strip cell MRFC configuration where strips of solid anode and cathode materials are interdigitated over strips of porous membrane electrolyte and on the same side of a nonconducting support film (Fig. 19.4) [63]. In this design the mixed fuel and oxidant stream follow over the anode and cathode strips and through the porous ionomer, which serves as an electrical connection between electrodes. Sung et al. demonstrated a microscale strip cell configuration using an aqueous alkaline electrolyte in the mixed-reactant stream rather than a porous stationary membrane [64]. The porous MEA-based design, as described by several authors, consists of two perforated gas diffusion electrodes that are positioned on either side of the porous membrane or sometimes liquid electrolyte to form a porous anode electrolyte cathode assembly (Fig. 19.3b) [60,61]. In these configurations, the mixed-reactant stream flows through the entire porous MEA. For both configurations, the mixed-reactant stream flows through the individual MRFC compared to a conventional membrane-based fuel cell, where individual reactant streams flow across electrodes. This flow-through design improves mass transport to the electrodes, eliminates water management issues, and facilitates the removal of reactant products such as carbonate species. The individual MRFC technologies may be easily scaled up into cheap fuel cell stacks with high volumetric energy density due to the elimination of gastight structures and the simplification of manifolding, which dramatically reduces the system volume and costs compared to membrane-based systems [60]. These design advantages have spurred research efforts to develop high-performance compact MR-DMFC-based power sources to compete with conventional DMFC-based power sources [60,61,63–68].

Unfortunately, at present, ideally selective and active electrocatalysts for MRFCs remain undiscovered. Indeed, those catalysts that demonstrate substantial selectivity are not as active towards fuel oxidation or oxygen reduction as are conventional unselective fuel cells catalysts. To address this challenge, researchers are exploring methanol-tolerant cathode catalysts for MR-DMFCs as well as for conventional, PEM-based DMFC designs, to alleviate the adverse effects of methanol crossover. Ruthenium–selenium (Ru–Se)-based chalcogenides are among the most promising methanol-resistant oxygen reduction catalysts [69–78]. Despite displaying excellent methanol tolerance and substantial oxygen reduction activity (although still quite lower than that of Pt), other studies indicate that Ru–Se-based catalysts may not be sufficiently stable for extended use in operating fuel cells [79]. Johnston et al. performed durability studies with Ru–Se-based cathodes in a $H_2$/air fuel cell and found evidence of Se migration from the cathode to the anode [79]. The stability of these chalcogenides may be improved by addition of late transition metals such as molybdenum (Mo) and tungsten (W) [75,80]. Transition metal (i.e. Co, Fe) porphyrins have also been investigated as methanol-resistant oxygen reduction catalysts. Iron tetramethyloxyphenylporphyrin (FeTMPP) has been reported as the most active of the transition metal porphyrin catalysts [61,66,81]. Shukla et al. investigated the performance of FeTMMP/C, CoTMMP/C, FeCoTMMP/C, and RuSe/C cathodes in a MR-DMFC and obtained the best results using the RuSe/C cathode [61]. Other authors have studied Pt transition metal (i.e., Ni, Fe, Co) alloys, similar to those used for DMFC anodes, as methanol-tolerant cathode catalysts [61,66,82]. While these catalysts exhibit enhanced oxygen reduction activities compared to pure Pt, prolonged exposure to acidic conditions leads to transition metal leaching and performance losses (see Section 19.3.2). Moreover, Pt transition metal alloy anode catalysts, such as Pt-Ru, have also shown instability in conventional DMFCs. Piela et al. reported Ru migration from the anode to the cathode, causing a significant decrease in cell voltage, up to $\sim$200 mV [83].

In addition to these electrocatalytic challenges, several MRFC system design issues remain. Inefficient electrode catalysis and high volumetric energy density in MRFC-based systems can lead to significant heat generation. Consequently, MRFC stacks must be designed with thermally conductive materials to act as a heat sink. Also, oxidant delivery is challenging because air is either dissolved into the electrolyte or entrained as bubbles. Unfortunately, oxygen solubility and diffusivity in water and/or electrolyte is low, thus limiting cathode performance. Furthermore, gas–liquid two-phase flows can lead to intermittent electrode reactions and unstable current densities.

Novel MRFC configurations are a promising alternative to conventional membrane-based fuel cells because of their lower cost, smaller size, and higher volumetric power density. However, future success of MRFC technologies clearly hinges on the development of electrocatalysts that demonstrate high activity, selectivity, and long-term stability under fuel cell operating conditions.

### 19.2.3 Membraneless Fuel Cells

*Microfluidics* can be defined as science and engineering of fluidic and transport phenomena within structures with at least one characteristic dimension in the range of tens to hundreds of micrometers [84,85]. At the microscale, fluid flow is typically laminar and characterized by low Reynolds numbers; viscous effects dominate over inertial effects and surface forces are more relevant than body forces [86]. Consequently, multiple streams may merge side by side in a single microchannel with only diffusive mixing occurring in between two streams [87]. This precise control of transport properties enables a wide range of applications, including drug discovery, protein crystallization, biomedical analysis, microfabrication, and energy conversion [87–89]. The Kenis group [77,90–97] and others [86,98–111] have exploited these microfluidic phenomena to develop a class of *membraneless* fuel cells that are also referred to as laminar flow–based fuel cells (LFFCs). The laminar nature of flow eliminates the need for a physical barrier, such as an expensive polymeric membrane, while still allowing for ionic transport between the anode and the cathode. Furthermore, these fuel cells benefit from high surface area to volume ratios and cheap component fabrication techniques, both of which facilitate the development of inexpensive high-energy-density microscale fuel cell systems [86].

The membraneless architecture overcomes some of the aforementioned membrane-related challenges.

1. *Fuel crossover can be minimized by adjusting cell dimensions and stream flow rates* [90,95,97]. For example, with operation of a membraneless fuel cell at a high Peclet number, (i.e., $Pe > 3000$), the streamwise convective velocity dominates the transverse diffusive velocity, thus restricting diffusive mixing to a narrow interfacial width at the center of the channel that broadens as a function of downstream position and mean velocity [90]. In addition, Jayashree et al. demonstrated that fuel crossover may be further reduced in membraneless fuel cells by hydrodynamically focusing the fuel into a thin stream on the anode by varying the stream flow rate ratios, thus widening the envelope of operating conditions [95]. Thus, compared to conventional membrane-based DMFCs, membraneless DMFCs exhibit higher open-circuit potentials and consequently improved fuel cell efficiencies [93]. Moreover, unlike MR-DMFCs, membraneless DMFCs do not require selective, but less active, methanol-tolerant cathode catalysts [77].
2. *Water management issues are absent in these all-aqueous membraneless fuel cells*. Thus challenges such as electrode dryout/flooding and stream humidification, which hamper both acidic and alkaline membrane-based fuel cells, can be avoided.
3. *The flowing streams also remove byproducts of electrode reactions.* This enables operation of membraneless alkaline DLFCs as the constantly refreshing electrolyte removes any formed carbonates from the electrode surface, and hinders the development of adverse pH gradients [93,96].

4. *The composition of the fuel, electrolyte, and oxidant streams can be specified independently.* This not only allows for operation of membraneless fuel cells with different fuels and different media (i.e., acidic, alkaline, mixed) but also enables optimization of reaction kinetics at the individual electrodes [91,92,96,99]. By placement of an external reference electrode (e.g., Ag/AgCl in 3 M NaCl) at the fuel cell outlet, anode and cathode performance characteristics can easily be independently monitored in an operating fuel cell [92]. More recently, Brushett et al. demonstrated the benefits of stream flexibility by investigating LFFC performance with five different fuels (formic acid, methanol, ethanol, hydrazine, and sodium borohydride) in either acidic or alkaline media [96]. Operation under alkaline conditions significantly improves methanol and ethanol oxidation kinetics and stabilizes sodium borohydride. Furthermore, the exact pH of the electrolyte solutions can be tailored to optimize anode performance via several in situ diagnostics techniques, such as impedance spectroscopy [96].

The unique properties of membraneless fuel cells enable the exploration of unusual configurations including mixed-media fuel cells and bio-fuel cells. Cohen et al. demonstrated a planar membraneless fuel cell utilizing dissolved hydrogen in alkaline media and dissolved oxygen in acidic media with an open-circuit potential in excess of 1.4 V, significantly higher than the 1.23 V obtained in conventional fuel cells [99]. Unfortunately, mixed-media fuel cells are unsustainable as power sources because of the exothermic neutralization of $OH^-$ and $H^+$ ions to water that occurs at the liquid–liquid interface reducing ionic strength and increasing ohmic resistances. Choban et al. showed that interfacial electrolyte consumption reduces the overall energy density of the mixed-media systems, to approximately one-fifth that of the corresponding all-acidic and all-alkaline systems [91]. The LFFC concept is also well suited for the development of biofuel cells that use enzymes, rather than traditional metallic catalysts (i.e., Pt, Pd), to catalyze the electrode reactions [112–121]. In an early example, Moore et al. presented a biofuel cell powered by flowing ethanol over a bioanode with an external Pt cathode [112]. A major challenge for biofuel cells is to maintain enzyme stability and activity, which is highly dependent on the local pH. Autonomous control of individual stream compositions enables the optimization of individual electrode environments within membraneless fuel cells. This, in turn, facilitates the development of stable and highly active bioelectrodes. This initial stabilization enables the design of advanced enzyme support structures (e.g., Nafion membranes modified by quaternary salts), which further stabilize the microenvironments surrounding enzymatic catalysts and improve electron transport to/from the enzyme from/to the current collector [119]. Consequently, substantial growth in the area of biofuel cell research can be expected in the near future.

Since the inception of microfluidic fuel cells in 2002, a wide array of membraneless architectures have been developed, including a commercial venture, INI Power Systems (Morrisville NC), based on technology developed in the Kenis group at the University of Illinois at Urbana-Champaign. The most common configuration is the

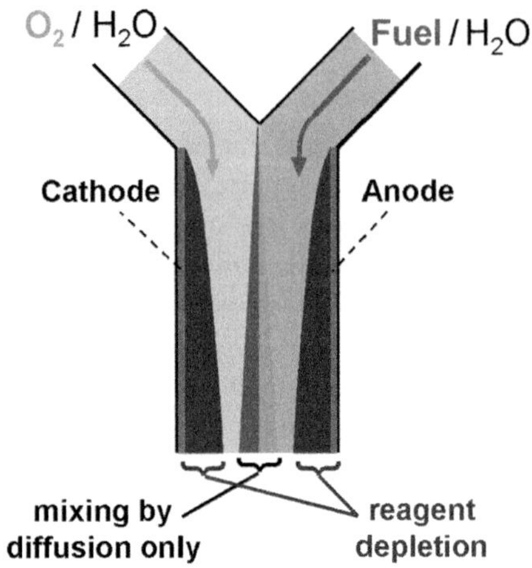


**FIGURE 19.5** Schematic of a Y-shaped laminar flow-based fuel cell (LFFC) developed by the Kenis group. The electrodes are the catalyst-coated sidewalls on opposite sides of the microchannel (red lines). The cell operates by flowing an anodic fuel stream (green) and a cathodic oxygen-saturated stream (yellow) in parallel. Though not shown, both streams also include a supporting electrolyte to enhance conductivity. The reaction depletion at the electrodes and diffusional mixing zones at the liquid-liquid interface are both highlighted (not drawn to scale). Reprinted with the permission of Professor Paul J.A. Kenis. (See insert for color representation.)

Y- or T-shaped microchannel design with two aqueous fuel and oxidant streams flowing in parallel in a laminar fashion and catalyst-coated electrodes on the opposing channel sidewalls [90]. In a seminal publication, Choban et al. demonstrated one of the earliest microfluidic fuel cells employing a Y-shaped design with aqueous anode and cathode streams of formic acid and dissolved oxygen in sulfuric acid ($H_2SO_4$) solution, respectively (Fig. 19.5) [90]. The fuel cell consisted of two precision-machined graphite plates horizontally aligned to form a Y channel and sealed on the top and bottom by a polydimethylsiloxane (PDMS) and polycarbonate layers, respectively. Anode (palladium-Pd) and cathode (Pt) nanoparticle catalysts were hand-painted onto the opposing channel sidewalls. The multilayer structure was held together using binder clips enabling rapid assembly and disassembly. This initial design demonstrated peak current and power densities of $\sim$14 mA/cm$^2$ and 5 mW/cm$^2$ due to mass transport limitations at the cathode [90]. Furthermore, fuel utilization per pass was only 0.04%, due primarily to the limiting cathode performance characteristics [94]. The cause of these cathode restrictions are twofold: (1) the low diffusivity of oxygen in solution ($\sim 2 \times 10^{-5}$ cm$^2$/s) and (2) the low dissolved oxygen saturation concentrations (1 – 4 mM), both of which cannot sufficiently replenish the depletion boundary layer that forms on the cathode.

A possible strategy to reduce these cathode limitations is the use of oxidants that are soluble at concentrations higher than those of dissolved oxygen, such as hydrogen peroxide, potassium permanganate, and hypochlorite bleach [90,103,105]. Use of high-energy-density liquid oxidants can be challenging. For example, hydrogen peroxide reduction is often coupled with decomposition that leads to vigorous gaseous oxygen evolution, which perturbs the laminar regime and often leads to two-phase behavior [103]. These adverse effects can be mitigated by (1) operating at low concentrations [122], (2) introducing a third stream to reduce interfacial interactions [106], (3) utilizing alternative catalysts (e.g., gold-Au) to minimize hydrogen peroxide decomposition [123], and/or (4) optimizing the electrode architecture (e.g., porous gas diffusion electrodes, grooved microstructure) to quickly utilize the gaseous oxygen [103,124]. More recently, Kjeang et al. presented an alkaline microfluidic fuel cell operating with formate and hypochlorite bleach that produced no gaseous products [105]. Note that for the microfluidic fuel cell, systems that utilize liquid oxidants must store and pump these additional fluids, which leads to an increase in parasitic losses and reduced system energy density.

The cathode limitations of early fuel cell designs have been overcome by the introduction of an air-breathing cathode into the membraneless architecture [77,93–96]. This passive configuration benefits from the relatively high concentration (~10 mM) and high diffusivity of oxygen ($\sim 2 \times 10^{-1}\ cm^2/s$) in the ambient environment. Jayashree et al. demonstrated a first membraneless fuel cell with an integrated air-breathing cathode (Fig. 19.6) [94]. The fuel cell consisted of a polymethylmethacrylate (PMMA) window that forms the microfluidic channel, a graphite plate coated with Pd nanoparticles as an anode, and a Toray carbon paper gas diffusion electrode covered with Pt nanoparticles as a cathode. The anode and cathode streams enter the cell through inlets bored into the graphite plate and travel through the microchannel in a vertically oriented laminar fashion. An aqueous mixture of formic acid and sulfuric acid is used for the anode stream. As oxygen is now supplied to the cathode directly

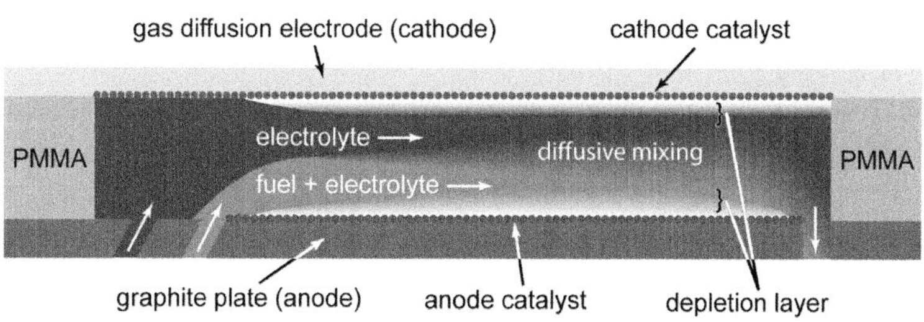


**FIGURE 19.6** Schematic of an air-breathing laminar flow-based fuel cell (LFFC) developed by the Kenis group and shown in Brushett et al. [96]. The electrodes are a catalyst-coated graphite plate as an anode and a catalyst-coated gas diffusion electrode on opposite sides of the microchannel (red lines). The cell operates by flowing an anodic fuel and electrolyte stream (green) and a cathodic electrolyte stream (blue) which serves as a barrier to fuel crossover. Oxygen diffuses to the cathode from the ambient environment. The reaction depletion at the electrodes and diffusional mixing regions at the liquid-liquid interface are both highlighted (not drawn to scale). Reprinted from Ref. 96, 2009 by permission of Elsevier. (See insert for color representation.)

from ambient environment, the cathode stream is only a conductive electrolyte (i.e., $H_2SO_4$) that serves as a barrier to fuel crossover. Implementation of the air-breathing cathode resulted in a 5-fold increase in the peak power density and a 10-fold increase in the maximum current density under near-identical operating conditions [94], compared to the performance of the first LFFC as reported by Choban et al. [90].

For all membraneless fuel cell designs, a tradeoff exists between maximizing power density and fuel utilization. For example, the *power density* of a membraneless fuel cell can be improved by reducing the electrode-to-electrode distance, increasing the stream flow rates, and/or increasing the fuel concentrations. Unfortunately, all of these improvements hamper both *fuel utilization* and system efficiency. By decreasing the electrode-to-electrode gap, the cell resistances are minimized and power density increases. However, an interelectrode gap that is too small will enhance fuel crossover, thus negatively affecting the *maximum* power density and also reducing fuel utilization. Moreover, in terms of balance of plant (BOP), parasitic losses will increase as a result of an increase in pressure drop in narrower channels. Increasing the stream flow rates can improve cell performance by minimizing both fuel crossover, as a result of shorter residence times, and the boundary-layer thickness, as a result of enhanced mass transport through a thinner depletion boundary layer. However, increasing the stream flow rates directly increases parasitic losses associated with pumping. Furthermore, the shortened residence time decreases fuel utilization as the fuel has less time to diffuse to and react at the anode. Increasing the fuel concentration improves cell performance by enhancing mass transport to the anode through the depletion boundary layer, due to the increased concentration gradients. But increasing fuel concentration increases the extent of fuel crossover, thus reversing the desired increase in power density. In addition, fuel utilization again decreases because of the fuel crossover, and because a significant fuel fraction must remain unreacted to maintain the high concentration gradients. On the other hand, fuel utilization may be increased by reducing fuel concentration and stream flow rates, both of which significantly lower cell power density.

Enhancing both fuel utilization and overall power density is a challenging task that requires different strategies, including novel operating methods and innovative cell designs. For example, Jayashree et al. optimized power density and fuel utilization of an air-breathing membraneless fuel cell by varying operational parameters such as stream flow rates and stream ratios, and structural parameters such as channel length and interelectrode distance [95]. By hydrodynamic focusing the fuel to a thin stream on the anode, the authors were able to increase the fuel utilization up to 38% without reducing stream flow rates or increasing fuel crossover. Using such a focusing technique, both fuel utilization and power density may be increased as it enables the use of higher concentrations of fuel and oxidant. Altering structural parameters of the microchannel in which the reactants flow can further improve performance and fuel utilization [108–110,125,126]. For example, Ahmed et al. developed a "trident-shaped" design that used electrolyte stream in the channel center to focus both the fuel and oxidant streams onto their respective electrodes [108]. In another interesting example, Yoon et al. investigated different active and passive methods of minimizing the depletion boundary layer, which limits cell performance: (1) using multiple outlets to remove the depleted regions, (2) using multiple inlets to replenish the depleted regions, and (3) using a herringbone structure to generate secondary transverse flow, enabling chaotic mixing within a single laminar stream, to replace the depleted layer with fluid of higher fuel concentration [126]. In all three examples, the performance improvements obtained from the design modifications must be balanced with either the increased system complexity and parasitic pumping losses associated with additional fluidic streams, or the possibility of increased fuel crossover associated with the chaotic advection. Another strategy to improve both cell performance and fuel utilization is to increase electrode surface area to further exploit the high surface area to volume ratio in the microscale fuel cells. By introducing porous high surface area electrode structures, instead of directly depositing catalysts on channel sidewalls, Kjeang and coworkers have developed several promising microfluidic fuel cell configurations that utilize so-called flow-through electrode architectures to improve both fuel utilization and power density [102–105,111]. Note that these flow-through configurations would suffer from significant parasitic pumping losses. For further reading, a more recent comprehensive review by Kjeang et al. provides an excellent summary of the present microfluidic fuel cell technologies. In this review, the authors highlight a range of fascinating novel configuration and compare the performance of each [86].

Transitioning from these laboratory-scale "proof of concept" configurations to commercially viable microfluidic fuel-cell-based power sources is a present challenge. In addition to general engineering challenges associated with designing conventional fuel cell system, several unique design constraints exist for membraneless fuel cells. As the membraneless fuel cell technologies are based on microscale phenomena, large-scale fuel cell systems must *scale out* rather than scale up to maintain the laminar flow regime. Indeed, while a variety of structural and operating parameters can be adjusted to improve the performance of microfluidic fuel cells, the system design is dependent on the explicit intended application with specific power and operational lifetime requirements as well as weight and volume restrictions. First, to obtain a certain absolute power level, a certain number of individual membraneless fuel cells will need to be arranged in an array; then to avoid having to integrate the capability to circulate the fluids within the BOP, a larger number of membraneless fuel cells can be run at lower power density but optimized fuel utilization in a single pass. Alternatively, one can run fewer cells at higher power density but low fuel utilization per pass (e.g., 25%) in a fuel and electrolyte recirculation scenario. The latter case would make more sense for a larger system that is desired to run for an extended period of time. As mentioned previously, varying the number and design of membraneless unit fuel cells in an array is relatively inexpensive, due to well-established micromanufacturing procedures and the lack of a membrane (although catalyst costs remain high). On the basis of this scaling-out concept, INI Power Systems have successfully developed a 20-W prototype direct methanol LFFC-based system [127]. Considering the remarkable growth of this fuel cell technology since the early 2000s, one can expect a number of microfluidic fuel-cell-based power sources to become commercially available in the next few decades.

## 19.3 ELECTROCHEMICAL COMBUSTION IN FUEL CELLS: CONCEPTS AND METHODS

The exploration and enhancement of electrocatalytic materials for the electrode reactions is a vital component to the continued advancement of fuel cell technologies. The high cost and limited availability of Pt necessitates the development of inexpensive alternative catalysts that either reduce or eliminate Pt content while maintaining high activity and stability under the harsh environment of an operating fuel cell [128]. Here, more recent advances in anode and cathode catalyst theory and development are highlighted.

### 19.3.1 Electrooxidation Reaction on the Anode

Among the most developed fuel cell technologies, zero-emission PEMFCs have been successfully demonstrated in stationary, extraterrestrial, and, most recently, automotive applications [2]. A major factor is the rapid hydrogen oxidation reaction (HOR) kinetics at the anode under acidic conditions [9]:

$$H_2 \rightarrow 2H^+ + 2e^- \qquad (E^\circ = 0.00\text{V vs. SHE}) \qquad (19.1)$$

If neat hydrogen is the fuel, Pt nanoparticles may be used as the anode catalyst [9,128,129]. If the hydrogen is contaminated by CO (e.g., in hydrogen reformate obtained from biomass or small organic fuels), the Pt can be alloyed with a secondary transition metal (i.e., Ru, Sn, Mo) to minimize the effects of the CO poisoning (see text below). As highlighted above, the present PEMFC technology remains prohibitively expensive mainly, due to the high costs and limited availability of Pt catalysts [3]. Consequently, significant research efforts, particularly in the automotive sector, have been focused on developing PEMFCs with ultralow Pt loadings. Gasteiger et al. note that a 5-fold reduction in Pt content, corresponding to an anode Pt loading of $\sim$0.05 mg/cm$^2$, would be required for use of PEMFC stacks in large-scale automotive applications [128]. Achieving such a reduction in catalyst loading requires (1) substantial improvements in the design and manufacture of porous high-surface-area electrodes to eliminate any mass transport losses [128] and (2) development of high-surface-area conductive catalyst supports to reduce particle aggregation, which improves Pt utilization, and minimize resistive losses within the catalyst layer (i.e., polymerized organic supports, graphite nanowires and nanotubes) [130–133].

In acidic DLFCs, direct electrooxidation of small organic fuels (i.e., methanol, ethanol, formic acid) is both slower and significantly more complex than hydrogen oxidation [9]. In acidic DMFCs, the methanol is directly electrooxidized into $CO_2$ in a multistep reaction, which produces six protons and six electrons:

$$CH_3OH + H_2O \rightarrow CO_2 + 6H^+ + 6e^- \qquad (E^\circ = 0.02\text{V vs. SHE}) \qquad (19.2)$$

As discussed earlier, bimetallic Pt–Ru catalysts demonstrate enhanced electrocatalytic activity in DLFC applications. On bimetallic Pt-Ru catalysts, the oxidation process occurs via a bifunctional mechanism where (1) methanol electroadsorbs to the catalytic Pt sites, (2) an oxygen species adsorbs on the Ru sites, (3) the oxygen species combines with the carbon-containing intermediate formed from methanol to yield $CO_2$, and (4) the $CO_2$ leaves the surface and releases the Pt catalytic sites for further action [9,61,134–137]. Kauranen et al. proposed the following sequence of elementary reaction pathway starting with the dissociative chemisorption of methanol to the Pt surface and subsequent sequential donation of electrons to the catalyst [136].

$$\mathrm{Pt} + \mathrm{CH_3OH} \xrightarrow{k_1} \mathrm{PtCH_2OH} + \mathrm{H}^+ + \mathrm{e}^- \tag{19.3}$$

$$\mathrm{Pt} + \mathrm{PtCH_2OH} \xrightarrow{k_2} \mathrm{Pt_2CHOH} + \mathrm{H}^+ + \mathrm{e}^- \tag{19.4}$$

$$\mathrm{Pt} + \mathrm{Pt_2CHOH} \xrightarrow{k_3} \mathrm{Pt_3COH} + \mathrm{H}^+ + \mathrm{e}^- \tag{19.5}$$

At this point $Pt_3COH$ is the dominant surface species as $k_1 < k_2 < k_3$. This species is subjected to a surface rearrangement leading to surface-bonded CO, in a linear or bridge-bonded form, and another electron donation:

$$\mathrm{Pt_3COH} \rightarrow \mathrm{Pt_3CO} + \mathrm{H}^+ + \mathrm{e}^- \tag{19.6}$$

On a pure Pt surface, for CO surface species to be removed, a water-splitting reaction must occur on an adjacent Pt surface site:

$$\mathrm{Pt} + \mathrm{H_2O} \rightarrow \mathrm{PtOH} + \mathrm{H}^+ + \mathrm{e}^- \tag{19.7}$$

Unfortunately, this last reaction step requires significant applied anodic potential, resulting in significant CO surface poisoning that hinders catalyst performance at higher fuel cell potentials. Alloying of secondary, and sometimes tertiary, transition metals to the Pt surface, a concept first introduced by Watanabe and Motoo, has been the most successful promotion method [138,139]. Adding a secondary transition metal, like ruthenium (Ru) to the Pt surface, lowers the required anode overpotential for water splitting

$$\mathrm{H_2O} + \mathrm{Ru} \rightarrow \mathrm{RuOH} + \mathrm{H}^+ + \mathrm{e}^- \tag{19.8}$$

thus facilitating the oxidation of the surface-adsorbed CO to gaseous $CO_2$ and the completion of the six-electron oxidation process:

$$\mathrm{PtCO} + \mathrm{RuOH} \rightarrow \mathrm{CO_2} + \mathrm{Ru} + \mathrm{Pt} + \mathrm{H}^+ + \mathrm{e}^- \tag{19.9}$$

The role of the Ru promoters on enhancing the catalytic of a Pt surface was confirmed by Maillard et al., who investigated CO oxidations on Pt surfaces decorated with Ru [137]. Both Pt-Ru and Pt-Sn electrocatalysts have demonstrated significantly enhanced methanol oxidation activity compared to pure Pt as the potential regions for chemisorption of methanolic and hydroxyl species overlap [61]. Consequently, the potential of the complete methanol oxidation shifts negatively by ~300 mV versus pure Pt, making the DMFC a practical fuel cell. Note that Herrero et al. confirmed that second possible reaction pathway, typically referred to as the dual pathway mechanism, in which methanol directly oxidizes to $CO_2$, without CO formation [140].

A complete electrooxidation of larger alcohols to $CO_2$ under acidic conditions, most prominently ethanol, remains a significant challenge because of the presence of strong C–C bonds. Consequently, most known catalysts follow unfavorable indirect oxidation mechanism resulting in large byproduct [141–145]. To date, the best known catalysts are Pt-based alloys that utilize Ru, tin (Sn) and indium (In) additives [144]. More recent studies highlight the use of tin oxide and rhodium (Rh) as the best oxidation catalyst for ethanol in acids [145]. For molecules larger than ethanol, the choice of catalysts is even more difficult due to the presence of multiple strong C–C bonds. For formic acid oxidation, Pd exhibits better catalytic activity than does Pt as it directly oxidizes formic acid to CO2 without a CO intermediate [146–149]:

$$HCOOH \rightarrow CO_2 + 2H^+ + 2e^- \quad (19.10)$$

Much like Pt, Pd anode catalyst activity also benefits from the incorporation of transition metal additives [148,149]. CO generation is possible via an unfavorable decomposition side reaction but this only appears in low coverage and does not hamper activity.

$$HCOOH \rightarrow CO + H_2O \quad (19.11)$$

However, Pd-based anode becomes significantly poisoned, while working in a fuel cell. The poisoning species that form on the surface, most likely as a result of Pd oxidation, remain unknown. Formic acid could be regarded as liquid hydrogen: the liquid state formed due to molecular hydrogen bound to $CO_2$ [129,150]. For commercial applications, such as in portable electronics, a 70% aqueous formic acid solution has proved to be an ideal fuel. Further information regarding formic acid oxidation mechanisms on both Pd and Pt may be found in the literature [151,152].

As mentioned previously, electrooxidation reactions of small organics are enhanced under alkaline conditions thus improving the performance of alkaline DLFCs [24,96]. Specifically, strongly adsorbed carbon-containing intermediates formed at the anode, namely, CO, $C_1$ and $C_2$ hydrocarbon residues, oxidize more readily by reacting with surface hydroxyl species adsorbed from the alkaline solution. The more facile fuel oxidation reaction kinetics also enable the introduction of inexpensive Pt-free catalysts (e.g., Ni–Co–Fe catalyst for ethanol oxidation), which are both more stable and more active under alkaline conditions compared to acidic conditions [48]. For further reading on this subject, Spendelow and

Wieckowski provide a detailed review of the electrocatalysis of small organics and oxygen species in alkaline media [24].

### 19.3.2 Electroreduction Reaction on the Cathode

Improving the cathodic oxygen reduction reaction (ORR) remains a key challenge in fuel cell development [153] as the performance and stability of the present catalyst of choice, Pt, is not ideal [154]. Sluggish reaction kinetics and high overpotentials associated with ORR on Pt hamper both fuel cell energetic efficiency and peak power production [155,156]:

$$O_2 + 4H^+ + 4e^- \leftrightarrow 2H_2O \qquad (E^\circ = 1.23\text{V vs. SHE}) \qquad (19.12)$$

The oxygen reduction reaction occurs at a theoretical potential of 1.23 V versus SHE under acidic conditions, but the large activation overpotential, up to 300 mV, that results from formation of oxide species on the Pt surface, reduces the practical ORR potential to near 0.8 V versus SHE. Pt surface oxides begin to destabilize at 0.8 V, which enables the overall ORR to proceed.

As mentioned above, for successful commercialization of fuel cell technologies, expensive Pt catalyst loadings must be reduced as much as possible while still maintaining acceptable power densities [128]. Alloying Pt with transition metals (M = Co, Ni, Fe, Cu, etc.) not only reduces catalyst costs by lowering Pt loading but also demonstrates higher oxygen reduction activity than does pure Pt in acidic media [155,157–161]. In the literature, several different factors have been proposed as the contributors to enhanced catalytic activity, including, but not limited to, changes in the geometric Pt structure (Pt–Pt bond distance), shifts in electronic structure (increased Pt *d*-band vacancy), and increased particle surface roughness and particle size effects [155,157,159–166]. Studies also indicate that under acidic conditions the Pt–M alloys can demonstrate instability and lose oxygen reduction activity as a result of leaching of the transition metals from the alloys [128,167]. Gasteiger et al. reported three major causes for the dissolution: (1) excess transition metal on the Pt surface, (2) incomplete alloying of the two metals, and (3) transition metal leaching for a well-alloyed Pt–M catalyst under acidic fuel cell operating conditions [128]. Also, Antolini et al. noted that reported stabilities of Pt–M alloys in acidic environments are often conflicting and vary widely depending not only on experimental parameters (i.e., temperature, pressure, pH) but also on material characteristics (i.e., degree of alloying) rather than on transition metal choice [167]. Furthermore, the choice of experimental apparatus can significantly impact the reported Pt–M stabilities as authors have observed high stabilities in PEM-based fuel cells and low stabilities in half-cell setups for identical materials [167]. In addition to reducing electrocatalytic activity, leaching transition metal and, to a lesser extent, Pt ions increase ohmic resistances and accelerate MEA degradation in membrane-based fuel cells [128,154,168–170]. The dissolution results in a core–shell configuration consisting of underlying *core* transition metal atoms (e.g., Co,

Fe), which cause stable enhancements to ORR activity of the protective Pt *shell* via geometric and electronic distortions [160,164,171–174]. The best core–shell catalysts display the reduction potential close to 0.9 V, significantly higher than from pure Pt [175]. The newest core-shell catalysts have Pd, Fe, copper (Cu) or Cu-Fe as cores [173,174,176,177].

Many Pt-free oxygen reduction catalysts have been explored, including Ru-based chalcogenides (see Section 19.2.2) [66,69–78], metal oxides [178,179], transition carbides [180], cobalt–polypyrrole–carbon composites [181], enzymes [120,121], and pyrolized porphyrins [66,81,182–185], but none have shown the necessary electrocatalytic activity, stability under acidic conditions, and durability to replace Pt-based catalysts in fuel cell systems. However, quite recently, Lefevre et al. reported on the development of Fe/N/C catalysts that demonstrate high oxygen reduction activity equivalent to that of Pt/C [184,185]. While very promising, the stability of the Fe/N/C catalyst in the harsh environment of the operating fuel cell remains inadequate as the authors reported substantial reductions in current density over 100 operating hours when fuel cell durability studies were performed [184]. Consequently, significant research efforts in academia and industry are devoted to the development and characterization of novel cathode catalysts that exhibit high activity and durability in the harsh acidic environment of an operating fuel cell.

## 19.4 NEW METHODS FOR STUDYING FUEL CELL CATALYSTS

The design and synthesis of novel catalytic materials is vital to the continued development of increasingly powerful yet cheaper fuel cell technologies. In the early 1990s, Watanabe, arguably the father of modern fuel cell electrocatalysis, has contributed pivotally to developing novel nanoscale catalytic structures for fuel cell applications [165]. This discovery, coupled with the increased availability of electron microscopy [165,186] and synchrotron light sources [187], has dramatically enhanced our abilities to design and characterize novel electrocatalysts. To take full advantage of these powerful tools, methods must be developed to rapidly screen and identify materials of interest [188]. Indeed, (1) the synthesis of new Pt-based catalysts [156], (2) the increasing role of Pt additives [156,189,190], and (3) the emergence of Pt-free catalysts [69,70,73,74,191–194], have proved to be essential in the development of increasingly better fuel cell systems. Fuel cell scientists may now use many of the methods developed since the 1970s or so, including both the ex situ ultra-high-vacuum methods [195,196] and in situ infrared (IR) spectroscopy, nuclear magnetic resonance (NMR), and synchrotron X-ray methods [194,197–205]. Nowadays, theory also contributes significantly to the study of fuel cell catalysis due to the giant leap in computational capabilities that has occurred since the late 1980s [206,207].

Two representative in situ synchrotron methods are X-ray absorption near-edge structure (XANES) and extended X-ray absorption fine structure (EXAFS) [187,198,200,201,208–212], in which scattered photoelectrons are collected and

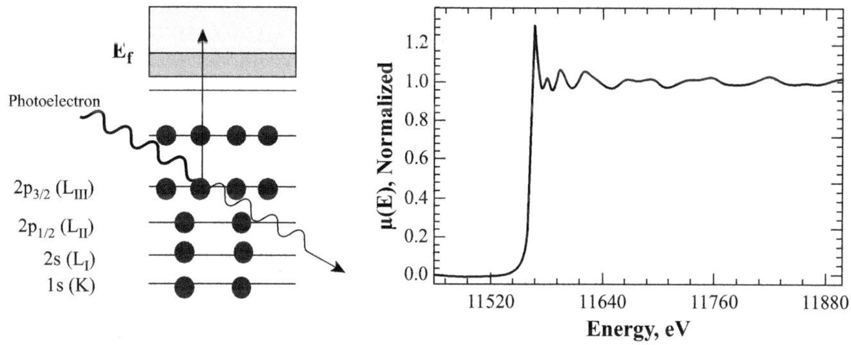


**FIGURE 19.7** General schematic of photoelectron scattering and resulting response as a function of energy region where the red line represents X-ray absorption near-edge spectroscopy (XANES) at $< 50$ eV and the blue line represents extended X-ray absorption fine structure (EXAFS) at $> 50$ eV. XANES studies provide information on absorber site symmetry, electronic configuration and geometric binding sites. EXAFS studies provide information on geometric structure, bond length, coordination number and *bulk* short range order. Courtesy of Professor Sanjeev Mukerjee, Department of Chemistry and Chemical Biology, Northeastern University.

analyzed to determine the energetic structure of catalytic nanoparticles (for EXAFS, see the blue trace in Fig. 19.7). EXAFS produces information on the center of symmetry for the X-ray photons, the electronic configuration of the scattering centers, surface crystallography, and nanoparticle bulk geometry, including bond length and coordination numbers. EXAFS yields data on the oxidation state of nanoparticle catalysts and their electronic distribution and, in particular, indicates the catalyst "*d*-band vacancies" as they depend on the catalyst composition [187].

Aside from broader use of synchrotron techniques, a substantial amount of the progress in the methodology of fuel cell catalysis research can be attributed to the development of two different methods [213]: metal/surface NMR [or electrochemical NMR (EC-NMR)], ideally suited to study nanoparticles [214,215], and broadband sum frequency generation spectroscopy (BB-SFG), ideally suited to investigate smooth solid/liquid interfaces, including single crystals [216,217]. These two methods are discussed in Sections 19.4.1 and 19.4.2.

### 19.4.1 Electrochemical Nuclear Magnetic Resonance (EC-NMR)

Electrochemical NMR, based on the use of such spins as $^{13}C$, $^{195}Pt$, or $^{77}Se$ [194,218–225], has been dedicated exclusively to examination of nanoparticle samples, specifically to determining the electronic state (semiconductor vs. metal) of different catalyst particles. For example, the relaxation time ($T_1$) of the semiconductor selenium ($^{77}Se$) at 80 K (left-hand black line) is ~400 s (Fig. 19.8) [194].

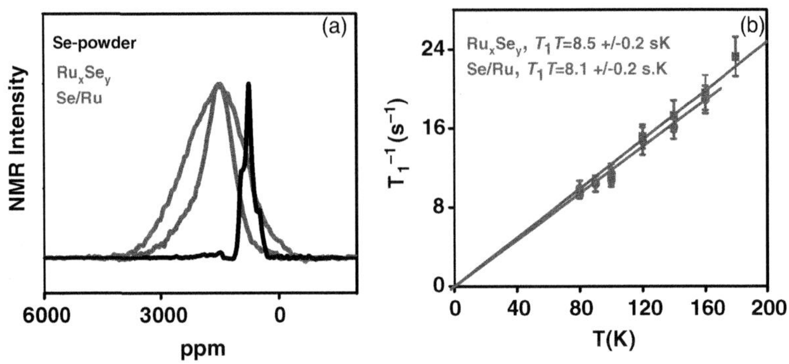


**FIGURE 19.8** Representative electrochemical nuclear magnetic resonance (EC-NMR) to demonstrate the presence of metallic Se deposited on Ru. (a) $^{77}$Se NMR spectra of Se powder (black line), Ru-Se (red line) and $Ru_xSe_y$ (blue line) catalysts obtained at 8.45 T and 80 K. (b) Temperature dependence of the spin-lattice relaxation rate in the two catalyst samples. Both have similar $T_1T$ values. Reprinted from Ref. 194. Copyright 2007 American Chemical Society.

However, the $T_1$ of Se in Se/Ru and $Ru_xSe_y$ at 80 K is much lower and is equal to 0.1 s (the blue and red lines in Fig. 19.10). The Korringa plot (Fig. 19.10b) demonstrates that the product of $T_1$ and $T$ (temperature) is constant, thus proving the metallic character of Se deposited on Ru. The data demonstrate semiconductor–metal transition of Se when one moves from the elemental (powder) Se to a Se chemisorbed states (Fig. 19.9). EC-NMR is also an excellent method for study of surface diffusion. The use of EC-NMRin the motion-sensitive mode has shown fast CO surface diffusion on Pt and Pt-Ru alloy nanoparticles [226–228], which is of relevance to unraveling the rate determining steps of fuel cell anode catalysts. Other mechanistic features of EC-NMR have been reviewed previously [218,220,225] .

### 19.4.2 Sum Frequency Generation (SFG)

A common feature of the sum frequency generation (SFG) methods is that out of two beams illuminating an interface of interest, a third (sum frequency) beam is produced carrying complete information of the vibration characteristics of the interface (Fig. 19.10). The two illuminating beams are usually in the infrared and visible ranges. However, since the reflected infrared beam is no longer used for data analysis, the thickness of the solution film between the optical window and the working electrode surface can be extended to 25 μm [216,217]; that is, in BB-SFG, all vibrations lying within the spectral region defined by the BBIR pulse spectrum are probed simultaneously [229]. Figure 19.11 shows BB-SFG spectra of CO chemisorbed on Pt. The amplitude of the CO spectra disappears when CO is electrooxidized from the surface [216]. Upon replacement of the polycrystalline Pt surface with a Pt(111) surface, a phase transition of surface-bounded CO was found using both SFG and second-harmonic generation (SHG). This detailed insight into

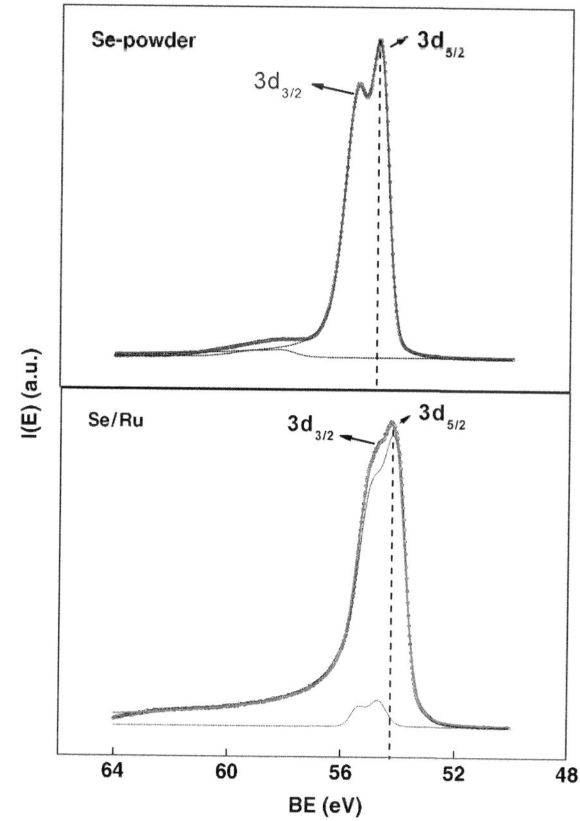


**FIGURE 19.9** Representation of the shift in binding energy from using elemental Se to Se in chemisorbed states, a semiconductor—metal transition. XPS spectra of the Se powder (upper) and the Ru-Se catalyst (lower), in the Se 3d region. The asymmetric line shape seen in the catalyst indicates Se becomes metallic, the shift indicates Ru → Se charge transfer. The smaller peaks observed in both spectra are due to traces of $SeO_2$ and elemental Se. Reprinted from Ref. 194. Copyright 2007 American Chemical Society.

surface chemistry of electrosorbed CO is particularly important in fuel cell research as CO is a major poison for anode catalysts. Indeed, CO poisoning is one of the key factors hampering progress in the development of fuel cell technologies to replace less efficient, less environmentally friendly power sources in a range of low-temperature applications.

## 19.5 NEW METHODS FOR STUDYING FUEL CELL ELECTRODES

Typically, once novel catalysts are shown to exhibit promising activity and stability, they are integrated into MEAs for further characterization in an actual fuel cell. However, confounding factors such as uneven performance across the electrodes,

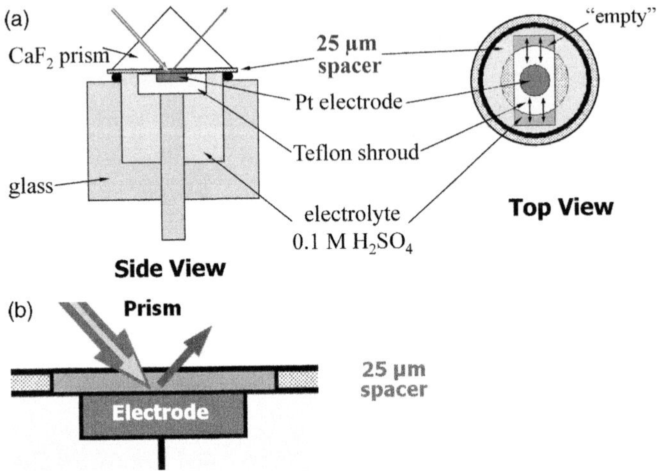


**FIGURE 19.10** Schematic of broadband sum frequency generation thin layer electrochemical cell (BB-SFG TLE). (a) The electrochemical cell has a 25 μm Teflon spacer that provides an electrolyte layer of known and controlled thickness. (b) A thin layer of electrolyte (e.g., 0.1 M $H_2SO_4$) rests in the spacer layer with a calcium flouride prism above to direct the impinging infrared and visible light and generated SFG signal. A Pt electrode is used to both reflect the SFG signal and perform the electrochemical analyses. A CCD detector acquires the broadband SFG spectra while a potentiostat acquires the electrochemical output. The potentiostat and CCD are synchronized with a common trigger. Reprinted from Ref. 217, 2006 by permission of Elsevier.

electrode flooding and dryout, and temperature gradients complicate independent characterization of the electrochemical, transport, and degradation processes taking place at each electrode in the fuel cell configurations typically used for this purpose. To overcome these challenges—*and thus to bridge the gap between fundamental catalyst characterization and fuel cell testing*—novel microfluidic analytical platforms have been developed to probe the performance of novel catalyst and electrode architectures for fuel cell applications [230–232]. Furthermore, imaging techniques for monitoring flooding and dryout phenomena at electrodes in a fuel cell have been developed, particularly neutron radiography, X-ray microtomography (microCT), and magnetic resonance imaging (MRI) [233–238].

### 19.5.1 Microfluidic Catalyst and Electrode Characterization Platforms

Developing a better understanding of the complex electrochemical, transport, and degradation processes that govern the performance of novel catalysts and electrodes *within an operating fuel cell* is key to designing robust fuel cells that are needed for commercial introduction [3]. However, detailed study of novel catalysts embedded in a MEA within an actual fuel cell is challenging because other factors such as water management, uneven performance across the electrodes, and temperature

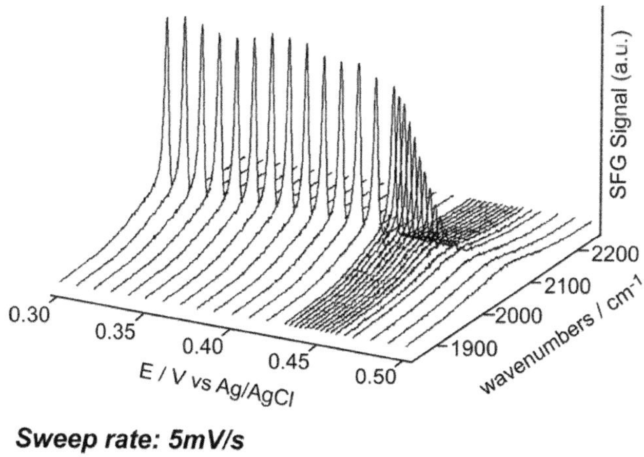


**FIGURE 19.11** BB-SFG spectra of CO chemisorbed on Pt electrode. A series of SFG spectra of CO adsorbed on a Pt electrode in CO-free 0.1 M $H_2SO_4$ electrolyte. The electrode potential is swept at 5 mV/s with SFG spectra obtained every 200 ms. While spectra were obtained at 1 mV intervals, to avoid congestion, spectra were averaged over 10 mV intervals at V < 0.43 V (the preoxidation region) and over 3.3 mV intervals at V > 0.43 V (the oxidation region). The amplitude of spectra disappears as CO is electro-oxidized from the surface at V > 0.43 V. Reprinted from Ref. 216, 2005 by permission of Elsevier.

gradients complicate elucidation of the individual processes taking place at the electrodes. Too many processes are interdependent of the same few parameters that can be varied.

To bridge the gap between fundamental characterization of catalyst structure and activity and characterization of performance and durability of the catalyst within an actual fuel cell, the Keris group has created a microfluidic platform that allows for investigation of electrochemical, transport, and degradation processes at the two electrodes *independently*, without factors such as water management complicating the experiment and data analysis [230,231]. Specifically, we have developed a microfluidic $H_2/O_2$ fuel cell that utilizes a flowing aqueous electrolyte stream in place of a stationary PEM or AAEM (Fig. 19.12). The convecting electrolyte stream minimizes water management issues, facilitates byproduct removal, and enables independent control of electrolyte parameters (i.e., no pH restrictions). This microfluidic $H_2/O_2$ fuel cell is an ideal platform for analytical investigations as the constantly replenishing electrolyte stream eliminates adverse fuel cell system limitations, namely, water management, enabling unfettered analysis of electrode performance. Furthermore, autonomous control over the convecting stream enables *in operandi* compositional flexibility without fuel cell disassembly. Moreover, by placing an external reference electrode in the exiting electrolyte stream, the performance of individual cell components can be independently analyzed [92]. Thus, this microfluidic platform enables a high number of experimental degrees of freedom, previously limited to the traditional three-electrode electrochemical cell, to be employed in a fuel cell.

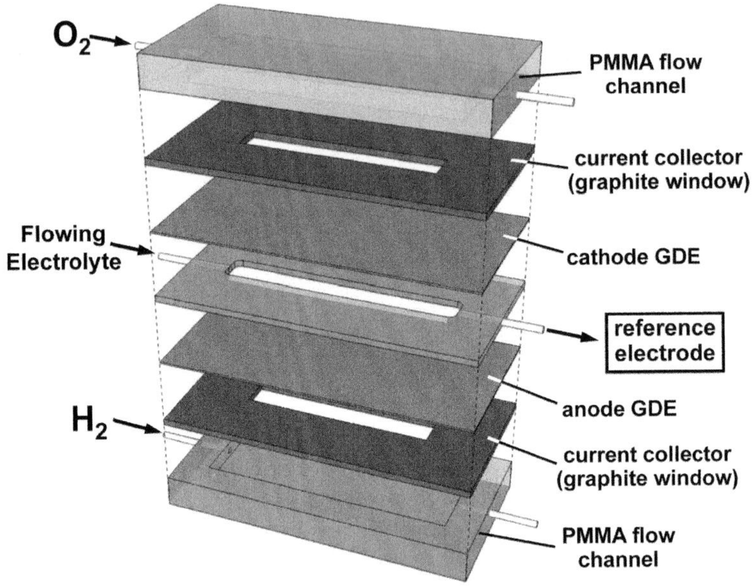


**FIGURE 19.12** Schematic of a microfluidic $H_2/O_2$ fuel cell that utilizes a flowing electrolyte stream that serves as an analytical platform for in operandi catalyst/electrode characterization. For analytical investigations, the dynamic stream (1) enables autonomous control over electrolyte parameters (i.e., pH, flow rate) and consequently the local electrode environments, as well as (2) allows for the independent in situ analyses of catalyst and/or electrode performance and degradation characteristics via an external reference electrode (e.g., Ag/AgCl). Reproduced from Ref. 211, 2009 by permission of The Electrochemical Society.

For example, the cathodic oxygen reduction reaction (ORR) presently requires high loading of expensive Pt catalysts to maintain acceptable power densities. While alternative Pt transition metal alloys are being investigated to improve catalytic activity and lower costs, such alloys can exhibit insufficient stability under acidic conditions [128]. Furthermore, degradation processes of other cell components (catalyst layers, catalyst–support materials, gas diffusion layers, membranes, etc.) within the fuel cell remain poorly understood. Probing these in situ degradation mechanisms is a significant challenge in an actual fuel cell, yet it becomes feasible in the microfluidic analytical platforms presented here.

Brushett et al. have demonstrated the utility of pH-flexible microfluidic platform as a catalyst/electrode characterization tool by analyzing the performance of Pt/C and Ag/C cathode catalysts in an operating fuel cell under varying alkaline conditions [231]. Furthermore, the performance and long-term stability of Pt and Pt-M alloys cathode catalysts have been investigated as a function of acidic conditions [232]. These preliminary investigations show the utility of the microfluidic analytical platform as a bridge between the capabilities and limitations of traditional electrochemical cells and the actual operational environment of fuel cells. The key is the flowing electrolyte stream, which enables precise control and analysis of the local electrode environments and facilitates the study of interfacial phenomena and the optimization of electrode/MEA design. In the future, this platform may to used to provide insight into the following research areas: (1) novel electrocatalytic and electrode support architecture development, (2) in situ degradation mechanisms and rates of different electrodes, (3) interfacial studies in MEA such the catalyst layer-membrane junction, and (4) the effects of various real-world contaminants on electrode and fuel cell performance. In all these areas, information garnered from such an in situ analytical platform would advance the development of more robust electrode configurations and thus more durable fuel cells.

### 19.5.2 Methods for Studying Water Management

Improving water management at the electrode–membrane interface remains a major challenge, in particular at the cathode of acidic PEM-based fuel cells. Here, the ORR produces water, which can flood the cathode, hampering oxygen transport to the electrocatalytic sites and reducing fuel cell performance. To address this challenge, complex fuel cell models are being developed to optimize electrode performance [239–241]. However, quantitative experimental data regarding the dynamic behavior of water within the gas diffusion layer (GDL) structure is critical to developing accurate water distribution models [239]. The three major imaging techniques for monitoring water transport in a fuel cell are neutron radiography, X-ray microtomography (microCT), and magnetic resonance imaging (MRI) [233,234,236–238,242]. Neutron radiography provides two-dimensional maps of liquid water in fuel cells and thus can neither determine the location of the liquid water, such as within the GDL or in the flow field, nor resolve water distribution throughout the three-dimensional GDL [236]. Consequently, when analyzing neutron radiographs, one assumes the cathode GDL to be full of liquid water, which may not be true under all operating conditions. MRI can be used to visualize liquid water transport across the PEM and in the gas flow field. However, due to the presence of magnetically inductive carbonaceous materials in a fuel cell, measurements often require custom-made architectures, such as graphite-coated Teflon flow fields, whose properties may differ from those of an actual fuel cell design [237,238]. Furthermore, analysis of water distribution within the carbon-based GDL is impossible because of the high conductivity. MicroCT, an X-ray absorption technique, offers unique insights into water transport and distribution in the porous GDL architecture within a fuel cell [235]. Furthermore, microCT enables detailed investigations of complex three-dimensional electrode architectures and interfaces between fuel cell components. Such studies may facilitate the development of novel catalyst support structures and also further understanding of component degradation mechanisms within the operating fuel cell.

## 19.6 CONCLUDING REMARKS

A clear need exists for novel approaches to producing and utilizing energy in more efficient ways, in light of society's ever-increasing demand as well as growing concerns with respect to climate change related to $CO_2$ emissions [13]. The development of low-temperature fuel cell technologies will continue to play an important role in many alternative energy conversion strategies, especially for portable electronics and automotive applications. To date, widespread commercialization of fuel cell technologies has yet to be achieved, due mainly to high costs (particularly of the precious metal catalysts), poor durability, and, to a lesser extent, system performance limitations [3]. To overcome the limitations of present technologies, novel fuel cell configurations must be designed and implemented to simultaneously reduce costs and improve performance. New AAEM-based fuel cell technologies offer advantages such as favorable reaction kinetics and cheaper constituent materials, especially less precious metal catalysts, but more work is needed with respect to extended fuel cell operation. Other options to overcome present performance and durability limitations include complete elimination of the conducting membranes and reliance on either catalyst selectivity, as in MRFCs, or microscale fluid flow phenomena, as in membraneless fuel cells or LFFCs.

Substantial work is still needed in catalyst and electrode development, as well as in further exploration of alternative fuel cell configuration before fuel cell technologies can become the power source of choice for a wide range of off-the-grid applications. In relation to the topics discussed in this chapter, specific areas in need of further research include

1. Electrode and fuel cells designs that use physicochemical characteristics of the microscale to their advantage need to be further explored [25,61,86]. This type of advanced engineering needs to go hand in hand with detailed scientific understanding of the electrocatalytic process at the molecular level.
2. Cheap, stable, and highly active novel catalytic materials need to be conceived, developed, and tested under realistic fuel cell conditions. Less explored fuel cell operation regimes, (e.g., alkaline) need to be revisited in light of the potential to use less precious metal catalysts while benefiting from better reaction kinetics.
3. The level of understanding of fuel cell electrode reactions, and the interplay of these reactions with the electrode structure at both mesoscale and microscale needs to be improved, presumably using a combination of theory and experiment [162,190,206,207,243–251].
4. New conducting polymers and better, more structurally defined MEAs need to be developed. Particularly promising in this regard is the recent development of anion exchange membranes for membrane-based alkaline fuel cells [25].

   Rapidly advancing market success of fuel cells, predominantly in cell phones, other small electronic devices, and stationary units will accelerate studies of inorganic energy materials, as well as studies toward more efficient, more durable electrodes and fuel cell configurations. The true fuel cell revolution in the automotive industry will come but more slowly, because of its much more stringent requirements on robustness and durability, especially in light of the vast range of ambient conditions—extreme cold, heat, dryness, and humidity—that it needs to be compatible with. Undoubtedly, though, the near future will continue to see fuel cells being introduced into the market for niche applications, and gradually also for mass-produced applications, such as certain types of portable electronics. This direction of progress and success are guaranteed by the already attained high-level knowledge of fuel cell processes and technology and the continued high level of investments in fuel cell research and technology worldwide that is spurred by global interest and a need for more sustainable power conversion technologies.

## ACKNOWLEDGMENTS

The authors gratefully acknowledge financial support from the National Science Foundation (Grant NSF CHE06-51083 and Career Grant CTS 05-47617 to PJAK), the Army Research Office (Grant MURI DAAD19-03-1-0169 to Case Western Reserve University), and the Department of Energy (LANL 53183-001-7, Office of Hydrogen, Fuel Cells & Infrastructure Technologies and Grant DEFG02005ER46260), INI Power Systems, Northrop Grumman, and SONY toward these reported research themes. FRB acknowledges support from the Graduate Engineering for Minorities (GEM) Fellowship.

## REFERENCES

1. L. Carrette, K. A. Friedrich, and U. Stimming, Fuel cells—fundamentals and applications, *ChemPhysChem* **1**, 5–39 (2001).
2. M. L. Perry and T. F. Fuller, A historical perspective of fuel cell technology in the 20th century, *J. Electrochem. Soc.* **149**(7), S59–S67 (2002).
3. M. Jacoby, Analytics for fuel cells, *Chem. Eng. News* **87**(13), 39–41 (2009).
4. T. E. Springer, T. Rockward, T. A. Zawodzinski, and S. Gottesfeld, Model for polymer electrolyte fuel cell operation on reformate feed: Effects of CO, $H_2$ dilution, and high fuel utilization, *J. Electrochem. Soc.* **148**, A11–A23 (2001).
5. T. A. Zawodzinski and D. Schiraldi, The future of polymers for fuel cell membranes, *Polym. Rev.* **46**(3), 215–217 (2006).
6. K. Miyatake, T. Takahiro, Y. Chikashige, H. Uchida, and M. Watanabe, Enhanced proton conduction in polymer electrolyte membranes with acid-functionalized polysilsesquioxane, *Angew. Chem. Int. Ed.* **46**, 6646–6649 (2007).
7. E. Higuchi, K. Okamoto, K. Miyatake, H. Uchida, and M. Watanabe, Gas diffusion electrodes for polymer electrolyte fuel cell using sulfonated polyimide, *Res. Chem. Intermed.* **32**(5), 533–542 (2006).
8. A. Guha, T. A. Zawodzinski, and D. A. Schiraldi, Evaluation of electrochemical performance for surface-modified carbons as catalyst support in polymer electrolyte membrane (PEM) fuel cells, *J. Power Sources* **172**, 530–541 (2007).

9. S. Srinivasan, *Fuel Cells. From Fundamentals to Applications*, Springer, New York, 2006.
10. W. Vielstich, A. Lamm, and H. Gasteiger, eds., *Fuel Cell Handbook: Fuel Cell Electrocatalysis*, Vols. **1–4**, Wiley-VCH, New York, 2003.
11. S. Litster and G. McLean, PEM fuel cell electrodes, *J. Power Sources* **130**, 61–76 (2004).
12. S. Gottesfeld, Polymer electrolyte and direct methanol fuel cells, in *Electrochemical Engineering*, D. D. Macdonald and P. Schmuki, eds., Wiley-VCH Verlag GmbH, Weinheim, 2007, pp. 544–661.
13. G. M. Whitesides and G. W. Crabtree, Perspective: Don't forget long-term fundamental research in energy, *Science* **315**, 796–798 (2007).
14. S. C. Thomas, X. M. Ren, S. Gottesfeld, and P. Zelenay, Direct methanol fuel cells: Progress in cell performance and cathode research, *Electrochim. Acta* **47**, 3741–3748 (2002).
15. X. M. Ren, P. Zelenay, S. Thomas, J. Davey, and S. Gottesfield, Recent advances in direct methanol fuel cells at Los Alamos National Laboratory, *J. Power Sources* **86**, 111–116 (2000).
16. A. S. Arico, S. Srinivasan, and V. Antonucci, DMFCs: From fundamental aspects to technology development, *Fuel Cells* **1**(2), 133–161 (2001).
17. R. Borup, J. Meyers, B. Pivovar, Y. S. Kim, R. Mukundan, N. Garland, D. Myers, M. Wilson, F. Garzon, D. Wood, P. Zelenay, K. More, K. Stroh, T. Zawodzinski, J. Boncella, J. E. McGrath, M. Inaba, K. Miyatake, M. Hori, K. Ota, Z. Ogumi, S. Miyata, A. Nishikata, Z. Siroma, Y. Uchimoto, K. Yasuda, K. I. Kimijima, and N. Iwashita, Scientific aspects of polymer electrolyte fuel cell durability and degradation, *Chem. Rev.* **107**(10), 3904–3951 (2007).
18. C. R. Buie, J. D. Posner, T. Fabian, C. A. Suk-Won, D. Kim, F. B. Prinz, J. K. Eaton, and J. G. Santiago, Water management in proton exchange membrane fuel cells using integrated electroosmotic pumping, *J. Power Sources* **161**(1), 191–202 (2006).
19. J. S. Yi, J. D. L. Yang, and C. King, Water management along the flow channels of PEM fuel cells, *AIChE J.* **50**(10), 2594–2603 (2004).
20. S. H. Ge, X. G. Li, and I. M. Hsing, Internally humidified polymer electrolyte fuel cells using water absorbing sponge, *Electrochim. Acta* **50**(9), 1909–1916 (2005).
21. A. Kuver and W. Vielstich, Investigation of methanol crossover and single electrode performance during PEMDMFC operation—a study using a solid polymer electrolyte membrane fuel cell system, *J. Power Sources* **74**(2), 211–218 (1998).
22. S. J. Hamrock, A. M. Herring, and T. A. Zawodzinski, Fuel cell chemistry and operation, *J. Power Sources* **172**, 1 (2007).
23. C. K. Dyer, Fuel cells for portable applications, *J. Power Sources* **106**(1–2), 31–34 (2002).
24. J. S. Spendelow and A. Wieckowski, Electrocatalysis of oxygen reduction and small alcohol oxidation in alkaline media, *Phys. Chem. Chem. Phys.* **9**(21), 2654–2675 (2007).
25. J. R. Varcoe and R. C. T. Slade, Prospects for alkaline anion-exchange membranes in low temperature fuel cells, *Fuel Cells* **5**(2), 187–200 (2005).
26. E. Gülzow, Alkaline fuel cells: A critical view, *J. Power Sources* **61**(1–2), 99–104 (1996).
27. E. Gülzow, Alkaline fuel cells, *Fuel Cells* **4**(4), 251–255 (2004).

28. G. F. McLean, T. Niet, S. Prince-Richard, and N. Djilali, An assessment of alkaline fuel cell technology, *Int. J. Hydrogen Energy* **27**(5), 507–526 (2002).

29. M. Cifrain and K. V. Kordesch, Advances, aging mechanism and lifetime in AFCs with circulating electrolytes, *J. Power Sources* **127**, 234–242 (2004).

30. K. V. Kordesch, 25 years of fuel cell development (1951–1976), *J. Electrochem. Soc.* **125**(3), C77–C91 (1978).

31. P. Gouerec, L. Poletto, J. Denizot, E. Sanchez-Cortezon, and J. H. Miners, The evolution of the performance of alkaline fuel cells with circulating electrolyte, *J. Power Sources* **129**(2), 193–204 (2004).

32. T. Burchardt, P. Gouerec, E. Sanchez-Cortezon, Z. Karichev, and J. H. Miners, Alkaline fuel cells: Contemporary advancement and limitations, *Fuel* **81**, 2151–2155 (2002).

33. B. Y. S. Lin, D. W. Kirk, and S. J. Thorpe, Performance of alkaline fuel cells: A possible future energy system? *J. Power Sources* **161**(1), 474–483 (2006).

34. G. A. Koscher and K. Kordesch, Alkaline methanol-air system—a historical survey and some new work, *J. Solid State Electrochem.* **7**(9), 632–636 (2003).

35. Y. Hiroyuki and F. Kenji, Anion exchange membrane and ionomer for alkaline membrane fuel cells (AMFCs), *Electrochem. Soc. Trans.* **16**(2), 257–262 (2008).

36. B. Xing and O. Savadogo, Hydrogen/oxygen polymer electrolyte membrane fuel cells (PEMFCs) based on alkaline-doped polybenzimidazole (PBI), *Electrochem. Commun.* **2**(10), 697–702 (2000).

37. J. R. Varcoe, Investigations of the ex situ ionic conductivities at 30 degrees celsius of metal-cation-free quaternary ammonium alkaline anion-exchange membranes in static atmospheres of different relative humidities, *Phys. Chem. Chem. Phys.* **9**(12), 1479–1486 (2007).

38. J. R. Varcoe and R. C. T. Slade, An electron-beam-grafted ETFE alkaline anion-exchange membrane in metal-cation-free solid-state alkaline fuel cells, *Electrochem. Commun.* **8**(5), 839–843 (2006).

39. J. R. Varcoe, R. C. T. Slade, and E. Lam How Yee, An alkaline polymer electrochemical interface: a breakthrough in application of alkaline anion-exchange membranes in fuel cells, *Chem. Commun.* **2006**(13), 1428–1429 (2006).

40. J. R. Varcoe, R. C. T. Slade, G. L. Wright, and Y. L. Chen, Steady-state dc and impedance investigations of $H_2/O_2$ alkaline membrane fuel cells with commercial Pt/C, Ag/C, and Au/C cathodes, *J. Phys. Chem. B* **110**(42), 21041–21049 (2006).

41. J. R. Varcoe, R. C. T. Slade, E. L. H. Yee, S. D. Poynton, and D. J. Driscoll, Investigations into the ex situ methanol, ethanol and ethylene glycol permeabilities of alkaline polymer electrolyte membranes, *J. Power Sources* **173**(1), 194–199 (2007).

42. J. R. Varcoe, R. C. T. Slade, E. L. H. Yee, S. D. Poynton, D. J. Driscoll, and D. C. Apperley, Poly(ethylene-co-tetrafluoroethylene)-derived radiation-grafted anion-exchange membrane with properties specifically tailored for application in metal-cation-free alkaline polymer electrolyte fuel cells, *Chem. Mater.* **19**(10), 2686–2693 (2007).

43. R. C. T. Slade and J. R. Varcoe, Investigations of conductivity in FEP-based radiation-grafted alkaline anion-exchange membranes, *Solid State Ion.* **176**(5–6), 585–597 (2005).

44. T. N. Danks, R. C. T. Slade, and J. R. Varcoe, Alkaline anion-exchange radiation-grafted membranes for possible electrochemical application in fuel cells, *J. Mater. Chem.* **13**(4), 712–721 (2003).

45. A. E. S. Sleightholme, J. R. Varcoe, and A. R. Kucernak, Oxygen reduction at the silver/hydroxide-exchange membrane interface, *Electrochem. Commun.* **10**(1), 151–155 (2008).

46. H. Herman, R. C. T. Slade, and J. R. Varcoe, The radiation-grafting of vinylbenzyl chloride onto poly (hexafluoropropylene-co-tetrafluoroethylene) films with subsequent conversion to alkaline anion-exchange membranes: optimisation of the experimental conditions and characterisation, *J. Membr. Sci.* **218**(1–2), 147–163 (2003).

47. S. F. Lu, J. Pan, A. B. Huang, L. Zhuang, and J. T. Lu, Alkaline polymer electrolyte fuel cells completely free from noble metal catalysts, *Proc. Natl. Acad. Sci. USA* **105**(52), 20611–20614 (2008).

48. C. Bianchini, P. Bert, G. Giambastiani, S. Catanorchi, M. Piana, R. Psaro, S. Santiccioli, A. Tampucci, and F. Vizza, New anode and cathode catalysts for highly efficient direct alcohol fuel cells based on first-row late-transition metals, *Proc. Int. Symp. Surface Imaging/Spectroscopy at the Solid/Liquid Interface,* Krakow, Poland, May 28–June 1, 2006.

49. K. Asazawa, K. Yamada, H. Tanaka, A. Oka, M. Taniguchi, and T. Kobayashi, A platinum-free zero-carbon-emission easy fuelling direct hydrazine fuel cell for vehicles Angew, *Chem. Int. Ed.* **46**(42), 8024–8027 (2007).

50. K. Scott, E. Yu, G. Vlachogiannopoulos, M. Shivare, and N. Duteanu, Performance of a direct methanol alkaline membrane fuel cell, *J. Power Sources* **175**(1), 452–457 (2008).

51. E. H. Yu and K. Scott, Development of direct methanol alkaline fuel cells using anion exchange membranes, *J. Power Sources* **137**(2), 248–256 (2004).

52. E. H. Yu and K. Scott, Direct methanol alkaline fuel cell with catalysed metal mesh anodes, *Electrochem. Commun.* **6**(4), 361–365 (2004).

53. E. H. Yu, K. Scott, R. W. Reeve, L. X. Yang, and R. G. Allen, Characterisation of platinised Ti mesh electrodes using electrochemical methods: Methanol oxidation in sodium hydroxide solutions, *Electrochim. Acta* **49**(15), 2443–2452 (2004).

54. K. Matsuoka, Y. Iriyama, T. Abe, M. Matsuoka, and Z. Ogumi, Alkaline direct alcohol fuel cells using an anion exchange membrane, *J. Power Sources* **150**, 27–31 (2005).

55. Y. S. Li, T. S. Zhao, and Z. X. Liang, Performance of alkaline electrolyte-membrane-based direct ethanol fuel cells, *J. Power Sources* **187**(2), 387–392 (2009).

56. L. Demarconnay, S. Brimaud, C. Coutanceau, and J. M. Leger, Ethylene glycol electro-oxidation in alkaline medium at multi-metallic Pt based catalysts, *J. Electroanal. Chem.* **601**(1–2), 169–180 (2007).

57. Y. Wang, L. Li, L. Hu, L. Zhuang, J. T. Lu, and B. Q. Xu, A feasibility analysis for alkaline membrane direct methanol fuel cell: Thermodynamic disadvantages versus kinetic advantages, *Electrochem. Commun.* **5**(8), 662–666 (2003).

58. L. A. Adams, S. D. Poynton, C. Tamain, R. C. T. Slade, and J. R. Varcoe, A carbon dioxide tolerant aqueous-electrolyte-free anion-exchange membrane alkaline fuel cell, *Chemsuschem* **1**(1–2), 79–81 (2008).

59. B. Gurau and E. S. Smotkin, Methanol crossover in direct methanol fuel cells: A link between power and energy density, *J. Power Sources* **112**(2), 339–352 (2002).

60. M. A. Priestnall, V. P. Kotzeva, D. J. Fish, and E. M. Nilsson, Compact mixed-reactant fuel cells, *J. Power Sources* **106**, 21–30 (2002).

61. A. K. Shukla, R. K. Raman, and K. Scott, Advances in mixed-reactant fuel cells, *Fuel Cells* **5**(4), 436–447 (2005).

62. G. Gruneberg, W. Wicke, and E. Justi, British Petent Office, ed. British Patent GB994448 (1961).

63. S. C. Barton, T. Patterson, E. Wang, T. F. Fuller, and A. C. West, Mixed-reactant, strip-cell direct methanol fuel cells, *J. Power Sources* **96**(2), 329–336 (2001).

64. W. Sung and J. W. Choi, A membraneless microscale fuel cell using non-noble catalysts in alkaline solution, *J. Power Sources* **172**(1), 198–208 (2007).

65. A. K. Shukla, C. L. Jackson, K. Scott, and G. Murgia, A solid-polymer electrolyte direct methanol fuel cell with a mixed reactant and air anode, *J. Power Sources* **111**(1), 43–51 (2002).

66. A. K. Shukla and R. K. Raman, Methanol-resistant oxygen-reduction catalysts for direct methanol fuel cells, *Annu. Rev. Mater. Res.* **33**, 155–168 (2003).

67. K. Scott, A. K. Shukla, C. L. Jackson, and W. R. A. Meuleman, A mixed-reactants solid-polymer-electrolyte direct methanol fuel cell, *J. Power Sources* **126**(1–2), 67–75 (2004).

68. H. Meng, M. Wu, X. X. Hu, M. Nie, Z. D. Wei, and P. K. Shen, Selective cathode catalysts for mixed-reactant alkaline alcohol fuel cells, *Fuel Cells* **6**(6), 447–450 (2006).

69. N. Alonso-Vante, Novel nanostructured materials based on transition-metal compounds for electrocatalysis, in *Catalysis and Electrocatalysis at Nanoparticle Surfaces*, A. Wieckowski, E. Savinova, and C. Vayenas, ed., Marcel Dekker, New York, 2003.

70. T. J. Schmidt, U. A. Paulus, H. A. Gasteiger, N. Alonso-Vante, and R. J. Behm, Oxygen reduction on $Ru_{1.92}Mo_{0.08}SeO_4$, Ru/carbon, and Pt/carbon in pure and methanol-containing electrolytes, *J. Electrochem. Soc.* **147**(7), 2620–2624 (2000).

71. V. I. Zaikovskii, K. S. Nagabhushana, V. V. Kriventsov, K. N. Loponov, S. V. Cherepanova, R. I. Kvon, H. Bonnemann, D. I. Kochubey, and E. R. Savinova, Synthesis and Structural characterization of Se-modified carbon-supported Ru nanoparticles for the oxygen reduction reaction, *J. Phys. Chem. B* **110**, 6881–6890 (2006).

72. D. Cao, A. Wieckowski, J. Inukai, and N. Alonso-Vante, Oxygen reduction reaction on ruthenium and rhodium nanoparticles modified with selenium and sulfur, *J. Electrochem. Soc.* **153**(5), A869–A874 (2006).

73. A. Lewera, J. Inukai, W. P. Zhou, D. Cao, H. T. Duong, N. Alonso-Vante, and A. Wieckowski, Chalcogenide oxygen reduction reaction catalysis: X-ray photoelectron spectroscopy with Ru, Ru/Se and Ru/S samples emersed from aqueous media, *Electrochim. Acta* **52**, 5759–5765 (2007).

74. I. V. Malakhov, S. G. Nikitenko, E. R. Savinova, D. I. Kochubey, and N. Alonso-Vante, In situ EXAFS study to probe active centers of Ru chalcogenide electrocatalysts during oxygen reduction reaction, *J. Phys. Chem. B* **106**(7), 1670–1676 (2002).

75. H. Cheng, W. Yuan, and K. Scott, A liquid-gas phase mixed-reactant fuel cell with a RuSeW cathode electrocatalyst, *J. Power Sources* **183**(2), 678–681 (2008).

76. A. Racz, P. Bele, C. Cremers, and U. Stimming, Ruthenium selenide catalysts for cathodic oxygen reduction in direct methanol fuel cells, *J. Appl. Electrochem.* **37**(12), 1455–1462 (2007).

77. D. T. Whipple, R. S. Jayashree, D. Egas, N. Alonso-Vante, and P. J. A. Kenis, Ruthenium cluster-like chalcogenide as a methanol tolerant cathode catalyst in air-breathing laminar flow fuel cells, *Electrochim. Acta* **54**(18), 4384–4388 (2009).

78. H. Tributsch, M. Bron, M. Hilgendorff, H. Schulenburg, I. Dorbandt, V. Eyert, P. Bogdanoff, and S. Fiechter, Methanol-resistant cathodic oxygen reduction catalysts for methanol fuel cells, *J. Appl. Electrochem.* **V31**(7), 739–748 (2001).

79. C. M. Johnston, J. H. Choi, P. K. Babu, A. Wieckowski, and P. Zelenay, *Performance and Durability of Chalcogen-Modified Ruthenium Catalysts for Oxygen Reduction: Hydrogen-Air MEA and RRDE Studies*, IL, Electrochemical Society, Inc., Chicago, 2008.

80. F. J. Rodriguez, P. J. Sebastian, O. Solorza, and R. Perez, Mo–Ru–W chalcogenide electrodes prepared by chemical synthesis and screen printing for fuel cell applications, *Int. J. Hydrogen Energy* **23**(11), 1031–1025 (1998).

81. S. L. Gojkovic, S. Gupta, and R. F. Savinell, Heat-treated iron(III) tetramethoxyphenyl porphyrin chloride supported on high-area carbon as an electrocatalyst for oxygen reduction—Part II. Kinetics of oxygen reduction, *J. Electroanal. Chem.* **462**(1), 63–72 (1999).

82. H. Yang, C. Coutanceau, J. M. Leger, N. Alonso-Vante, and C. Lamy, Methanol tolerant oxygen reduction on carbon-supported Pt-Ni alloy nanoparticles, *J. Electroanal. Chem.* **576**(2), 305–313 (2005).

83. P. Piela, P. Eickes, E. Brosha, F. Garzon, and P. Zelenay, Ruthenium crossover in the direct methanol fuel cell with a Pt-Ru anode, *J. Electrochem. Soc.* **151**, A2053–A2059 (2004).

84. G. M. Whitesides, The origins and the future of microfluidics, *Nature* **442**(7101), 368–373 (2006).

85. T. M. Squires and S. R. Quake, Microfluidics: Fluid physics at the nanoliter scale, *Rev. Mod. Phys.* **77**(3), 977–1026 (2005).

86. E. Kjeang, N. Djilali, and D. Sinton, Microfluidic fuel cells: A review, *J. Power Sources* **186**(2), 353–369 (2009).

87. P. J. A. Kenis, R. F. Ismagilov, and G. M. Whitesides, Microfabrication inside capillaries using multiphase laminar flow patterning, *Science* **285**(5424), 83–85 (1999).

88. M. W. Toepke, S. H. Brewer, D. M. Vu, K. D. Rector, J. E. Morgan, R. B. Gennis, P. J. A. Kenis, and R. B. Dyer, Microfluidic flow-flash: Method for investigating protein dynamics, *Anal. Chem.* **79**(1), 122–128 (2007).

89. S. L. Perry, G. W. Roberts, J. D. Tice, R. B. Gennis, and P. J. A. Kenis, Microfluidic generation of lipidic mesophases for membrane protein crystallization, *Crystal Growth Design* **9**(6), 2566–2569 (2009).

90. E. R. Choban, L. J. Markoski, A. Wieckowski, and P. J. A. Kenis, Microfluidic fuel cell based on laminar flow, *J. Power Sources* **128**, 54–60 (2004).

91. E. R. Choban, J. S. Spendelow, L. Gancs, A. Wieckowski, and P. J. A. Kenis, Membraneless laminar flow-based micro fuel cells operating in alkaline, acidic, and acidic/alkaline media, *Electrochim. Acta* **50**(27), 5390–5398 (2005).

92. E. R. Choban, P. Waszczuk, and P. J. A. Kenis, Characterization of limiting factors in laminar flow-based membraneless microfuel cells, *Electrochem. Solid State Lett.* **8**(7), A348–A352 (2005).

93. R. S. Jayashree, D. Egas, J. S. Spendelow, D. Natarajan, L. J. Markoski, and P. J. A. Kenis, Air-breathing laminar flow-based direct methanol fuel cell with alkaline electrolyte, *Electrochem. Solid State Lett.* **9**(5), A252–A256 (2006).

94. R. S. Jayashree, L. Gancs, E. R. Choban, A. Primak, D. Natarajan, L. J. Markoski, and P. J. A. Kenis, Air-breathing laminar flow-based microfluidic fuel cell, *J. Am. Chem. Soc.* **127**(48), 16758–16759 (2005).

95. R. S. Jayashree, S. K. Yoon, F. R. Brushett, D. Natarajan, L. J. Markoski, and P. J. A. Kenis, On the performance of membraneless laminar flow based microfluidic fuel cells, manuscript in preparation, 2009.

96. F. R. Brushett, R. S. Jayashree, W. P. Zhou, and P. J. A. Kenis, Investigation of fuel and media flexible laminar flow-based fuel cells, *Electrochim. Acta* **54**, 7099–7105 (2009).

97. L. J. Markoski, P. J. A. Kenis, and E. R. Choban, Fuel Cells Comprising Laminar Flow Induced Dynamic Conducting Interfaces, Electronic Devices Comprising Such Cells and Method Employing Same, United Stated Petent Office, ed., US Patent (2007).

98. J. L. Cohen, D. A. Westly, A. Pechenik, and H. D. Abruna, Fabrication and preliminary testing of a planar membraneless microchannel fuel cell, *J. Power Sources* **139**(1–2), 96–105 (2005).

99. J. L. Cohen, D. J. Volpe, D. A. Westly, A. Pechenik, and H. D. Abruna, A dual electrolyte $H_2/O_2$ planar membraneless microchannel fuel cell system with open circuit potentials in excess of 1.4 V, *Langmuir* **21**(8), 3544–3550 (2005).

100. S. M. Mitrovski, L. C. C. Elliott, and R. G. Nuzzo, Microfluidic devices for energy conversion: Planar integration and performance of a passive, fully immersed $H_2$-$O_2$ fuel cell, *Langmuir* **20**(17), 6974–6976 (2004).

101. A. Bazylak, D. Sinton, and N. Djilali, Improved fuel utilization in microfluidic fuel cells: A computational study, *J. Power Sources* **143**(1–2), 57–66 (2005).

102. E. Kjeang, B. T. Proctor, A. G. Brolo, D. A. Harrington, N. Djilali, and D. Sinton, High-performance microfluidic vanadium redox fuel cell, *Electrochim. Acta* **52**(15), 4942–4946 (2007).

103. E. Kjeang, A. G. Brolo, D. A. Harrington, N. Djilali, and D. Sinton, Hydrogen peroxide as an oxidant for microfluidic fuel cells, *J. Electrochem. Soc.* **154**(12), B1220–B1226 (2007).

104. E. Kjeang, J. McKechnie, D. Sinton, and N. Djilali, Planar and three-dimensional microfluidic fuel cell architectures based on graphite rod electrodes, *J. Power Sources* **168**(2), 379–390 (2007).

105. E. Kjeang, R. Michel, D. A. Harrington, D. Sinton, and N. Djilali, An alkaline microfluidic fuel cell based on formate and hypochlorite bleach, *Electrochim. Acta* **54**(2), 698–705 (2008).

106. M. H. Sun, G. V. Casquillas, S. S. Guo, J. Shi, H. Ji, Q. Ouyang, and Y. Chen, Characterization of microfluidic fuel cell based on multiple laminar flow, *Microelectron. Eng.* 1182–1185 (2007).

107. R. Ferrigno, A. D. Stroock, T. D. Clark, M. Mayer, and G. M. Whitesides, Membraneless vanadium redox fuel cell using laminar flow, *J. Am. Chem. Soc.* **124**(44), 12930–12931 (2002).

108. D. H. Ahmed, H. B. Park, and H. J. Sung, Optimum geometrical design for improved fuel utilization in membraneless micro fuel cell, *J. Power Sources* **185**(1), 143–152 (2008).

109. H. B. Park, D. H. Ahmed, K. H. Lee, and H. J. Sung, An H-shaped design for membraneless micro fuel cells, *Electrochim. Acta* **54**(18), 4416–4425 (2009).

110. K. S. Salloum, J. R. Hayes, C. A. Friesen, and J. D. Posner, Sequential flow membraneless microfluidic fuel cell with porous electrodes, *J. Power Sources* **180**(1), 243–252 (2008).

111. E. Kjeang, R. Michel, D. A. Harrington, N. Djilali, and D. Sinton, A microfluidic fuel cell with flow-through porous electrodes, *J. Am. Chem. Soc.* **130**(12), 4000–4006 (2008).

112. C. M. Moore, S. D. Minteer, and R. S. Martin, Microchip-based ethanol/oxygen biofuel cell, *Lab on a Chip* **5**(2), 218–225 (2005).
113. F. Sato, M. Togo, M. K. Islam, T. Matsue, J. Kosuge, N. Fukasaku, S. Kurosawa, and M. Nishizawa, Enzyme-based glucose fuel cell using vitamin K-3-immobilized polymer as an electron mediator, *Electrochem. Commun.* **7**(7), 643–647 (2005).
114. M. Togo, A. Takamura, T. Asai, H. Kaji, and M. Nishizawa, An enzyme-based microfluidic biofuel cell using vitamin K-3-mediated glucose oxidation, *Electrochim. Acta* **52**(14), 4669–4674 (2007).
115. M. Togo, A. Takamura, T. Asai, H. Kaji, and M. Nishizawa, Structural studies of enzyme-based microfluidic biofuel cells, *J. Power Sources* **178**(1), 53–58 (2008).
116. S. Topcagic and S. D. Minteer, Development of a membraneless ethanol/oxygen biofuel cell, *Electrochim. Acta* **51**(11), 2168–2172 (2006).
117. N. Mano, J. L. Fernandez, Y. Kim, W. Shin, A. J. Bard, and A. Heller, Oxygen is electroreduced to water on a "wired" enzyme electrode at a lesser overpotential than on platinum, *J. Am. Chem. Soc.* **125**(50), 15290–15291 (2003).
118. V. Soukharev, N. Mano, and A. Heller, A four-electron $O_2$ electroreduction biocatalyst superior to platinum and a biofuel cell operating at 0.88 V, *J. Am. Chem. Soc.* **126**(27), 8368–8369 (2004).
119. N. L. Akers, C. M. Moore, and S. D. Minteer, Development of alcohol/$O_2$ biofuel cells using salt-extracted tetrabutylammonium bromide/Nafion membranes to immobilize dehydrogenase enzymes, *Electrochim. Acta* **50**(12), 2521–2525 (2005).
120. R. A. Bullen, T. C. Arnot, J. B. Lakeman, and F. C. Walsh, Biofuel cells and their development, *Biosens. Bioelectron.* **21**(11), 2015–2045 (2006).
121. J. A. Cracknell, K. A. Vincent, and F. A. Armstrong, Enzymes as working or inspirational electrocatalysts for fuel cells and electrolysis, *Chem. Rev.* **108**(7), 2439–2461 (2008).
122. A. Li, S. H. Chan, and N. T. Nguyen, A laser-micromachined polymeric membraneless fuel cell, *J. Micromech. Microeng.* **17**(6), 1107–1113 (2007).
123. L. F. Gu, N. Luo, and G. H. Miley, Cathode electrocatalyst selection and deposition for a direct borohydride/hydrogen peroxide fuel cell, *J. Power Sources* **173**(1), 77–85 (2007).
124. A. D. Stroock, S. K. W. Dertinger, A. Ajdari, I. Mezic, H. A. Stone, and G. M. Whitesides, Chaotic mixer for microchannels, *Science* **295**(5555), 647–651 (2002).
125. J. Phirani and S. Basu, Analyses of fuel utilization in microfluidic fuel cell, *J. Power Sources* **175**(1), 261–265 (2008).
126. S. K. Yoon, G. W. Fichtl, and P. J. A. Kenis, Active control of the depletion boundary layers in microfluidic electrochemical reactors, *Lab on a Chip* **6**(12), 1516–1524 (2006).
127. INI Power Systems, `www.inipower.com`. 2008 (cited July 31, 2009).
128. H. A. Gasteiger, S. S. Kocha, B. Sompalli, and F. T. Wagner, Activity benchmarks and requirements for Pt, Pt-alloy, and non-Pt oxygen reduction catalysts for PEMFCs, *Appl. Catal. B Environ.* **56**(1–2), 9–35 (2005).
129. R. Masel, Energy technology—hydrogen quick and clean, *Nature* **442**, 521–522 (2006).
130. M. K. Debe, A. K. Schmoeckel, G. D. Vernstrom, and R. Atanasoski, High Voltage stability of nanostructured thin film catalysts for PEM fuel cells, *J. Power Sources* **161**, 1002–1011 (2006).
131. A. A. Argun, J. N. Ashcraft, and P. T. Hammond, Highly conductive, methanol resistant polyelectrolyte multilayers, *Adv. Mater.* **20**(8), 1539 (2008).

132. K. Lee, J. Zhang, H. Wang, and D. P. Wilkinson, Progress in the synthesis of carbon nanotube and nanofiber-supported Pt electrocatalysts for PEM fuel cell catalysis, *J. Appl. Electrochem.* **36**(5), 507–522 (2006).

133. Y. Y. Shao, J. Liu, Y. Wang, and Y. H. Lin, Novel catalyst support materials for PEM fuel cells: Current status and future prospects, *J. Mater. Chem.* **19**(1), 46–59 (2009).

134. A. V. Tripkovic, K. D. Popovic, B. N. Grgur, B. Blizanac, P. N. Ross, and N. M. Markovic, Methanol electrooxidation on supported Pt and PtRu catalysts in acid and alkaline solutions, *Electrochim. Acta* **47**(22–23), 3707–3714 (2002).

135. A. V. Tripkovic, K. D. Popovic, J. D. Lovic, V. M. Jovanovic, and A. Kowal, Methanol oxidation at platinum electrodes in alkaline solution: Comparison between supported catalysts and model systems, *J. Electroanal. Chem.* **572**(1), 119–128 (2004).

136. P. S. Kauranen, E. Skou, and J. Munk, Kinetics of methanol oxidation on carbon-supported Pt and Pt + Ru catalysts, *J. Electroanal. Chem.* **404**(1), 1–13 (1996).

137. F. Maillard, G. Q. Lu, A. Wieckowski, and U. Stimming, Ru-decorated Pt surfaces as model fuel cell electrocatalysts for CO electrooxidation, *J. Phys. Chem. B* **109**, 16230–16243 (2005).

138. M. Watanabe and S. Motoo, Electrocatalysis by Ad-atoms Part II. Enhancement of the oxidation of methanol on platinum by ruthenium Ad-atoms, *J. Electroanal. Chem.* **60**, 267–273 (1975).

139. M. Watanabe and S. Motoo, Electrocatalysis by Ad-atoms Part III. Enhancement of the oxidation of carbon monoxide on platinum by ruthenium Ad-atoms, *J. Electroanal. Chem.* **60**, 275–283 (1975).

140. E. Herrero, W. Chrzanowski, and A. Wieckowski, Dual path mechanism in methanol electrooxidation on a platinum-electrode, *J. Phys. Chem.* **99**(25), 10423–10424 (1995).

141. H. Hitmi, E. M. Belgsir, J. M. Leger, C. Lamy, and R. O. Lezna, A kinetic analysis of the electrooxidation of ethanol at a platinum-electrode in acid medium, *Electrochim. Acta* **39**(3), 407–415 (1994).

142. S. Rousseau, C. Coutanceau, C. Lamy, and J. M. Leger, Direct ethanol fuel cell (DEFC): Electrical performances and reaction products distribution under operating conditions with different platinum-based anodes, *J. Power Sources* **158**(1), 18–24 (2006).

143. E. Antolini, Catalysts for direct ethanol fuel cells, *J. Power Sources* **170**(1), 1–12 (2007).

144. J. Mann, N. Yao, and A. B. Bocarsly, Characterization and analysis of new catalysts for a direct ethanol fuel cell, *Langmuir* **22**, 10432–10436 (2006).

145. A. Kowal, M. Li, M. Shao, K. Sasaki, M. B. Vukmirovic, J. Zhang, N. S. Marinkovic, P. Liu, A. I. Frenkel, and R. R. Adzic, Ternary Pt/Rh/$SnO_2$ electrocatalysts for oxidizing ethanol to $CO_2$, *Nature Mater.* **8**(4), 325–330 (2009).

146. G. Q. Lu, A. Crown, and A. Wieckowski, Formic acid decomposition on polycrystalline platinum and palladium electrodes, *J. Phys. Chem. B* **103**, 9700–9711 (1999).

147. C. Rice, R. I. Ha, R. I. Masel, P. Waszczuk, A. Wieckowski, and T. Barnard, Direct formic acid fuel cells, *J. Power Sources* **111**(1), 83–89 (2002).

148. C. Rice, S. Ha, R. I. Masel, and A. Wieckowski, Catalysts for direct formic acid fuel cells, *J. Power Sources* **115**(2), 229–235 (2003).

149. R. S. Jayashree, J. S. Spendelow, J. Yeom, C. Rastogi, M. A. Shannon, and P. J. A. Kenis, Characterization and application of electrodeposited Pt, Pt/Pd, and Pd catalyst structures for direct formic acid micro fuel cells, *Electrochim. Acta* **50**(24), 4674–4682 (2005).

150. S. Ha, Z. Dunbar, and R. I. Masel, Characterization of a high performing passive direct formic acid fuel cell, *J. Power Sources* **158**, 129–136 (2006).

151. G. Samjeske and M. Osawa, Current oscillations during formic acid oxidation on a Pt electrode: Insight into the mechanism by time-resolved IR spectroscopy, *Angew. Chem. Int. Ed.* **44**(35), 5694–5698 (2005).

152. Y. X. Chen, M. Heinen, Z. Jusys, and R. B. Behm, Kinetics and mechanism of the electrooxidation of formic acid—spectroelectrochemical studies in a flow cell, *Angew. Chem. Int. Ed.* **45**(6), 981–985 (2006).

153. K. Yamamoto, T. Imaoka, W. J. Chun, E. Osamu, H. Katoh, M. Takenaga, and A. Sonoi, Size-specific catalytic activity of platinum clusters enhances oxygen reduction reactions, *Nature Chem.* **1**(1), 307–402 (2009).

154. Y. Shao-Horn, W. C. Sheng, S. Chen, P. J. Ferreira, E. F. Holby, and D. Morgan, Instability of supported platinum nanoparticles in low-temperature fuel cells, *Top. Catal.* **46**(3–4), 285–305 (2007).

155. J. K. Nørskov, J. Rossmeisl, A. Logadottir, L. Lindqvist, J. R. Kitchin, T. Bligaard, and H. Jonsson, Origin of the overpotential for oxygen reduction at a fuel-cell cathode, *J. Phys. Chem. B* **108**(46), 17886–17892 (2004).

156. R. R. Adzic, Recent Advances in the Kinetics of Oxygen Reduction, in *Electrocatalysis*, J. Lipkowski and P. N. Ross, ed., Wiley-VCH, New York, 1998, pp. 197–243.

157. T. Toda, H. Igarashi, H. Uchida, and M. Watanabe, Enhancement of the electroreduction of oxygen on Pt alloys with Fe, Ni, and Co, *J. Electrochem. Soc.* **146**(10), 3750–3756 (1999).

158. P. Mani, R. Srivastava, and P. Strasser, Dealloyed Pt-Cu core-shell nanoparticle electrocatalysts for use in PEM fuel cell cathodes, *J. Phys. Chem. C* **112**(7), 2770–2778 (2008).

159. S. Mukerjee and S. Srinivasan, Enhanced electrocatalysis of oxygen reduction on platinum alloys in proton-exchange membrane fuel cells, *J. Electroanal. Chem.* **357**(1–2), 201–224 (1993).

160. S. Chen, P. J. Ferreira, W. C. Sheng, N. Yabuuchi, L. F. Allard, and Y. Shao-Horn, Enhanced activity for oxygen reduction reaction on "$Pt_3Co$" nanoparticles: Direct evidence of percolated and sandwich-segregation structures, *J. Am. Chem. Soc.* **130**(42), 13818–13819 (2008).

161. V. R. Stamenkovic, B. Fowler, B. S. Mun, G. F. Wang, P. N. Ross, C. A. Lucas, and N. M. Markovic, Improved oxygen reduction activity on $Pt_3Ni(111)$ via increased surface site availability, *Science* **315**(5811), 493–497 (2007).

162. V. Stamenkovic, B. S. Mun, K. J. J. Mayrhofer, P. N. Ross, N. M. Markovic, J. Rossmeisl, J. Greeley, and J. K. Nørskov, Changing the activity of electrocatalysts for oxygen reduction by tuning the surface electronic structure, *Angew. Chem. Int. Ed.* **45**, 2897–2901 (2006).

163. V. Stamenkovic, T. J. Schmidt, P. N. Ross, and N. M. Markovic, Surface segregation effects in electrocatalysis: kinetics of oxygen reduction reaction on polycrystalline $Pt_3Ni$ alloy surfaces, *J. Electroanal. Chem.* **554**, 191–199 (2003).

164. V. R. Stamenkovic, B. S. Mun, K. J. J. Mayrhofer, P. N. Ross, and N. M. Markovic, Effect of surface composition on electronic structure, stability, and electrocatalytic properties of Pt-transition metal alloys: Pt-skin versus Pt-skeleton surfaces, *J. Am. Chem. Soc.* **128**(27), 8813–8819 (2006).

165. M. Watanabe, K. Tsurumi, T. Mizukami, T. Nakamura, and P. Stonehart, Activity and stability of ordered and disordered Co-Pt alloys for phosphoric acid fuel cells, *J. Electrochem. Soc.* **141**, 2659–2668 (1994).

166. B. C. Beard and P. N. Ross, The structure and activity of Pt-Co alloys as oxygen reduction electrocatalysts, *J. Electrochem. Soc.* **137**(11), 3368–3374 (1990).

167. E. Antolini, J. R. C. Salgado, and E. R. Gonzalez, The stability of Pt-M (M = first row transition metal) alloy catalysts and its effect on the activity in low temperature fuel cells—a literature review and tests on a Pt-Co catalyst, *J. Power Sources* **160**(2), 957–968 (2006).

168. J. Aragane, T. Murahashi, and T. Odaka, Change of Pt distribution in the active components of phosphoric-acid fuel cell, *J. Electrochem. Soc.* **135**(4), 844–850 (1988).

169. K. Yasuda, A. Taniguchi, T. Akita, T. Ioroi, and Z. Siroma, Characteristics of a platinum black catalyst layer with regard to platinum dissolution phenomena in a membrane electrode assembly, *J. Electrochem. Soc.* **153**(8), A1599–A1603 (2006).

170. K. Yasuda, A. Taniguchi, T. Akita, T. Ioroi, and Z. Siroma, Platinum dissolution and deposition in the polymer electrolyte membrane of a PEM fuel cell as studied by potential cycling, *Phys. Chem. Chem. Phys.* **8**(6), 746–752 (2006).

171. H. T. Duong, M. A. Rigsby, W. P. Zhou, and A. Wieckowski, Oxygen reduction catalysis of the $Pt_3Co$ alloy in alkaline and acidic media studied by X-ray photoelectron Spectroscopy and electrochemical methods, *J. Phys. Chem. C* **111**(36), 13460–13465 (2007).

172. M. B. Vukmirovic, J. Zhang, K. Sasaki, A. U. Nilekar, F. Uribe, M. Mavrikakis, and R. R. Adzic, Platinum monolayer electrocatalysts for oxygen reduction, *Electrochim. Acta* **52**, 2257–2263 (2007).

173. M. H. Shao, T. Huang, P. Liu, J. Zhang, K. Sasaki, M. B. Vukmirovic, and R. R. Adzic, Palladium monolayer and palladium alloy electrocatalysts for oxygen reduction, *Langmuir* **22**(25), 10409–10415 (2006).

174. M. H. Shao, K. Sasaki, P. Liu, and R. R. Adzic, $Pd_3Fe$ and Pt monolayer-modified $Pd_3Fe$ electrocatalysts for oxygen reduction, *Z. Phys. Chem. Int. J. Res. Phys. Chem. Chem. Phys.* **221**, 1175–1190 (2007).

175. S. Koh, Y. C. Mani, P. Srivastava, and P. Strasser, Activity of ordered and disordered Pt-Co alloy phases for the electroreduction of oxygen in catalysts with multiple coexisting phases, *J. Power Sources* **172**, 50–56 (2007).

176. S. Koh and P. Strasser, Electrocatalysis on bimetallic surfaces: modifying catalytic reactivity for oxygen reduction by voltammetric surface dealloying, *J. Am. Chem. Soc.* **129**, 12624 (2007).

177. Z. C. Liu, S. Koh, C. F. Yu, and P. Strasser, Synthesis, dealloying, and ORR electrocatalysis of PDDA-stabilized Cu-rich Pt alloy nanoparticles, *J. Electrochem. Soc.* **154**, B1192–B1199 (2007).

178. L. Jorissen, Bifunctional oxygen/air electrodes, *J. Power Sources* **155**(1), 23–32 (2006).

179. J. Prakash, D. A. Tryk, and E. B. Yeager, Kinetic investigations of oxygen reduction and evolution reactions on lead ruthenate catalysts, *J. Electrochem. Soc.* **146**(11), 4145–4151 (1999).

180. H. Meng and P. K. Shen, Novel Pt-free catalyst for oxygen electroreduction, *Electrochem. Commun* **8**(4), 588–594 (2006).

181. R. Bashyam and P. Zelenay, A class of non-precious metal composite catalysts for fuel cells, *Nature* **443**(7107), 63–66 (2006).
182. J. Fournier, G. Lalande, R. Cote, D. Guay, and J. P. Dodelet, Activation of various Fe-based precursors on carbon black and graphite supports to obtain catalysts for the reduction of oxygen in fuel cells, *J. Electrochem. Soc.* **144**(1), 218–226 (1997).
183. F. Jaouen and J. P. Dodelet, Average turn-over frequency of $O_2$ electro-reduction for Fe/N/C and Co/N/C catalysts in PEFCs, *Electrochim. Acta* **52**(19), 5975–5984 (2007).
184. M. Lefevre, E. Proietti, F. Jaouen, and J. P. Dodelet, Iron-based catalysts with improved oxygen reduction activity in polymer electrolyte fuel cells, *Science* **324**(5923), 71–74 (2009).
185. H. A. Gasteiger and N. M. Markovic, Just a dream—or future reality? *Science* **324**(5923), 48–49 (2009).
186. K. Du and F. Ernst, Quantitative assessment of nanoparticle size distributions from HRTEM images, *Int. J. Mater. Res.* **7**, 928–933 (2006).
187. J. McBreen and S. Mukerjee, In-situ X-ray absorption studies of a Pt-Ru electrocatalyst, *J. Electrochem. Soc* **142**(10), 3399–3404 (1995).
188. K. M. Brace, B. E. Hayden, A. E. Russell, and J. R. Owen, A parallel optical screen for the rapid combinatorial electrochromic analysis of electrochemical materials, *Ad. Mater.* **18**, 3253 (2006).
189. K. A. Friedrich, K. P. Geyzers, A. J. Dickinson, and U. Stimming, Fundamental aspects in electrocatalysis: From the reactivity of single-crystals to fuel cell electrocatalysis, *J. Electroanal. Chem.* **524–525**, 261–272 (2002).
190. J. Greeley, J. Rossmeisl, A. Hellman, and J. K. Nørskov, Theoretical trends in particle size effects for the oxygen reduction reaction, *Z. Phys. Chem. Int. J. Res. Phys. Chem. Chem. Phys.* **221**, 1209–1220 (2007).
191. F. Dassenoy, W. Vogel, and N. Alonso-Vante, Structural studies and stability of cluster-like RuxSey electrocatalysts, *J. Phys. Chem. B* **106**(47), 12152–12157 (2002).
192. O. Solorza-Feria, K. Ellmer, M. Giersig, and N. Alonso-Vante, Novel low-temperature synthesis of semiconducting transition metal chalcogenide electrocatalyst for multi-electron charge transfer: Molecular oxygen reduction, *Electrochim. Acta* **39**(11–12), 1647–1653 (1994).
193. A. Kolary-Zurowska, A. Zieleniak, K. Miecznikowski, B. Baranowska, A. Lewera, S. Fiechter, P. Bogdanoff, I. Dorbandt, R. Marassi, and P. J. Kulesza, Activation of methanol-tolerant carbon-supported $RuSe_x$ electrocatalytic nanoparticles towards more efficient oxygen reduction, *J. Solid State Electrochem.* **11**(7), 915–921 (2007).
194. P. K. Babu, A. Lewera, J. H. Chung, R. Hunger, W. Jaegermann, N. Alonso-Vante, A. Wieckowski, and E. Oldfield, Selenium becomes metallic in Ru-Se fuel cell catalysts: An EC-NMR and XPS investigation, *J. Am. Chem. Soc.* **129**, 15140–15141 (2007).
195. A. T. Hubbard, Electrochemistry at well-defined surfaces, *Acc. Chem. Res.* **13**, 177 (1980).
196. A. T. Hubbard, Electrochemistry of metals, *Heterog. Chem. Rev.* **1**, 3–39 (1994).
197. A. Bewick, In situ infrared spectroscopy of the electrode/electrolyte solution interphase, *J. Electroanal. Chem.* **150**, 481–493 (1983).

198. L. Blum, H. D. Abruna, J. White, J. G. Gordon II, G. L. Borges, M. G. Samant, and O. R. Merloy, Study of underpotentially deposited copper on gold by fluorescence detected surface EXAFS, *J. Chem. Phys.* **85**, 6732–6738 (1986).

199. S. Mukerjee, R. C. Urian, S. J. Lee, E. A. Ticianelli, and J. McBreen, Electrocatalysis of CO tolerance by carbon-supported PtMo electrocatalysts in PEMFCs, *J. Electrochem. Soc.* **151**, A1094–A1103 (2004).

200. S. Murkerjee, In situ X-ray absorption spectroscopy of carbon supported Pt and Pt alloy electrocatalysts: Correlation of electrocatalytic activity with particle size and alloying, in *Catalysis and Electrocatalysis at Nanoparticle Surfaces*, A. Wieckowski, E. Savinova, and C. Vayenas, eds., Marcel Dekker, New York, Basel, 2003.

201. F. J. Scott, S. Mukerjee, and D. E. Ramaker, CO coverage/oxidation correlated with PtRu electrocatalyst particle morphology in 0.3 M methanol by in situ XAS, *J. Electrochem. Soc.* **154**, A396–A406 (2007).

202. S. J. Lee, S. Mukerjee, E. A. Ticianelli, and J. McBreen, Electrocatalysis of CO tolerance in hydrogen oxidation reaction in PEM fuel cells, *Electrochim. Acta* **44**, 3283–3293 (1999).

203. A. Lewera, W. P. Zhou, C. Vericat, J. H. Chung, R. Haasch, A. Wieckowski, and P. S. Bagus, X-ray photoelectron spectroscopy and reactivity measurements of Ru/Pt nanoparticles supported on a gold disk, *Electrochim. Acta* **51**, 3950–3956 (2006).

204. C. Vericat, M. Wakisaka, R. Haasch, P. S. Bagus, and A. Wieckowski, Binding energy of ruthenium submonolayers deposited on a Pt(111) electrode, *J. Solid State Electrochem.* **8**, 794–803 (2004).

205. W. P. Zhou, A. Lewera, R. Larsen, R. I. Masel, P. S. Bagus, and A. Wieckowski, Size effects in electronic and catalytic properties of unsupported palladium nanoparticles in electrooxidation of formic acid, *J. Phys. Chem. B* **110**(27), 13393–13398 (2006).

206. B. Hammer, Y. Morikawa, and J. K. Nørskov, CO chemisorption at metal surfaces and overlayers, *Phys. Rev. Lett.* **76**, 2141–2144 (1996).

207. M. Mavrikakis, B. Hammer, and J. K. Nørskov, Effect of strain on the reactivity of metal surfaces, *Phys. Rev. Lett.* **81**, 2819–2822 (1998).

208. K. M. Robinson and W. E. O'Grady, X-Ray surface diffraction studies of the restructuring and electrodeposition of Pb monolayers on Au(100) single crystals, *Faraday Discuss.* **95**, 55–64 (1993).

209. W. E. O'Grady, P. L. Hagans, K. I. Pandya, and D. L. Maricle, Structure of Pt/Ru catalysts using X-ray absorption near edge structure studies, *Langmuir* **17**, 3047–3050 (2001).

210. C. Roth, N. Benker, T. Buhrmester, M. Mazurek, M. Loster, H. Fuess, D. C. Koningsberger, and D. E. Ramaker, Determination of O[H] and CO coverage and adsorption sites on PtRu electrodes in an operating PEM fuel cell, *J. Am. Chem. Soc.* **127**, 14607–14615 (2005).

211. S. Stoupin, E. H. Chung, S. Chattopadhyay, C. U. Segre, and E. S. Smotkin, Pt and Ru X-ray absorption spectroscopy of PtRu anode catalysts in operating direct methanol fuel cells, *J. Phys. Chem. B* **110**, 9932–9938 (2006).

212. J. Inukai, D. Cao, A. Wieckowski, K.-C. Chang, A. Menzel, V. Komanicky, and H. You, In-situ synchrotron X-ray spectroscopy of ruthenium nanoparticles modified with selenium for an oxygen reduction reaction, *J. Phys. Chem. C* **111**(45), 16889–16894 (2007).

213. A. Wieckowski, Low temperature fuel cell technologies: Materials and new device concepts, 2007 (cited; Available from `http://www.hydrogen.edu.pl/`).

214. K. W. H. Chan and A. Wieckowski, Probing adsorbates on Pt electrode surfaces by the use of 13C spin-echo NMR: Studies of CO generated from methanol electrosorption, *J. Electrochem. Soc.* **137**, 367–372 (1990).

215. P. J. Slezak and A. Wieckowski, On interfacing surface electrochemistry with solid-state NMR: Characterization of surface CO on polycrystalline platinum, *J. Magn. Reson. A* **102**(2), 166–172 (1993).

216. G. Q. Lu, A. Lagutchev, D. D. Dlott, and A. Wieckowski, Quantitative vibrational sum-frequency generation spectroscopy of thin layer electrochemistry: CO on a Pt electrode, *Surf. Sci.* **585**, 3–16 (2005).

217. A. Lagutchev, G. Q. Lu, T. Takeshita, D. D. Dlott, and A. Wieckowski Vibrational sum frequency generation studies of the $(2 \times 2) \rightarrow (\sqrt{19} \times \sqrt{19})$ phase transition of CO on Pt(111) electrodes, *J. Chem. Phys.* **125**, 154705 (2006).

218. Y. Y. Tong, C. Belrose, A. Wieckowski, and E. Oldfield, First observation of platinum-195 nuclear magnetic resonance in commercial graphite-supported platinum electrodes in an electrochemical environment, *J. Am. Chem. Soc.* **119**(48), 11709–11710 (1997).

219. Y. Y. Tong, H. S. Kim, P. K. Babu, P. Waszczuk, A. Wieckowski, and E. Oldfield, An NMR investigation of CO tolerance in a Pt/Ru fuel cell catalyst, *J. Am. Chem. Soc.* **124**, 468–473 (2002).

220. Y. Y. Tong, E. Oldfield, and A. Wieckowski, Exploring metal/solution electrochemical interfaces by solid-state nuclear magnetic resonance, *Anal. Chem.* **70**, 518A–522A (1998).

221. Y. Y. Tong, E. Oldfield, and A. Wieckowski, Diffusion on a nanoparticle surface as revealed by electrochemical NMR, *Faraday Discuss.* **121**, 323–330 (2002).

222. Y. Y. Tong, C. Rice, N. Godbout, A. Wieckowski, and E. Oldfield, Correlation between the knight shift of chemisorbed CO and the Fermi level local density of states at clean platinum catalyst surfaces, *J. Am. Chem. Soc.* **121**, 2996–3003 (1999).

223. Y. Y. Tong, C. Rice, A. Wieckowski, and E. Oldfield, A detailed NMR-based model for CO on Pt catalysts in an electrochemical environment, *J. Am. Chem. Soc.* **122**, 1123–1129 (2000).

224. Y. Y. Tong, C. Rice, A. Wieckowski, and E. Oldfield, Pt-195 NMR of platinum electrocatalysts: Friedel-Heine invariance and correlations between platinum knight shifts, healing length, and adsorbate electronegativity, *J. Am. Chem. Soc.* **122**(48), 11921–11924 (2000).

225. Y. Y. Tong, A. Wieckowski, and E. Oldfield, NMR of electrocatalysts, *J. Phys. Chem. B* **106**(10), 2434–2446 (2002).

226. T. Kobayashi, P. K. Babu, J. H. Chung, E. Oldfield, and A. Wieckowski, Coverage dependence of CO diffusion on Pt nanoparticle surfaces: An EC-NMR study, *J. Phys. Chem. C* **111**, 7078–7083 (2007).

227. T. Kobayashi, P. K. Babu, L. Gancs, J. H. Chung, E. Oldfield, and A. Wieckowski, An NMR determination of CO diffusion on platinum electrocatalysts, *J. Am. Chem. Soc.* **127**, 14164–14165 (2005).

228. P. K. Babu, J. H. Chung, E. Oldfield, and A. Wieckowski, CO surface diffusion on platinum fuel cell catalysts by electrochemical NMR, *Electrochim. Acta* **53**(23), 6672–6679 (2008).

229. G. Q. Lu, A. Lagutchev, T. Takeshita, R. L. Behrens, D. D. Dlott, and A. Wieckowski, Broad-band sum frequency generation studies of surface intermediates involved in fuel cell electrocatalysis, in *Fuel Cell Catalysis: A Surface Science Approach*, M. Koper, ed., Wiley, Hoboken, NJ, 2009.

230. R. S. Jayashree, M. Mitchell, D. Natarajan, L. J. Markoski, and P. J. A. Kenis, Microfluidic hydrogen fuel cell with a liquid electrolyte, *Langmuir* **23**, 6871–6874 (2007).

231. F. R. Brushett, W. P. Zhou, R. S. Jayashree, and P. J. A. Kenis, Alkaline microfluidic hydrogen-oxygen fuel cell as a cathode characterization platform, *J. Electrochem. Soc.* **156**(5), B565–B571 (2009).

232. F. R. Brushett, H. T. Duong, J. W. D. Ng, A. Wieckowski, and P. J. A. Kenis, Investigation of Pt, $Pt_3Co$ and $Pt_3Co/Mo$ cathodes for the oxygen reduction reaction in an acidic microfluidic $H_2/O_2$ fuel cell, manuscript in preparation, 2009.

233. S. R. Stock, Recent advances in X-ray microtomography applied to materials, *Int. Mater. Rev.* **53**(3), 129–181 (2008).

234. P. K. Sinha, P. Halleck, and C. Y. Wang, Quantification of liquid water saturation in a PEM fuel cell diffusion medium using x-ray microtomography, *Electrochem. Solid State Lett.* **9**(7), A344–A348 (2006).

235. P. K. Sinha, P. P. Mukherjee, and C. Y. Wang, Impact of GDL structure and wettability on water management in polymer electrolyte fuel cells, *J. Mater. Chem.* **17**(30), 3089–3103 (2007).

236. R. Satija, D. L. Jacobson, M. Arif, and S. A. Werner, In situ neutron imaging technique for evaluation of water management systems in operating PEM fuel cells, *J. Power Sources* **129**(2), 238–245 (2004).

237. Z. Dunbar and R. I. Masel, Quantitative MRI study of water distribution during operation of a PEM fuel cell using Teflon (R) flow fields, *J. Power Sources* **171**(2), 678–687 (2007).

238. Z. W. Dunbar and R. I. Masel, Magnetic resonance imaging investigation of water accumulation and transport in graphite flow fields in a polymer electrolyte membrane fuel cell: Do defects control transport? *J. Power Sources* **182**(1), 76–82 (2008).

239. L. Ma, D. B. Ingham, M. Pourkashanian, and E. Carcadea, Review of the computational fluid dynamics modeling of fuel cells, *J. Fuel Cell Sci. Technol.* **2**(4), 246–257 (2005).

240. D. Cheddie and N. Munroe, Review and comparison of approaches to proton exchange membrane fuel cell modeling, *J. Power Sources* **147**(1–2), 72–84 (2005).

241. C. Y. Wang, Fundamental models for fuel cell engineering, *Chem. Rev.* **104**(10), 4727–4765 (2004).

242. K. Yoshizawa, K. Ikezoe, Y. Thsaki, D. Kramer, E. H. Lehmann, and G. G. Scherer, Analysis of gas diffusion layer and flow-field design in a PEMFC using neutron radiography, *J. Electrochem. Soc.* **155**(3), B223–B227 (2008).

243. B. Hammer and J. K. Nørskov, Theoretical surface science and catalysis—calculations and concepts, *Adv. Catal.* **45**, 71–129 (2000).

244. R. A. van Santen and M. Neurock, *Molecular Heterogeneous Catalysis: A Conceptual and Computational Approach*, Wiley-VCH, Weinhem, 2006.

245. T. Bligaard and J. K. Nørskov, Ligand effects in heterogeneous catalysis and electrochemistry, *Electrochim. Acta* **52**, 5512–5516 (2007).

246. J. C. Davies, J. Bonde, A. Logadottir, J. K. Nørskov, and I. Chorkendorff, The ligand effect: CO desorption from Pt/Ru catalysts, *Fuel Cells* **5**, 429–435 (2005).

247. B. Hinnemann and J. K. Nørskov, Catalysis by enzymes: The biological ammonia synthesis, *Top. Catal.* **37**, 55–70 (2006).

248. J. R. Kitchin, J. K. Nørskov, M. A. Barteau, and J. G. Chen, Role of strain and ligand effects in the modification of the electronic and chemical properties of bimetallic surfaces, *Phys. Rev. Lett.* **93** (2004).

249. J. K. Nørskov, M. Scheffler, and H. Toulhoat, Density functional theory in surface science and heterogeneous catalysis, *MRS Bull.* **31**, 669–674 (2006).

250. M. T. de Groot, T. H. Evers, M. Merkx, and M. T. M. Koper, Electron transfer and ligand binding to cytochrome c' immobilized on self-assembled monolayers, *Langmuir* **23**(2), 729–736 (2007).

251. M. T. de Groot, M. Merkx, and M. T. M. Koper, Reorganization of immobilized horse and yeast cytochrome c induced by pH changes or nitric oxide binding, *Langmuir* **23**, 3832–3839 (2007).

# INDEX




*Fuel Cell Science: Theory, Fundamentals, and Biocatalysis*,
Edited by Andrzej Wieckowski and Jens K. Nørskov

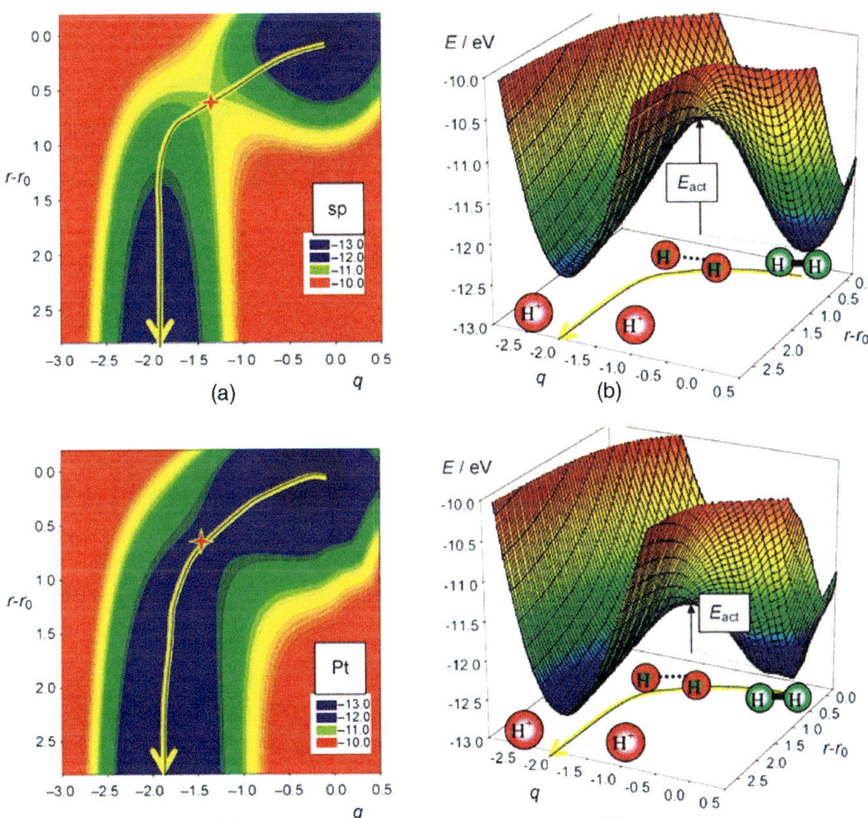


**FIGURE 1.21** Potential-energy surfaces for oxidation of $H_2$ as a function of bond length $r$ and solvent coordinate $q$. The equilibrium bond length is denoted by $r_0$; $r$ is given in atomic units (1 au = 0.529 Å). (a,b) *sp* metal; (c,d) platinum. The red crosses denote the saddle points [98].

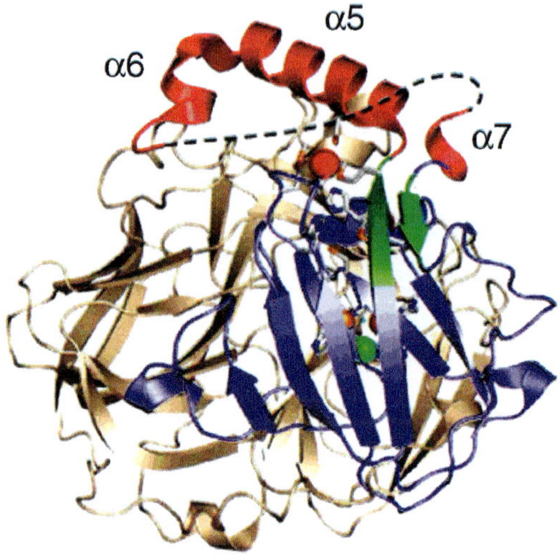


**FIGURE 5.4** The structure of copper efflux oxidase (to generate the Δα5–7 deletion mutant, the amino acid residues in the α-helical region from Pro357 to His406 (red ribbon region) were deleted and replaced by the dipeptide Gly–Gly) [53].

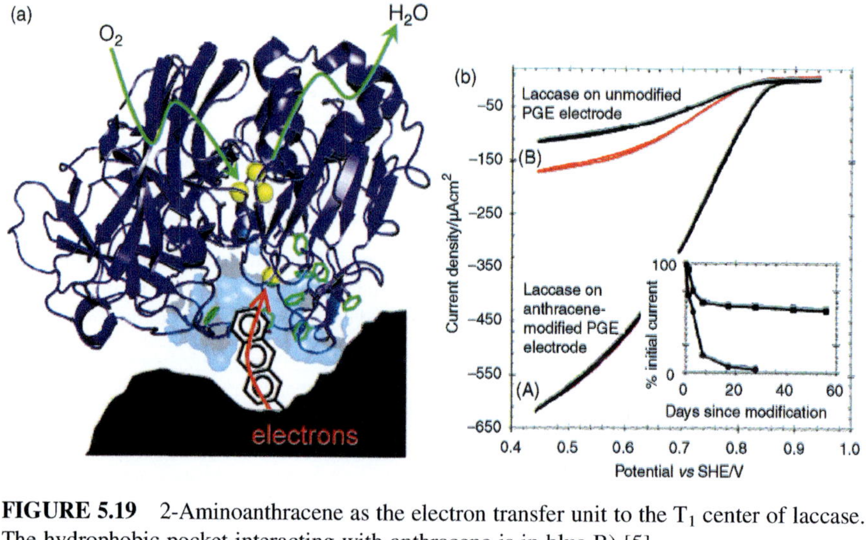


**FIGURE 5.19** 2-Aminoanthracene as the electron transfer unit to the $T_1$ center of laccase. The hydrophobic pocket interacting with anthracene is in blue B) [5].

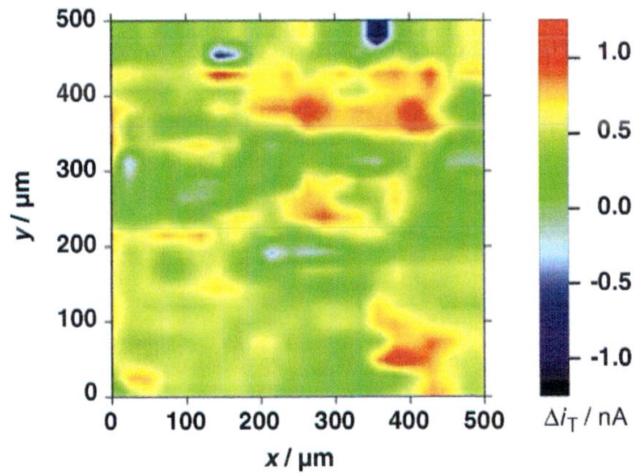


**FIGURE 5.22** The structure of copper efflux oxidase (to generate the $\Delta\alpha$5–7 deletion mutant, the amino acid residues in the $\alpha$-helical region from Pro357 to His406 (red ribbon region) were deleted and replaced by the dipeptide Gly–Gly) [53].

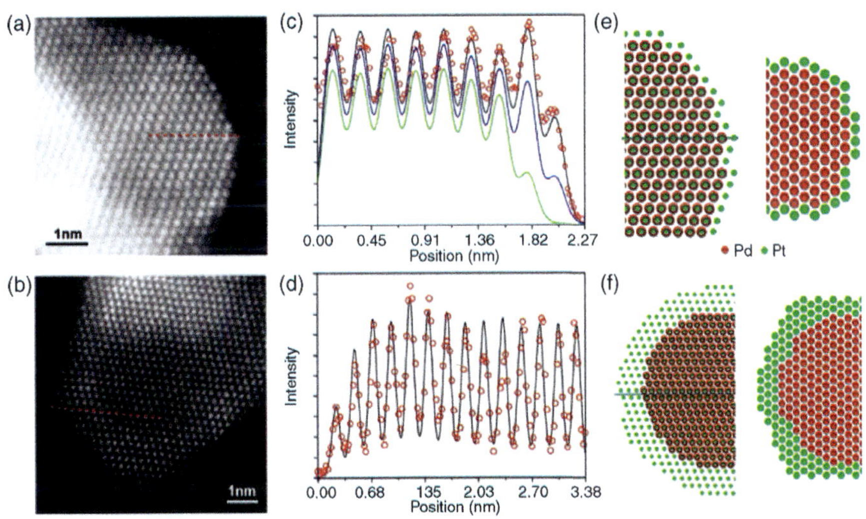


**FIGURE 6.3** (a,b) HAADF-STEM images of the Pd(core)–Pt(shell) nanoparticles obtained for the $Pd_CPt_1$ and $Pd_CPt_4$ samples, respectively. We note that the nanoparticles overlap with others at the left side in (a), and at the top side in (b). (c,d) Intensity profiles from the scan lines in (a) and (b) (open circles), and the best fits (black lines), based on the structure models shown in (e) and (f) for Pd particles with one and three to four Pt surface layers, respectively. The calculations are convoluted with a Gaussian point spread function. Two additional lines in (c) show the calculations made with all the Pt surface atoms being removed (green line), or replaced by Pd (blue line). (e) Projection of the one-Pt-layer core–shell structure model along the [101] direction (left), and the atomic arrangement (right) of the (1–1–1) plane at the position indicated by the line in the left. (f) Projection of the structure model with four Pt layers on the STEM image plane (left) and the arrangements of atoms in the vertical columns along the scan line (right) [19].

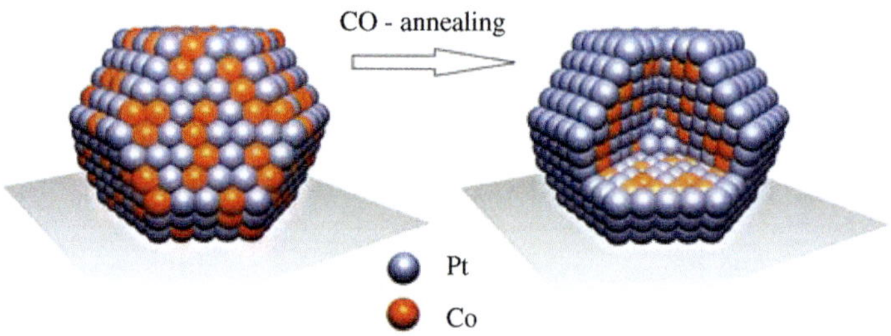


**FIGURE 9.8** Scheme of induced surface segregation. (Taken from Ref. 114.)

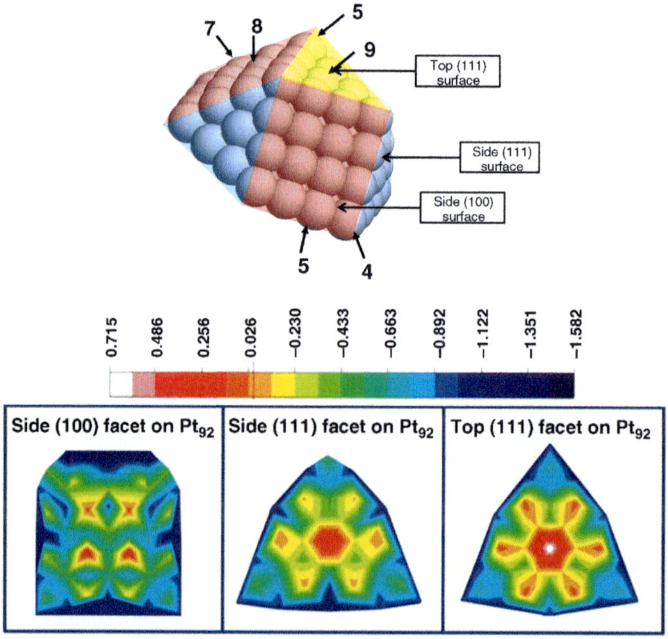


**FIGURE 10.7** Hemispherical cuboctahedral $Pt_{92}$ nanoparticle with distinct facets and distinct local coordination structures of surface atoms, indicated by their nearest-neighbor atom numbers (top). Contour plots of oxygen adsorption energies for the distinct surface facets of hemispherical cuboctahedral $Pt_{92}$ nanoparticle (bottom). (Reprinted with permission from *J. Phys. Chem. C*, 2009, 113 (42), pp 17989–17996. Copyright 2009 American Chemical Society.)

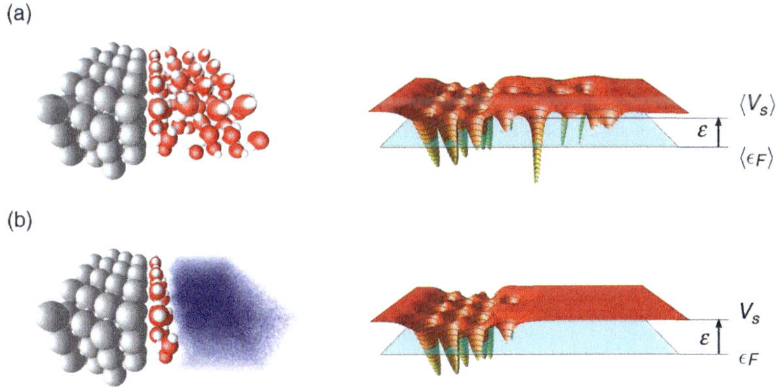


**FIGURE 13.1** Determination of the absolute electrode potential $\mathcal{E}$ using (a) an explicit solvent model, and (b) an implicit solvent model. Within the explicit model, the potential $\mathcal{E}$ is calculated from the dynamical averages of the solvent electrostatic energy $\langle \mathcal{V}_S \rangle$ and of the metal Fermi energy $\langle \epsilon_F \rangle$.

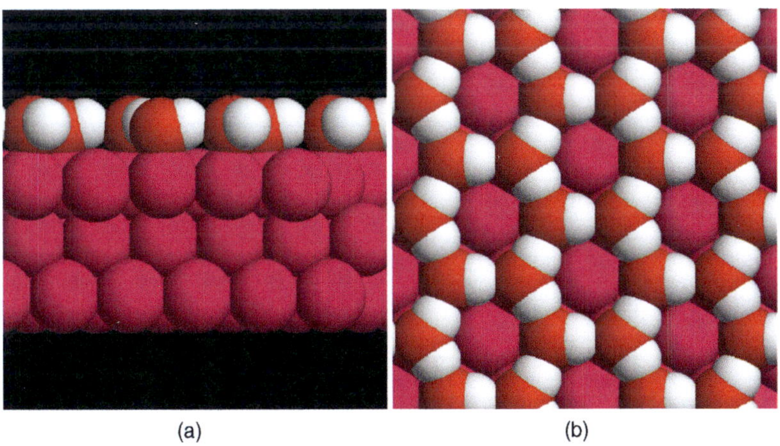


**FIGURE 16.1** The hexagonal half-dissociated water layer on Pt (111): sideview (a) and topview (b). All hydrogen bonds in the plane of the surface are saturated. White spheres indicate hydrogen; red, oxygen, and pink, Pt.

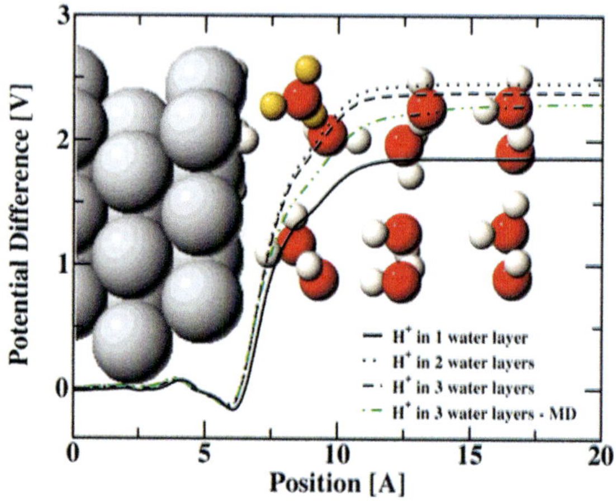


**FIGURE 16.2** The charged solid–liquid interface: charged Pt (111) slab with three water layers outside and one solvated hydronium ion per unit cell. The electrode potential, due to the charged interface averaged parallel to the surface, is shown for systems with one, two, and three relaxed water layers along with results from a DFT molecular dynamics simulation of a proton solvated in three water layers. (Reproduced from Ref. 17.)

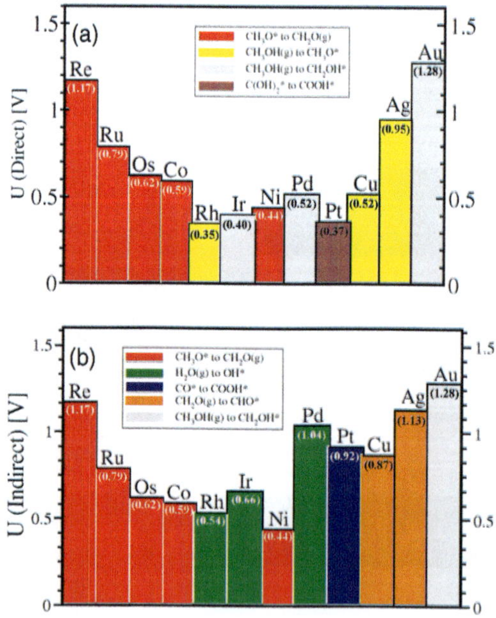


**FIGURE 16.5** (a) The lowest potential needed to ensure that all steps along the *direct* mechanism path are downhill in free energy on 12 close-packed metal surfaces. For Os, the final step ($CO^* + OH^* \rightarrow CO_2$) has a $\Delta G$ within 0.1 eV of that of the potential-determining step. For Pt, initial methanol activation ($CH_3OH \rightarrow CH_2OH$) has a $\Delta G$ within 0.1 eV of that of the potential-determining step. (b) The lowest potential needed to ensure that all steps along the *indirect* mechanism path are downhill in free energy on 12 close-packed metal surfaces. For Os, the final step ($CO^* + OH^* \rightarrow CO_2$) has a $\Delta G$ within 0.1 eV of that of the potential-determining step. The color code is related to the nature of the potential-determining step, as described in the legend in the figure.

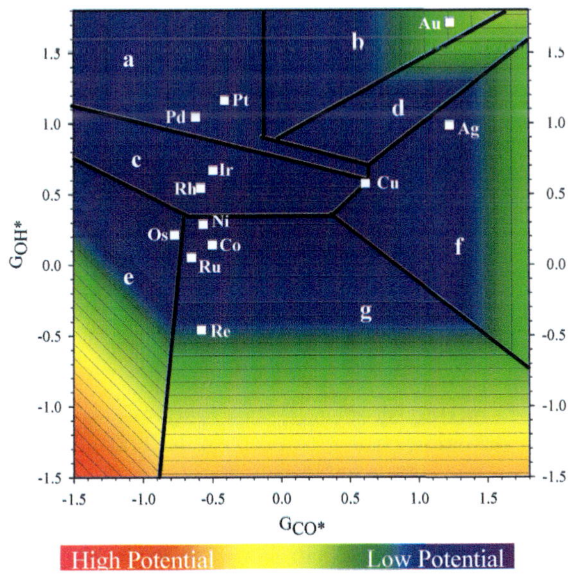


**FIGURE 16.6** The onset potential and potential-determining steps (PDSs) for the *indirect* mechanism of methanol electrooxidation on close-packed hcp and fcc surfaces plotted using $G_{CO^*}$ and $G_{OH^*}$ as descriptors. These two adsorption free energies describe the overall reaction landscape and can be used to estimate the potential through the linear relations, as described in Reference 43. The PDSs (without showing proton/electron transfer explicitly) for each region are as follows: (a) $CO^* \rightarrow COOH^*$; (b) $H_3COH \rightarrow H_2COH^*$; (c) $H_2O \rightarrow OH^*$; (d) $H_3COH \rightarrow H_3CO^*$;(e) $CO^* + OH^* \rightarrow CO_2$; (f) $H_2CO \rightarrow CHO^*$; (g) $H_3CO^* \rightarrow H_2CO$. Isopotential lines are included for reference. Each line represents a difference of 0.1 V. Because of scatter in the correlations used, the potential-determining region for the individual metal does not always match the actual PDS for that metal, which is given in Figure 16.5b.

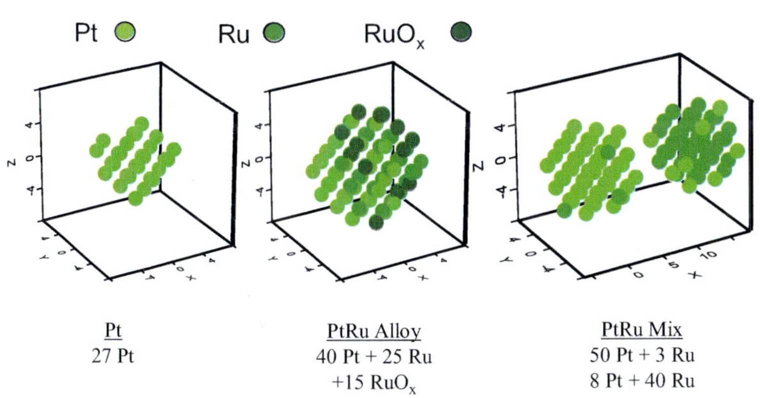


**FIGURE 17.9** Structural models estimated from EXAFS analysis for the indicated electrocatalysts prepared as described in the text. The PtRu mix catalyst was prepared by mechanically mixing Pt/C and Ru/C powder, with some atoms of the "other" element thereby being deposited on each cluster surface. (From Ref. 68.)

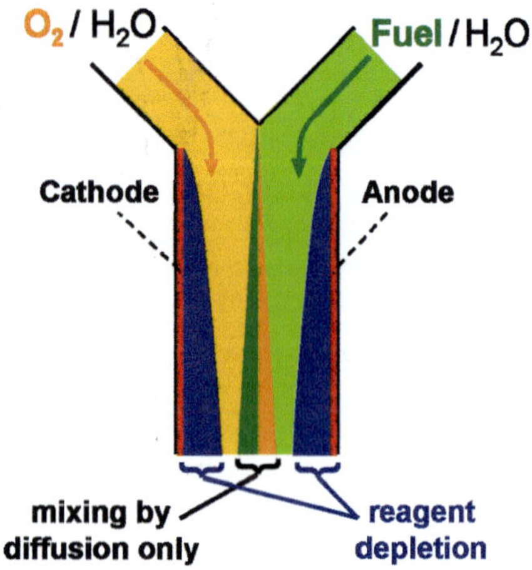


**FIGURE 19.5** Schematic of a Y-shaped laminar flow-based fuel cell (LFFC) developed by the Kenis group. The electrodes are the catalyst-coated sidewalls on opposite sides of the microchannel (red lines). The cell operates by flowing an anodic fuel stream (green) and a cathodic oxygen-saturated stream (yellow) in parallel. Though not shown, both streams also include a supporting electrolyte to enhance conductivity. The reaction depletion at the electrodes and diffusional mixing zones at the liquid-liquid interface are both highlighted (not drawn to scale). Image reprinted with the permission of Professor Paul J.A. Kenis.

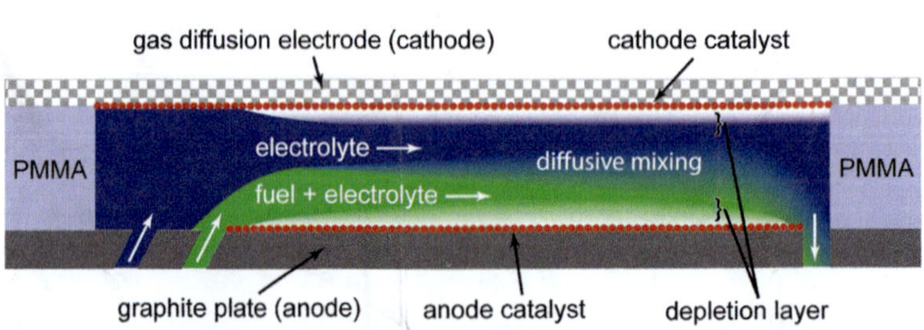


**FIGURE 19.6** Schematic of an air-breathing laminar flow-based fuel cell (LFFC) developed by the Kenis group and shown in Brushett et al. [96]. The electrodes are a catalyst-coated graphite plate as an anode and a catalyst-coated gas diffusion electrode on opposite sides of the microchannel (red lines). The cell operates by flowing an anodic fuel and electrolyte stream (green) and a cathodic electrolyte stream (blue) which serves as a barrier to fuel crossover. Oxygen diffuses to the cathode from the ambient environment. The reaction depletion at the electrodes and diffusional mixing regions at the liquid-liquid interface are both highlighted (not drawn to scale). Reprinted from Ref. 96, 2009 by permission of Elsevier.